Distance Formula

Distance from (x_1, y_1) to (x_2, y_2)

$$\sqrt{(x_1 - x_2)^2 + (y_1 - y_2)^2}$$

Midpoint Formula

Midpoint of the segment joining

(x_1, y_1) and (x_2, y_2) is $\left(\frac{x_1 + x_2}{2}, \frac{y_1 + y_2}{2}\right)$

Slope Formula

Slope of line through (x_1, y_1) and (x_2, y_2)

$$m = \frac{y_1 - y_2}{x_1 - x_2}$$

Point-slope Equation

Equation of line through (x_0, y_0) with slope m

$$y - y_0 = m(x - x_0)$$

Tests for Symmetry

With respect to the x-axis:
equation unchanged if y is replaced by $-y$
With respect to y-axis:
equation unchanged if x is replaced by $-x$
With respect to origin:
equation is unchanged if x is replaced by $-x$
and y is replaced by $-y$

Polar Coordinates

$x = r\cos\theta \qquad y = r\sin\theta$

$\tan\theta = \frac{y}{x} \qquad r^2 = x^2 + y^2$

Translations

Replace	*Result*
x by $x - h$	shift curve horizontally $\lvert h\rvert$ units, right if $h > 0$ left if $h < 0$
y by $y - k$	shift curve vertically $\lvert k\rvert$ units, up if $k > 0$ down if $k < 0$

Conic Sections

Circle with center at (h, k), radius r

$$(x - h)^2 + (y - k)^2 = r^2$$

Parabola with vertex (h, k)

$$(y - k)^2 = 4p(x - h)$$
$$(x - h)^2 = 4p(y - k)$$

Ellipse

$$\frac{(x - h)^2}{a^2} + \frac{(y - k)^2}{b^2} = 1$$

Hyperbola

$$\frac{(x - h)^2}{a^2} - \frac{(y - k)^2}{b^2} = 1$$

$$\frac{(y - k)^2}{a^2} - \frac{(x - h)^2}{b^2} = 1$$

COLLEGE ALGEBRA and TRIGONOMETRY

COLLEGE ALGEBRA and TRIGONOMETRY

Edward D. Gaughan
New Mexico State University

Carl E. Hall
University of Texas at El Paso

Brooks/Cole Publishing Company
Monterey, California

Contemporary Undergraduate Mathematics Series
Consulting Editor: Robert J. Wisner

Brooks/Cole Publishing Company
A Division of Wadsworth, Inc.

Printed in the United States of America
10 9 8 7 6 5 4 3 2

Library of Congress Cataloging in Publication Data

Gaughan, Edward.
College algebra and trigonometry.

(Contemporary undergraduate mathematics series)
Includes index.
1. Algebra. 2. Trigonometry. I. Hall, Carl E., 1937– . II. Title. III. Series.
QA154.2.G384 1984 512′.13 83-15470
ISBN 0-534-02777-6

Sponsoring Editor: *Craig F. Barth*
Production Editor: *Jane Stanley*
Manuscript Editor: *Charles Hibbard*
Interior Design: *Jamie Sue Brooks*
Cover Design: *Stanley Rice*
Cover Illustration: *David Aguero*
Art Coordinator: *Rebecca Ann Tait*
Interior Illustration: *Lori Heckelman*
Typesetting: *Bi-Comp, Inc., York, Pennsylvania*
Cover Printing: *Phoenix Color Corporation, Long Island City, New York*
Printing and Binding: *The Maple Press Company, York, Pennsylvania*

Preface

The purpose of this book is to provide a firm grounding in those topics in algebra and trigonometry that a student needs to master before taking a course in calculus.

The approach of this text is modern, with a liberal emphasis on applications. We have presented problems for which old methods of solution will not work, so that students will feel encouraged to try new ideas. We have striven throughout to develop students' understanding of mathematical concepts, rather than to rely upon rote learning. This book contains a diverse selection of problems so that the underlying concepts can be demonstrated through a variety of situations. The first discussion of applied problems comes in Chapter 1 and this approach is maintained throughout the book.

You do not need a calculator to use this book, but some special exercises are offered for those who do have a calculator at their disposal. They are identified with the heading For Your Calculator. The intent of these exercises is to offer situations in which the calculator can be used to gain more insight into the ideas presented in that section.

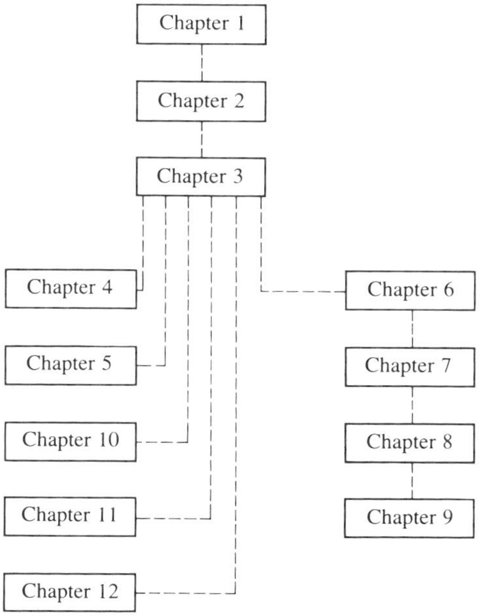

Readers with a strong background in algebra will want to begin with Chapter 3 and then concentrate their efforts on the new ideas in the following chapters. On the other hand, readers with a less thorough background in algebra should review by working through Chapters 1 and 2.

Chapter 3 is a mixture of some review and some new material and is essential to an understanding of all the material that follows. The diagram on the preceding page is intended to indicate the dependence and interdependence of Chapters 4 through 12. There are some minor exceptions to the diagram; for example, Chapters 9 and 12 are independent, except that DeMoivre's theorem is covered in Chapter 9 and is offered as an example in Chapter 12.

To the Instructor

This book is written with both the student and the instructor in mind. The next few paragraphs are intended to acquaint you, the instructor, with the features of the book that are designed to assist you in its use.

How do we usually teach mathematics? We begin with some motivation, present the new material to be taught, illustrate the new concepts with many examples, give the students exercises to practice on, and then test for mastery. This book has been prepared with this procedure in mind.

Most chapters begin with a paragraph or so offering a problem or setting that provides motivation for the content of the chapter. Further motivating problems are offered in some of the sections of each chapter.

As new concepts are introduced or old ones extended, the development is liberally supplemented with examples. The examples run the gamut from very elementary examples illustrating basic principles to more complex examples integrating new concepts into the overall picture.

The exercise sets are tied very closely to the text and examples. As an aid to you the exercises are paired. To illustrate this pairing, look at Exercises 19 and 20 in Section 1.6. Both ask the student to factor a polynomial that is the square of a binomial. Notice the similarity of even and odd exercises in that section. The reason for this pairing is to give you great flexibility in using the exercises. Answers to most odd-numbered exercises are in the back of the book. If you wish to assign problems with answers available to the student, you may assign only odd-numbered problems with the assurance that all types of problems will be covered. On the other hand, if you prefer that students not have answers to the exercises, you may assign only even-numbered problems with the same assurance. Further, if a student gets stuck on an even-numbered exercise, the answer to the paired odd-numbered exercise, which is very similar, is available in the back of the book. Of course, as most of us do, you may want to use your own variation on the even-odd scheme. The exceptions to the pairing are some of the word problems and some of the more difficult problems at the end of some exercise sets.

As an additional feature, certain exercise sets include a set of problems labeled For Your Calculator. We have included these sets where they add to the understanding of the content of the section.

At the end of each chapter you will find a review section that has a complete set of answers at the back of the book. These sections are intended to offer the student an opportunity to test his or her knowledge of the material in the chapter.

Most proofs and verifications of trigonometric identities are not included in the answer section. We have, however, included suggestions for the identities in Exercises 7.6 and for the chapter review and chapter test in Chapter 7.

After the chapter review you will find a chapter test, again with answers in the back of the book. It is not our intention that these actually be used as chapter tests; rather, they are designed to show students, in brief form, the skills they should have mastered in the chapter. Our experience is that students appreciate such sample tests and find them a useful aid in studying.

Calculator Notes

First of all, it should be made clear that a calculator is **not** required to study the material in this book, although it would be very useful in certain sections.

The areas where a calculator may be helpful in this book may be grouped into two general categories: sections in which the calculator is useful for computations and special sections labeled For Your Calculator that are designed to help in the understanding of the concept being studied.

The extent to which the calculator is used in a particular section will depend on the content and on the emphasis of the instructor on such usage. For example, Section 11.6, Irrational Roots of Polynomial Equations, requires quite a bit of computation to find roots. Here it is reasonable to use a calculator in doing the homework.

The sections labeled For Your Calculator are designed to utilize the calculator to give a different insight into the concepts of that particular section or to acquaint the student with some of the capabilities of the calculator. For example, Section 1.6 shows how to use the nested form of a polynomial to find its values with a calculator; in Section 6.1 the various modes of angle measurement available on the calculator are discussed.

Two types of logic are commonly used on calculators: reverse Polish notation (RPN) and algebraic (ALG) logic. Those calculators using RPN are characterized by a key marked [Enter] or [Save]. In the first chapter we give the key sequences for both types of logic; in later chapters we give only the key sequence for the algebraic type of calculator.

Since there are so many types of calculators available, we suggest you study your manual for more details on the operation of your calculator. Have fun!

Acknowledgments

We wish to express our appreciation to the reviewers of the book: John Cross, University of Northern Iowa; Mark Hale, Jr., University of Florida; Lotus D. Herschberger, Illinois State University; Jimmie J. McKim, University of Central Arkansas; William T. Pelletier, Wayne State University; Philip A. Schmidt, Berea College; and Carroll G. Wells, Western Kentucky University.

We also wish to express our appreciation to Robert J. Wisner of New Mexico State University for his careful reading of the manuscript and his many valuable suggestions, to Craig Barth of Brooks/Cole Publishing Company for his encouragement and assistance in this project, and to Jane Stanley for her cheerful patience throughout the trying stages of production. Many thanks must also go to our typists who spent many long hours doing an excellent job of preparing the manuscript.

Edward D. Gaughan
Carl E. Hall

Contents

1 Real Numbers and Polynomials 1

1.1 Real Numbers and the Number Line 1
1.2 The Order Relation for Real Numbers 8
1.3 Equations and Inequalities 11
1.4 Problem Solving 18
1.5 Polynomials 23
1.6 Factoring 28
Chapter Review 32
Chapter Test 34

2 Fractions, Radicals, and Quadratic Equations 35

2.1 Fractions and Quotients of Polynomials 35
2.2 Integer Exponents and Scientific Notation 42
2.3 Roots, Radicals, and Rational Exponents 48
2.4 Complex Numbers 54
2.5 Quadratic Equations 58
2.6 Other Equations and Inequalities 65
Chapter Review 69
Chapter Test 72

3 Functions, Graphs, and Linear Equations 73

3.1 Functions 73
3.2 Relations, Functions, and Graphs 78
3.3 Linear Functions and Equations 86
3.4 Symmetry and Translations of Graphs 93
3.5 Algebra of Functions 99
3.6 1–1 Functions and Inverse Functions 104
3.7 Applications 108
Chapter Review 112
Chapter Test 115

4 Polynomial Functions, Rational Functions, and Conic Sections 116

4.1 Quadratic Functions 116
4.2 Conic Sections: The Parabola and the Circle 123

4.3 Conic Sections: The Ellipse 130
4.4 Conic Sections: The Hyperbola 134
4.5 Polynomial Functions 139
4.6 Rational Functions 142
4.7 Partial Fractions 148
4.8 Inequalities in Two Variables 152
Chapter Review 157
Chapter Test 158

5 Exponential and Logarithmic Functions 159

5.1 Exponential Functions 159
5.2 Using Exponential Functions 165
5.3 Logarithmic Functions 169
5.4 Computations with Logarithms 175
5.5 Logarithmic and Exponential Equations 178
5.6 Interpolation with Logarithms 182
Chapter Review 186
Chapter Test 187

6 The Trigonometric Functions 188

6.1 Angles and Their Measure 188
6.2 The Trigonometric Functions 195
6.3 Special Angles 204
6.4 Trigonometric Functions of Real Numbers 210
6.5 Use of Reference Angle and Reference Number 217
6.6 Graphs of the Trigonometric Functions 225
6.7 Amplitude, Period, and Phase Shift in Graphing 232
6.8 Inverse Trigonometric Functions 238
Chapter Review 244
Chapter Test 247

7 Analytic Trigonometry 248

7.1 Equations and Identities 248
7.2 The Fundamental Identities 254
7.3 The Addition Formulas 260
7.4 Double-Angle and Half-Angle Formulas 266
7.5 Miscellaneous Formulas 273
7.6 Identities and Equations Revisited 277
Chapter Review 282
Chapter Test 284

8

Solution of Triangles 286

8.1 Solving Right Triangles 286
8.2 Law of Sines 291
8.3 Law of Cosines 296
8.4 Vector Algebra 301
8.5 Vector Applications 308
Chapter Review 314
Chapter Test 315

9

Complex Numbers and Polar Coordinates 316

9.1 Geometric Representation and Trigonometric Form 316
9.2 Multiplication and Division in Trigonometric Form 321
9.3 De Moivre's Theorem and nth Roots of Complex Numbers 327
9.4 Polar Coordinates 333
9.5 Graphs in Polar Coordinates 339
Chapter Review 343
Chapter Test 345

10

Systems of Equations, Inequalities, and Matrices 346

10.1 Systems of Equations in Two Unknowns 346
10.2 Systems of Inequalities 354
10.3 Linear Programming 360
10.4 Systems of Linear Equations in Three Unknowns 365
10.5 Matrices 371
10.6 Gaussian Elimination 378
10.7 Matrices and Matrix Inversion 384
10.8 Determinants 390
10.9 Cramer's Rule 395
Chapter Review 398
Chapter Test 400

11

Polynomial Equations 402

11.1 Remainder and Factor Theorems 402
11.2 Synthetic Division 406
11.3 Roots of Polynomial Equations 410
11.4 Locating Roots 414
11.5 Rational Roots of Polynomial Equations 419
11.6 Irrational Roots of Polynomial Equations 422
Chapter Review 425
Chapter Test 426

12 Sequences, Series, and Induction 428

12.1 Sequences 428
12.2 Series 433
12.3 Mathematical Induction 437
12.4 Binomial Theorem 441
Chapter Review 445
Chapter Test 447

Appendix I Answers 448

Appendix II Tables 511

Table I Squares, Square Roots, and Prime Factors 512
Table II Common Logarithms 513
Table III Exponential Functions 515
Table IV Natural Logarithms of Numbers 516
Table V Values of the Trigonometric Functions 517

Index 525

COLLEGE ALGEBRA and TRIGONOMETRY

1 Real Numbers and Polynomials

1.1 Real Numbers and the Number Line

Since the study of algebra is a study in depth of the properties of the system of real numbers, it is only appropriate that we begin with a look at the system of real numbers.

Natural numbers
Integers

The set of **natural numbers** is the set consisting of the numbers 1, 2, 3, 4, and so on, and is a subset of the set of **integers,** which consists of the numbers 0, 1, −1, 2, −2, 3, −3, and so on. The set of all **rational numbers** consists of all numbers that can be written in the form a/b, where a and b are integers with $b \neq 0$. Those real numbers that are not rational numbers are called **irrational numbers.** The numbers $\sqrt{2}$, π, and $1 + \sqrt[5]{3}$, just to mention a few, are irrational numbers. In this book we use R to represent the set of real numbers, Q to represent the set of rational numbers, and Z to represent the set of integers.

Rational numbers
Irrational numbers

Number line

There is a one-to-one correspondence between the set of real numbers and the set of points on a line, and we shall make good use of that correspondence in the study of algebra and trigonometry. To illustrate the correspondence more explicitly, draw a line and designate a point as the origin, marked by 0. Select a point to the right of the origin and mark that point 1. Call the length of the segment joining 0 and 1 one unit. Now mark the point one unit to the right of 1 with the number 2 and so on. Then mark the point one unit to the left of 0 with the number −1, the point one unit to the left of −1 with the number −2, and so on. Once the integers have been located on the line, it is easy to locate the rational numbers. The points not marked with either integers or rational numbers correspond to irrational numbers (see Figure 1.1).

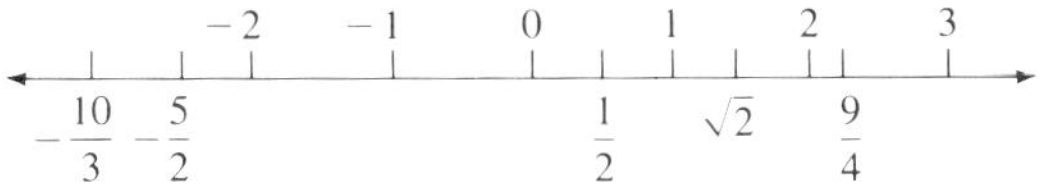

Figure 1.1

The system of real numbers consists of the set R of real numbers, along with two operations, addition and multiplication. The operations of addition and multiplication satisfy the following axioms.

For all real numbers x, y, and z the following are true:

Axioms for the system of real numbers

Axiom 1 $(x + y) + z = x + (y + z)$ $\quad (xy)z = x(yz)$ — Associative axiom

Axiom 2 $x + y = y + x \qquad xy = yx$ — Commutative axiom

Axiom 3 $x(y + z) = xy + xz$ — Distributive axiom

Axiom 4	There is a number 0 such that $0 + x = x$ for all x.	Additive identity
Axiom 5	There is a number 1 such that $1 \cdot x = x$ for all x.	Multiplicative identity
Axiom 6	For each x there is a unique real number called $-x$ such that $x + (-x) = 0$.	Additive inverse
Axiom 7	If $x \neq 0$ there is a unique number called x^{-1} such that $x \cdot x^{-1} = 1$.	Multiplicative inverse

Field Axioms 1 through 7 are called the field axioms, because any algebraic system that has two operations satisfying these axioms is called a **field.** The system of rational numbers, along with the operations of multiplication and addition, forms a field; as you will see later, the system of complex numbers also forms a field.

What follows is a list of theorems that are useful in computations, along with some examples to illustrate the uses. Some proofs are included, and the rest are left as exercises.

Theorem 1 $-(-x) = x$

Proof The number $-(-x)$ is the unique real number that one adds to $-x$ to get 0 (see Axiom 6). But

$$-x + x = x + (-x) = 0$$

so $-(-x)$ is indeed x.

Example 1 $-(-5) = 5$ ■

Theorem 2 $(-x) + (-y) = -(x + y)$

Example 2 $(-2) + (-5) = -(2 + 5) = -7$ ■

The sum $(-2) + (-5)$ is usually written $-2 - 5$. So Theorem 2 may be written

$$-x - y = -(x + y)$$

Example 3 $-7 - 9 = -16$ ■

When computing with numbers grouped by parentheses and brackets, it is a good idea to work from the inside out.

Example 4

$$\begin{aligned}[(4 + 2) + 5] - [2 + (3 + 4)] &= [6 + 5] - [2 + 7] \\ &= 11 - 9 \\ &= 2\end{aligned}$$ ■

Example 5

$$\begin{aligned}[2 - (9 - 5)] - [(6 - 7) - 4] &= [2 - 4] - [-1 - 4] \\ &= -2 - [-5] \\ &= -2 + 5 \\ &= 3\end{aligned}$$ ■

Theorem 3 $a \cdot 0 = 0$ for each real number a.

Proof Since $0 + 0 = 0$, we have

$$a \cdot 0 = a(0 + 0) = a \cdot 0 + a \cdot 0$$

Adding $-(a \cdot 0)$ to both sides yields

$$a \cdot 0 + [-(a \cdot 0)] = a \cdot 0 + a \cdot 0 + [-(a \cdot 0)]$$

or

$$0 = a \cdot 0$$

Theorem 4 For all real numbers a and b, $-(ab) = (a)(-b) = (-a)(b)$.

Theorem 5 For all real numbers a and b, $ab = (-a)(-b)$.

Example 6 $(-4)(5) = -20$ ■

Example 7 $(-4)(-5) = 20$ ■

Example 8 $(-2)(-3)(6)(-4) = (2 \cdot 3)(-24) = -144$ ■

Example 9 $(-4)(8 - 11) = (-4)(8) + (-4)(-11) = -32 + 44 = 12$ ■

Multiplicative inverse

Axiom 7 guarantees that every nonzero real number x has a **multiplicative inverse** or **reciprocal,** denoted by x^{-1} or $1/x$, such that $x \cdot x^{-1} = 1$.

Example 10 $3 \cdot \dfrac{1}{3} = 1$ ■

Example 11 $(\sqrt{2})(\sqrt{2})^{-1} = 1$ ■

Theorem 6 If $x \neq 0$, then $(x^{-1})^{-1} = x$ or $x = \dfrac{1}{\frac{1}{x}}$.

Theorem 7 If $a \cdot b = 0$, then either $a = 0$ or $b = 0$.

Proof If $a = 0$, fine. If not, then a has a multiplicative inverse a^{-1}. Multiply the equation $a \cdot b = 0$ by a^{-1} to obtain

$$a^{-1} \cdot (a \cdot b) = a^{-1} \cdot 0$$
$$(a^{-1} \cdot a) \cdot b = 0 \qquad (a^{-1} \cdot 0 = 0 \text{ by Theorem 3})$$
$$1 \cdot b = 0$$
$$b = 0$$

So either $a = 0$ or, if $a \neq 0$, then $b = 0$.

Theorem 8 If $a \neq 0$ and $b \neq 0$, then $ab \neq 0$ and $\dfrac{1}{ab} = \dfrac{1}{a} \cdot \dfrac{1}{b}$.

Proof Since neither a nor b is zero, $ab \neq 0$ by Theorem 7. To prove that

$$\frac{1}{ab} = \frac{1}{a} \cdot \frac{1}{b}$$

we must show that

$$(ab)\left(\frac{1}{a} \cdot \frac{1}{b}\right) = 1 \quad \text{(See Axiom 7.)}$$

Using the associative and commutative laws,

$$(ab)\left(\frac{1}{a} \cdot \frac{1}{b}\right) = \left(a \cdot \frac{1}{a}\right)\left(b \cdot \frac{1}{b}\right) = 1 \cdot 1 = 1$$

Example 12 $\dfrac{1}{3} \cdot \dfrac{1}{7} = \dfrac{1}{21}$ ■

Division The multiplicative inverse of a nonzero real number is used in the process of **division.** If $b \neq 0$, to "divide a by b means to find a solution x to the equation $bx = a$. Since

$$b(a)\left(\frac{1}{b}\right) = a(b)\left(\frac{1}{b}\right) = a$$

Quotient the solution is $x = a(1/b)$. In keeping with the customary notation and language, we call x the **quotient** of a and b and write

$$a \div b = a\left(\frac{1}{b}\right) = \frac{a}{b} = a \cdot b^{-1}$$

Example 13 $4 \div \dfrac{1}{2} = 4 \cdot \left(\dfrac{1}{2}\right)^{-1} = 4 \cdot 2 = 8$ ■

Notice that if $b \neq 0$ and $d \neq 0$, then

$$\frac{a}{b} \cdot \frac{c}{d} = (a)\left(\frac{1}{b}\right)(c)\left(\frac{1}{d}\right) = (ac)\left(\frac{1}{bd}\right) = \frac{ac}{bd}$$

Theorem 9 If $b \neq 0$ and $d \neq 0$, then $\dfrac{a}{b} \cdot \dfrac{c}{d} = \dfrac{ac}{bd}$.

Theorem 10 If $a \neq 0$ and $b \neq 0$, then $\left(\dfrac{a}{b}\right)^{-1} = \dfrac{b}{a}$.

Example 14 $12 \div \dfrac{3}{4} = 12 \cdot \left(\dfrac{3}{4}\right)^{-1} = 12 \cdot \dfrac{4}{3} = 16$ ■

The fact that we have been careful to develop some of the rules of arithmetic doesn't mean that your previous knowledge is being ignored. Rather, we want you to see how your experiences with numbers fit into the structure of the real numbers.

Theorem 11 If $a \neq 0$, then $-\frac{1}{a} = \frac{1}{-a}$.

Theorem 12 If $b \neq 0$, then

$$-\frac{a}{b} = \frac{-a}{b} = \frac{a}{-b} = -\frac{-a}{-b}$$

and

$$\frac{a}{b} = -\frac{-a}{b} = -\frac{a}{-b} = \frac{-a}{-b}$$

Example 15 $(-16) \div \left(-\frac{4}{3}\right) = (-16)\left(-\frac{3}{4}\right) = (16)\left(\frac{3}{4}\right) = 12$ ■

The distributive axiom is very useful in simplifying algebraic expressions. There will be more on this later, but for now let us examine a few simple examples.

Example 16 $2x + 5x = (2 + 5)x = 7x$ ■

Example 17

$$\begin{aligned} 2x - 3(x + 7) + 4(3 - x) &= 2x - 3x - 21 + 12 - 4x \\ &= (2 - 3 - 4)x - 9 \\ &= -5x - 9 \end{aligned}$$ ■

Collecting like terms The process illustrated in Examples 16 and 17 is called **collecting like terms.**

Exercises 1.1

Find each sum or difference in Exercises 1–18.

1. $(-2) + (-6)$
2. $(-9) + (-7)$
3. $-5 - 2$
4. $-8 - 5$
5. $-5 + 8$
6. $-7 + 16$
7. $42 - (12 - 45)$
8. $22 - (67 - 12)$
9. $-\frac{3}{4} - \left(\frac{5}{4} - \frac{1}{4}\right)$
10. $-\frac{1}{2} - \left(\frac{3}{2} - \frac{1}{2}\right)$
11. $3.4 - (3.2 - 4.7)$
12. $5.6 - (6.2 - 4.5)$
13. $(4 - 16) - (4 - 18)$
14. $(5 - 12) - (5 - 16)$
15. $[(1 + 4) - 5] - [3 - (1 + 5)]$
16. $[(3 + 2) - 7] - [9 - (2 + 6)]$
17. $[6.7 + (3.2 - 1.4)] - [2.3 - (3.5 + 1.2)]$
18. $[-7.9 + (1.2 - 5.6)] - [1.4 - (9.5 + 2.3)]$

Find each product or quotient in Exercises 19–42.

19. $(-2)(6)$
20. $(-6)(2)$
21. $(-9)(-7)$
22. $(-6)(-8)$
23. $(-2)(-5)(-1)$
24. $(-5)(-6)(-1)$

25. $-(-6)(7)$
26. $-(5)(-3)$
27. $-[-(-4)(-5)]$
28. $-[-(-8)(-5)]$
29. $6 \div (-3)$
30. $8 \div (-4)$
31. $(-15) \div (-3)$
32. $(-24) \div (-4)$
33. $\left(-\frac{1}{2}\right) \div \left(\frac{1}{3}\right)$
34. $\left(-\frac{2}{3}\right) \div \left(\frac{5}{3}\right)$
35. $(9 \div 3)(6)$
36. $(12 \div 3)(5)$
37. $4 \div (5 \div 6)$
38. $3 \div (5 \div 9)$
39. $\frac{2}{3} \div \frac{3}{4}$
40. $\frac{5}{6} \div \frac{12}{7}$
41. $\left(\frac{2}{3}\right)\left(\frac{7}{8}\right)\left(\frac{9}{7}\right)$
42. $\left(\frac{4}{5}\right)\left(\frac{3}{2}\right)\left(\frac{10}{9}\right)$

In Exercises 43–54 collect like terms.

43. $7v - 9v + 3v$
44. $6t - 2t + 3t$
45. $-4x + 7x - 3x$
46. $-7x + 9x - 2x$
47. $-5t + 2t - 8t$
48. $-9r + 7r - 5r$
49. $2s + 5 - (s - 1)$
50. $4c + 7 - (c - 2)$
51. $2w - 4(2w + 6)$
52. $8y - 3(2y + 5)$
53. $3(2s + 3t) - t$
54. $4(3r + 2w) - w$

55. If $x + 2$ is the additive inverse of 3, what is x?
56. If $x - 4$ is the additive inverse of 3, what is x?
57. What number is its own additive inverse? In other words, for what x is $x = -x$?
58. What number is its own multiplicative inverse? That is, for what x is $x = x^{-1}$?
59. Prove the cancellation law of addition: If a, b, and c are real numbers with $a + c = b + c$, then $a = b$. (Hint: add $-c$ to both $a + c$ and $b + c$.)
60. Prove the cancellation law of multiplication: If a, b, and c are real numbers with $c \neq 0$, and $ac = bc$, then $a = b$. (Hint: multiply both ac and bc by c^{-1}.)
61. Prove Theorem 2 by showing that $(-x) + (-y)$ is the additive inverse of $x + y$.
62. Prove Theorem 4 by showing that $(-a)(b)$ and $(a)(-b)$ are both the additive inverse of ab. Use the distributive law and Theorem 3.
63. Use Theorem 1 and Theorem 4 to prove Theorem 5.
64. Prove Theorem 10 by showing that $(a/b)(b/a) = 1$.
65. Prove Theorem 11 by showing that $(-a)\left(-\frac{1}{a}\right) = 1$. You may want to use Theorem 5.
66. Show by example that the operation of division is not associative, that is,

$$(a \div b) \div c = a \div (b \div c)$$

is *not* always true.
67. Show by example that the operation of subtraction is not associative, that is,

$$(a - b) - c = a - (b - c)$$

is *not* always true.

For Your Calculator

Your calculator has a key for addition, [+], a key for subtraction, [−], and a key to change the sign of a number, usually [+/−] or [chs]. The following exercises show how addition and subtraction of signed numbers are done by the calculator.

Example 1 $(-5.36) + (-2.1)$
Key sequence:

RPN 5.36 [chs] [enter] 2.1 [chs] [+] answer −7.46
Alg 5.36 [chs] [+] 2.1 [chs] [=] answer −7.46

Alternate approach:

RPN 5.36 [enter] 2.1 [+][chs] answer −7.46
Alg 5.36 [+] 2.1 [=][chs] answer −7.46

The alternate approach illustrates that

$(-a) + (-b) = -(a + b)$ ■

Example 2 2.45 − 3.76
Key sequence:

RPN 2.45 [enter] 3.76 [−] answer −1.31
Alg 2.45 [−] 3.76 [=] answer −1.31

Alternate approach:

RPN 2.45 [enter] 3.76 [chs][+] answer −1.31
Alg 2.45 [+] 3.76 [chs][=] answer −1.31

The alternate approach illustrates that

$a - b = a + (-b)$ ■

Use your calculator to find the sum or difference in Exercises 68–83.

68. $2.56 + (-9.7)$ **69.** $4.67 + (-6.5)$ **70.** $5.6 - (-6.7)$ **71.** $9.8 - (-2.3)$
72. $10.1 - 15.6$ **73.** $9.5 - 12.3$ **74.** $-2.3 - 4.5$ **75.** $-6.7 - 2.8$
76. $-(2.678 + 1.334)$ **77.** $-(1.908 + 2.405)$ **78.** $-(2.001 - 4.505)$ **79.** $-(1.002 - 5.404)$

Your calculator has a key for multiplication, [×], a key for division, [÷], and a key for finding reciprocals, [1/x]. Example 3 shows how to perform division using the [÷] key or the [1/x] key.

Example 3 Compute 16 ÷ 8.
Key sequence:

RPN 16 [enter] 8 [÷] answer 2
Alg 16 [÷] 8 [=] answer 2

Alternate approach:

RPN 16 [enter] 8 [1/x][×] answer 2
Alg 16 [×] 8 [1/x][=] answer 2

The alternate approach uses the fact that $a \div b = a(1/b)$. ■

Compute each of the quotients by both methods, using the [1/x] *key and the* [÷] *key.*

80. $(-6.4) \div (-.4)$ **81.** $(-4.5) \div (-.2)$ **82.** $(-8.2) \div (-2)$ **83.** $(-9.6) \div (-4)$

1.2 The Order Relation for Real Numbers

In addition to the arithmetic operations of addition and multiplication, there is an order relation on the system of real numbers. Those numbers corresponding to points to the right of 0 on the number line are called **positive** numbers and those to the left are called **negative** numbers.

Positive
Negative
Less than

A real number x is **less than** a real number y if there is a positive real number z such that $x + z = y$. In Figure 1.2 we see that this definition is equivalent to "x lies to the left of y on the number line." If x is less than y we write $x < y$.

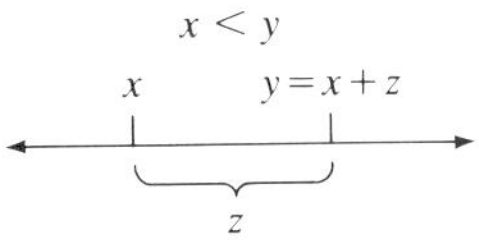

Figure 1.2

In this notation x is positive if $0 < x$ and negative if $x < 0$.

In many cases the order relation between numbers is easiest to recognize by looking at the number line.

Example 1 $5 < 8$ because 5 lies to the left of 8 on the line. ■

Example 2 $-6 < -5$ because -6 lies to the left of -5 on the line. ■

Greater than

We say that x is **greater than** y if and only if y is less than x, and we write $x > y$. In other words, $x > y$ and $y < x$ are just different ways of saying the same thing.

Example 3 $8 > 5 \qquad -5 > -6$ ■

The next two results follow from the geometric interpretation of the order relation on the number line.

Theorem 13 For any two real numbers x and y, exactly one of the following is true:

$$x < y \qquad x > y \qquad x = y$$

Theorem 14 If $x < y$ and $y < z$ then $x < z$.

Theorem 13 is often referred to as the trichotomy law, and Theorem 14 is called the law of transitivity.

The next four theorems relate the order relation to the arithmetic operations.

Arithmetic operations and the order relation

Theorem 15 If $x < y$, then $x + z < y + z$.

Theorem 16 If $x < y$ and $z > 0$, then $xz < yz$.

Theorem 17 If $x < y$ and $z < 0$, then $xz > yz$.

Theorem 18 If $0 < x < y$, then $0 < \frac{1}{y} < \frac{1}{x}$.

Example 4 If $x + 5 < 12$, then $x + 5 - 5 < 12 - 5$ or $x < 7$. ■

Example 5 If $4x < 28$, then since $0 < 4$, we have $0 < \frac{1}{4}$, and hence $\frac{1}{4}(4x) < \frac{1}{4}(28)$ or $x < 7$. ■

Example 6 To compare $-\frac{4}{5}$ and $-\frac{1}{5}$, write $-\frac{4}{5} = \frac{-4}{5}$ and $-\frac{1}{5} = \frac{-1}{5}$. Then $-4 < -1$ and $\frac{1}{5} > 0$, hence $-\frac{4}{5} = \frac{-4}{5} < \frac{-1}{5} = -\frac{1}{5}$ ■

Example 6 gives a way to compare two fractions: write the fractions with the same positive denominator; the fraction with the larger numerator is then the larger number.

Example 7 If $-4x + 1 > 13$, then $-4x > 12$ and $-\frac{1}{4} = \frac{-1}{4} < \frac{0}{4} = 0$, hence

$$-\frac{1}{4}(-4x) < -\frac{1}{4}(12)$$

or

$$x < -3$$ ■

It is convenient to have a slightly different version of the order relation from time to time as given below:

Less than or equal, greater than or equal notation

a is less than or equal b, written $a \leq b$, if $a < b$ or $a = b$.

a is greater than or equal b, written $a \geq b$, if $a > b$ or $a = b$.

The segment joining the origin 0 and the point corresponding to 1 on the number line determines the unit length on the number line. For example, the point corresponding to 6 is located 6 units to the right of the origin, and the number corresponding to -6 is located 6 units to the left of the origin. This means that real numbers determine **directed distances** on the number line. The absolute value of a number is the undirected distance of that number from the origin.

Absolute value

If a is any real number, the **absolute value** of a, written $|a|$, is defined to be the undirected distance from 0 to the point corresponding to a on the number line. So

$$|a| = \begin{cases} a & \text{if } a \geq 0 \\ -a & \text{if } a < 0 \end{cases}$$

Example 8 $|12| = 12$ since $12 > 0$

$|-5| = -(-5) = 5$ since $-5 < 0$ ■

Example 9 $5 - 2|6 - 15| - 6 = 5 - 2(9) - 6 = 5 - 18 - 6 = -19$ ■

Example 10 If $a < b$, then $a - b < 0$ and so $|a - b| = b - a$. ■

Exercises 1.2

In Exercises 1–8 relate the two numbers with the symbol <.

1. 6, 8
2. 7, 11
3. −2, −5
4. −7, −11
5. $-\frac{2}{7}, -\frac{1}{3}$
6. $-\frac{3}{7}, -\frac{2}{5}$
7. $\frac{4}{5}, \frac{5}{6}$
8. $\frac{2}{3}, \frac{3}{4}$

In Exercises 9–18 arrange the numbers in increasing order.

9. −1, 7, 3
10. −2, 4, −6
11. −2.3, −3.4, −5
12. −5.4, −9.2, −7
13. 1.1, 1.01, 1.001
14. 2.2, 2.02, 2.002
15. −2.02, −2.002, −2.2
16. −8.8, −8.08, −8.008
17. $\frac{2}{3}, -\frac{3}{4}, \frac{7}{12}, -\frac{4}{5}$
18. $\frac{1}{9}, -\frac{2}{7}, -\frac{3}{5}, \frac{2}{3}$

In Exercises 19–28 fill in the missing symbol that makes the statement true.

19. If $x < y$, then $x - 2$ ____ $y - 2$.
20. If $x < y$, then $x + 7$ ____ $y + 7$.
21. If $x < y$, then $-2x$ ____ $-2y$.
22. If $x < y$, then $-3x$ ____ $-3y$.
23. If $x > y$, then $4.2x$ ____ $4.2y$.
24. If $x > y$, then $.2x$ ____ $.2y$.
25. If $x > 2$, then $\frac{1}{x}$ ____ $\frac{1}{2}$.
26. If $x > 5$, then $\frac{1}{5}$ ____ $\frac{1}{x}$.
27. If $x < -2$, then $\frac{1}{x}$ ____ $-\frac{1}{2}$.
28. If $x < -5$, then $-\frac{1}{5}$ ____ $\frac{1}{x}$.

In Exercises 29–46 write the number without using absolute value symbols.

29. $|-7|$
30. $|-12|$
31. $|-2.3|$
32. $|-4.5|$
33. $|17 - 9|$
34. $|16 - 3|$
35. $|4 - 19|$
36. $|8 - 19|$
37. $\left|-\frac{3}{4}\right|$
38. $\left|-\frac{2}{3}\right|$
39. $\left|-\frac{1}{2} - \frac{3}{4}\right|$
40. $\left|\frac{2}{5} - \frac{3}{4}\right|$
41. $4 - 2|4 - 7| + 1$
42. $5 + 2|9 - 7| - 6$
43. $\frac{|2 - 14|}{3} - \frac{|15 - 9|}{|-2|}$
44. $\frac{|-12|\,|9 - 6|}{|7 - 3|} - \frac{|15 - 5|}{7 + |-3|}$
45. $|x + 9|$ if $x < -9$
46. $|y - 7|$ if $y > 7$

47. One of the following statements is true for all real numbers x and y.

$|x + y| = |x| + |y| \qquad |x + y| \le |x| + |y|$
$|x + y| \ge |x| + |y|$

By trying different values of x and y, determine which is the true statement.

48. Use the fact that $0 < 1$ to prove that if $x > 0$ then $x^{-1} > 0$.

49. Prove Theorem 18.

50. Show that $|ab| = |a|\,|b|$.

51. Show that $|a| = \sqrt{a^2}$.

52. If $0 < a < b$, show that $a^2 < b^2$.

1.3 Equations and Inequalities

Equation An **equation** is a statement concerning numbers and involving the phrase "is equal to." It can be a very simple statement such as $2 + 3 = 5$ or it can involve one or more letters representing numbers. These letters are usually referred to as **variables** or **unknowns.** A number is called a **solution** or **root** of an equation if substituting that number for the unknown yields a true statement. If an equation contains more than one unknown, a solution would require an assignment of a number to each unknown to yield a true statement. A solution is said to **satisfy** the equation, and finding the solution or solutions of an equation is called **solving** the equation. The set of solutions of an equation is called the **solution set,** and two equations are **equivalent** if they have the same solution set. Our strategy in this section will be to solve equations by transforming them into equivalent equations whose solutions are apparent.

Unknowns

Root

Solution set

There are two fundamental principles that will be used in dealing with equations:

The equations $A = B$ and $A + C = B + C$ are equivalent.

The equations $A = B$ and $AC = BC$ are equivalent if $C \neq 0$.

Example 1 Solve the equation $3x + 7 = 5x - 8$.

Add $-5x - 7$ to both members:

$$3x + 7 - 5x - 7 = 5x - 8 - 5x - 7$$
$$-2x = -15$$

Divide both members by -2:

$$x = \frac{15}{2}$$

It is apparent that the only solution to the last equation is the number $\frac{15}{2}$, which is thus the solution to the original equation. ■

Sometimes preliminary steps are necessary before using the principles given above.

Example 2 Solve the equation $2x - 3(x + 7) = x - 4(x + 2)$.
First, use the distributive law to remove parentheses and collect like terms:

$$\begin{aligned} 2x - 3x - 21 &= x - 4x - 8 \\ -x - 21 &= -3x - 8 \end{aligned}$$

Add $3x + 21$ to both members:

$$2x = 13$$

Multiply by $\frac{1}{2}$:

$$x = \frac{13}{2}$$ ■

Perhaps it would be a good idea to summarize the techniques used in Examples 1 and 2.

Step 1 Remove parentheses and collect like terms.

Step 2 Use addition to isolate the unknown.

Step 3 Divide by the coefficient of the unknown.

If you keep in mind the object of the game—to transform the equation into an equation whose solution is apparent—the steps used and the order in which they are used are unimportant. Just be sure that an equivalent equation results from each operation.

Example 3 Solve the equation $\frac{x}{4} + \frac{3}{2} = \frac{x}{5} + \frac{1}{2}(x + 1)$.
Multiply by 20 to clear fractions:

$$20\left(\frac{x}{4}\right) + 20\left(\frac{3}{2}\right) = 20\left(\frac{x}{5}\right) + 20\left(\frac{1}{2}\right)(x + 1).$$

$$5x + 30 = 4x + 10x + 10$$

Collect like terms:

$$5x + 30 = 14x + 10$$

Add $-14x - 30$:

$$\begin{aligned} 5x - 14x &= 10 - 30 \\ -9x &= -20 \end{aligned}$$

Divide by -9:

$$x = \frac{20}{9}$$ ■

Linear equations

So far we have been solving **linear equations,** which are equations that can be written in the form $ax + b = c$. The techniques used, however, can be applied to some equations that are not linear.

Example 4 Solve the equation $\frac{5}{x} - 3 = \frac{7}{x} - 4$.

Multiply by x:

$$5 - 3x = 7 - 4x$$

The second of the fundamental principles given above states that $A = B$ and $AC = BC$ are equivalent *if* $C \neq 0$. Here we multiply by x, so the equations are equivalent only if $x \neq 0$. Continue:

$$5 - 3x = 7 - 4x$$

Add $4x - 5$:

$$4x - 3x = 7 - 5$$
$$x = 2$$

■

Example 5 Solve the equation $2x + 3y = 7$ for y.
Treat y as the unknown and x as a number.

$$2x + 3y = 7$$

Add $-2x$:

$$3y = 7 - 2x$$

Divide by 3:

$$y = \frac{7 - 2x}{3}$$

■

Example 6 Solve the equation $|4x - 6| = 2$.
The number x is a solution to the equation if either

$$4x - 6 = 2 \quad \text{or} \quad 4x - 6 = -2$$

Thus there are two solutions to the equation, found by solving the two equations

$$4x - 6 = 2 \quad \text{and} \quad 4x - 6 = -2$$
$$4x = 8 \qquad\qquad 4x = 4$$
$$x = 2 \qquad\qquad x = 1$$

The two solutions to the equation are 1 and 2. ■

The principles used in solving inequalities are very similar to those used in solving equations.

Equivalent inequalities

The inequalities $A < B$ and $A + C < B + C$ are equivalent.

The inequalities $A < B$ and $AC < BC$ are equivalent if $C > 0$.

The inequalities $A < B$ and $AC > BC$ are equivalent if $C < 0$.

Recall that $A \leq B$ means that $A < B$ or $A = B$. Consequently the principles given above may be rewritten to include the possibility of equality by using the principles about equations.

The inequalities $A \leq B$ and $A + C \leq B + C$ are equivalent.

The inequalities $A \leq B$ and $AC \leq BC$ are equivalent if $C > 0$.

The inequalities $A \leq B$ and $AC \geq BC$ are equivalent if $C < 0$.

Let us illustrate these principles with some numerical examples.

Example 7 $2 < 3$, hence $2 + 6 = 8 < 9 = 3 + 6$ ■

Example 8 $2 < 3$ and $4 > 0$, hence $2 \cdot 4 = 8 < 12 = 3 \cdot 4$ ■

Example 9 $2 < 3$ and $-4 < 0$, so $-4 \cdot 2 = -8 > -12 = -4 \cdot 3$ ■

Example 10 Solve the inequality $\dfrac{2x + 3}{6} \leq \dfrac{x + 3}{2}$.

Multiply by 6:

$$2x + 3 \leq 3(x + 3)$$
$$2x + 3 \leq 3x + 9$$

Add $-3x - 3$:

$$2x - 3x \leq 9 - 3$$
$$-x \leq 6$$

Multiply by -1:

$$x \geq -6 \qquad \text{(Remember, this reverses inequality.)}$$ ■

Example 11 Solve the inequalities $-5 < 3x - 2 < 4$.

In reality there are two inequalities, but they can be solved simultaneously:

$$-5 < 3x - 2 < 4$$

Add 2:

$$-5 + 2 < 3x < 4 + 2$$
$$-3 < 3x < 6$$

Divide by 3:

$$-1 < x < 2$$ ■

Example 12 Solve the inequality $3x + 6 < 2x - 7$.
Add $-6 - 2x$:

$$\begin{aligned} 3x - 2x &< -7 - 6 \\ x &< -13 \end{aligned}$$

■

Set-builder notation

Since the solution set of an inequality usually contains more than one number, it is often convenient to use **set-builder notation** to describe the solution set.

If $P(x)$ is a statement concerning the object x, then

$$\{x: P(x)\}$$

is the set of all objects x such that $P(x)$ is a true statement.

Example 13 Let $P(x)$ be the statement "x is a positive integer less than 6." Then

$$\begin{aligned} \{x: P(x)\} &= \{x: x \text{ is a positive integer less than } 6\} \\ &= \{1, 2, 3, 4, 5\} \end{aligned}$$

■

Example 14 $\{x: x < -13\}$ is the set of all real numbers less than -13 and is the solution set for the inequality in Example 12. ■

Example 15 Solve the inequality $|2 - 3x| > 5$.
A solution to this inequality is any solution to either

$$\begin{aligned} 2 - 3x &< -5 \quad \text{or} \quad & 2 - 3x &> 5 \\ -3x &< -7 & -3x &> 3 \\ x &> \frac{7}{3} & x &< -1 \end{aligned}$$

The solution set is $\{x: x > \frac{7}{3} \text{ or } x < -1\}$. ■

Another way of describing the solution set of an inequality is to show the set as a subset of the number line.

Example 16 Show the set $\{x: x > \frac{7}{3} \text{ or } x < -1\}$ on the number line.

The parenthesis, either (or), is used to indicate that the endpoint is not included. ■

Example 17 Show the set $\{x: -2 < x \leq 5\}$ on the number line.

The parenthesis, (, indicates that -2 is not included, and the bracket,], indicates that 5 *is* included. ■

Recall that the absolute value of a number is the undirected distance from the origin to the point on the number line corresponding to that number. Thus the inequality $|a| > 5$ is true if and only if either $a < -5$ or $a > 5$. Similarly, $|a| < 5$ is true if and only if a is between -5 and 5, that is, if $-5 < a < 5$.

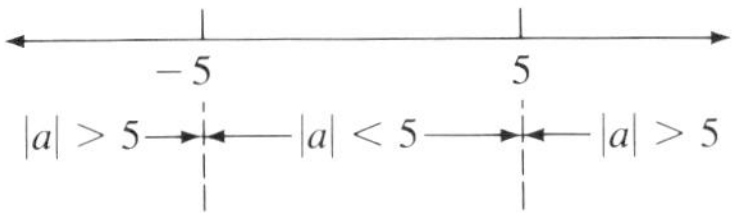

Example 18 Solve the inequality $|x| < 4$.
The inequality $|x| < 4$ is true if and only if the point on the number line corresponding to x is less than 4 units from the origin; that is, if and only if $-4 < x < 4$. The solution set is $\{x\colon -4 < x < 4\}$. ■

Example 19 Solve the inequality $|5 - 3x| \le 7$.
This is equivalent to the two inequalities

$$-7 \le 5 - 3x \le 7$$

$$-12 \le -3x \le 2$$

$$4 \ge x \ge -\frac{2}{3} \qquad \textit{(Dividing by } -3 \textit{ reverses the inequalities.)}$$

The solution set is $\{x\colon -\frac{2}{3} \le x \le 4\}$, or graphically

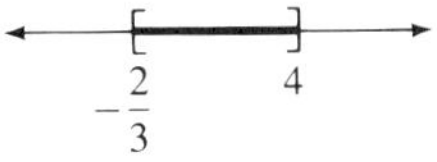

■

Intervals on the number line such as those in Examples 18 and 19 occur often enough to deserve notation. Note how the bracket and parenthesis are used to indicate whether an endpoint is included or excluded.

For any real numbers a and b with $a < b$, we define

$(a, b) = \{x\colon a < x < b\}$

$(a, b] = \{x\colon a < x \le b\}$

$[a, b) = \{x\colon a \le x < b\}$

$[a, b] = \{x\colon a \le x \le b\}$

Exercises 1.3

In Exercises 1–26 solve the equation.

1. $22 - 8x = -15$

2. $14 - 7x = -16$

3. $1.2x + 3.6 = 2.4$

4. $1.5a + 4.5 = -3$

5. $2(x - 3) = x + 5$

6. $4(x - 4) = 3x + 1$

7. $3(x + 1) - 2x = 4(x - 1) - 2$

8. $4(x + 1) - 3x = 3(x - 1) - 3$

9. $3(x + 1) - 9 = 2(x + 1) - 6$

10. $4(x - 2) + 8 = 3(x - 2) + 5$

11. $15 - (3x - 2) = 14 - (2x + 7)$

12. $-3 - (9 - 2x) = 6 - (14 - 3x)$

13. $-2(3x + 7) - 3 = 8(2x - 1) - 9$

14. $-7(3x + 1) - 5 = 2(9x + 1) - 14$

15. $\frac{3}{4}(x + 2) - \frac{9}{4} = \frac{1}{2}(x + 2) - \frac{3}{2}$

16. $\frac{2}{3}(x - 1) + \frac{4}{3} = \frac{1}{2}(x - 1) + 5$

17. $\frac{3}{20}(x - 1) - \frac{1}{4} = \frac{6}{5} - \frac{1}{10}(2x + 2)$

18. $\frac{1}{7}(x - 1) - \frac{1}{14} = \frac{3}{2} + \frac{2}{7}(3x - 1)$

19. $.3(2x - 1) + .6 = .3(2x + 1) - .4x$

20. $.4(3x - 2) + .1 = .2(x - 5) + .3$

21. $\frac{3}{x} - 6 = \frac{2}{x} + 1$

22. $\frac{2}{x} + 2 = \frac{4}{x} - 3$

23. $|3x - 7| = 9$

24. $|4x - 6| = 10$

25. $\left|5 - \frac{r}{6}\right| = 1$

26. $\left|3 - \frac{s}{4}\right| = 1$

Solve for the indicated unknown in Exercises 27–30.

27. $2x - 3s = 5$, for s

28. $2r - 5t = 8$, for r

29. $S = 5 + (n - 1)t$, for n

30. $Q = (n + 2)d - 6$, for n

In Exercises 31–48 find the solution set of the inequality or inequalities.

31. $3x - 10 < 5x + 2$

32. $2x - 9 < 4x + 1$

33. $x + 12 \geq 4x - 2$

34. $x + 4 \geq 3x - 1$

35. $\frac{3s + 9}{4} \leq \frac{2s - 7}{3}$

36. $\frac{5t + 7}{3} \leq \frac{2t - 5}{8}$

37. $3(x - 4) < 5(x - 6)$

38. $5(u - 3) < 3(u - 1)$

39. $\frac{x}{3} - 2 > \frac{5x + 9}{2}$

40. $\frac{y}{2} - 3 > \frac{5y + 9}{3}$

41. $4(m + 1) - 2(m - 1) \geq 0$

42. $3(n + 1) - 2(n - 1) \geq 0$

43. $2 < 3x + 6 < 5$

44. $3 < 2x + 5 < 8$

45. $2 - x < 3 < 8 - x$

46. $4 - r < 6 < 8 - r$

47. $x + 5 \leq 2x + 3 \leq x + 9$

48. $w + 2 \leq 2w + 1 \leq w + 6$

In Exercises 49–56 solve the inequality and show the solution set on the number line.

49. $|2x - 1| \leq 3$

50. $|2x - 5| \leq 4$

51. $|3x + 16| < 8$

52. $|5x + 17| < 7$

53. $|5x + 1| \geq 4$

54. $|3x + 2| \geq 7$

55. $\left|\frac{x}{3} - 5\right| > 4$

56. $\left|\frac{s}{4} - 7\right| > 9$

In Exercises 57–60 solve the inequalities for the indicated letter.

57. $S < (n - 2)t + 6$, for n, $t > 0$

58. $S < (2 - n)t - 6$, for n, $t > 0$

59. $-v < k + gt < v$, for t, $g < 0$

60. $-v < k - gt < v$, for t, $g > 0$

1.4 Problem Solving

The first and often the most important step in solving a problem is to translate the pertinent facts into a mathematical statement, usually either an equation or an inequality. In this section we examine some strategies for problem solving, that is, strategies for representing problems with equations or inequalities and thus finding solutions.

George is in charge of a motorcycle shop and receives a shipment of 24 helmets without an itemized invoice. He knows that the total charge for the shipment was \$926 and the freight (included in the total cost of \$926) was \$4.40. Can you help George compute the cost per helmet, not including freight?

This problem is not a difficult one to solve, but let us use it to illustrate the general techniques to be used in problem solving.

Step 1 First read the problem *very* carefully; then read it again. *Determine the unknown that you wish to find to solve the problem.* Here it is the cost per helmet. Use the letter c for this unknown quantity:

c = cost per helmet

Step 2 Determine what conditions this unknown quantity must satisfy and formulate these conditions in mathematical language; that is, *write the equations or inequalities the unknown must satisfy.* Since 24 helmets were ordered, $24c$ is the cost of the helmets, and $24c + 4.40$ is the cost of the helmets plus the freight, that is, the total cost. In the language of mathematics

$$\begin{aligned} \text{cost of helmets} + \text{freight} &= \text{total cost} \\ 24c \qquad + \quad 4.40 &= 926 \end{aligned}$$

Step 3 *Solve the equation or inequalities.*

$$\begin{aligned} 24c + 4.40 &= 926 \\ 24c &= 921.60 \\ c &= 38.40 \end{aligned}$$

Step 4 *Check and interpret the solution.* Take the number (or numbers in some cases) you found in Step 3 and see if it is a solution to the problem. Don't just check it in the equation; go back to the original problem. Based on the information in Step 3, the cost per helmet must be 38.40. To check,

$$\begin{array}{lcl} \text{cost of helmets} & + \text{ freight} & = \text{total cost} \\ 24(38.40) & + \ 4.40 & \\ 921.60 & + \ 4.40 & \\ & 926.00 & = 926 \end{array}$$

Our solution is correct; the cost per helmet is \$38.40.

You drive from Flintrock Springs to Roadrunner State Park along I-690 at 88 km/hr and return along Route 49 to Flintrock Springs at 72 km/hr. If each trip takes the same amount of time and the total distance is 80 km, how long does each trip take?

Step 1 The time for the trip each way is the unknown quantity to be determined. Let

t = time in hours for the trip one way

Step 2 Distance equals rate times time, so

$88t$ = distance along I-690

$72t$ = distance along Route 49

and

$$\begin{array}{cccc} \text{distance along I-690} & \text{plus} & \text{distance along Rt. 49} & = \text{total distance} \\ 88t & + & 72t & = \quad 80 \end{array}$$

Step 3 Solve the equation:

$$\begin{aligned} 88t + 72t &= 80 \\ 160t &= 80 \\ t &= \frac{1}{2} \end{aligned}$$

Step 4 Check and interpret the solution. From Step 3, $t = 1/2$. So

$$\begin{array}{cccc} \text{distance along I-690} & + & \text{distance along Rt. 49} & = \text{total distance} \\ 88\left(\frac{1}{2}\right) & + & 72\left(\frac{1}{2}\right) & \\ 44 & + & 36 & \\ & & 80 & = \quad 80 \end{array}$$

All is well! The time for each part of the trip is 1/2 hour.

Some of the problems may seem a bit silly or impractical, but the point is to give you an opportunity to practice setting up equations and inequalities for simple problems before tackling more difficult situations. People don't attempt to learn to drive during rush hours on very busy streets. Rather, the fundamental skills are usually learned and practiced under simple circumstances, on nearly vacant roads or streets. Then as the basic skills are acquired, more difficult situations can be approached. The same is true of mathematics.

There may be more than one unknown quantity in a problem. But if a choice of an unknown is made for one of the desired quantities, usually there is enough information to describe the other quantity in terms of that unknown. Follow the next example.

Example 1 The perimeter of a rectangle is 24 cm and the length is twice the width. What are the dimensions?

Step 1 There are two unknown quantities: the length and the width. Let

w = width of the rectangle

Since the length is twice the width,

$2w$ = length of the rectangle

Step 2 The perimeter of a rectangle is the sum of the lengths of all four sides:

$$\begin{array}{ccccccccc} \text{length} & + & \text{length} & + & \text{width} & + & \text{width} & = & \text{perimeter} \\ 2w & + & 2w & + & w & + & w & = & 24 \end{array}$$

Step 3

$$\begin{aligned} 2w + 2w + w + w &= 24 \\ 6w &= 24 \\ w &= 4 \text{ cm} \end{aligned}$$

Step 4 The width of the rectangle is 4 cm and its length (twice the width) is 8 cm. This checks because the perimeter of such a rectangle is 24 cm. ■

Example 2 Suppose you wish to mix peanuts worth 80¢ per pound with cashews worth \$1.20 per pound to make 80 pounds of a mixture worth 96¢ per pound. How many pounds of each should you use?

Step 1 Let

x = number of pounds of peanuts to use

Then

$80 - x$ = number of pounds of cashews to use

Step 2

$$\begin{array}{ccccc} \text{value of peanuts} & + & \text{value of cashews} & = & \text{value of mixture} \\ .80x & + & 1.20(80 - x) & = & .96(80) \end{array}$$

Step 3

$$\begin{aligned} .8x + 1.2(80 - x) &= .96(80) \\ .8x + 96 - 1.2x &= 76.80 \\ -.4x &= -19.20 \\ x &= 48 \text{ pounds} \\ 80 - x &= 32 \text{ pounds} \end{aligned}$$

Step 4 You should use 48 pounds of peanuts with a value of

$48(.8) = 38.40$

and 32 pounds of cashews with a value of

$32(1.2) = 38.40.$

The total value is then

$$38.40 + 38.40 = 76.80$$

as desired. ■

Example 3 Eve has scores of 90, 88, and 72 on three tests. What score must she earn on the fourth test to have an average of at least 90?

Step 1 Let

x = score on the fourth test

Step 2 Eve's average will be

$$\frac{90 + 88 + 72 + x}{4}$$

and the requirement is that this be at least 90. The inequality is

$$\frac{90 + 88 + 72 + x}{4} \geq 90$$

Step 3 Solve the inequality:

$$\frac{90 + 88 + 72 + x}{4} \geq 90$$

$$\begin{aligned} 90 + 88 + 72 + x &\geq 360 \\ 250 + x &\geq 360 \\ x &\geq 110 \end{aligned}$$

Step 4 If Eve scores 110 on the fourth test, her average will be

$$\frac{90 + 88 + 72 + 110}{4} = \frac{360}{4} = 90$$

If she scores higher than 110, her average will be higher than 90. ■

We have outlined the basic steps to follow in attacking problems. The beauty of the mathematician's approach should now be apparent. If we can write an equation or inequality that accurately describes the problem, then we have stripped the problem down to its essence and can proceed to seek the solution without being bothered by unnecessary details such as the color of the motorcycle helmets or the state in which we find I-690 or Route 49.

Exercises 1.4

Solve the problems.

1. David and Pat share a paper route and David delivers twice as many papers as does Pat. If they deliver 81 papers together, how many papers does David deliver?

2. Hatch has a population 3 times that of Rincon. If the difference in populations is 1,524, what is the population of each?

3. The length of a rectangle is 10 less than 5 times

the width. The perimeter is 11 times the width. Find the dimensions.

4. Mary weighs 10 kg more than Jean and together they weigh 80 kg. What is the weight of each?

5. Carol is 3 years younger than Mona and the sum of their ages is 71. How old is Carol?

6. Lea and Angie play 100 sets of tennis and Lea wins 14 more sets than Angie wins. How many sets did Lea win?

7. Drusilla receives a bonus check from her employer for $455. She knows that 30% has been withheld for taxes. What was the amount of the bonus before taxes?

8. Mrs. Hay is 3 times as old as her son, and the difference in their ages is 28. How old is Mrs. Hay?

9. Mike and Jerry raise rabbits, with Jerry owning $\frac{3}{5}$ of the rabbits and Mike owning $\frac{2}{5}$ of the rabbits. If Jerry owns 3 more rabbits than Mike, how many do they have together?

10. Evelyn started with an annual salary of $9,000 with an annual increase of $600. At the same time Susan started with an annual salary of $7,200 with an annual raise of $900. When will their salaries be the same?

11. Records are on sale for $1.25 off list price. At the regular price 4 records cost the same as 6 records at the sale price. What is the sale price?

12. Bertha sells orange juice for 30¢ per glass and Coke for 20¢ per glass. By the end of the day she has sold 400 drinks and taken in $95. How many glasses of orange juice did she sell?

13. Eve cuts a piece of ribbon into two pieces of equal length and then cuts off $\frac{1}{3}$ of one of the pieces. If the remaining piece of the ribbon is 8 in. long, how long was the original piece of ribbon?

14. Jack tells his wife Betty to sell half of his books. Before she can accomplish this, he changes his mind and keeps 50 of the books she was going to sell. She complains that this means that he is only going to give up one-third of his books. How many books does Jack have?

15. Warren has $6,000 invested, part at 6% and part at 8%. If his interest earned for the year is $440, how much does he have invested at each rate?

16. Anne and Alfred have $9,100 invested, part at 8% and part at 6%. If the income from each investment is the same, how much is invested at each rate?

17. Joan has $2,000 invested, part at 5% and part at 7%. Her interest income for the year is $125. How much does she have invested at each rate?

18. Jack bought a sailboat and paid 25% down. The remainder plus 9% interest is due at the end of 2 years. If the interest due at the end of 2 years is $3,240, how much did the boat cost?

19. Craig has borrowed money from two banks. The first loan is at 9% and the second loan is at 10%. If the amount of the first loan is $400 more than the amount of the second loan, and the interest due at the end of the year is $720, how much did Craig borrow?

20. James earns $40 each day he works and forfeits $3 for each day he does not work. If he nets $456 at the end of a 20-day period, how many days did he work?

21. Barbara's pay for 5 hr and Karen's pay for 7 hr totaled $70.50. If each had worked 2 more hours, their pay would have been $93.50. What is the hourly rate for each?

22. How much nonfat dry milk that is 21% protein should be mixed with whole milk that is 9% protein to produce 2 liters of milk that is 12% protein?

23. Eve has 50 coins, all nickels and dimes, amounting to $3.50. How many nickels does she have?

24. How much water should be added to 8 liters of 40% acid solution to yield a 16% acid solution?

25. A cooling system has a capacity of 24 qt. If the solution in the system is 40% antifreeze, how much antifreeze must be added to obtain a solution that is 50% antifreeze? (Be careful! The cooling system will hold only 24 qt, so to add a quart of antifreeze you must drain out a quart of the 40% solution.)

26. A butcher has ground round worth $2.00/lb and ground chuck worth $1.20/lb. How many pounds of each should he mix together to obtain 20 lb of hamburger worth $1.52 per pound?

27. Many people like to keep their thermostats between 68° and 72° Fahrenheit during the winter months. What is the corresponding range on the Celsius scale? (The formula for F, degrees Fahrenheit, in terms of C, degrees Celsius, is $F = \frac{9}{5}C + 32$.)

28. If the minimum speed on the highway is 40 mi/hr and the maximum speed is 55 mi/hr, what is the range of time necessary to make a trip of 275 mi?

1.5 Polynomials

The relationship between C, temperature measured in °C, and F, temperature measured in °F, is given by the formula $F = \frac{9}{5}C + 32$. The systolic blood pressure, the pressure during the contraction of the heart, is approximated by the formula $P = 0.5A + 110$, where A is the subject's age and P is the pressure.

Monomial
Polynomial

Expressions like $\frac{9}{5}C + 32$ and $0.5A + 110$ occur in many real-life situations. These are examples of polynomials. A **monomial** is a number, a variable, or a product of numbers and variables. A **polynomial** is a monomial or a sum of monomials. The monomials whose sum forms the polynomial are called the **terms** of the polynomial. A polynomial with two terms is called a **binomial,** and one with three terms is called a **trinomial.** The expression $r^3 + r^2 - 3r + 4$ is a polynomial with terms r^3, r^2, $-3r$, and 4. The term 4, which does not involve any variable, is called the **constant term.** In the term $-3r$, the number -3 is called the **coefficient** of r. This polynomial may be referred to as a polynomial **in** r.

Terms
Constant term
Coefficient

A polynomial may be considered to be a polynomial in any of the letters occurring in any of the terms. For example, the polynomial

$$ax^2 + bxy + cy^2 + a^2$$

may be considered to be a polynomial in x, a polynomial in y, a polynomial in a, a polynomial in b, a polynomial in c, or a polynomial in several of the variables—for example, in x and y.

Degree

The **degree** of a monomial in x is the exponent on the variable x. The **degree** of a monomial in several variables is the sum of the exponents on the variables. The **degree** of a polynomial is the degree of the term with highest degree.

Example 1 $3x^2y$ is a monomial, the coefficient of x^2y is 3, the degree in x is 2, the degree in y is 1 and the degree in x and y is 3. ■

Example 2 $4x^2 + 6x - 7$ is a polynomial, -7 is the constant term, and the degree of the polynomial is 2. ■

Example 3 $3x^4y + 2x^2 + y^3$ is of degree 4 in x, degree 3 in y, and degree 5 in x and y. ■

P(x) notation

It is often convenient to specify a polynomial with a letter. We will use the symbolism $P(x)$ to denote a polynomial in x and $P(x, y)$ to denote a polynomial in x and y. There is nothing magic about the letter P; we could equally well use $Q(a, b, c)$ to denote a polynomial in a, b, and c.

Value of a polynomial

If $P(x)$ is a polynomial and a is a number or variable, then $P(a)$ is the **value** of P at a; that is, it is the number or expression obtained by replacing the variable x by a in the polynomial. Similar notation is used for polynomials in several variables.

Example 4 If $P(x) = 2x^2 - 3x + 5$, then

$$P(2) = 2(2)^2 - 3(2) + 5 = 7$$
$$P(0) = 2(0)^2 - 3(0) + 5 = 5$$
$$P(a) = 2a^2 - 3a + 5$$
■

Example 5 If $R(a, b) = -a^2b + ab^2 - 14$, then

$$R(0, -1) = -(0)^2(-1) + (0)(-1)^2 - 14 = -14$$
$$R(2, -3) = -(2)^2(-3) + 2(-3)^2 - 14 = 16$$ ■

The distributive law is useful in a variety of ways in dealing with polynomials.

Example 6 $2x(x + 5) = 2x(x) + 2x(5) = 2x^2 + 10x$ ■

Example 7
$$\begin{aligned} -4x(2x + 3y) &= -4x(2x) + (-4x)(3y) \\ &= -8x^2 - 12xy \end{aligned}$$ ■

Example 8
$$\begin{aligned} (x + 2)(x - 2) &= x(x - 2) + 2(x - 2) \\ &= x^2 - 2x + 2x - 4 \\ &= x^2 - 4 \end{aligned}$$ ■

Special product The pattern in Example 8 illustrates what is called a **special product.**

Example 9
$$\begin{aligned} (x - y)(x + y) &= x(x + y) - y(x + y) \\ &= x^2 + xy - xy - y^2 \\ &= x^2 - y^2 \end{aligned}$$ ■

Example 10
$$\begin{aligned} (a - 9)(a + 9) &= a^2 - 9^2 \quad \textit{(from Example 9)} \\ &= a^2 - 81 \end{aligned}$$ ■

Example 11
$$\begin{aligned} (x + y)(x + y) &= x(x + y) + y(x + y) \\ &= x^2 + xy + xy + y^2 \\ &= x^2 + 2xy + y^2 \end{aligned}$$ ■

Example 12
$$\begin{aligned} (2x - 3y)(2x - 3y) &= (2x)^2 + 2(2x)(-3y) + (-3y)^2 \quad \textit{(from Example 11)} \\ &= 4x^2 - 12xy + 9y^2 \end{aligned}$$ ■

The types of special products in Examples 9 and 11 deserve to be highlighted. They are very useful in computing products and also in factoring polynomials.

1. $(x + y)(x - y) = x^2 - y^2$
2. $(x + y)^2 = (x + y)(x + y) = x^2 + 2xy + y^2$
3. $(x - y)^2 = (x - y)(x - y) = x^2 - 2xy + y^2$

Example 13 $(3x - 6y)(3x + 6y) = 9x^2 - 36y^2$ *(See 1 above.)* ■

Example 14 $(5x + y)^2 = 25x^2 + 10xy + y^2$ *(See 2 above.)* ■

Example 15 $(2s - 7t)^2 = 4s^2 - 28st + 49t^2$ *(See 3 above.)* ■

We can add one more special product to the list to deal with products like $(x + a)(x + b)$:

4. $(x + a)(x + b) = x^2 + (a + b)x + ab$

Example 16 $$\begin{aligned}(x + 2)(x + 4) &= x^2 + (2 + 4)x + (2)(4) \\ &= x^2 + 6x + 8\end{aligned}$$ ■

Example 17 $$\begin{aligned}(2x - 3y)(4x + 5y) &= 2x(4x + 5y) - 3y(4x + 5y) \\ &= 8x^2 + 10xy - 12xy - 15y^2 \\ &= 8x^2 - 2xy - 15y^2\end{aligned}$$ ■

Sometimes it is easier to carry out all the multiplications and then collect terms than to put the product into one of the four special forms given above. This is illustrated in Example 17.

There is a method called the FOIL method that can be useful in remembering how to find the product of two binomials. We illustrate with the next example.

Example 18 Find the product $(3x - 2)(4x + 5)$ by the FOIL method.

$$(3x - 2)(4x + 5) = \overset{\text{F}}{(3x)(4x)} + \overset{\text{O}}{(3x)(5)} + \overset{\text{I}}{(-2)(4x)} + \overset{\text{L}}{(-2)(5)}$$

Note that F represents the product of the *first* terms of the two binomials, O represents the product of the *outer* terms, I represents the product of the *inner* terms, and L represents the product of the *last* terms of the binomials. To continue,

$$\begin{aligned}(3x - 2)(4x + 5) &= (3x)(4x) + (3x)(5) + (-2)(4x) + (-2)(5) \\ &= 12x^2 + 15x - 8x - 10 \\ &= 12x^2 + 7x - 10\end{aligned}$$ ■

Example 19 Find the product $(4x - 7y)(2x + 3y)$.

$$\begin{aligned}(4x - 7y)(2x + 3y) &= \overset{\text{F}}{(4x)(2x)} + \overset{\text{O}}{(4x)(3y)} + \overset{\text{I}}{(-7y)(2x)} + \overset{\text{L}}{(-7y)(3y)} \\ &= 8x^2 + 12xy - 14xy - 21y^2 \\ &= 8x^2 - 2xy - 21y^2\end{aligned}$$ ■

In computing some products, you must use the distributive law several times.

Example 20 $$\begin{aligned}(x + y)^3 &= (x + y)(x + y)^2 \\ &= (x + y)(x^2 + 2xy + y^2) \\ &= x(x^2 + 2xy + y^2) + y(x^2 + 2xy + y^2) \\ &= x^3 + 3x^2y + 3xy^2 + y^3\end{aligned}$$ ■

Example 21 $(x - y)^3 = (x + (-y))^3$
$= x^3 + 3x^2(-y) + 3x(-y)^2 + (-y)^3$
$= x^3 - 3x^2y + 3xy^2 - y^3$ ■

Exponents In Examples 20 and 21 we used a property of **exponents** that we should review here. If x is any real number and n is a positive integer, then

$$x^n = x \cdot x \cdots x \quad (n \text{ factors})$$

Example 22 $4^3 = 4 \cdot 4 \cdot 4 = 64$ ■

Example 23 $(-2)^5 = (-2)(-2)(-2)(-2)(-2) = -32$ ■

If m and n are both positive integers and x is any real number, then

Multiplication with exponents

$$x^n x^m = \underbrace{(x \cdot x \cdots x)}_{n \text{ factors}}\underbrace{(x \cdot x \cdots x)}_{m \text{ factors}} = \underbrace{x \cdot x \cdots x}_{m+n \text{ factors}} = x^{m+n}$$

Example 24 $x^6x^5 = x^{6+5} = x^{11}$ ■

Example 25 $xy^2x^2y^3 = x(x^2)(y^2y^3)$
$= x^{1+2}\, y^{2+3}$
$= x^3y^5$ ■

Example 26 $3x^2y(2xy^2 - xy + y) = 3x^2y(2xy^2) + 3x^2y(-xy) + 3x^2y(y)$
$= 6x^3y^3 - 3x^3y^2 + 3x^2y^2$ ■

Exercises 1.5

Determine the degree of the polynomials in Exercises 1–6.

1. $2u^2 - 3u + 1$ **2.** $4y - 4y^3 + y^2$ **3.** $ab + 4a^3 - b^2$
4. $m^2n + 2m + n^3$ **5.** $3 - 5y^7$ **6.** $12 - s^4$

Given $R(x) = -x^3 + 7x^2 - 12x + 1$, *find*

7. $R(3)$ **8.** $R(4)$ **9.** $R(-3)$ **10.** $R(-4)$ **11.** $R(-a)$ **12.** $R(2a)$

Given $Q(x, y) = x^2 - 2xy - y^2$, *find*

13. $Q(2, -4)$ **14.** $Q(-2, 4)$ **15.** $Q(0, 6)$ **16.** $Q(6, 0)$

Simplify the polynomials in Exercises 17–24 by collecting like terms.

17. $-(2x + 6) + 6(2 - x)$

18. $-(5 + 3a) + 4(3 - a)$

19. $(a + b - 3c) - 2(a + 3b - c)$

20. $(m + n - 4p) - 3(m + 3n - 3p)$

21. $a - 3(a - b) - 2(b - a)$

22. $-3y - 2(x + y) - 4(2y - x)$

23. $s + t - [(s - 3t) - (s + 4t)]$

24. $u - v - [(v - 3u) - (v + 4u)]$

In Exercises 25–78 write each product as a polynomial.

25. $x(x - 2y)$

26. $y(2x - y)$

27. $3u(u + v)$

28. $4r(r + t)$

29. $-xy(x - y)$

30. $-rt(t - r)$

31. $(r - 4t)(r + 4t)$

32. $(u + 6v)(u - 6v)$

33. $(9 - p)(9 + p)$

34. $(4 + x)(4 - x)$

35. $(2t + 3s)(3s - 2t)$

36. $(3n + 4m)(4m - 3n)$

37. $(x - 9)^2$

38. $(t - 2)^2$

39. $(2x + 7y)^2$

40. $(5u + 3v)^2$

41. $(5t - s)^2$

42. $(6m - n)^2$

43. $(4t + 3r)(3r + 4t)$

44. $(7u + 4v)(4v + 7u)$

45. $(w + 4)(w + 7)$

46. $(r + 2)(r + 7)$

47. $(b - 5)(b + 1)$

48. $(h - 6)(h + 3)$

49. $(3r - 7t)(2r + t)$

50. $(3x - y)(2x + 5y)$

51. $(4t + 5s)(2t + 7s)$

52. $(5a + 6b)(2a + 7b)$

53. $(4n - 3m)(9n - 2m)$

54. $(5a - 3b)(3a - 8b)$

55. $(w - 2)^3$

56. $(u - 3)^3$

57. $(x + 2y)^3$

58. $(s + 3t)^3$

59. x^4x^7

60. a^6a^8

61. $xy^3(x^2y - xy)$

62. $s^4t(st^2 - st)$

63. $3m^3n^2(mn - 2n)$

64. $6x^5y(-2xy^2 + 4y)$

65. $(a - 2)(a^2 + 2a + 4)$

66. $(s - 3)(s^2 + 3s + 9)$

67. $(m + 4)(m^2 - 4m + 16)$

68. $(u + 3)(u^2 - 3u + 9)$

69. $(x^2 - 1)(x^2 + 1)$

70. $(a^2 - 2)(u^2 + 2)$

71. $(xy^2 - 5)(xy^2 + 5)$

72. $(a^2b + 4)(a^2b - 4)$

73. $-2[x^2 - (x - y)^2]$

74. $-3[(r - s)^2 - s^2]$

75. $(a - b)^2 - (a + b)^2$

76. $(2s - r)^2 - (2r + s)^2$

77. $(x + y - 2z)^2$

78. $(a - 2b + c)^2$

79. If $P(x) = 3x^2 - 4x + 9$, write $P(x + 2)$ as a polynomial.

80. If $Q(y) = 4y^2 - 3y + 8$, write $Q(x - 2)$ as a polynomial.

81. If $R(x) = 3x^2 - 2x$, write $R(2 + h) - R(2)$ as a polynomial.

82. If $R(x) = 3x^2 - 2x$, write $R(2 - h) - R(2)$ as a polynomial.

1.6 Factoring

Have you ever tried to disassemble something that someone else had put together? It's often not easy without some instructions. However, if the gadget is something with which you have had previous experience, the task can be much easier. For example, an experienced automobile mechanic could probably disassemble any carburetor without having seen that particular type before. The situation in mathematics is similar.

In Section 1.5 we learned to write a product of polynomials as a polynomial; in this section we will **factor** polynomials, that is, write polynomials as products. If you are skilled in the process of computing products, factoring will not be a difficult task.

Removing common factors

The simplest type of factoring is that in which all terms of the polynomial have a common factor. For example, each term of the polynomial $3x + 6y$ is divisible by 3, and we would factor it as $3x + 6y = 3(x + 2y)$.

Example 1 $4a^2bc^3 - 8ab^2c = 4abc(ac^2 - 2b)$ ■

Example 2 $-y^2 + xy^2z + z^2y^2 = y^2(-1 + xz + z^2)$

or

$$= -y^2(1 - xz - z^2)$$ ■

The second type of factoring we will consider follows directly from the special products given in Section 1.5.

1. $(x + y)(x - y) = x^2 - y^2$	3. $(x - y)^2 = x^2 - 2xy + y^2$
2. $(x + y)^2 = x^2 + 2xy + y^2$	4. $(x + a)(x + b) = x^2 + (a + b)x + ab$

Factoring the difference of two squares

The first special product shows how to factor the **difference of two squares:**

$$x^2 - y^2 = (x - y)(x + y)$$

Example 3 $x^2 - 36 = x^2 - 6^2 = (x - 6)(x + 6)$ ■

Example 4 $9y^2 - 4z^2 = (3y)^2 - (2z)^2 = (3y - 2z)(3y + 2z)$ ■

Example 5 $3s^2 - 48t^2 = 3(s^2 - 16t^2) = 3(s - 4t)(s + 4t)$ ■

Factoring perfect squares

The second special product allows us to recognize **perfect squares:**

$$x^2 + 2xy + y^2 = (x + y)^2$$

The third special product is also a perfect square:

$$x^2 - 2xy + y^2 = (x - y)^2$$

Example 6 $b^2 + 10b + 25 = b^2 + 2(5)(b) + (5)^2$
$= (b + 5)^2$ ■

Example 7 $4u^2 - 12uv + 9v^2 = (2u)^2 - 2(2u)(3v) + (3v)^2$
$= (2u - 3v)^2$ ■

Example 8 $5w^2 - 10w + 5 = 5(w^2 - 2w + 1) = 5(w - 1)^2$ ■

Factoring general trinomials

Special products of type 4 are a bit more difficult to recognize. This product is:

$$x^2 + (a + b)x + ab = (x + a)(x + b)$$

Example 9 Factor $x^2 + 5x + 6$.
We seek a and b such that $a + b = 5$ and $ab = 6$. A little trial and error yields $a = 3$, $b = 2$, and so

$$x^2 + 5x + 6 = (x + 2)(x + 3)$$ ■

Example 10 Factor $x^2 - 2x - 15$.
The task is to find a and b such that $a + b = -2$ and $ab = -15$. Again, trial and error yields $a = -5$, $b = 3$. The factorization is

$$x^2 - 2x - 15 = (x - 5)(x + 3)$$ ■

If the polynomial is more complicated, the same trial-and-error method is used, although it may take a bit more time.

Example 11 Factor $12x^2 - 4 + 2x$.

Step 1 Arrange the terms in decreasing (or increasing) powers of the variable:

$$12x^2 + 2x - 4$$

Step 2 Remove any common factors:

$$2(6x^2 + x - 2)$$

Step 3 Consider factorizations of the first term, $(6x)(x)$ or $(3x)(2x)$:

$$2(3x + ?)(2x + ?) \qquad 2(6x + ?)(x + ?)$$

Step 4 Consider factorizations of the last term, $(-2)(1)$ or $(2)(-1)$:

$$2(3x + 2)(2x - 1) \qquad 2(6x + 2)(x - 1)$$
$$2(3x - 2)(2x + 1) \qquad 2(6x - 2)(x + 1)$$
$$2(3x - 1)(2x + 2) \qquad 2(6x - 1)(x + 2)$$
$$2(3x + 1)(2x - 2) \qquad 2(6x + 1)(x - 2)$$

Step 5 Find the factorization that works (if there is one):

$$2(3x + 2)(2x - 1) = 2(6x^2 + x - 2) = 12x^2 + 2x - 4$$ ■

Example 12 Factor $11ab - 20a^2 + 3b^2$.

Step 1 Arrange in descending powers of b (you could use descending powers of a if you wished):

$$3b^2 + 11ab - 20a^2$$

Step 2 There are no common factors.

Step 3 Consider factorizations of the first term:

$$(3b + ?)(b + ?)$$

Step 4 Consider factorizations of the last term. There are lots of factorizations of $20a^2$; we should try to shorten the search. If we try $(3b + 20a)(b - a)$, we obtain a middle term of $17ab$. The product $(3b - 20a)(b + a)$ has a middle term of $-17ab$, which is the same except for the sign. This means, for example, that since the product $(3b + 10a)(b - 2a)$ has a middle term of $4ab$, there is no reason to try $(3b - 10a)(b + 2a)$, because that product will have a middle term of $-4ab$.

Step 5 Find the factorization that works. In this case:

$$3b^2 + 11ab - 20a^2 = (3b - 4a)(b + 5a)$$ ■

Two other special products are of sufficient interest to mention.

Sum and difference of two cubes

5. $x^3 - y^3 = (x - y)(x^2 + xy + y^2)$
6. $x^3 + y^3 = (x + y)(x^2 - xy + y^2)$

Example 13 $s^3 - 8 = s^3 - 2^3 = (s - 2)(s^2 + 2s + 4)$ ■

Example 14 $27t^3 + s^3 = (3t)^3 + s^3 = (3t + s)(9t^2 - 3st + s^2)$ ■

Factoring by grouping

Some polynomials may be factored by "grouping."

Example 15

$$\begin{aligned} x^2 + ax + 2a - 4 &= x^2 - 4 + a(x + 2) \\ &= (x - 2)(x + 2) + a(x + 2) \\ &= (x - 2 + a)(x + 2) \end{aligned}$$ ■

Example 16

$$\begin{aligned} x^3 - 4x^2 + 8x - 32 &= x^2(x - 4) + 8(x - 4) \\ &= (x^2 + 8)(x - 4) \end{aligned}$$ ■

Notice that

$$(x - 1 - \sqrt{3})(x - 1 + \sqrt{3}) = x^2 - 2x - 2$$

which means that our methods would not succeed in finding the factorization of $x^2 - 2x - 2$. Although $x^2 - 2x - 2$ can be factored, it cannot be factored using only integers. We say it is **irreducible** over the set of integers.

Exercises 1.6

Factor the polynomials.

1. $5y^2 - 25y$
2. $4x^2 - 16x$
3. $7xyz - 14x^2y$
4. $8abc - 24a^2c$
5. $-3x - 9x^2$
6. $-5t - 15t^2$
7. $3xy - 6xy^2$
8. $4st - 12st^2$
9. $a^2b - 3ab + 5a^2b^2$
10. $xy^2 - 6xy - 9x^2y^2$
11. $15a^2b^3 - 12a^3b^2$
12. $18a^2b^3 - 45a^2b^2$
13. $x^2 - 25$
14. $a^2 - 121$
15. $49 - s^2$
16. $64 - p^2$
17. $5n^2 - 80m^2$
18. $3r^2 - 48t^2$
19. $w^2 - 18w + 81$
20. $y^2 - 14y + 49$
21. $u^2 - 16uv + 64v^2$
22. $n^2 - 20mn + 100m^2$
23. $p^2 + 12p + 36$
24. $q^2 + 18q + 81$
25. $2p^2 + 20p + 50$
26. $5z^2 + 40z + 80$
27. $w^2 + 6w - 16$
28. $v^2 + 4v - 12$
29. $w^2 - 6w - 27$
30. $b^2 - 5b - 14$
31. $4r^2 + 8r - 32$
32. $5t^2 - 30t - 35$
33. $2b^2 - 14b + 24$
34. $3a^2 + 21a + 30$
35. $5m^2 - 5m - 360$
36. $6n^2 - 6n - 336$
37. $5x^2 + 11xy + 2y^2$
38. $2x^2 + 7xy + 3y^2$
39. $3a^2 - 7a + 2$
40. $7y^2 - 22y + 3$
41. $5x^2 - 9xy - 2y^2$
42. $2r^2 - 5rt - 3t^2$
43. $3u^2 - 16uv - 12v^2$
44. $2m^2 - 5mn - 12n^2$
45. $10z^2 + 19z + 6$
46. $15p^2 + 16p + 4$
47. $4r^2 - 10r + 6$
48. $15x^2 + 9x - 6$
49. $a^3 + 27$
50. $s^3 + 64$
51. $8z^3 - 27t^3$
52. $64u^3 - 27v^3$
53. $ax + ay + x + y$
54. $mn + mp + p + n$
55. $x^3 + 7x^2 - 4x - 28$
56. $a^3 + 5a^2 - 4a - 20$

For Your Calculator

Factoring can be used to write a polynomial in what is called nested form. In nested form

$$\begin{aligned} P(x) &= 3x^3 - 4x^2 + 7x - 9 \\ &= (3x^2 - 4x + 7)x - 9 \\ &= [(3x - 4)x + 7]x - 9 \end{aligned}$$

The nested form is convenient for computing values of $P(x)$ on the calculator. To compute $P(2)$, the key sequences are as follows:

Alg	2 [×] 3 [−] 4 [=] [×] 2 [+] 7 [=] [×] 2 [−] 9 [=]	answer 13
RPN	3 [enter] 2 [×] 4 [−] 2 [×] 7 [+] 2 [×] 9 [−]	answer 13

Write the polynomials in nested form and calculate the values indicated.

57. $P(x) = x^3 - 4x^2 + 2x - 6$; find $P(3)$.
58. $Q(x) = x^3 + 4x^2 - 2x + 7$; find $Q(2)$.
59. $Q(x) = 5x^4 - 3x^2 + 3x - 14$; find $Q(1.4)$.
60. $S(a) = 6a^4 - 2a^2 + 4a - 12$; find $S(2.1)$.
61. $T(s) = 1.2s^3 + 2.7s^2 - 1.1s + .5$; find $T(.3)$.
62. $S(x) = x^5 + x^4 + x^3 + x^2 + x + 1$; find $S(.4)$.
63. $S(x) = x^5 - x^4 + x^3 - x^2 + x + 1$; find $S(.4)$.

Chapter Review

Compute each sum or difference.

1. $2 + (-8) - 6$

2. $2 + |-2|$

3. $3 - (6 - 5)$

4. $2.5 - [(1.6 - 7) + (-4.5 - 7.1)]$

5. $14 - [(4 - 7) - (2 - 5)]$

6. $\frac{1}{2} - \left(\frac{3}{4} - \frac{7}{8}\right)$

Write each product or quotient as a single number.

7. $4(-7)$

8. $(-5)(-7)$

9. $(-9)(-2)(-4)$

10. $(-9)(4 - 8)$

11. $(-1.5)(-2.4)(1.6)$

12. $-(-(14 - 22))$

13. $6 \div (-2)$

14. $(-8) \div (-4)$

15. $\frac{1}{2} \div \left(-\frac{1}{4}\right)$

16. $5 \div \left(-\frac{1}{10}\right)$

17. $\frac{3}{4} \div \frac{2}{3}$

18. $-\frac{1}{2} \div \left(-\frac{2}{3}\right)$

Collect like terms.

19. $2a + 6a$

20. $8u - 2u - 5u$

21. $-6y + 8y - 9y$

22. $2s - 3(s + 2s)$

23. $5t - (6t - 2t)$

24. $8p - (p - 1)$

25. $5r - 2(r + 1)$

26. $2 - 8m + 2(m + 4)$

Arrange the numbers in increasing order.

27. $-6, -9, 1, 0, 4$

28. $-3.4, -2.1, -3.8, 0$

29. $-0.01, -0.002, 1, 2, 6$

30. $-\frac{3}{4}, -\frac{7}{9}, -\frac{5}{6}$

31. $\frac{6}{7}, \frac{9}{11}, \frac{11}{13}, \frac{13}{15}$

32. $2, -2, |-2|, -2.5, |-2.5|$

Solve the equations.

33. $5t + 7 = 4t - 1$

34. $r + \frac{1}{2} = -\frac{3}{4}$

35. $.06x = 3.6$

36. $-\frac{3}{4}x = 42$

37. $22 + 3x = 16 - 5x$

38. $0.09n + 1.2 = 0.03n - 4.8$

39. $\frac{3}{4}s - \frac{1}{2} = \frac{3}{2} + \frac{3}{2}s$

40. $3x - (x + 1) = 2(x + 1) + 3x$

41. $5(x - 2) - 6 = 3(x - 1)$

42. $0.7(x - 1) + 0.3 = 0.06(x + 4)$

43. $\frac{1}{3}(x - 2) + \frac{1}{4} = \frac{3}{4}(x + 1)$

44. $\frac{5}{x} + 14 = \frac{2}{x} - 7$

45. $2n - 3m = pq$, for m.

46. $|2x - 3| = 7$

47. $s = 1 + (s - 2)t$, for s.

48. $s = 1 + (s - 2)t$, for t.

49. $|5 - \frac{3}{x}| = 4$

Solve each inequality and show the solution set on the number line.

50. $5x + 3 < 7x - 5$

51. $4x - 9 \leq -6x + 1$

52. $2(x + 1) - 6 > -4x + 7$

53. $\frac{3x - 1}{6} \geq -5$

54. $-3 < 3x + 1 < 5$

55. $-9 \leq 6 - 3x \leq 15$

56. $|2x + 6| < 4$

57. $|6 - 4x| \geq 18$

58. $2w < 4w - 1 < 8 + 2w$

59. Rose has twice as many letters to type as does Jill. Together they have 30 letters. How many does each have to type?

60. Joe plays golf with Chuck and together their sum for 18 holes is 164. If Joe took 2 fewer strokes than Chuck, what was Joe's score?

61. Nancy scores 145, 133, 152, 160, and 148 in 5 successive games. What score must she make in her next game to have an average between 150 and 160?

62. Carla and Bryan ride their bicycles for 12 mi. Then they walk an additional 4 mi at a rate 4 mi/hr slower than they cycled. If each part of the trip took the same amount of time, find the rate at which they walked.

63. The perimeter of a triangle is 36 cm. One side is 1 cm longer than the shortest side and the remaining side is 1 cm shorter than twice the length of the shortest side. Find the dimensions.

Given $P(x) = 3x^2 - 6x + 7$, *find*

64. $P(3)$

65. $P(2)$

66. $P(-2)$

67. $P(-3)$

Simplify the polynomials.

68. $(x - 2y)(x + 2y)$

69. $(3a - 5b)(3a + 5b)$

70. $(s - 2t)^2$

71. $(m + 3n)^2$

72. $(x - 2y)(x + 3y)$

73. $(a + 2b)(a - 5b)$

74. $(3a + 6)(2a - 1)$

75. $(2m + 5)(3m - 2)$

76. $(x - 1)^2 - (x + 1)^2$

77. $(a + 2)^2 - (a - 2)^2$

Factor.

78. $a^2 - 81$

79. $5m^2 - 20n^2$

80. $3n^2 - 48m^2$

81. $a^2 - 8a + 16$

82. $s^2 - 10s + 25$

83. $3s^2 + 21s + 30$

84. $2x^2 - 14x + 24$

85. $5a^2 + 11ab + 2b^2$

86. $2t^2 + 7st + 3s^2$

87. $7q^2 + 33q - 10$

88. $3v^2 - 16v - 12$

89. $a^3 + 4a^2 - 9a - 36$

90. $a^3 - 64$

91. $3x^3 + 24y^3$

92. $2ax - 3ay + 14x - 21y$

Chapter Test

Compute each sum or difference.

1. $5 + (-2) + (-3)$ **2.** $-4 - 6 + 3$ **3.** $(4 - 6) - (-2 - 7)$

Compute each product or quotient.

4. $(-6.2)(-1.5)$ **5.** $(-15) \div (-3)$ **6.** $\left(-\frac{1}{5}\right) \div \frac{2}{3}$

7. $(-3)(-2.4)(5)(-1)$ **8.** $(15 \div 5) \div \left(-\frac{1}{6}\right)$

Simplify each expression.

9. $5x + 7 - 3(2x - 3)$ **10.** $\frac{1}{2}t - \frac{3}{4}t + 2t$

Arrange in increasing order.

11. $1.2, -2.2, -1.7, 3.4$ **12.** $\frac{3}{5}, -\frac{2}{5}, \frac{4}{7}, -\frac{3}{7}$

Solve.

13. $3(x + 5) - 6 = 5x + 7$ **14.** $2(3x + 1) + 5 = 4(3 - x)$

15. $2x + 1 < 5$ **16.** $2rp - 5 = 14$, for p

17. $3x - 6 \leq 5x + 9$ **18.** $|2x + 6| \geq 12$

19. How much pure silver must be added to 110 oz of 60% silver alloy to obtain a 75% alloy?

20. If $P(x) = 2x^3 + 4x - 5$, find $P(-2)$ and $P(3)$.

Compute the products and simplify.

21. $(2x - 1)(2x + 1)$ **22.** $(3x - 2y)^2$

23. $(3x + 4y)(2x - 5y)$ **24.** $2x(x - 1)(3x + 5)$

Factor.

25. $5x^2 - 20y^2$ **26.** $a^2 + 22a + 121$

27. $x^2 - 9x - 52$ **28.** $12a^2 - 5ab - 2b^2$

29. $27z^3 - 8$

2 Fractions, Radicals, and Quadratic Equations

2.1 Fractions and Quotients of Polynomials

Before considering quotients of polynomials, we review some facts about fractions.

If b is any nonzero real number, then there is a unique real number c such that $bc = 1$. We write $c = b^{-1}$ or $c = 1/b$.

If a is any real number and b is any nonzero real number, we define a/b to be $a(1/b)$. We refer to a/b as a **fraction** with a the **numerator** and b the **denominator.** The number a/b is also called the **quotient** of a and b because it is the solution to the division problem $a \div b = c$. In symbols

$$\frac{a}{b} = a\left(\frac{1}{b}\right) = a \div b$$

Equality for fractions

If a, b, c, and d are any real numbers with $b \neq 0$ and $d \neq 0$, then

$$\frac{a}{b} = \frac{c}{d} \quad \text{if and only if} \quad ad = bc.$$

Cancellation law

This result gives an easy way to check on the equality of two fractions and also gives what is called the **cancellation law.** Since $a(bx) = b(ax)$ for all a, b, and x, we have

$$\frac{ax}{bx} = \frac{a}{b} \quad \text{if} \quad x \neq 0 \quad \text{and} \quad b \neq 0$$

Example 1 $\frac{2}{3} = \frac{14}{21}$ since $2(21) = 42 = 3(14)$

$$\frac{2}{3} = \frac{7 \cdot 2}{7 \cdot 3} = \frac{14}{21}$$ ■

In the second part of Example 1 the cancellation law was used in reverse when numerator and denominator were multiplied by the common factor 7. This is useful in adding fractions.

Example 2 $\frac{x}{y} = \frac{4x}{4y}$ if $y \neq 0$ ■

Example 3 $\frac{689}{3445} = \frac{689(1)}{689(5)} = \frac{1}{5}$ ■

Example 3 shows how to use the cancellation law to reduce fractions by removing or "cancelling" common factors. But you must use the cancellation law with care! It allows you to cancel *common factors* from numerator and denominator, not common terms. For example,

WARNING You can cancel common factors, but not common terms.

$$\frac{\not{x} + y}{\not{x} + 2y} = \frac{\not{y}}{2\not{y}} = \frac{1}{2}$$

is *not* a proper use of the cancellation law.

Example 4

$$\frac{4y^2 + 4y - 24}{5y - 10} = \frac{4(y^2 + y - 6)}{5(y - 2)} = \frac{4(y + 3)(y - 2)}{5(y - 2)}$$
$$= \frac{4(y + 3)}{5} \text{ if } y \neq 2.$$ ■

Signs of a fraction

For any real numbers a and b with $b \neq 0$,

$$\frac{a}{b} = \frac{-a}{-b} = -\frac{-a}{b} = -\frac{a}{-b}$$

and

$$-\frac{a}{b} = \frac{-a}{b} = \frac{a}{-b} = -\frac{-a}{-b}$$

There is a helpful rule of thumb to remember these relationships. If we associate with each fraction three choices of sign—that of the numerator, that of the denominator, and that of the fraction—we can change any two of these without changing the sign of the number. But if we change one of these signs or three of them, the sign of the number is changed.

Example 5

$$\frac{2}{x - y} = \frac{-2}{-(x - y)} = \frac{-2}{y - x} \text{ if } x \neq y$$
$$\frac{2}{x - y} = -\frac{-2}{x - y} = -\frac{2}{y - x} \text{ if } x \neq y$$ ■

If two fractions have the same denominator, the sum is given by the formula

Sum of two fractions

$$\frac{a}{b} + \frac{c}{b} = \frac{a + c}{b} \quad \text{if} \quad b \neq 0$$

Common denominator

To find the sum if the denominators are different, it is necessary to use a **common denominator.**

Example 6 $\frac{2}{5} + \frac{2}{15} = \frac{2 \cdot 3}{5 \cdot 3} + \frac{2}{15} = \frac{6}{15} + \frac{2}{15} = \frac{8}{15}$ ■

Example 7 $\dfrac{2}{x} + \dfrac{3}{y} = \dfrac{2y}{xy} + \dfrac{3x}{xy} = \dfrac{2y + 3x}{xy}$ ■

The product of the denominators will work, as in Example 7, but sometimes a simpler denominator can be used, as in Example 6. The number $75 = 5 \cdot 15$ could have been used in Example 6, but 15 served the purpose just as well. Factoring is useful in finding common denominators.

Example 8 Find a common denominator for the fractions

$$\frac{3x}{x^2 - y^2}, \quad \frac{5xy}{2x^2 - 4xy + 2y^2}$$

Factoring each denominator:

$$\frac{3x}{x^2 - y^2} = \frac{3x}{(x - y)(x + y)}$$

$$\frac{5xy}{2x^2 - 4xy + 2y^2} = \frac{5xy}{2(x - y)(x - y)}$$

Any common denominator must contain all factors occurring in both denominators, and in the common denominator each factor must occur the largest number of times it occurs in any one denominator. The common denominator so chosen is

$$2(x + y)(x - y)(x - y)$$ ■

Example 9 Write the following sum as a single fraction:

$$\frac{3x}{x^2 - y^2} + \frac{5xy}{2x^2 - 4xy + 2y^2}$$

Using the common denominator from Example 8,

$$\begin{aligned}\frac{3x}{x^2 - y^2} + \frac{5xy}{2x^2 - 4xy + 2y^2} &= \frac{3x}{(x + y)(x - y)} + \frac{5xy}{2(x - y)(x - y)} \\ &= \frac{3x(2)(x - y)}{2(x + y)(x - y)(x - y)} + \frac{5xy(x + y)}{2(x + y)(x - y)(x - y)} \\ &= \frac{6x^2 - 6xy}{2(x + y)(x - y)^2} + \frac{5x^2y + 5xy^2}{2(x + y)(x - y)^2} \\ &= \frac{6x^2 - 6xy + 5x^2y + 5xy^2}{2(x + y)(x - y)^2}\end{aligned}$$ ■

Product of fractions

If a, b, c, and d are real numbers with $b \neq 0$ and $d \neq 0$, then

$$\frac{a}{b} \cdot \frac{c}{d} = \frac{ac}{bd}$$

Example 10 $\dfrac{8a^2b}{2c} \cdot \dfrac{c}{5ab^2} = \dfrac{8a^2bc}{10ab^2c} = \dfrac{2abc(4a)}{2abc(5b)} = \dfrac{4a}{5b}$ ■

Example 11 $\left(\frac{5xy}{16w}\right)\left(-\frac{4xw}{3y}\right) = \left(\frac{5xy}{16w}\right)\left(\frac{-4xw}{3y}\right) = \frac{-20x^2wy}{48wy} = \frac{-5x^2}{12}$ ■

As mentioned earlier in this section, when we divide a by b, the result can be expressed in several ways:

$$a \div b = a \cdot b^{-1} = a\left(\frac{1}{b}\right) = \frac{a}{b}$$

Dividing fractions In particular, to find the quotient of two numbers, invert the divisor and multiply.

Example 12 $\frac{3}{4} \div \frac{2}{3} = \frac{3}{4} \cdot \frac{3}{2} = \frac{9}{8}$ ■

Example 13
$$\frac{x^2 - y^2}{z} \div \frac{x - y}{3z} = \frac{x^2 - y^2}{z} \cdot \frac{3z}{x - y} = \frac{(x - y)(x + y)(3z)}{(x - y)(z)}$$
$$= 3(x + y)$$ ■

Fractions that have complicated numerators and denominators can sometimes be simplified by rewriting them in terms of division.

Example 14
$$\frac{\frac{2}{3}}{\frac{5}{12}} = \frac{2}{3} \div \frac{5}{12} = \frac{2}{3} \cdot \frac{12}{5} = \frac{24}{15} = \frac{8}{5}$$ ■

Another way to simplify complicated fractions is to multiply numerator and denominator by factors designed to eliminate the fractions in the numerator and denominator.

Example 15
$$\frac{x + \frac{1}{x}}{x - \frac{1}{x}} = \frac{x\left(x + \frac{1}{x}\right)}{x\left(x - \frac{1}{x}\right)} = \frac{x^2 + 1}{x^2 - 1}$$ ■

The techniques considered in this section are very helpful in solving equations that contain fractions.

Example 16 Solve the equation $\frac{x - 4}{x + 4} = \frac{x + 2}{x - 2}$.

The two fractions are equal if and only if

$$(x - 4)(x - 2) = (x + 4)(x + 2)$$
$$x^2 - 6x + 8 = x^2 + 6x + 8$$
$$12x = 0$$
$$x = 0$$ ■

Note that neither $x = -4$ nor $x = 2$ could have been a solution.

Example 17 Solve the equation $\frac{1}{5} + \frac{a}{2a - 1} = \frac{8}{4a - 2}$.

Multiply by $5(2)(2a - 1)$ to clear fractions:

$$5(2)(2a - 1)\left(\frac{1}{5} + \frac{a}{2a - 1}\right) = 5(2)(2a - 1)\left(\frac{8}{4a - 2}\right)$$

$$2(2a - 1) + 5(2)(a) = 5(8)$$

$$4a - 2 + 10a = 40$$

$$14a = 42$$

$$a = 3$$

■

Many rate problems lead to fractions. If a job can be done in a hours, then b/a is the fraction of the job that can be done in b hours. The next example illustrates this idea.

Example 18 A man has two plows and wishes to plow a field. Using the larger plow, it takes 6 hr to do the job and using the smaller plow it takes 8 hr. He decides to hire a helper and use both plows. How long will it take to plow the field?

Let x = number of hours required to plow the field using both plows. Then

$\frac{x}{6}$ = fraction of the field plowed with the large plow since the large plow covers 1/6 of the field in 1 hr.

$\frac{x}{8}$ = fraction of the field plowed with the smaller plow since the smaller plow covers 1/8 of the field in 1 hr.

The equation is

$$\frac{x}{6} + \frac{x}{8} = 1$$

(The sum of the fractions must equal 1 since the entire field is to be plowed.)

$$\frac{4x}{24} + \frac{3x}{24} = 1$$

$$\frac{7x}{24} = 1$$

$$x = \frac{24}{7}$$

(approximately 3 hr and 26 min) ■

To focus a camera, a photographer adjusts the distance from the lens to the film to get the clearest image. If s is the object distance (the distance from the lens to the object to be photographed) and s' is the image distance (the distance from the lens to the image on the film), then the clearest picture occurs when

$$\frac{1}{s} + \frac{1}{s'} = \frac{1}{f}$$

where f is the focal length of the lens.

Example 19 If the object distance is 10 m and the lens has a focal length of 50 mm, what distance from the lens to the film will give the sharpest picture?

We first convert the object distance to millimeters (mm), obtaining an object distance of 10,000 mm. In the equation above

$$s = 10{,}000 \quad \text{and} \quad f = 50$$

so

$$\frac{1}{10{,}000} + \frac{1}{s'} = \frac{1}{50}$$

Clearing fractions, we obtain

$$\begin{aligned} s' + 10{,}000 &= 200s' \\ 10{,}000 &= 199s' \\ \frac{10{,}000}{199} &= s' \\ 50.25 \text{ mm} &= s' \end{aligned}$$

■

Exercises 2.1

Simplify the fractions in Exercises 1–18 by removing common factors from numerator and denominator. Assume all denominators are nonzero.

1. $\frac{8a^2b^2}{4ab}$

2. $\frac{9x^2y^2}{3y}$

3. $\frac{45a^2b^3c^2}{15a^4bc}$

4. $\frac{36x^2y^3z^2}{9x^3yz^2}$

5. $\frac{xy + xz}{-x}$

6. $\frac{mn - mp}{-m}$

7. $\frac{x^2 - 6x - 16}{2x - 16}$

8. $\frac{a^2 + 4a - 21}{3a + 21}$

9. $\frac{x - 2y}{x^2 - 4y^2}$

10. $\frac{a + 4b}{a^2 - 16b^2}$

11. $\frac{x^2 + xy - 2y^2}{x - y}$

12. $\frac{a^2 - 8ab + 12b^2}{a - 2b}$

13. $\frac{21 - 3x}{x^2 - 49}$

14. $\frac{20 - 4x}{x^2 - 25}$

15. $\frac{x^2 - 5x - 6}{2x + 2}$

16. $\frac{m^2 - 4m - 21}{3m + 9}$

17. $\frac{x^2 - 7x + 12}{x^2 - x - 6}$

18. $\frac{a^2 - 5a - 24}{a^2 + 4a + 3}$

In Exercises 19–34 write each sum or difference as a single fraction and simplify. Assume all denominators are nonzero.

19. $\frac{3x}{x - y} + \frac{3y}{y - x}$

20. $\frac{2s}{s - t} + \frac{2t}{t - s}$

21. $\frac{5}{3m - 2n} + \frac{5}{2n - 3m}$

22. $\frac{3}{6u - 5v} + \frac{3}{5v - 6u}$

23. $\frac{3}{x + 2} - \frac{2}{x - 3}$

24. $\frac{5}{a - 4} - \frac{3}{a - 2}$

25. $\dfrac{y+5}{3} + \dfrac{y}{y-2}$

26. $\dfrac{6}{x-1} + \dfrac{x-8}{2}$

27. $\dfrac{5}{2x-y} - \dfrac{4}{x-2y}$

28. $\dfrac{2}{2a+3b} - \dfrac{4}{a-2b}$

29. $m + \dfrac{5}{m-2}$

30. $x + \dfrac{4}{x+1}$

31. $\dfrac{4}{x^2-2xy+y^2} - \dfrac{5}{x^2-y^2}$

32. $\dfrac{6}{a^2+2ab+b^2} - \dfrac{8}{a^2-b^2}$

33. $\dfrac{4}{a} - \dfrac{1}{a^2-b^2} + \dfrac{1}{a-b}$

34. $\dfrac{2}{x^2-y^2} + \dfrac{3}{x-y} - \dfrac{4}{x}$

In Exercises 35–58 write each product or quotient as a fraction and simplify. Assume all denominators and divisors are nonzero.

35. $\dfrac{12x^2y^2z^2}{5w} \cdot \dfrac{-9w}{4xy} \cdot \dfrac{5w}{3xy}$

36. $\dfrac{15a^2b^2c}{3r} \cdot \dfrac{-7r}{5ab} \cdot \dfrac{3r}{3ab}$

37. $\dfrac{3y+9}{2} \cdot \dfrac{3}{2y+6}$

38. $\dfrac{a+3}{3} \cdot \dfrac{5b}{2a+6}$

39. $\dfrac{9a-3}{a^2-1} \cdot \dfrac{a+1}{9a^2-1}$

40. $\dfrac{6x-2}{x^2-4} \cdot \dfrac{x+2}{9x^2-1}$

41. $\dfrac{3x+3y}{5x+10y} \cdot \dfrac{x+2y}{3x-6y} \cdot \dfrac{4x-8y}{6x+6y}$

42. $\dfrac{2a-2b}{6a+18b} \cdot \dfrac{a-3b}{a-b} \cdot \dfrac{3a+9b}{2a-6b}$

43. $\dfrac{3xyz}{8y^2} \div \dfrac{-9z^2}{4y^2}$

44. $\dfrac{5rst}{12r^2} \div \dfrac{-15t^3}{6r^2}$

45. $\dfrac{2xyz}{x^2-2xy+y^2} \div \dfrac{4z}{y-x}$

46. $\dfrac{3ab}{a^2-2ab+b^2} \div \dfrac{6b}{b-a}$

47. $\dfrac{y^2-10y+21}{4y^2-12y+9} \div \dfrac{y-7}{3-2y}$

48. $\dfrac{a^2-8a+15}{9y^2-12y+4} \div \dfrac{a-5}{2-3y}$

49. $\left(\dfrac{a^2b}{c} \div \dfrac{ab^2}{2c^2}\right) \div \dfrac{c}{ab}$

50. $\left(\dfrac{xy^2}{z^2} \div \dfrac{x^2}{2z}\right) \div \dfrac{z}{xy}$

51. $\left(\dfrac{2r}{r+s} \div \dfrac{2s}{r^2-s^2}\right) \div \dfrac{r}{s}$

52. $\left(\dfrac{3x}{x^2-y^2} \div \dfrac{2y}{x+y}\right) \div \dfrac{x}{y}$

53. $\dfrac{x-\dfrac{1}{6}}{x+\dfrac{1}{6}}$

54. $\dfrac{a-\dfrac{1}{4}}{a+\dfrac{1}{4}}$

55. $\dfrac{1+\dfrac{1}{x}}{1-\dfrac{1}{x}}$

56. $\dfrac{1-\dfrac{1}{y}}{1+\dfrac{1}{y}}$

57. $\dfrac{\dfrac{1}{2}x-\dfrac{1}{3}y}{\dfrac{1}{3}x-\dfrac{1}{2}y}$

58. $\dfrac{\dfrac{1}{4}a-\dfrac{1}{3}b}{\dfrac{1}{3}a+\dfrac{1}{4}b}$

Solve the equations in Exercises 59–68.

59. $\frac{1-2x}{3x+1} = \frac{2x-4}{4-3x}$

60. $\frac{2x+1}{x-7} = \frac{2x+14}{x-8}$

61. $\frac{4n}{n-2} - \frac{1}{3} = \frac{16}{3n-6}$

62. $\frac{2x}{x+3} + \frac{5}{2} = \frac{24}{2x+6}$

63. $\frac{2}{m+4} + \frac{5}{m^2-16} = \frac{1}{m-4}$

64. $\frac{6}{a^2-25} - \frac{1}{a+5} = \frac{3}{a-5}$

65. $\frac{2}{x} - \frac{5}{x} = 3$

66. $\frac{7}{u} - \frac{9}{u} = 4$

67. $\frac{1}{5} + \frac{x}{x-2} = \frac{3x-10}{5x-10}$

68. $\frac{3}{4} + \frac{2x}{x+3} = \frac{13x-5}{4x+12}$

69. Veronica has a farm, which she irrigates from a storage reservoir. The inlet pipe can fill the reservoir in 15 hr and the outlet pipe can drain it in 18 hr. If both pipes are open, how long will it take to fill Veronica's lake?

70. John can clean the boathouse in 20 min and Keith can do the job in 30 min. Mark has agreed to let John use his sailboat after the boathouse is cleaned if the boat is returned by 8:00 P.M. If John and Keith begin cleaning the boathouse at 5:00 P.M., how much time will John have for a sail in Mark's boat?

71. Working together, Mike and Chicho can pick 100 lb of cotton in $1\frac{1}{2}$ hr. Chicho can pick 100 lb of cotton by himself in $2\frac{1}{4}$ hr. How long will it take Mike to pick 100 lb of cotton?

72. Jessie can type a manuscript in 18 hr and Ross can type it in 12 hr. If both work together, how long will it take them to type the manuscript?

73. David's lawn mower cuts a swath twice as wide as Pat's mower does. If David can mow his yard in 1 hr, how long will it take David and Pat together to mow David's yard?

74. Printing press A can do a job in 6 hr and press B can do the same job in 4 hr. How long will it take to do the job if both presses are used?

75. If a camera has a focal length of 50 mm and the object distance is 100 m, what distance from the lens to the film gives the sharpest picture?

76. Solve the equation $(1/s) + (1/s') = 1/f$ for s'. Recall that s' is the distance from the lens to the film for a good picture. What happens to s' as the object distance s gets large? If you were to construct a camera with the distance from the lens to the film fixed, what distance might you choose for s'?

2.2 Integer Exponents and Scientific Notation

Recall that if x is a real number and n is a positive integer,

$$x^n = x \cdot x \cdot x \cdots x \quad (n \text{ factors})$$

The number n is called an **exponent.** We extend the definition of x^n to include integer exponents as follows:

Zero exponent 1. $x^0 = 1$ if $x \neq 0$

Negative exponents 2. $x^{-n} = \frac{1}{x^n}$ if $x \neq 0$

Listed below are the properties of exponents that are particularly useful in computations:

For all real numbers x and y and integers m and n,

Properties of exponents

1. $x^n \cdot x^m = x^{n+m}$
2. $\dfrac{x^n}{x^m} = x^{n-m}$
3. $(x^n)^m = x^{nm}$
4. $(xy)^n = x^n y^n$
5. $\left(\dfrac{x}{y}\right)^n = \dfrac{x^n}{y^n}$

In Properties 1–5 it is understood that no expression is defined if division by zero is involved. For example, $x^4/x^2 = x^{4-2} = x^2$ is not a valid statement if $x = 0$, even though x^4 and x^2 are defined for $x = 0$.

Example 1 $x^2 \cdot x^3 = x^{2+3} = x^5$ ■

Example 2 $x^4 \cdot x^{-5} = x^{4-5} = x^{-1} = \dfrac{1}{x}$ if $x \neq 0$ ■

Example 3 $\dfrac{p^6}{p^4} = p^{6-4} = p^2$ if $p \neq 0$ ■

Example 4 $\dfrac{u^4}{u^7} = u^{4-7} = u^{-3} = \dfrac{1}{u^3}$ if $u \neq 0$ ■

Property 2 is sometimes stated in two versions:

2a. $\dfrac{x^n}{x^m} = x^{n-m}$ if $n \geq m$.

2b. $\dfrac{x^n}{x^m} = \dfrac{1}{x^{m-n}}$ if $n < m$.

There is a rule of thumb that is useful in dealing with negative exponents. *You may move a factor from numerator to denominator or vice versa by changing the sign of the exponent. Note*: This may be done with *factors* of the numerator or denominator, but *not* with terms of a sum in numerator or denominator. We illustrate this principle in the next three examples.

Example 5 $\dfrac{1}{y^{-2}} = \dfrac{y^2}{1} = y^2$; $\quad \dfrac{x^2}{x^{-3}} = x^2x^3 = x^5$; $\quad \dfrac{a^{-2}}{b^{-3}} = \dfrac{b^3}{a^2}$ ■

Example 6 Write using only positive exponents:

$$\frac{x^{-2}y^3}{a^4b^{-5}}$$

Using the principle stated above:

$$\frac{x^{-2}y^3}{a^4b^{-5}} = \frac{b^5y^3}{x^2a^4}$$ ■

Example 7 Write using only positive exponents:

$$\frac{2s^{-3}t^5r^{-2}}{-3a^3b^{-5}c^{-1}}$$

Using the principle stated above:

$$\frac{2s^{-3}t^5r^{-2}}{-3a^3b^{-5}c^{-1}} = \frac{2b^5t^5c}{-3a^3s^3r^2}$$ ■

Example 8 $(a^4)^3 = a^{4\cdot 3} = a^{12}$ ■

Example 9 $(s^{-2})^3 = s^{-6} = \frac{1}{s^6}$ ■

Example 10 $\left(\frac{a}{b}\right)^4 = \frac{a^4}{b^4}$ ■

Example 11 $\left(\frac{a^2}{a^{-3}}\right)^2 = \frac{a^4}{a^{-6}} = a^{4-(-6)} = a^{10}$ ■

Example 12 $(a^{-2}b^3)^{-1} = (a^{-2})^{-1} \cdot (b^3)^{-1} = a^2b^{-3} = \frac{a^2}{b^3}$ ■

Example 13 $\left(\frac{a}{b}\right)^{-1} = \frac{a^{-1}}{b^{-1}} = \frac{b}{a}$ ■

Example 14 $\left(\frac{a}{b}\right)^{-2} = \frac{a^{-2}}{b^{-2}} = \frac{b^2}{a^2} = \left(\frac{b}{a}\right)^2$ ■

Examples 13 and 14 illustrate the following principle:

If the sign of the exponent on a fraction is changed, the result is to invert the fraction.

Example 15 Remove parentheses and write using only positive exponents:

$$\left(\frac{x^2y^3}{6z^{-2}}\right)^{-2}$$

Method 1 Change the sign of the exponent on the fraction and invert the fraction:

$$\left(\frac{x^2y^3}{6z^{-2}}\right)^{-2} = \left(\frac{6z^{-2}}{x^2y^3}\right)^2 = \frac{6^2z^{-4}}{x^4y^6} = \frac{36}{x^4y^6z^4}$$

Method 2 Use properties of exponents first:

$$\left(\frac{x^2y^3}{6z^{-2}}\right)^{-2} = \frac{x^{-4}y^{-6}}{6^{-2}z^4} = \frac{36}{x^4y^6z^4}$$

Method 3 Rewrite fraction using positive exponents first, then change sign on exponent of fraction and invert:

$$\left(\frac{x^2y^3}{6z^{-2}}\right)^{-2} = \left(\frac{x^2y^3z^2}{6}\right)^{-2} = \left(\frac{6}{x^2y^3z^2}\right)^2 = \frac{36}{x^4y^6z^4}$$ ■

Example 16 Write as a fraction using only positive exponents:

$$a^{-1} + b^{-1} = \frac{1}{a} + \frac{1}{b} = \frac{a+b}{ab}$$

$$3 + 3^{-2} = 3 + \frac{1}{3^2} = 3 + \frac{1}{9} = \frac{27}{9} + \frac{1}{9} = \frac{28}{9}$$ ■

Exponents are very useful in many ways, one of them being scientific notation.

Scientific notation

A number is written in **scientific notation** if it is expressed as a product of a number between 1 and 10 and an integer power of 10.

Number	Scientific Notation
1,000,000	1×10^6
0.00005	5×10^{-5}
2,300,000	2.3×10^6
0.034	3.4×10^{-2}

Note that there is an easy rule of thumb to make the transformation from decimal notation to scientific notation. If the number is between 0 and 1, the exponent of 10 is negative and is the number of places it is necessary to move the decimal point to the right to obtain a number between 1 and 10. If the number is greater than 10, the exponent of 10 is positive and is the number of places it is necessary to move the decimal point to the left to obtain a number between 1 and 10.

Example 17

$$105 = 1.05 \times 10^2$$
$$0.0105 = 1.05 \times 10^{-2}$$
$$0.0000105 = 1.05 \times 10^{-5}$$
$$105{,}000 = 1.05 \times 10^5$$ ■

Example 18 The distance to Barnard's star is 6.1 light-years, and a light-year is about 5.8×10^{12} mi. If a space ship travels 200,000 mi/hr, how long would it take to travel to the vicinity of Barnard's star? Using the formula time = distance/rate with

distance = $(6.1)(5.8 \times 10^{12})$

rate = 2×10^5

the time for the trip will be

$$t = \frac{(6.1)(5.8 \times 10^{12})}{2 \times 10^5}$$
$$= 17.7 \times 10^7 \text{ hr} = 1.77 \times 10^8 \text{ hr}$$

In years, the time would be

$$\frac{1.77 \times 10^8}{(3.65 \times 10^2)(2.4 \times 10)} = 2 \times 10^4 \text{ yr} \qquad \textit{(rounded off)}$$

The trip would take about 20,000 years! ■

Exercises 2.2

Simplify and write the expressions in Exercises 1–40 using positive exponents.

1. $a^2b^3a^2b^4$ **2.** $x^2y^4x^6y^2$ **3.** $\frac{x^3}{x^2}$

4. $\frac{a^4}{a^2}$ **5.** $\frac{m^2}{m^6}$ **6.** $\frac{n^3}{n^5}$

7. $(x^2)^4$ **8.** $(a^3)^2$ **9.** $(a^2b)^2$

10. $(xy^2)^2$ **11.** $\left(\frac{x^2}{y}\right)^2$ **12.** $\left(\frac{a^2}{b}\right)^2$

13. $\frac{a^0b^{-1}c^2}{a^{-2}bc}$ **14.** $\frac{x^{-3}y^0z^2}{x^2y^{-2}z}$ **15.** $\left(\frac{x^2y^3}{z}\right)^{-2}$

16. $\left(\frac{a^2b^4}{c}\right)^{-2}$ **17.** $\left(\frac{2x^2y^{-2}z^3}{-3xz^2}\right)^{-4}$ **18.** $\left(\frac{-2a^{-3}b^2c^{-4}}{3a^2bc}\right)^{-2}$

19. $2^{-3} + 3^{-2}$ **20.** $2^{-3} - 3^{-2}$ **21.** $(-2)^{-2}x^2y^{-3}$

22. $(-3)^{-3}x^{-4}y^2$ **23.** $17^0 - 0^{17}$ **24.** $5^0 + 0^5$

25. $x + x^{-1}$ **26.** $y - y^{-1}$ **27.** $(x + x^{-1})^{-1}$

28. $(y - y^{-1})^{-1}$ **29.** $a - a^{-1}b$ **30.** $x + x^{-1}y$

31. $ab^{-1} - ba^{-1}$ **32.** $r^{-1}s + s^{-1}r$ **33.** $2^{-1} + 3^{-1}$

34. $4^{-1} + 5^{-1}$ **35.** $\frac{s^{-1} + t^{-1}}{s^{-1}t^{-1}}$ **36.** $\frac{x^{-1} - y^{-1}}{x^{-1}y^{-1}}$

37. $\left(\frac{a}{b} - 1\right)^{-1}$ **38.** $\left(\frac{x}{y} + 1\right)^{-1}$ **39.** $\frac{x}{y^{-1}} + \frac{y}{x^{-1}}$

40. $\frac{a}{b^{-1}} - \frac{b}{a^{-1}}$

Write each of the numbers in Exercises 41–52 in scientific notation.

41. 155 **42.** 375 **43.** 15.5 **44.** 37.5

45. .00045 **46.** .0057 **47.** .0000043 **48.** .00000032

49. 14×10^3 **50.** 12×10^6 **51.** 23×10^{-3} **52.** 154×10^{-2}

Write each of the numbers in Exercises 53–64 in decimal notation.

53. 1.3×10^3 **54.** 2.5×10^4 **55.** 2.5×10^{-4} **56.** 1.3×10^{-5}

57. 8.45×10^2 **58.** 2.34×10^2 **59.** 25×10^{-1} **60.** 48×10^{-1}

61. $.24 \times 10^4$ **62.** $.47 \times 10^3$ **63.** $.08 \times 10^{-4}$ **64.** $.03 \times 10^{-2}$

Perform the calculations in Exercises 65–70.

65. $\dfrac{(2.5 \times 10^{2})(2 \times 10^{-1})}{(5 \times 10^{-3})}$

66. $\dfrac{(4 \times 10^{6})(3 \times 10^{3})}{(6 \times 10^{-5})}$

67. $\dfrac{(36 \times 10^{-1})(5.2 \times 10^{-3})}{(6 \times 10^{-2})(1.3 \times 10^{-2})}$

68. $\dfrac{(24 \times 10^{-2})(3.4 \times 10^{3})}{(8 \times 10^{-3})(1.7 \times 10^{-2})}$

69. $\dfrac{(63{,}000)(.02)}{(180)(.004)}$

70. $\dfrac{(15{,}000)(.06)}{(.0005)(300)}$

For Your Calculator

Most scientific calculators have a key for scientific notation, usually [ee] or [exp]. To enter a number such as 1.3×10^{6}, you use the following key sequence:

1.3 [exp] 6

The display should read

1.3 6

For the number 1.3×10^{-6}, the key sequence is

1.3 [exp] 6 [chs]

The display should read

1.3 −6

The key sequence for computing

$$\frac{(1.4 \times 10^{-7})(2.6 \times 10^{8})}{(5.6 \times 10^{9})(1.3 \times 10^{6})}$$

in scientific notation is

Alg 1.4 [ee] 7 [chs] [×] 2.6 [ee] 8 [÷] 5.6 [ee] 9 [÷] 1.3 [ee] 6 [=]

The display would be

5 −15

This is the number 5×10^{-15}.

The average distance from the sun to each planet is given below in kilometers. Compute the time required for light from the sun to reach each planet. The velocity of light is approximately 3×10^{5} km/sec.

71. Mercury: 57.6×10^{6} **72.** Venus: 1.08×10^{8} **73.** Mars: 2.26×10^{8} **74.** Jupiter: 1.42×10^{9}

75. Uranus: 2.85×10^{9} **76.** Neptune: 4.47×10^{9} **77.** Pluto: 5.86×10^{9} **78.** Earth: 1.49×10^{8}

79. Saturn: 1.44×10^{9}

2.3 Roots, Radicals, and Rational Exponents

One would like to extend the definition of exponent to include rational exponents and have Properties 1–5 in Section 2.2 still hold. For example, $3^{1/2}$ should be a number that satisfies the relation

$$(3^{1/2})^2 = 3^{(1/2)\cdot 2} = 3^1 = 3$$

so $3^{1/2}$ should be one of the square roots of 3. This is the rationale for the definition of x^r, where r is rational. But first a few words about roots and radicals.

nth Root The number x is an **nth root** of a if $x^n = a$. For example, 2 is a 5th root of 32 since $2^5 = 32$. If n is equal to 2, we speak of a **square root,** and for $n = 3$ the term is **cube root.**

Example 1 Both 3 and -3 are square roots of 9, because $3^2 = 9$ and $(-3)^2 = 9$. ■

Example 2 The only (real) cube root of -27 is -3, because $(-3)^3 = -27$ and -3 is the only real number with that property. ■

Although the number -27 has only one *real* cube root, it does have some cube roots that are complex numbers. This point is discussed in detail in Chapter 9.

If n is odd, then a number a has only one real nth root, which we denote by

$$\sqrt[n]{a}$$

If n is even and a is positive, then a has two real nth roots; we call the positive nth root the **principal** nth root of a and denote it by

$$\sqrt[n]{a}$$

Example 3 $\sqrt{4} = 2$ (It is customary to write $\sqrt{}$ rather than $\sqrt[2]{}$ for the square root.) ■

Example 4 $\sqrt[3]{8} = 2$ ■

Example 5 $\sqrt[5]{-3125} = -5$ ■

Radicals The symbol $\sqrt[n]{}$ is called a **radical** sign, and expressions of the form $\sqrt[n]{x}$ are often referred to as **radicals.** From the definition of the nth root,

$$(\sqrt[n]{a})^n = a$$

assuming that $\sqrt[n]{a}$ is defined.

Example 6 $(\sqrt[3]{34})^3 = 34$ ■

Example 7 $(\sqrt{47})^4 = [(\sqrt{47})^2]^2 = (47)^2 = 2209$ ■

Example 8 $(\sqrt[3]{-8})^6 = [(\sqrt[3]{-8})^3]^2 = (-8)^2 = 64$ ■

The situation with $\sqrt[n]{a^n}$ is a bit different.

Example 9 $\sqrt{(-4)^2} = \sqrt{16} = 4$ ■

If n is even, then x^n is nonnegative and $\sqrt[n]{x^n}$ is the principal nth root, so

$$\sqrt[n]{x^n} = |x| \quad \text{if} \quad n \quad \text{is even}$$

and

$$\sqrt[n]{x^n} = x \quad \text{if} \quad n \quad \text{is odd}$$

Example 10 $\sqrt[5]{2^5} = 2$ ■

Example 11 $\sqrt[3]{(-3)^3} = -3$ ■

Example 12 $\sqrt[6]{(-2)^6} = |-2| = 2$ ■

And now back to the discussion of rational exponents. We want to define $a^{1/n}$ so that it obeys the laws of exponents, in particular

$$(a^{1/n})^n = a^{(1/n)(n)} = a^1 = a$$

So we would want $a^{1/n}$ to be an nth root of a.

1. If n is even and a is a positive real number, we define $a^{1/n}$ to be the principal nth root of a, that is,

$$a^{1/n} = \sqrt[n]{a}$$

Expressing nth roots with exponents

2. If n is odd and a is any real number, we define $a^{1/n}$ to be the nth root (there is only one real nth root in this case) of a, that is,

$$a^{1/n} = \sqrt[n]{a}$$

3. $0^{1/n} = 0$ for any positive integer n.
4. If n is even and a is negative, $a^{1/n}$ is not defined.

Example 13 $(16)^{1/2} = \sqrt{16} = 4$ ■

Example 14 $(-27)^{1/3} = \sqrt[3]{-27} = -3$ ■

Definition for rational exponents

If a is a positive real number or 0, and m and n are positive integers, we define

$$a^{m/n} = (a^{1/n})^m$$

If a is negative, we define $a^{m/n}$ only if n is odd, and then as above,

$$a^{m/n} = (a^{1/n})^m$$

If r is a negative rational number, then we define

$$a^r = \frac{1}{a^{-r}}$$

provided a^r is defined and nonzero.

Example 15 $(9)^{5/2} = (9^{1/2})^5 = 3^5 = 243$ ■

Example 16 $(9^5)^{1/2} = \sqrt{59{,}049} = 243$ ■

Examples 15 and 16 illustrate the fact that

$$a^{m/n} = \sqrt[n]{a^m} = (\sqrt[n]{a})^m$$

provided $\sqrt[n]{a}$ is defined. The proof is Exercise 121 of this section.

The properties of exponents used before with integer exponents also hold for rational exponents and are used in Examples 17 and 18.

Example 17 $a^{1/2}a^{1/4} = a^{1/2+1/4} = a^{3/4}$ ■

Example 18 $\left(\frac{x^{1/2}y^{3/4}}{z^{1/6}}\right)^8 = \frac{(x^{1/2})^8(y^{3/4})^8}{(z^{1/6})^8} = \frac{x^4y^6}{z^{4/3}}$ ■

Two properties of radicals are easily derived from the properties of exponents or from the definition of nth roots.

1. $\sqrt[n]{ab} = \sqrt[n]{a}\,\sqrt[n]{b}$
2. $\sqrt[n]{\frac{a}{b}} = \frac{\sqrt[n]{a}}{\sqrt[n]{b}}$

Both rules apply as long as all the radicals involved are defined.

Example 19 $\sqrt{8} = \sqrt{4}\,\sqrt{2} = 2\sqrt{2}$ ■

Simplifying radicals These properties of radicals are useful in **simplifying** radical expressions. To simplify a radical means to make the expression inside the radical as simple as possible. Observe the following examples.

Example 20 $\sqrt[5]{64x^6y^7} = \sqrt[5]{32x^5y^5}\,\sqrt[5]{2xy^2} = 2xy\,\sqrt[5]{2xy^2}$ ■

Example 21 $\sqrt{45x^3y^2} = \sqrt{9x^2y^2}\,\sqrt{5x} = 3xy\,\sqrt{5x}$ ■

The properties can also be used in combining radicals and dealing with fractions inside radicals.

Example 22 $\sqrt{3}\,\sqrt{27} = \sqrt{81} = 9$ ■

Example 23 $\sqrt{\dfrac{9}{16}} = \dfrac{\sqrt{9}}{\sqrt{16}} = \dfrac{3}{4}$ ■

Example 24 $\sqrt{\dfrac{1}{2}} = \sqrt{\dfrac{2}{4}} = \dfrac{\sqrt{2}}{2}$ ■

There are times when it is very useful to use rational exponents in place of radical notation to simplify expressions.

Example 25 $\sqrt{x} + \dfrac{1}{x} = \dfrac{x\sqrt{x} + 1}{x} = \dfrac{x^{3/2} + 1}{x}$ ■

Example 26 Simplify

$$\frac{(\sqrt{x + 1})(2) - \dfrac{1}{\sqrt{x + 1}}}{x + 1} \qquad (x > -1)$$

Multiply numerator and denominator by $\sqrt{x + 1}$:

$$\frac{(\sqrt{x + 1})(\sqrt{x + 1})(2) - \dfrac{\sqrt{x + 1}}{\sqrt{x + 1}}}{(\sqrt{x + 1})(x + 1)}$$

$$= \frac{2(x + 1) - 1}{(x + 1)^{3/2}} = \frac{2x + 1}{(x + 1)^{3/2}}$$ ■

Rationalizing the denominator

If the denominator of an expression contains a radical, it is sometimes desirable to rewrite the expression with a radical-free denominator. Such a process is called **rationalizing the denominator.** If the denominator is just a single term, rather than a sum, the process is fairly simple. See the next two examples.

Example 27 $\dfrac{1}{\sqrt{2x^3}} = \dfrac{\sqrt{2x}}{\sqrt{2x^3}\,\sqrt{2x}} = \dfrac{\sqrt{2x}}{\sqrt{4x^4}} = \dfrac{\sqrt{2x}}{2x^2} \qquad (x > 0)$ ■

Notice the strategy: since the radical in the denominator is a square root, multiply numerator and denominator by what is needed (in this case, $\sqrt{2x}$) to make the quantity inside the radical in the denominator a perfect square. In this way the radical is eliminated. In the next example the radical is a cube root, hence the multiplier is chosen to make the expression inside the radical a cube.

Example 28 $\dfrac{1}{\sqrt[3]{5}} = \dfrac{\sqrt[3]{25}}{\sqrt[3]{5}\,\sqrt[3]{25}} = \dfrac{\sqrt[3]{25}}{\sqrt[3]{5^3}} = \dfrac{\sqrt[3]{25}}{5}$ ■

It is sometimes possible to simplify sums of radicals by using properties of radicals and the distributive law.

Example 29 $2\sqrt{3} + 6\sqrt{3} = 8\sqrt{3}$ ■

Example 30 $\sqrt{8} + \sqrt{2} = 2\sqrt{2} + \sqrt{2} = 3\sqrt{2}$ ■

Example 31 $\sqrt{2}(\sqrt{2} + \sqrt{8}) = \sqrt{2}\sqrt{2} + \sqrt{2}\sqrt{8} = 2 + 4 = 6$ ■

Example 32 $(\sqrt{x} + \sqrt{y})(\sqrt{x} - \sqrt{y}) = (\sqrt{x})^2 - (\sqrt{y})^2 = x - y$ if $x \geq 0$ and $y \geq 0$. ■

The result in Example 32 is also useful in rationalizing denominators when the denominator is a sum.

Example 33
$$\frac{1}{\sqrt{x} + \sqrt{y}} = \frac{(\sqrt{x} - \sqrt{y})}{(\sqrt{x} + \sqrt{y})(\sqrt{x} - \sqrt{y})} = \frac{\sqrt{x} - \sqrt{y}}{x - y}$$
if $x \neq y$ and $x \geq 0$ and $y \geq 0$. ■

Example 34
$$\frac{5}{\sqrt{3} - \sqrt{2}} = \frac{5(\sqrt{3} + \sqrt{2})}{(\sqrt{3} - \sqrt{2})(\sqrt{3} + \sqrt{2})} = \frac{5\sqrt{3} + 5\sqrt{2}}{3 - 2} = 5\sqrt{3} + 5\sqrt{2}$$ ■

Exercises 2.3

In Exercises 1–16 evaluate the radical.

1. $\sqrt{25}$ **2.** $\sqrt{49}$ **3.** $\sqrt[3]{64}$ **4.** $\sqrt[3]{216}$

5. $\sqrt[4]{16}$ **6.** $\sqrt[4]{256}$ **7.** $(\sqrt{67})^2$ **8.** $(\sqrt{14})^2$

9. $(\sqrt[3]{-3})^3$ **10.** $(\sqrt[3]{-7})^3$ **11.** $\sqrt{(-12)^2}$ **12.** $\sqrt{(-5)^2}$

13. $\sqrt{4^3}$ **14.** $\sqrt[3]{8^2}$ **15.** $\sqrt[5]{(-3)^5}$ **16.** $\sqrt[3]{(-2)^3}$

Write the expressions in Exercises 17–24 using rational exponents. Assume that all variables represent positive real numbers.

17. $\sqrt{6}$ **18.** $\sqrt{7}$ **19.** $\sqrt[3]{x^5}$ **20.** $\sqrt[3]{y^5}$

21. $\sqrt{a^3b^5}$ **22.** $\sqrt{xy^2}$ **23.** $\sqrt{x + y}$ **24.** $\sqrt{a^2 + b^2}$

Write the expressions in Exercises 25–32 in radical form. Assume all variables are positive real numbers.

25. $(xy)^{3/4}$ **26.** $(ab)^{2/3}$ **27.** $(a + b)^{5/2}$ **28.** $(x + y)^{3/2}$

29. $m^{-3/5}$ **30.** $n^{-3/2}$ **31.** $(a^3b^2)^{5/6}$ **32.** $(m^2n^3)^{5/12}$

Evaluate the expressions in Exercises 33–44.

33. $16^{3/4}$ **34.** $8^{2/3}$ **35.** $25^{3/2}$ **36.** $(-27)^{2/3}$

37. $(-64)^{2/3}$ **38.** $81^{3/4}$ **39.** $125^{4/3}$ **40.** $256^{3/4}$

41. $2^{1/2}8^{1/2}$ **42.** $5^{1/3}25^{1/3}$ **43.** $4^{1/3}16^{1/3}$ **44.** $3^{1/2}27^{1/2}$

Simplify the expressions in Exercises 45–56. Assume all variables represent positive real numbers.

45. $a^{1/5}a^{1/2}$ **46.** $x^{1/3}x^{1/4}$ **47.** $t^{2/3}t^{1/2}$ **48.** $s^{3/4}s^{1/3}$

49. $\dfrac{x^{1/2}}{x^{1/3}}$ **50.** $\dfrac{u^{1/3}}{u^{1/2}}$ **51.** $\left(\dfrac{x^{1/2}y^3}{x\,y^{1/2}}\right)^3$ **52.** $\left(\dfrac{ab^{1/3}}{a^{1/3}\,b}\right)^2$

53. $\left(\dfrac{x^2y^3}{w^4}\right)^{5/6}$ **54.** $\left(\dfrac{a^2b^5}{c^4}\right)^{3/10}$ **55.** $\left(\dfrac{a^{2/3}b^{1/3}}{c^{1/2}}\right)^{3/2}$ **56.** $\left(\dfrac{x^{1/5}y^{2/5}}{z^{1/2}}\right)^{5/2}$

In Exercises 57–76 simplify or evaluate the radical. Assume all variables represent positive real numbers.

57. $\sqrt{250}$ **58.** $\sqrt{640}$ **59.** $\sqrt[3]{8{,}000}$ **60.** $\sqrt[3]{27{,}000}$

61. $\sqrt[3]{81}$ **62.** $\sqrt[3]{40}$ **63.** $\sqrt{16x^2}$ **64.** $\sqrt{25z^2}$

65. $-\sqrt{169a^4}$ **66.** $-\sqrt{121b^6}$ **67.** $\sqrt{32x^4y^8}$ **68.** $\sqrt{81m^4n^5}$

69. $\sqrt{50}\,\sqrt{2}$ **70.** $\sqrt{5}\,\sqrt{20}$ **71.** $\sqrt[3]{16}\,\sqrt[3]{4}$ **72.** $\sqrt[3]{6}\,\sqrt[3]{36}$

73. $\sqrt{.0009}$ **74.** $\sqrt{.0036}$ **75.** $\sqrt{25 \times 10^{-4}}$ **76.** $\sqrt{36 \times 10^{-6}}$

In Exercises 77–98 simplify and collect like terms where possible. Assume all variables represent positive real numbers.

77. $\sqrt{28} + 2\sqrt{7}$ **78.** $\sqrt{20} + 2\sqrt{5}$ **79.** $\sqrt{8a^3} + \sqrt{18a^3}$

80. $\sqrt{27b^3} - \sqrt{75b^3}$ **81.** $\sqrt{12x} + \sqrt{27x} - \sqrt{3x}$ **82.** $\sqrt{8x} + \sqrt{18x} - \sqrt{2x}$

83. $\sqrt[3]{16} + \sqrt[3]{2} - \sqrt[3]{54}$ **84.** $\sqrt[3]{81} - \sqrt[3]{24} + \sqrt[3]{3}$ **85.** $\dfrac{2\sqrt{3}}{3} - \dfrac{1}{\sqrt{3}}$

86. $\dfrac{2}{\sqrt{5}} - \dfrac{2\sqrt{5}}{5}$ **87.** $\dfrac{1}{\sqrt{8}} - \dfrac{1}{\sqrt{2}}$ **88.** $\dfrac{1}{\sqrt{27}} + \dfrac{1}{\sqrt{3}}$

89. $\sqrt{3}(\sqrt{3} + \sqrt{15})$ **90.** $\sqrt{2}(\sqrt{10} + \sqrt{2})$ **91.** $\sqrt[3]{2}(\sqrt[3]{4} + \sqrt[3]{6})$

92. $\sqrt[3]{3}(\sqrt[3]{9} + \sqrt[3]{6})$ **93.** $(2 + \sqrt{3})(2 - \sqrt{3})$ **94.** $(3 - \sqrt{2})(3 + \sqrt{2})$

95. $(\sqrt{2} + \sqrt{3})(\sqrt{2} - 4\sqrt{3})$ **96.** $(\sqrt{5} - 2\sqrt{3})(\sqrt{5} + \sqrt{3})$ **97.** $(2 - \sqrt{x - 2})^2$

98. $(3 + \sqrt{x - 3})^2$

In Exercises 99–114 rationalize the denominator in each expression. Assume all radicals are defined and denominators are nonzero.

99. $\sqrt{\dfrac{2}{7}}$ **100.** $\sqrt{\dfrac{3}{5}}$ **101.** $\sqrt[3]{\dfrac{1}{9}}$

102. $\sqrt[3]{\dfrac{1}{25}}$ **103.** $\sqrt{\dfrac{5}{6a^2b^3}}$ **104.** $\sqrt{\dfrac{7}{12x^3y^4}}$

105. $\sqrt[3]{\dfrac{3}{25m^4n^2}}$ **106.** $\sqrt[3]{\dfrac{5}{9a^2b^4}}$ **107.** $\dfrac{1}{1 + \sqrt{2}}$

108. $\dfrac{1}{1 + \sqrt{3}}$ **109.** $\dfrac{7\sqrt{3}}{2\sqrt{3} - 1}$ **110.** $\dfrac{2\sqrt{2}}{3\sqrt{2} + 1}$

111. $\dfrac{\sqrt{5}}{\sqrt{5} - \sqrt{10}}$ **112.** $\dfrac{\sqrt{3}}{\sqrt{3} - \sqrt{12}}$ **113.** $\dfrac{\sqrt{x - 1}}{1 - \sqrt{x - 1}}$

114. $\dfrac{\sqrt{a - 1}}{1 + \sqrt{a - 1}}$

In Exercises 115–118 simplify the expression. Assume the quantities inside the radicals are positive if necessary.

115. $\dfrac{(\sqrt{x^2+1})(2) - \dfrac{x}{\sqrt{x^2+1}}(2x)}{x^2+1}$

116. $\dfrac{(\sqrt{x^3+2})(4) - \dfrac{3x^2}{\sqrt{x^3+2}}(2x)}{x^3+2}$

117. $\dfrac{(x+5)^{1/3}(6) - (x+5)^{-2/3}(2x)}{(x+5)^{2/3}}$

118. $\dfrac{(x-3)^{1/3}(9) - (x-3)^{-2/3}(3x)}{(x-3)^{2/3}}$

119. Show that for all $n \geq 0$, $\sqrt{n+1} - \sqrt{n} = \dfrac{1}{\sqrt{n+1}+\sqrt{n}}$. (Hint: write $\sqrt{n+1} - \sqrt{n}$ as a fraction and rationalize the *numerator*.)

120. Show that for all $n \geq 0$, $\sqrt{n^2+2n} - n = \dfrac{2n}{\sqrt{n^2+2n}+n}$.

121. Show that $a^{m/n} = \sqrt[n]{a^m}$.

2.4 Complex Numbers

The system of natural numbers evolved from our need to count objects, and we can well imagine how early mathematicians could have justified the need for rational numbers. However, irrational numbers such as π and $\sqrt{2}$ could not be long ignored, and soon the concept of the set of real numbers was generally accepted, although the idea was not very well formulated for some time. Still, the basic philosophy held by many was that numbers were to be used for describing quantities such as lengths, areas, and volumes.

As we have seen, for any real number x, $x^2 \geq 0$. For this reason, we did not give meaning to such expressions as $\sqrt{-1}$ and $\sqrt{-4}$. We now need to enlarge our number system to remedy this situation. People who were convinced that the real numbers were sufficient to do all mathematics related to the real world saddled the new numbers with the unfortunate name of **imaginary** numbers. Do not be frightened by this choice of words; indeed, these numbers have many applications in the sciences.

We will content ourselves with an informal introduction to the system of complex numbers, with a word of assurance that a formal and very precise definition of this system is available in more advanced texts.

Complex numbers

The system of **complex numbers** is the set

$$C = \{a + bi\colon a,\ b \quad \text{real numbers}\}$$

together with addition, multiplication, and equality as defined below. Just in case you can't wait, we will discover that $i^2 = -1$; in other words, the set C will contain a solution to the equation $x^2 + 1 = 0$.

Equality
Real and imaginary parts
Addition

1. Two complex numbers $a + bi$ and $c + di$ are **equal** if and only if $a = c$ and $b = d$. In the complex number $a + bi$, a is called the **real part** and b is called the **imaginary part.**
2. We define addition as follows:

$$(a + bi) + (c + di) = (a + c) + (b + d)i$$

Example 1 $(2 + 3i) + (1 + 4i) = (2 + 1) + (3 + 4)i$
$= 3 + 7i$ ■

Example 2 $(3 - 4i) + (-8 + 4i) = (3 - 8) + (-4 + 4)i$
$= -5 + 0i$
$= -5$ ■

The notation in the result of Example 2 is convenient, and we agree to write

$a + 0i$ as a
$0 + bi$ as bi

It is clear that $0 + 0i = 0$ is the additive identity for the complex numbers, and that the additive inverse of $a + bi$ is $-a + (-b)i$ or, more simply, $-a - bi$.

Example 3 $(a + bi) + (0 + 0i) = a + bi$ ■

Example 4 $(3 + 2i) + (-3 - 2i) = 0 + 0i = 0$ ■

Example 5 $(8 - 4i) - (3 + 2i) = (8 - 4i) + (-3 - 2i)$
$= 5 - 6i$ ■

Multiplication of complex numbers is motivated by the need for all the field axioms to hold—the distributive law in particular—and the desire for i^2 to be -1. We present the formula below, but in practice you can multiply two complex numbers as if they were binomials in the variable i and then replace i^2 by -1 wherever it occurs.

Multiplication of complex numbers

3. We define multiplication as follows:

$$(a + bi)(c + di) = (ac - bd) + (ad + bc)i$$

Example 6 $(2 + 3i)(1 - i) = 2(1 - i) + 3i(1 - i)$
$= 2 - 2i + 3i - 3i^2$
$= 2 + i - 3(-1)$
$= 5 + i$ ■

Example 7 $i(1 - 2i) = i - 2i^2 = 2 + i$ ■

Example 8 $(1 - i)^2 = 1 - 2i + i^2 = 1 - 2i - 1 = -2i$ ■

Example 9 $3(4 + 3i) = 12 + 9i$ ■

Example 10 $(2i)(2i) = 4i^2 = -4$ ■

Example 10 tells us that $2i$ is a square root of -4. In general, we can use complex numbers to denote the square roots of negative numbers.

Example 11 $\sqrt{-16} = 4i$ ■

Example 12 $\sqrt{-25} = 5i$ ■

Example 13 $\sqrt{-47} = \sqrt{47}\, i$ since $(\sqrt{47}\, i)^2 = 47i^2 = -47$ ■

Division is best accomplished by a method reminiscent of the strategy for rationalizing the denominator of a radical expression. Remember that for real numbers, a/b means ab^{-1}, where b^{-1} is defined by the relation $b \cdot b^{-1} = 1$.

Example 14 Show that $(a + bi)\left(\dfrac{a}{a^2 + b^2} - \dfrac{b}{a^2 + b^2}\, i\right) = 1$.

$$\begin{aligned}(a + bi)\left(\frac{a}{a^2 + b^2} - \frac{b}{a^2 + b^2}\, i\right) &= a\left(\frac{a}{a^2 + b^2} - \frac{b}{a^2 + b^2}\, i\right) \\ &\quad + bi\left(\frac{a}{a^2 + b^2} - \frac{b}{a^2 + b^2}\, i\right) \\ &= \frac{a^2 + b^2}{a^2 + b^2} + \frac{ab - ab}{a^2 + b^2}\, i \\ &= 1\end{aligned}$$ ■

In Example 14 we found that the complex number $a + bi$ has a multiplicative inverse if a and b are not both zero. This assures us that division is possible in the system of complex numbers, but the operation appears a bit cumbersome. The next example illustrates a much easier method for finding the quotient of two complex numbers.

Example 15 Write the quotient $\dfrac{2 + 3i}{1 + i}$ in the form $a + bi$.

Quotients of complex numbers

Multiply numerator and denominator by $1 - i$:

$$\frac{2 + 3i}{1 + i} = \frac{(2 + 3i)(1 - i)}{(1 + i)(1 - i)} = \frac{5 + i}{1 - i^2} = \frac{5}{2} + \frac{1}{2}\, i$$ ■

Standard form
Conjugate

When a complex number is written in the form $a + bi$, a and b real numbers, it is said to be in **standard form.** If $a + bi$ is a complex number, the **conjugate** of $a + bi$ is the complex number $a - bi$, written

$$\overline{a + bi} = a - bi$$

Notice that $(a + bi)(a - bi) = a^2 - b^2i^2 = a^2 + b^2$. This is the property of a complex number and its conjugate that was used in Example 15 to write a quotient in standard form.

Example 16 Write $\dfrac{6 - i}{3 + 2i}$ in standard form.

Multiply numerator and denominator by the conjugate of $3 + 2i$, namely $3 - 2i$:

$$\frac{6-i}{3+2i} = \frac{(6-i)(3-2i)}{(3+2i)(3-2i)} = \frac{16-15i}{3^2+2^2}$$
$$= \frac{16-15i}{13} = \frac{16}{13} - \frac{15}{13}i \qquad \blacksquare$$

By the way, we will often find it convenient to leave a complex number in the form $\frac{(16-15i)}{13}$ rather than convert it to $\left(\frac{16}{13}\right) - \left(\frac{15}{13}\right)i$.

Example 17 $\frac{1}{i} = \frac{1 \cdot i}{i^2} = \frac{i}{-1} = -i$ $\blacksquare$

Example 18 $\frac{i}{2-i} = \frac{i(2+i)}{(2-i)(2+i)} = \frac{-1+2i}{5}$ $\blacksquare$

Exercises 2.4

In Exercises 1–8 write the expression as a complex number.

1. $\sqrt{-25}$
2. $\sqrt{-49}$
3. $-\sqrt{-4}$
4. $-\sqrt{-25}$
5. $3-\sqrt{-25}$
6. $2-\sqrt{-32}$
7. $36-\sqrt{-36}$
8. $-81-\sqrt{-81}$

In Exercises 9–52 write each sum, difference, product, or quotient in the form a + bi, a and b real numbers.

9. $(2+i)+(6+3i)$
10. $(4+i)+(4+5i)$
11. $(4-2i)+(4+2i)$
12. $(5+4i)+(5-2i)$
13. $(-3-2i)-(-3+5i)$
14. $(-6-4i)-(-6-3i)$
15. $3i+(7-6i)$
16. $2i+(7-6i)$
17. $(1+i)(2-i)$
18. $(2+i)(1-i)$
19. $(1-i)(1+i)$
20. $(3-i)(3+i)$
21. $(4+2i)(5-i)$
22. $(2+3i)(3-i)$
23. $(6-2i)(6+2i)$
24. $(2-6i)(2+6i)$
25. $(4i)(-4i)$
26. $(-3i)(3i)$
27. $5i+i^2$
28. i^2-2i
29. $(2-i)^2+(2+i)^2$
30. $(1-i)^2-(1+i)^2$
31. i^2+i^3
32. i^3-i^2
33. $(\sqrt{2}-\sqrt{2}\,i)^2$
34. $(\sqrt{3}+\sqrt{3}\,i)^2$
35. $\left(\frac{\sqrt{3}}{2}-\frac{1}{2}i\right)^3$
36. $\left(\frac{\sqrt{3}}{2}+\frac{1}{2}i\right)^3$
37. $\left(-\frac{1}{2}+\frac{\sqrt{3}}{2}i\right)^3$
38. $\left(-\frac{1}{2}-\frac{\sqrt{3}}{2}i\right)^3$
39. $i^4+i^3+i^2+i+1$
40. $i^4-i^3+i^2-i+1$
41. $\frac{1}{2-i}$
42. $\frac{1}{3+i}$
43. $\frac{i}{2+3i}$
44. $\frac{-i}{3-3i}$
45. $\frac{4-5i}{2i}$
46. $\frac{6-3i}{-2i}$

47. $\dfrac{2i}{1 - i} - \dfrac{2i}{1 + i}$

48. $\dfrac{i}{2 + 2i} - \dfrac{-2i}{2 - 2i}$

49. $\dfrac{1 + i}{1 - i}$

50. $\dfrac{1 - i}{1 + i}$

51. $\dfrac{-2 + 2i}{1 + i}$

52. $\dfrac{-1 - i}{2 - 2i}$

If $P(x) = x^2 - 2x + 3$, *find*:

53. $P(-i)$

54. $P(2i)$

55. $P(2 + i)$

56. $P(-2 - i)$

57. Is $-2 - i$ a root of the equation $x^2 - 4x + 5 = 0$?

58. Is $2 + i$ a root of the equation $x^2 - 4x + 5 = 0$?

59. If u and v are complex numbers with $\overline{u}$ and $\overline{v}$ their conjugates, show that the conjugate of uv is $\overline{u} \cdot \overline{v}$, that is, $(\overline{uv}) = \overline{u} \cdot \overline{v}$.

60. Show that the conjugate of $u + v$ is $\overline{u} + \overline{v}$.

61. If $P(x)$ is a polynomial with real coefficients and u is a complex number such that $P(u) = 0$, show that $P(\overline{u}) = 0$. Use Exercises 59 and 60.

62. If a and b are complex numbers and $ab = 0$, show that either $a = 0$ or $b = 0$.

2.5 Quadratic Equations

Quadratic equation

A **quadratic equation** is any equation equivalent to one of the form

$$ax^2 + bx + c = 0$$

where $a \neq 0$. In this section we examine several methods for solving quadratic equations.

One of the simplest examples of a quadratic equation is

$$x^2 - 1 = 0$$

This equation is equivalent to

$$x^2 = 1$$

and since the number 1 has two square roots, the solution set for this equation is $\{-1, 1\}$. Similarly, the equation

$$(x - 5)^2 = 4$$

has solutions given by the equations

$$\begin{aligned} x - 5 &= 2 \quad \text{and} \quad & x - 5 &= -2 \\ x &= 7 & x &= 3 \end{aligned}$$

One method of solution of a quadratic equation is to use algebra to write the equation in one of the forms of the two examples given above. This involves writing a quadratic polynomial as a perfect square.

Recall that

$$(x + a)^2 = x^2 + 2ax + a^2$$
$$(x - a)^2 = x^2 - 2ax + a^2$$

Completing the square

We shall use these two relationships to carry out a process called **completing the square.**

The question is, given a polynomial such as

$$x^2 - 6x$$

what can one add to yield a polynomial that is a perfect square? Looking at the examples above, we see that we need to add the square of one-half of the coefficient of the x term. In this case, $\left(\frac{1}{2} \cdot 6\right)^2 = 9$:

$$x^2 - 6x + 9 = (x - 3)^2$$

Example 1 Complete the square for $x^2 + 8x$.

$$\begin{aligned} x^2 + 8x + \left(\frac{1}{2} \cdot 8\right)^2 &= x^2 + 8x + 16 \\ &= (x + 4)^2 \end{aligned}$$

■

Example 2 Complete the square for $x^2 - 14x$.

$$\begin{aligned} x^2 - 14x + \left(\frac{1}{2} \cdot 14\right)^2 &= x^2 - 14x + 49 \\ &= (x - 7)^2 \end{aligned}$$

■

Let us see how to put this process into practice in solving quadratic equations.

Example 3 Solve the equation $x^2 - 6x + 4 = 0$ by completing the square.

$$\begin{aligned} x^2 - 6x + 4 &= 0 \\ x^2 - 6x &= -4 \\ x^2 - 6x + 9 &= -4 + 9 \quad \textit{(completing the square by adding 9 to both sides)} \\ (x - 3)^2 &= 5 \end{aligned}$$

The number 5 has two square roots, so we have

$$\begin{aligned} x - 3 &= \sqrt{5} & \text{or} \quad x - 3 &= -\sqrt{5} \\ x &= 3 + \sqrt{5} & x &= 3 - \sqrt{5} \end{aligned}$$

■

Example 4 Solve $x^2 - 16x - 36 = 0$.

$$\begin{aligned} x^2 - 16x - 36 &= 0 \\ x^2 - 16x &= 36 \\ x^2 - 16x + 64 &= 36 + 64 \quad \textit{(completing the square)} \\ (x - 8)^2 &= 100 \\ x - 8 = 10 \quad \text{or} \quad x - 8 &= -10 \\ x = 18 \quad \text{or} \quad x &= -2 \end{aligned}$$

■

We have developed a step-by-step procedure for solving a quadratic equation; or, in other words, an **algorithm** for solving a quadratic equation. This algorithm can be stated as a formula called the **quadratic formula.**

Consider the quadratic equation

$$ax^2 + bx + c = 0, \qquad a \neq 0$$

This equation is equivalent to

$$x^2 + \frac{b}{a}x = -\frac{c}{a}$$

Completing the square

$$x^2 + \frac{b}{a}x + \left(\frac{b}{2a}\right)^2 = -\frac{c}{a} + \left(\frac{b}{2a}\right)^2$$

$$\left(x + \frac{b}{2a}\right)^2 = -\frac{c}{a} + \left(\frac{b}{2a}\right)^2$$

The solutions now come from taking the two square roots to obtain

$$x = -\frac{b}{2a} + \sqrt{-\frac{c}{a} + \left(\frac{b}{2a}\right)^2}$$

$$x = -\frac{b}{2a} - \sqrt{-\frac{c}{a} + \left(\frac{b}{2a}\right)^2}$$

Simplifying the radical and combining the two solutions into one form, we obtain the **quadratic formula:**

Quadratic formula

$$x = \frac{-b \pm \sqrt{b^2 - 4ac}}{2a}$$

Example 5 Use the quadratic formula to solve the equation

$$2x^2 + 3x - 1 = 0$$

Here $a = 2$, $b = 3$, and $c = -1$. By the quadratic formula, the solutions are

$$x = \frac{-3 + \sqrt{3^2 - 4(2)(-1)}}{2(2)} = \frac{-3 + \sqrt{17}}{4}$$

and

$$x = \frac{-3 - \sqrt{17}}{4}$$

■

Example 6 Solve the equation $x^2 - 6x + 9 = 0$.
We have

$$a = 1 \qquad b = -6 \qquad c = 9$$

so

$$x = \frac{-(-6) \pm \sqrt{(-6)^2 - 4(1)(9)}}{2(1)} = \frac{6 \pm 0}{2} = 3$$

■

Up to now all quadratic equations have had two roots, but in Example 6 there is only one solution. In this case, the root is called a **double root.**

Example 7 Solve the equation $2x^2 + 5 = 6x$.
Rewrite the equation as

$$2x^2 - 6x + 5 = 0$$

Then

$$a = 2 \qquad b = -6 \qquad c = 5$$

and

$$x = \frac{-(-6) \pm \sqrt{(-6)^2 - 4(2)(5)}}{2(2)} = \frac{6 \pm \sqrt{-4}}{4} = \frac{3}{2} \pm \frac{1}{2}i$$

■

The roots of the quadratic equation

$$ax^2 + bx + c = 0$$

are given by the formula

$$x = \frac{-b \pm \sqrt{b^2 - 4ac}}{2a}$$

The nature of the roots

The **nature of the roots** is determined by the quantity

$$b^2 - 4ac$$

Discriminant

which is called the **discriminant** of the equation.

1. If $b^2 - 4ac = 0$ the equation has only one root, a double root.
2. If $b^2 - 4ac > 0$ the equation has two distinct real roots.
3. If $b^2 - 4ac < 0$ the equation has two distinct complex roots that are conjugates of each other.

Sometimes quadratic equations can be solved by factoring. The basic principle to be used is:

If a and b are real (or complex) numbers and $ab = 0$, then either $a = 0$ or $b = 0$.

Example 8 Solve the equation $2x^2 + x - 1 = 0$ by factoring.
Factoring yields the equation

$$(2x - 1)(x + 1) = 0$$

The roots come from the equations

$$2x - 1 = 0 \quad \text{and} \quad x + 1 = 0$$

$$x = \frac{1}{2} \qquad\qquad x = -1$$

■

You have three methods at your disposal for solving quadratic equations: completing the square, using the quadratic formula, and factoring. You should use the method that is easiest for you in each particular instance. For example, if the polynomial is easy to factor, that is probably the quickest way to find the solutions. On the other hand, it may be easy to complete the square and find the solutions. And finally, you always have recourse to the quadratic formula. With some practice, you will be able to find the best method for each problem.

WARNING
Some roots of the equation may not be solutions to the problem.

When a quadratic equation arises in the solution of a problem, the presence of two roots may pose a dilemma. While both roots may be solutions to the equation, often only one root can give a solution to the problem. This is illustrated in the next two examples.

Example 9 The price of eggs has gone up 10¢ per dozen since last month. For \$6.00 you now purchase 2 dozen fewer eggs at the higher price than at last month's price. What is the current price per dozen eggs?
Let x = current price for a dozen eggs. Then $x - 10$ is the price of eggs last month. For \$6.00 you purchase $600/x$ dozen eggs this month; last month you purchased $600/(x - 10)$ dozen eggs for the same amount. Since you get 2 dozen fewer eggs for your money this month, the equation is

$$\frac{600}{x - 10} = \frac{600}{x} + 2$$

Multiply by $x(x - 10)$:

$$\begin{aligned} 600x &= 600(x - 10) + 2x(x - 10) \\ 600x &= 600x - 6000 + 2x^2 - 20x \\ 0 &= 2x^2 - 20x - 6000 \\ 0 &= x^2 - 10x - 3000 \\ 0 &= (x - 60)(x + 50) \end{aligned}$$

The roots are $x = 60$ and $x = -50$. We reject -50 as a solution because the price of eggs must be positive. Thus the price of eggs this month must be 60¢ per dozen. So, at 60¢ per dozen you get 10 dozen for \$6.00 and at $60 - 10 = 50$¢ per dozen you get 12 dozen for \$6.00, 2 dozen more. ■

Example 10 A rectangular fish pool is 4 ft wide and 9 ft long and is surrounded by a walk of uniform width. If the area of the walk is 68 ft^2, what are the dimensions of the walk? Let x = the width of the walk. A sketch is often useful in a situation like this one.

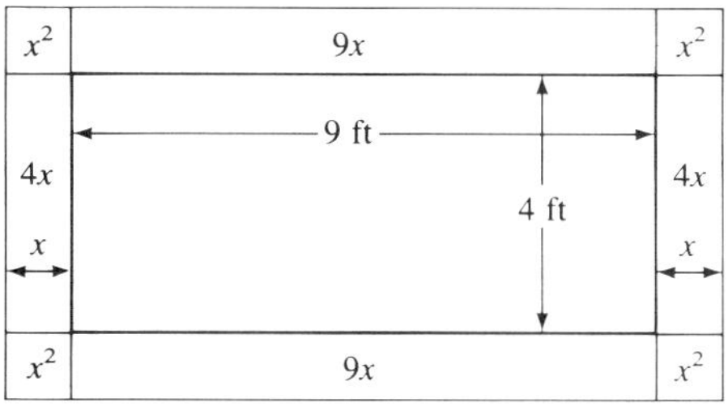

The area of the walk is given by the formula

$$\begin{aligned} A &= x^2 + x^2 + x^2 + x^2 + 9x + 9x + 4x + 4x \\ &= 4x^2 + 26x \end{aligned}$$

The area is 68 ft^2, so we want to solve the equation

$$\begin{aligned} 4x^2 + 26x &= 68 \\ 4x^2 + 26x - 68 &= 0 \\ 2x^2 + 13x - 34 &= 0 \\ (2x + 17)(x - 2) &= 0 \\ x = -\frac{17}{2} \qquad x &= 2 \end{aligned}$$

The root $-\frac{17}{2}$ is rejected because the dimensions cannot be negative. The walk is 2 ft wide. Check it for yourself. ■

Exercises 2.5

Solve the quadratic equations by taking the square root.

1. $x^2 = 16$ **2.** $x^2 = 25$ **3.** $x^2 = \frac{1}{9}$

4. $x^2 = \frac{1}{4}$ **5.** $(x - 1)^2 = 16$ **6.** $(x + 2)^2 = 25$

Solve the quadratic equations by completing the square.

7. $x^2 + 12 = 8x$ **8.** $a^2 - 12 = 4a$ **9.** $x^2 - 2x = 48$

10. $m^2 + 8m = -15$ **11.** $z^2 + 6z - 4 = 0$ **12.** $m^2 - 2m - 5 = 0$

13. $u^2 - 8u + 12 = -4$ **14.** $v^2 + 6v - 5 = -14$

15. $s^2 - 10s + 5 = 0$ **16.** $t^2 - 6t - 3 = 0$

17. $a^2 - 3a + 7 = 0$ **18.** $b^2 - 5b + 10 = 0$

19. $6x^2 - 5x + 1 = 0$ **20.** $8a^2 - 6a + 1 = 0$

Use the quadratic formula to solve the equations in Exercises 21–40.

21. $x^2 = x$

22. $x^2 = 5x$

23. $x^2 - 3x + 2 = 0$

24. $x^2 + x - 2 = 0$

25. $7x^2 - 11x + 4 = 0$

26. $3x^2 - 15 = 4x$

27. $x^2 + 1 = 2x$

28. $x^2 + 16 = 8x$

29. $3x^2 - 15x - 42 = 0$

30. $4x^2 - 8x - 12 = 0$

31. $x^2 - 6x + 4 = 0$

32. $x^2 + 2x - 4 = 0$

33. $2x^2 = 10x + 9$

34. $3x^2 = 2x + 6$

35. $4x^2 + 3x + 2 = 0$

36. $5a^2 + 8a + 7 = 0$

37. $2t^2 - 4t = 3$

38. $5w - 2w^2 = 1$

39. $3x^2 + 7x = 6$

40. $2t^2 - 4t = 3$

Solve the equations in Exercises 41–52 by factoring.

41. $2s^2 - 3s = s^2 - 5s$

42. $5y^2 + 4y = 4y^2 - 3y$

43. $x^2 - 7x + 6 = 0$

44. $y^2 - 5y + 6 = 0$

45. $m^2 - 4m - 12 = 0$

46. $n^2 - 2n - 8 = 0$

47. $p^2 + 3p - 10 = 0$

48. $s^2 + 9s - 22 = 0$

49. $5a^2 + 7a + 2 = 0$

50. $2a^2 + 11a + 5 = 0$

51. $2a(a + 1) + a^2 = 2a + 12$

52. $a(2a + 1) + 2a^2 = a + 16$

In Exercises 53–58 solve for the variable indicated.

53. $a^2 + 2ab + b^2 = 0$, for a

54. $m^2 - 4mn + 4n^2 = 0$, for m

55. $2x^2 - 5xy - 12y^2 = 0$, for y

56. $3x^2 + 14xy - 5y^2 = 0$, for y

57. $2x^2 + 14xz - 8z^2 = 0$, for x

58. $3x^2 + 9xw - 6w^2 = 0$, for x

59. A box is made by cutting 2-in. squares from each corner of a square of cardboard and folding up the edges. Find the dimensions of the square if the volume of the box is 242 in^3.

60. A piece of wire 60 cm long is bent into the shape of a right triangle. If the length of the hypotenuse is 26 cm, find the lengths of the two legs of the triangle.

61. The exterior dimensions of a building are 10 m by 12 m. A footing of uniform width is to be poured with a surface area of 21 m^2. What is the width of the footing?

62. A room that is 4 ft longer than it is wide requires 96 ft^2 of carpeting. What are the dimensions of the room?

63. The length of a rectangle is 3 times the width. If the width is increased by 2 m and the length is increased by 9 m, the area is doubled. What are the dimensions of the rectangle?

64. If one side of a square is increased by 2 m and the other dimension is decreased by 2 m, the resulting rectangle has an area of 21 m^2. What are the dimensions of the square?

65. Find two consecutive integers such that the sum of their squares is 85.

66. Find two consecutive positive integers such that the difference of their squares is 17.

67. Find two consecutive integers such that the square of the first plus twice the square of the second is 134.

68. Find three consecutive positive integers such that twice the square of the first plus the square of the second is twice the square of the third.

69. Find two consecutive odd integers whose product is 143.

70. Find two consecutive even integers whose product is 288.

2.6 Other Equations and Inequalities

Some equations, although not quadratic, may be solved by the methods applicable to quadratic equations.

Example 1 Solve $x^4 + x^2 - 2 = 0$ (which is equivalent to $(x^2)^2 + (x^2) - 2 = 0$). This equation is quadratic in x^2. Using the quadratic formula,

$$x^2 = \frac{-1 \pm \sqrt{1 - 4(1)(-2)}}{2} = \frac{-1 \pm 3}{2}$$

Thus x is a root if either

$$x^2 = -2 \quad \text{or} \quad x^2 = 1$$

The roots are $\sqrt{2}\, i$, $-\sqrt{2}\, i$, -1, and 1. ∎

Example 2 Solve $x - 5\sqrt{x} + 6 = 0$ (or $(\sqrt{x})^2 - 5\sqrt{x} + 6 = 0$). This equation is quadratic in $\sqrt{x}$. Factoring yields

$$(\sqrt{x} - 3)(\sqrt{x} - 2) = 0.$$

Thus x is a root if either

$$\sqrt{x} = 3 \quad \text{or} \quad \sqrt{x} = 2$$

that is, if either

$$x = 9 \quad \text{or} \quad x = 4$$ ∎

The equation in Example 2 can be solved by a different method.

Example 3 Solve $x - 5\sqrt{x} + 6 = 0$. Rewrite the equation with the radical alone on one side:

$$x + 6 = 5\sqrt{x}$$

Square both members:

$$\begin{aligned}(x + 6)^2 &= (5\sqrt{x})^2 \\ x^2 + 12x + 36 &= 25x \\ x^2 - 13x + 36 &= 0 \\ (x - 9)(x - 4) &= 0\end{aligned}$$

The roots are $x = 9$ and $x = 4$. ∎

WARNING Check solutions in the original equation.

There is only one drawback to the method used in Example 3. The equation $A = B$ is not necessarily equivalent to the equation $A^2 = B^2$. For example, the equation $x = -3$ has only one solution, yet the equation $x^2 = (-3)^2 = 9$ has two solutions, $x = 3$ and $x = -3$. So when both members of an equation are squared in the process of seeking the solutions, all alleged solutions should be checked in the *original* equation.

Example 4 Solve the equation $\sqrt{3x + 7} = x - 1$.
Square both members:

$$3x + 7 = (x - 1)^2$$
$$0 = x^2 - 5x - 6$$
$$0 = (x - 6)(x + 1)$$

It would appear that the roots are $x = 6$ and $x = -1$, but they must be checked in the original equation.
Try $x = 6$:

$$\sqrt{3(6) + 7} = \sqrt{25} = 5$$
$$6 - 1 = 5 \qquad \textit{OK}$$

Try $x = -1$:

$$\sqrt{3(-1) + 7} = \sqrt{4} = 2$$
$$-1 - 1 = -2 \qquad \textit{NO}$$

The only solution is $x = 6$. ∎

Example 5 $\sqrt{10 - x} + 3 = \sqrt{3x + 7}$
Square both members:

$$10 - x + 6\sqrt{10 - x} + 9 = 3x + 7$$
$$6\sqrt{10 - x} = 4x - 12$$
$$3\sqrt{10 - x} = 2x - 6$$

Square again:

$$9(10 - x) = 4x^2 - 24x + 36$$
$$4x^2 - 15x - 54 = 0$$
$$(x - 6)(4x + 9) = 0$$
$$x = 6 \qquad x = -\frac{9}{4}$$

Try $x = 6$:

$$\sqrt{10 - 6} + 3 = 2 + 3 = 5$$
$$\sqrt{3 \cdot 6 + 7} = \sqrt{25} = 5 \qquad \textit{OK}$$

Try $x = -\frac{9}{4}$:

$$\sqrt{10 - \left(-\frac{9}{4}\right)} + 3 = \frac{7}{2} + 3 = \frac{13}{2}$$
$$\sqrt{3\left(-\frac{9}{4}\right) + 7} = \frac{1}{2} \neq \frac{13}{2} \qquad \textit{NO}$$

The only solution is $x = 6$. ∎

Factoring is quite useful in solving nonlinear inequalities.

Example 6 The XYZ Manufacturing Company predicts that the profit P from producing x items per day is given by the formula

$$P(x) = 28x - 132 - x^2$$

What range of production guarantees a positive profit?

To answer the question, one must solve the inequality

$$28x - 132 - x^2 > 0$$

Rewrite the inequality:

$$x^2 - 28x + 132 < 0$$

Factor:

$$(x - 22)(x - 6) < 0$$

The number x will be a solution to the inequality if and only if $x - 22$ and $x - 6$ have opposite signs. See the graph below.

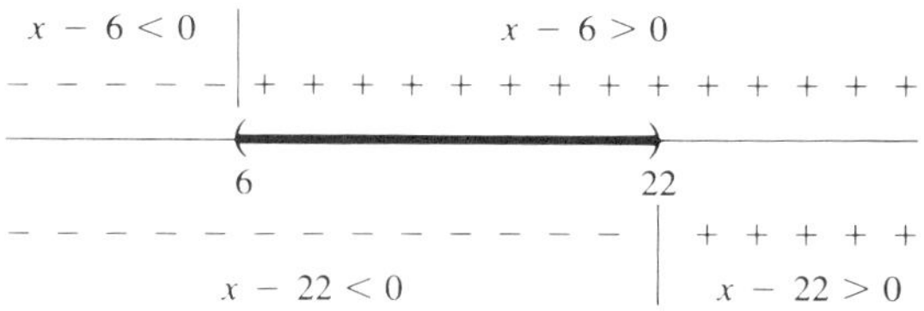

If $x < 6$, then $x - 6$ and $x - 22$ are both negative, and the product is positive. If $x > 22$, then both $x - 6$ and $x - 22$ are positive, and the product is positive. Only for $6 < x < 22$ do the factors $x - 22$ and $x - 6$ have opposite signs. Hence the solution to the inequality

$$(x - 22)(x - 6) < 0$$

is

$$\{x\colon 6 < x < 22\}$$

The company will make a profit if it produces more than 6 units per day and fewer than 22 units per day. ■

Example 7 Solve the inequality $x^2 - 12 \geq x$.

Rewrite and factor:

$$x^2 - x - 12 \geq 0$$
$$(x - 4)(x + 3) \geq 0$$

The number x will be a solution if and only if $x - 4$ and $x + 3$ have the *same* sign or if one of them is zero. See the graph below.

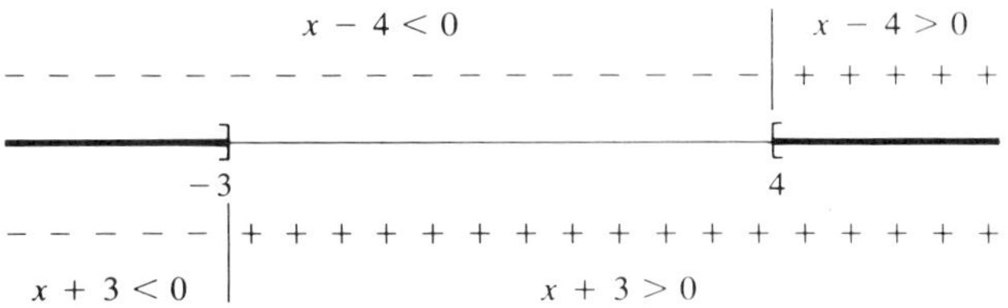

The solution set is $\{x\colon x \leq -3 \text{ or } x \geq 4\}$. ■

Example 8 Solve the inequality $\dfrac{x}{x+1} \geq 2$.

The temptation is strong to multiply both members by $x + 1$ to clear the fractions. However, the inequality is reversed if $x + 1 < 0$. To avoid this difficulty, multiply by $(x + 1)^2$, which is always nonnegative:

$$(x+1)^2 \frac{x}{x+1} \geq 2(x+1)^2$$

$$(x+1)x \geq 2(x^2 + 2x + 1)$$

$$0 \geq x^2 + 3x + 2$$

$$0 \geq (x+1)(x+2)$$

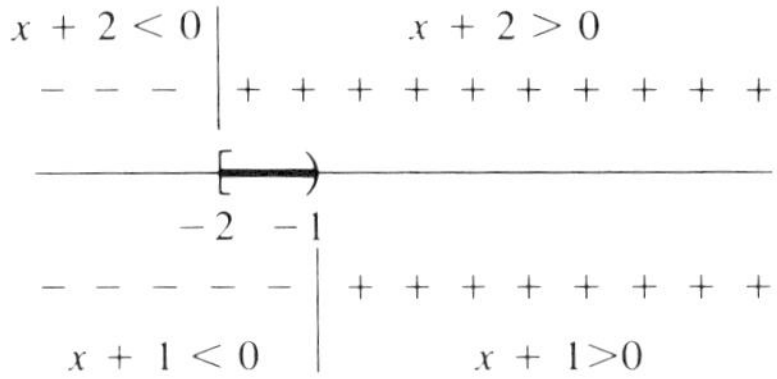

The solution set is $\{x\colon -2 \leq x < -1\}$. Notice that -1 is not included in the solution set because of the factor $x + 1$ in the denominator in the original inequality. ■

Exercises 2.6

Find the solution sets of the equations in Exercises 1–28.

1. $x - 2\sqrt{x} - 48 = 0$

2. $a - 12 = 4\sqrt{a}$

3. $3m = 13\sqrt{m} - 4$

4. $4m = 13\sqrt{m} - 3$

5. $4(x-1)^2 - 6 = 5(x-1)$

6. $10(a-1)^2 + 7(a-1) = 12$

7. $x^4 - 2x^2 + 1 = 0$

8. $y^4 - 6y^2 + 9 = 0$

9. $x^4 - 1 = 0$

10. $y^4 - 16 = 0$

11. $\sqrt{2x+1} = 3$

12. $\sqrt{3x-2} = 5$

13. $\sqrt{x-2}\,\sqrt{x+3} = 6$

14. $\sqrt{a-2}\,\sqrt{a+13} = 10$

15. $\sqrt{x+1} = x + 1$

16. $\sqrt{2x+1} = x + 1$

17. $\sqrt{3-x} - \sqrt{2+x} = 0$

18. $\sqrt{4+x} - \sqrt{2x-1} = 0$

19. $\sqrt{3x+8} = 2 + \sqrt{3x+2}$

20. $\sqrt{4x+11} = 3 + \sqrt{4x+1}$

21. $\sqrt{5-2x} + 3 = \sqrt{2-2x}$

22. $\sqrt{4-3x} + 5 = \sqrt{1-3x}$

23. $\sqrt[3]{2x+1} = 3$

24. $\sqrt[3]{3x-1} = 2$

25. $x - 6 = \sqrt{22 - x^2}$

26. $x + 4 = \sqrt{18 - x^2}$

27. $\sqrt{x+3} = 12 + \sqrt{x+1}$

28. $\sqrt{x+4} = 4 + \sqrt{x-2}$

Find the solution sets for the inequalities in Exercises 29–54. Show the sets on the number line.

29. $x(x+1) < 0$

30. $x(x-1) < 0$

31. $(x-1)(x+5) > 0$

32. $(x-9)(x+2) > 0$

33. $x^2 + 9x < -20$

34. $x^2 - 9x < -20$

35. $-9 - 6x < x^2$

36. $-25 + 10x < x^2$

37. $4x^2 \leq 4x - 1$

38. $9x^2 + 6x \leq -1$

39. $x + 12 < x^2$

40. $2x + 15 < x^2$

41. $2x - x^2 \geq 0$

42. $5x + x^2 \geq 0$

43. $(x+2)^2 \geq 9$

44. $(x-1)^2 \geq 16$

45. $x^3 - 9x > 0$

46. $x^3 - 25x > 0$

47. $x(x-1)(x+3) \leq 0$

48. $x(x+2)(x-3) \leq 0$

49. $\dfrac{x-1}{x+4} \leq 2$

50. $\dfrac{x+1}{x-4} \leq 2$

51. $\dfrac{x-2}{x+5} > 0$

52. $\dfrac{x+2}{x-6} > 0$

53. $\dfrac{2}{x-2} \geq \dfrac{4}{x}$

54. $\dfrac{4}{x+3} \geq \dfrac{9}{x}$

The height of a rocket after t sec is given by the formula $h = -16t^2 + 256t$.

55. At what times is the height greater than 624 ft?

56. At what times is the height less than 448 ft?

Chapter Review

Simplify the fractions in Exercises 1–8.

1. $\dfrac{16xy}{4x^2}$

2. $\dfrac{12ab}{3a^3}$

3. $\dfrac{x-3y}{x^2-9y^2}$

4. $\dfrac{s+2t}{s^2-4t^2}$

5. $\dfrac{x^2+7x+10}{x+2}$

6. $\dfrac{a^2+5a+6}{a+3}$

7. $\dfrac{m^2-5m-24}{m^2+4m+3}$

8. $\dfrac{u^2-4u+3}{u^2-2u+1}$

In Exercises 9–17 write each sum, product, or quotient as a single fraction and simplify.

9. $\frac{1}{x} - x$

10. $\frac{a}{a-1} + \frac{a}{a+1}$

11. $\frac{1}{x^2 - y^2} - \frac{x-y}{x^2 - 2xy + y^2}$

12. $\frac{3a+9}{4} \cdot \frac{8}{a+3}$

13. $\frac{4x-6}{3} \cdot \frac{9}{2x-3}$

14. $\frac{6a^2bc}{5xy} \cdot \frac{10x^2}{3ab^2}$

15. $\frac{8mn^2}{3ab} \cdot \frac{9b^2}{2n^2m}$

16. $\frac{a-2}{a^2 - 7a + 12} \div \frac{a^2 - 4a + 4}{3a - 12}$

17. $\frac{x-3}{x^2 - 5x + 4} \div \frac{x^2 - 9}{3x - 12}$

Solve the equations in Exercises 18–25.

18. $\frac{x-1}{x+1} = \frac{4}{5}$

19. $\frac{3s+1}{2s-1} = \frac{7}{9}$

20. $\frac{4a+7}{2a-1} = \frac{6a-5}{3a+1}$

21. $\frac{s}{3} + \frac{s}{4} = 7$

22. $1 + \frac{9}{x} = 28$

23. $\frac{4}{t} - \frac{6}{t} = 4$

24. $\frac{1}{x-5} + \frac{3}{2x-10} = -5$

25. $\frac{2}{x-1} + \frac{1}{x^2-1} = \frac{7}{x+1}$

In Exercises 26–34 simplify the expressions using only positive exponents. Assume all variables are nonzero.

26. x^2x^4

27. $(a^2b^3)^4$

28. $\frac{x^3}{x^2}$

29. $\frac{a^{-2}b^3c^{-2}}{ab^{-1}c^{-1}}$

30. $\left(\frac{-3a^{-2}b^4}{-2ab^{-2}}\right)^{-3}$

31. $x - x^{-2}$

32. $2^{-2} - (-2)^{-2}$

33. $(a^2b^3)^{-2}(a^{-2}b)^{-1}$

34. $\left(\frac{a}{b} - 2\right)^{-1}$

Write in scientific notation.

35. 0.00014

36. 1,300,000

37. 23.4

Write in decimal notation.

38. 4.4×10^{-5}

39. 2.3×10^3

40. 1.1×10^{-2}

Simplify the expressions in Exercises 41–55, rationalizing the denominators where necessary. Assume all radical expressions are defined.

41. $\left(\frac{x^{1/2}x^{1/4}}{x}\right)^4$

42. $(m^{2/3}n^{5/6})^{-3}$

43. $(x^2y^3)^{-1/3}$

44. $\sqrt[3]{\frac{16x^2}{3a}}$

45. $\sqrt[3]{27x^6y^4}$

46. $\sqrt{32a^3b^2}$

47. $\sqrt{12x^2y} + \sqrt{27x^2y}$

48. $\sqrt[3]{40x^{-6}}$

49. $\sqrt[3]{0.0008}$

50. $\sqrt[3]{2}(\sqrt[3]{4} + \sqrt[3]{16})$

51. $\frac{1}{\sqrt{27}} - \frac{2}{\sqrt{12}}$

52. $\sqrt{50}\,\sqrt{8}$

53. $\sqrt{25 \times 10^{-6}}$ **54.** $\dfrac{5}{2 + \sqrt{3}}$ **55.** $\dfrac{2}{\sqrt{x} + \sqrt{y}}$

Write in the form $a + bi$, where a and b are real numbers.

56. $6 + \sqrt{-64}$ **57.** $\sqrt{-8} + \sqrt{-2}$ **58.** $(6 + 3i) - (1 + i)$

59. $(1 + i)^2$ **60.** $(2 - 3i)^2 + (1 - 2i)^2$ **61.** $\dfrac{2}{1 + i}$

62. $\dfrac{1 + i}{1 + 2i}$ **63.** $\dfrac{32 + 4i}{8 - i}$

64. If $P(x) = x^2 - 2x + 3$, compute $P(i)$.
65. If $P(x) = x^2 + 5x - 1$, compute $P(1 - i)$.
66. If $P(x) = x^2 - 2\sqrt{3}\,x - 2$, compute $P(\sqrt{3} + \sqrt{5})$.

Find the solution sets for the equations in Exercises 67–86.

67. $x(x + 7)(3x + 4) = 0$ **68.** $27x^2 = 3$

69. $x^2 = 14 - 5x$ **70.** $x^2 + 12 = 7x$

71. $2x^2 + 5x = 7$ **72.** $12x^2 - 2 = -5x$

73. $(x + 3)^2 = 7$ **74.** $x^2 + 5x - 2 = 0$

75. $x^2 + 5x + 7 = 0$ **76.** $4x^2 - 7x + 9 = 0$

77. $\dfrac{8}{x} + x = 6$ **78.** $x(x - 2) + 3x = 16$

79. $x - 3 = \dfrac{10}{x}$ **80.** $.15x^2 - .01x = .02$

81. $\sqrt{x + 2} + 4 = x$ **82.** $\sqrt{x - 5} - \sqrt{x} = 1$

83. $x - 2\sqrt{x} - 8 = 0$ **84.** $x - 5\sqrt{x} + 2 = 26$

85. $\sqrt{x + 3} - \sqrt{x} = 1$ **86.** $x^4 - x^2 - 2 = 0$

For Exercises 87–92, show the solution set on the line.

87. $x^2 - 5x - 6 < 0$ **88.** $x^2 - 2x \geq -1$ **89.** $x^2 < 14 - 5x$

90. $\dfrac{x^2}{2} \leq 4 - x$ **91.** $(x + 3)^2 \leq 7$ **92.** $\dfrac{6}{x - 1} \leq x$

93. Ron is a master locksmith and Jerry is his apprentice. It is generally agreed that Ron can do a job in half the time it takes Jerry to do the same job. Goldie is having trouble with burglars and wishes to have all the locks in her house changed. Ron estimates that he and Jerry working together could do the job in $2\frac{1}{2}$ hr. If Ron worked alone, how long would it take him to change Goldie's locks?

94. Peter Piper is a pepper picker who finds that he can pick green peppers twice as fast as he can pick speckled peppers. In 1 hr he can pick 6 more pecks of green peppers than of speckled peppers. How long does it take for Peter Piper to pick a peck of speckled peppers?

95. Conrad put an advertisement that cost \$1.50 per square inch in a newspaper. If the ad was 1.5 in. longer than it was wide and the cost was \$20.25, what were the dimensions of Conrad's ad?

Chapter Test

1. Simplify: **a.** $\dfrac{-14a^3b^2}{2a^2b}$ **b.** $\dfrac{t^2 - 5t - 14}{7 - t}$

2. Write as a single fraction and simplify:

   a. $\dfrac{4}{2x - 1} + \dfrac{5}{3 - 6x}$ **b.** $\dfrac{6a^2 + 19a - 20}{a^2 - 18a + 81} \cdot \dfrac{a^2 - 3a - 54}{a^2 - 16}$

   c. $\dfrac{t^2 - t}{24 - 6t^2} \div \dfrac{5t^2 - 3t - 2}{2t - t^2}$

3. Simplify using only positive exponents. Assume all variables are nonzero real numbers.

   a. $(a^2b^{-4})^2$ **b.** $\left(\dfrac{-3s^2r^{-2}t^3}{6srt^{-1}}\right)^{-2}$ **c.** $(-3)^{-3} - (-2)^{-2}$

4. Write .00012 in scientific notation.
5. Write 2.3×10^4 in decimal notation.
6. Simplify using only positive exponents. Assume all variables are positive real numbers.

   $\left(\dfrac{x^2y^{1/2}}{x^{1/2}y^2}\right)^{-1/3}$

7. Simplify the expressions, rationalizing the denominator if necessary. Assume all radicals and fractions are defined.

   a. $\sqrt{32a^4b^5c^3}$ **b.** $\sqrt{3r^3s} - \sqrt{12r^3s} + \sqrt{75r^3s}$

   c. $\dfrac{4}{\sqrt{3} - \sqrt{7}}$ **d.** $\dfrac{6}{2\sqrt{a} - 3\sqrt{b}}$

8. Write in the form $a + bi$, a and b real numbers.

   a. $(2 - 3i) + (4 - 5i)$ **b.** $(2i)(4 - i)(4 + i)$

   c. $\dfrac{2 - 3i}{3 + 2i}$ **d.** $\sqrt{-8} + \sqrt{-9}$

9. Solve the equations.

   a. $4x^2 = 25$ **b.** $(9x - 1)^2 = 36$

   c. $x^2 - 3x + 5 = 0$ **d.** $x^2 - 6x - 16 = 0$

   e. $-3 + 5x = -2x^2$ **f.** $x + \dfrac{4}{x} = 5$

   g. $\sqrt{s + 7} - \sqrt{s} = 1$

10. Show the solution sets on the line.

    a. $x^2 - 9x \leq 22$ **b.** $\dfrac{4}{x + 1} < 8$

11. The perimeter of a rectangle is 30 cm and the area is 36 cm². Find the dimensions.
12. A bathtub can be filled with hot water in 10 min and with cold water in 6 min. How long will it take to fill the tub with both faucets on?

3 Functions, Graphs, and Linear Equations

Probably the largest purchase that any of us will make during our lifetime is that of a home or condominium. Of prime importance at such a time is the answer to the question "What will be the monthly payments?" The answer to this question may well be "It depends." What this means is that the amount of the monthly payment depends on the purchase price, the amount of the down payment, the number of months of the loan, the rate of interest, and possibly some other factors. In light of the new variable-rate mortgages, the dependency will play an important role throughout the life of the loan. This type of relationship between various quantities is quite common and is the subject of our consideration in this chapter.

3.1 Functions

The word *function* is one of those words that has both a precise mathematical meaning and a nontechnical meaning, which are closely related. Let us examine the intuitive meaning of function by means of the following examples:

1. A clerk at a store enters the price of each item being purchased into the register; the machine computes the sum, adds the sales tax, and displays the amount owed.
2. The loan officer at a bank uses a table to determine the amount of the monthly payments of a car loan, using the amount borrowed, the interest rate, and the number of months of the loan.
3. A service station attendant gauges the capacity of your cooling system and uses a graph on the back of the antifreeze container to determine how much antifreeze to put in your radiator.

Each of these examples involves a function. Notice the basic ingredients: an input consisting of one or more numbers and an output determined by some means. In the first case the clerk used a machine, in the second case the bank officer used a table, and in the third case the station attendant used a graph. These are but a few of the many ways in which the output of a function can be determined.

To describe a function we need to identify the following: a set of possible inputs—the *domain*—and a means of finding the output. It is important that for each possible input there be exactly one output. If you and I make identical purchases in the above-mentioned store and are charged different amounts, something is wrong with the system.

Function
Domain
Range
F(x) notation
Image

Let A and B be sets. A **function** from A to B is a correspondence that assigns to each input from A exactly one member of B. The set A is called the **domain** of the function, and the set of all possible outputs is called the **range** of the function. If F is a function from A to B and x is an element of A (a permissible input for F), then the element of B corresponding to x is written $\boldsymbol{F(x)}$. $F(x)$ is also referred to as the **image** of x under F, or the **value** of F at x.

Another commonly used method for defining a function is to give a formula or rule for finding the output. In the example below, recall that R is the set of all real numbers.

Example 1 Define F from R to R by means of the formula $F(x) = x^2 + 1$. If the input is x, the output is $x^2 + 1$. For example,

$$F(0) = 1 \qquad F(-2) = 5 \qquad F(x + 1) = (x + 1)^2 + 1$$ ■

Finding the range of a function can be a difficult task. The next example illustrates how one can go about determining the range of a function.

Example 2 Find the range of the function in Example 1.
The function F is defined by $F(x) = x^2 + 1$. A number y will be in the range of F if y is a possible output for F, that is, if there is x in the domain of F such that $y = F(x)$. In other words, can we solve the equation $y = x^2 + 1$ for x? The equation $x^2 + 1 = y$ has a real number solution for x as long as $y \geq 1$. This means that the range of F is $\{y: y \geq 1\}$. ■

If a function is defined by a formula and the domain is not specified, it will be understood that the domain is the set of all real numbers for which the formula makes sense. Rather than write "let F be the function defined by the formula $F(x) = \ldots$" we will simply speak of "the function $F(x) = \ldots$."

Example 3 Find the domain and range of the function $F(x) = \sqrt{1 - x}$.
The function F will be defined as long as $1 - x \geq 0$, that is, if $x \leq 1$. The domain is $\{x: x \leq 1\}$. To find the range of F, we wish to determine the values of y for which the equation $y = \sqrt{1 - x}$ has a solution x in $\{x: x \leq 1\}$. Squaring both members, we obtain $y^2 = 1 - x$, which is equivalent to $x = 1 - y^2$. Thus the equation $y^2 = 1 - x$ has a solution $x \leq 1$ for all y. But since $y = \sqrt{1 - x}$, the positive square root, the range is $\{y: y \geq 0\}$. ■

Example 4 The table shown is a portion of a table published by the Internal Revenue Service for the purpose of computing earned income credit. The input comes from line 3 or 4 of the worksheet and the output is E, the earned income credit. From the table,

$$E(124) = 13 \qquad E(250) = 23$$ ■

If line 3 or 4 of the worksheet is—		Your earned income credit is—
Over	But not over	
\$0	\$50	\$3
50	100	8
100	150	13
150	200	18
200	250	23
250	300	28
300	350	33
350	400	38
400	450	43
450	500	48
500	550	53
550	600	58
600	650	63
650	700	68
700	750	73
750	800	78
800	850	83
850	900	88
900	950	93
950	1,000	98

Certain functions are special enough to deserve special names.

Example 5 *Identity function* The function $f(x) = x$ is called the **identity function.** The output and the input are the same. ■

Example 6 The function $g(x) = 6$ is an example of what is called a *constant* function. ■

Constant function A **constant function** is one in which the output is always the same regardless of the input.

1-1 function A function f is **1-1 (one-to-one)** if and only if $f(a) = f(b)$ implies that $a = b$. In other words, f is *not* 1-1 if there are a and b with $a \neq b$ and $f(a) = f(b)$.

Example 7 The function $f(x) = x^2 + 1$ is *not* 1-1 because $f(2) = -5$ and $f(-2) = 5$ although $2 \neq -2$. ■

Example 8 The function $f(s) = 3s - 4$ is 1-1. To verify this, assume

$$f(a) = f(b)$$

that is

$$3a - 4 = 3b - 4$$
$$3a = 3b$$
$$a = b$$

Thus f is 1-1. ■

In Example 8 we used the letter s to denote the input for the function rather than the letter x as we had earlier. The point of this discussion is that *any* letter may be used to represent the input of a function in giving the formula to define the function. It is customary, however, to use letters toward the end of the alphabet for the input.

Functions may be used to describe many situations, and sometimes where several variables are involved it is desirable to find a function that gives the value of one of the variables in terms of the other(s).

Example 9 A ladder 15 ft long is leaning against a building. Find a function h that gives the height of the top of the ladder above the ground in terms of the distance s of the bottom of the ladder from the wall.

From the sketch,

$$s^2 + h^2 = 15^2 = 225$$

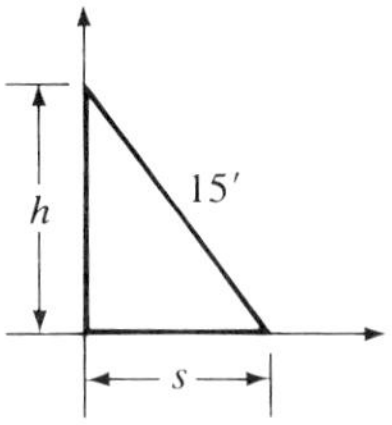

Solving for h:

$$h = \sqrt{225 - s^2}$$

or in functional notation,

$$h(s) = \sqrt{225 - s^2}$$ ■

Sometimes, as in the last example, it will be convenient to leave the function h written in the form $h = \sqrt{225 - s^2}$ rather than using the $h(s)$ form.

Example 10 A right circular cone is such that the radius is always half the height. Write the volume V as a function of the radius r.

The formula for the volume of the cone is $V = \frac{1}{3}\pi r^2 h$, where r is the radius and h is the height of the cone. Since $r = \frac{1}{2}h$, then $h = 2r$. Substituting this expression for h in the formula for V we obtain

$$V = \frac{1}{3}\pi r^2(2r) = \frac{2\pi}{3}r^3$$

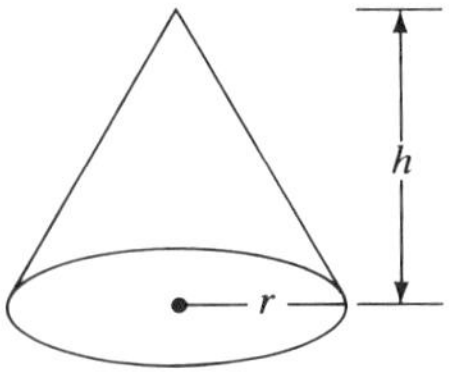

■

Exercises 3.1

The domain of each of the functions in Exercises 1–6 is the set $A = \{0, 1, 2, 3, 4\}$. Find the range of each function.

1. $f(x) = x^2 - 2$ **2.** $h(x) = 2x - 5$ **3.** $g(t) = \dfrac{3}{2t + 1}$

4. $H(s) = s^2 - 6s + 1$ **5.** $P(y) = 3 - 2y$ **6.** $S(t) = \dfrac{t}{t - 5}$

If $F(x) = 2x^2 - 3x - 1$, find

7. $F(-1)$ **8.** $F(-2)$ **9.** $F(2x)$ **10.** $F(x + 1)$

If $G(a) = 2a\left(a^2 - \dfrac{1}{a}\right)$, find

11. $G(-2)$ **12.** $G(-1)$ **13.** $G(a + 1)$ **14.** $G(3a)$

If $k(t) = |2 - 3t|$, find

15. $k(3)$ **16.** $k(-3)$ **17.** $k(4) - k(-4)$ **18.** $k(3) + k(-3)$

If $L(x) = 5 - 3x$, find

19. $\dfrac{L(4) - L(2)}{2}$ **20.** $\dfrac{L(5) - L(2)}{3}$

The table gives the amount of income tax a single person claiming one exemption must pay. Find the tax $T(a)$ owed for a taxable income a.

Taxable Income		
Over	But not over	Tax
3,400	3,450	18
3,450	3,500	25
3,500	3,550	32

21. $T(3{,}516)$ **22.** $T(3{,}479)$

23. $T(3{,}500)$ **24.** $T(3{,}450)$

In Exercises 25–32 find the domain of the function.

25. $H(x) = 2x - 3x^2 + 4$ **26.** $K(a) = 16 - (4 - a)^2$

27. $h(x) = \dfrac{x}{2x - 1}$ **28.** $p(s) = \dfrac{2s + 1}{s - 3}$

29. $T(a) = \sqrt{2 - 3a}$ **30.** $Q(y) = \sqrt{5 - 5y}$

31. $f(x) = \dfrac{3x}{x^2 - 9}$ **32.** $g(x) = \dfrac{x}{x^2 - 16}$

In Exercises 33–36 find the range of the function.

33. $f(x) = \sqrt{x + 1}$

34. $g(x) = \sqrt{x - 4}$

35. $h(t) = 1 - (t + 1)^2$

36. $h(s) = 4 - (s - 3)^2$

In Exercises 37–40 determine if the function is 1-1. Justify your conclusion.

37. $k(s) = \dfrac{s}{4 - s}$

38. $p(t) = 2 - 3t + t^2$

39. $s(t) = t^2 + 2t - 1$

40. $h(x) = \dfrac{1}{x - 2}$

41. A square of side length x is cut from each corner of a piece of tin that is 14 in. by 18 in. The sides are then folded up to form a box. Write the volume V as a function of x.

42. A wire 25 cm long is bent to form a rectangle. Find the area A as a function of the width w of the rectangle.

43. The ratio of the height to the radius of a cone is 1:3. Write the volume V as a function of the radius r.

44. A rectangle is inscribed in a semicircle of radius 4. Write the area A of the rectangle as a function of the width w of the rectangle.

45. Define $f(x)$ for $x \geq 0$ as follows: $f(x) = x^2$ if x is irrational; $f(x) = 2x - 1$ if x is rational. Is f a 1-1 function? Explain.

46. Define $g(x)$ for all x as follows: $g(x) = 2x$ if x is an integer; $g(x) = \dfrac{1}{x + 1}$ if x is not an integer. Is g a 1-1 function? Explain.

For Your Calculator

In Exercises 47–50 the function is defined by a key sequence for a calculator using algebraic logic. Write a formula for the function and find the domain of the function.

47. [√] [+] [1] [=]

48. [1/x] [−] [2] [=] [√]

49. [+] [1] [=] [√]

50. [−] [2] [=] [√] [1/x]

In Exercises 51–54 write a key sequence for the given function. Find the domain of the function.

51. $f(x) = \sqrt{x^2 - 1}$

52. $g(x) = \sqrt{x^2 - 4}$

53. $F(x) = \dfrac{1}{\sqrt{4 - x^2}}$

54. $G(x) = \dfrac{1}{\sqrt{9 - x^2}}$

3.2 Relations, Functions, and Graphs

One of the ways to describe a function is by means of a graph. In this section we explore the relationship between a function and its graph and the notion of a relation.

Rectangular coordinate system

To construct a **rectangular** (or **Cartesian**) **coordinate system,** draw two lines intersecting at right angles, label the point of intersection (or *origin*) as O and mark a scale on both lines. The horizontal line is called the x axis and the vertical line is the y axis. The coordinate axes divide the plane into four regions or *quadrants* (see Figure 3.1).

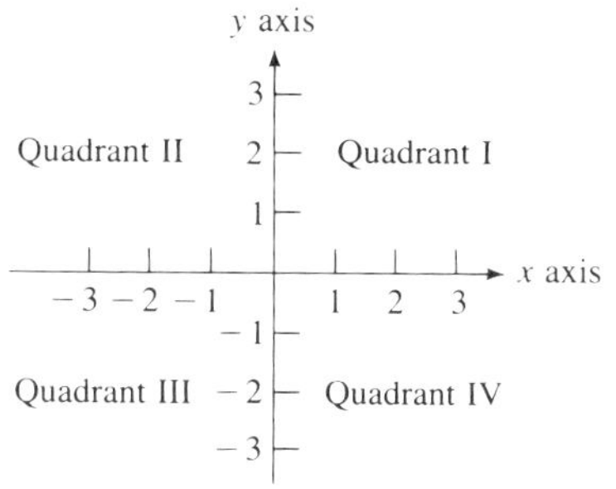

Figure 3.1

Ordered pair
x Coordinate
y Coordinate

An **ordered pair** is a pair of numbers (x, y) with one number x designated as the **first** or ***x* coordinate** and the other number y designated as the **second** or ***y* coordinate** of the pair. To each point P we assign the ordered pair (x, y), where x is the directed distance from the y axis to P and y is the directed distance from the x axis to P (see Figure 3.2). It is common usage to say "P has coordinates (x, y)." Sometimes the x coordinate is called the **abscissa** and the y coordinate is called the **ordinate.**

Abscissa
Ordinate

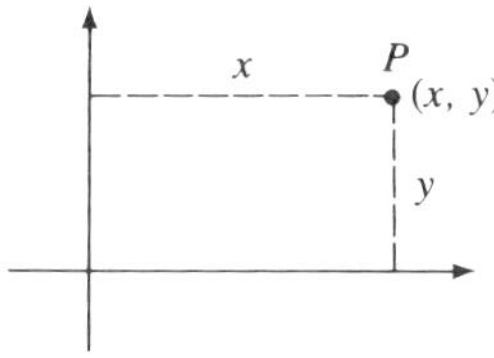

Figure 3.2

Conversely, given an ordered pair (x, y), we may use that pair to determine a point by using x as the directed distance from the y axis and y as the directed distance from the x axis (Figure 3.2).

We now have a 1-1 correspondence between points in the plane and ordered pairs of real numbers. Because of this relationship we will often speak of the point (x, y) rather than the point corresponding to the pair (x, y). To illustrate this correspondence, see Figure 3.3 for a few points identified by ordered pairs.

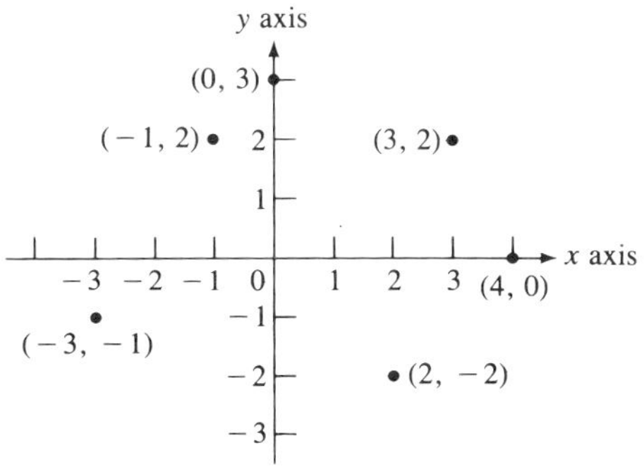

Figure 3.3

The Pythagorean theorem may be used to derive the formula for the distance between two points in the plane. Suppose P_1 and P_2 are two points, with coordinates (x_1, y_1) and (x_2, y_2), respectively. Suppose also that $x_1 \neq x_2$ and $y_1 \neq y_2$. Let P_3 be the point with coordinates (x_2, y_1) and draw the triangle with vertices P_1, P_2, and P_3 (see Figure 3.4).

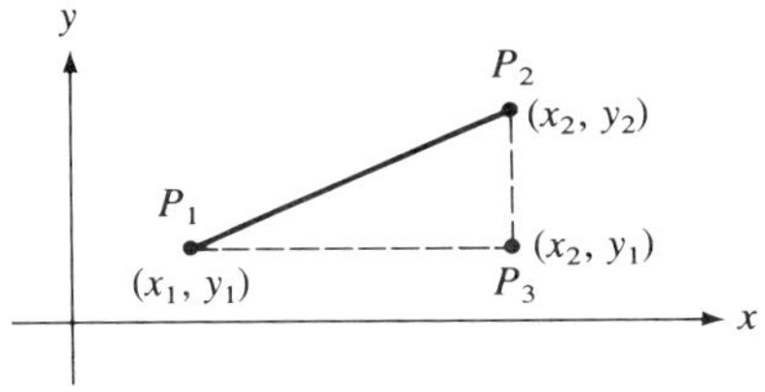

Figure 3.4

The distance from P_1 to P_2 is the length of the hypotenuse of the right triangle, and if we let $\overline{P_iP_j}$ denote the length of the segment joining P_i and P_j, the Pythagorean theorem tells us that

$$(\overline{P_1P_2})^2 = (\overline{P_1P_3})^2 + (\overline{P_2P_3})^2$$

The numbers $\overline{P_1P_3}$ and $\overline{P_2P_3}$ are easy to compute:

$$\overline{P_1P_3} = |x_1 - x_2| \quad \text{and} \quad \overline{P_2P_3} = |y_1 - y_2|$$

So we have the formula

Distance formula

$$\overline{P_1P_2} = \sqrt{(x_1 - x_2)^2 + (y_1 - y_2)^2}$$

for the distance between the points (x_1, y_1) and (x_2, y_2). Note that the formula still works when either $x_1 = x_2$ or $y_1 = y_2$.

The coordinates of the midpoint of a line segment can be expressed in terms of the coordinates of the endpoints of the segment, a relationship we will now derive. Let A and B be points with coordinates (x_1, y_1) and (x_2, y_2), respectively, and let M be the midpoint of the segment (see Figure 3.5). The three vertical lines through A, M, and B are parallel, hence the ratios

$$\frac{\overline{AM}}{\overline{BM}} \quad \text{and} \quad \frac{\overline{am}}{\overline{bm}}$$

are equal. But $\overline{AM} = \overline{BM}$, hence $\overline{am} = \overline{bm}$. Therefore the x coordinate of m (and hence of M) is $x_1 + \frac{1}{2}(x_2 - x_1) = \frac{1}{2}(x_1 + x_2)$. By a similar analysis we find that the y coordinate of M is $\frac{1}{2}(y_1 + y_2)$.

Midpoint formula

The midpoint of the line segment with endpoints (x_1, y_1) and (x_2, y_2) is the point $\left(\frac{1}{2}(x_1 + x_2), \frac{1}{2}(y_1 + y_2)\right)$.

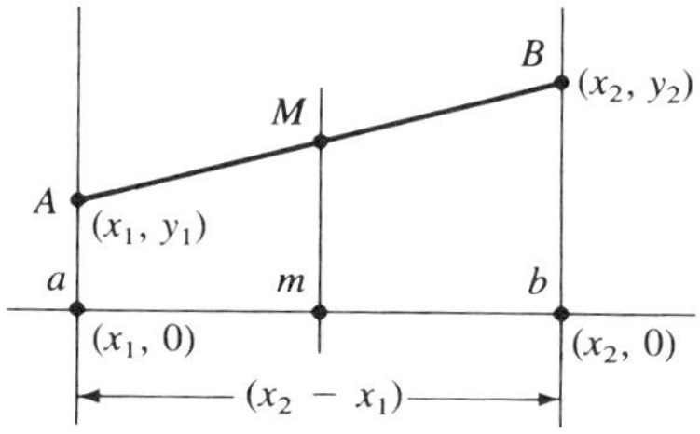

Figure 3.5

Example 1 The distance from (2, 3) to (5, −1) is

$$d = \sqrt{(2-5)^2 + [3-(-1)]^2} = 5$$

The midpoint of the segment is $\left(\frac{2+5}{2}, \frac{3-1}{2}\right) = \left(\frac{7}{2}, 1\right)$ ■

Circle A **circle** consists of all those points at a fixed distance r, called the **radius,** from a fixed point (h, k), called the **center.** Thus a point (x, y) is on the circle of radius r with center (h, k) if and only if

$$\sqrt{(x-h)^2 + (y-k)^2} = r \quad (\textit{distance from } (x, y) \textit{ to } (h, k))$$

or (squaring both members)

$$(x-h)^2 + (y-k)^2 = r^2$$

Example 2 Write the condition that (x, y) be on the circle of radius 2 with center at (0, 0). In this case, $h = k = 0$ and $r = 2$. The condition is

$$x^2 + y^2 = 4$$ ■

In Example 2 we have an equation $x^2 + y^2 = 4$ in two unknowns, x and y. The solution set consists of all pairs (x, y) such that $x^2 + y^2 = 4$; in this case the set is the circle of radius 2 with center at (0, 0). Because of this, the equation $x^2 + y^2 = 4$ is called the equation of the circle with radius 2 and center at (0, 0).

Equation of the circle To generalize: the **equation of the circle** of radius r with center at (h, k) is $(x-h)^2 + (y-k)^2 = r^2$. In other words, a point (x, y) is on the circle if and only if the pair (x, y) is in the solution set of the equation of the circle.

The equation $x^2 + y^2 = 4$ states a relationship between the coordinates of a point on the circle. We wish to generalize this idea of relation as it pertains to ordered pairs.

Relation *Domain* *Range* A **relation** is a set of ordered pairs. If S is a relation, the **domain** of S is the set of all first coordinates of members of S and the **range** of S is the set of all second coordinates of members of S.

Example 3 The set $S = \{(x, y): x^2 + y^2 = 4\}$ is a relation.
The domain of S is $\{x: -2 \le x \le 2\}$ and the range of S is $\{y: -2 \le y \le 2\}$.

■

Since a function F pairs a member x of the domain with a member $F(x)$ of the range, the notion of a relation offers an alternative means of defining a function.

Alternative definition of function A **function** is a relation with the property that if (u, v) and (u, w) belong to the relation, then $v = w$. Thus if F is a function and x belongs to the domain of F, there is exactly one member y in the range of F such that (x, y) belongs to F. The usual notation for y is $F(x)$.

Example 4 The relation $\{(x, y): x + 2y = 3\}$ is a function.
To verify this, suppose that (u, v) and (u, w) belong to the relation. Then $u + 2v = 3$ and $u + 2w = 3$. Thus

$$u + 2v = u + 2w$$
$$2v = 2w$$
$$v = w$$

The relation is a function. ■

Example 5 The relation $\{(x, y): |x - y| = 1\}$ is *not* a function, because (2, 1) and (2, 3) both belong to the relation. ■

Graph of a relation If S is a relation (that is, a set of ordered pairs), then the corresponding set of points in the plane is called the **graph** of S.

Example 6 The graph of the relation described by the equation $x^2 + y^2 = 4$ is the circle of radius 2 with center at (0, 0). ■

Graph of a function The **graph of a function** f consists of all points (x, y) for which x belongs to the domain of f and $y = f(x)$.

Example 7 Sketch the graph of the function $f(x) = \sqrt{x}$.
It is understood that the domain of f is the set of all real numbers greater than or equal to zero. We cannot possibly compute the coordinates of every point on the graph, so we will find some representative points and draw a smooth curve through them. Recall that a point (x, y) is on the graph if and only if x is in the domain of f and $y = f(x) = \sqrt{x}$.

x	$y = f(x) = \sqrt{x}$
0	0
1	1
2	1.4 (approx.)
3	1.7 (approx.)
4	2
5	2.2 (approx.)

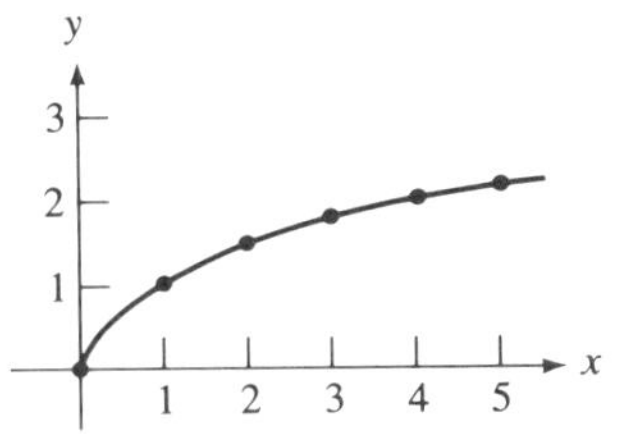

■

Example 8 Graph the function $k(x) = -2$.
This is a constant function, and the graph consists of all points with second coordinate equal to -2. This is a horizontal line passing through $(0, -2)$.

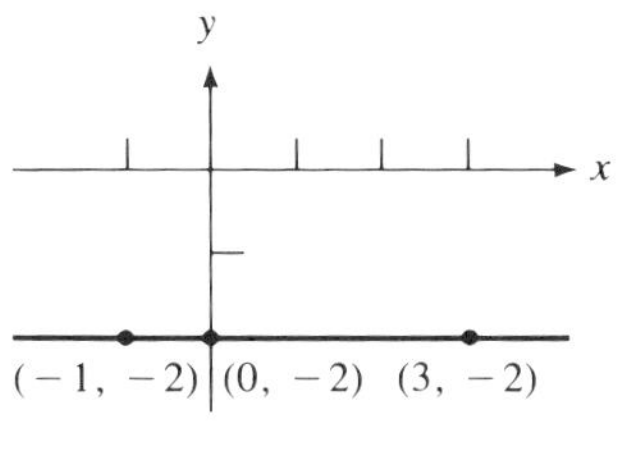

■

If F is a function and x is in the domain of F, there is exactly one point on the graph of F with first coordinate x, namely the point $(x, F(x))$. This means that a relation is a function if and only if each vertical line intersects the graph at most once. This is called the **vertical line test.** In Figure 3.6 the graph in (a) is *not* the graph of a function, whereas the graphs in (b) and (c) pass the vertical line test and are therefore the graphs of functions.

Vertical line test

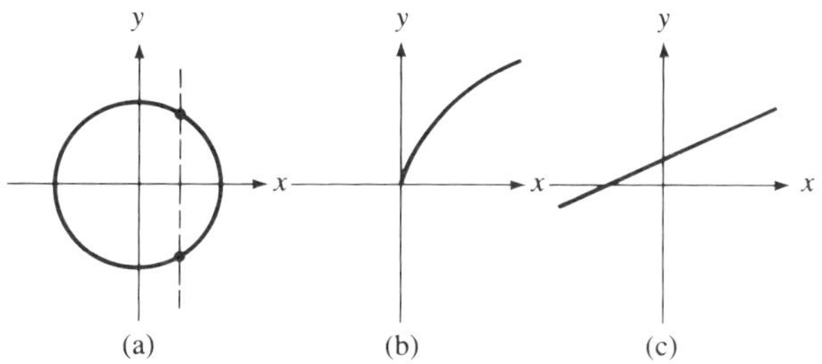

Figure 3.6

Example 9 The graph of the relation $\{(x, y): x^2 + y^2 = 4, y \geq 0\}$ is the upper half of the circle given by the equation $x^2 + y^2 = 4$, hence the graph passes the vertical line test. The relation is a function, and solving the equation $x^2 + y^2 = 4$ for y with $y \geq 0$ we obtain the function $y = \sqrt{4 - x^2}$.

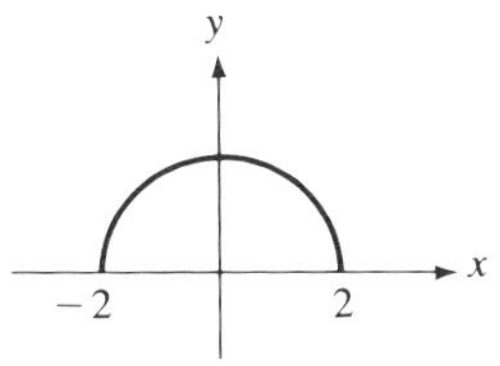

■

A function F defined by a formula of the type $F(x) = ax + b$ is called a *linear* function. As you will see in the next section, the graph of a linear function is a line. Consequently, to graph a linear function, locate two points on the line and draw the line through those two points.

Example 10 Graph the linear function $F(x) = 2x - 1$.
Find any two points on the graph. If $x = 0$, $y = F(0) = -1$, and if $x = 1$, $y = F(1) = 1$. Draw the line through $(0, -1)$ and $(1, 1)$.

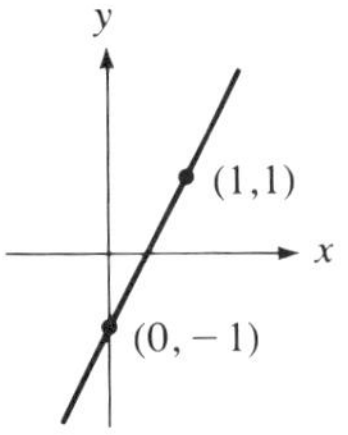

■

Because of the relationship between the x and y coordinates of points on the graph of a function F, it is common to use the terminology $y = F(x)$ to refer to the function. For example, in Example 10 one would speak of the function $y = 2x - 1$. When appropriate, this terminology will be used in this book.

Exercises 3.2

Plot the points in Exercises 1–10 relative to a coordinate system.

1. $(1, 1)$ **2.** $(2, 2)$ **3.** $(2, -3)$ **4.** $(3, -2)$ **5.** $(-4, -2)$
6. $(-2, -4)$ **7.** $(-5, 1)$ **8.** $(-1, 5)$ **9.** $(0, 4)$ **10.** $(-2, 0)$

Identify the quadrant to which the point belongs in Exercises 11–18.

11. $(2, 6)$ **12.** $(3, -4)$ **13.** $(-2, 3)$ **14.** $(1, 8)$
15. $(1, -3)$ **16.** $(-4, -5)$ **17.** $(-6, -7)$ **18.** $(-5, 2)$

In Exercises 19–26 find the midpoint and length of the line segment joining the points.

19. $(-4, -2)$, $(1, 3)$ **20.** $(-1, 3)$, $(1, -4)$ **21.** $(2, 3)$, $(4, 3)$
22. $(-2, -3)$, $(2, 4)$ **23.** $(3, 5)$, $(3, 7)$ **24.** $(1, 4)$, $(5, 4)$
25. $(-4, 5)$, $(2, -3)$ **26.** $(4, 1)$, $(4, -2)$

In Exercises 27–32, write the equation of the circle with the given radius and center.

27. Center $(0, 0)$, radius 5. **28.** Center $(-2, 3)$, radius 3
29. Center $(-4, 5)$, radius 2 **30.** Center $(0, 0)$, radius 4
31. Center $(2, -3)$, radius 1 **32.** Center $(3, -2)$, radius 2

Determine which of the following relations are functions. Justify your conclusion.

33. $\{(x, y): 2x - y = 7\}$ **34.** $\{(x, y): x^2 - 2y = 1\}$
35. $\{(x, y): y^2 - 2x = 1\}$ **36.** $\{(x, y): 2y - x = 8\}$
37. $\{(x, y): x^2 + y^2 = 1, y \le 0\}$ **38.** $\{(x, y): x^2 + y^2 = 1, x \ge 0\}$
39. $\{(x, y): |2x + y| = 6\}$ **40.** $\{(x, y): |2x| + y = 6\}$

In Exercises 41–46 sketch the graph of the function by plotting points corresponding to $x = -3$, -2, -1, 0, 1, 2, and 3 and drawing a smooth curve through those points.

41. $y = 2x - 3$
42. $y = \sqrt[3]{x}$
43. $f(x) = 8 - x^2$
44. $g(x) = 4 - x$
45. $y = 2x$
46. $h(x) = x^2 - 6$

The graph of each of the functions in Exercises 47–58 is a line. Draw the graph by finding two points on the line and drawing the line through them.

47. $y = 3x - 1$
48. $y = 3x + 1$
49. $f(x) = 2 - 3x$
50. $f(x) = 3 - 2x$
51. $f(x) = -3$
52. $g(x) = 2$
53. $h(x) = -2x$
54. $k(x) = 3x$
55. $L(x) = \frac{1}{2}x$
56. $S(x) = -\frac{1}{3}x$
57. $y = \frac{2}{3}x + 1$
58. $y = -\frac{2}{3}x - 2$

Each of the relations in Exercises 59–64 is a function. Write the function in the form $y = f(x)$.

59. $\{(x, y): x^2 + y^2 = 16, y \le 0\}$
60. $\{(x, y): xy = 1, x \ne 0\}$
61. $\{(x, y): x^2 - y^2 = 25, y \ge 0\}$
62. $\{(x, y): y^2 = x + 4, y \ge 0\}$
63. $\{(x, y): 2x - 5y = 14\}$
64. $\left\{(x, y): x = \dfrac{y+2}{y-3}\right\}$

In Exercises 65–68 find the area of the figure with the given vertices. The figures in Exercises 65 and 66 are right triangles; in 67 and 68, rectangles.

65. (1, 3), (5, 5), (2, 11)
66. (−2, 5), (1, 11), (2, 3)
67. (−2, 1), (1, 3), (−1, 6), (−4, 4)
68. (4, −2), (1, −4), (2, 1), (−1, −1)

Graph the functions defined in Exercises 69–70.

69. $f(x) = \begin{cases} x + 1 & x \ge 0 \\ 1 & 0 > x \end{cases}$

70. $g(x) = \begin{cases} 2x - 1 & x \le 1 \\ 1 & 1 < x \end{cases}$

71. Find the equation of the locus of points equidistant from (2, 4) and (−4, 2).

72. A circle passes through the point (2, 3) and has its center at (4, −5). Write the equation of the circle.

For Your Calculator

For the functions in Exercises 73–78 use your calculator to compute the values of the functions for $x = 0.2$, 0.4, 0.6, etc. up to $x = 2.0$. If the function is defined for $x = 0$, include the value at 0 also. Plot the points on graph paper, using a large scale, and sketch the curve.

73. $y = x^{1/4}$
74. $y = x^{2/3}$
75. $f(x) = x^{-1/3}$
76. $L(x) = x^{-1/2}$
77. $y = x^{-3/2}$
78. $y = x^{-2/3}$

3.3 Linear Functions and Equations

A **linear** function is one that can be written in the form $f(x) = ax + b$. If $a = 0$, the function $f(x) = b$ is called a **constant** function. In this section we will see that the graph of a linear function is a line, so we begin with a discussion of lines.

Slope of a line segment

Let P_1 and P_2 be any two points with coordinates (x_1, y_1) and (x_2, y_2), with $x_1 \neq x_2$, as shown in Figure 3.7. We define the **slope of the line segment** P_1P_2 to be the ratio

$$\frac{y_2 - y_1}{x_2 - x_1} \quad (\textit{Remember, } x_1 \neq x_2.)$$

Notice that if $x_1 \neq x_2$, then

$$\frac{y_1 - y_2}{x_1 - x_2} = \frac{y_2 - y_1}{x_2 - x_1}$$

so we obtain the same value for the slope regardless of which point we label as P_1 or P_2.

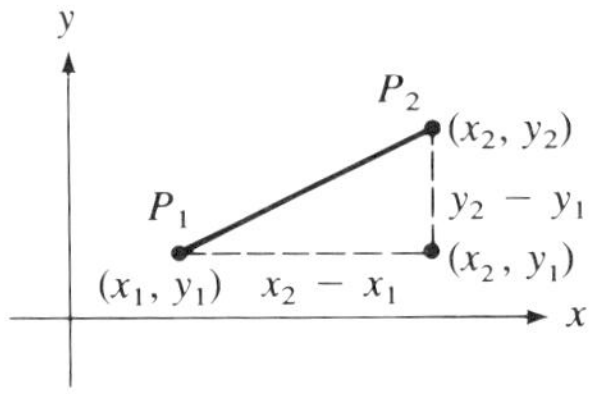

Figure 3.7

Example 1 The slope of the line segment joining (2, 3) and (5, 1) is

$$\text{slope} = \frac{3 - 1}{2 - 5} = -\frac{2}{3}$$

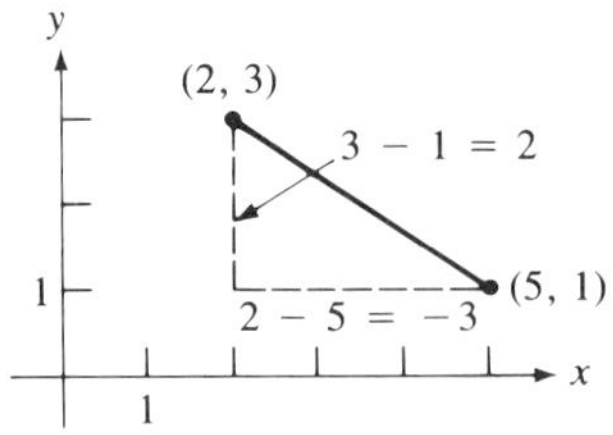

■

Example 2 The slope of the line segment joining (2, −5) and (5, −1) is

$$\text{slope} = \frac{-5 - (-1)}{2 - 5} = \frac{-4}{-3} = \frac{4}{3}$$

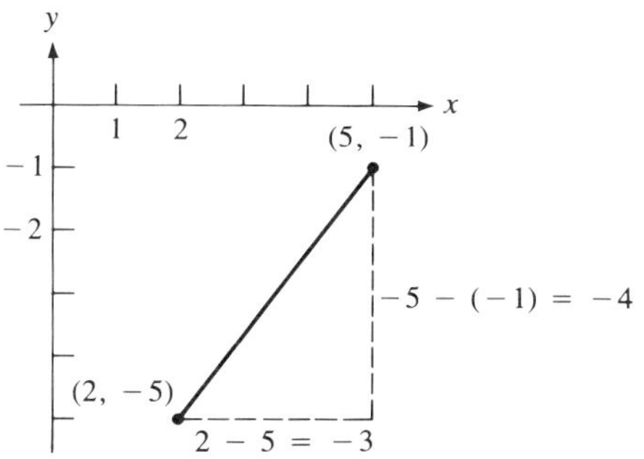

■

Suppose that L is a line not parallel to the y axis. Pick two points P_1 and P_2 on L as in Figure 3.8 and compute the slope of the segment P_1P_2. If we pick two other points Q_1 and Q_2 on L, then the triangles $P_1P_2P_3$ and $Q_1Q_2Q_3$ are similar, and hence segments P_1P_2 and Q_1Q_2 will have the same slope because the ratios of corresponding sides are equal. This number, which is independent of the points chosen, is called the **slope** of the line L.

Slope of a line

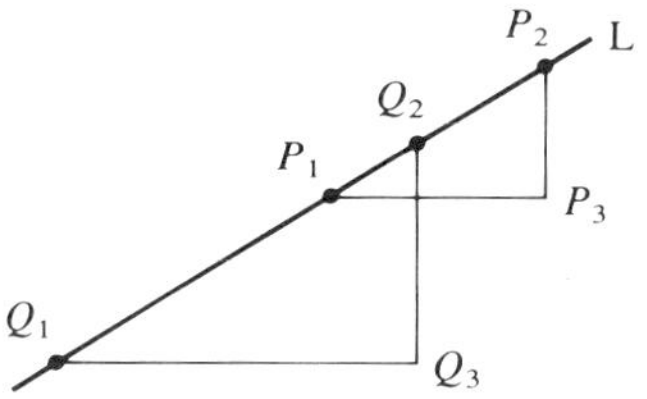

Figure 3.8

If the line is vertical it is impossible to find two points on the line with different x coordinates. This means that we cannot compute the slope of a vertical line; we therefore agree that the slope is **undefined** for a vertical line.

Suppose L is a nonvertical line with slope m passing through (x_1, y_1). If (x, y) is on L and $x \neq x_1$, then by the slope formula,

$$m = \frac{y - y_1}{x - x_1}$$

or

$$y - y_1 = m(x - x_1) \qquad (*)$$

On the other hand, suppose (x, y) satisfies the equation (*). Since L is nonvertical, there is a w such that (x, w) is on L, and by the discussion above, $w - y_1 = m(x - x_1)$. So we have both

$$y - y_1 = m(x - x_1)$$

and

$$w - y_1 = m(x - x_1)$$

Hence $w = y$ and the point (x, y) is on L.

Point-slope form for the equation of a line

> To summarize: a point (x, y) is on the line L with slope m passing through (x_1, y_1) if and only if $y - y_1 = m(x - x_1)$. The equation $y - y_1 = m(x - x_1)$ is called the **point-slope form** of the equation of a line.

Example 3 Write the equation of the line through $(-1, 4)$ with slope $-\frac{2}{3}$.

Using the point-slope form,

$$y - 4 = -\frac{2}{3}[x - (-1)]$$

or, after simplification,

$$3y = -2x + 10$$

■

Example 4 Write the equation of the line passing through $(2, -5)$ and $(5, -1)$.

From Example 2, the slope is $\frac{4}{3}$. Using the point-slope form with slope $\frac{4}{3}$ and point $(2, -5)$ we obtain

$$y - (-5) = \frac{4}{3}(x - 2)$$

or

$$3y = 4x - 23$$

We could use either point; using $(5, -1)$ we obtain

$$y - (-1) = \frac{4}{3}(x - 5)$$

or

$$3y = 4x - 23$$

■

Equation of a vertical line

If the line is vertical (and the point-slope form can't be used), then all points on the line have the same x coordinate and the equation of the line is $x = k$, where k is a real number.

Example 5 Graph the line through $(2, 5)$ and $(2,0)$ and write the equation of the line. The line is the vertical line through $(2, 5)$ (see the sketch). Since all points on a vertical line have the same x coordinate, the equation is $x = 2$.

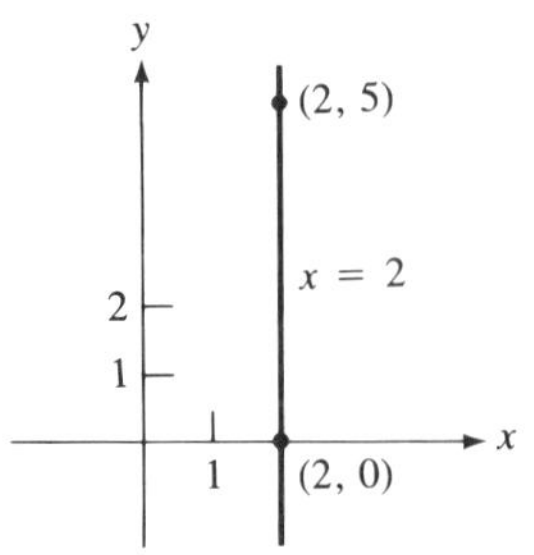

■

A linear equation in x and y is any equation that can be written in the form $Ax + By = C$, where A and B are not both zero. If $B \neq 0$, we can rewrite the equation in the form $y - C/B = -(A/B)(x - 0)$, which is the equation of the line through $(0, C/B)$ with slope $-A/B$. If $B = 0$ the equation can be written in the form $x = C/A$, which is the equation of a vertical line. We summarize below.

Linear equations and lines

The equation of a line is a linear equation in x and y. Any linear equation in x and y is the equation of a line.

From the point-slope form of the line we can see that a linear equation of the form $y = mx + b$ is the equation of a line with slope m passing through $(0, b)$.

Slope-intercept form

This is called the **slope-intercept form** of the equation of the line. The point $(0, b)$ is called the *y intercept*, that is, the point where the line crosses the y axis.

Example 6 Graph the line given by the linear equation $3x + 4y = 12$. What is the slope?
The points (2, 1.5) and (4, 0) are on the line. Writing the equation in the form $y = -\frac{3}{4}x + 3$, we see that the slope is $-\frac{3}{4}$.

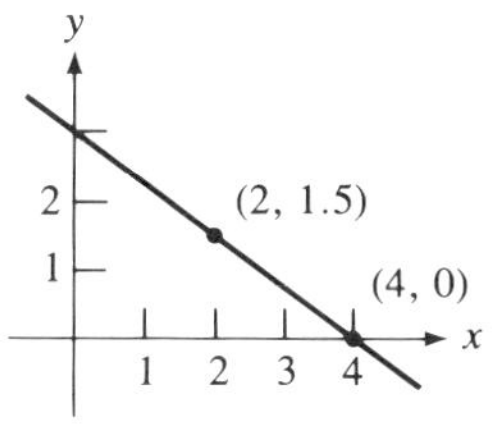

■

Intercepts

The points where the line crosses the coordinate axes, the **intercepts,** are found by setting one coordinate equal to zero and solving for the other coordinate.

Example 7 Find the intercepts and graph the line $3x - 9 = 2y$.
If $x = 0$, then $-9 = 2y$, and $y = -4.5$. If $y = 0$, then $3x - 9 = 0$, and $x = 3$. The intercepts are $(0, -4.5)$ and $(3, 0)$.

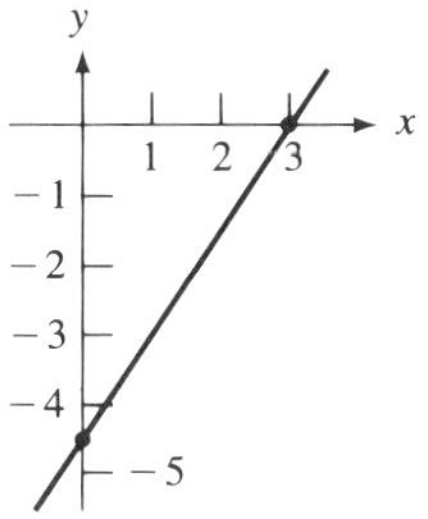

■

If $B \neq 0$, the graph of the linear equation $Ax + By = C$ is a nonvertical line, and hence passes the vertical line test. This means that the equation $Ax + By = C$ defines a function. On the other hand, every linear function $f(x) = mx + b$ defines a linear equation $y = mx + b$, which is the equation of a nonvertical line.

To summarize: the graph of a linear function is a nonvertical line, and every nonvertical line is the graph of a linear function.

Example 8 Graph the function $g(x) = 3 - 2x$.
This is a linear function, so the graph is a line. If $x = 0$, then $y = g(0) = 3$, and if $x = 1$, then $y = g(1) = 1$. Draw the line through (0, 3) and (1, 1).

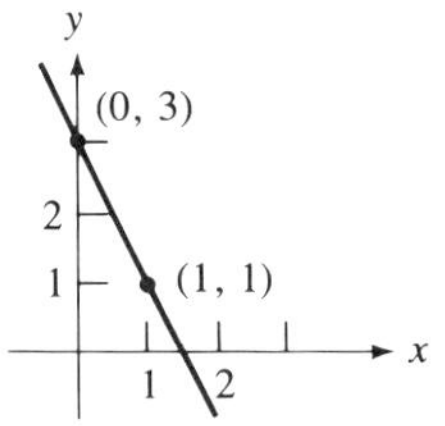

■

Consider two nonvertical lines with the same slope, $y = mx + b_1$ and $y = mx + b_2$. Since the lines are distinct, $b_1 \neq b_2$. The lines are either parallel or intersect at some point (h, k). But if there were a point of intersection (h, k), then we would have $k = mh + b_1$ and $k = mh + b_2$, which is impossible because $b_1 \neq b_2$. Thus lines with the same slope must be parallel. The converse can be proven using similar triangles.

Slopes of parallel lines

Two nonvertical lines are parallel if and only if they have the same slope.

The slope of the line $y = mx + b$ is m, so any line parallel to that line will have an equation of the form $y = mx + k$ for some value of k. Likewise the lines $Ax + By = C$ and $Ax + By = D$ both have slope $-A/B$ if $B \neq 0$, or are both vertical if $B = 0$. These two observations are intended to make it easy to find the equation of a line parallel to a given line.

Example 9 Write the equation of the line parallel to $y = 3x - 1$ passing through $(2, -4)$.
For each real number k, the line $y = 3x + k$ is parallel to the line $y = 3x - 1$; we need only find the value of k for which $y = 3x + k$ passes through $(2, -4)$.

$$-4 = 3(2) + k$$
$$-10 = k$$

The desired equation is $y = 3x - 10$. ■

Example 10 Write the equation of the line through $(1, -2)$ that is parallel to the line $3x - 5y = 6$.
The line $3x - 5y = C$ will be parallel to the line $3x - 5y = 6$ for any value of C. We wish to find C so that the line passes through $(1, -2)$.

$$3(1) - 5(-2) = C$$
$$13 = C$$

The desired line is $3x - 5y = 13$. ■

Suppose that L_1 and L_2 are perpendicular lines, neither vertical, which intersect at (h, k). Let the slopes be m_1 and m_2, respectively. Draw the vertical line $x = h + 1$ that intersects the two lines at the points shown in Figure 3.9.

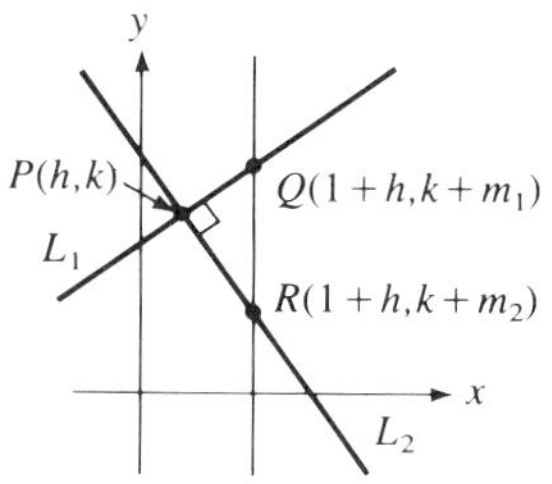

Figure 3.9

Since PQR is a right triangle,

$$\overline{PQ}^2 + \overline{PR}^2 = \overline{QR}^2$$

or

$$(1 + m_1^2) + (1 + m_2^2) = (m_1 - m_2)^2$$

Simplifying this equation, we obtain

$$m_1 m_2 = -1$$

The converse is also true.

> Two nonvertical lines are perpendicular if and only if the product of their slopes is -1.

The slope of the line $y = mx + b$ is m, and if $m \neq 0$ the line perpendicular to that line has slope $-1/m$.

Example 11 Write the equation of the line that is perpendicular to the line $y = \frac{3}{2}x - 6$ and passes through the point $(-4, 1)$.

The slope must be $-\frac{2}{3}$; using the point-slope form, the equation is

$$y - 1 = -\frac{2}{3}(x + 4)$$

or

$$y = -\frac{2}{3}x - \frac{5}{3}$$

■

Exercises 3.3

In Exercises 1–12 find the slope of the line segment joining the two points (if the slope is defined).

1. $(-2, 3), (1, 6)$
2. $(-1, 4), (3, 8)$
3. $(-2, -3), (4, -1)$
4. $(-4, -5), (2, -3)$
5. $(-2, 3), (1, -2)$
6. $(-1, 4), (2, -4)$
7. $(-5, 1), (-2, -1)$
8. $(-6, 3), (-1, -4)$
9. $(4, 5), (5, 5)$
10. $(2, -3), (3, -3)$
11. $(2, 4), (2, 3)$
12. $(-4, 5), (-4, -4)$

In Exercises 13–22 write the equation of the line with the given slope that passes through the given point.

13. Slope 4; $(-2, 4)$
14. Slope 3; $(-1, 3)$
15. Slope -2; $(6, -2)$
16. Slope -3; $(-1, 4)$
17. Slope 0; $(1, 1)$
18. Slope 0; $(2, 3)$
19. Slope undefined; $(-2, 3)$
20. Slope undefined; $(3, -2)$
21. Slope $\frac{1}{3}$; $(-3, -4)$
22. Slope $\frac{2}{3}$; $(-4, -3)$

In Exercises 23–34 write the equation of the line through the given points.

23. $(0, 3), (4, 0)$
24. $(-2, 0), (0, 3)$
25. $(1, 4), (3, 8)$
26. $(2, 5), (4, 9)$
27. $(-5, 1), (-2, -1)$
28. $(-6, 3), (-1, -4)$
29. $(-2, 3), (1, 6)$
30. $(-1, 4), (3, 8)$
31. $(3, 4), (3, 6)$
32. $(4, 3), (6, 3)$
33. $(5, 6), (-5, 6)$
34. $(-2, 5), (-2, 2)$

In Exercises 35–46 find the slope (if defined) and the intercept(s) and graph the line.

35. $y = 2x - 7$
36. $y = 5x - 9$
37. $x = 3y + 7$
38. $x = 4y + 6$
39. $4x - 5y = 20$
40. $6x - 7y = 42$
41. $4y + 5x = 40$
42. $2y + 9x = 27$
43. $y = 5y - 6$
44. $y = -2y + 1$
45. $2x = 3x - 7$
46. $4x = 9 - 7x$

In Exercises 47–54 write the equation satisfying the given conditions.

47. Parallel to $5x - 3y = 17$; passing through $(1, -2)$
48. Parallel to $3x + 5y = 19$; passing through $(-2, 3)$
49. Perpendicular to the line $3x - y = 1$; passing through $(9, -1)$
50. Perpendicular to the line $2x + y = 1$; passing through $(-8, 3)$
51. Vertical line through $(6, -3)$
52. Vertical line through $(4, -3)$
53. Horizontal line through $(-6, 7)$
54. Horizontal line through $(9, -5)$

55. The equation $\frac{x}{a} + \frac{y}{b} = 1$ is called the **intercept form** for the equation of a line. Show that $\frac{x}{a} + \frac{y}{b} = 1$ is the equation of the line with intercepts $(a, 0)$ and $(0, b)$.

56. Find the point A with coordinates (x, y) such that the line through A and $(0, 3)$ is perpendicular to the line through $(0, 3)$ and $(2, 0)$.

57. The points $(3, -1)$, $(1, 1)$, and $(2, 4)$ are three vertices of a parallelogram. Find the fourth vertex. How many possible answers are there?

58. If $A \neq 0$ and $B \neq 0$, show that the lines $Ax + By + C = 0$ and $Bx - Ay + D = 0$ are perpendicular. (Hint: find the slopes.)

59. Show that the points (x_1, y_1), (x_2, y_2), and (x_3, y_3) are collinear, that is, lie on the same line, if and only if

$$(y_1 - y_2)(x_3 - x_2) = (y_3 - y_2)(x_1 - x_2)$$

(Hint: find the slopes of the line segments involved.)

60. A median of a triangle is a line segment joining a vertex with the midpoint of the opposite side. If a triangle has vertices (x_1, y_1), (x_2, y_2), and (x_3, y_3), show that the point

$$\left(\frac{x_1 + x_2 + x_3}{3}, \frac{y_1 + y_2 + y_3}{3}\right)$$

lies on each median.

61. A function is said to be an affine function if $f(au + bv) = af(u) + bf(v)$ whenever $a + b = 1$. Show that a linear function is an affine function.

3.4 Symmetry and Translation of Graphs

In this section we will examine several concepts that are very helpful in sketching the graph of a function or an equation. We begin with an example that illustrates the notion of *symmetry*.

Example 1 Sketch the graph of the function $f(x) = x^2$.
To sketch this function we plot some points and draw a smooth curve through them.

x	$f(x)$
−2	4
−1.5	2.25
−1	1
−0.5	0.25
−0.3	0.09
0	0
0.3	0.09
0.5	0.25
1	1
1.5	2.25
2	4

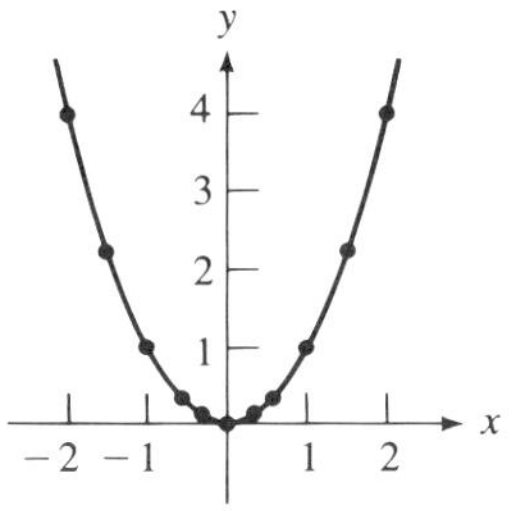

■

There is something special about the graph in Example 1 that should be noted. The portion of the curve in the second quadrant is a mirror image of the portion of the curve in the first quadrant. This happens because for every point (x, y) on the curve, the point $(-x, y)$ is also on the curve. We say that the curve is symmetric with respect to the y axis.

Symmetry with respect to the y axis

A curve is **symmetric with respect to the y axis** if for each point (x, y) on the curve, the point $(-x, y)$ is also on the curve.

If a curve is described by an equation, the curve is symmetric with respect to the y axis if the equation is unchanged when x is replaced by $-x$. The graph of a function is symmetric with respect to the y axis if the value of the function at x is the same as the value at $-x$. Such a function is called an even function.

Even function

A function f is called an **even function** if $f(x) = f(-x)$ for all x in the domain of f.

From our observations above, the graph of an even function is symmetric with respect to the y axis.

Example 2 The function $f(x) = |x|$ is an even function, because $f(-x) = |-x| = |x| = f(x)$. This means that the graph is symmetric with respect to the y axis. Thus we can graph the function for $x \geq 0$ and then draw the mirror image to obtain the graph for $x < 0$. For $x \geq 0$, $f(x) = |x| = x$.

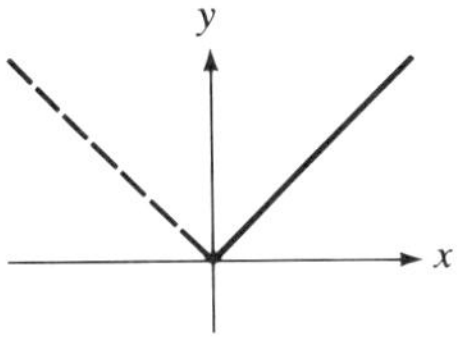

■

Line of symmetry

If the graph of an equation or of a function is symmetric with respect to the y axis, the line $x = 0$ (the y axis) is called a **line of symmetry** for the curve.

Example 3 Graph the function $y = x^3$.
In anticipation of the symmetry this curve may possess, notice that changing the sign of x changes the sign of y, i.e., if $y = x^3$ then $(-x)^3 = -x^3 = -y$. This means that if (x, y) is on the curve, then $(-x, -y)$ is also on the curve.
Each choice of x therefore gives two points on the curve, (x, y) and, by the remarks above, $(-x, -y)$.

x	$y = x^3$
1.5	3.375
1	1
0.5	0.125
0	0
−1.5	−3.375
−1	−1
−0.5	−0.125

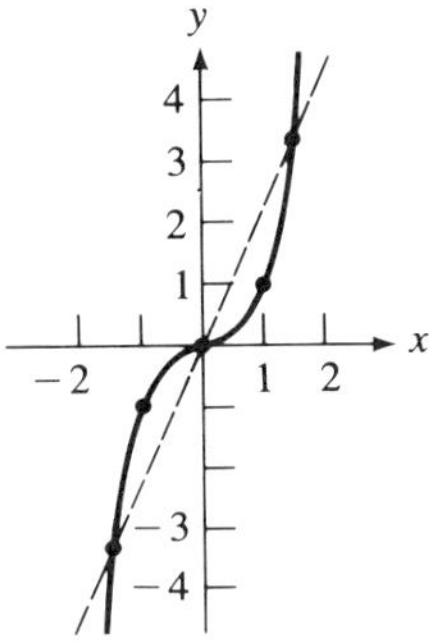

■

Notice how a line through the origin that intersects the curve $y = x^3$ at (x, y) also intersects the curve at the point $(-x, -y)$. We say that such a curve is symmetric with respect to the origin.

Symmetry with respect to the origin

A curve is **symmetric with respect to the origin** if for each point (x, y) on the curve, the point $(-x, -y)$ is also on the curve.

If a curve is described by an equation, the curve will be symmetric with respect to the origin if an equivalent equation is obtained by replacing x by $-x$

and y by $-y$. The graph of a function will be symmetric with respect to the origin if changing the sign of x results only in a change in sign of the value of the function. Such a function is called an odd function.

Odd function

A function f is said to be an **odd function** if $f(-x) = -f(x)$ for each x in the domain of f.

Example 4 The function $y = \sqrt[3]{x}$ is an odd function because $f(-x) = \sqrt[3]{-x} = -\sqrt[3]{x} = -f(x)$.
Plotting a few points gives the graph shown. Notice the symmetry with respect to the origin.

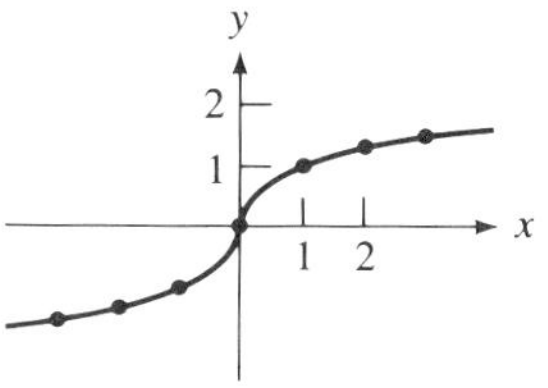

■

Example 5 Graph the equation $x = \sqrt{y^2 + 1}$.
Notice that changing the sign of y leaves the equation unchanged. Thus if (x, y) is on the curve, so is the point $(x, -y)$. Using this fact and plotting some points gives the curve shown.

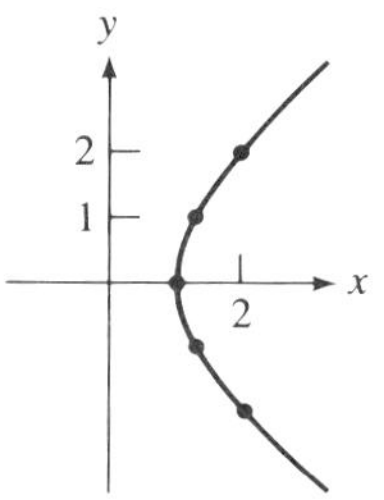

■

Following the pattern set before, the graph in Example 5 is symmetric with respect to the x axis.

Symmetry with respect to the x axis

A curve is **symmetric with respect to the x axis** if for each point (x, y) on the curve, the point $(x, -y)$ is also on the curve.

A curve given by an equation is symmetric with respect to the x axis if replacing y by $-y$ leaves the equation unchanged. Notice that a curve that is symmetric with respect to the x axis does not pass the vertical line test, and hence cannot be the graph of a function.

Example 6 The graph of $x = y^2$ is symmetric with respect to the x axis because $(-y)^2 = y^2$.

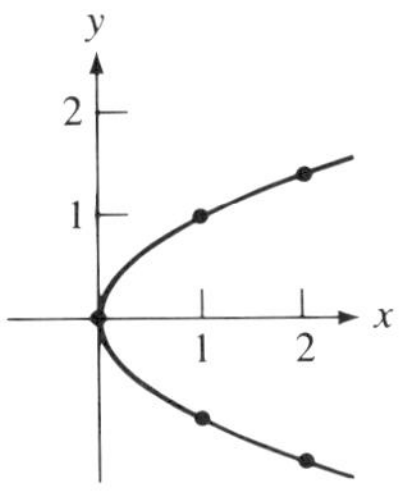

■

The graphs of some functions may be found by means of a translation of the graph of a familiar function.

Example 7 Graph the function $y = |x| - 2$.
This graph can be found by subtracting 2 from the y coordinate of each point on the graph of $y = |x|$, or, in other words, shifting the graph of $y = |x|$ down 2 units. The line $x = 0$ is still a line of symmetry.

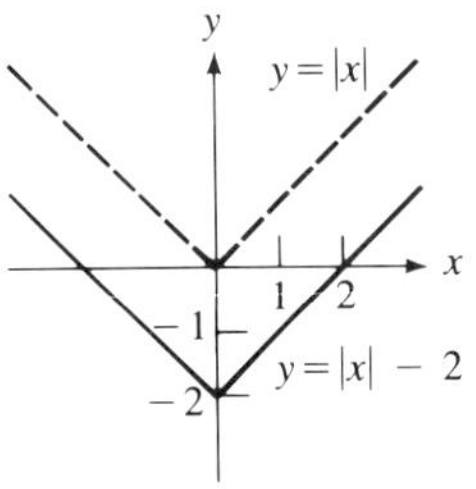

■

Example 8 Graph the function $y = \sqrt[3]{x + 3}$.
Notice that if the point (x, y) is on the graph of the function $y = \sqrt[3]{x}$, then the point $(x - 3, y)$ is on the graph of $y = \sqrt[3]{x + 3}$. For example, (8, 2) is on the graph of $y = \sqrt[3]{x}$, and (5, 2) is on the graph of $y = \sqrt[3]{x + 3}$. Thus the graph of $y = \sqrt[3]{x + 3}$ is the graph of $y = \sqrt[3]{x}$ shifted 3 units to the left.

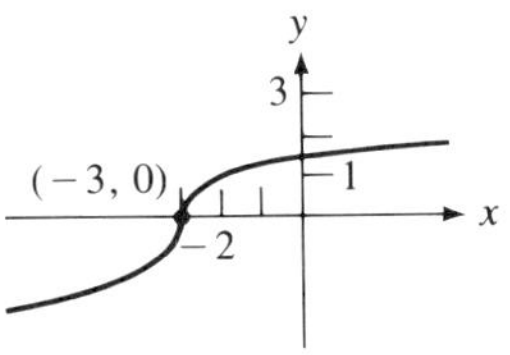

■

The following table summarizes the observations made in Examples 7 and 8.

Replace in the Equation of the Curve:	Result
x by $x - h$	Shift the curve horizontally $\lvert h\rvert$ units, right if $h > 0$ and left if $h < 0$.
y by $y - k$	Shift the curve vertically $\lvert k\rvert$ units, up if $k > 0$ and down if $k < 0$.

Example 9 Graph the following functions on the same coordinate axes:

$$y = \sqrt{x} \qquad y = \sqrt{x + 2} \qquad y = 3 + \sqrt{x} \qquad y - 3 = \sqrt{x + 2}$$

The graph of $y = \sqrt{x}$ was given in Example 7 in Section 3.2. The graph of $y = \sqrt{x + 2}$ is found by shifting that graph 2 units to the left. If we rewrite the equation $y = 3 + \sqrt{x}$ in the form $y - 3 = \sqrt{x}$, we see that the graph is the graph of $y = \sqrt{x}$ shifted 3 units up. Finally, the graph of $y - 3 = \sqrt{x + 2}$ is the graph of $y = \sqrt{x}$ shifted 2 units to the left and 3 units up.

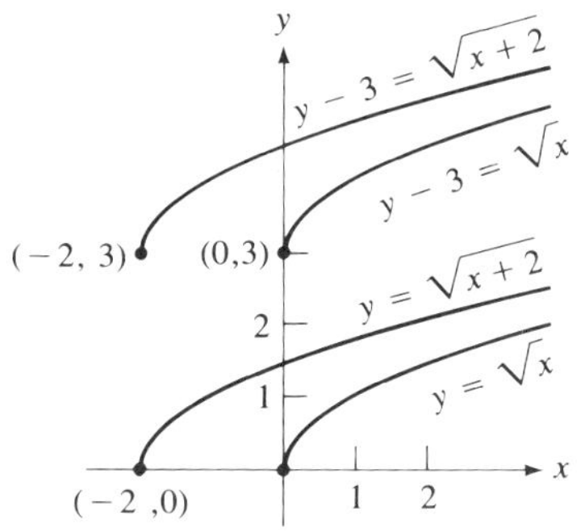

■

Example 10 Graph the function $y = |x - 1| - 2$.
If we rewrite the equation in the form

$$y + 2 = |x - 1|$$

we see that the graph is a translation of $y = |x|$. In this case $h = 1$ and $k = -2$. We translate 1 unit to the right and 2 units down. The line of symmetry $x = 0$ for the graph of $y = |x|$ is translated 1 unit to the right to give the line of symmetry $x = 1$ for the graph of $y = |x - 1| - 2$.

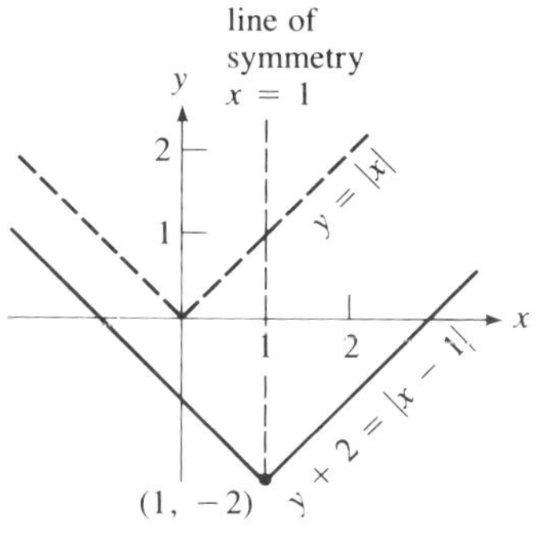

■

Exercises 3.4

In Exercises 1–16 identify the symmetries, if any, that the graph of the relation possesses.

1. $y = -3x^2 + 1$

2. $y^3 + x^4 = 7$

3. $y - \sqrt{x^2} = 9$

4. $y = |x - 4|$

5. $x^2 - y^2 = 9$

6. $y^3 = x - 8x^3$

7. $y = \sqrt[3]{-3x}$

8. $\sqrt{y} - x^4 = 0$

9. $x = \sqrt[3]{y} - y$

10. $x = y^2 + |y|$

11. $xy - y = y^2$

12. $2x^2 + 4y^2 = 8$

13. $x = -3y^2 - 5$

14. $y^{1/3} + x^{1/3} = 0$

15. $x = (y - 1)^2 + y^2$

16. $xy = 16$

In Exercises 17–20 complete the graph, assuming the curve is symmetric with respect to (a) the y axis, (b) the x axis, and (c) the origin. You will have three graphs for each exercise.

17.

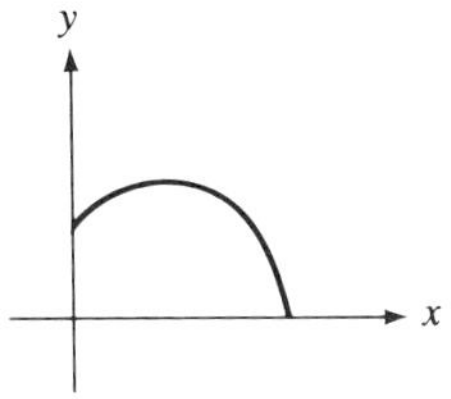

18.

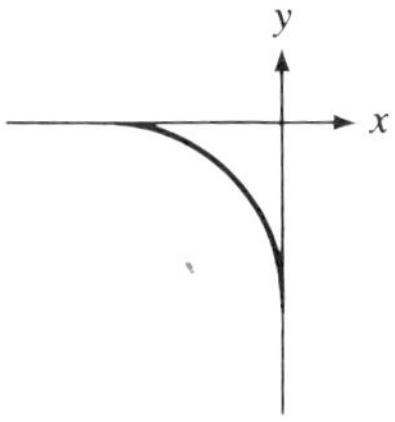

19.

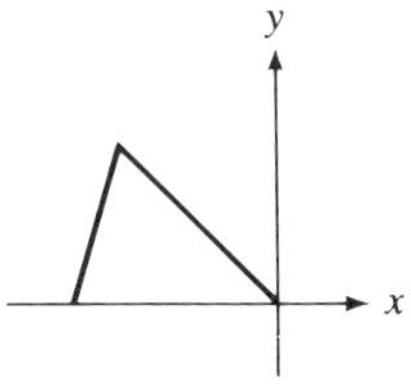

20.

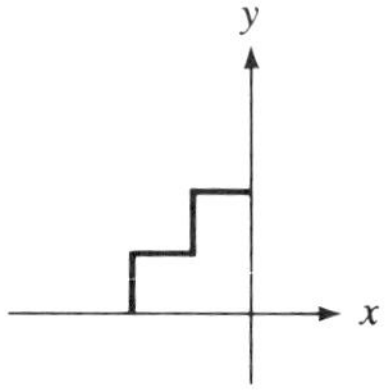

In Exercises 21–34 use the graphs shown in the examples in this section to sketch the graph.

21. $y = x^2 - 3$

22. $y = x^2 + 2$

23. $y = (x - 1)^2$

24. $y = (x + 2)^2$

25. $y = x^3 + 2$

26. $y = x^3 - 2$

27. $y = (x + 2)^3$

28. $y = (x - 1)^3$

29. $x - 2 = y^2 + 1$

30. $x + 3 = y^2 + 1$

31. $y = |x - 2| + 1$

32. $y = |x + 1| - 2$

33. $y = \sqrt[3]{x - 2} + 1$

34. $y = \sqrt[3]{x + 1} - 2$

35. If f is an odd function with domain R, prove that $f(0) = 0$.

36. If a curve is symmetric with respect to both the x axis and the y axis, prove that the curve is symmetric with respect to the origin.

37. Prove that any curve that possesses two of the three symmetries discussed in this section also possesses the third symmetry.

38. Let f be any function with domain R. Prove that the function g defined by $g(x) = f(x) + f(-x)$ is an even function. Prove that the function h defined by $h(x) = f(x) - f(-x)$ is an odd function.

For Your Calculator

Some scientific calculators will accept negative values for y if x is an integer and $\boxed{y^x}$ is pressed, but all will refuse to compute y^x if y is negative and x is not an integer. This happens because y^x for nonintegral values of x is computed using logarithms, a topic covered in Chapter 5. In order to compute the values of certain functions with your calculator you will need to use the fact that certain functions are odd and certain functions are even. For example, to compute $(-8)^{1/3}$ one would compute $8^{1/3}$ first and then use the fact that $(-8)^{1/3} = -(8)^{1/3}$.

Use the calculator to find the value of the functions as given below.

39. $f(x) = x^3 - x^5 + \sqrt[3]{x} + \sqrt[5]{x}$, $x = -2.1$
40. $f(x) = x^5 - x^3 - \sqrt[3]{x} - \sqrt[5]{x}$, $x = -1.2$
41. $g(x) = -2x^3 - 3x^2 + x - 2\sqrt[3]{x}$, $x = -2.02$
42. $h(x) = x^{5/3} + x^{1/3} - 5x + 2x^2$, $x = -.002$

3.5 Algebra of Functions

Sometimes a problem may be best described by the use of several functions. For example, the area of a window that is made in the shape of a square surmounted by a semicircle is the sum of the area of the square and the area of the semicircle. If x represents the length of a side of the square, the area function A is given by the formula

$$A = x^2 + \frac{\pi}{2}\left(\frac{x}{2}\right)^2$$

Here is an example of a function, A, that is the sum of two functions: one that gives the area of the square and one that gives the area of the semicircle.

As illustrated in the preceding paragraph, some of the simplest ways to combine functions to create new functions is to use the arithmetic operations of addition, subtraction, multiplication, and division.

If f and g are functions, we define for all x in the intersection of the domains of f and g,

$f + g$ 1. $(f + g)(x) = f(x) + g(x)$
$f - g$ 2. $(f - g)(x) = f(x) - g(x)$
fg 3. $(fg)(x) = f(x)g(x)$
$\frac{f}{g}$ 4. $\left(\frac{f}{g}\right)(x) = \frac{f(x)}{g(x)}$ if $g(x) \neq 0$

Notice that the domain of $f + g$, $f - g$, and fg is the set of all x for which *both* f and g are defined, and the domain of f/g is the set of all x for which both f and g are defined and $g(x) \neq 0$.

Example 1 If $f(x) = \sqrt{x + 1}$ and $g(x) = 2x$, find $f + g$, $f - g$, fg, and f/g and the domain of each.

The domain of f is $\{x: x \geq -1\}$, and the domain of g is R, the set of all real numbers. Thus the domain of $f + g$, $f - g$, and fg is the same as the domain of f, namely $\{x: x \geq -1\}$. The domain of f/g is $\{x: x \neq 0 \text{ and } x \geq -1\}$.

$$(f + g)(x) = \sqrt{x + 1} + 2x$$
$$(f - g)(x) = \sqrt{x + 1} - 2x$$
$$(fg)(x) = 2x\sqrt{x + 1}$$
$$\left(\frac{f}{g}\right)(x) = \frac{\sqrt{x + 1}}{2x}$$

■

If the function f is multiplied by the constant function that has the value a for each x, the result is usually written af. In other words,

$$(af)(x) = a \cdot f(x)$$

The graph of f may be used to graph the function af by multiplying the y coordinate of each point on the graph by a and drawing a curve through the corresponding points. Note that if $a < 0$, the curve is reflected in the x axis.

Example 2 Graph the function $f(x) = -|x|$.
First graph the function $g(x) = |x|$, shown in the sketch as a dashed graph; then reflect the curve in the x axis.

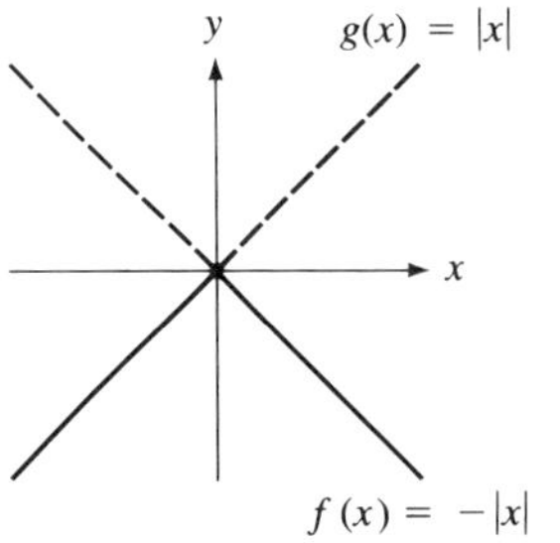

■

Example 3 Graph $y = 2x^2$ and $y = -\frac{1}{2}x^2$ on the same axes with the graph of $y = x^2$.
To graph $y = 2x^2$, first graph $y = x^2$, then multiply the y coordinate (or ordinate as it is often called) by 2 and plot the corresponding point on the graph of $y = 2x^2$. To graph $y = -\frac{1}{2}x^2$, graph $y = -x^2$ (found by reflecting the graph of $y = x^2$ about the x axis) and multiply each ordinate by $\frac{1}{2}$ to locate points on the graph of $y = -\frac{1}{2}x^2$.

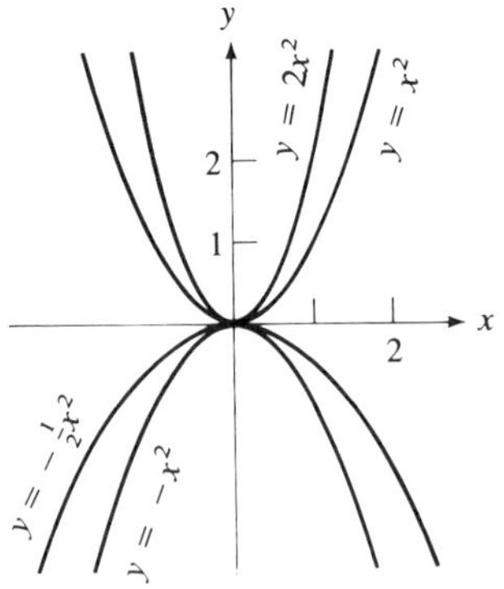

■

Addition of ordinates The sum of two functions may be graphed by the **addition of ordinates.** A point (x, y) will be on the graph of $f + g$ if and only if $y = f(x) + g(x)$. To find the point (x, y) on the graph of $f + g$, find the corresponding points (x, y_1) on the graph of f and (x, y_2) on the graph of g and add the ordinates y_1 and y_2 to obtain y. This can be done using a compass or ruler.

Example 4 Graph by addition of ordinates the function $y = x - 1 + \sqrt{x}$.
Graph $y = x - 1$ and $y = \sqrt{x}$ (shown as dashed curves). Then add ordinates to obtain the graph.

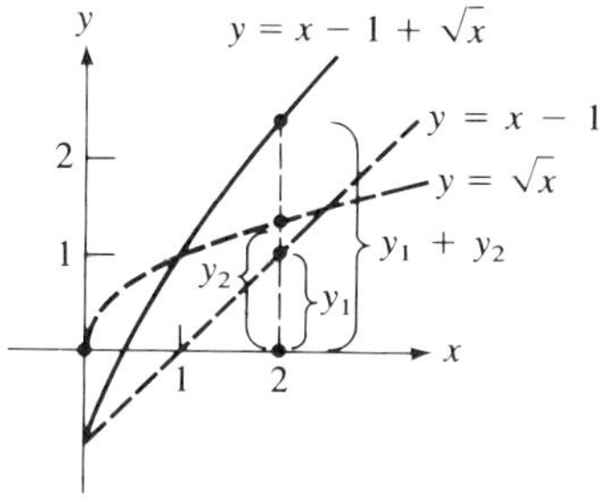

■

When your bill is computed at the grocery store it is the usual practice to find the total price of the merchandise and use that number as the input for another function, the one that computes the cost plus sales tax. This method of combining two functions is very common and has a special name.

Composition of functions

If f and g are functions with the range of f contained in the domain of g (this allows the output of f to be an admissible input for g), we define the **composition of g by *f*,** written $g \circ f$, as follows:

$$(g \circ f)(x) = g(f(x))$$

In other words, the composition of g by f is just the function f followed by the function g. The only restriction is that each output of f must be in the domain of g.

Example 5 If $f(x) = x^2 - 2x$ and $g(x) = x - 1$, then

$$\begin{aligned} g \circ f(x) &= g[f(x)] = g(x^2 - 2x) = x^2 - 2x - 1 \\ f \circ g(x) &= f[g(x)] = f(x - 1) = (x - 1)^2 - 2(x - 1) \\ &= x^2 - 4x + 3 \end{aligned}$$

■

Notice from Example 5 that $f \circ g$ and $g \circ f$ are not always the same.

If we use the notation $x \xrightarrow{f} f(x)$ to depict the function f, then the composition $g \circ f$ is depicted as

$$x \xrightarrow{f} f(x) \xrightarrow{g} g[f(x)]$$

This notation is convenient when we wish to write a function as the composition of simpler functions.

Example 6 Write $f(x) = \sqrt{x + 1}$ as the composition of functions, each involving only one calculation.

Think of the steps you would go through to compute a value of f. First input x, add 1, and then find the square root. Using the symbolism introduced above,

$$x \xrightarrow{\text{(add 1)}} x + 1 \xrightarrow{\text{(square root)}} \sqrt{x+1}$$

If $A(x) = x + 1$ and $S(x) = \sqrt{x}$, then $f = S \circ A$. ■

Example 7 Write $f(x) = \dfrac{1}{x^3 + 1}$ as the composition of functions, each involving only one calculation.

To evaluate $f(x)$, first cube x (raise to the 3rd power), then add 1, then invert. Symbolically,

$$x \xrightarrow{\text{cube}} x^3 \xrightarrow{\text{add 1}} x^3 + 1 \xrightarrow{\text{invert}} \frac{1}{x^3 + 1}$$

If $C(x) = x^3$, $A(x) = x + 1$, and $I(x) = 1/x$, then

$$f(x) = I(A(C(x)))$$

or

$$f = I \circ A \circ C$$ ■

Exercises 3.5

In Exercises 1–12 write a formula for $f + g$, $f - g$, fg, and f/g and give the domain of each.

1. $f(x) = 2x - 3$, $g(x) = x^2 - 3x - 28$

2. $f(x) = 3 - 4x$, $g(x) = x^2 + 5x - 36$

3. $f(x) = \sqrt{2x - 3}$, $g(x) = \sqrt{3x + 1}$

4. $f(x) = \sqrt{3x + 2}$, $g(x) = \sqrt{4x + 3}$

5. $f(x) = |x + 7|$, $g(x) = |4 - 3x|$

6. $f(x) = |3x - 6|$, $g(x) = |4 - 5x|$

7. $f(x) = x^3 - 3x^2 + 7$, $g(x) = \dfrac{1}{x}$

8. $f(x) = 3x^2 - x^3 + 1$, $g(x) = \dfrac{1}{x}$

9. $f(x) = 2$, $g(x) = 2x - 4$

10. $f(x) = -3$, $g(x) = 3x + 7$

11. $f(x) = x - 1$, $g(x) = x^2 - 1$

12. $f(x) = x + 2$, $g(x) = x^2 - 4$

In Exercises 13–26 use the techniques illustrated in Examples 3 and 4 to graph the function.

13. $f(x) = 2|x|$

14. $f(x) = 3|x|$

15. $y = \dfrac{1}{2}x^3$

16. $y = \dfrac{1}{3}x^3$

17. $y = -\sqrt[3]{x}$

18. $y = -\sqrt{x}$

19. $y = |x| + x$

20. $y = |x| - x$

21. $y = x^2 - x$

22. $y = x^2 + x$

23. $y = |x + 1| + |x - 1|$

24. $y = |x - 2| + |x + 2|$

25. $y = x + \sqrt{x - 1}$

26. $y = x - \sqrt{x - 1}$

In Exercises 27–34 give a formula for $F \circ G$ and $G \circ F$.

27. $F(x) = 2x^2 - 8$, $G(x) = 3x - 1$

28. $F(x) = 3x + 7$, $G(x) = x^2 - 5x$

29. $F(x) = \dfrac{1}{x}$, $G(x) = 3x + 5$

30. $F(x) = \dfrac{2}{x}$, $G(x) = 4 - 3x$

31. $F(x) = \frac{1}{x}$, $G(x) = \frac{1}{x - 1}$

32. $F(x) = \frac{2}{x}$, $G(x) = \frac{1}{x - 2}$

33. $F(x) = \frac{x}{x - 2}$, $G(x) = \frac{2x}{x - 1}$

34. $F(x) = \frac{x - 2}{x + 3}$, $G(x) = \frac{2 + 3x}{1 - x}$

In Exercises 35–40 give a formula for $f \circ g \circ h$.

35. $f(x) = x^2$, $g(x) = x + 2$, $h(x) = 3x$

36. $f(x) = 3x$, $g(x) = x + 2$, $h(x) = x^2$

37. $f(x) = \frac{1}{x}$, $g(x) = x - 4$, $h(x) = x^2$

38. $f(x) = x^2$, $g(x) = \frac{1}{x}$, $h(x) = x - 4$

39. $f(x) = \sqrt{x}$, $g(x) = x - 5$, $h(x) = x^2$

40. $f(x) = x^2$, $g(x) = x - 5$, $h(x) = \sqrt{x}$

In Exercises 41–48 write the function as a composition of functions, each requiring only one computation.

41. $f(x) = 2x - 8$

42. $g(x) = 4 - 3x$

43. $h(x) = \frac{1}{2}(x + 9)$

44. $k(x) = \frac{2}{3}(x - 5)$

45. $L(x) = 2(x - 5)^2$

46. $T(x) = 4(x + 2)^2$

47. $H(x) = \frac{1}{\sqrt{x + 7}} + 4$

48. $R(x) = \frac{1}{\sqrt{x - 4}} - 6$

49. If $f(x) = mx + b$ and $g(x) = kx + c$ are two linear functions, show that the composition of f and g is also a linear function.

50. If f and g are linear functions and their graphs are perpendicular lines, what can you say about the slope of the graph of $f \circ g$?

51. Prove that the product of two even functions is even. What can you say about the product of two odd functions? Justify your conclusion.

52. If f and g are 1-1, show that $f \circ g$ is 1-1.

53. Is the product of an even and an odd function an even function? Is it an odd function? Justify your conclusion.

54. If f and g are odd functions, show that $f \circ g$ is an odd function.

55. Show that every function can be written as a sum of an odd function and an even function. (Hint: see Exercise 38 in Section 3.4.)

For Your Calculator

Write each function as a composition of functions, each of which is one function on the calculator, then write the key sequence for computing values of the function.

56. $f(x) = \sqrt{x - 4} + 15$

57. $h(x) = \sqrt{2x - 5} - 8$

58. $G(x) = \frac{3}{x + 7} + 5$

59. $F(x) = \frac{-2}{x - 4} + 4$

60. $f(t) = \frac{-3}{\sqrt{2 - 3t}} + 14$

61. $g(r) = \frac{-4}{\sqrt{4 - 2r}} - 6$

62. $f(s) = |4s - 3| - 5$
(Hint: $|x| = \sqrt{x^2}$)

63. $g(y) = |5y - 2| + 1$

64. $f(x) = x^2 - 4x + 7$ (Hint: complete the square to write the function as $f(x) = (x - 2)^2 + 3$.)

65. $g(x) = x^2 + 6x - 4$

66. $L(r) = 2r^2 - 8r + 1$

3.6 1-1 Functions and Inverse Functions

In Section 3.4 we graphed the functions $f(x) = x^3$ and $g(x) = \sqrt[3]{x}$. There is an interesting relationship between these two functions in that each "undoes" what the other does. To illustrate:

$$f \circ g(x) = f(\sqrt[3]{x}) = (\sqrt[3]{x})^3 = x$$
$$g \circ f(x) = g(x^3) = \sqrt[3]{x^3} = x$$

When two functions are related in this fashion they are said to be inverses of each other. Not all functions have inverses; we will see shortly how to determine those that do.

Inverse function

If f is a function with domain A and range B, then a function g with domain B and range A is called an **inverse** of f if

$f \circ g(x) = x$ for all $x \varepsilon B$ *(Recall that $x \varepsilon B$ means x belongs to B.)*

and

$g \circ f(x) = x$ for all $x \varepsilon A$

f^{-1}

If a function f has an inverse, the inverse is unique (see Exercise 41). It is denoted by f^{-1}.

Under what conditions does a function have an inverse? If a function f has input x and output y then the inverse function must take input y and give output x, i.e., it must undo what f does. If we think of f as a set of ordered pairs (x, y), then the inverse of f must consist of the ordered pairs (y, x) found by reversing the coordinates of pairs in f. The question to be answered is, "When does this process yield a function?"

Existence of inverse for a 1-1 function

Theorem A function f has an inverse if and only if f is 1-1.

Proof Suppose f is 1-1. Define $f^{-1} = \{(y, x): (x, y) \varepsilon f\}$. We need to prove that f^{-1} is a function. Suppose (s, w) and (s, u) belong to f^{-1}. Then, by the definition of f^{-1}, (u, s) and (w, s) belong to f, i.e., $f(u) = s = f(w)$. But f is 1-1, hence $u = w$. This shows that f^{-1} is a function. Suppose f has an inverse f^{-1} and $f(u) = f(w)$. Then

$$u = f^{-1}[f(u)] = f^{-1}[f(w)] = w$$

This means that f is 1-1.

From the fact that $f^{-1} = \{(y, x): (x, y) \varepsilon f\}$ if f is 1-1, the following properties are apparent.

Properties of the inverse function

1. The domain of f^{-1} is the range of f.
2. The range of f^{-1} is the domain of f.
3. For each a in the domain of f, $f^{-1}[f(a)] = a$.
4. For each b in the range of f, $f[f^{-1}(b)] = b$.

Example 1 Find the domain of $f(x) = \dfrac{x-1}{2x+1}$. Verify that $g(x) = -\dfrac{x+1}{2x-1}$ is the inverse of f. Find the range of f.

The domain of f is $\left\{x: x \neq -\frac{1}{2}\right\}$. To verify that g is the inverse of f,

$$f[g(x)] = \frac{-\dfrac{x+1}{2x-1} - 1}{2\left(-\dfrac{x+1}{2x-1}\right) + 1} = \frac{-x-1-2x+1}{-2x-2+2x-1} = x$$

$$g[f(x)] = -\frac{\dfrac{x-1}{2x+1} + 1}{2\left(\dfrac{x-1}{2x+1}\right) - 1} = -\frac{x-1+2x+1}{2x-2-2x-1} = x$$

Since g is the inverse of f, the range of f is the domain of g, $\left\{x: x \neq \frac{1}{2}\right\}$. ■

For some functions, the inverse function is easy to find. It is simply the function that undoes what the original function does. Some examples follow.

1. If $f(x) = x + a$, then $f^{-1}(x) = x - a$.
2. If $f(x) = kx$, $k \neq 0$, then $f^{-1}(x) = (1/k)\,x$
3. If $F(x) = 1/x$, then $F^{-1}(x) = 1/x = F(x)$
4. If $G(x) = x^n$, where n is an odd positive integer, then $G^{-1}(x) = \sqrt[n]{x}$. If n is even, G is not 1-1 unless the domain is restricted.

The above examples are helpful in finding the inverse of a function that is the composition of functions such as those in the list. See the next two examples.

Example 2 Find the inverse of $f(x) = 2x + 3$.

Write f as the composition of functions involving only one calculation:

$$x \xrightarrow{\text{multiply by 2}} 2x \xrightarrow{\text{add 3}} 2x + 3$$

Replace each function by its inverse and compose in the opposite order:

$$\frac{x-3}{2} \xleftarrow{\text{divide by 2}} x - 3 \xleftarrow{\text{subtract 3}} x$$

The inverse function is $f^{-1}(x) = \dfrac{x-3}{2}$. ■

Example 3 Verify that $f(x) = \dfrac{1}{x^3+1}$ is 1-1 and find its inverse.

If $f(a) = f(b)$, then

$$\frac{1}{a^3+1} = \frac{1}{b^3+1}$$

hence $a^3 + 1 = b^3 + 1$. This implies that $a = b$ and consequently f is 1-1. To find the inverse, write f as a composition and proceed as in Example 1:

$$x \xrightarrow{\text{cube } x} x^3 \xrightarrow{\text{add 1}} x^3 + 1 \xrightarrow{\text{invert}} \frac{1}{x^3+1}$$

Replacing each function by its inverse and reversing the order,

$$\sqrt[3]{\frac{1}{x} - 1} \xleftarrow{\text{cube root}} \frac{1}{x} - 1 \xleftarrow{\text{subtract 1}} \frac{1}{x} \xleftarrow{\text{invert}} x$$

The inverse is $f^{-1}(x) = \sqrt[3]{\frac{1}{x} - 1}$. ■

If the function is given by a formula $y = f(x)$, another method of finding the inverse is given below:

Step 1 From the equation $y = f(x)$, solve for x in terms of y to obtain $x = f^{-1}(y)$.

Step 2 Interchange x and y to obtain $y = f^{-1}(x)$. This step is included because we are accustomed to using x for the input and y for the output of a function.

Example 4 Find the inverse of $f(x) = 4 - 3x$.
Write the equation $y = 4 - 3x$. Solve for x to obtain $x = -\frac{1}{3}(y - 4)$. Interchange x and y to obtain

$$y = f^{-1}(x) = -\frac{1}{3}(x - 4)$$ ■

Some functions, such as $f(x) = x^2$, are not 1-1 and hence do not have inverses. We may, however, restrict the domain to make the function 1-1, and under those circumstances the function will have an inverse.

Example 5 Restrict the domain of $f(x) = x^2$ so that the resulting function is 1-1.
From the graph of $y = x^2$ we see that either the branch of the curve in the first quadrant or the branch of the curve in the second quadrant will give a 1-1 function. Let us choose the branch in the second quadrant. The domain is $\{x: x \leq 0\}$.

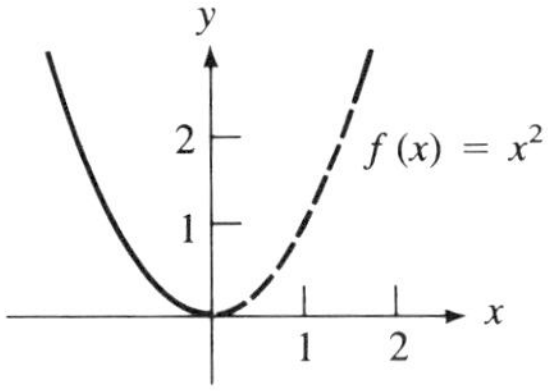

■

Example 6 Find the inverse of the function defined in Example 5.
From Example 5, the function is $f(x) = x^2$, $x \leq 0$. To find the inverse, write the equation $y = f(x)$ and solve for x:

$$y = x^2$$
$$-\sqrt{y} = x \quad (\textit{Recall that } x \leq 0.)$$

Interchange x and y to obtain

$$y = -\sqrt{x}$$ ■

If f is a 1-1 function, a point (a, b) is on the graph of f if and only if (b, a) is on the graph of f^{-1}. This means that the graphs of f and f^{-1} are symmetric with respect to the line $y = x$ (see Figure 3.10 for the graphs of $f(x) = x^2$, $x \leq 0$, and $f^{-1}(x) = -\sqrt{x}$).

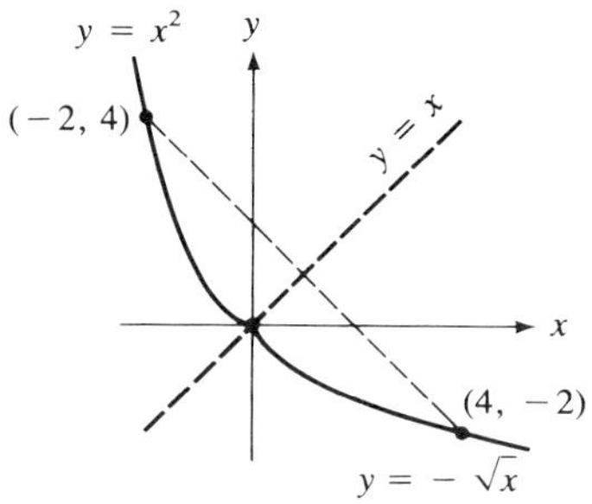

Figure 3.10

Example 7 Show that $f(x) = \sqrt{x - 2}$, $x \geq 2$, is 1-1, find the inverse, and graph both f and f^{-1} on the same set of axes.

Suppose $f(a) = f(b)$, that is, $\sqrt{a - 2} = \sqrt{b - 2}$. Then $a - 2 = b - 2$ and hence $a = b$. So f is indeed 1-1. To find f^{-1}, set $y = \sqrt{x - 2}$ and solve for x to obtain

$$x = y^2 + 2$$

Switching x and y yields

$$y = f^{-1}(x) = x^2 + 2 \qquad x \geq 0$$

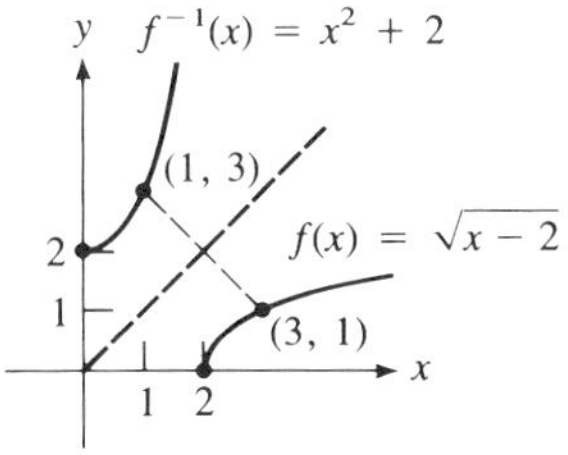

■

Exercises 3.6

In Exercises 1–12 determine whether the function is 1-1, and if it is, find the domain and range of the inverse.

1. $y = \sqrt{x + 1} - 2$

2. $y = \sqrt{x - 4} + 5$

3. $y = (x - 1)^2 - 4$

4. $y = (x + 1)^2 + 9$

5. $f(x) = \dfrac{1}{x - 3}$

6. $f(x) = \dfrac{1}{2 - x}$

7. $g(x) = |x + 2|$

8. $g(x) = |x - 5|$

9. $h(t) = t^2 + 3$, $t \geq 0$

10. $h(s) = s^2 - 5$, $s \geq 0$

11. $f(x) = (x - 1)^2 + 1$, $x \geq 0$

12. $g(r) = (r + 1)^2 + 1$, $r \leq 0$

In Exercises 13–24 write the function as a composition of functions, each requiring only one computation, then find the inverse as a composition of inverses as is done in Examples 2 and 3. Each function is 1-1.

13. $y = 2 - 4x$

14. $y = \frac{1}{2} - \frac{2}{3}x$

15. $y = \frac{3}{2} - \frac{2}{5}x$

16. $y = 3 - 6x$

17. $f(x) = 3x^2 - 4, x \geq 0$

18. $g(x) = 4x^2 + 3, x \geq 0$

19. $h(t) = 2(t - 3)^2 + 1, t \leq 3$

20. $k(s) = 4(s + 2)^2 - 5, s \leq -2$

21. $y = \sqrt{x - 2} + 1$

22. $y = \sqrt{x + 3} - 6$

23. $y = \frac{5}{2x - 3}$

24. $y = \frac{-4}{3x + 6}$

In Exercises 25–32 the given function is 1-1. Find the inverse by the method used in Examples 4 and 6.

25. $f(x) = 9 - x^2, 1 \leq x \leq 2$

26. $t(x) = 9 - x^2, 2 \leq x \leq 3$

27. $g(t) = 2(t - 4)^2 + 5, 4 \leq t \leq 6$

28. $h(s) = 4(s - 3)^2 - 2, 3 \leq s \leq 6$

29. $h(r) = \frac{r + 2}{r - 5}$

30. $k(n) = \frac{n - 4}{n + 4}$

31. $f(m) = \frac{2 - 3m}{m - 2}$

32. $h(p) = \frac{3 - 2p}{p + 1}$

In Exercises 33 and 34 the function is 1-1. Find the inverse and graph both the function and the inverse on the same coordinate axes.

33. $y = 4 - x^2, -2 \leq x \leq 0$

34. $y = 9 - x^2, -3 \leq x \leq 0$

35. Find a so that the function $f(x) = (x + 2)^2 - 4$, $-3 \leq x \leq a$, is 1-1.

36. Find b so that the function $g(x) = 6 - (x + 1)^2$, $-2 \leq x \leq b$, is 1-1.

37. Find c so that the function $f(x) = 6 - (x - 4)^2$, $-c \leq x \leq c$, is 1-1.

38. Find a so that the function $f(x) = 15 + (x - 5)^2$, $-a \leq x \leq a$, is 1-1.

39. Choose a so that the function $f(x) = |2x + a|$ is 1-1 for $-2 \leq x \leq 2$.

40. Choose a so that the function $g(x) = |3x - a|$ is 1-1 for $-4 \leq x \leq 4$.

41. If f and g are both inverses for a function h, prove that $f(x) = g(x)$ for all x in the range of h. This proves that the inverse of a 1-1 function is unique.

3.7 Applications

Many relationships are linear, and in this section we examine ways of expressing those and other relationships.

Direct variation

When a relation between variables x and y can be expressed in the form

$$y = kx$$

Constant of variation

then we say that y is **directly proportional to** x or y **varies directly as** x. The number k is called the **constant of proportionality** or **constant of variation.**

Example 1 The pressure P exerted by a liquid at a point is directly proportional to the depth d below the surface. If the pressure is 50 lb/in.^{2} at a depth of 8 ft, find the pressure at 10 ft.

The formula is

$$P = kd$$

If $d = 8$, $P = 50$:

$$50 = k(8)$$
$$6.25 = k$$

The formula is

$$P = 6.25d$$

At $d = 10$

$$P = 62.5 \text{ lb/in.}^2$$ ■

If x and y are related by the relationship

$$y = kx^n$$

we say that **y varies directly as the nth power of x, or y is directly proportional to the nth power of x.** As before, k is called the constant of proportionality.

Example 2 The area A of a circle of radius r is directly proportional to the second power of r or the square of r:

$$A = \pi r^2$$

The constant of variation is π. ■

If x and y are related by the equation

$$y = \frac{k}{x^n}$$

Inverse variation

we say that **y varies inversely as the nth power of x, or y is inversely proportional to the nth power of x.**

Example 3 The resistance R of a wire to electrical current is inversely proportional to the square of the diameter d of the wire:

$$R = \frac{k}{d^2}$$ ■

Example 4 The resistance of a wire of diameter 50 mils is 4 ohms (Ω). Find the resistance of a wire of the same material of diameter 100 mils.

The formula is

$$R = \frac{k}{d^2}$$

where R is the resistance in ohms and d is the diameter in mils.
When $d = 50$, $R = 4$, so

$$4 = \frac{k}{(50)^2}$$

$$k = 10{,}000$$

When $d = 100$,

$$R = \frac{10{,}000}{(100)^2} = 1\ \Omega$$ ■

Joint variation When relationships involve both inverse and direct variation, we will use the term **joint variation.**

Example 5 The pressure P of an enclosed gas is directly proportional to the temperature T of the gas and inversely proportional to the volume V occupied by the gas. This is known as Boyle's Law. The formula is

$$P = k\frac{T}{V}$$ ■

Example 6 If y is given by the formula

$$y = \frac{2x}{w^2v}$$

we say that y is directly proportional to x, inversely proportional to the square of w, and inversely proportional to v. ■

Most long distance calls cost a certain amount for the first 3 minutes plus a certain amount for each additional minute or fraction thereof. For example, an operator-assisted call from Las Cruces, New Mexico to Chicago, Illinois costs \$2.15 for the first 3 minutes plus \$0.35 for each additional minute at the weekday full rate. This means that the cost C of such a call can be expressed as a linear function of n, the number of additional minutes, where any fraction of a minute is counted as a full minute:

$$C = 2.15 + 0.35n$$

Example 7 A 5-minute call from Las Cruces to Los Angeles costs \$2.73, and an 8-minute call costs \$3.75. What is the cost for the first 3 minutes and what is the cost for each additional minute?

The cost C can be expressed in the form

$$C = T + rn$$

where T is the cost for the first 3 minutes and r is the cost per additional minute.

We find r in the same way that we find the slope of a linear function.

n	C
$5 - 3 = 2$	2.73
$8 - 3 = 5$	3.75

So

$$r = \frac{3.75 - 2.73}{5 - 2} = \$0.34$$

The cost for each additional minute is \$0.34. To find T, substitute $C = 2.73$ and $n = 2$ in the formula

$$C = T + 0.34n$$
$$2.73 = T + 0.34(2)$$
$$T = 2.05$$

The minimum charge for a call of 3 minutes or less is \$2.05. ■

Exercises 3.7

In Exercises 1–10 write an equation relating the variables.

1. The illumination I from a source of light varies inversely as the square of distance d from the source.
2. The amount A of air space available for each person in a room is inversely proportional to the number N of persons in the room.
3. The time T a commuter spends traveling to work is inversely proportional to the commuter's average speed.
4. Neglecting air resistance, the distance d a body falls from rest varies directly as the square of the time t.
5. The distance D on a map is directly proportional to the distance d on the ground.
6. The amount A of force necessary to apply a torque of 100 ft · lb is inversely proportional to the length L of the wrench.
7. The number W of watts used by an appliance is proportional to the voltage E and to the current I.
8. The lift force F on a wing varies directly as the area A of the wing and square of the velocity V.
9. The heat loss in an electrical transmission line is proportional to the square of the current I.
10. The period (time for one complete oscillation) T of a pendulum is directly proportional to the square root of the length L of the pendulum.

In Exercises 11–20 write an equation relating the variables, find the constant of variation, and compute any requested values.

11. The amount A a spring stretches under a force F is proportional to the stretching force. A force of 4 lb stretches the spring 6 ft.
12. The weight of a person on the moon is directly proportional to the weight of that person on earth. An astronaut weighs 150 lb on earth and 25 lb on the moon. How much does a person who weighs 200 lb on earth weigh on the moon?

13. The distance in meters that a skydiver falls before opening the parachute is proportional to the square of the time of fall in seconds. The skydiver falls 45 m in 3 sec.

14. The gravitational force between two objects varies inversely as the square of the distance between them. When the two objects are 5 cm apart, the force is 8 dynes.

15. Deductions for a retirement fund are directly proportional to the worker's pay. If $46.20 is deducted from a paycheck of $840, how much would be deducted from a paycheck of $1,200?

16. The resistance of a wire of fixed length is inversely proportional to the square of the diameter of the wire. If a wire of diameter 0.012 in. has a resistance of 10 Ω, what is the resistance of a wire of diameter 0.036 in.?

17. The maximum range of a projectile varies directly as the square of the initial velocity. If the maximum range is 15,000 ft for an initial velocity of 300 ft/sec, what is the maximum range for an initial velocity of 450 ft/sec?

18. The pressure of an enclosed gas is directly proportional to the temperature and inversely proportional to the volume occupied. A certain gas has a pressure of 15 lb/in.^{2} at 275°K when confined in a container of 100 in.^{3} If the temperature is increased to 300°K and the container is reduced to 20 in.^{3}, what will the pressure be?

19. The kinetic energy of a moving object is directly proportional to the mass of the object and to the square of its velocity. If a 5-g object traveling at 20 cm/sec has a kinetic energy of 1000 ergs, what is the kinetic energy of an 8-g object traveling at 40 cm/sec?

20. The amount withheld for Social Security (until a certain maximum is reached) is directly proportional to the amount of the pay. For a paycheck of $800 the amount withheld is $48.40. How much would be withheld from a check of $1,200?

In Exercises 21–24 the relationship described can be represented by a linear function.

21. Your electric bill includes a base charge for the first 200 kilowatt hours (kWh) plus a charge for each additional kilowatt hour used. The cost for 1,000 kWh is $35, and the cost for 1,400 kWh is $47. What is the base charge? What is the cost per kilowatt hour over 200? What would the cost be for 1,550 kWh?

22. A television repair service charges a fixed fee for each service call plus a charge for each hour or fraction thereof on the job. The charge for a job requiring 2 hr is $52 and the charge for a job requiring 4 hr is $82. What is the fixed fee? What is the hourly charge? How much would it cost for a job that required 3 hr?

23. A car rents for a fixed amount per day plus a charge for each mile driven. The cost for 1 day for 100 mi is $41 and the cost for 1 day for 250 mi is $69.50. What is the charge per mile? What is the charge per day?

24. The temperature of dry air is a linear function of altitude. At an altitude of 1,000 m the temperature is 10°C, and at an altitude of 2,500 m the temperature is −5°C. What is the temperature at sea level (0 m)?

Chapter Review

Give the domain of the function.

1. $G(x) = 2x^2 + \dfrac{1}{x}$

2. $f(x) = |2 - 3x|$

3. $H(x) = \dfrac{3x + 2}{x^2 - 9}$

4. $g(x) = \dfrac{x + 5}{\sqrt{2x - 1}}$

5. $k(x) = \dfrac{1}{(x - 4)^2}$

Find the range of each of the functions.

6. $h(x) = x^2 - 2$

7. $k(x) = \frac{1}{x}$

8. $h(x) = 1 - \sqrt{x - 2}$

9. $g(x) = |2 - 3x|$

Determine which of the functions are 1-1 and justify your conclusions.

10. $k(t) = \frac{1 + t}{1 - t}$

11. $R(s) = 2 - s^2, -1 < s < 1$

12. $h(r) = r + \frac{1}{r}$

13. $g(n) = n(n - 1) + 1$

If $G(x) = 2x^3 - 4x^2 + 6x - 7$, *find*

14. $G(-2)$ **15.** $G(3)$ **16.** $G(\sqrt{2})$ **17.** $G(x + 1)$

If $g(s) = |3s + 4|$, *find*

18. $g(-2)$ **19.** $g(3)$ **20.** $g(t - 1)$

21. A sheet of paper is 8 in. wide and 10 in. long. If a margin of t in. is used at the top and bottom and a margin of $t + 1$ in. is used at the edges, write the area A of the printed material as a function of t.

22. A point P is on the curve $y = \sqrt{x - 1}$. Write the distance d from P to the origin as a function of the x coordinate of the point.

Plot each pair of points and compute the distance between them.

23. (2, 3), (3, 4) **24.** (−2, 3), (−3, −1) **25.** (0, 1), (2, −3)

26. (1, −2), (4, 0) **27.** (−4, 0), (0, 5) **28.** (−2, 3), (3, −2)

Determine which of the relations are functions and justify your conclusions.

29. $\{(x, y): x^2 - 3y = 7\}$

30. $\{(x, y): |2x - 3y| = 8\}$

31. $\{(x, y): |2x| - 3y = 8\}$

32. $\{(x, y): |2x - 3y| = 8, x \leq 4\}$

Find the slope of the line segment joining the two points and write the equation of the line passing through the two points.

33. (2, 3), (4, 6) **34.** (−2, −1), (6, 0) **35.** (2, −3), (9, 0)

36. (0, 6), (4, 0) **37.** (2, 3), (−2, 3) **38.** (−2, 4), (−2, −2)

Write the equation of the line with the given slope passing through the given point.

39. Slope −1; (0, 4)

40. Slope 2; (−1, 4)

41. Slope 5; (−2, −2)

42. Slope 0; (4, 5)

43. Slope undefined, (1, 2)

44. Slope $-\frac{3}{4}$; (6, 9)

Write the equation of the line satisfying the given conditions.

45. Parallel to $3x - 5y = 8$; passing through (2, −3)

46. Perpendicular to $3x - 5y = 8$; passing through (2, −3)

47. Vertical line passing through (5, −6)

48. Horizontal line passing through (5, −6)

Find the slope and intercepts and graph the line.

49. $3x - 4y = 12$

50. $2x + 5y = 10$

51. $x = 3y - 7$

52. $y = 3x - 9$

Identify the symmetries (if any) of the graphs of the following relations.

53. $x^2 - y^2 = 16$

54. $y^3 - x = 45$

55. $y^2 = 3 - x$

56. $|x| - 3y = 1$

Use translation to graph each of the following.

57. $y = x^2 + \frac{1}{2}$

58. $y = 2 - x^3$

59. $y = |x + 1| - 2$

60. $y = (x - 1)^2 + 1$

61. $y = \sqrt{x - 2} + 1$

62. $y = \sqrt[3]{x + 2} - 4$

Use addition of ordinates to graph the following.

63. $y = x - |x|$

64. $y = x + \frac{1}{2}|x|$

If $f(t) = t^2 - 4$ *and* $g(t) = t + \frac{2}{t}$, *find*

65. $f \circ g$

66. $g \circ f$

67. $f + g$

68. $f - g$

69. g/f

70. fg

71. *Find the domain of each of the functions in Exercises 65–70.*

Write each of the following functions as a composition of functions, each requiring only one computation.

72. $f(x) = x^2 - 5$

73. $h(s) = \frac{2}{3}s + 5$

74. $g(x) = |x - 3| + 2$

75. $f(r) = 2\sqrt{r + 1} + 4$

Determine whether the function is 1-1, *and if it is, find the inverse.*

76. $y = (x - 3)^2 + 1,\ 3 \le x \le 6$

77. $y = 2x - 3$

78. $y = \frac{4}{3 - x}$

79. $y = \sqrt{4 - x^2},\ -2 \le x \le 2$

Write an equation relating the variables.

80. The amount S of sales tax varies directly as the amount A of the purchase.

81. The property tax T is directly proportional to the value V of the property.

82. The momentum M of an object is directly proportional to the velocity v and directly proportional to the mass m of the object.

83. The current I produced by a given voltage is inversely proportional to the resistance R.

Write an equation relating the variables and find the requested values.

84. The amount of sales tax varies directly as the amount of the purchase. The tax on \$40.00 is \$1.70. What is the tax on \$50.00?

85. The property tax varies directly as the value of the property. If the tax on a home valued at \$50,000 is \$620, what is the tax on a home valued at \$72,000?

86. Griselda's Garage charges a fixed amount for the oil filter plus a fixed amount for each quart of oil used in changing the oil in an automobile. A change with 5 qt costs \$10.70 and a change with 6 qt costs \$11.85. What is the charge for the filter, and what is the charge for each quart of oil?

87. A person's caloric needs are a linear function of outdoor temperature. At 21°C an adult needs 3,000 Cal/day and at 0°C an adult needs 3,630 Cal/day. What are an adult's caloric needs at 16°C?

Chapter Test

1. Find the domain and range of $f(x) = \sqrt{4 - x^2}$.

For Problems 2 and 3 use $f(x) = 2x - 1$ *and* $g(x) = 2x^2$.

2. Find $f(3)$, $g(-2)$, $f(1/x)$, and $g(2 - x)$.

3. Give formulas for $f \circ g$ and $g \circ f$.

4. The volume V of a rectangular parallelepiped (box) is $V = lwh$, where l and w are the length and width of the base, respectively, and h is the height. Write V as a function of w if the base is square and the height is twice the width of the base.

In Problems 5 and 6 find the distance between the points, the slope of the segment joining them, and the equation of the line through the two points.

5. $(2, -6)$, $(-4, 5)$

6. $(3, 8)$, $(-1, -1)$

In Problems 7 and 8 find the slope and the intercepts and graph the line.

7. $2x - 3y = 6$

8. $5x + y = 9$

9. Find the equation of the line parallel to $5x = 2y + 6$ that passes through $(-3, 4)$.

10. Find the equation of the line perpendicular to the line $2y - 7 = 3x$ that passes through $(4, -5)$.

In Problems 11 and 12 identify the symmetries (if any) of the graph of the given relation.

11. $y = -2x^2 + |x| + 9$

12. $y^2 = x^3 + 5x - 6$

13. Graph $y = 2 - (x + 1)^2$.

In Problems 14–16 determine which functions are 1-1 and find the inverse function if there is one.

14. $f(x) = x^2 + 1, \ -1 \le x \le 1$

15. $g(x) = 4x - 1, \ 1 \le x \le 2$

16. $H(x) = \dfrac{-4}{2x - 3}$

17. A car rental agency rents cars for a certain amount per day plus a charge for each mile driven. One day's rental for 150 mi is \$53.50 and one day's rental for 250 mi is \$72.50. What is the daily charge and what is the charge per mile?

4 Polynomial Functions, Rational Functions, and Conic Sections

When a projectile is fired at a certain angle with a known initial velocity, the motion can be described by a *quadratic* function. For example, if a projectile is fired with an angle of elevation of 80°32′ and an initial velocity of 243 ft/sec, the path of the projectile is described by the function $y = 6x - 0.1x^2$, where x is the horizontal distance in feet from the point of firing and y is the height in feet of the projectile above the point of firing. In this chapter we will examine, among other things, functions of this type.

4.1 Quadratic Functions

In Chapter 3 we examined the graph of the function $f(x) = x^2$ by plotting some points and drawing a smooth curve through them. The resulting curve is called a **parabola.** As you will see later, the graph of every quadratic function is a parabola.

Quadratic function

A **quadratic function** is any function that can be written in the form

$$f(x) = ax^2 + bx + c \quad \text{where} \quad a \neq 0$$

The graph of the quadratic function $f(x) = ax^2$ can be sketched by slightly changing the shape of the graph of the parabola $y = x^2$.

Example 1 Sketch the graphs of $f(x) = 2x^2$ and $g(x) = \frac{1}{2}x^2$ on the same axes.

The curves are found by multiplying the ordinates of $y = x^2$ (the dashed curve) by 2 and by $\frac{1}{2}$, respectively.

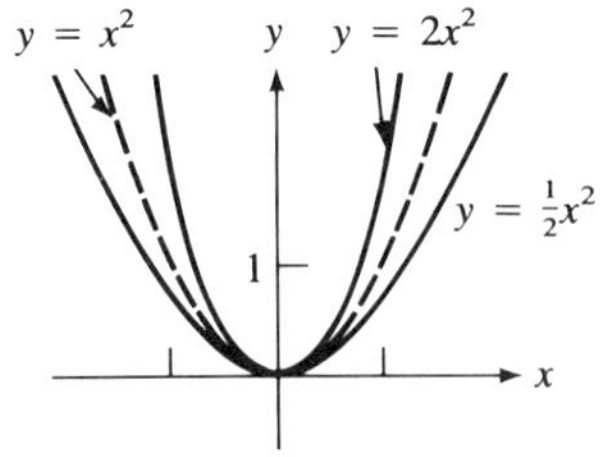

Note that the graph of $f(x) = 2x^2$ is "narrower" than the graph of $g(x) = \frac{1}{2}x^2$. ■

Example 2 Sketch the graph of $h(x) = -2x^2$.

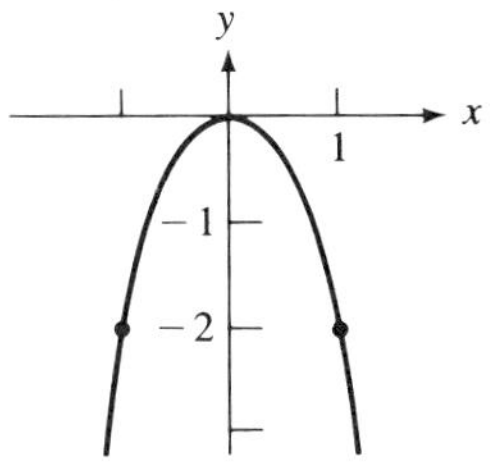

The graph of $h(x) = -2x^2$ is the same shape as the graph of $f(x) = 2x^2$, but it opens downward rather than upward. ■

Suppose that we now have a quadratic function written in the form $f(x) = a(x - h)^2 + k$. If we rewrite this in the language of Section 3.4 we would have $y - k = a(x - h)^2$. This means that the graph of $f(x) = a(x - h)^2 + k$ can be found by translating the graph of $y = ax^2$. We summarize below.

> The graph of $f(x) = a(x - h)^2 + k$ is found by shifting the graph of $y = ax^2$ by $|h|$ units, to the right if $h > 0$ and to the left if $h < 0$, and by $|k|$ units, up if $k > 0$ and down if $k < 0$.

Example 3 Graph the function $y = 2(x - 1)^2 + 3$.
In this case, $h = 1$ and $k = 3$. We shift the graph of $y = 2x^2$ to the right 1 unit and up 3 units.

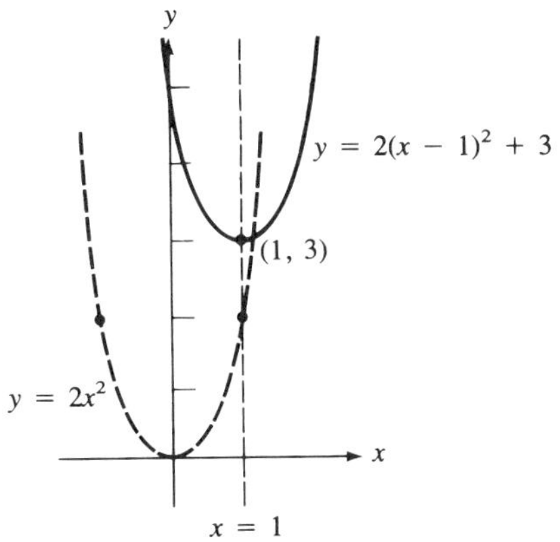

■

Example 4 Graph the function $y = (x + 2)^2 - 1$.
Here $h = -2$ and $k = -1$. Shift the graph of $y = x^2$ to the left 2 units and down 1 unit.

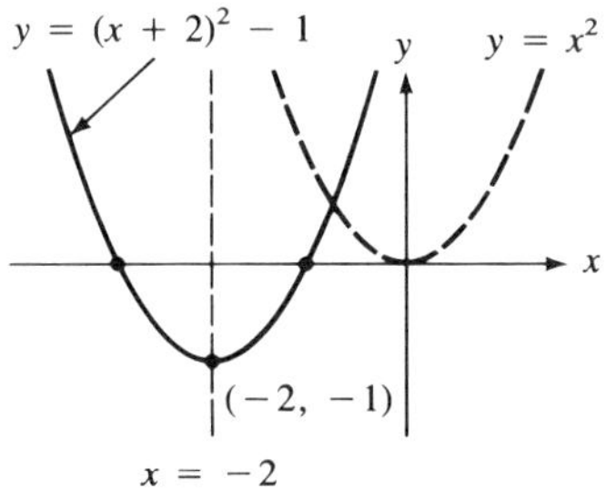

■

Each of the parabolas in Examples 1–4 has certain noteworthy properties:

Vertex
1. Each parabola has an extreme point called the **vertex.** In Examples 1, 3, and 4 the vertex is the lowest point on the curve, and in Example 2 the vertex is the highest point on the curve.

Line of symmetry
2. Each parabola has a **line of symmetry** that passes through the vertex. In Examples 1 and 2 the line of symmetry is the y axis, the line $x = 0$. In Example 3 the line of symmetry is the line $x = 1$, and in Example 4 the line of symmetry is the line $x = -2$.

Every quadratic polynomial $ax^2 + bx + c$ may be written in the form $a(x - h)^2 + k$ by completing the square. We therefore summarize the properties of the parabola when the equation is written in the form $y = a(x - h)^2 + k$.

The graph of the function $y = a(x - h)^2 + k$ is a parabola with the following properties.

Properties of the parabola

1. The parabola opens upward if $a > 0$ and downward if $a < 0$.
2. The vertex is at (h, k).
3. The line of symmetry is the line $x = h$.

Example 5 Graph the function $f(x) = x^2 - 2x$, showing the line of symmetry and the vertex.

Completing the square yields

$$f(x) = x^2 - 2x = x^2 - 2x + 1 - 1 = (x - 1)^2 - 1$$

The parabola opens upward, the vertex is at $(1, -1)$, and the line of symmetry is $x = 1$.

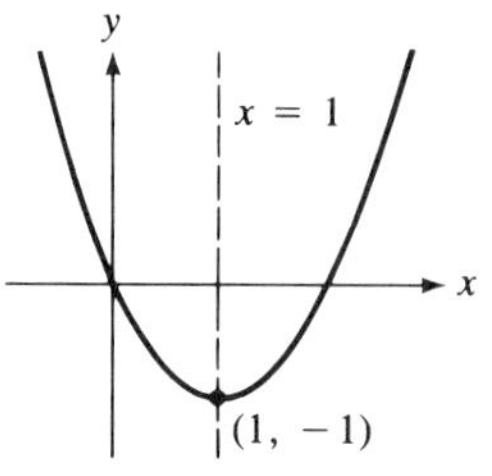

■

Completing the square may be used to find maximum or minimum values of quadratic functions.

Example 6 A piece of galvanized metal 28 cm wide is to be formed into a gutter by folding up the edges at right angles. What dimensions will maximize the capacity of the gutter?

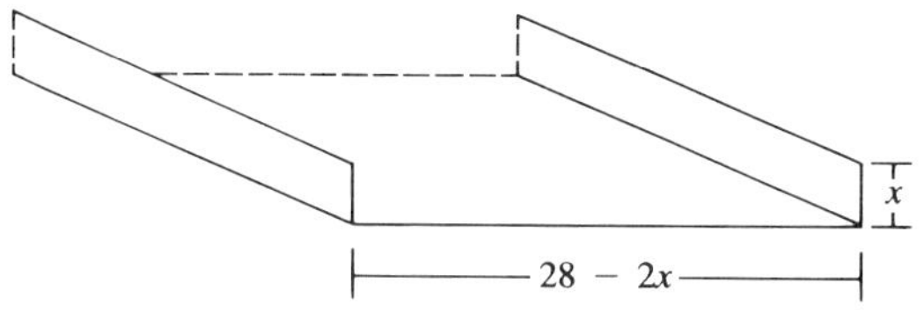

Let x be the width of the piece of metal to be turned up. To maximize the capacity of the gutter we must maximize the area of the cross section. That area A is given by the formula

$$A = x(28 - 2x)$$

To maximize A, complete the square:

$$\begin{aligned} A &= x(28 - 2x) = -2x^2 + 28x = -2(x^2 - 14x) \\ &= -2(x^2 - 14x + 49) + 2(49) \\ &= -2(x - 7)^2 + 98 \end{aligned}$$

The graph of A is a portion of a parabola. The largest value of A occurs at the vertex, (7, 98).

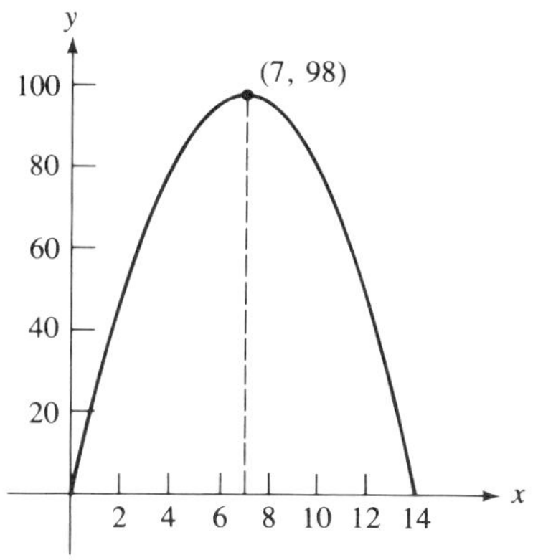

The maximum area of the cross section is 98 when $x = 7$. The gutter should be formed as shown in the sketch.

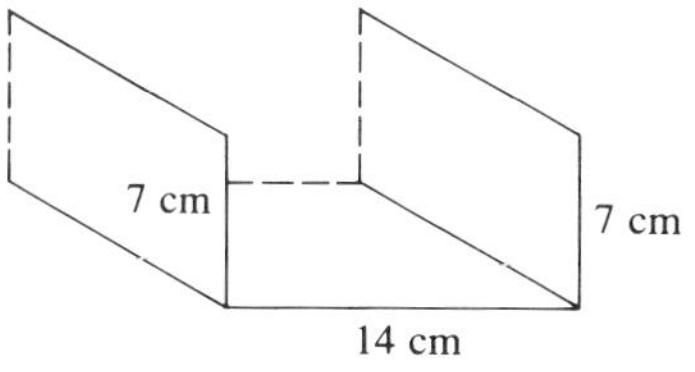

■

Example 7 An apartment manager rents each of his 40 apartments at \$360 per month. He thinks that for each \$20 increase in rent he will have on the average 2 vacant apartments per month. What rent should he charge to maximize his monthly income?

Let n be the number of \$20 increases to be put into effect. Then the manager's rent per apartment will be $\$360 + 20n$, and the number of apartments rented will be $40 - 2n$. The monthly income I will be given by the quadratic function

$$\begin{aligned} I &= (360 + 20n)(40 - 2n) \\ &= 14{,}400 + 800n - 720n - 40n^2 \end{aligned}$$

Collecting terms and completing the square gives

$$\begin{aligned} I &= -40(n^2 - 2n + 1) + 40 + 14{,}400 \\ &= -40(n - 1)^2 + 14{,}440 \end{aligned}$$

To maximize I, choose $n = 1$, that is, raise the rent to \$380 per apartment. The income per month is then \$14,440, only a \$40 increase over the income at \$360 per month. Hardly worth the fuss! ■

Consider the function

$$f(x) = ax^2 + bx + c \quad \text{where} \quad a \neq 0$$

Zero of a function

The real roots of the equation $ax^2 + bx + c = 0$ are those numbers r such that $f(r) = 0$. The number r is called a **zero** of f and the graph of f passes through the point $(r, 0)$. We know that a quadratic equation with real coefficients may have two distinct real roots, a real double root, or two imaginary roots. These possibilities correspond to the parabola intersecting the x axis at two points, at only one point (tangent to the x axis), or not intersecting the x axis at all. Figure 4.1 illustrates these possibilities.

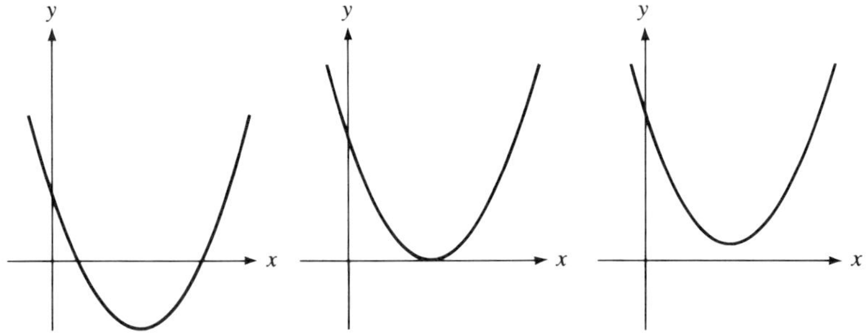

Two intersections, two real and distinct roots | One intersection, one double root | No intersection, two imaginary roots

Figure 4.1

Example 8 Find the vertex, line of symmetry, and x intercepts for the parabola

$$y = -x^2 - x + 12$$

Completing the square yields

$$y = -x^2 - x + 12$$
$$= -\left(x + \frac{1}{2}\right)^2 + 12\frac{1}{4}$$

The vertex is at $\left(-\frac{1}{2}, 12\frac{1}{4}\right)$, and the line of symmetry is $x = -\frac{1}{2}$. The x intercepts are found by solving the equation

$$-x^2 - x + 12 = 0$$
$$x^2 + x - 12 = 0$$
$$(x - 3)(x + 4) = 0$$
$$x = 3 \qquad x = -4$$

The intercepts are (3, 0) and (−4, 0). Notice that symmetry guarantees that the line of symmetry is halfway between the x intercepts.

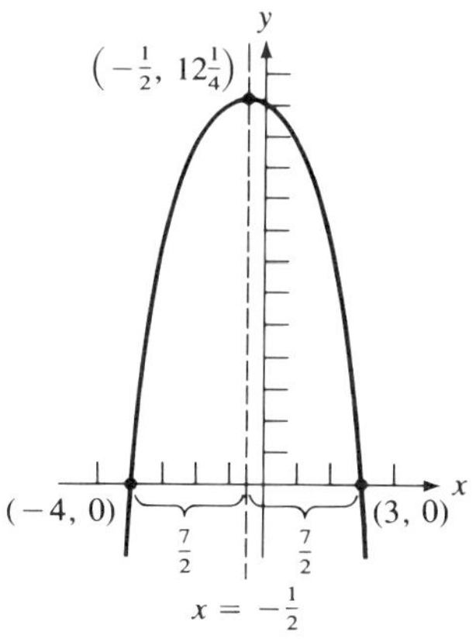

■

Example 9 Find the equation of the parabola with vertex at (−1, 5) that passes through (2, −13).
The equation will have the form

$$y = a(x + 1)^2 + 5$$

because the vertex is at (−1, 5). Substituting $x = 2$ and $y = -13$ yields

$$-13 = 9a + 5 \quad \text{or} \quad a = -2$$

The desired equation is

$$y = -2(x + 1)^2 + 5$$

■

Example 10 Find the equation of the parabola with x intercepts (−4, 0) and (2, 0) and with highest point having a y coordinate of 3.
The line of symmetry will intersect the x axis halfway between the x intercepts

at the point $(-1, 0)$. The line of symmetry passes through the vertex, and the vertex is thus at $(-1, 3)$. The equation has the form

$$y = a(x + 1)^2 + 3$$

Substituting $x = 2$ and $y = 0$ yields $a = -\frac{1}{3}$. The equation is

$$y = -\frac{1}{3}(x + 1)^2 + 3$$

■

Exercises 4.1

In Exercises 1–22 graph the function, showing the vertex and line of symmetry.

1. $y = -\frac{1}{2}x^2$
2. $y = -\frac{1}{3}x^2$
3. $y = 3x^2$
4. $y = 1.5x^2$
5. $y = 1 - x^2$
6. $y = 2 - x^2$
7. $y = 2(x + 1)^2$
8. $y = 1.5(x + 2)^2$
9. $y = (x - 1)^2 - 1$
10. $y = (x + 1)^2 - 1$
11. $y = 4 - (x + 1)^2$
12. $y = 5 - (x - 2)^2$
13. $y = x^2 - 2x + 4$
14. $y = x^2 + 2x + 4$
15. $f(x) = -x^2 + 2x - 4$
16. $g(x) = -x^2 - 2x - 4$
17. $g(x) = -x^2 + 2x$
18. $h(x) = -x^2 + 4x$
19. $y = 2x^2 - 12x + 24$
20. $y = 2x^2 + 12x + 24$
21. $f(x) = 6x - 3x^2 - 4$
22. $f(x) = 12x - 3x^2 - 11$

In Exercises 23–30 write the equation of a parabola with the given properties.

23. Vertex at $(-2, 3)$; passing through $(-1, 5)$
24. Vertex at $(1, 4)$; passing through $(2, 7)$
25. x Intercepts $(-2, 0)$ and $(4, 0)$; highest point has y coordinate 3
26. x Intercepts $(-3, 0)$ and $(7, 0)$; lowest point has y coordinate -2
27. Passes through $(2, 4)$ and $(5, 4)$ and has only one x intercept
28. Passes through $(1, -2)$ and $(4, -2)$ and has only one x intercept
29. Line of symmetry $x = -2$; passes through $(0, 0)$ and $(-1, -3)$
30. Line of symmetry $x = -4$; passes through $(0, 0)$ and $(1, -9)$.
31. The height of a rocket after t sec is given by the function $H(t) = 256t - 16t^2$. Find the maximum height. When does it occur?
32. Jean arranges a special concert for a club. Tickets will cost \$4.00 each for 100 tickets or fewer, but for each additional 50 tickets over 100 sold the price of each ticket is reduced 20¢. What number of tickets sold will give the most revenue?
33. G & G Leasing Company leases automobiles to corporations. The company charges \$1,000 per car per year, but for each car more than 10 cars in the lease contract they discount the lease cost of each car by 2%. What number of cars leased to a single corporation yields the most income to G & G?
34. Prove that $x^2 - 2x \geq -1$ for each real number x. (Hint: find the vertex of the graph of the function $f(x) = x^2 - 2x$.)
35. Prove that $4x^2 + 4x \geq -1$ for each real number.
36. Buddy decides to build a chicken pen. He buys a roll of chicken wire that is 72 ft long and 4 ft wide and decides to construct a rectangular pen that has an 8-ft-high fence on all four sides. However, he can utilize an existing 4-ft-high fence as the bottom half of one side of the pen. What dimensions for the pen yield the maximum area?
37. If an orange grove is planted 32 trees to the acre, the average yield per acre is expected to be 400 oranges. For each additional tree planted per acre, the average yield per tree is decreased by 10 oranges. How many trees to the acre will produce the maximum yield per acre?
38. A rectangular pen is to be built with a single partition. If 36 yards of fencing is available, what dimensions maximize the area of the pen?

39. A playing field is to be built in the shape of a rectangle with a semicircle at each end. The perimeter of the field must be 400 m. Find the dimensions of the field that maximize the area of the rectangular part.

40. A window is made in the shape of a rectangle with a semicircle on the top. If the perimeter is to be 8π ft, find the dimensions that maximize the area of the window.

41. A piece of wire 8 in. long is cut into two pieces. One piece is formed into a circle, the other into a square. How can the wire be cut so that the sum of the areas of the square and the circle is minimized?

4.2 Conic Sections: The Parabola and the Circle

The most general second-degree equation in x and y is

$$ax^2 + bxy + cy^2 + dx + ey + f = 0$$

where at least one of a, b, and c is nonzero. The curves that result from the solution sets of such equations are called **conic sections.** The conic sections are the circle, the parabola, the ellipse, and the hyperbola and are named conic sections because they may all be obtained as the intersections of a plane and a cone, as shown in Figure 4.2. Included in the category of conic sections are what are called the **degenerate** conics, which can be a point, a line, or two lines.

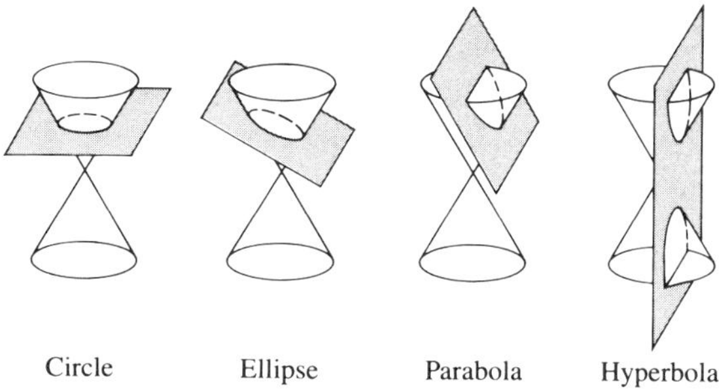

Figure 4.2

In this and the next section, we will graph the conics that are the graphs of equations of the form

$$ax^2 + cy^2 + dx + ey + f = 0$$

that is, those second-degree equations in x and y in which the xy term is missing.

The circle has already been discussed in Chapter 3, but a review seems appropriate here.

Circle A **circle** consists of all points at a fixed distance r, called the **radius,** from a fixed point (h, k), called the **center.**

Equation of the circle

Using the distance formula, we obtain

$$(x - h)^2 + (y - k)^2 = r^2$$

for the equation of the circle with radius r and center at (h, k).

Example 1 The equation of the circle with center at $(3, -4)$ and radius 6 is

$$(x - 3)^2 + (y + 4)^2 = 36$$

■

Example 2 The equation

$$(x - 1)^2 + (y + 8)^2 = 9$$

is the equation of the circle with center at $(1, -8)$ and radius 3. ■

Example 3 Show that the equation

$$x^2 + y^2 + 4x - 3y = 5$$

is the equation of a circle.

In order to write the equation in the form $(x - h)^2 + (y - k)^2 = r^2$, complete the square for both the x and y terms:

$$x^2 + 4x \qquad + y^2 - 3y \qquad = 5$$

$$x^2 + 4x + 4 + y^2 - 3y + \frac{9}{4} = 5 + 4 + \frac{9}{4}$$

$$(x + 2)^2 + \left(y - \frac{3}{2}\right)^2 = \frac{45}{4}$$

This is the equation of a circle with center at $\left(-2, \frac{3}{2}\right)$ and radius $3\sqrt{5}/2$. ■

Example 4 Find the equation of the circle passing through the points $(1, 3)$, $(5, 3)$, and $(5, 9)$.

Plot the points on a coordinate system. Remember that the perpendicular bisector of any chord of a circle passes through the center. The perpendicular bisector of the segment connecting $(1, 3)$ and $(5, 3)$ is the line $x = 3$. The perpendicular bisector of the segment connecting $(5, 3)$ and $(5, 9)$ is the line $y = 6$ (see the sketch). The center must be at $(3, 6)$, and hence the radius is the distance from $(3, 6)$ to $(1, 3)$ (or any of the three points for that matter.) We find $r = \sqrt{13}$, and the equation is

$$(x - 3)^2 + (y - 6)^2 = 13$$

You should check to see that each of the three points satisfies the equation.

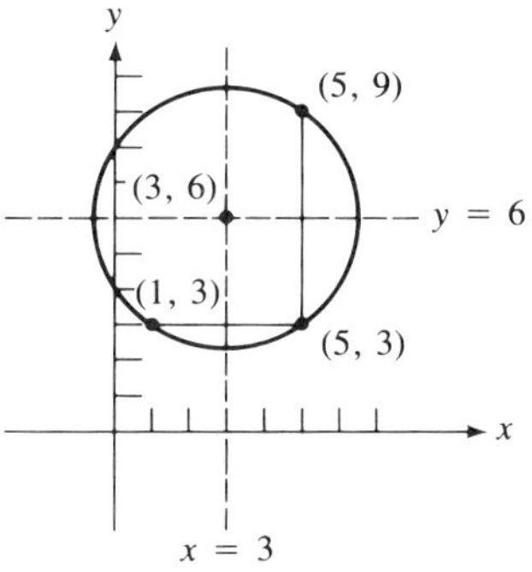

■

The method used in Example 3 shows that any equation of the form

$$ax^2 + ay^2 + by + cx + d = 0$$

may be written in the form

$$(x - h)^2 + (y - k)^2 = m$$

assuming $a \neq 0$. If $m > 0$ the graph is a circle, if $m = 0$ the graph is the single point (h, k), and if $m < 0$ there are no points on the graph.

In the last section we examined the parabola as the graph of a quadratic function, but it also has a geometric interpretation.

Geometric definition of the parabola

A parabola is the set of all points such that the distance from a fixed point, called the **focus,** and the distance from a fixed line, called the **directrix,** are equal.

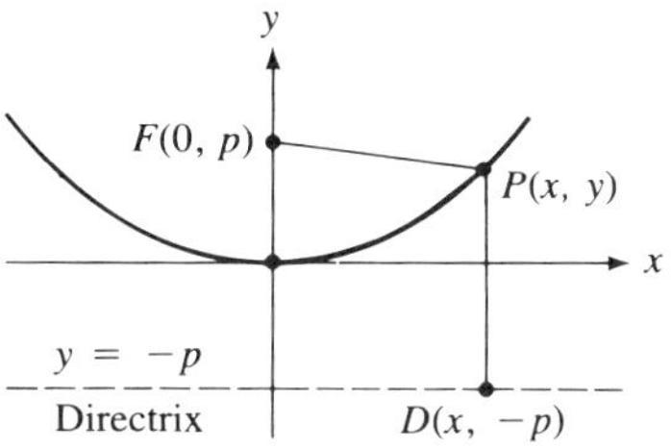

Figure 4.3

A point P with coordinates (x, y) is on the parabola if $\overline{PF} = \overline{PD}$, where F is the focus and $\overline{PD}$ is the distance from P to the directrix (see Figure 4.3). Using the distance formula we obtain

$$\sqrt{(x - 0)^2 + (y - p)^2} = \sqrt{(x - x)^2 + (y + p)^2}$$

Squaring both sides and simplifying yields the equation

$$x^2 = 4py$$

Dividing both sides by $4p$ yields the quadratic function

$$y = \frac{1}{4p} x^2$$

a function whose graph is now very familiar.

If the roles of x and y are interchanged, we obtain the equation for a parabola with the focus at $(p, 0)$ and with directrix $x = -p$:

$$y^2 = 4px$$

See Figure 4.4 for the case $p > 0$. The vertex V is at $(0, 0)$ and the x axis is the axis of symmetry.

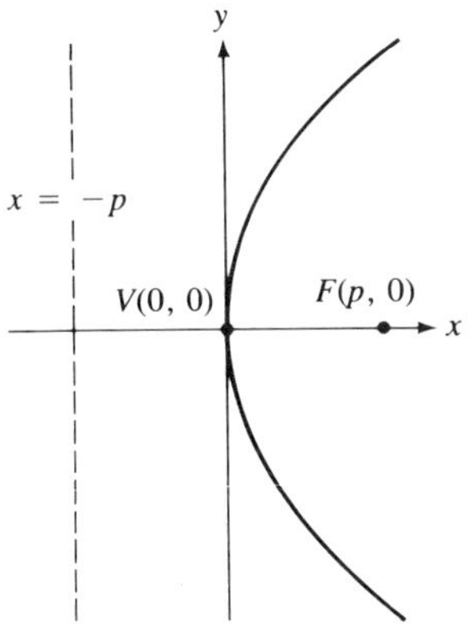

Figure 4.4

Using the notions of translation from Chapter 3, we may obtain the equation and graph of any parabola with directrix parallel to either of the coordinate axes.

General equation for the parabola

> The equation $(x - h)^2 = 4p(y - k)$ is the equation of the parabola with focus at $(h, p + k)$, directrix $y = -p + k$, and vertex at (h, k). The line of symmetry is the line $x = h$. Notice that the vertex lies on the line of symmetry halfway between the focus and the directrix. The parabola opens upward if $p > 0$ and downward if $p < 0$ (see Figure 4.5).

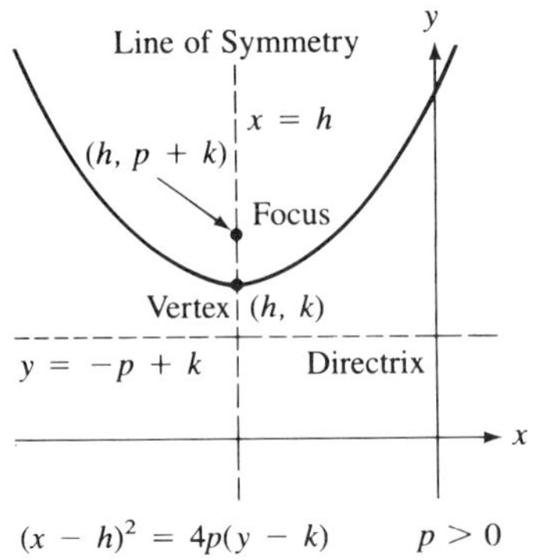

Figure 4.5

The equation $(y - k)^2 = 4p(x - h)$ is the equation of the parabola with focus at $(p + h, k)$, directrix $x = -p + h$, and vertex at (h, k). The line of symmetry is the line $y = k$. Again note that the vertex is on the line of symmetry halfway between the focus and the directrix. The parabola opens to the right if $p > 0$ and opens to the left if $p < 0$ (see Figure 4.6).

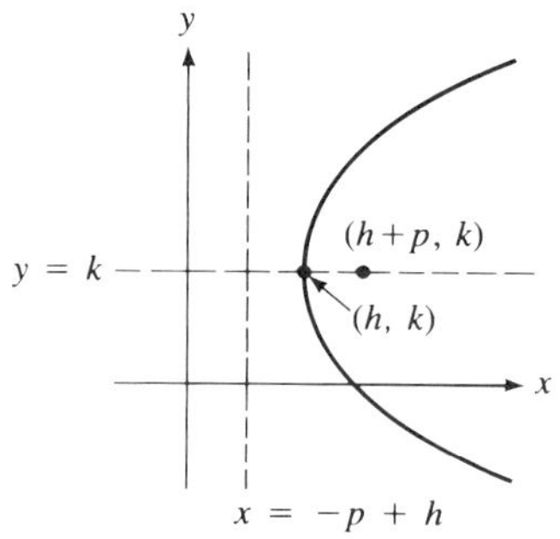

$(y - k)^2 = 4p(x - h) \qquad p > 0$

Figure 4.6

Example 5 Graph the parabola $y^2 = x$, showing the focus and directrix.

Here we have the equation $y^2 = 4px$ with $p = \frac{1}{4}$.

The focus is at $\left(\frac{1}{4}, 0\right)$ and the directrix is $x = -\frac{1}{4}$.

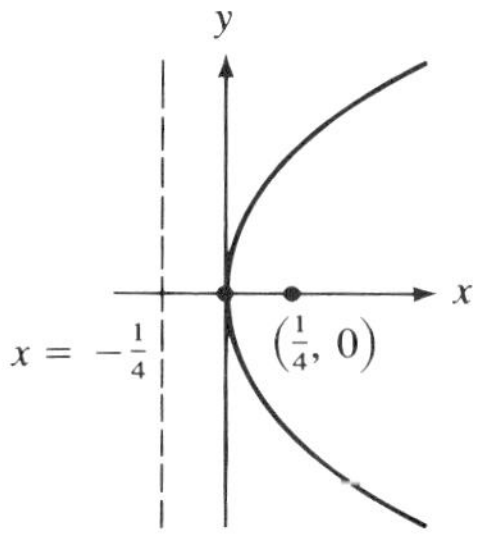

■

Example 6 Graph the parabola $x = 2y - y^2$, showing the vertex, line of symmetry, focus, and directrix.

Completing the square for the y terms we obtain

$$\begin{aligned} x &= 2y - y^2 \\ &= -(y^2 - 2y + 1) + 1 \\ (y - 1)^2 &= -(x - 1) \end{aligned}$$

This is the graph of $y^2 = -x$ translated 1 unit to the right and 1 unit up. The vertex is at $(1, 1)$, and the line of symmetry is the line $y = 1$. Here we have

$p = -\frac{1}{4}$, so the focus is at $\left(1 - \frac{1}{4}, 1\right) = \left(\frac{3}{4}, 1\right)$ and the directrix is the line $x = 1 + \frac{1}{4} = \frac{5}{4}$.

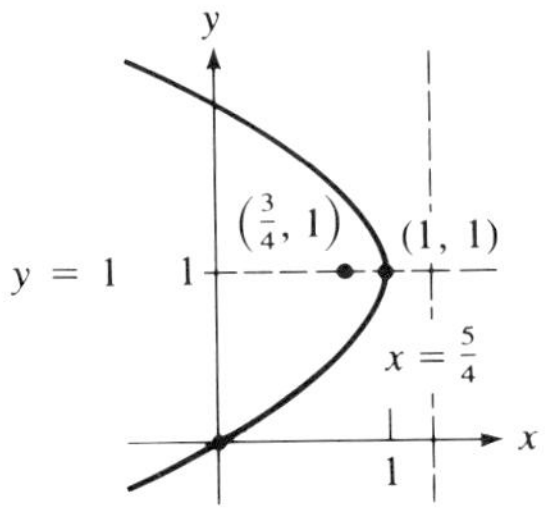

■

Example 7 Write the equation of the parabola with focus at $(-2, 3)$ and directrix $y = 5$. The distance from the focus to the directrix is $2p$ and from the graph we see that $2p = 2$, $p = 1$. The parabola opens downward and the vertex is at $(-2, 4)$. The equation is

$$(x + 2)^2 = -4(1)(y - 4)$$

or

$$(x + 2)^2 = -4(y - 4)$$

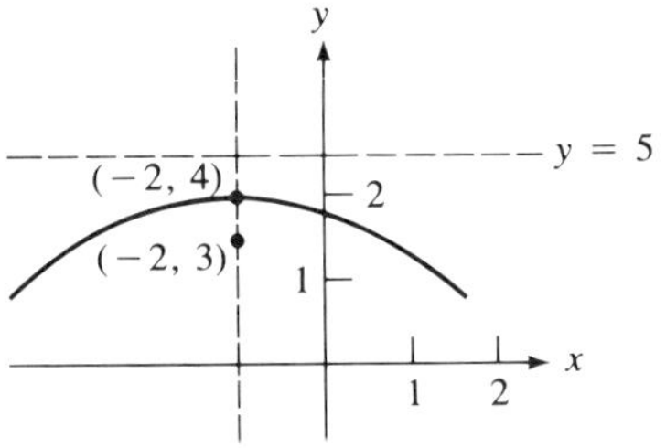

■

Example 8 Write the equation of the parabola with vertex at $(2, -1)$ and focus at $\left(2, -\frac{7}{8}\right)$. Since the vertex is at $(2, -1)$ and the focus is at $\left(2, -\frac{7}{8}\right)$, the parabola opens upward. The distance from the vertex to the focus is p, so $p = \frac{1}{8}$. The directrix is the line $y = -\frac{9}{8}$, and the equation is

$$(x - 2)^2 = 4\left(\frac{1}{8}\right)(y + 1)$$

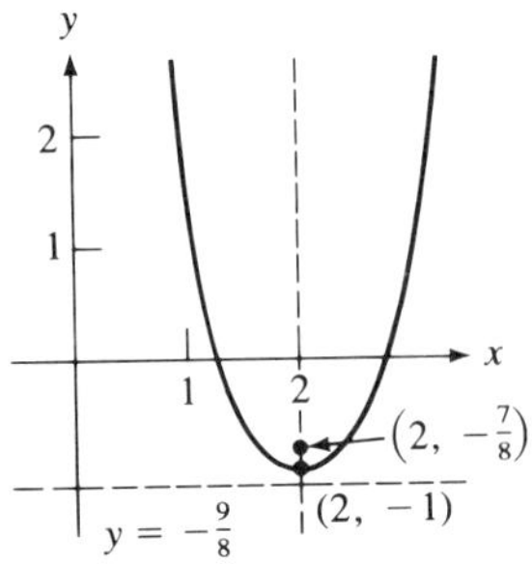

■

When a parabola is rotated about its line of symmetry, a curved surface called a paraboloid of revolution is formed. Parabolic mirrors are ground so that the reflecting surface is part of a parabolic surface. Such mirrors are used in astronomical telescopes because all rays of light that come into the telescope parallel to the axis of symmetry are reflected to the focus of the parabola.

Example 9 A parabolic mirror is 32 in. wide and 10 in. deep. How far from the vertex of the mirror is the focus?

Place a cross section of the mirror on a coordinate system with the vertex at the origin. Then the parabola passes through $(-16, 10)$ and $(16, 10)$. Since the vertex is at the origin, the equation of the parabola is $y = ax^2$, and substituting $x = 16$ and $y = 10$ yields $a = \frac{10}{256} = \frac{5}{128}$. Recall that the focus of the parabola $y = (1/4p)x^2$ is at the point $(0, p)$. Solving

$$\frac{1}{4p} = \frac{10}{256}$$

we obtain

$$p = \frac{256}{40} = 6.4 \text{ in.}$$

The focus is 6.4 in. above the vertex.

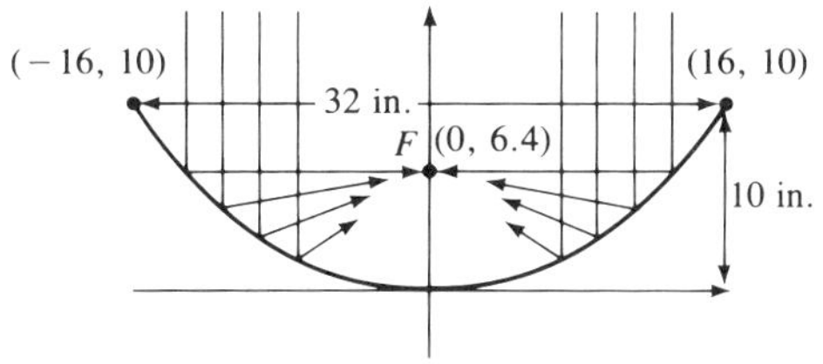

■

Exercises 4.2

Find the center and radius of each circle.

1. $x^2 + y^2 + 14x - 6y = 3$

2. $x^2 + y^2 - 8y = 9$

3. $2x^2 + 2y^2 = 4y - 7x - 8$

4. $3x^2 + 3y^2 - 12x = 6 - 8y$

Find the equation of the circle passing through the given points. Remember, the perpendicular bisector of a chord (a segment connecting two points on the circle) passes through the center.

5. $(0, 0), (6, 0), (6, 4)$

6. $(0, 0), (8, 0), (8, 4)$

7. $(0, 0), (0, 6), (2, 6)$

8. $(0, 0), (0, 4), (1, 4)$

9. $(2, 4), (2, 7), (4, 4)$

10. $(-1, 5), (-1, 8), (-5, 5)$

Write an equation for the parabola with the given information.

11. Focus (2, 0); directrix $x = -2$

12. Focus (4, 0); directrix $x = -4$

13. Focus (0, 3); vertex (0, 0)

14. Focus (0, −3); vertex (0, 0)

15. Focus (−2, 4); directrix $y = 8$

16. Focus (3, −5); directrix $y = -1$

17. Vertex (2, 6); focus (2, 9)

18. Vertex (2, 6); focus (4, 6)

Graph the parabola, showing vertex, focus, and directrix.

19. $x = \frac{1}{8}y^2$

20. $16x = y^2$

21. $y^2 - 4y + 12 = 8x$

22. $y^2 - 2y - 31 = 8x$

23. $x^2 + 2x + 8y = 15$

24. $y^2 - 8x - 2y + 17 = 0$

25. A parabolic mirror is 64 in. wide and 8 in. deep. How far is the focus from the vertex?

26. A parabolic mirror is to be made 48 in. wide and with a focus 4 in. from the vertex. How deep should it be?

27. The path of a projectile which is given an initial velocity and then is subject only to the force of gravity is a parabola. If the path of an artillery shell is given by $y = x - 0.00001x^2$, where y is the height and x is the horizontal distance from the gun, both measured in feet, what is the maximum height? How far from the gun does the shell hit the ground?

28. If the path of a projectile is given by the function $y = x - 0.002x^2$, what is its maximum height? How far away from the point of firing does it hit the ground?

4.3 Conic Sections: The Ellipse

The ellipse looks somewhat like a squashed circle, and indeed, the circle may be considered a special case of the ellipse.

Geometric definition of the ellipse

The **ellipse** is the set of points such that for each point the sum of its distances from two fixed points is constant. The two fixed points are called the **foci** of the ellipse.

A point P with coordinates (x, y) is on the ellipse if $\overline{PF_1} + \overline{PF_2} = 2a$, where F_1 and F_2 are the foci and a is a constant (see Figure 4.7). The reason for using $2a$ will be apparent shortly.

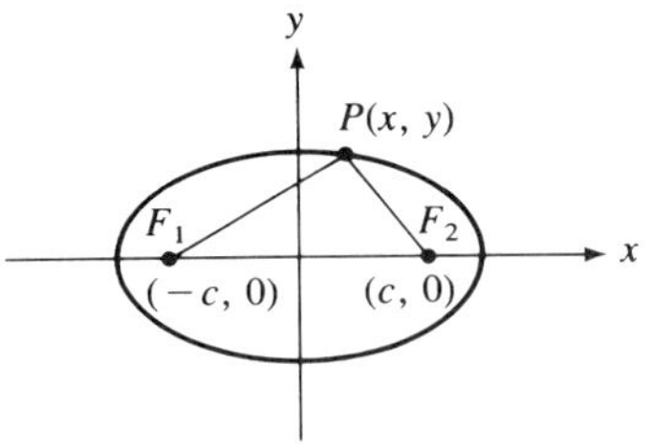

Figure 4.7

Using the distance formula we have

$$\sqrt{(x - c)^2 + y^2} + \sqrt{(x + c)^2 + y^2} = 2a$$

Eliminating the radicals and collecting terms yields the equation

$$\frac{x^2}{a^2} + \frac{y^2}{a^2 - c^2} = 1$$

If the foci are at $(-c, 0)$ and $(c, 0)$ the distance between them is $2c$; hence the sum of the distances $\overline{PF}_1 + \overline{PF}_2 = 2a$ must be larger than $2c$ (see Figure 4.7). If $2a = 2c$, the only points satisfying the condition are those points on the segment F_1F_2. Since $a > c$, we have $a^2 - c^2 > 0$, and it is convenient to write $a^2 - c^2 = b^2$, where $b > 0$.

Looking at the equation

$$\frac{x^2}{a^2} + \frac{y^2}{b^2} = 1$$

we see that the curve will be symmetric with respect to the x axis, with respect to the y axis, and with respect to the origin, and that the points $(-a, 0)$, $(a, 0)$, $(0, -b)$, and $(0, b)$ will be on the curve.

Example 1 Graph the ellipse $\dfrac{x^2}{16} + \dfrac{y^2}{4} = 1$.

Plotting a few points in the first quadrant and using the symmetry yields the curve shown. The relationship $c^2 = a^2 - b^2 = 16 - 4 = 12$ gives $c = \sqrt{12} = 2\sqrt{3}$, and the foci are at $(2\sqrt{3}, 0)$ and $(-2\sqrt{3}, 0)$.

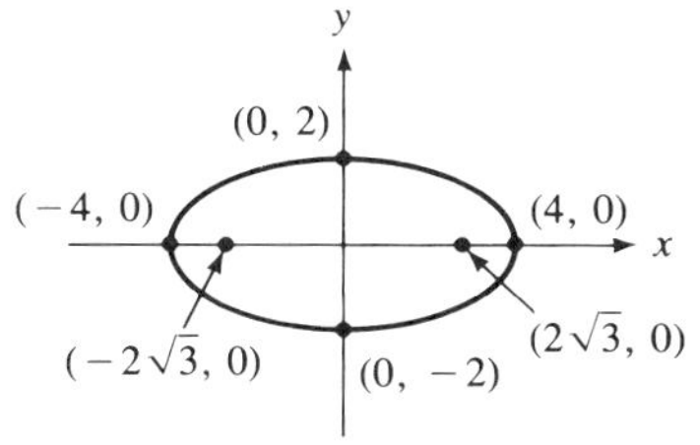

■

The shape of the ellipse

$$\frac{x^2}{a^2} + \frac{y^2}{b^2} = 1 \quad \text{where} \quad a > b$$

is similar to that shown in Figure 4.8. The segment PR is called the **major axis,** and the segment QS is called the **minor axis.** The points P, Q, R, and S are called the **vertices,** and the point O where the major and minor axes intersect is called the **center.**

If we interchange x and y we obtain an ellipse with the major axis vertical and the minor axis horizontal, as shown in Figure 4.9. Since the vertices are at the ends of the major and minor axes, finding those vertices allows one to tell easily whether the major axis is vertical or horizontal.

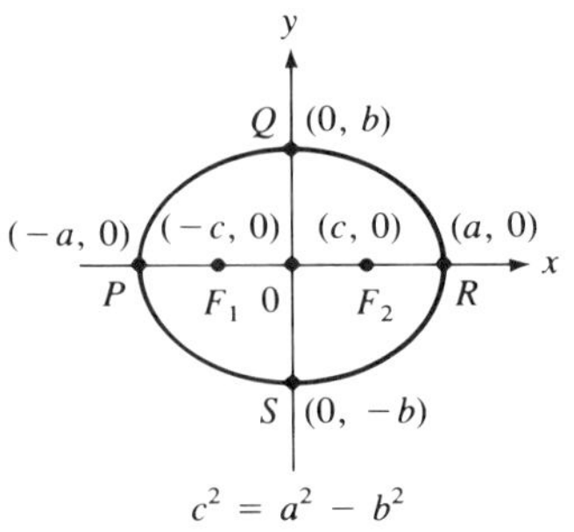

Figure 4.8

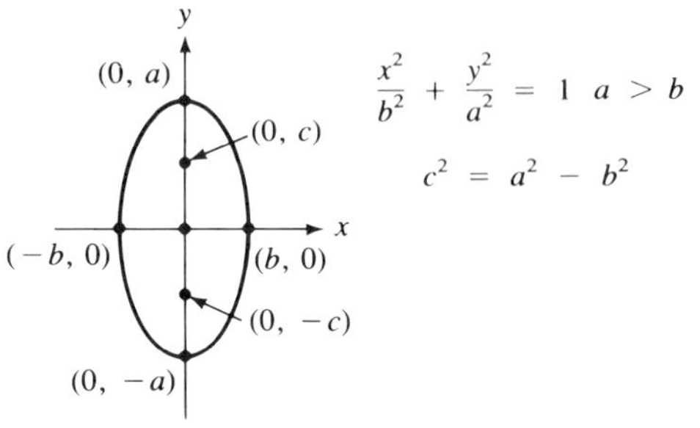

Figure 4.9

Example 2 Graph $\frac{x^2}{4} + \frac{y^2}{25} = 1$, identifying vertices, major and minor axes, center, and foci.

The vertices are found by setting $x = 0$ and solving for y, then setting $y = 0$ and solving for x. The vertices are (2, 0), (−2, 0), (0, 5), and (0, −5). From $c^2 = 25 - 4$, the foci are $(0, \sqrt{21})$ and $(0, -\sqrt{21})$. The major axis is vertical (the longer axis), and the minor axis is horizontal. The center is at the origin.

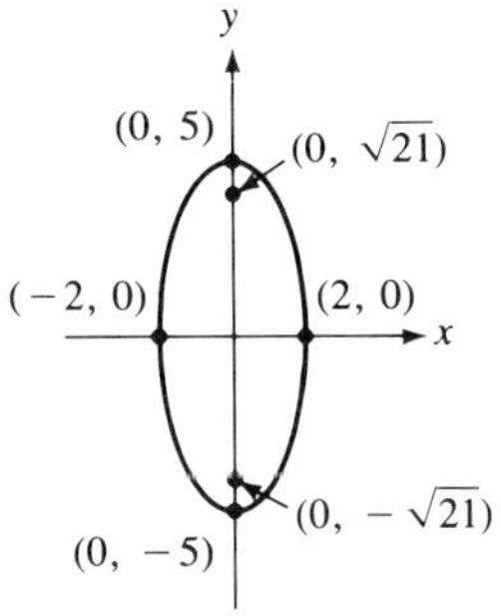

■

If an ellipse is located with its center at the origin and its major and minor axes on the coordinate axes, it is said to be in **standard position.** Any equation of the form

$$px^2 + qy^2 = r \qquad p \neq 0, q \neq 0, r \neq 0$$

where p, q, and r have the same sign, may be written in the form

$$\frac{x^2}{a^2} + \frac{y^2}{b^2} = 1$$

and is therefore the equation of an ellipse in standard position. If $p = q$ the curve is a circle, which may be considered a special case of the ellipse. As with the circle and the parabola, the techniques of completing the square and translation can be used to identify ellipses that are not in standard position.

Example 3 Complete the square to write the equation

$$x^2 + 2x + 4y^2 - 16y + 13 = 0$$

in the form

$$\frac{(x-h)^2}{a^2} + \frac{(y-k)^2}{b^2} = 1$$

The first goal is to complete the square for both x and y:

$$(x^2 + 2x \qquad) + 4(y^2 - 4y \qquad) = -13$$
$$(x^2 + 2x + 1) + 4(y^2 - 4y + 4) = -13 + 1 + 16$$
$$(x + 1)^2 + 4(y - 2)^2 = 4$$

Divide by 4:

$$\frac{(x+1)^2}{4} + \frac{(y-2)^2}{1} = 1$$

■

Example 4 Graph the ellipse $x^2 + 2x + 4y^2 - 16y + 13 = 0$, showing the lines of symmetry, vertices, center, and foci.

From Example 3, the equation can be written in the form

$$\frac{(x+1)^2}{4} + \frac{(y-2)^2}{1} = 1$$

This ellipse is the graph of the ellipse $(x^2/4) + (y^2/1) = 1$ translated 2 units up and 1 unit to the left. In this case $a = 2$, $b = 1$, and $c = \sqrt{3}$, from the equation $c^2 = a^2 - b^2$. The center is at $(-1, 2)$, the lines of symmetry are $x = -1$ and $y = 2$, the vertices are $(-3, 2)$, $(1, 2)$, $(-1, 1)$, and $(-1, 3)$, and the foci are $(-1 + \sqrt{3}, 2)$ and $(-1 - \sqrt{3}, 2)$.

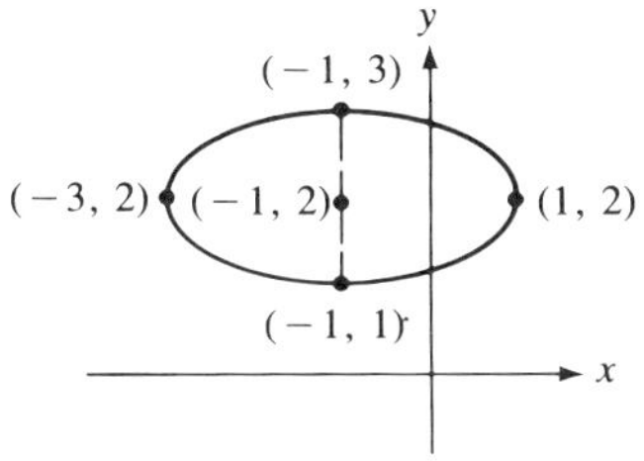

■

Elliptical cams or gears are used to change rotary motion into horizontal motion. The next example illustrates this.

Example 5 An elliptical cam rotates about its center and drives a pushrod as shown in the sketch. If the equation of the ellipse is $3x^2 + 7y^2 = 21$, how far does the pushrod move as it goes from one extreme position to another?

Place a set of coordinate axes as shown. The end point of the pushrod is at $(\sqrt{7}, 0)$, one of the vertices of the ellipse. After the cam rotates, the end of the pushrod is at $(\sqrt{3}, 0)$; the distance moved is therefore $\sqrt{7} - \sqrt{3}$.

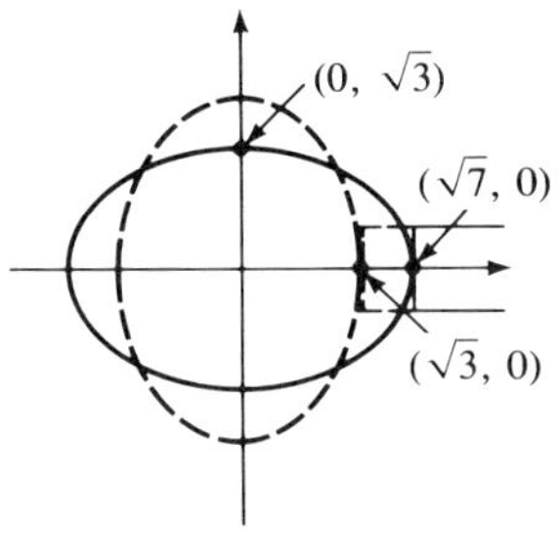

■

Exercises 4.3

Graph the ellipses, showing vertices and foci.

1. $\frac{x^2}{36} + \frac{y^2}{4} = 1$

2. $9x^2 + 4y^2 = 36$

3. $4y^2 + x^2 = 36$

4. $4x^2 + y^2 = 100$

5. $4x^2 + 9y^2 = 25$

6. $9x^2 + 4y^2 = 49$

7. $4x^2 + y^2 + 16x - 4y = 80$

8. $x^2 + 4y^2 + 6x - 16y = -9$

9. $25x^2 + 9y^2 - 200x + 90y = -400$

10. $16x^2 + 9y^2 - 64x + 90y = 17$

Write the equation of the ellipse with the given information.

11. Foci at $(-3, 0)$ and $(3, 0)$; sum of the distances 12

12. Foci at $(4, 0)$ and $(-4, 0)$; sum of the distances 10

13. Foci at $(0, 3)$ and $(0, -3)$; sum of the distances 10

14. Foci at $(0, 1)$ and $(0, -1)$; sum of the distances 5

15. Vertices $(0, 4)$, $(0, -4)$, $(5, 0)$, and $(-5, 0)$

16. Vertices $(0, -2)$, $(0, 2)$, $(6, 0)$, and $(-6, 0)$

17. Foci at $(0, 5)$ and $(0, -5)$; length of major axis 14

18. Foci at $(-2, 0)$ and $(2, 0)$; length of major axis 12

19. Foci at $(-3, 4)$ and $(-3, -4)$; sum of distances 12

20. Foci at $(4, -1)$ and $(-4, -1)$; sum of distances 10

21. Vertices $(2, 4)$, $(2, -6)$, $(5, -1)$, and $(-1, -1)$

22. Vertices $(5, -2)$, $(5, 8)$, $(9, 3)$, and $(1, 3)$

23. An elliptical cam rotates about its center and drives a pushrod. If the equation of the cam is $9x^2 + y^2 = 1$, what is the distance the pushrod moves as it goes from one extreme position to another?

24. The equation of an elliptical cam is $4x^2 + 9y^2 = 1$. If it rotates about its center and drives a pushrod, what is the maximum distance the pushrod moves?

25. An arch is in the form of a semiellipse. It is 24 feet wide at the base and 10 feet high at the center. How wide is the arch at a height of 5 feet above the base?

26. An arch is in the form of a semiellipse. It is 24 feet wide at the base and 18 feet wide at a point 5 feet above the base. What is the height of the arch?

4.4 Conic Sections: The Hyperbola

Geometric definition of the hyperbola

The **hyperbola** is the set of all points such that for each point the difference between its distances from two fixed points is constant. The two fixed points are called the **foci** of the hyperbola.

A point P with coordinates (x, y) is on the hyperbola if and only if

$$|\overline{PF}_1 - \overline{PF}_2| = 2a$$

where F_1 and F_2 are the foci and a is a constant (see Figure 4.10).

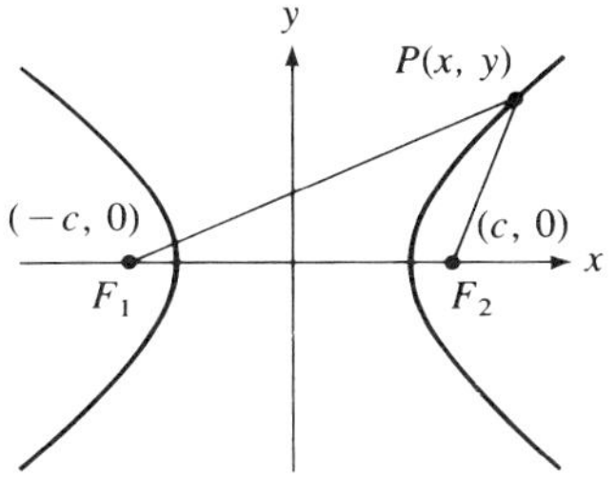

Figure 4.10

Using the distance formula, we obtain

$$|\sqrt{(x - c)^2 + y^2} - \sqrt{(x + c)^2 + y^2}| = 2a$$

Eliminating the radicals and collecting terms yields the equation

$$\frac{x^2}{a^2} + \frac{y^2}{a^2 - c^2} = 1$$

From the triangle in Figure 4.10 we see that

$$\overline{PF}_1 - \overline{PF}_2 = 2a \quad \text{and} \quad \overline{F_1F}_2 = 2c$$

Hence, because the sum of the lengths of two sides of a triangle is greater than the length of the third side,

$$\overline{PF}_2 + 2a = \overline{PF}_1 \leq \overline{PF}_2 + 2c$$

Therefore $a \leq c$. If $a = c$, the only points on the hyperbola are the point on the x axis to the right of F_2 or to the left of F_1. Thus $a < c$ and hence $a^2 < c^2$, and we write $c^2 - a^2 = b^2$.

The equation of the hyperbola with foci at $(-c, 0)$ and $(c, 0)$ with the difference of the distances equal to $2a$ is

$$\frac{x^2}{a^2} - \frac{y^2}{b^2} = 1 \quad \text{where} \quad b^2 = c^2 - a^2$$

The graph is shown in Figure 4.11.

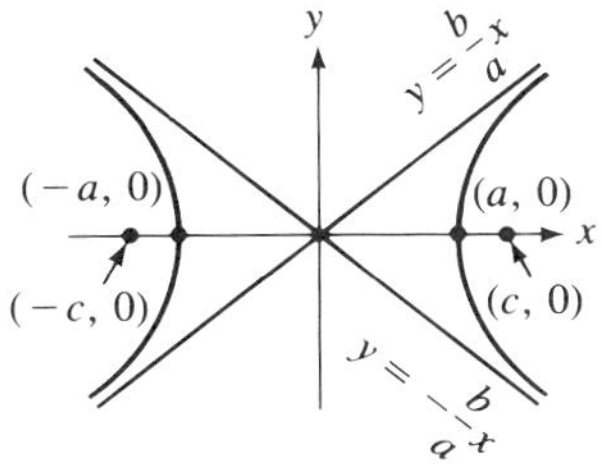

Figure 4.11

The x intercepts $(-a, 0)$ and $(a, 0)$ are called the **vertices.** The curve is symmetric with respect to the x axis, the y axis, and the origin. The center is the point $(0, 0)$. Solving for y, we obtain the two equations

$$y = -\frac{b}{a}x\sqrt{1 - \frac{a^2}{x^2}} \quad \text{and} \quad y = \frac{b}{a}x\sqrt{1 - \frac{a^2}{x^2}}$$

For large values of x we find that y can be approximated by the two equations

$$y = -\frac{b}{a}x \quad \text{and} \quad y = \frac{b}{a}x$$

These two lines are called the **asymptotes** of the hyperbola and are very useful in drawing the graph. Notice in Figure 4.11 how the points on the hyperbola come closer and closer to the asymptotes as one moves away from the center of the hyperbola. Perhaps the easiest way to find the equations for the asymptotes of the hyperbola $(x^2/a^2) - (y^2/b^2) = 1$ is to solve the related equation $(x^2/a^2) - (y^2/b^2) = 0$.

Example 1 Graph the hyperbola $4x^2 - 25y^2 = 100$, showing foci, vertices, and asymptotes. Rewrite the equation in the form

$$\frac{x^2}{25} - \frac{y^2}{4} = 1$$

We have $a = 5$, $b = 2$, and, from the equation $a^2 + b^2 = c^2$, $c = \sqrt{29}$. The vertices are $(-5, 0)$ and $(5, 0)$, the foci are $(-\sqrt{29}, 0)$ and $(\sqrt{29}, 0)$, and the asymptotes are $y = \frac{2}{5}x$ and $y = -\frac{2}{5}x$.

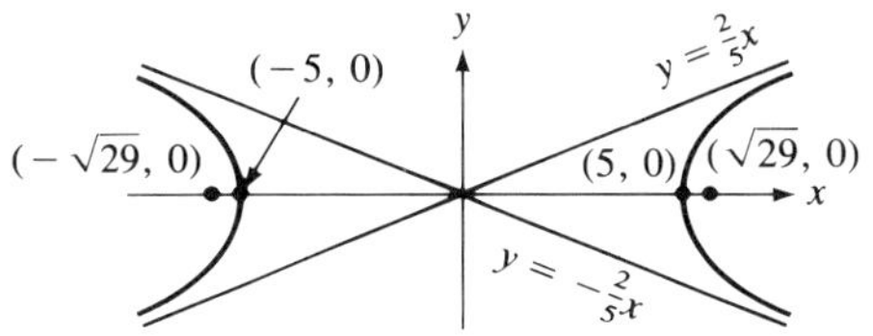

■

If we interchange x and y we obtain a hyperbola with foci and vertices on the y axis. See Figure 4.12 for a typical hyperbola with equation of the form

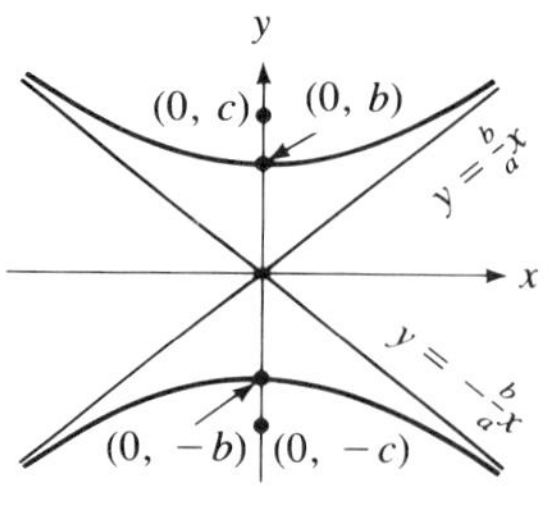

Figure 4.12

$$\frac{y^2}{b^2} - \frac{x^2}{a^2} = 1$$

The vertices are $(0, -b)$ and $(0, b)$, the foci are $(0, -c)$ and $(0, c)$, where $c^2 = a^2 + b^2$, and the asymptotes are $y = (b/a)x$ and $y = -(b/a)x$.

Example 2 Graph the hyperbola $\frac{y^2}{16} - \frac{x^2}{4} = 1$, showing vertices, foci, and asymptotes.

In this case the hyperbola will be similar to that in Figure 4.12, with $b = 4$, $a = 2$, and $c = \sqrt{20}$. The foci are $(0, -\sqrt{20})$ and $(0, \sqrt{20})$, the vertices are $(0, -4)$ and $(0, 4)$, and the asymptotes are $y = 2x$ and $y = -2x$.

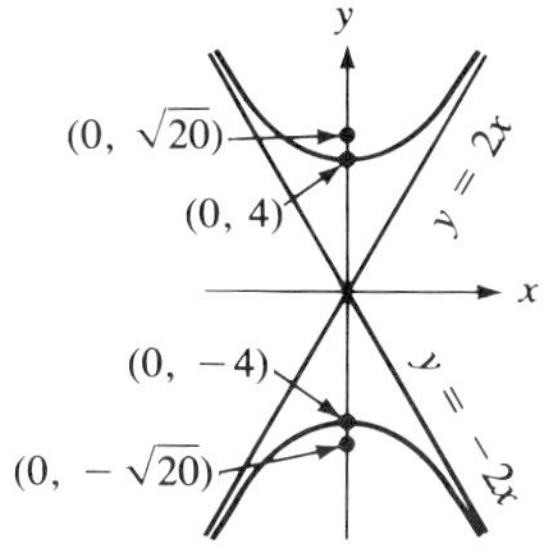

■

If a hyperbola is located with its center at the origin and foci on one of the coordinate axes, the equation can be written in one of the forms

$$\frac{y^2}{b^2} - \frac{x^2}{a^2} = 1 \quad \text{or} \quad \frac{x^2}{a^2} - \frac{y^2}{b^2} = 1$$

and the hyperbola is said to be in **standard position.** Just as with the ellipse, the techniques of completing the square and translating can be used to identify hyperbolas not in standard position.

Example 3 Complete the square to write $9x^2 - 18x - 4y^2 - 16y = 43$ in the form $\frac{(x-h)^2}{a^2} - \frac{(y-k)^2}{b^2} = 1$.

$$9x^2 - 18x - 4y^2 - 16y = 43$$
$$9(x^2 - 2x) - 4(y^2 + 4y) = 43$$
$$9(x^2 - 2x + 1) - 4(y^2 + 4y + 4) = 43 + 9 - 16$$
$$9(x-1)^2 - 4(y+2)^2 = 36$$
$$\frac{(x-1)^2}{4} - \frac{(y+2)^2}{9} = 1$$

■

Example 4 Graph the conic $9x^2 - 18x - 4y^2 - 16y = 43$.

From Example 3, the equation may be written in the form

$$\frac{(x-1)^2}{4} - \frac{(y+2)^2}{9} = 1$$

This is the equation of a hyperbola obtained by translating the hyperbola $(x^2/4) - (y^2/9) = 1$ down 2 units and to the right 1 unit. Here we have $a = 2$, $b = 3$, and $c = \sqrt{13}$. The center is at $(1, -2)$, the vertices are at $(-1, -2)$ and $(3, -2)$, the foci are at $(1 - \sqrt{13}, -2)$ and $(1 + \sqrt{13}, -2)$, and the asymptotes are $y + 2 = \frac{3}{2}(x - 1)$ and $y + 2 = -\frac{3}{2}(x - 1)$. As before, the asymptotes may be found by solving the equation

$$\frac{(x-1)^2}{4} - \frac{(y+2)^2}{9} = 0$$

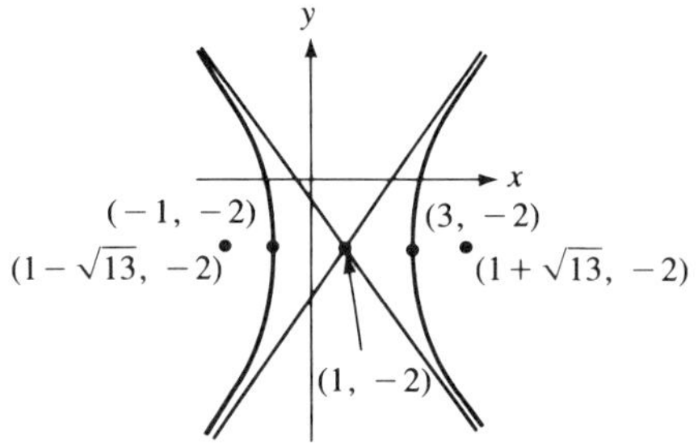

■

Exercises 4.4

In Exercises 1–12 graph the hyperbola, showing vertices, foci, and asymptotes.

1. $\frac{x^2}{16} - \frac{y^2}{25} = 1$

2. $\frac{x^2}{25} - \frac{y^2}{4} = 1$

3. $\frac{y^2}{49} - \frac{x^2}{16} = 1$

4. $\frac{y^2}{16} - \frac{x^2}{49} = 1$

5. $9x^2 - 4y^2 = 36$

6. $4x^2 - 9y^2 = 36$

7. $4x^2 - y^2 = 1$

8. $4y^2 - x^2 = 1$

9. $4x^2 - y^2 + 16x - 4y = 88$

10. $x^2 - 4y^2 + 6x - 16y = -9$

11. $9x^2 - y^2 + 36x - 2y = 1$

12. $x^2 - 16y^2 - 4x + 32y = 76$

In Exercises 13–22 write the equation of the hyperbola with the given information.

13. Foci at $(0, 4)$, $(0, -4)$; difference of the distances 6

14. Foci at $(0, 3)$, $(0, -3)$; difference of the distances 2

15. Foci at $(5, 0)$, $(-5, 0)$; difference of the distances 6

16. Foci at $(6, 0)$, $(-6, 0)$; difference of the distances 8

17. Foci at $(5, 0)$, $(-5, 0)$; vertices at $(4, 0)$, $(-4, 0)$

18. Foci at $(6, 0)$, $(-6, 0)$; vertices at $(2, 0)$, $(-2, 0)$

19. Foci at $(0, 5)$, $(0, -5)$; asymptotes $y = 2x$ and $y = -2x$

20. Foci at $(2, 0)$, $(-2, 0)$; asymptotes $y = 2x$ and $y = -2x$

21. Foci at $(2, 3)$, $(6, 3)$; asymptotes with slope $-\frac{1}{2}$ and $\frac{1}{2}$

22. Foci at $(2, 3)$, $(2, 9)$; asymptotes with slope $-\frac{2}{3}$ and $\frac{2}{3}$

For Your Calculator

In Exercises 23–26 find the y coordinates of the points on the hyperbola in the first quadrant that have x coordinates of 10, 100, 1,000, and 10,000. Compare with the y coordinates of the points on the asymptote that have the same x coordinates.

23. $\frac{x^2}{25} - \frac{y^2}{4} = 1$

24. $x^2 - 4y^2 = 1$

25. $9y^2 - 16x^2 = 1$

26. $16y^2 - 25x^2 = 1$

4.5 Polynomial Functions

A **polynomial function** is any function that can be written in the form

$$f(x) = a_0 + a_1x + \cdots + a_nx^n$$

If $a_n \neq 0$, the degree of the polynomial is n. For $n > 2$, advanced techniques are needed to analyze the graph of the polynomial function, but in this section we examine a few special examples and some techniques that will be useful later.

Example 1 Graph the function $f(x) = x^3$.
This function was graphed in Example 3, Section 3.4. Just as a reminder, this is an odd function; that is, $f(-x) = -f(x)$. The graph will therefore be symmetric with respect to the origin.

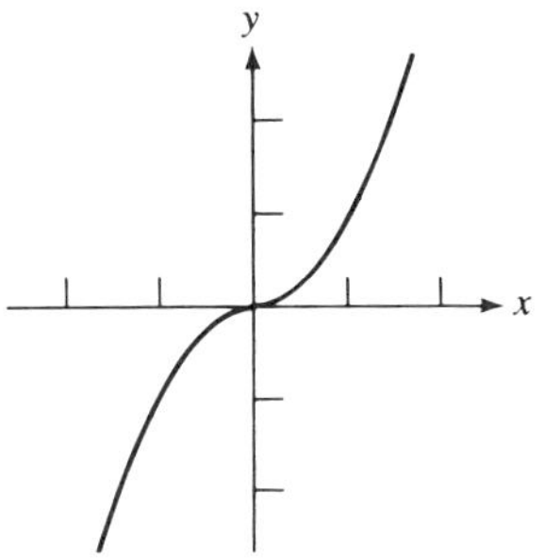

■

Drawing from our experience in Section 3.4 we can see that any function of the form $f(x) = a(x - h)^3 + k$ may be graphed by shifting the graph of $y = ax^3$ so that the point of symmetry (corresponding to (0, 0) for $y = ax^3$) is at the point (h, k). This means a shift of $|h|$ units, left if $h < 0$ and right if $h > 0$, and $|k|$ units, up if $k > 0$ and down if $k < 0$.

Example 2 Graph the function $y = 2(x + 1)^3 + 2$.
First graph $y = 2x^3$ and note that it is a bit "thinner" than the graph of $y = x^3$, just as with the parabola. The graph of $y = 2x^3$ is shown here as a dashed curve. Now shift the curve 2 units up and 1 unit to the left. The point $(-1, 2)$ is now the point of symmetry.

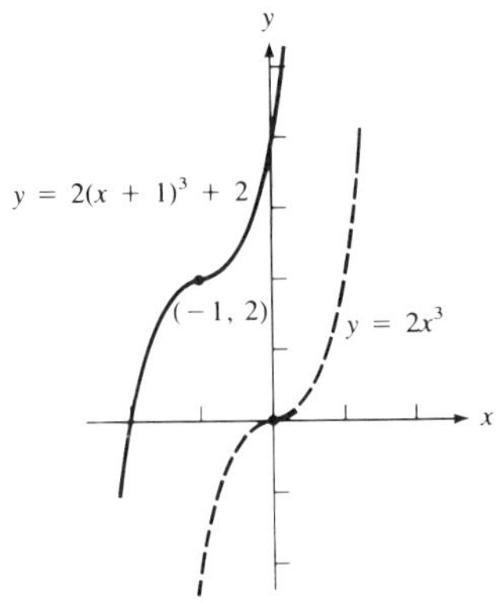

■

Example 3 Graph the function $g(x) = x^4$.
This is an even function, i.e., $g(x) = g(-x)$, which means that the graph is symmetric with respect to the y axis.

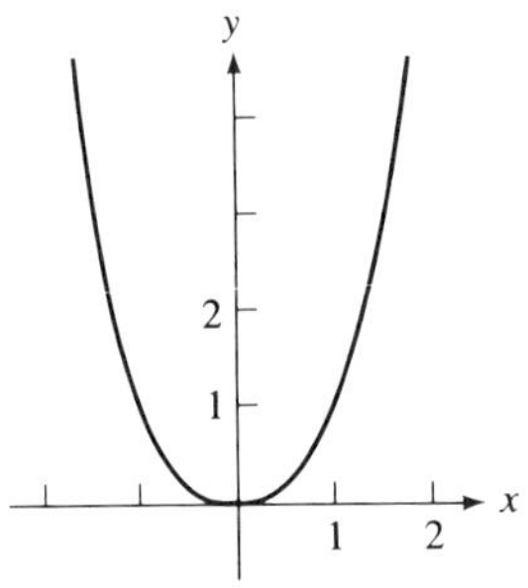

■

To graph a function of the form $f(x) = a(x - h)^4 + k$, follow the same procedure we used in graphing the quadratic and cubic polynomials in the same form.

Example 4 Graph the function $f(x) = -(x + 1)^4 - 2$.
First graph $y = -x^4$. That curve is shown here as a dashed curve. Now shift that curve 2 units down and 1 unit to the left. The line of symmetry for the curve is the line $x = -1$.

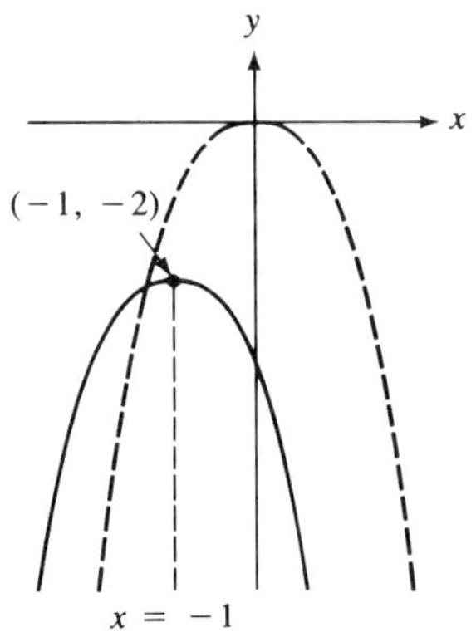

■

If the polynomial is in factored form or can be factored easily, the factorization can be useful in sketching the graph.

Example 5 Sketch the graph of $y = (x - 1)(x + 2)(x - 3)$.
Make a chart showing the signs of the factors.

$x - 1$ − − − − − − − − − − − − − − 0 + + + + + + +
$x + 2$ − − − − − 0 + + + + + + + + + + + + + + +
$x - 3$ − − − − − − − − − − − − − − − − − − − 0 + + +

If an odd number of the factors are negative, then y is negative. If an even number of the factors are negative, y is positive. The chart below shows the sign of y over various intervals.

y − − − − − + + + + + + + − − − − + + + +

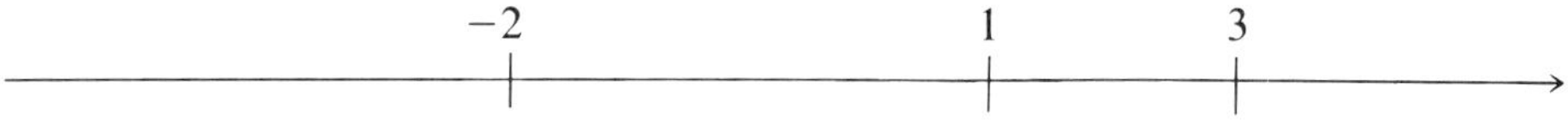

After plotting a few points and using this information, we obtain the sketch shown.

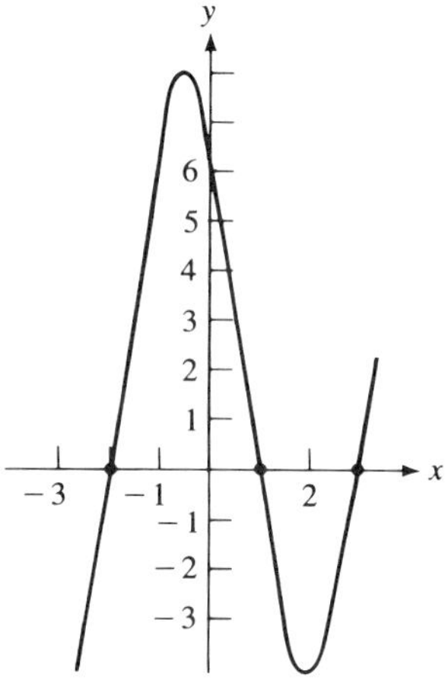

There are several important points on this graph that you are not able at this time to locate. Somewhere between $x = -2$ and $x = 1$ there is a point at which the curve stops rising and starts downhill. Likewise, somewhere between $x = 1$ and $x = 3$ there is a point at which the curve stops falling and begins to rise. In the study of calculus, techniques are developed to find those points. ■

Exercises 4.5

In Exercises 1–16 graph the function.

1. $y = -\frac{1}{2}x^3$

2. $y = -\frac{1}{3}x^3$

3. $y = \frac{1}{3}x^3$

4. $y = \frac{1}{2}x^3$

5. $y = \frac{1}{2}(x - 1)^3$

6. $y = \frac{1}{3}(x - 2)^3$

7. $y = 2(x + 2)^3 - 3$

8. $y = 2(x - 3)^3 + 2$

9. $y = -(2 - x)^3 + 1$

10. $y = -(3 - x)^3 - 4$

11. $y = -\frac{1}{2}x^4$

12. $y = -\frac{1}{3}x^4$

13. $y = -\frac{1}{2}(x - 4)^4 + 1$

14. $y = -\frac{1}{3}(x + 3)^4 - 2$

15. $y = x^5 - 1$

16. $y = x^5 + 2$

In Exercises 17–24 the polynomials are either factored or can be easily factored by grouping. Use the factorization to graph the functions.

17. $f(x) = (x - 4)(x + 4)(x - 1)$

18. $g(x) = (x + 1)(x - 1)(x + 4)$

19. $h(x) = x^2(x + 2)(x - 5)$

20. $k(x) = x^2(x - 2)(x + 5)$

21. $f(x) = x^3 + 3x^2 - 9x - 27$

22. $g(x) = x^3 + 4x^2 - x - 4$

23. $y = x^3 - x^2 - 12x$

24. $y = x^3 + 3x^2 - 10x$

4.6 Rational Functions

A function that can be written in the form

$$f(x) = \frac{P(x)}{Q(x)}$$

where $P(x)$ and $Q(x)$ are polynomials, is called a **rational function.** The domain of such a function is the set of all x such that $Q(x) \neq 0$. In this section we graph some rational functions and examine the asymptotes for such functions.

Example 1 Graph the function $y = \frac{1}{x}$.

First of all, note that an equivalent equation is obtained if x is replaced by $-x$ and y by $-y$. Thus the curve will be symmetric with respect to the origin. Note that as x gets larger and larger, $y = 1/x$ gets closer and closer to zero. Also, as x approaches 0 from the right (through positive values), $y = 1/x$ gets larger and larger. Using this information and plotting a few points yields the graph shown.

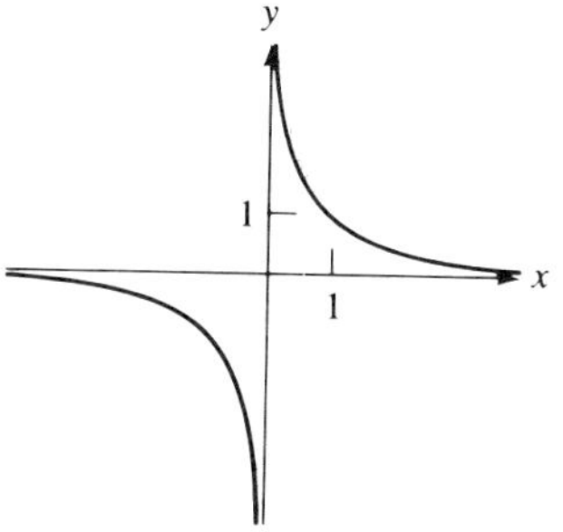

■

The equation $y = 1/x$ may be rewritten in the form $xy = 1$, which is a second-degree equation in x and y. Consequently the graph must be one of the conic sections, and it is clear that the only candidate is the hyperbola. Indeed, the equation $xy = 1$ is the equation of the hyperbola with foci at $(-\sqrt{2}, -\sqrt{2})$ and $(\sqrt{2}, \sqrt{2})$ and vertices at $(1, 1)$ and $(-1, -1)$. The x and y axes are asymptotes for this hyperbola, as the graph indicates. We now need a more general definition of asymptote.

Asymptote for a curve

As a point P moves away from the origin along a branch of a curve, if the distance from that point to a given line tends to 0, the line is called an **asymptote** for the curve.

The use of the term asymptote with reference to the hyperbola certainly fits this description. In order to discuss asymptotes and the behavior of curves as one moves away from the origin, some new language is very useful. If we wish to discuss some behavior as x increases without bound, we will use the terminology "x tends to ∞." The same language can be used when y is increasing without bound. If x is decreasing without bound, i.e., moving farther and farther to the left through negative values, we will use the terminology "x tends to $-\infty$." One thing should be made clear: ∞ and $-\infty$ are not numbers; they are just symbols used to convey a message. We refer to ∞ as infinity and $-\infty$ as minus infinity.

Vertical and horizontal asymptotes

The curve $y = 1/x$ has two asymptotes, the x axis and the y axis. For obvious reasons, the x axis is called a **horizontal** asymptote and the y axis is called a **vertical** asymptote. Horizontal and vertical asymptotes are very useful in graphing rational functions.

How do vertical asymptotes occur? The line $x = a$ is a vertical asymptote if, as x tends to a, y tends to either ∞ or $-\infty$. This will happen for $y = P(x)/Q(x)$ (reduced to lowest terms) when $Q(a) = 0$.

Condition for vertical asymptotes

The rational function $f(x) = P(x)/Q(x)$ (reduced to lowest terms) has a vertical asymptote $x = a$ if and only if $Q(a) = 0$.

Example 2 Graph the function $y = \dfrac{1}{(x-2)^2}$.

The graph will have a vertical asymptote at $x = 2$. Also notice that as x tends to 2 from either the left or the right, y tends to ∞; in particular, y is always positive. Also as x tends to ∞, y tends to 0, so the x axis is a horizontal asymptote.

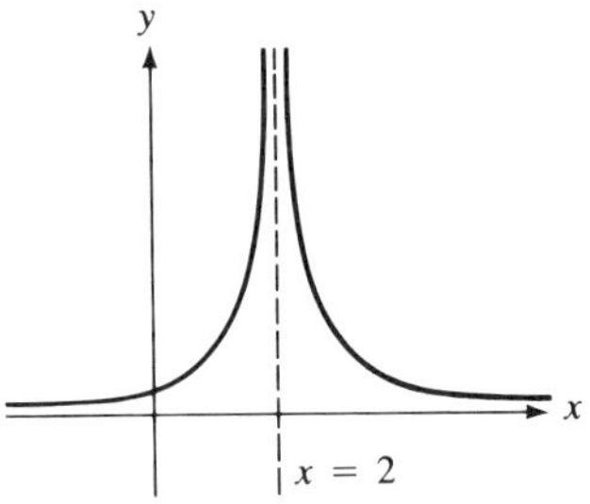

Notice that the line $x = 2$ appears to be a line of symmetry for the curve. That happens because the line $x = 0$, the y axis, is a line of symmetry for the curve $y = 1/x^2$ and the graph of $y = 1/(x - 2)^2$ is found by translating the graph of $y = 1/x^2$ to the right 2 units. ■

The horizontal asymptotes for some functions are easy to determine. That was the case with the functions in Examples 1 and 2. In the next three examples, we illustrate a technique that can be used to find horizontal asymptotes.

Example 3 Find the horizontal asymptote(s) for the graph of the function $y = \dfrac{3x - 2}{5x + 9}$.

As the function is now written, it is not easy to determine what happens to y as x tends to ∞. But if we divide numerator and denominator by x, we obtain an equation that gives the same values for y as long as $x \neq 0$:

$$y = \frac{3 - \dfrac{2}{x}}{5 + \dfrac{9}{x}}$$

Now we see that as x tends to either ∞ or $-\infty$, the numerator tends to 3 and the denominator tends to 5; hence y tends to $\dfrac{3}{5}$. The line $y = \dfrac{3}{5}$ is the only horizontal asymptote. ■

Example 4 Find the horizontal asymptote(s) for the graph of

$$y = \frac{3x^2 - 2x + 7}{5x^3 - 7x + 19}$$

If we divide numerator and denominator by x^3 we obtain

$$y = \frac{\dfrac{3}{x} - \dfrac{2}{x^2} + \dfrac{7}{x^3}}{5 - \dfrac{7}{x^2} + \dfrac{19}{x^3}}$$

Now as x tends to ∞, the numerator tends to 0 and the denominator tends to 5. Consequently y tends to 0. The only horizontal asymptote is the line $y = 0$. ■

Notice that in both Examples 3 and 4 we divided numerator and denominator by the highest power of x in the denominator. We continue this strategy in Example 5.

Example 5 Find the horizontal asymptotes for the graph of

$$y = \frac{5x^3 - 7x + 19}{3x^2 - 2x + 7}$$

Divide numerator and denominator by x^2 to obtain

$$y = \frac{5x - \frac{7}{x} + \frac{19}{x^2}}{3 - \frac{2}{x} + \frac{7}{x^2}}$$

Now it is easy to see that as x tends to either ∞ or $-\infty$ the denominator tends to 3, whereas the numerator tends to either ∞ or $-\infty$. This graph does not have a horizontal asymptote. ■

The procedure used in Examples 3, 4, and 5 can be analyzed to establish the following.

Finding horizontal asymptotes

The graph of $y = P(x)/Q(x)$ ($P(x)$ and $Q(x)$ polynomials) has

1. the x axis as the only horizontal asymptote if the degree of P is less than the degree of Q (as in Example 4)
2. the horizontal asymptote $y = a/b$, where a is the coefficient of the highest power of x in P and b is the coefficient of the highest power of x in Q, if P and Q have the same degree (as in Example 3)
3. no horizontal asymptote if the degree of P is larger than the degree of Q (as in Example 5)

Example 6 Graph the function $y = \dfrac{(x-1)^2}{3x^2 - 6x}$, showing vertical and horizontal asymptotes. Rewriting the numerator we obtain

$$y = \frac{x^2 - 2x + 1}{3x^2 - 6x}$$

The graph has a horizontal asymptote $y = \frac{1}{3}$. Factoring the denominator yields

$$y = \frac{x^2 - 2x + 1}{3x(x - 2)}$$

The denominator is 0 at $x = 0$ and $x = 2$. The vertical asymptotes are $x = 0$ and $x = 2$. As with polynomials, the factorization of numerator and denominator will allow us to determine when the function is negative and when it is positive. In this case, the numerator, being a square, is always nonnegative. The sign of the fraction is determined by the denominator.

x	− − − − − − −	+ + + + +	+ + + + + + + + + + +
$x - 2$	− − − − − − −	− − − − −	+ + + + + + + + + + +
$x(x - 2)$	+ + + + + + +	− − − − −	+ + + + + + + + + + + +
	0	2	

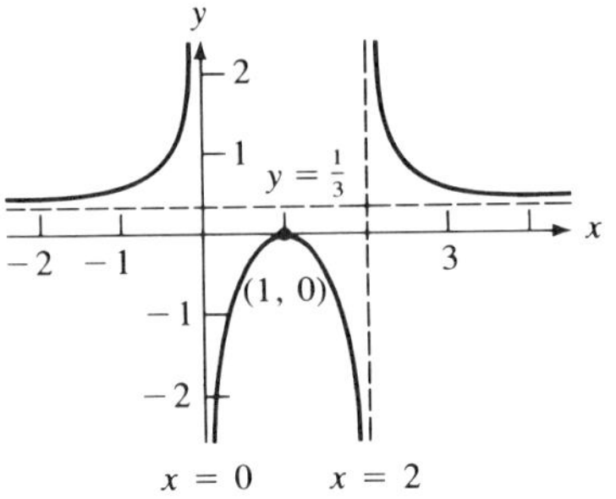

■

Example 7 Graph the function $y = \dfrac{x - 2}{x^2 - 4x + 3}$, showing horizontal and vertical asymptotes.

Since the degree of the numerator is less than the degree of the denominator, the only horizontal asymptote is the x axis. Factoring the denominator yields the equation

$$y = \frac{x - 2}{(x - 3)(x - 1)}$$

The vertical asymptotes are $x = 3$ and $x = 1$. From the factored form we can determine the sign of y.

$x - 1$	− − −	+ + +	+ + + +	+ + +
$x - 2$	− − −	− − −	+ + + +	+ + +
$x - 3$	− − −	− − −	− − − −	+ + +
y	− − −	+ + +	− − − −	+ + +
	1	2	3	

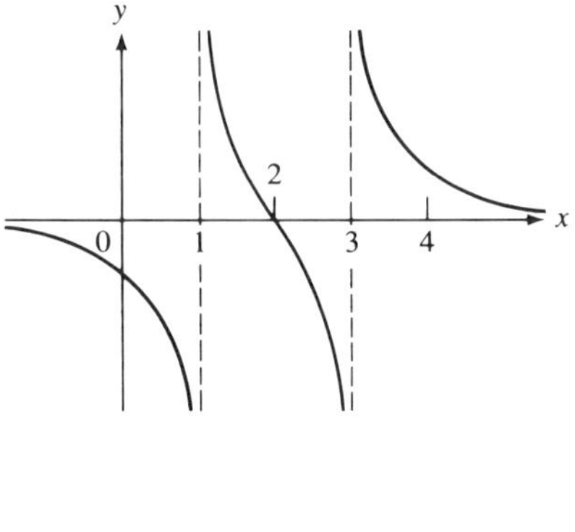

■

Exercises 4.6

For each function in Exercises 1–16 find horizontal and vertical asymptotes (if any).

1. $y = \dfrac{-2}{x + 3}$
2. $y = \dfrac{3}{x - 2}$
3. $y = \dfrac{3x - 4}{3 - 4x}$
4. $y = \dfrac{2x + 3}{2 - 3x}$
5. $y = \dfrac{3 + 5x}{2 - 3x} + 7$
6. $y = \dfrac{2 + 4x}{3 - 5x} + 3$
7. $y = \dfrac{3x - 9}{x^2 - 4}$
8. $y = \dfrac{4x + 7}{x^2 - 9}$
9. $y = \dfrac{x^2 - 6}{x - 6}$
10. $y = \dfrac{x^2 - 4}{x - 4}$
11. $y = \dfrac{2x^2 - 9}{3x^2 - 3}$
12. $y = \dfrac{3x^2 - 4}{2x^2 - 2}$
13. $y = \dfrac{5x^3 - 2x^2 + 8}{6x^3 + 6x}$
14. $y = \dfrac{7x^3 - 4x + 17}{3x^3 + 15x}$
15. $y = \dfrac{4x^2 - 7x - 3x^3}{13x + 7x^3}$
16. $y = \dfrac{7x - 5x^2 + 15x^3}{17x + 4x^3}$

In Exercises 17–24 graph each function, showing asymptotes.

17. $y = \dfrac{1}{x + 4}$
18. $y = \dfrac{1}{x + 3}$
19. $y = \dfrac{-2}{x - 3} + 4$
20. $y = \dfrac{-2}{x - 4} + 2$
21. $y = \dfrac{(x - 1)^2}{2(x^2 - 2x - 3)}$
22. $y = \dfrac{(x + 1)^2}{3(x^2 + 2x - 3)}$
23. $y = \dfrac{x + 1}{3(x^2 + 2x - 8)}$
24. $y = \dfrac{x - 4}{-2(x^2 - 8x + 7)}$

For Your Calculator

Each of the functions below has a graph with a horizontal asymptote. Compute $f(x)$ for large values of x and attempt to find the asymptote.

25. $f(x) = \sqrt{x + 1} - \sqrt{x}$
26. $f(x) = \sqrt{x^2 + 2x} - \sqrt{x^2}$
27. $f(x) = \dfrac{x^2}{x - 1} - \dfrac{x^2}{x + 1}$
28. $f(x) = \sqrt{x^4 + 5x^2 + 3} - x^2$

4.7 Partial Fractions

In the study of calculus it is often useful to write a rational function as a sum of simpler rational functions. In this section we will see how to write a rational function as a sum of what are called **partial fractions.**

Suppose $P(x)$ and $Q(x)$ are polynomials with the degree of P less than the degree of Q. Then the rational function $P(x)/Q(x)$ can be written as a sum of partial fractions as follows.

Partial fraction decomposition with no repeated factors

Case I $Q(x) = a(x - a_1)(x - a_2) \cdots (x - a_n)$, where the a_i are distinct, that is, $Q(x)$ has no repeated factors. Then

$$\frac{P(x)}{Q(x)} = \frac{A_1}{x - a_1} + \frac{A_2}{x - a_2} + \cdots + \frac{A_n}{x - a_n}$$

where $A_1, A_2, \ldots, A_n$ are constants to be determined.

Example 1 Find the partial fraction decomposition for the function $\frac{4x}{x^2 - 1}$.

The denominator factors as follows:

$$x^2 - 1 = (x - 1)(x + 1)$$

so the decomposition is

$$\frac{4x}{x^2 - 1} = \frac{A}{x - 1} + \frac{B}{x + 1}$$

To determine A and B, combine the terms on the right and compare numerators:

$$\frac{4x}{x^2 - 1} = \frac{A(x + 1) + B(x - 1)}{(x - 1)(x + 1)}$$

Since the denominators are the same, the numerators must be equal for all x:

$$4x = A(x + 1) + B(x - 1)$$

We substitute values for x that make the factors $x - 1$ and $x + 1$ equal to 0, that is $x = 1$ and $x = -1$. If $x = 1$, then

$$4 = 2A$$

and

$$2 = A$$

If $x = -1$,

$$-4 = -2B$$

and

$$2 = B$$

The decomposition is

$$\frac{4x}{x^2 - 1} = \frac{2}{x - 1} + \frac{2}{x + 1}$$ ■

If the denominator Q of a rational function can be factored as a product of linear factors with no repeated factors, the technique used in Example 1 works quite well, that is, substituting values for x that make the linear factors 0.

Example 2 Write $\dfrac{x^2 + 2x - 1}{x(x - 2)(2x + 1)}$ as a sum of partial fractions.

Since the denominator is already in factored form and no factors are repeated, the form is

$$\frac{x^2 + 2x - 1}{x(x - 2)(2x + 1)} = \frac{A}{x} + \frac{B}{x - 2} + \frac{C}{2x + 1}$$

Combining the fractions on the right and comparing numerators, we obtain

$$x^2 + 2x - 1 = A(x - 2)(2x + 1) + Bx(2x + 1) + Cx(x - 2)$$

Substituting $x = 0$, we obtain

$$-1 = -2A \qquad A = \frac{1}{2}$$

Substituting $x = 2$, we obtain

$$7 = 10B \qquad B = \frac{7}{10}$$

Substituting $x = -\frac{1}{2}$, we obtain

$$-\frac{7}{4} = \frac{5}{4}C \qquad C = -\frac{7}{5}$$

The desired sum is

$$\frac{x^2 + 2x - 1}{x(x - 2)(2x + 1)} = \frac{\frac{1}{2}}{x} + \frac{\frac{7}{10}}{x - 2} + \frac{-\frac{7}{5}}{2x + 1}$$ ■

Partial fraction decomposition for repeated linear factors

Case II Q can be factored as a product of linear factors with one or more repeated linear factors. The terms that correspond to nonrepeated linear factors will be just as in Case I. If a linear factor $x - a$ occurs k times in the factorization of Q, then the partial fraction decomposition will have terms of the form

$$\frac{A_1}{x - a} + \frac{A_2}{(x - a)^2} + \cdots + \frac{A_k}{(x - a)^k}$$

where the constants $A_1, A_2, \ldots, A_k$ are to be determined.

Example 3 Find the partial fraction decomposition for the function $\dfrac{2x^2 - 3}{x(x - 2)^2}$.

The factor x is not repeated so the decomposition will include a term A/x. However, the factor $x - 2$ occurs twice, so we will need two terms:

$$\frac{B}{x - 2} \quad \text{and} \quad \frac{C}{(x - 2)^2}$$

We obtain

$$\frac{2x^2 - 3}{x(x - 2)^2} = \frac{A}{x} + \frac{B}{x - 2} + \frac{C}{(x - 2)^2}$$

Writing the right-hand member with denominator $x(x - 2)^2$ and comparing numerators we obtain

$$2x^2 - 3 = A(x - 2)^2 + Bx(x - 2) + Cx$$

If we set $x = 0$, we obtain

$$-3 = 4A \qquad A = -\frac{3}{4}$$

If we set $x = 2$, we obtain

$$5 = 2C \qquad C = \frac{5}{2}$$

Since $x - 2$ is a repeated factor, we must use another value for x to determine B. Set $x = 3$; then

$$\begin{aligned} 15 &= A + 3B + 3C \\ &= -\frac{3}{4} + 3B + \frac{15}{2} \\ B &= \frac{11}{4} \end{aligned}$$

The sum is

$$\frac{2x^2 - 3}{x(x - 2)^2} = \frac{-\frac{3}{4}}{x} + \frac{\frac{11}{4}}{x - 2} + \frac{\frac{5}{2}}{(x - 2)^2}$$ ■

Partial fraction decomposition for nonlinear factors

> Case III If the factorization of Q contains an irreducible factor $ax^2 + bx + c$, there will be in the decomposition a term of the form
>
> $$\frac{Ax + B}{ax^2 + bx + c}$$
>
> If the factor is repeated, then additional terms are included for the different powers of the factor, as with the repeated linear factors.

Example 4 Find the partial fraction decomposition of the function $\dfrac{x^2 - 2x + 3}{(x - 2)(x^2 + x + 3)}$.

The factor $x^2 + x + 3$ is irreducible, that is, it cannot be factored using real numbers. The decomposition is therefore

$$\frac{x^2 - 2x + 3}{(x - 2)(x^2 + x + 3)} = \frac{A}{x - 2} + \frac{Bx + C}{x^2 + x + 3}$$

Combining the terms on the right with a common denominator and comparing numerators, we obtain

$$x^2 - 2x + 3 = A(x^2 + x + 3) + (Bx + C)(x - 2)$$

If we set $x = 2$, we obtain

$$3 = 9A \qquad A = \frac{1}{3}$$

If we set $x = 0$, we obtain

$$\begin{aligned} 3 &= 3A - 2C \\ &= 1 - 2C \qquad C = -1 \end{aligned}$$

And finally, $x = 1$ yields

$$\begin{aligned} 2 &= 5A - B - C \\ &= \frac{5}{3} - B + 1 \qquad B = \frac{2}{3} \end{aligned}$$

The decomposition is

$$\frac{x^2 - 2x + 3}{(x - 2)(x^2 + x + 3)} = \frac{\frac{1}{3}}{x - 2} + \frac{\frac{2}{3}x - 1}{x^2 + x + 3}$$

■

Exercises 4.7

Write each rational function as a sum of partial fractions.

1. $y = \dfrac{5x - 5}{(x - 3)(x + 2)}$

2. $y = \dfrac{x - 12}{(x + 3)(x - 2)}$

3. $y = \dfrac{6 - x}{2x^2 - 3x}$

4. $y = \dfrac{-4}{3x^2 - 2x}$

5. $y = \dfrac{4x - 11}{2x^2 + 7x - 4}$

6. $y = \dfrac{2x + 28}{2x^2 - 7x - 4}$

7. $y = \dfrac{2x^2 - 5x - 1}{x(x - 1)(x + 1)}$

8. $y = \dfrac{3x^2 - 10x - 8}{x(x - 2)(x + 2)}$

9. $y = \dfrac{x^2 - 11x + 6}{(x^2 - 1)(x - 2)}$

10. $y = \dfrac{12 + 6x - x^2}{(x^2 - 4)(x + 3)}$

11. $y = \dfrac{-4x^2 + x}{(x - 4)(x + 1)(x + 2)}$

12. $y = \dfrac{-x^2 + 9x + 16}{(x + 3)(x - 1)(2x + 1)}$

13. $y = \dfrac{4x^2 - 1}{x^3 - x^2}$

14. $y = \dfrac{3x^2 - 2}{x^3 + x}$

15. $y = \dfrac{4x^2 + 5x + 3}{(x - 1)^2(x + 2)}$

16. $y = \dfrac{-9x + 23}{(x - 2)^2(x + 3)}$

17. $y = \dfrac{4x^2 - x + 7}{(x - 1)(x^2 + x + 3)}$

18. $y = \dfrac{-2x^2 - 2x + 15}{(x + 1)(x^2 + 2x + 6)}$

19. $y = \dfrac{5x^2 + 12x + 4}{(x - 2)(2x^2 + x + 2)}$

20. $y = \dfrac{4x^2 - 10x + 2}{(x - 6)(3x^2 - x + 4)}$

4.8 Inequalities in Two Variables

Recall that a relation is a set of ordered pairs, and that most of the examples of relations considered in this chapter and Chapter 3 were given by equations. The graph of the relation was the solution set for that equation. Not all situations, however, can be described by equations. Sometimes an inequality is needed. In this section we examine the solution sets of certain types of inequalities.

In order to illustrate the general principle to be used in graphing inequalities, we look at a linear inequality. Consider the inequality

$$y \geq 3x - 2$$

We begin by graphing the line $y = 3x - 2$, since any point on the line will be a solution to the inequality (see Figure 4.13). Now any point (x, y) on the line is a solution to $y = 3x - 2$; hence any point directly above that point is a solution to $y > 3x - 2$, since the x coordinate hasn't changed, while the y coordinate has increased. This means that the solution set for the inequality $y \geq 3x - 2$ consists of all points on the line $y = 3x - 2$ and all points above the line $y = 3x - 2$ (see Figure 4.14). In light of this fact, the solution set for the inequality

$$y < 3x - 2$$

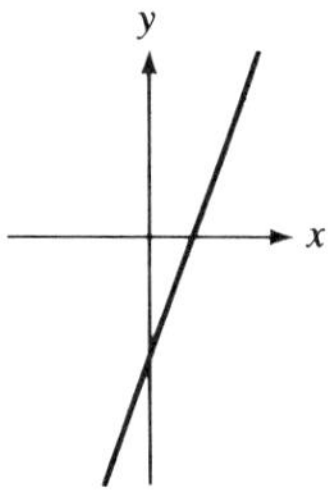

Figure 4.13

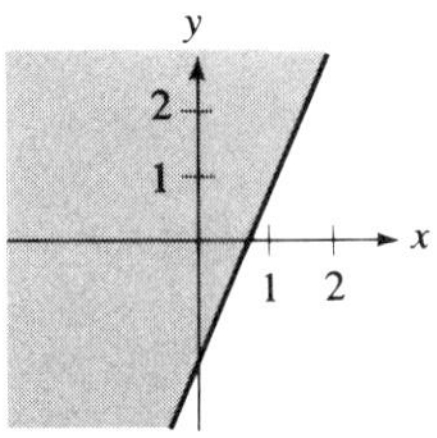

Figure 4.14

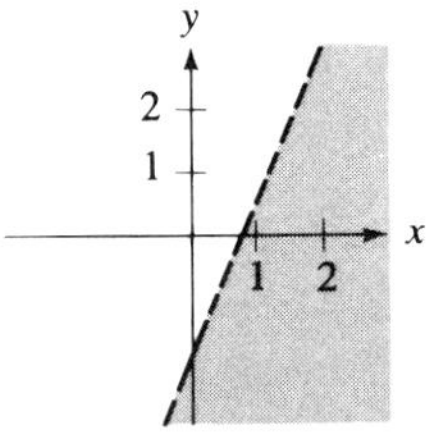

Figure 4.15

consists of all the points *below* the line $y = 3x - 2$. In Figure 4.15 the line is shown as a broken line to indicate that points on the line are *not* in the solution set.

The type of reasoning used in the example above allows us to give the following description of the solution set of a linear inequality.

Solution set for a linear inequality

The solution set for a linear inequality $ax + by < c$ consists of all points on one side of the line $ax + by = c$. If the linear inequality is of the form $ax + by \leq c$, the solution set consists of all points on one side of the line $ax + by = c$ and also all points on the line $ax + by = c$.

If the inequality has the form $ax + by > c$, simply multiply by -1 to see that the principle above applies.

Example 1 Graph the inequality $2x - 3y \leq 6$.
First graph the line $2x - 3y = 6$. To determine which side of the line makes up the solution set, choose any point on one side and check to see whether it is a solution to the inequality. In this case (0, 0) is an easy one to try and it is indeed a solution to the inequality. The solution set is shown in the sketch.

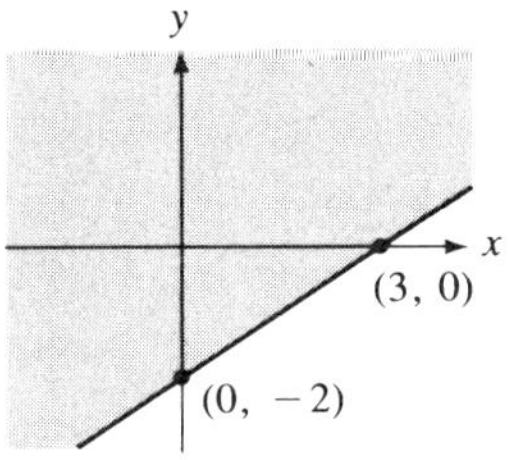

■

Example 2 Graph the inequality $3x - 4y > 0$.
The solution set consists of all points on one side of the line $3x - 4y = 0$, but the line is not included. In the sketch the line is shown as a broken line to indicate this. Since (0, 0) is on the line, another test point must be chosen. Try $(-1, 1)$. This point is *not* a solution, so the solution set consists of those points on the side of the line that does not contain $(-1, 1)$ (see the sketch).

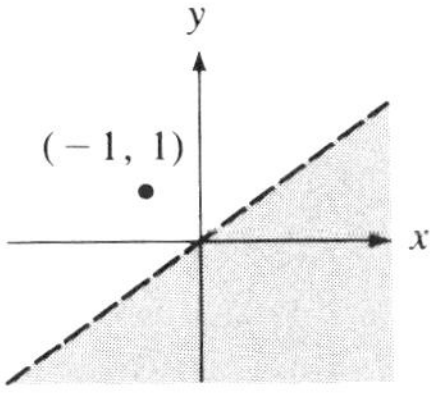

■

The principle used in graphing linear inequalities may be extended to nonlinear inequalities. We illustrate this by examining the conic sections and related inequalities.

Example 3 Graph the inequality $x^2 + y^2 < 25$.
The quantity $x^2 + y^2$ is the square of the distance from (0, 0) to (x, y). If $x^2 + y^2 < 25$, then the square of the distance from (0, 0) to (x, y) is less than 25,

that is, the distance is less than 5. This means that (x, y) is *inside* the circle of radius 5 with center at (0, 0). ■

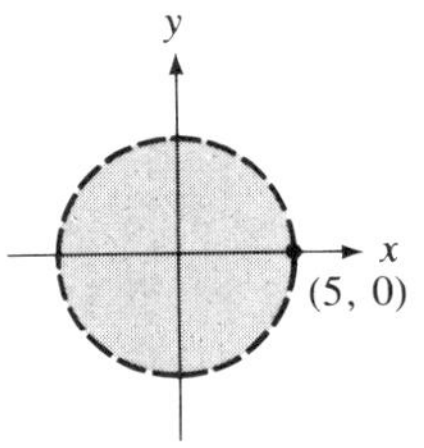

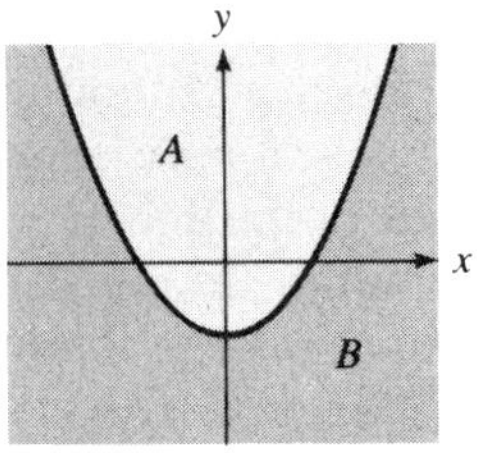

Figure 4.16

Three of the conic sections–the circle, the ellipse, and the parabola–have the property that they partition the plane into two regions in a natural way. In the case of the circle and the ellipse, one region is the set of points inside the circle or ellipse, and the other region is the set of points outside the circle or ellipse. The parabola partitions the plane into two regions as shown in Figure 4.16 above, with A and B marking the two regions.

Suppose we seek to graph the solution set for an inequality, and the corresponding equality is the equation of one of the conic sections described above: the circle, ellipse or parabola. Then, just as with the linear inequality, the solution set is one of the two regions determined by the curve.

Example 4 Graph the inequality $y < x^2 - 5$.
First graph the parabola $y = x^2 - 5$. Then pick a point in one of the regions—for example, (0, 0). The coordinates of (0, 0) do *not* satisfy the inequality, so the solution set is the other region determined by the parabola.

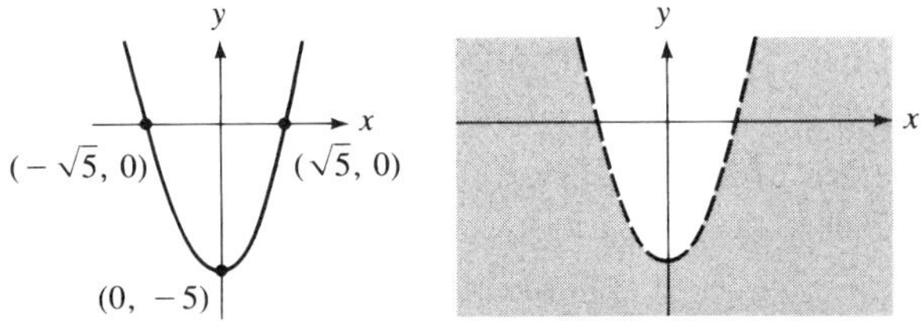

■

The situation with the hyperbola is slightly different, because that curve is made up of *two* separate pieces. The hyperbola partitions the plane into *three* regions as shown in Figure 4.17, with the regions labeled A, B, and C.

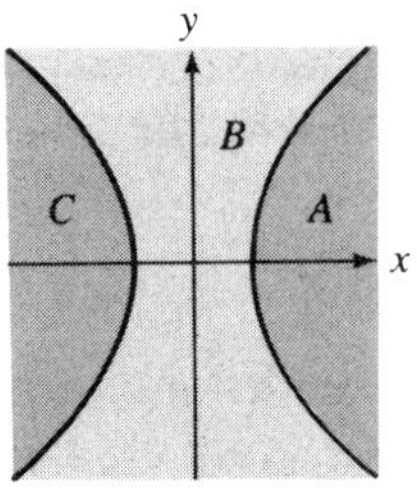

Figure 4.17 ■

Example 5 Graph the inequality $x^2 - y^2 > 1$.
First, graph the hyperbola $x^2 - y^2 = 1$. Then test each of the three regions determined by the hyperbola. It turns out that (2, 0) and (−2, 0) are solutions, whereas (0, 0) is not a solution. The graph is shown in the sketch.

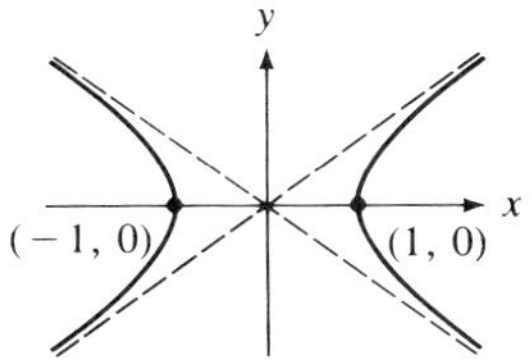

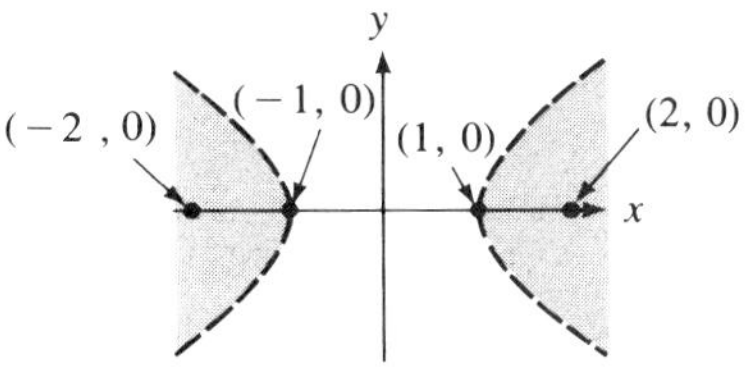

■

Example 6 The graph of $x^2 - y^2 < 1$ is shown in the sketch.

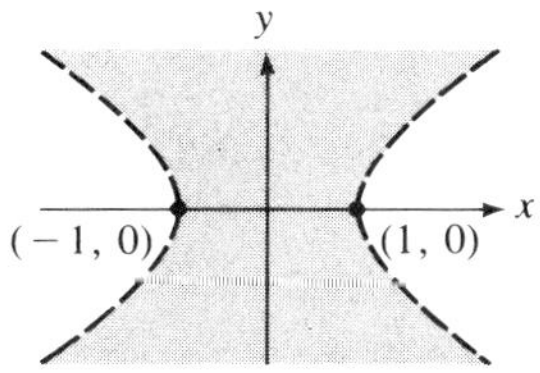

■

In general, most inequalities in two variables may be graphed by the techniques used so far. The related equation is the equation of a curve that divides the plane into a number of regions, one or more of which make up the solution set of the inequality.

Example 7 Graph the inequality $y \leq (x - 1)^3 + 2$.
First, graph the equation $y = (x - 1)^3 + 2$. The curve is drawn by translating the graph of $y = x^3$ up 2 units and to the right 1 unit. This curve partitions the plane into two regions; after checking a point we find that the region on the right is the solution set.

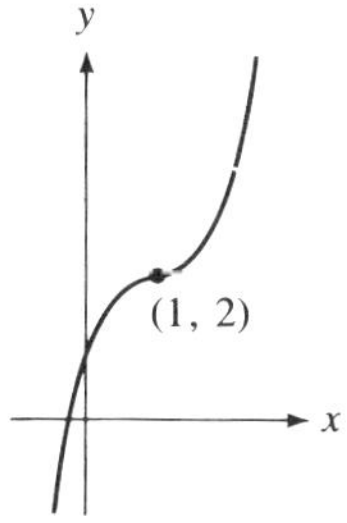

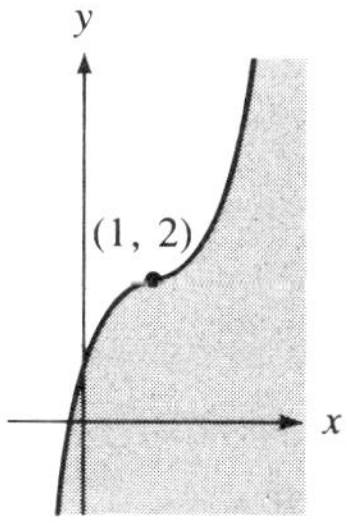

■

Sometimes the inequality will exclude certain values of x or y, and this exclusion may partition the plane in addition to the partition given by the curve.

Example 8 Graph the inequality $y \leq \frac{1}{x}$.

First, we graph the equation $y = 1/x$. At first glance it appears that the curve divides the plane into three regions. However, the additional restriction that

$x \neq 0$ makes the y axis (the line $x = 0$) also a natural boundary. The plane is thus partitioned into *four* regions, indicated by A, B, C, and D in the first sketch.

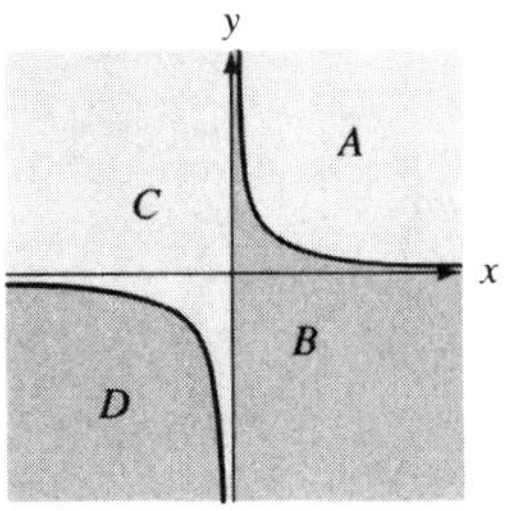

Choose a point in each region and test to see if the inequality is satisfied. Testing the points (2, 2), (−2, −2), (−1, 0), and (1, 0) yields the solution set shown in the second sketch.

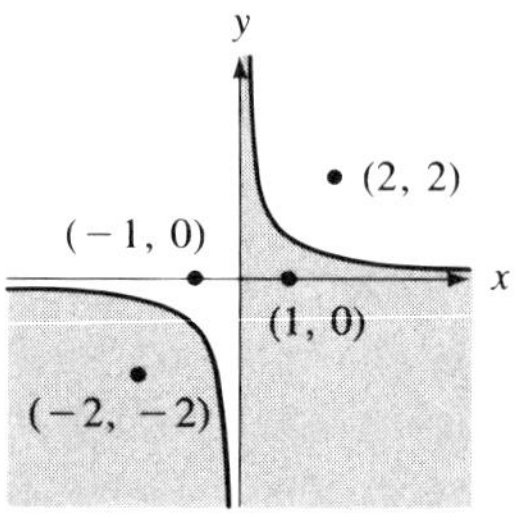

■

Exercises 4.8

Graph the solution set for the inequalities.

1. $x + y \leq 1$
2. $x + y \leq 2$
3. $x - y < 2$
4. $x - y > -1$
5. $2x + 3y < 12$
6. $3x + 2y < -6$
7. $y < x$
8. $y > x$
9. $y > x + 2$
10. $y > x + 3$
11. $x - y \leq 0$
12. $y - x \leq 0$
13. $x > 3$
14. $x > -2$
15. $y < 4$
16. $y < -3$
17. $2 - x \geq 1$
18. $4 - x \geq 3$
19. $x^2 < 1 - y^2$
20. $y^2 < 4 - x^2$
21. $x^2 - 9 \geq -y^2$
22. $y^2 - 25 > x^2$
23. $2x^2 \leq 8 - y^2$
24. $2y^2 \leq 8 - x^2$
25. $9x^2 > 36 - 4y^2$
26. $9y^2 > 36 - 4x^2$
27. $y < 3x^2 - 5$
28. $y + 1 < x^2$
29. $x \geq y^2 + 3$
30. $x \geq y^2 - 3$
31. $x^2 < y^2 + 2$
32. $y^2 < x^2 - 1$
33. $4x^2 < -4y^2 + 4y$
34. $2y^2 - 8x < -2x^2$
35. $y \geq -2x^3$
36. $y > 3x^3$
37. $y + 2 \leq (x - 1)^4$
38. $y - 1 \leq (x + 2)^4$
39. $y > \dfrac{-2}{x}$
40. $y > \dfrac{4}{x}$
41. $y < \dfrac{1}{(x - 1)^2}$
42. $y > \dfrac{2}{(x + 2)^2}$

Chapter Review

Find the line of symmetry and the vertex and graph the function.

1. $y = \frac{2}{3}x^2$ **2.** $y = -2x^2 - 4x$ **3.** $y = 1 - x^2$

4. $F(x) = x^2 - 6x + 15$ **5.** $G(x) = -2x^2 - x - 6$

6. An object moves in a straight line in such a way that the velocity v at the end of t sec is given by the formula $v = t^2 - 2t + 2$. Find the least velocity and the time at which it occurs.

7. An apartment owner can keep each of his 50 apartments rented at \$150 per month. For each \$5 increase in rent he will have 1 more apartment vacant each month. What rent will give the maximum income?

8. Find the center and radius of the circle $x^2 + y^2 - 9y + 6x = 0$.

9. Write the equation of the parabola with focus $(-3, 0)$ and directrix $x = 3$. Graph the parabola.

10. Write the equation of the parabola with focus $(3, 8)$ and vertex $(3, 6)$. Graph the parabola.

11. Graph the ellipse $9x^2 + y^2 - 18x + 6y = 18$, showing vertices and foci.

12. Write the equation of the ellipse with foci $(-2, 3)$ and $(-2, 8)$ and sum of the distances 8.

13. Graph the hyperbola $9x^2 - y^2 - 18x + 6y = 36$, showing foci, vertices, and asymptotes.

14. Write the equation of the hyperbola with foci $(-2, 3)$ and $(-2, 8)$ and difference of the distances 3.

15. Write the equation of the hyperbola with foci $(0, 5)$ and $(0, -5)$ and asymptotes $y = 4x$ and $y = -4x$.

16. Write the equation of the hyperbola with vertices $(3, 4)$ and $(3, -8)$ and foci $(3, 5)$ and $(3, -9)$.

Graph the following polynomial functions.

17. $y = (x + 2)^3 - 1$ **18.** $y = -(x + 4)^3 - 4$

19. $y = -(x - \frac{1}{2})^4 - \frac{3}{2}$ **20.** $y = (x + \frac{3}{2})^4 - \frac{1}{2}$

21. $y = x(x + 1)(x - 3)$ **22.** $y = x^2(x + 1)(x - 2)$

23. $y = x^3 - 4x^2 - 12x$ **24.** $y = 14x - 9x^2 + x^3$

Graph the following rational functions, showing horizontal and vertical asymptotes.

25. $y = \dfrac{1}{(x - 4)^2} + 6$ **26.** $y = \dfrac{3}{x + 2} - 6$

27. $y = \dfrac{(x - 3)^2}{x^2 - 6x + 5}$ **28.** $y = \dfrac{x + 2}{x^2 + 4x - 5}$

Find the partial fraction decomposition for each of the following.

29. $\dfrac{2x - 15}{(2x - 1)(x + 3)}$ **30.** $\dfrac{14 - x}{6 + 7x - 3x^2}$

31. $\dfrac{3x^2 - 4x - 5}{x^2(x - 5)}$ **32.** $\dfrac{x^2 + 7x - 10}{x(x^2 - 3x + 5)}$

Sketch the solution set for the following inequalities.

33. $y < 4 - x$
34. $y + 5 \geq 7$
35. $3x - 4y < 12$
36. $3y + 9 \leq 0$
37. $x + 2y \leq 9$
38. $x - 3y \leq 6$
39. $y > x^2$
40. $y^2 \geq 25 - x^2$
41. $y^2 \leq 36 - 4x^2$
42. $x^2 \leq y^2 + 1$
43. $y + 4x + 4 < -2x^2$
44. $y^2 > x^2 + 1$
45. $y \geq (x - 5)^3 + 1$
46. $y < -(x - 6)^4 - 2$
47. $y < \dfrac{1}{x + 1} + 4$
48. $y > \dfrac{1}{(x - 6)^2} - 2$

Chapter Test

1. Find the vertex and line of symmetry and graph the function $f(x) = 4x^2 - 16x + 12$.
2. Find the center and radius of the circle $x^2 + y^2 - 2x + 4y = 4$.
3. Write the equation of the parabola with vertex at (2, 0) and focus at (2, 4).
4. Write the equation of the ellipse with foci at (0, −2) and (0, 2) and length of the major axis 8.
5. Graph the ellipse $x^2 + 4y^2 - 6x + 8y = 3$, showing vertices and foci.
6. Write the equation of the hyperbola with foci (−2, 3) and (−2, 9) and vertices (−2, 5) and (−2, 7).
7. Graph the hyperbola $x^2 - 4y^2 + 6x - 8y = 11$, showing vertices, foci, and asymptotes.
8. Identify the conic section for the following equations:
   a. $x^2 - y^2 - 4y + 7y = 9$
   b. $2x^2 + x - 6y = 14 - 2y^2$
   c. $2x^2 + x = 7y - 9$
9. Graph the function $y = -\dfrac{1}{2}(x - 1)^3 + 7$.
10. Graph the function $y = x(x + 1)(x - 2)$.
11. Graph the function $y = \dfrac{1}{(x - 2)^2} + 4$, showing asymptotes.
12. Find the partial fraction decomposition of $\dfrac{x + 13}{(x - 2)(x + 3)}$.

Graph the solution set for the inequalities.

13. $4x - 3y \leq 8$
14. $x^2 > y$
15. $x^2 - y^2 \leq 9$
16. $x^2 + 2y^2 > 16$
17. $y \geq 2(x - 1)^3$
18. $y < \dfrac{1}{(x + 1)^2} - 5$

5 Exponential and Logarithmic Functions

In 1908 a New Mexico cowboy named George McJunkin discovered some strange bones near Folsom, New Mexico. Further investigation revealed bones of an extinct relative of the bison—larger and with horns like a Texas long-horn—that had vanished more than 10,000 years ago. A stone point between two bison ribs indicated that human beings had inhabited the region at that time.

In 1932 excavators found near Clovis, New Mexico bones of extinct animals accompanied by stone points that differed from those found at Folsom.

If a method could be found to date the artifacts, archaeologists would have a partial answer to the question "When did human beings first reach the North American continent?"

During its lifetime a living organism absorbs a certain amount of a radioactive element called carbon-14. When the organism dies, the carbon-14 decays at a predictable rate. By measuring the amount of carbon-14 left in the remains of a once-living organism, a scientist can estimate how long it has been since the organism died. By this method, the Folsom culture has been dated at about 11,000 years ago and the Clovis culture at about 12,000 years ago.

The decay of a radioactive material such as carbon-14 is best described by means of an *exponential* function. Such functions are discussed in this chapter.

5.1 Exponential Functions

If b is any positive real number, then b^r is defined for any rational exponent r. Just as a reminder, for $b > 0$:

Review of exponents

If n is a positive integer,

$$b^n = b \cdot b \cdot b \cdots b \quad (n \text{ factors})$$

If $n = 0$,

$$b^n = b^0 = 1$$

If n is a negative integer,

$$b^n = \frac{1}{b^{-n}}$$

If n is a positive integer,

$$b^{1/n} = \sqrt[n]{b}$$

If $r = \dfrac{m}{n}$, m and n integers with n positive,

$$b^r = (b^{1/n})^m$$

If $b > 1$, then for any positive rational number r, $b^r > 1$. Thus if $b > 1$ and $s < r$, then $r - s > 0$, and hence

$$b^r = b^s(b^{r-s}) > b^s \cdot 1 = b^s$$

This result is important in the discussion to follow.

If $b > 1$ and r and s are rational numbers with $s < r$, then

$$b^s < b^r$$

The purpose of this discussion is to define b^x when x is an irrational number. From Chapter 1 we know that any irrational number can be approximated by rational numbers to any desired degree of accuracy, and this is the key to our definition of b^x for irrational x.

The following theorem can be proven using the completeness of the system of real numbers and the fact that we can find a rational number as close to any real number as we wish.

Theorem Given any real number b greater than 1 and any irrational number x, there is *exactly* one number z such that $b^s < z < b^r$ for every pair of rational numbers r and s such that $s < x < r$.

Definition of b^x for irrational x

In light of this theorem, we define b^x, $b > 1$, to be that unique real number with the property that

$$b^s < b^x < b^r$$

for every pair of rational numbers r and s such that

$$s < x < r$$

If $0 < b < 1$, define

$$b^x = \left(\frac{1}{b}\right)^{-x}$$

Perhaps a few words of philosophy are in order at this stage. Defining b^x for irrational x is no more mysterious than defining $\sqrt{2}$. The number is defined in terms of the properties it possesses. You cannot give a decimal representation of $\sqrt{2}$ because it is not a rational number. You can, however, compute rational approximations to $\sqrt{2}$ by any one of a number of means, the easiest being to use your handheld calculator. Similarly, your handheld calculator may have a key marked $\boxed{y^x}$ which will approximate y^x with a rational number. Of course the catch here is that we only know how to input rational numbers into the calculator, hence to compute $\sqrt{3}^{\sqrt{2}}$ we would use rational approximations for both $\sqrt{3}$ and $\sqrt{2}$.

We have now defined b^x for any real number x and any positive real number b. This idea can be used to define a new type of function.

Exponential function

If $b > 0$, define $f(x) = b^x$. This is an example of what we will call an **exponential function.**

An exponential function may best be understood by considering its graph. The graph of $f(x) = 2^x$ is shown in Figure 5.1. The graph is found by plotting a few points for integer values of x and drawing a smooth curve through those points.

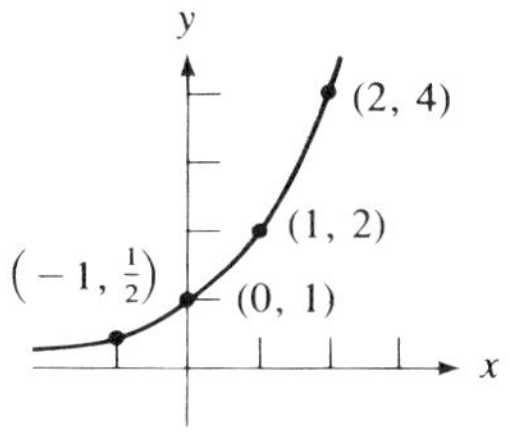

Figure 5.1

One of the first things to note follows from an earlier observation that if $b > 1$ and $s < r$, s and r rational numbers, then $b^s < b^r$. Hence for any real numbers u and v, if $b > 1$ and $u < v$, then $b^u < b^v$. This means that the exponential function $f(x) = b^x$, $b > 1$, is what we call an *increasing* function.

Increasing function

A function f is said to be an **increasing function** if whenever $x < y$, then $f(x) < f(y)$.

As you may note from Figure 5.1, an increasing function is 1-1. To see this, assume that $f(u) = f(v)$. Then either $u = v$, $u < v$, or $u > v$. If $u < v$ then $f(u) < f(v)$, contrary to $f(u) = f(v)$. Likewise, if $u > v$, then $f(u) > f(v)$, again contrary to $f(u) = f(v)$. The only remaining possibility is $u = v$. Thus f is 1-1.

If $b > 1$, the exponential function $f(x) = b^x$ has the following properties.

Properties of exponential functions

1. f is an increasing function, that is, $u < v$ implies $f(u) < f(v)$.
2. f is a 1-1 function.
3. If $x < 0$, then $0 < f(x) < 1$.
4. $f(0) = 1$.
5. If $x > 0$, then $f(x) > 1$.
6. As x tends to $-\infty$, $f(x)$ tends to 0. This means that the x axis is an asymptote for the graph of $y = b^x$.

Example 1 Graph the function $f(x) = 3^x$.

Plotting the points corresponding to $x = -2, -1, 0, 1$, and 2 and drawing a smooth curve through them yields the curve shown.

x	3^x
-2	$3^{-2} = \frac{1}{9}$
-1	$3^{-1} = \frac{1}{3}$
0	$3^0 = 1$
1	$3^1 = 3$
2	$3^2 = 9$

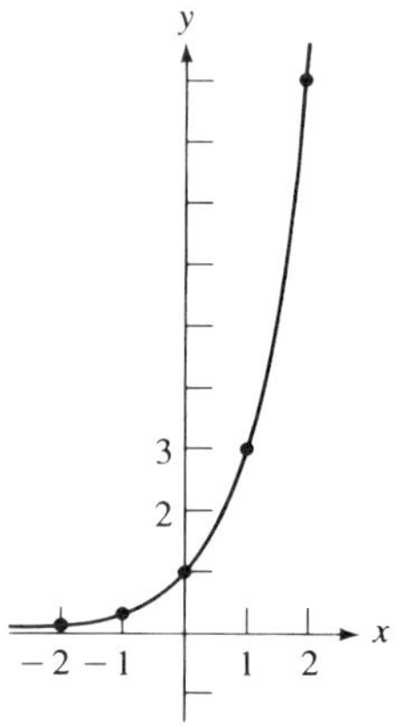

■

If $0 < b < 1$ and s and r are rational numbers with $s < r$, then $b^s > b^r$. Hence $b^u > b^v$ for any real numbers u and v with $u < v$. This means that the exponential function $f(x) = b^x$, $0 < b < 1$, is what is called a *decreasing* function.

Decreasing function

A function f is a **decreasing function** if whenever $x < y$, $f(x) > f(y)$.

It is easy to show that a decreasing function is a 1-1 function. The proof is very similar to the proof that an increasing function is 1-1.

If $0 < b < 1$, the exponential function $f(x) = b^x$ has the following properties.

Properties of exponential functions

1. f is a decreasing function, that is, $u < v$ implies $f(u) > f(v)$.
2. f is a 1-1 function.
3. If $x < 0$, then $f(x) > 1$.
4. $f(0) = 1$.
5. If $x > 0$, then $0 < f(x) < 1$.
6. As x tends to ∞, $f(x)$ tends to 0. The x axis is an asymptote for the graph of $y = b^x$.

Example 2 Graph the function $y = \left(\frac{1}{2}\right)^x$.

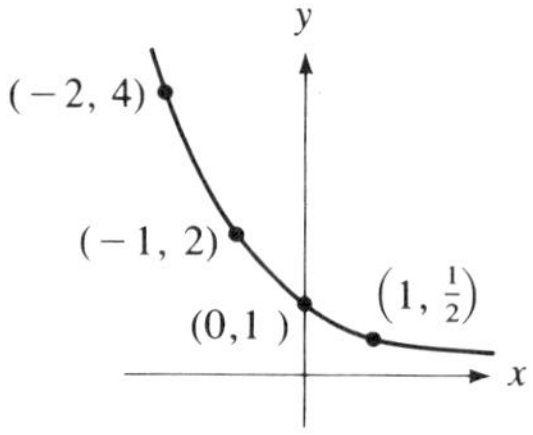

■

The properties of exponents listed in Chapter 2 also hold for real exponents; we list them here for completeness.

If a and b are positive real numbers and x and y are any real numbers, then

Properties of exponents

1. $a^x a^y = a^{x+y}$
2. $\dfrac{a^x}{a^y} = a^{x-y}$
3. $(a^x)^y = a^{xy}$
4. $(ab)^x = a^x b^x$
5. $\left(\dfrac{a}{b}\right)^x = \dfrac{a^x}{b^x}$

Example 3 Graph the function $y = 2^{x+1}$.

The graph of $y = 2^{x+1}$ is found by shifting the graph of $y = 2^x$ left 1 unit. We can use properties of exponents to obtain the graph of $y = 2^{x+1}$ in another way. Note that

$$y = 2^{x+1} = 2^1 2^x = 2(2^x)$$

Thus the graph of $y = 2^{x+1}$ can be obtained by multiplying the ordinate of each point on the graph of $y = 2^x$ by the factor 2.

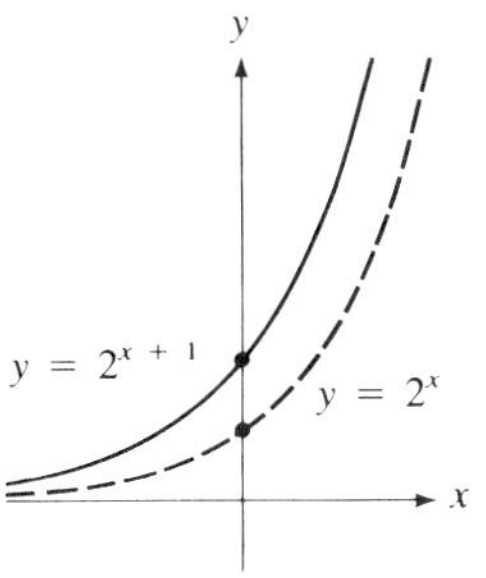

■

Example 4 Find the value of k such that the point $\left(2, \frac{1}{2}\right)$ is on the graph of $y = 2^{kx}$.

Requiring that the point $\left(2, \frac{1}{2}\right)$ be on the graph of $y = 2^{kx}$ gives the equation

$$\frac{1}{2} = 2^{2k}$$

when we substitute $x = 2$ and $y = \frac{1}{2}$. But $\frac{1}{2} = 2^{-1}$, so the equation becomes

$$2^{-1} = 2^{2k}$$

Since the exponential function is 1-1, it must be true that

$$-1 = 2k, \qquad k = -\frac{1}{2}$$

■

Example 5 Graph the function $y = 1 - 2^x$.
The graph is obtained by shifting the graph of $y = -2^x$ up 1 unit. The graph of $y = -2^x$ is found by reflecting the graph of $y = 2^x$ about the x axis.

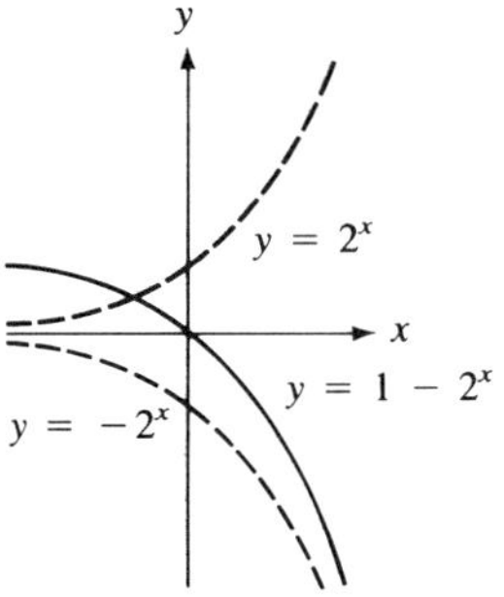

■

Exercises 5.1

In Exercises 1–16 evaluate the given function for the given value of x. Use properties of exponents rather than your calculator.

1. $y = 2^{x-1}$; $x = -2$

2. $y = 3^{x+1}$; $x = -5$

3. $f(x) = \left(\frac{1}{4}\right)^{2x}$; $x = 2$

4. $g(x) = \left(\frac{1}{3}\right)^{3x}$; $x = 2$

5. $F(x) = \left(\frac{1}{7}\right)^{-2x}$; $x = 1$

6. $G(x) = \left(\frac{1}{5}\right)^{-3x}$; $x = 1$

7. $y = \left(\frac{2}{3}\right)^{3x}$; $x = 2$

8. $y = \left(\frac{5}{2}\right)^{2x}$; $x = 2$

9. $y = 2 - 3(2^x)$; $x = -1$

10. $y - 4 + 2(3^x)$; $x = \;\; 2$

11. $y = \frac{2^x}{3^{x+1}}$; $x = -2$

12. $y = \frac{5^{-x}}{2^{1+x}}$; $x = -2$

13. $y = (3^x)^2$; $x = -2$

14. $y = (2^x)^3$; $x = -4$

15. $y = \left(\frac{1}{2}\right)^x\left(\frac{2}{3}\right)^{2x-1}$; $x = 2$

16. $y = \left(\frac{3}{4}\right)^{2x}\left(\frac{2}{3}\right)^{3x-2}$; $x = 2$

17. Find the value of k such that the graph of $y = 3^{kx}$ passes through the point $(-4, 9)$.

18. Find the value of k such that the graph of $y = 2^{kx}$ passes through the point $(-4, 8)$.

19. Find the value of a such that the graph of $y = a^x$ passes through the point $(2, 2)$.

20. Find the value of a such that the graph of $y = a^x$ passes through the point $(3, 3)$.

Sketch the graphs of the functions in Exercises 21–34.

21. $f(x) = 2^{-x}$

22. $f(x) = -2^{-x}$

23. $g(x) = (1.5)^x$

24. $h(x) = (1.2)^x$

25. $y = \left(\frac{4}{3}\right)^{-x}$

26. $y = \left(\frac{3}{4}\right)^{-x}$

27. $y = -3^{-x}$

28. $y = 3^{-x}$

29. $y = 4 - 2^x$

30. $y = 3 - 3^x$

31. $y = 3^{x-1}$

32. $y = 2^{x-1}$

33. $y = 3(2^{x+2})$

34. $y = 2(3^{x+2})$

5.2 Using Exponential Functions

Many growth and decay situations can best be described by exponential functions. A common example is that of depreciation.

Depreciation is the decline in value of an asset, and the process of recording this depreciation is the allocation of these costs over a period of years. There are various ways of determining depreciation; one is the method of *declining balances*.

Example 1 Compute the annual depreciation and remaining balance for a \$10,000 asset with an annual rate of 40% by the method of declining balances. Show the amounts for the first 3 years and give a formula for $V(n)$, the remaining value after n years.

Year	Depreciation	Remaining Value
1	(10,000)(0.4) = 4,000	6,000 = (10,000)(0.6)
2	(6,000)(0.4) = 2,400	3,600 = $(10{,}000)(0.6)^2$
3	(3,600)(0.4) = 1,440	2,160 = $(10{,}000)(0.6)^3$

At the end of the nth year the remaining value $V(n)$ is given by the formula

$$V(n) = 10{,}000(0.6)^n$$ ■

Example 2 How many years will it take for the asset in Example 1 to depreciate to a salvage value of \$100?

For the present, trial and error is our only method of solution.

$V(8) = \$167.96$

$V(9) = \$100.78$

Nine years of depreciation reduces the value to approximately \$100. ■

Money grows at compound interest in somewhat the same manner that the value of an asset depreciates under the method of declining balances.

Example 3 If \$5,000 is invested in an account at 9% interest compounded annually, how much is in the account at the end of 3 years?

The term *compound interest* here means that the interest is computed at the end of each interest period and is added to the account. The amount in the account thus increases each interest period, and interest is paid on the interest itself. In this case the interest is compounded annually, meaning that the interest period is 1 year. Thus at the end of each year the amount in the account is increased by 9%, the interest earned for the year. The table shows the amounts in the account at the end of each of the 3 years.

Year	Amount in the Account
1	$5{,}000(1.09) = 5{,}450$
2	$5{,}450(1.09) = 5{,}000(1.09)^2 = 5{,}940.50$
3	$5{,}000(1.09)^3 = 6{,}475.15$

■

The computations in Example 3 can be used to obtain a formula for finding the value of an investment at compound interest:

$$A = P(1 + r)^n$$

where A is the amount in the account, P is the amount invested, r is the rate per interest period, and n is the number of interest periods.

Example 4 Determine the value of an investment of \$5,000 after 3 years if the amount is invested at 9% compounded semiannually, quarterly, and monthly.
If interest is compounded semiannually there are 6 interest periods in 3 years, and the rate for each interest period is 4.5%, half of the annual rate:

$$A = 5{,}000(1.045)^6 = 6{,}511.30$$

If the interest is compounded quarterly there are 12 interest periods, and the rate per interest period is 2.25%, one-fourth of the annual rate:

$$A = 5{,}000(1.0225)^{12} = 6{,}530.25$$

Finally, if the interest is compounded monthly, there are 36 interest periods and the rate per period is 0.75%, one-twelfth of the annual rate:

$$A = 5{,}000(1.0075)^{36} = 6{,}543.23$$

■

Notice in Example 4 that as interest is compounded more often the return is increased. How far can this go? In an effort to compete for the saver's dollar, savings institutions have at times offered daily compounding and even continuous compounding. How would continuous compound interest be computed?

Example 5 Compute the value of an investment of \$1 after 1 year at 100% annual interest compounded monthly, daily, and hourly.

$$\left(1 + \frac{1}{12}\right)^{12} = 2.613 \qquad \text{monthly}$$

$$\left(1 + \frac{1}{365}\right)^{365} = 2.71457 \qquad \text{daily}$$

$$\left(1 + \frac{1}{8{,}760}\right)^{8{,}760} = 2.71812 \qquad \text{hourly}$$

■

The number e

If the process in Example 5 is continued, computing interest compounded every minute, every second, every microsecond, every nanosecond, etc., the numbers thus obtained come closer and closer to a number universally denoted by the letter e. To nine decimal places,

$$e = 2.718281828\ldots$$

Despite what the decimal approximation might suggest, e is an irrational number, not a repeating decimal. This number is used to compute continuous compound interest. Actually it is also used in computing daily compound interest, because there is little difference between interest compounded daily and interest compounded continuously.

If an amount P is invested at an annual rate r (expressed as a decimal) compounded continuously for t years, the amount $A(t)$ in the account at the end of t years is given by the exponential function

$$A(t) = Pe^{rt}$$

where e is the number described above.

Example 6 If \$5,000 is invested at 8% annual interest rate compounded continuously, how much will be in the account at the end of 1.5 years?

Use the formula $A(t) = Pe^{rt}$ with $P = 5{,}000$, $r = 0.08$, and $t = 1.5$:

$$\begin{aligned} A(1.5) &= 5{,}000(e^{0.12}) && (0.12 = (0.08)(1.5)) \\ &= 5{,}000(1.1275) && \text{(From Table III)} \\ &= 5{,}637.50 \end{aligned}$$

■

Any population that grows or decays at a more or less continuous rate may be described by using an exponential function of the form $A = Pe^{rt}$, where P is the size of the population when $t = 0$ and the constant r depends on the situation. Human population growth may be considered continuous, because births and deaths do not occur at any specified time.

Because the number e occurs so often in applications, a table of values of both e^x and e^{-x} for certain values of x is given in Table III. Also, all scientific calculators and most business calculators have a key that produces values of e^x. In many cases this key is labeled $\boxed{e^x}$, but for some machines you will need to use the $\boxed{\text{INV}}$ key followed by the $\boxed{\text{ln}}$ key. The reason for this will become apparent in Section 5.3. Values from the table may be used to obtain the graphs of $y = e^x$ and $y = e^{-x}$ shown in Figure 5.2.

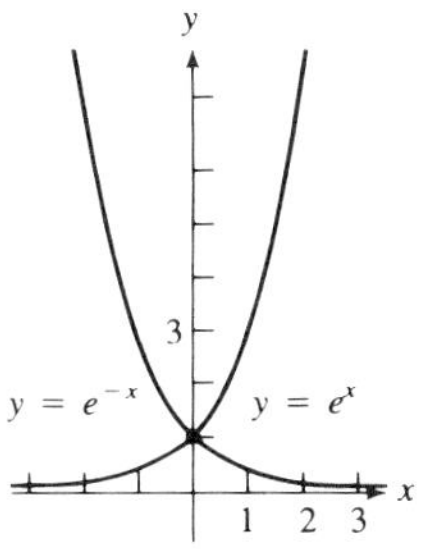

Figure 5.2

Example 7 The population $N(t)$ of a certain country is approximated by the function $N(t) = 350{,}000e^{0.04t}$, where t is the number of years after 1980. Use the table in Table III (or your calculator) to estimate the population in 1985 and 1990. Using the table:

In 1985

$$\begin{aligned} N(5) = 350{,}000e^{0.04(5)} &= 350{,}000e^{0.2} \\ &= 350{,}000(1.2214) \\ &= 427{,}490 \end{aligned}$$

In 1990

$$\begin{aligned} N(10) = 350{,}000e^{0.04(10)} &= 350{,}000e^{0.4} \\ &= 350{,}000(1.4918) \\ &= 522{,}130 \end{aligned}$$

■

Since these are estimations, there is little reason to carry the computations down to the very last person. The answers in Example 7 should be rounded off to 427,500 and 522,100, respectively.

Exercises 5.2

Use Table III to graph the functions in Exercises 1–4.

1. $P(x) = 2e^x$
2. $P(x) = \frac{1}{2}e^x$
3. $y = x + e^x$
4. $y = x - e^x$

In Exercises 5–10 find the remaining value of the asset if it depreciates at the indicated rate. Round off to the nearest dollar.

5. \$1,000 at 25% for 4 years
6. \$1,000 at 25% for 3 years
7. \$4,000 at 30% for 6 years
8. \$4,000 at 30% for 5 years
9. \$1,000 at 75% for 2 years
10. \$1,000 at 75% for 3 years
11. If an asset worth \$5,000 depreciates at a rate of 20% each year, how long does it take to depreciate to a value of less than \$1,500? (Use trial and error methods.)
12. If an asset worth \$4,000 depreciates at a rate of 30% each year, how long does it take to depreciate to a value of less than \$1,000?
13. After 3 years at 9% compounded annually, what is the value of an initial investment of \$5,000?
14. After 4 years at 9% compounded annually, what is the value of an initial investment of \$6,500?
15. If interest on an investment of \$3,000 is compounded continuously for 3 years at 10%, how much will the investment be worth?
16. If interest on an investment of \$3,000 is compounded continuously at 15% for 4 years, how much will the investment be worth?
17. John is paying rent of \$250 per month for his apartment. If the annual rate of inflation is 8% and his rent increases at the same rate, how much will his rent be in 4 years?
18. If the annual rate of inflation is 6% (see Problem 17), how much rent will John have to pay 4 years from now?

The population of a certain country is approximated by the function

$$N(t) = 250{,}000e^{0.05t}$$

where t is measured in years after 1980.

19. Find the population in 1985.
20. Find the population in 1990.
21. What is the annual percent of increase in population from 1984 to 1985?
22. Show that a decreasing function is 1-1.
23. If f is a decreasing function, prove that f^{-1} is a decreasing function.

For Your Calculator

Your scientific calculator has an exponential function key, usually labeled $\boxed{y^x}$. In this set of exercises use your calculator to see how exponential functions grow or decay.

If a natural resource is being used up at a rate of 15% per year, the percent of the resource left after t years is given by the formula

$$P = (1 - 0.15)^t = (0.85)^t$$

24. If a natural resource is to be used up at a rate of 16% per year, what percent of the present supply will be left after 20 yr?
25. With an annual rate of inflation of 8%, how much will an automobile that now sells for \$5,000 cost in 10 yr?
26. If you invest \$1,000 at 9% interest compounded annually, how much will you have in 20 yr?
27. Double-digit inflation refers to inflation at an annual rate of more than 10%. Suppose you had invested the \$1,000 from Problem 26 in property that increased in value at the rate of inflation. If the annual rate of inflation were 10% for 20 yr, how much would your investment be worth at the end of 20 yr? Compare this to the answer to Problem 26.
28. If a candy bar sells for 25¢ today and its price increases at the same rate as inflation, what will be the price in 20 yr at a 12% annual rate of inflation?

5.3 Logarithmic Functions

Look at the graph of $y = 2^x$ in Figure 5.3. Notice that:

1. The domain of the function is the set of all real numbers.
2. The range is the set of all positive real numbers.
3. The function is 1-1.

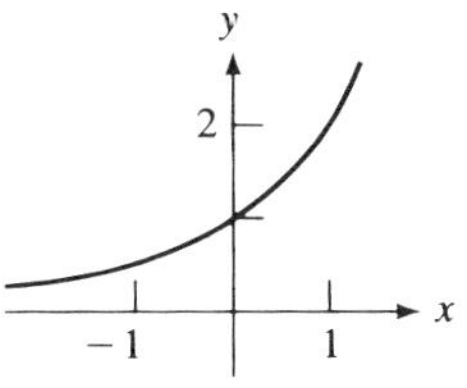

Figure 5.3

The observations about $y = 2^x$ carry over to any exponential function $f(x) = b^x$, where b is positive and $b \neq 1$.

1. The domain of the exponential function is the set of all real numbers.
2. The range of the exponential function ($b \neq 1$) is the set of all positive real numbers.
3. The exponential function is 1-1 ($b \neq 1$).

Because the exponential function is 1-1, it has an inverse. The inverse function is called a *logarithmic* or *logarithm* function.

Definition of $\log_b x$

Let b be any positive real number different from 1. The inverse of the exponential function $f(x) = b^x$ is called the *logarithmic function to the base b* and is denoted by

$$L(x) = \log_b x$$

To help in understanding the logarithmic function, we show the graph of $y = \log_2 x$ in Figure 5.4 on the same set of axes with the graph of $y = 2^x$.

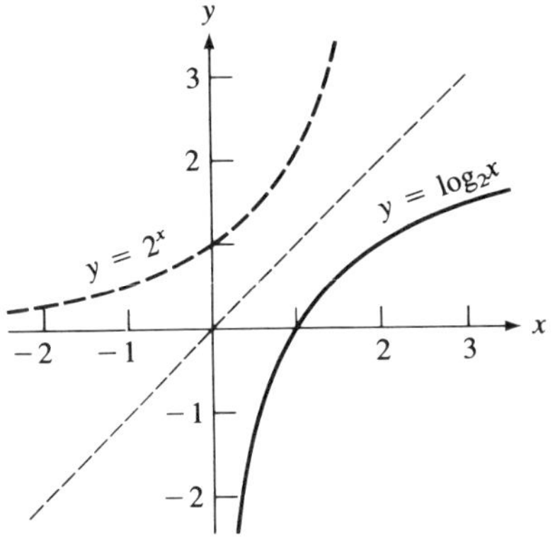

Figure 5.4

The graph of $y = \log_2 x$ is found by reflecting the graph of $y = 2^x$ about the line $y = x$, because the two functions are inverses of each other.

The shape of the graph of $y = \log_2 x$ is typical of the graph of any logarithmic function $y = \log_b x$ for $b > 1$. From the graph of the logarithmic function and the properties of the exponential function we can obtain the following properties of the logarithmic function.

For $b > 1$, the function $f(x) = \log_b x$ has the following properties.

Properties of the logarithmic function

1. f is an increasing function and hence a 1-1 function.
2. If $0 < x < 1$, then $f(x) < 0$.
3. If $x > 1$, then $f(x) > 0$.
4. $f(1) = 0$.
5. As x tends to 0, $f(x)$ tends to $-\infty$, that is, the y axis is a vertical asymptote.
6. The domain of f is the set of all positive real numbers and the range is the set of all real numbers.

It is important to remember the relationship between the logarithmic function and the corresponding exponential function. This relationship is summarized below.

7. $\log_b b^x = x$.
8. $b^{\log_b x} = x$.
9. $y = \log_b x$ if and only if $x = b^y$.

Example 1 $\log_2 8 = 3$, since $2^3 = 8$. ■

Example 2 $\log_{10}(0.0001) = -4$, since $0.0001 = 10^{-4}$. ■

Example 3 $\log_9 3 = \frac{1}{2}$, since $3 = 9^{1/2}$. ■

Natural logarithm

Common logarithm

The inverse of the exponential function $y = e^x$ is often referred to as the **natural logarithm** and is denoted by $y = \ln x$. The table in Table IV gives some values of $\ln x$. The inverse of the function $y = 10^x$ is called the **common logarithm** and is usually denoted by $y = \log x$. Values of $\log x$ can be found in the table in Table II. Almost all of the numbers given in the tables are approximations, and, strictly speaking, we should not use the symbol for equality when we write

$$\log 2 = 0.301$$

However, we will follow custom and use equality, even though "approximately equal" is more accurate.

Since the functions $y = e^x$ and $y = \ln x$ are inverses of each other, one obtains values of e^x on some calculators by depressing the [INV] key (for inverse function) and then the [ln] key. Likewise, values of 10^x may be obtained by depressing the [INV] key and then the [log] key.

Because the logarithmic function is the inverse of the exponential function, we can expect to find similarities between the properties of the two functions. We list thc properties below.

Properties of Exponential Functions	Properties of Logarithmic Functions
1. $b^s b^t = b^{s+t}$	1. $\log_b(uv) = \log_b u + \log_b v$
2. $\dfrac{b^s}{b^t} = b^{s-t}$	2. $\log_b\left(\dfrac{u}{v}\right) = \log_b u - \log_b v$
3. $(b^t)^s = b^{ts}$	3. $\log_b x^u = u \log_b x$

We will derive Property 1 of logarithmic functions and leave the rest to the exercises.

Suppose that $u = b^m$ and $v = b^n$; then $m = \log_b u$ and $n = \log_b v$. But $uv = b^m b^n = b^{m+n}$, hence $\log_b uv = m + n = \log_b u + \log_b v$.

Example 4 $\log_b 15 = \log_b 5 + \log_b 3$ ■

Example 5 $\log 50 = \log 5 + \log 10 = \log 5 + 1$ (*Remember*, log *means* $\log_{10}$.) ■

Example 6 $\log_b \frac{2}{3} = \log_b 2 - \log_b 3$ ■

Example 7 $\log_b 8 = \log_b 2^3 = 3 \log_b 2$ ■

Example 8 If $\log 2 = 0.301$ and $\log 3 = 0.477$, then

$$\log 6 = \log 2 + \log 3 = 0.301 + 0.477 = 0.778$$
$$\begin{aligned}\log 18 = \log 9 + \log 2 &= \log 3^2 + \log 2 \\ &= 2 \log 3 + \log 2 \\ &= 1.255\end{aligned}$$
■

Example 9 If $\log 5 = 0.699$, then

$$\log 50 = \log 10 + \log 5 = 1 + 0.699 = 1.699$$
$$\log 500 = \log 100 + \log 5 = 2 + 0.699 = 2.699$$
$$\log 0.005 = \log 0.001 + \log 5 = -3 + 0.699 = -2.301$$
$$\log \frac{2}{5} = \log 2 - \log 5 = 0.301 - 0.699 = -0.398$$
$$\begin{aligned}\log \frac{2}{5} &= \log 0.4 = \log 0.1 + \log 4 \\ &= -1 + 2 \log 2 = -1 + 0.602 = -0.398\end{aligned}$$
■

Example 10 $e^{\ln 2} = 2$ (*Remember*, $\ln x = \log_e x$.) ■

Example 11 $\ln e^3 = 3$ ■

Example 12 $\log 10^4 = 4$ ($\log x = \log_{10} x$) ■

Example 13 $10^{-\log 4} = 10^{\log(1/4)} = \frac{1}{4}$ ■

The *intensity* of a sound wave is the average power the sound wave transports per unit area. Because of the large range of intensities over which the human ear is sensitive, a logarithmic scale is used to measure *intensity level*. The intensity level L of a sound wave with intensity I is given by the formula

$$L(I) = 10 \log \frac{I}{I_0}$$

where I_0 is an arbitrary reference intensity which is taken as 10^{-6} watts per square centimeter (W/cm^2), corresponding roughly to the intensity of the faintest sound that can be heard. The human ear is almost too sensitive for its own good. Certain tribesmen in Africa can hear a whisper across a clearing the size of a football field. Continued noise at a high intensity level for long periods of time can do permanent damage to our hearing. For example, a jackhammer at 100 decibels (dB, the unit of intensity level) for 2 hours without any hearing protection will cause some loss of hearing.

Example 14 A normal conversation has an intensity level of 60 dB and a jackhammer has an intensity level of 100 dB. How many times more powerful than ordinary conversation is the sound of the jackhammer?

Let I_1 denote the intensity of conversation and I_2 the intensity of the sound of the jackhammer. Then

$$60 = 10 \log \frac{I_1}{I_0}$$

and

$$100 = 10 \log \frac{I_2}{I_0}$$

Then

$$\frac{I_1}{I_0} = 10^6$$

and

$$\frac{I_2}{I_0} = 10^{10}$$

Hence

$$I_2 = 10^4 I_1$$

Thus the intensity of the jackhammer sound is 10,000 times that of normal conversation. ■

Exercises 5.3

In Exercises 1–4 graph both functions on the same set of axes.

1. $y = \log_3 x,\ y = 3^x$

2. $y = \log_4 x,\ y = 4^x$

3. $y = \log x,\ y = 10^x$

4. $y = \ln x,\ y = e^x$

In Exercises 5–16 evaluate the expression.

5. $\log_2 \frac{1}{4}$

6. $\log_5 \left(\frac{1}{25}\right)$

7. $\log_3 \sqrt{3}$

8. $\log_2 \sqrt{2}$

9. $\log_4 64$

10. $\log_3 27$

11. $\log_2 \left(\frac{1}{\sqrt{2}}\right)$

12. $\log_3 \left(\frac{1}{\sqrt{3}}\right)$

13. $2^{\log_2 5}$

14. $3^{\log_3 6}$

15. $9^{\log_3 6}$

16. $4^{\log_2 5}$

In Exercises 17–32 use log 2 = 0.301, log 3 = 0.477, *and* log 5 = 0.699 *to compute the logarithms.*

17. $\log \frac{5}{2}$ **18.** $\log \frac{5}{4}$ **19.** $\log \frac{15}{2}$ **20.** $\log \frac{6}{5}$

21. $\log \frac{1}{3}$ **22.** $\log \frac{1}{5}$ **23.** log 32 **24.** log 25

25. log 30 **26.** log 150 **27.** log 0.002 **28.** log 0.005

29. log 300 **30.** log 6,000 **31.** log 0.012 **32.** log 0.015

In Exercises 33–38 rewrite the expression in terms of $\log x$, $\log y$, *and* $\log z$. *Assume that* x, y, *and* z *are positive real numbers.*

33. $\log \frac{1}{xy}$ **34.** $\log \frac{x}{yz}$ **35.** $\log \frac{z^2}{x^3y}$

36. $\log \frac{z^4}{y^2x^3}$ **37.** $\log \frac{\sqrt[3]{xy^2}}{\sqrt{z}}$ **38.** $\log \frac{\sqrt[3]{y}}{\sqrt{xz^2}}$

39. Sound at an intensity level of 120 dB is painful to the ear. How many times more powerful is such a sound than the sound of ordinary conversation at 60 dB?

40. A whisper is rated at 20 dB and thunder at 110 dB. How many times more powerful is thunder than a whisper?

For Your Calculator

Your calculator doesn't have all the values of the log and ln functions stored in its memory; rather, it has certain algorithms to follow in computing values of those functions. For values of x between -0.3 and 0.3, the approximation

$$\ln(1 + x) \approx x - \frac{x^2}{2} + \frac{x^3}{3} - \frac{x^4}{4}$$

is fairly good. *Use your calculator and the approximation given above to compute the logarithms below and then compare those results to those found from your calculator using the* [ln] *key.*

41. ln 1.1 **42.** ln 1.01 **43.** ln 1.03 **44.** ln 1.05

45. ln 1.15 **46.** ln 1.25 **47.** ln 0.8 **48.** ln 0.85

49. ln 0.75

When your calculator responds to the [y^x] key, it usually does the computation using natural logarithms and the fact that

$$e^{x \ln y} = y^x$$

For this reason, negative or zero values for y are not accepted, even though such powers as $(-3)^5$ are perfectly valid. Some calculators, however, will compute y^x for negative values of y if x is an integer. You should check yours to see. In Exercises 50–53 write the expression in the form y^x, where y is an integer.

50. $e^{-2 \ln 3}$ **51.** $e^{3 \ln 4}$ **52.** $e^{(1/2) \ln 2}$ **53.** $e^{(1/4) \ln 5}$

5.4 Computations with Logarithms

Before the advent of high-speed computers and handheld calculators, logarithms were used extensively for purposes of calculation. However, with the availability of inexpensive four-function calculators, there is little excuse to use logarithms for multiplication and division, and even exponentiation can be handled with scientific calculators that sell for less than \$30. In this section we discuss *common* logarithms—that is, logarithms of base 10—and show how a table of logarithms can be used to carry out rather complicated computations.

Any positive real number can be written in scientific notation as the product of a number between 1 and 10 and a power of 10. Since the table of common logarithms in Table II only gives the logarithms of numbers between 1 and 10, we use scientific notation and properties of logarithms to find logarithms of numbers not in the table.

Example 1 From the table, $\log 9.9 = 0.9956$ ■

Example 2 From the table, $\log 1.07 = 0.0294$ ■

Example 3 From the table, $\log 9.9 = 0.9956$. So

$$\begin{aligned}\log 99 = \log(9.9 \times 10) &= \log 9.9 + \log 10\\ &= 0.9956 + 1\\ &= 1.9956\end{aligned}$$ ■

Example 4

$$\begin{aligned}\log 0.099 &= \log(9.9 \times 10^{-2})\\ &= \log 9.9 + \log 10^{-2}\\ &= 0.9956 - 2\\ &= -1.0044\end{aligned}$$ ■

If a positive real number r is written in scientific notation, $r = a \times 10^n$, where $1 \leq a < 10$ and n is an integer, then

$$\log r = \log a + n$$

Mantissa and characteristic

The number $\log a$ is called the **mantissa** and the integer n is called the **characteristic** of $\log r$. Notice that the mantissa is *always* a nonnegative number less than 1, because it is the logarithm of a number between 1 and 10.

Example 5 $\log 99 = 1.9956$; the characteristic is 1 and the mantissa is 0.9956. ■

Example 6 Find N so that $\log N = 3.8069$.
The mantissa of $\log N$ is 0.8069 and the characteristic is 3. We look in the table for a number with logarithm equal to the mantissa, 0.8069, and use the characteristic to place the decimal point. From the table,

$$\log 6.41 = 0.8069$$

so

$$\begin{aligned}\log 6{,}410 &= \log(6.41 \times 10^3)\\ &= \log 6.41 + \log 10^3\\ &= 0.8069 + 3\\ &= 3.8069\end{aligned}$$

The answer is $N = 6{,}410$. ■

Example 7 Find N so that $\log N = -2.1818$.
To find N we want to look up the mantissa in Table II. Consequently, we need to write -2.1818 in a different form—as a positive number between 0 and 1 (the mantissa) plus an integer (the characteristic). To accomplish this, add and subtract 3:

$$-2.1818 = 3 - 2.1818 - 3 = 0.8182 - 3$$

From the table, $\log 6.58 = 0.8182$, so

$$\begin{aligned}\log 0.00658 &= \log 6.58 + \log 10^{-3}\\ &= 0.8182 - 3\end{aligned}$$

Therefore, $N = 0.00658$. ■

Notice how the mantissa determines the digits in the number and the characteristic locates the decimal point.

Example 8 Find N so that $\log N = -0.2596$.
Add and subtract 1:

$$\begin{aligned}-0.2596 &= 1 - 0.2596 - 1\\ &= 0.7404 - 1\end{aligned}$$

From the table, $\log 5.5 = 0.7404$, so

$$\begin{aligned}\log 0.55 &= 0.7404 - 1 = -0.2596\\ N &= 0.55\end{aligned}$$

■

If you have only a four-function calculator, you may use the table of common logarithms to find roots and powers and carry out more complicated computations.

Example 9 Use logarithms to compute $(9.43)^{1.8}$.
Let $N = (9.43)^{1.8}$. Then

$$\begin{aligned}\log N &= \log(9.43)^{1.8}\\ &= 1.8 \log(9.43)\\ &= 1.8(0.9745)\\ &= 1.7541\\ &= 0.7541 + 1\end{aligned}$$

From the tables, log 5.68 = 0.7543 is the closest, so

$$N = 56.8$$ ■

If you wish more accuracy in a computation, as in Example 9, you may want to use more accurate tables or use *interpolation* to determine more decimal places. See Section 5.6 for a discussion of interpolation if you are interested.

Example 10 Compute $N = \dfrac{\sqrt[3]{45}(2.02)^{3.2}}{0.394}$.

$$\begin{aligned} \log N &= \log \sqrt[3]{45} + \log(2.02)^{3.2} - \log(0.394) \\ &= \frac{1}{3}\log 45 + 3.2\log(2.02) - \log(0.394) \\ &= 0.5511 + 0.9773 - (0.5955 - 1) \\ &= 1.9329 \end{aligned}$$

From the tables, log 8.57 = 0.9330, so

$$N = 85.7$$ ■

Exercises 5.4

In Exercises 1–12 find the common logarithm of the given number.

1. 5.68 **2.** 6.58 **3.** 0.000568 **4.** 0.000658

5. 40,700 **6.** 30,900 **7.** 1.01 **8.** 2.02

9. 0.999 **10.** 0.888 **11.** 0.09865 **12.** 0.08675

In Exercises 13–24 find N.

13. $\log N = 0.7024$ **14.** $\log N = 0.6222$

15. $\log N = 2.7024$ **16.** $\log N = 3.6222$

17. $\log N = -0.3778$ **18.** $\log N = -1.2976$

19. $\log N = 2.8$ **20.** $\log N = 4.9777$

21. $\log N = -3.2$ **22.** $\log N = -3.0223$

23. $\log N = -2.4056$ **24.** $\log N = -2.4045$

In Exercises 25–34 use logarithms to perform the computations. Round off to the nearest entry in the logarithm table.

25. $(1.07)^{14}$ **26.** $(1.06)^{12}$ **27.** $\dfrac{\sqrt[7]{97}\sqrt{56}}{\sqrt[8]{10.4}}$

28. $\dfrac{\sqrt[7]{45}\sqrt{57}}{\sqrt[8]{13.2}}$ **29.** $(1.11)^{-6}$ **30.** $(1.12)^{-5}$

31. $\sqrt{\sqrt[5]{16.8}\sqrt[3]{45}}$ **32.** $\sqrt{\sqrt[5]{18.6}\sqrt[3]{24}}$ **33.** $(2.45)^{1.1}(2.22)^{0.04}$

34. $(4.55)^{0.02}(12.4)^{1.2}$

5.5 Logarithmic and Exponential Equations

In solving logarithmic equations and exponential equations we use the basic properties of logarithmic and exponential functions plus the tables or a calculator to find the solutions. There are no set rules for such equations, but in this section we use some examples to illustrate the techniques.

Example 1 *Solving exponential equations*

Solve the equation $2^x = 8$.
We apply $\log_2$ to both members of the equation:

$$\log_2 2^x = \log_2 8 = \log_2 2^3$$
$$x = 3$$ ■

Example 2 Solve the equation $2(3^{x-1}) = 5^x$.
Take the common logarithm of both members of the equation:

$$\log[2(3^{x-1})] = \log 5^x$$
$$\log 2 + \log 3^{x-1} = \log 5^x$$
$$\log 2 + (x - 1)\log 3 = x \log 5$$
$$x = \frac{\log 3 - \log 2}{\log 3 - \log 5} = -\ 0.7937$$ ■

Example 3 *Solving logarithmic equations*

Solve the equation $\log 7 - \log 35 = \log x$.
Using properties of logarithms,

$$\log 7 - \log 35 = \log x$$
$$\log \frac{7}{35} = \log x$$

Since the function $y = \log x$ is 1-1,

$$\frac{7}{35} = x$$ ■

Example 4 Solve the equation $\log_2(x^2 - 6x) = 4$.
Since $\log_2(x^2 - 6x) = 4$, from the definition of $\log_2$,

$$x^2 - 6x = 2^4 = 16$$
$$x^2 - 6x - 16 = 0$$
$$(x - 8)(x + 2) = 0$$
$$x = 8 \qquad x = -2$$

To check:

$$x = 8, \qquad \log_2(8^2 - 6 \cdot 8) = \log_2 16 = \log_2 2^4 = 4$$
$$x = -2, \qquad \log_2[(-2)^2 - 6(-2)] = \log_2 16 = 4$$ ■

Example 5 Solve the equation $\ln x + \ln(x + 1) = \ln 6$.
Using properties of logarithms:

$$\ln x + \ln(x + 1) = \ln 6$$
$$\ln x(x + 1) = \ln 6$$
$$x(x + 1) = 6$$
$$x^2 + x - 6 = 0$$
$$(x - 2)(x + 3) = 0$$
$$x = 2 \qquad x = -3$$

To check:

$x = 2,\quad \ln 2 + \ln 3 = \ln 6$

$x = -3,\quad$ is *not* a solution, because $\ln(-3)$ and $\ln(-2)$ are *not* defined!

■

Note what happened in Example 5 and always check solutions in the *original* equation, not a version of the equation found by using properties of logarithms.

In Section 5.1 it was shown that if an asset with value A is depreciating at a rate r (expressed as a decimal) annually, the value $V(n)$ after n years is given by the formula

$$V(n) = A(1 - r)^n$$

Example 6 If the present value of an asset is \$5,000 and the annual depreciation rate is 25%, how long will it take for the asset to depreciate to a value of less than \$100?

Using the formula $V(n) = A(1 - r)^n$ with $V = 100$, $A = 5{,}000$ and $r = 0.25$, we obtain the equation

$$100 = 5{,}000(0.75)^n$$
$$0.02 = (0.75)^n$$

Taking the common logarithm of both sides of the equation,

$$\log 0.02 = \log(0.75)^n$$
$$= n \log 0.75$$
$$n = \frac{\log 0.02}{\log 0.75} = 13.59$$

At the end of the 14th year the value will have declined to less than \$100.

■

Example 7 The population of a certain country may be estimated by the function $P(t) = 250{,}000{,}000e^{0.025t}$, where t is the number of years after 1970. How long will it take for the population to double?

We wish to find t such that

$$500{,}000{,}000 = 250{,}000{,}000e^{0.025t}$$

or

$$2 = e^{0.025t}$$

Taking the natural logarithm of both sides of the equation we obtain

$$\ln 2 = \ln e^{0.025t} = 0.025t$$

$$t = \frac{\ln 2}{0.025} = \frac{0.693}{0.025} = 27.72$$

The doubling time is 27.72 yr. ■

As discussed in Section 5.2, continuous growth or decay may be described by a function of the form $A(t) = A_0e^{kt}$, where A_0 is the size of the population when $t = 0$ and k is determined by the situation and the units of measure used. Radioactive decay is an example of a continuous process. The decay of carbon-14 at a predictable rate has enabled scientists to date certain artifacts of ancient man and learn more about the beginnings of man. The term **half-life** is used to measure the rate at which a radioactive substance decays. To say that the half-life of carbon-14 is about 5,700 years means that if one begins with a certain amount of carbon-14, at the end of 5,700 years half of the material will have decayed.

Half-life

Example 8 If the half-life of carbon-14 is 5,700 years, write a formula for the amount $A(t)$ of carbon-14 remaining after t years if one begins with A grams of carbon-14. The function desired is of the form

$$A(t) = Ae^{kt}$$

When $t = 5{,}700$, $A(5{,}700) = \frac{1}{2}A$, yielding the equation

$$\frac{1}{2}A = Ae^{5{,}700k}$$

$$\frac{1}{2} = e^{5{,}700k}$$

$$\ln\frac{1}{2} = \ln e^{5{,}700k} = 5{,}700k$$

$$-\ln 2 = 5{,}700k$$

$$k = -\frac{\ln 2}{5{,}700}$$

The desired function is

$$A(t) = Ae^{-\left(\frac{\ln 2}{5{,}700}\right)t}$$

■

Example 9 A piece of wood contains 25% of the carbon-14 that it contained when it was a part of a living tree. How long ago was the tree cut?
Using the function derived in Example 8, we obtain the equation

$$0.25A = Ae^{-\left(\frac{\ln 2}{5{,}700}\right)t}$$

$$0.25 = e^{-\left(\frac{\ln 2}{5{,}700}\right)t}$$

$$\ln 0.25 = -\frac{\ln 2}{5{,}700}t$$

$$-2\ln 2 = -(\ln 2)\left(\frac{1}{5{,}700}\right)t$$

$$t = 5700(2) = 11{,}400$$

The tree was cut approximately 11,400 years ago. ■

Example 10 The population of Transylvania decreased from 100,000 in 1950 to 80,000 in 1960. Find a function that estimates the population $P(t)$, where t is the number of years after 1950. Estimate the population in 2000.

We are looking for a function of the form

$$P(t) = 100{,}000e^{kt} \quad (\text{When } t = 0,\ P = 100{,}000)$$

Knowing $P(10)$, the population in 1960, will allow us to find k.

$$80{,}000 = 100{,}000e^{10k}$$

$$\frac{4}{5} = e^{10k}$$

$$\ln\frac{4}{5} = 10k$$

$$\ln 4 - \ln 5 = 10k$$

$$-0.2231 = 10k$$

$$k = -0.022 \text{ (rounded off)}$$

The desired function is

$$P(t) = 100{,}000e^{-0.022t}$$

When $t = 50$, that is, in 2000,

$$\begin{aligned} P(50) &= 100{,}000e^{(-0.022)(50)} \\ &= 100{,}000(0.3329) \\ &= 33{,}290 \end{aligned}$$

■

Exercises 5.5

In Exercises 1–24 solve the given equation.

1. $4^x = 256$

2. $3^x = 729$

3. $9^{2x-1} = 27$

4. $4^{3x-1} = 128$

5. $2^{x^2-2x} = \frac{1}{2}$

6. $3^{x^2+2x} = \frac{1}{3}$

7. $7^{2x} = 3^{4x+1}$

8. $8^{3x} = 5^{2x+4}$

9. $(1.3)^{-x} = 0.5$

10. $(1.2)^{-x} = 0.25$

11. $\dfrac{(1.04)^x - 1}{0.04} = 6.63$

12. $\dfrac{(1.05)^x - 1}{0.05} = 4.31$

13. $\log 2x = 2$

14. $\log 4x = -1$

15. $\log(x^2 - 2x) = \log 3$

16. $\log(x^2 - 3x) = 1$

17. $\log(7x - 12) = 2 \log x$

18. $\log(5x - 6) = 2 \log x$

19. $\log_2 x + \log_2(x + 2) = 3$

20. $\log_3 x + \log_3(x + 6) = 3$

21. $\log_2 x - \log_2(x - 8) = 1$

22. $\log_3 3x - \log_3(x - 8) = 2$

23. $\frac{1}{2} \log_5(x - 1) - \log_5 x = 0$

24. $\frac{1}{2} \log_7(x - 1) - \log_7 x = 0$

25. An automobile is purchased for \$7,500 and depreciated at an annual rate of 20%. What is the value at the end of 5 yr? When will it reach a salvage value of \$100?

26. A machine is purchased for a factory at a cost of \$8,800. After 5 yr the depreciated value is \$2,088. What is the rate of depreciation?

27. Find the doubling time for a population given by the function $P(t) = 350{,}000e^{0.02t}$.

28. Find the doubling time for an investment of \$4,000 at 9% compounded continuously.

29. When U.S. Savings Bonds were first introduced, an investment of \$18.75 would yield a return of \$25 10 years later. If the interest was compounded annually, what was the annual rate?

30. If the interest on U.S. Savings Bonds had been compounded continuously, what would have been the interest rate?

31. If \$18.75 were invested today at 12% interest compounded semiannually, how much would the investment be worth in 10 yr?

32. A bone is found to contain 40% of the carbon-14 that it contained when it was part of a living organism. How long ago did the organism die?

33. Assuming that the population of the United States was 200,000,000 in the year 1980, what will be the population in the year 2000 if the population is given by the function $P(t) = 200{,}000{,}000e^{(0.011)t}$, where t is measured in years after 1980? This corresponds to an annual growth rate of 1.1%.

34. What will be the population of the United States in the year 2000 if the annual rate is 1.2%; that is, if the function is $P(t) = 200{,}000{,}000e^{(0.012)t}$?

35. If your apartment rent is \$100 per month and the rent increases 3.6% per year (cost of living increase) for 5 yrs, how much rent will you be paying at the end of 5 yrs?

36. Suppose your rent today is \$100 per month, and over the past 3 years the rent had been increasing at the rate of 0.3% per month. What was the rent 3 years ago?

37. The population of a certain country was 450,000 in 1980 and will be 495,000 in 1985. Find a function of the form $P(t) = 450{,}000e^{kt}$ that gives the population as a function of t, the time in years from 1980. What will be the population in the year 1990?

5.6 Interpolation with Logarithms

Linear interpolation

In Section 5.4 we found certain logarithms that were not *exactly* the numbers given in the table. At that time we chose to round off to the nearest number in the table. However, there is a technique called **linear interpolation** which can be used to obtain more accuracy than can be obtained directly from the table.

Suppose we wish to use logarithms to compute

$$N = \frac{12\sqrt[3]{7}\sqrt{3}}{10\sqrt{14}}$$

Then

$$\begin{aligned}\log N &= \log 12 + \frac{1}{3}\log 7 + \frac{1}{2}\log 3 - \frac{1}{2}\log 14 - \log 10\\ &= 1.0792 + \frac{1}{3}(0.8451) + \frac{1}{2}(0.4771) - \frac{1}{2}(1.1461) - 1\\ &= 0.0264\end{aligned}$$

We do not find 0.0264 in the log table, but we do find 0.0253 and 0.0294; in fact,

$$\log 1.06 = 0.0253$$
$$\log 1.07 = 0.0294$$

so N is between 1.06 and 1.07. We will use linear interpolation to approximate N more closely.

We know that the graph of the logarithm function passes through the points (1.06, 0.0253) and (1.07, 0.0294). The idea behind linear interpolation is to replace the logarithm function by a linear function whose graph passes through the same two points and find the approximation to N from the linear function (see Figure 5.5). The point P on the graph of the logarithm function has coordinates (N, 0.0264), but we use the point (N', 0.0264) on the line to give the approximation N' to the value of N. The coordinate N' is found by equating ratios:

$$.01\left\{\begin{array}{l} x\left\{\begin{array}{l}\log 1.06 = 0.0253\\ \log N \;\;\, = 0.0264\end{array}\right\} 0.0011 \\ \quad\;\; \log 1.07 = 0.0294\end{array}\right\} 0.0041$$

We determine x from the proportion

$$\frac{x}{0.01} = \frac{0.0011}{0.0041}$$
$$x = 0.0027$$

which we round off to

$$x = 0.003$$

Our approximation is

$$N = 1.06 + 0.003 = 1.063$$

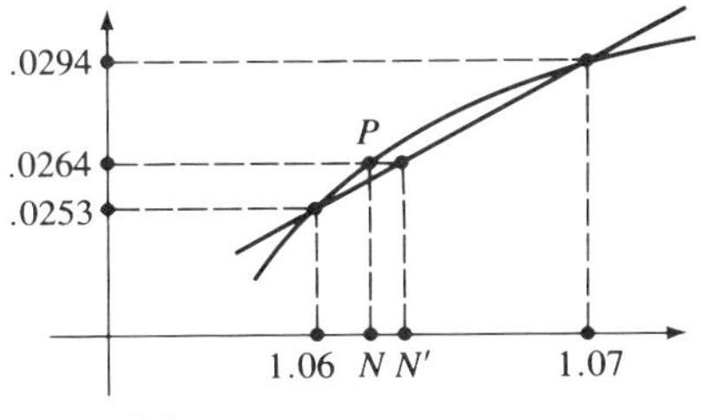

Figure 5.5

Example 1 Use logarithms and interpolation to compute $N = \dfrac{6\sqrt{1.376}}{\sqrt[3]{18}}$.

$$\log N = \log 6 + \frac{1}{2} \log 1.376 - \frac{1}{3} \log 18$$

Since log 1.376 is not in the tables, we must use interpolation to find that number.

$$0.01 \left\{ \begin{array}{l} 0.006 \left\{ \begin{array}{l} \log 1.37 \;\; = 0.1367 \\ \log 1.376 = \quad ? \end{array} \right\} x \\ \qquad\quad \log 1.38 \;\; = 0.1399 \end{array} \right\} 0.0032$$

From the proportion

$$\frac{0.006}{0.01} = \frac{x}{0.0032}$$

we obtain

$$x = 0.0019 \quad (\textit{Rounded off})$$

so

$$\log 1.376 = 0.1367 + 0.0019 = 0.1386$$

Now

$$\begin{aligned} \log N &= \log 6 + \frac{1}{2} \log 1.376 - \frac{1}{3} \log 18 \\ &= 0.7782 + \frac{1}{2}(0.1386) - \frac{1}{3}(1.2553) \\ &- 0.4291 \end{aligned}$$

From the table,

$$\log 2.68 = 0.4281$$
$$\log 2.69 = 0.4298$$

Using interpolation,

$$.01 \left\{ \begin{array}{l} x \left\{ \begin{array}{l} \log 2.68 = 0.4281 \\ \log N \;\;\; = 0.4291 \end{array} \right\} 0.001 \\ \quad \log 2.69 = 0.4298 \end{array} \right\} 0.0017$$

From the proportion

$$\frac{x}{0.01} = \frac{0.001}{0.0017}$$

$$x = 0.006 \quad (\textit{Rounded off})$$

So

$$N = 2.68 + 0.006 = 2.686$$

■

Example 2 Use logarithms and interpolation to compute $\dfrac{\sqrt[4]{16.5}}{\sqrt[3]{12}\,\sqrt[5]{89}}$.

$$\log N = \frac{1}{4}\log 16.5 - \frac{1}{3}\log 12 - \frac{1}{5}\log 89$$
$$= -0.4452$$

Adding and subtracting 1 gives

$$\log N = 0.5548 - 1$$

From the table

$$\log 3.58 = 0.5539$$
$$\log 3.59 = 0.5551$$

Using interpolation,

$$.01\left\{\begin{array}{l} x\left\{\begin{array}{l}\log 3.58 = 0.5539\\ \log N \quad = 0.5548\end{array}\right\}0.0009 \\ \quad\ \log 3.59 = 0.5551\end{array}\right\}0.0012$$

From the proportion,

$$\frac{x}{0.01} = \frac{0.0009}{0.0012}$$
$$x = 0.0075$$

So

$$\log 3.5875 = 0.5548$$

and

$$N = 0.35875$$

■

Exercises 5.6

In Exercises 1–12 find the logarithm for the given number.

1. 5.755 **2.** 7.234 **3.** 0.0007655 **4.** 0.0002214
5. 10,050 **6.** 20,450 **7.** 2.005 **8.** 4.006
9. 0.3752 **10.** 0.1875 **11.** 234.5 **12.** 561.5

In Exercises 13–24 find N.

13. $\log N = 0.7046$ **14.** $\log N = 0.4649$ **15.** $\log N = 0.602$
16. $\log N = 0.7385$ **17.** $\log N = -0.3405$ **18.** $\log N = -0.252$
19. $\log N = 3.8107$ **20.** $\log N = 3.925$ **21.** $\log N = -2.452$
22. $\log N = -2.0062$ **23.** $\log N = 4.7103$ **24.** $\log N = 5.153$

In Exercises 25–30 use logarithms to do the computations.

25. $(1.003)^{-12}$ **26.** $(1.075)^{12}$ **27.** $\sqrt[5]{2.34}\,\sqrt[4]{24.7}$

28. $\sqrt[3]{24}\,\sqrt[7]{4.56}$ **29.** $\sqrt{2+\sqrt{2+\sqrt{2}}}$ **30.** $\sqrt{3+\sqrt{3+\sqrt{3}}}$

Chapter Review

Sketch a portion of the graph.

1. $F(x) = 2^{2x}$ **2.** $G(x) = \frac{1}{2}(2^{-x})$

3. Find the remaining value of an asset worth \$5,000 after it depreciates 4 yr at an annual rate of 10%.

4. If \$5,000 is invested at 8% compounded annually, how much will the investment be worth after 3 yr?

Write each of the following as an integer.

5. $\log 0.001$ **6.** $\log_3 243$ **7.** $\log_2 \frac{1}{64}$

8. $6^{\log_6 5}$ **9.** $36^{\log_6 4}$ **10.** $25^{\log_5 8}$

Use $\log 3 = 0.477$ *and* $\log 2 = 0.301$ *to compute the following.*

11. $\log 6$ **12.** $\log 18$ **13.** $\log \frac{2}{3}$

14. $\log 12$ **15.** $\log \frac{3}{4}$ **16.** $\log \sqrt{2}$

Use logarithms to perform the computations.

17. $\sqrt[5]{2}\,\sqrt[3]{3}$ **18.** $(1.005)^{20}$ **19.** $\dfrac{\sqrt{168}\,\sqrt[4]{8}}{\sqrt[3]{67}}$

Solve the equations.

20. $3^t = 2^{t+1}$ **21.** $2^{3t} = 64$ **22.** $2^{x^2-3x} = 16$

23. $\log 4x = 2$ **24.** $\log_2(6x - x^2) = 3$

25. $\log_2(x - 3) - \log_2(2x) = -3$

26. The half-life of radon is 3.85 days. If a sample of 100 g is left to decay, how much remains at the end of 30.8 days?

27. If interest is compounded annually, at what annual rate will an investment double in 10 years?

28. The population of mainland China was estimated to be 583 million in 1953 and 823 million in 1975. If the same rate of growth continues, what will be the population in the year 2000? Assume the growth is exponential. (Note: you should use your calculator for exercise 28.)

Chapter Test

Sketch a portion of the graph.

1. $F(x) = \frac{3}{5}(2^x)$

2. $F(x) = \frac{1}{9}(3^{-x})$

3. Find the remaining value of an asset initially valued at \$6,000 after it has depreciated for 5 yr at an annual rate of 50%.

4. Find the value of an investment of \$4,000 after 4 yr if it earns interest at 8% compounded semiannually.

Write each of the following as an integer.

5. $\log_2 128$

6. $\log_3 81$

7. $\log_4 \frac{1}{64}$

8. $\log 0.001$

9. $10^{\log 6}$

10. $6^{\log_6 5}$

11. $3^{2\log_3 4}$

12. $9^{2\log_3 7}$

Use $\ln 2 = 0.693$ *and* $\ln 3 = 1.1$ *to compute the following.*

13. $\ln 6$

14. $\ln 12$

15. $\ln \frac{2}{3}$

16. $\ln \sqrt{96}$

17. $\ln 1.5$

18. $\ln \sqrt[5]{36}$

Use logarithms to compute the following.

19. $\sqrt{45}\,\sqrt[5]{17}$

20. $(1.055)^{24}$

21. $(0.56)^{-14}$

Solve the equations.

22. $4^{2x} = \frac{1}{2}$

23. $3^{x^2-1} = 3^{x+1}$

24. $5^x = 7$

25. $\log 4x - \log x = 3$

26. $\log_5(x^2 - 7) = \log_5(6x)$

27. $\log_2(x^2 + 1) - \log_2(5x) = -1$

28. The half-life of polonium is 140 days. If we begin with 100 g of polonium, how much will remain after 280 days?

6 The Trigonometric Functions

China, Texas is located more than 800 miles southeast (more east than south) of El Paso, Texas. A traveler setting out to go from El Paso to China with only these directions would probably get wet in the Gulf of Mexico and never see China. The concept of angle measurement is exactly what is needed. In this chapter we examine angles, angle measurement, and the trigonometric functions.

6.1 Angles and Their Measure

We can take many different approaches to the definition of angle. This is primarily due to the difficulty involved in making the term mathematically precise. We will take the point of view that the concept is intuitively familiar and adopt the following dynamic approach. If desired, "segment" can be replaced with "segment or ray."

Angle Consider a line segment L in the plane. An **angle** is formed if the segment is rotated about one of its endpoints v, as in Figure 6.1.

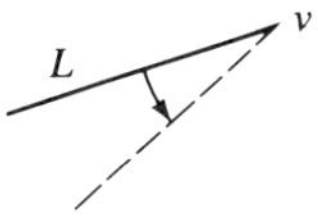

Figure 6.1

The point v is called the **vertex** of the angle, L is the **initial side** of the angle, and L is rotated to the **terminal side** of the angle. If a cartesian coordinate system is introduced and the angle is oriented with vertex at the origin and initial side on the positive x axis, then the angle is said to be in **standard position.** An angle is **positive** if the direction of rotation is counterclockwise and **negative** if the direction of rotation is clockwise. A quadrantal angle is one whose terminal side coincides with one of the coordinate axes. **Coterminal angles** have the same terminal sides.

Standard position

Coterminal angles

Example 1 In the illustration shown,
$\alpha, \beta, \gamma, \delta, \tau$ are in standard position.
α, δ, τ are positive angles.
β, γ are negative angles.
δ, γ, τ are coterminal angles.

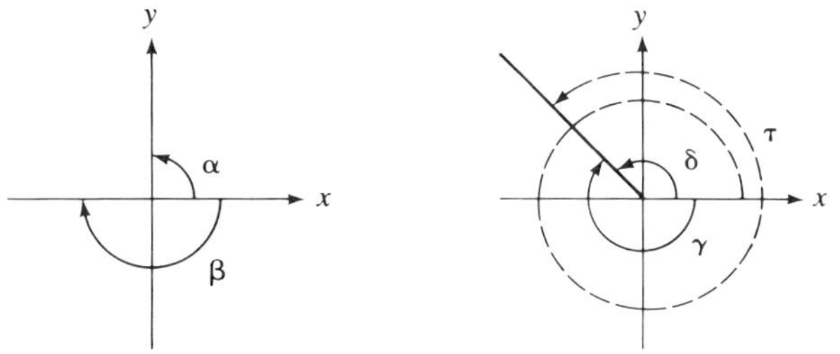

■

There are several standard units of measure for angles. The reader is likely to be most familiar with degree-minute-second measure. If the initial side of the angle is rotated through one complete revolution so that the initial side and terminal side coincide, the measure of 360 degrees is assigned to the angle.

Degree *Minute* *Second*

Thus one **degree** measures $\frac{1}{360}$ of one revolution. To obtain even finer measurement, degrees are divided into 60 equal parts called **minutes,** and minutes are divided into 60 equal parts called **seconds.** The notation for degrees is °, for minutes ′, and for seconds ″. For example, 37°23′31.3″ is read "37 degrees, 23 minutes, and 31.3 seconds."

Acute *Right* *Obtuse*

A positive angle is **acute** if its measure is less than 90°, **right** if its measure equals 90°, and **obtuse** if its measure is greater than 90° but less than 180°. Note that 90° measures an angle obtained by $\frac{1}{4}$ of a revolution, which means that the coordinate axes intersect at right angles. We identify an angle with its measure, so that if the measure of α is 30°, we write $\alpha = 30°$.

Example 2 In the illustration shown,

a. $\alpha = 90°$, $\beta = 45°$, $\gamma = -135°$, $\delta = 360°$

b. $\alpha = 30°$, $\beta = 60°$, $\gamma = -300°$

c. $\alpha = 150°$, $\beta = 225°$, $\gamma = -180°$

d. $\alpha = 750°$

e. $\alpha = -1{,}200°$

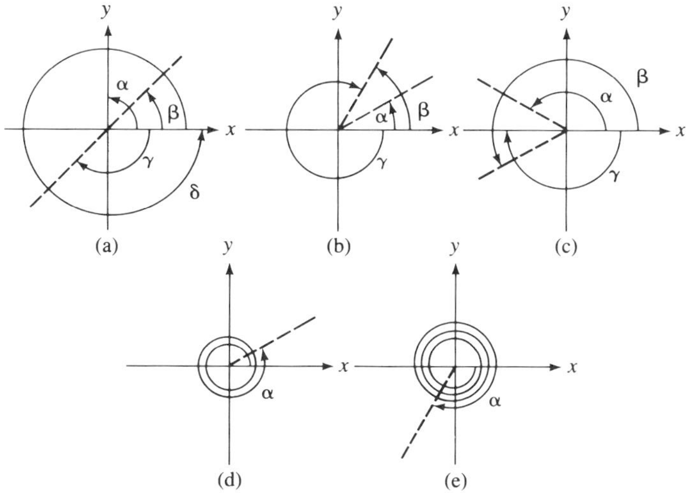

■

The radian is another widely used unit of measure for angles. Recall that the circumference c of a circle is 2π times the radius of the circle. In particular, on the unit circle U, $c = 2\pi$. If an angle α is in standard position with initial side of unit length, then the **radian** measure of the angle is defined to be the length of the arc subtended on U by the angle (see Figure 6.2).

Radian

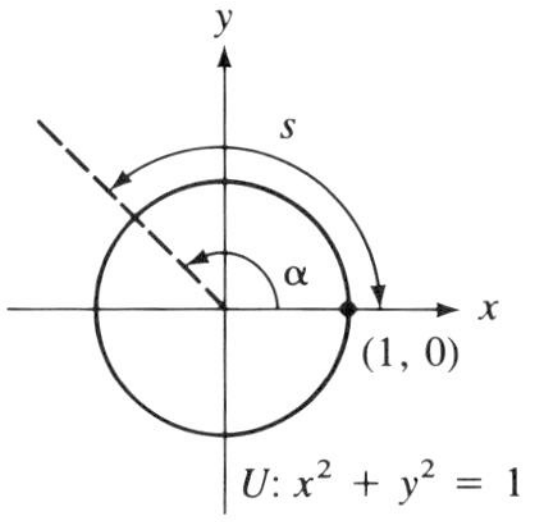

Figure 6.2 The arc length s is equal to the radian measure of α.

Thus the measure of 2π radians is assigned to 1 complete revolution and 1 radian measures a fraction $1/2\pi$ of 1 revolution. Decimal representation is used to obtain finer measurement, so that, for example, we might speak of an angle of measure 10.0178 radians. It is convenient to omit a notation for radian measure, so that if the measure of an angle is given without specifying the unit, it is understood that the unit is the radian. Notice that $\frac{1}{4}$ of a complete revolution forms an angle of measure $\pi/2$ (as well as 90°).

Example 3 In the illustration in Example 2,

a. $\alpha = \frac{\pi}{2}, \beta = \frac{\pi}{4}, \gamma = -\frac{3\pi}{4}, \delta = 2\pi$

b. $\alpha = \frac{\pi}{6}, \beta = \frac{\pi}{3}, \gamma = -\frac{5\pi}{3}$

c. $\alpha = \frac{5\pi}{6}, \beta = \frac{5\pi}{4}, \gamma = -\pi$

d. $\alpha = \frac{25\pi}{6}$

e. $\alpha = -\frac{20\pi}{3}$ ■

Measure conversion

It is not difficult to convert between radian and degree measure. Since 180° and π radians both measure half of 1 revolution, it seems easiest to use the relation $180° = \pi$ radians. Thus to convert from degrees to radians, we multiply by (π radians)/(180 degrees), and to convert from radians to degrees we multiply by (180 degrees)/(π radians). In order to remember this, notice in Example 4 that the units cancel. We point out that

$$1 \text{ radian} = (1 \text{ radian})\left(\frac{180 \text{ degrees}}{\pi \text{ radians}}\right) = \left(\frac{180}{\pi}\right)^{\circ} \simeq 57.3° = 57°18'.$$

Example 4

a. $30° = (30 \text{ \sout{degrees}})\left(\dfrac{\pi \text{ radians}}{180 \text{ \sout{degrees}}}\right) = \dfrac{30\pi}{180} \text{ radians} = \dfrac{\pi}{6}$

b. $\dfrac{5\pi}{12} \text{ radians} = \left(\dfrac{5\pi}{12} \text{ \sout{radians}}\right)\left(\dfrac{180 \text{ degrees}}{\pi \text{ \sout{radians}}}\right) = 75°$

c. $5 = (5)\left(\dfrac{180}{\pi}\right) \text{ degrees} \simeq 286.48° = 286°28.8' = 286°28'48''$

d. $-315° = -(315)\left(\dfrac{\pi}{180}\right) \text{ radians} \simeq -5.498$

e. $265°42'28'' \simeq 265°42.47' \simeq 265° + 0.71° = 265.71°$

$$= (265.71)\left(\frac{\pi}{180}\right) \text{ radians} \simeq 4.64$$

For part **(c)** we convert $(0.48)°$ to minutes by noting that

$$(0.48 \text{ \sout{degrees}})\left(\frac{60 \text{ minutes}}{\text{\sout{degree}}}\right) = [(0.48)(60)]' = 28.8'$$

Similarly,

$$(0.8 \text{ \sout{minutes}})\left(\frac{60 \text{ seconds}}{\text{\sout{minute}}}\right) = 48''$$

If minutes and seconds are involved in the expression of an angle in degree measure, we convert to decimal form before converting to radian measure. Thus for part **(e)** we convert $28''$ to minutes by noting that

$$(28 \text{ \sout{seconds}})\left(\frac{\text{minute}}{60 \text{ \sout{seconds}}}\right) = \frac{28}{60} \text{ minutes} \simeq 0.47'$$

Similarly,

$$(42.47 \text{ \sout{minutes}})\left(\frac{\text{degree}}{60 \text{ \sout{minutes}}}\right) = \left(\frac{42.47}{60}\right)^{\circ} \simeq 0.71°$$ ■

In the following sections it will be helpful to know the radian-degree equivalents for certain special angles. Remember especially that

$$30° = \frac{\pi}{6} \text{ radians}$$

$$45° = \frac{\pi}{4} \text{ radians}$$

$$60° = \frac{\pi}{3} \text{ radians}$$

$$90° = \frac{\pi}{2} \text{ radians}$$

From the definition of radian measure of an angle θ, we know that arc length s subtended by θ on a circle of radius 1 is given by $s = \theta$.

WARNING
Be sure to express θ in radian measure

Consider Figure 6.3. Since the ratio of arc lengths is equal to the ratio of the radii ($\theta/s = 1/r$), we see that the arc length s on a circle of radius r is given by $s = r\theta$ (θ in radian measure). The area of the shaded circular sector determined by the angle θ is given by $A = \frac{1}{2}r^2\theta$ (θ in radian measure).

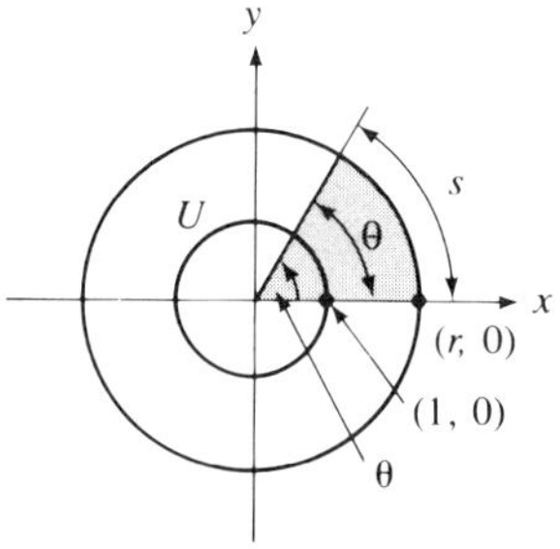

Figure 6.3

Example 5 On a circle of radius 5 meters find the length of arc and the area of the circular sector subtended by an angle of measure 150°.

In radian measure, $150° = (150)(2\pi/180) = 5\pi/6$.

Thus

$$s = r\theta = (5)\left(\frac{5\pi}{6}\right) = \frac{25\pi}{6} \simeq 13.1 \text{ meters}$$

Also

$$A = \frac{1}{2}r^2\theta = \frac{1}{2}(5)^2\left(\frac{5\pi}{6}\right) = \frac{125\pi}{12} \simeq 32.7 \text{ square meters}$$ ■

Example 6 An angle θ subtends an arc of 1 meter on a circle with radius 75 centimeters. What is the measure of θ?

Since $s = r\theta$, the radian measure of θ is $\theta = s/r = \frac{100}{75} = \frac{4}{3}$.
In degree measure,

$$\theta = \left(\frac{4}{3}\right)\left(\frac{180}{\pi}\right) = \frac{240}{\pi} \simeq 76.4°$$ ■

mil

We conclude this section with a note that other units of measure for angles have desirable properties for special applications. One of these is the **mil** used by military artillery gunners. One mil is $\frac{1}{6{,}400}$ of a revolution. The mil is convenient because 1 mil subtends an arc of approximately 1 yard on a circle of radius 1,000 yards.

Exercises 6.1

In Exercises 1–14 give an approximate sketch for the angle in standard position.

1. $210°$ **2.** $240°$ **3.** $-330°$ **4.** $-300°$

5. $-840°$ **6.** $-930°$ **7.** $\frac{5\pi}{6}$ **8.** $\frac{11\pi}{6}$

9. $-\frac{3\pi}{4}$ **10.** $-\frac{7\pi}{4}$ **11.** -11π **12.** $-\frac{31\pi}{3}$

13. $\frac{9\pi}{2}$ **14.** $-\frac{11\pi}{2}$

In Exercises 15–24 indicate those pairs of angles that are coterminal and list the adjectives that apply to each angle: positive, negative, acute, right, quadrantal, and obtuse.

15. $720°, 0°, 7\pi, \frac{5\pi}{2}, \frac{13\pi}{6}$

16. $-180°, \frac{3\pi}{4}, 90°, -270°, 3\pi$

17. $\frac{7\pi}{6}, \frac{13\pi}{6}, -\frac{5\pi}{6}, \frac{\pi}{6}, -\frac{20\pi}{6}$

18. $\frac{4\pi}{3}, -\frac{\pi}{3}, -\pi, -\frac{2\pi}{3}, \frac{5\pi}{3}$

19. $-90°, 450°, -450°, \frac{3\pi}{2}, 4\pi$

20. $-\frac{5\pi}{4}, 45°, \frac{3\pi}{4}, -405°, \frac{9\pi}{4}$

21. $-12°, 168°, \frac{14\pi}{15}, 348°, -\frac{14\pi}{15}$

22. $\frac{2\pi}{5}, -288°, -\frac{3\pi}{5}, \frac{12\pi}{5}, \frac{8\pi}{5}$

23. $\frac{3\pi}{10}, 18°, -306°, -\frac{17\pi}{10}, -54°$

24. $\frac{19\pi}{2}, -90°, \frac{5\pi}{2}, -\frac{3\pi}{2}, \frac{13\pi}{2}$

In Exercises 25–36 convert the degree measure to radian measure.

25. $45°$ **26.** $135°$ **27.** $210°$ **28.** $150°$ **29.** $-120°$

30. $-15°$ **31.** $1{,}800°$ **32.** $-2{,}520°$ **33.** $-150°$ **34.** $18°$

35. $300°$ **36.** $72°$

In Exercises 37–48 convert the radian measure to degree measure.

37. $\frac{7\pi}{6}$ **38.** $-\pi$ **39.** $\frac{8\pi}{3}$ **40.** $\frac{11\pi}{6}$ **41.** $-\frac{7\pi}{4}$

42. $-\frac{7\pi}{3}$ **43.** $\frac{2}{5}$ **44.** $\frac{4}{7}$ **45.** 3 **46.** 4

47. 9 **48.** 10

In Exercises 49–52 convert the angle measure from degrees to degrees, minutes, seconds.

49. $35.125°$ **50.** $160.755°$ **51.** $250.41°$ **52.** $325.87°$

In Exercises 53–56 convert the angle measure from degrees, minutes, seconds to radians.

53. 43°32′28″ **54.** 110°42′25″ **55.** 215°15′45″ **56.** 305°24′38″

In Exercises 57–60 convert the radian measure to degrees, minutes, seconds.

57. 3 **58.** 4 **59.** 9 **60.** 10

61. On a circle of radius 40 cm, what is the length of the arc and the area of the circular sector subtended by an angle of measure $\frac{1}{4}$ radian?

62. If an angle θ subtends an arc of 70 cm on a circle of radius 20 cm, what is the measure of θ and the area of the circular sector subtended by θ?

63. The minute hand of a watch is 10 cm long. How far does the tip of the hand move in 30 min?

64. A 10-in.-diameter circular saw has an angular velocity of 3,450 revolutions per minute. How far does the tip of a sawtooth travel in 30 sec?

65. Show that a mil subtends an arc of approximately 1 yard on a circle of radius 1,000 yards.

66. At 3,500 yards a gunner can subtend an angle of 10 mils. What is the length of the arc?

67. A 26-in. bicycle wheel travels 50 yards. Through what angle has the wheel turned?

68. Obtain a good map of Texas and describe the direction to travel from El Paso to China, Texas.

69. The velocity v of a point moving on a circle is related to the radius r and the angular velocity ω by the formula $v = r\omega$, where ω is measured in radians per unit time. Calculate the velocity of a point on the tip of a 10-in. saw rotating at 3,450 revolutions per minute.

For Your Calculator

Some calculators will accept angle measure in degrees, minutes, and seconds and then convert to degrees in decimal form and vice versa by means of functions stored in the machine. Also, most scientific calculators will convert from degree measure to radian measure and vice versa. You should read the manual that comes with your calculator to determine the methods to be used. The examples below illustrate for one model.

Press	Display	Meaning
45.5	45.5	45.5°
→DMS	45.30	45°30′
45.125	45.125	45.125°
→DMS	45.0730	45°07′30″
45.15	45.15	45°15′
←DMS	45.25	45.25°

In Exercises 70–75 use your calculator to convert the indicated measure to the requested mode.

70. 36°45′ to degrees decimal

71. 28°3′45″ to degrees decimal

72. 36.3333° to degrees, minutes, seconds

73. 178.875° to degrees, minutes, seconds

74. 4.567 radians to degrees

75. 47.85° to radians

76. At 5.23 mi a gunner can subtend an angle of 37.7 mils. What is the length of the arc?

77. A 27-in. bicycle wheel travels 15 m. What is the degree measure of the angle through which the wheel has turned?

6.2 The Trigonometric Functions

The word *trigonometry* has both Latin and Greek origins and loosely translates into *triangle measure*. Historically, the study of the subject was directed to the solution of triangles, and the trigonometric functions were developed for this purpose. As mathematical knowledge has progressed these functions have become important in numerous applications, many of which have no reference to angle or triangle. In this section we begin the development of these functions with the traditional approach. However, we quickly (Section 6.4) extend our analysis to trigonometric functions of real numbers.

Consider an angle θ in standard position and choose two points (x_1, y_1) and (x_2, y_2) on the terminal side of θ. For convenience, an illustration for an acute angle θ is provided in Figure 6.4.

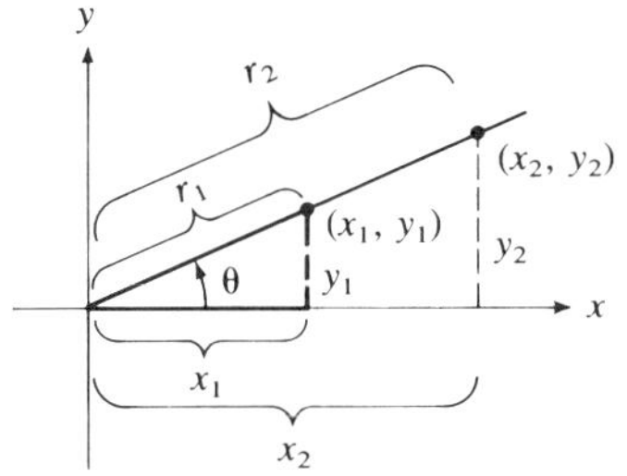

Figure 6.4

The two triangles shown are similar, and consequently $x_1/r_1 = x_2/r_2$ and $y_1/r_1 = y_2/r_2$. We see then that these ratios are the same for every point on the terminal side of the angle, and hence depend only on the angle. These observations permit us to make the following definitions.

Let θ be an angle in standard position and $(x, y) \neq (0, 0)$ be any point on the terminal side of θ, as in Figure 6.5. The distance r from the origin to (x, y) is $r = \sqrt{x^2 + y^2}$.

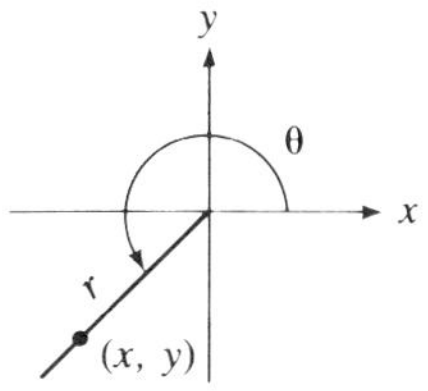

Figure 6.5

We define the six trigonometric functions of θ as follows.

The trigonometric functions

$$\text{sine } \theta = \frac{y}{r}$$

$$\text{cosine } \theta = \frac{x}{r}$$

$$\text{tangent } \theta = \frac{y}{x} \quad \text{if} \quad x \neq 0$$

$$\text{cotangent } \theta = \frac{x}{y} \quad \text{if} \quad y \neq 0$$

$$\text{secant } \theta = \frac{r}{x} \quad \text{if} \quad x \neq 0$$

$$\text{cosecant } \theta = \frac{r}{y} \quad \text{if} \quad y \neq 0$$

sin θ
cos θ
tan θ
cot θ
sec θ
csc θ

Due to the frequency of use of the trigonometric functions it is customary to assign each a three-letter abbreviation: sine θ = **sin θ,** cosine θ = **cos θ,** tangent θ = **tan θ,** cotangent θ = **cot θ,** secant θ = **sec θ,** and cosecant θ = **csc θ.** Note that these functions are not only independent of choice of point on the terminal side of θ, but also of choice of measure for the angle, be it degrees, radians, or what have you.

In practice it is easier to remember how the last four of the trigonometric functions are related to the first two than to remember the definitions of these functions in terms of x, y, and r. We display these relationships.

$$\tan \theta = \frac{\sin \theta}{\cos \theta}$$

$$\cot \theta = \frac{\cos \theta}{\sin \theta}$$

$$\sec \theta = \frac{1}{\cos \theta}$$

$$\csc \theta = \frac{1}{\sin \theta}$$

Example 1 In the illustration shown,

a. $\sin \alpha = \frac{2}{2} = 1 \qquad \cos \alpha = \frac{0}{2} = 0$

$\csc \alpha = \frac{1}{\sin \alpha} = 1 \qquad \cot \alpha = \frac{\cos \alpha}{\sin \alpha} = 0$

$\tan \alpha = \frac{\sin \alpha}{\cos \alpha}$ and $\sec \alpha = \frac{1}{\cos \alpha}$ are not defined.

b. $\sin \beta = \dfrac{0}{4} = 0 \qquad \cos \beta = \dfrac{-4}{4} = -1$

$$\tan \beta = \frac{\sin \beta}{\cos \beta} = \frac{0}{-1} = 0 \qquad \sec \beta = \frac{1}{\cos \beta} = -1$$

$\cot \beta$ and $\csc \beta$ are not defined.

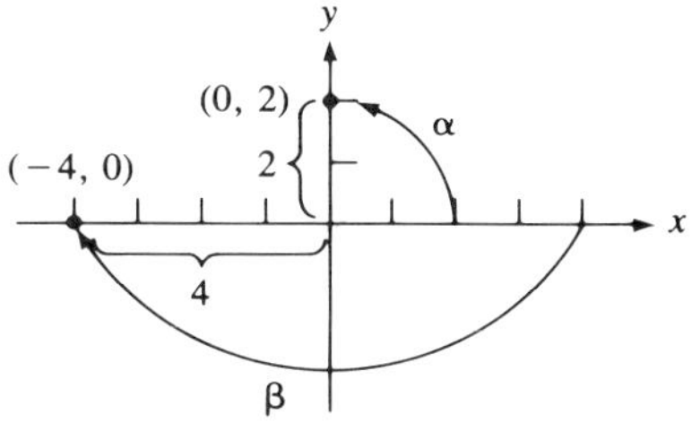

■

Example 1 illustrates that not all of the trigonometric functions are defined for all angles. As usual, we follow the convention that the domain of each function is as large as possible. If $P(x, y)$ is on the terminal side of θ, then $\tan \theta$ and $\sec \theta$ are undefined if $x = 0$, i.e., if P is on the y axis. Also, $\cot \theta$ and $\csc \theta$ are undefined if $y = 0$, i.e., if P is on the x axis. Since $r \neq 0$, $\sin \theta$ and $\cos \theta$ are defined for every angle θ.

Example 2 The line $y = 3x$ determines angles α and β as shown. Find the values of all six trigonometric functions of α and β.

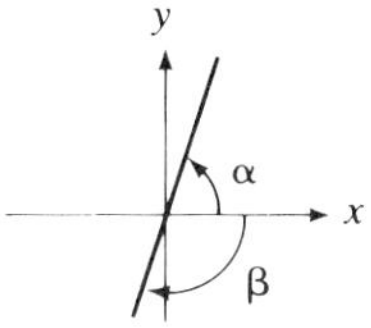

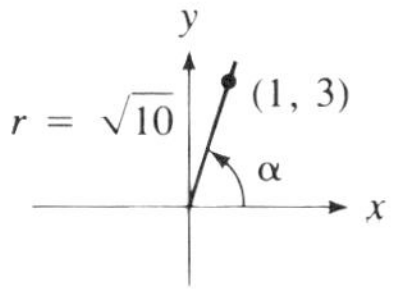

The point (1, 3) is on the terminal side of α. Thus,

$$\sin \alpha = \frac{3}{\sqrt{10}} \qquad \cos \alpha = \frac{1}{\sqrt{10}}$$

$$\csc \alpha = \frac{1}{\sin \alpha} = \frac{\sqrt{10}}{3} \qquad \sec \alpha = \frac{1}{\cos \alpha} = \sqrt{10}$$

$$\tan \alpha = \frac{\sin \alpha}{\cos \alpha} = 3 \qquad \cot \alpha = \frac{\cos \alpha}{\sin \alpha} = \frac{1}{3}$$

The point $(-1, -3)$ is on the terminal side of β. Thus,

$$\sin\beta = \frac{-3}{\sqrt{10}} \qquad \cos\beta = \frac{-1}{\sqrt{10}}$$

$$\csc\beta = -\frac{\sqrt{10}}{3} \qquad \sec\beta = -\sqrt{10}$$

$$\tan\beta = 3 \qquad \cot\beta = \frac{1}{3}$$

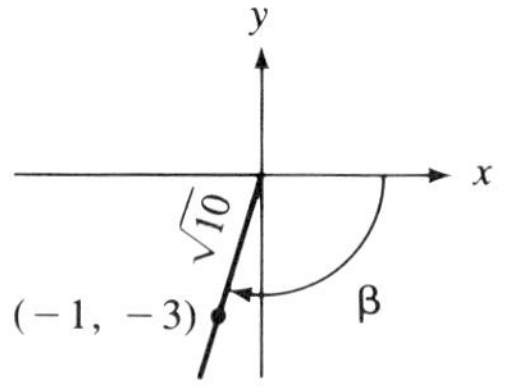

Notice in Example 2 that $\tan\alpha = 3 = \tan\beta$ and 3 is the slope of the line $y = 3x$. In general, if the line $y = mx$ makes an angle α with the positive x axis, then $\tan\alpha = m$, the slope of the line.

Example 3 Let θ be an angle in standard position such that $\sin\theta = \frac{1}{2}$ and $\cos\theta = \frac{\sqrt{3}}{2}$. Find x and y if (x, y) is a point on the terminal side of θ such that $r = 5$.

Since $\sin\theta = y/r$ and $\cos\theta = x/r$ we see that

$$y = r\sin\theta = 5\left(\frac{1}{2}\right) = \frac{5}{2} \quad \text{and} \quad x = r\cos\theta = 5\left(\frac{\sqrt{3}}{2}\right) = \frac{5\sqrt{3}}{2}$$

To analyze range values for the trigonometric functions, let θ be an arbitrary angle in standard position and (x, y) be on the terminal side of θ. Then $|x| \le r$ and $|y| \le r$, hence $-1 \le \cos\theta \le 1$, $-1 \le \sin\theta \le 1$, $\sec\theta \ge 1$ or $\sec\theta \le -1$, and $\csc\theta \ge 1$ or $\csc\theta \le -1$. It is not as clear, but the range for $\tan\theta$ and $\cot\theta$ consists of all real numbers.

There are many relationships among the trigonometric functions other than those mentioned above. Two of these are immediate and merit mention at this time.

$$\tan\theta = \frac{1}{\cot\theta}$$

$$\sin^2\theta + \cos^2\theta = 1$$

The first of these is simply the statement that $y/x = 1/(x/y)$. For the second we note that $(\sin\theta)^2 + (\cos\theta)^2 = 1$ since

$$\left(\frac{y}{r}\right)^2 + \left(\frac{x}{r}\right)^2 = \frac{y^2 + x^2}{r^2} = \frac{r^2}{r^2} = 1$$

$\sin^k \theta$ To avoid excessive grouping of symbols it is customary to write powers of trigonometric functions in the form $\mathbf{sin^k\ \theta}$ rather than $(\sin\theta)^k$. There is one exception, when $k = -1$; this is discussed in Section 6.8. Thus

$$1 = (\sin\theta)^2 + (\cos\theta)^2 = \sin^2\theta + \cos^2\theta$$

Example 4 Let θ be an acute angle such that $\sin\theta = \frac{1}{2}$. Find $\cos\theta$, $\tan\theta$, and $\cot\theta$.

Since $\cos^2\theta = 1 - \sin^2\theta$, we see that $\cos^2\theta = 1 - \left(\frac{1}{2}\right)^2 = \frac{3}{4}$. Thus $\cos\theta = \pm\sqrt{3}/2$. The cosine function is positive for acute angles since $x > 0$, so that $\cos\theta = \frac{1}{2}\sqrt{3}$. Then

$$\tan\theta = \frac{\sin\theta}{\cos\theta} = \frac{1}{2} \div \frac{\sqrt{3}}{2} = \frac{1}{\sqrt{3}} \quad \text{and} \quad \cot\theta = \frac{1}{\tan\theta} = \sqrt{3}$$ ■

Notice that if α and β are coterminal angles, then $\sin\alpha = \sin\beta$, $\cos\alpha = \cos\beta$, and so on. Thus, for example $\sin\theta = \sin(\theta + 2k\pi)$, $k \in Z$. A similar result holds for the remaining trigonometric functions. This means that the trigonometric functions are examples of periodic functions. A function f is **periodic** if there is a real number p such that $f(x + p) = f(x)$ for all x in the domain of f, $p \neq 0$. For this reason, to determine values of the trigonometric functions for arbitrary angles it is sufficient to know the values of the trigonometric functions for those angles θ such that $0 \leq \theta < 2\pi$. The following example illustrates the mechanics. We shall see later that for the tangent and cotangent functions, the interval $0 \leq \theta < \pi$ is sufficient and that these two functions have period π.

Periodic

Example 5 Given that $\cos 60° = \frac{1}{2}$, find $\cos 420°$, $\cos(-300°)$, and $\cos\left(-\frac{23\pi}{3}\right)$.

$$\cos 420° = \cos(60° + 360°) = \cos 60° = \frac{1}{2}$$

$$\cos(-300°) = \cos(60° - 360°) = \cos 60° = \frac{1}{2}$$

$$\cos\left(-\frac{23\pi}{3}\right) = \cos\left(\frac{\pi}{3} - 8\pi\right) = \cos\frac{\pi}{3} = \frac{1}{2}$$ ■

In general it is not possible to determine exact values for the trigonometric functions, because the values depend on the rectangular coordinates of a point on the terminal side of the angle. It is possible to obtain exact values for the quadrantal angles and some special angles. Using advanced techniques, tables have been developed to give approximate values for trigonometric functions of angles. Many handheld calculators are also designed to give approximate values. Special angles and general techniques for determining approximate trigonometric function values are developed in the following sections. For now, we offer Table 6.1, which displays the trigonometric function values for quadrantal

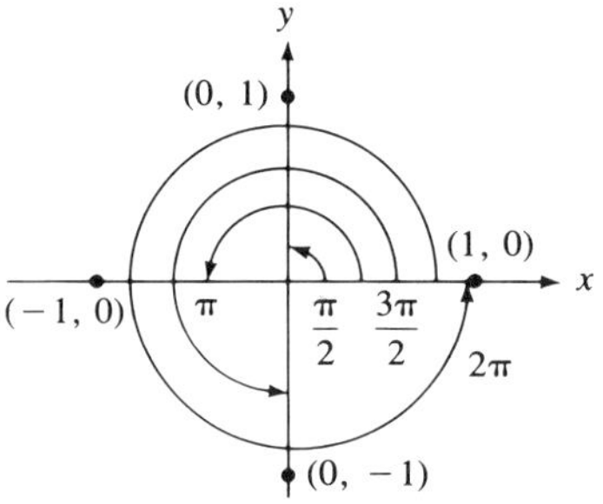

Figure 6.6

θ deg	0°	90°	180°	270°	360°
θ rad	0	$\frac{\pi}{2}$	π	$\frac{3\pi}{2}$	2π
$\sin\theta$	0	1	0	−1	0
$\cos\theta$	1	0	−1	0	1
$\tan\theta$	0	*	0	*	0
$\cot\theta$	*	0	*	0	*
$\sec\theta$	1	*	−1	*	1
$\csc\theta$	*	1	*	−1	*

Table 6.1

angles θ such that $0° \leq \theta \leq 360°$. An asterisk in the table indicates that the particular function is not defined for that angle. For convenience, we select points on the terminal side of θ such that $r = 1$. See Figure 6.6 and Table 6.1.

Example 6 Use Table 6.1 to find $\cos\left(-\frac{\pi}{2}\right)$, $\sec(-270°)$, $\cot\frac{5\pi}{2}$, and $\sec 1{,}260°$.

The angles $-\pi/2$ and $3\pi/2$ are coterminal and hence

$$\cos\left(-\frac{\pi}{2}\right) = \cos\frac{3\pi}{2} = 0$$

Similarly,

$$\sec(-270°) = \sec 90° \text{ is undefined}$$

$$\cot\frac{5\pi}{2} = \cot\left(\frac{\pi}{2} + 2\pi\right) = \cot\frac{\pi}{2} = 0$$

$$\sec 1{,}260° = \sec(180° + 3\cdot 360°) = \sec 180° = -1$$

■

We conclude this section with observations concerning the algebraic signs of the trigonometric functions relative to the quadrant in which the terminal side of the angle is located. It is important that one be able immediately to determine the sign of each of the six trigonometric functions for each quadrant. Since $r > 0$, $\sin\theta > 0$ if $y > 0$. This is true in quadrants I and II. Cos $\theta > 0$ if $x > 0$, and this is true in quadrants I and IV. Using these results and the definitions of the trigonometric functions, we obtain the diagram shown in Figure 6.7.

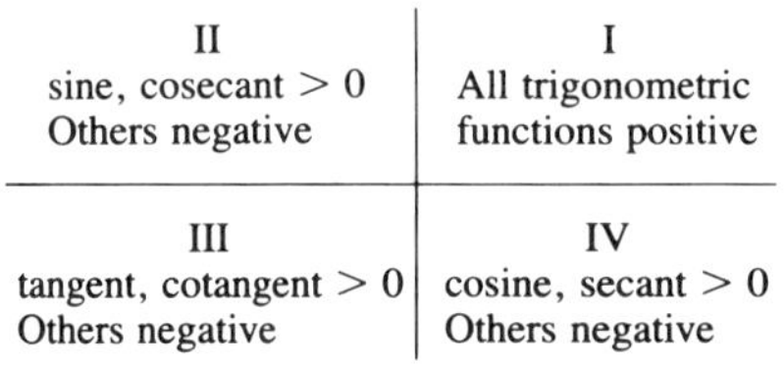

Figure 6.7

Example 7 **a.** $\sin 4 < 0$ and $\cos 4 < 0$ because the terminal side of 4 is in quadrant III.
b. If $\sin \theta < 0$ and $\cos \theta > 0$, then the terminal side of θ must fall in quadrant IV.
c. If $\tan \theta > 0$ and $0° < \theta < 360°$, then $0° < \theta < 90°$ or $180° < \theta < 270°$.
d. If $\theta = -\dfrac{11\pi}{6}$, then all trigonometric function values at θ are positive, because the terminal side of θ is located in quadrant I.
e. $\tan\left(-\dfrac{2\pi}{3}\right) > 0$ and $\sin\left(-\dfrac{2\pi}{3}\right) < 0$, because the terminal side of $-\dfrac{2\pi}{3}$ is located in quadrant III. ■

Exercises 6.2

In Exercises 1–12, given the point (x, y) on the terminal side of θ, find the values of the trigonometric functions at θ.

1. $(3, 4)$ **2.** $(4, 3)$ **3.** $(-2, -5)$ **4.** $\left(-\frac{8}{5}, \frac{2}{3}\right)$

5. $\left(-\frac{1}{2}, \frac{1}{3}\right)$ **6.** $(-4, -1)$ **7.** $\left(\frac{1}{2}\sqrt{3}, -\frac{1}{2}\right)$ **8.** $(-2, 0)$

9. $(0, 3)$ **10.** $\left(\frac{1}{2}, -\frac{1}{2}\sqrt{3}\right)$ **11.** $(3, -3)$ **12.** $(-2, 2)$

In Exercises 13–20, if the terminal side of θ is in the given quadrant and lies on the given line, find the values of the trigonometric functions at θ.

13. I; $y = x$ **14.** IV; $y = -x$ **15.** II; $y = -3x$

16. I; $y = \frac{1}{2}x$ **17.** III; $y = \sqrt{3}\,x$ **18.** III; $y = \sqrt{5}\,x$

19. IV; $y = -\frac{2}{3}x$ **20.** II; $y = -\frac{4}{5}x$

In Exercises 21–26 if the terminal side of θ is in the given quadrant and is parallel to the given line, find the values of the trigonometric functions at θ.

21. II; $3x + 2y = 5$ **22.** III; $4y - 3x = 5$

23. III; $x = \frac{1}{2}y + 7$ **24.** II; $x = -7y + 21$

25. IV; $3y + 6x + 2 = 0$ **26.** IV; $3x + 5y + 1 = 0$

In Exercises 27–34 find (x, y), given the two trigonometric function values and the value of r.

27. $\sin \theta = 1$; $\cos \theta = 0$; $r = 5$ **28.** $\cos \theta = -1$; $\sin \theta = 0$; $r = 7$

29. $\cos \theta = -\frac{1}{2}$; $\tan \theta = \sqrt{3}$; $r = 3$ **30.** $\cos \theta = -\frac{\sqrt{3}}{2}$; $\tan \theta = -\frac{1}{\sqrt{3}}$; $r = 2$

31. $\tan \theta = -1$; $\sin \theta = \frac{1}{\sqrt{2}}$; $r = 4$ **32.** $\sin \theta = -\frac{1}{\sqrt{2}}$; $\tan \theta = -1$; $r = 1$

33. $\cot \theta = \frac{12}{5}$; $\cos \theta = -\frac{12}{13}$; $r = 26$ **34.** $\cot \theta = -\frac{3}{4}$; $\cos \theta = -\frac{3}{5}$; $r = 3$

In Exercises 35–44, if the terminal side of θ is in the given quadrant and the given trigonometric function has the specified value at θ, find the values of the other trigonometric functions at θ.

35. II; $\cos\theta = -\dfrac{24}{25}$

36. I; $\sin\theta = \dfrac{7}{25}$

37. III; $\sec\theta = -\dfrac{13}{5}$

38. IV; $\csc\theta = -\dfrac{4}{\sqrt{15}}$

39. I; $\csc\theta = \dfrac{\sqrt{13}}{2}$

40. IV; $\sec\theta = \dfrac{13}{12}$

41. IV; $\sin\theta = -\dfrac{\sqrt{6}}{3}$

42. IV; $\cos\theta = \dfrac{2}{\sqrt{13}}$

43. III; $\csc\theta = -4$

44. II; $\sec\theta = -\dfrac{3}{\sqrt{3}}$

In Exercises 45–54 use periodicity to find the values of the trigonometric functions at the specified angle, given the following trigonometric values.

$\sin 30° = \dfrac{1}{2}$ $\quad \sin 45° = \dfrac{1}{\sqrt{2}}$ $\quad \sin 60° = \dfrac{\sqrt{3}}{2}$

$\cos\dfrac{2\pi}{3} = -\dfrac{1}{2}$ $\quad \cos\dfrac{3\pi}{4} = -\dfrac{1}{\sqrt{2}}$ $\quad \cos\dfrac{5\pi}{6} = -\dfrac{\sqrt{3}}{2}$

$\tan 300° = -\sqrt{3}$ $\quad \tan\dfrac{11\pi}{6} = -\dfrac{1}{\sqrt{3}}$ $\quad \tan\dfrac{7\pi}{4} = -1$

45. $\sin 390°$

46. $\sin(-690°)$

47. $\cos\dfrac{14\pi}{3}$

48. $\tan\dfrac{23\pi}{6}$

49. $\cos\dfrac{17\pi}{6}$

50. $\sin 405°$

51. $\sin(-675°)$

52. $\cos\dfrac{19\pi}{4}$

53. $\tan\left(-\dfrac{9\pi}{4}\right)$

54. $\sin 1{,}140°$

In Exercises 55–64 use Table 6.1 and periodicity to find the values of the trigonometric functions at the specified angle.

55. $\sec 7\pi$

56. $\csc\dfrac{7\pi}{2}$

57. $\tan 20\pi$

58. $\cot\dfrac{15\pi}{2}$

59. $\cos(-11\pi)$

60. $\sin\left(-\dfrac{5\pi}{2}\right)$

61. $\cot 630°$

62. $\cos 15\pi$

63. $\csc(-990°)$

64. $\sec(-5\pi)$

In Exercises 65–74, given the two trigonometric function values at θ, find the values of the remaining trigonometric functions at θ.

65. $\sin\theta = 0$; $\cos\theta = -1$

66. $\cos\theta = \dfrac{4}{5}$; $\csc\theta = -\dfrac{5}{3}$

67. $\cos\theta = -\frac{3}{5}$; $\csc\theta = -\frac{5}{4}$

68. $\sec\theta = \frac{25}{24}$; $\tan\theta = -\frac{7}{24}$

69. $\tan\theta = \frac{3}{2}$; $\csc\theta = -\frac{\sqrt{13}}{3}$

70. $\cot\theta = \frac{1}{\sqrt{15}}$; $\sin\theta = -\frac{\sqrt{15}}{4}$

71. $\cot\theta = -\sqrt{\frac{1}{2}}$; $\sec\theta = \frac{3}{\sqrt{3}}$

72. $\cos\theta = 0$; $\sin\theta = -1$

73. $\tan\theta = -\frac{2}{3}$; $\cos\theta = -\frac{3}{\sqrt{13}}$

74. $\csc\theta = -\frac{\sqrt{13}}{2}$; $\cot\theta = -\frac{3}{2}$

In Exercises 75–84 find the algebraic signs for the trigonometric functions at the given angle.

75. $\frac{3\pi}{4}$

76. $\frac{5\pi}{4}$

77. $-190°$

78. $-280°$

79. $\frac{21\pi}{12}$

80. $\frac{31\pi}{20}$

81. 11

82. 2

83. $-\frac{3\pi}{4}$

84. $-\frac{10\pi}{24}$

85. Give three positive and three negative angles for which $\sec\theta$ is not defined.

86. Give three positive and three negative angles for which $\csc\theta$ is not defined.

87. Give three positive and three negative angles for which $\tan\theta$ is not defined.

88. Give three positive and three negative angles for which $\cot\theta$ is not defined.

89. Given $\sin\alpha = \frac{2}{3}$, compute $\sin^4\alpha$.

90. Given $\tan\alpha = -2$, find $\tan^5\alpha$.

91. Give an example of an angle α such that $\sin^2\alpha + \cos^2\alpha \neq (\sin\alpha + \cos\alpha)^2$.

92. Given $\sec\alpha = -2$, find $\sec^{-2}\alpha$.

93. If $\cot\alpha = -\frac{1}{2}$, find $\cot^{-2}\alpha$.

94. If possible, give an example of an angle α such that $\sin^3\alpha = -27$.

95. A line intersects the positive x axis at an angle α such that $\cos\alpha = -\frac{3}{5}$. Write the equation of the line if it passes through $(-1, 2)$.

For Your Calculator

Most scientific calculators will compute the values of the trigonometric functions in either the degree or radian mode. Consult your manual for the proper means of changing modes on your calculator. The example below shows the procedure for getting the values of sin 45° and sin 45.

[deg] [sin] 45	answer	0.7071	
[rad] [sin] 45	answer	0.8509	*(45 rad = 2578.3°)*

In Exercises 96–101 use your calculator to evaluate the trigonometric functions. Remember that if no unit is specified it is understood that the unit is the radian.

96. sin 90°; sin 90

97. cos 60°; cos 60

98. tan 1.57°; tan 1.57

99. cos 100°; cos 100

100. tan 225°; tan 225

101. sin 6.02°; sin 6.02

102. There is a real number x, $170 \leq x \leq 180$, such that $\sin x = \sin x°$. Approximate x to three decimal places.

6.3 Special Angles

As we have seen, it is usually difficult to determine trigonometric function values. For some "special angles," however, the task is made simple by using results from plane geometry.

30°-60°-90° Triangle

In the equilateral triangle shown in Figure 6.8, drop the perpendicular from C to AB. Then D bisects the segment AB and angle $ACD = 30° =$ angle DCB. We see then that in a **30°-60°-90° triangle** the length of the side opposite the 30° angle must be 1/2 the length of the side opposite the right angle (hypotenuse). In Figure 6.9 observe that $b = \frac{1}{2}c$ and, from the Pythagorean theorem, we can compute

$$\begin{aligned} a &= \sqrt{c^2 - b^2} = \sqrt{c^2 - \frac{1}{4}c^2} \\ &= \sqrt{\frac{3}{4}c^2} = \frac{c}{2}\sqrt{3} = b\sqrt{3} \end{aligned}$$

In particular, if $c = 2$, note that $b = 1$ and $a = \sqrt{3}$.

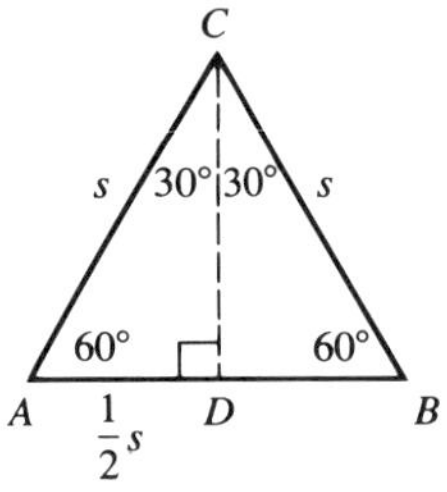

Figure 6.8

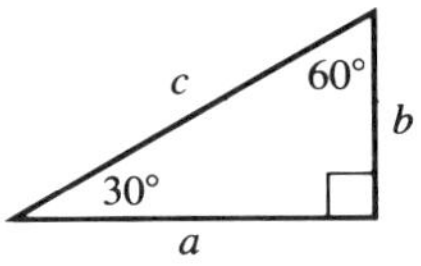

Figure 6.9

Example 1 In the illustration shown,

$$\alpha = 30° = \frac{\pi}{6} \qquad \beta = 120° = \frac{2\pi}{3}$$

$$\gamma = 240° = \frac{4\pi}{3} \quad \text{and} \quad \delta = -120° = -\frac{2\pi}{3}$$

Each of these has been constructed with terminal side of length $r = 2$. Using the 30°-60°-90° triangles it is easy to see then that:

a. $\sin 30° = \frac{y}{r} = \frac{1}{2} \qquad \cos 30° = \frac{x}{r} = \frac{\sqrt{3}}{2}$

$\tan 30° = \frac{\sin 30°}{\cos 30°} = \frac{y}{x} = \frac{1}{\sqrt{3}}$

b. $\sin \frac{2\pi}{3} = \frac{\sqrt{3}}{2} \qquad \cos \frac{2\pi}{3} = -\frac{1}{2} \qquad \tan \frac{2\pi}{3} = -\sqrt{3}$

c. $\sin 240° = -\dfrac{\sqrt{3}}{2} = \sin(-120°)$ $\quad\cos\dfrac{4\pi}{3} = -\dfrac{1}{2} = \cos\left(-\dfrac{2\pi}{3}\right)$

$\tan\dfrac{4\pi}{3} = \sqrt{3} = \tan\left(-\dfrac{2\pi}{3}\right)$

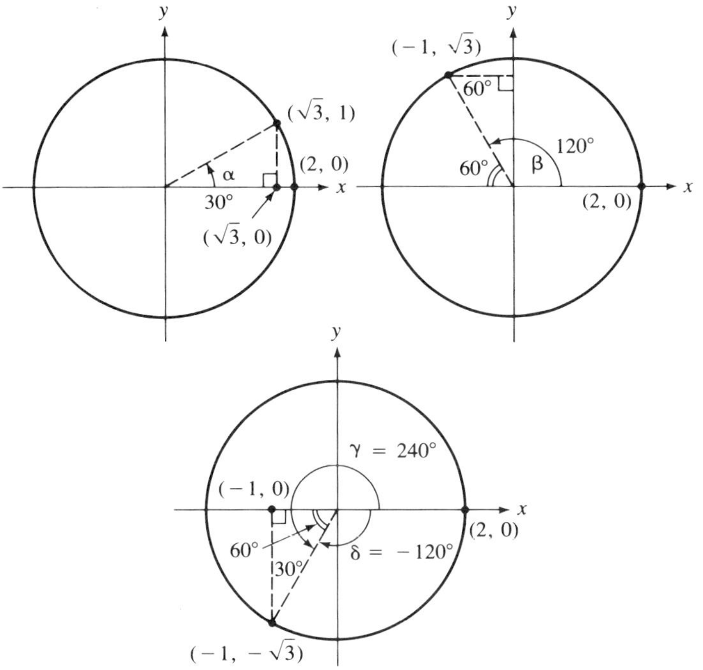

■

Example 2 Find the values of the six trigonometric functions for $\theta = -\dfrac{7\pi}{6}$.

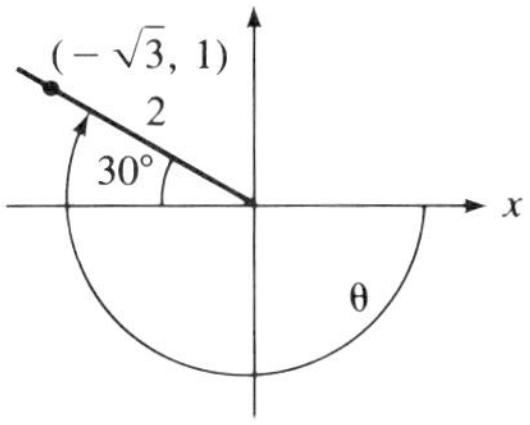

Begin by locating the angle in standard position. Then, using the 30°-60°-90° triangle, the point $(-\sqrt{3}, 1)$ is on the terminal side of θ. We quickly obtain $\sin(-7\pi/6) = \dfrac{1}{2}$, $\cos(-7\pi/6) = -\sqrt{3}/2$. From these two, $\tan(-7\pi/6) = -1/\sqrt{3}$, $\cot(-7\pi/6) = -\sqrt{3}$, $\sec(-7\pi/6) = -2/\sqrt{3}$, and $\csc(-7\pi/6) = 2$. ■

Notice in Example 2 that we computed the values of $\tan\theta$, $\cot\theta$, $\sec\theta$, and $\csc\theta$ from the values of $\sin\theta$ and $\cos\theta$. Since it is routine to do this we will begin to restrict the discussion of values of trigonometric functions to $\sin\theta$,

$\cos\theta$, and $\tan\theta$. Tan θ will be included so that the others may be obtained using only reciprocal relationships.

45°-45°-90° Triangle In a **45°-45°-90° triangle** the lengths of the two short sides are equal. In Figure 6.10 observe that $c = \sqrt{a^2 + a^2} = \sqrt{2a^2} = a\sqrt{2}$. In particular, if $a = 1$ then $c = \sqrt{2}$.

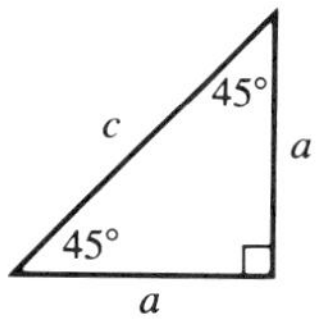

Figure 6.10

Example 3 In the illustration shown,

$$\alpha = 45° = \frac{\pi}{4} \qquad \beta = 225° = \frac{5\pi}{4} \quad \text{and } \gamma = -135° = -\frac{3\pi}{4}$$

Each of these angles has terminal side on the line $y = x$, so it is convenient to choose terminal sides of length $r = \sqrt{2}$. We see that

a. $\sin 45° = \dfrac{1}{\sqrt{2}} \qquad \cos 45° = \dfrac{1}{\sqrt{2}} \qquad \tan 45° = 1$

b. $\sin\dfrac{5\pi}{4} = -\dfrac{1}{\sqrt{2}} \qquad \cos\dfrac{5\pi}{4} = -\dfrac{1}{\sqrt{2}} \qquad \tan\dfrac{5\pi}{4} = 1$

c. $\sin\left(-\dfrac{3\pi}{4}\right) = -\dfrac{1}{\sqrt{2}} \qquad \cos\left(-\dfrac{3\pi}{4}\right) = -\dfrac{1}{\sqrt{2}} \qquad \tan(-135°) = 1$

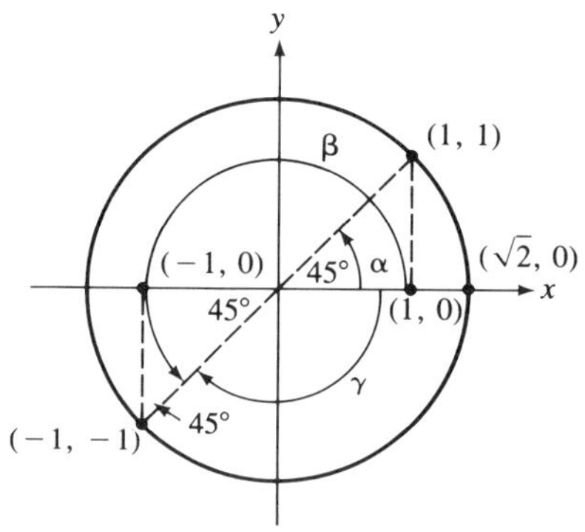

■

Example 4 Find the values of the trigonometric functions for $\theta = \dfrac{15\pi}{4}$.

Observe that

$$\frac{15\pi}{4} = \frac{8\pi}{4} + \frac{7\pi}{4} = 2\pi + \frac{7\pi}{4}$$

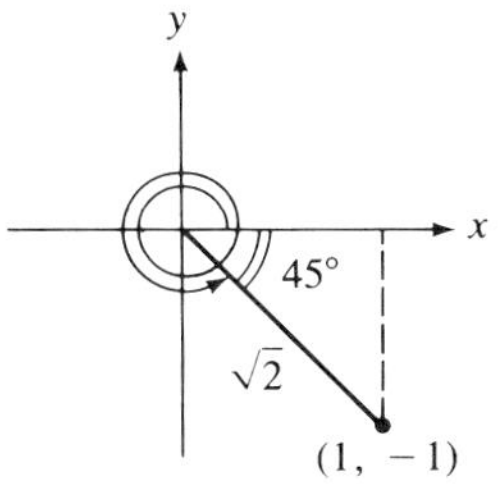

In standard position this angle lacks only $\pi/4$ of being two complete revolutions. Using the point $(1, -1)$ on the terminal side of θ, $r = \sqrt{2}$. Then

$$\sin\frac{15\pi}{4} = -\frac{1}{\sqrt{2}} \qquad \cos\frac{15\pi}{4} = \frac{1}{\sqrt{2}} \qquad \tan\frac{15\pi}{4} = -1$$

Thus

$$\csc\frac{15\pi}{4} = -\sqrt{2} \qquad \sec\frac{15\pi}{4} = \sqrt{2} \qquad \cot\frac{15\pi}{4} = -1$$ ■

By using the 30°-60°-90° and 45°-45°-90° triangles and the techniques of the above examples, Table 6.2 may be obtained. For completeness we have repeated the quadrantal angles.

θ deg	0°	30°	45°	60°	90°	180°	270°
θ rad	0	$\frac{\pi}{6}$	$\frac{\pi}{4}$	$\frac{\pi}{3}$	$\frac{\pi}{2}$	π	$\frac{3\pi}{2}$
$\sin\theta$	0	$\frac{1}{2}$	$\frac{1}{\sqrt{2}}$	$\frac{\sqrt{3}}{2}$	1	0	-1
$\cos\theta$	1	$\frac{\sqrt{3}}{2}$	$\frac{1}{\sqrt{2}}$	$\frac{1}{2}$	0	-1	0
$\tan\theta$	0	$\frac{1}{\sqrt{3}}$	1	$\sqrt{3}$	*	0	*

Table 6.2

A handheld calculator will show a decimal such as 0.8660254 for sin 60° rather than the $\sqrt{3}/2$ value shown in Table 6.2. It is important to realize that $\sqrt{3}/2$ is an exact value, whereas 0.8660254 is an approximate value of sin 60°.

Exact vs. approximate

Consider an acute angle θ in standard position and let $P(x, y)$ be a point on the terminal side of θ. Form the right triangle containing θ as in Figure 6.11. It is helpful to think of the trigonometric functions of θ without reference to the coordinate system. To do this use the words **side opposite** to describe the location of leg y of the triangle relative to θ and **side adjacent** to describe the location of leg x of the triangle relative to θ. As usual, r is called the **hypotenuse.** This language is applicable for any acute angle of a right triangle.

Side opp
Side adj

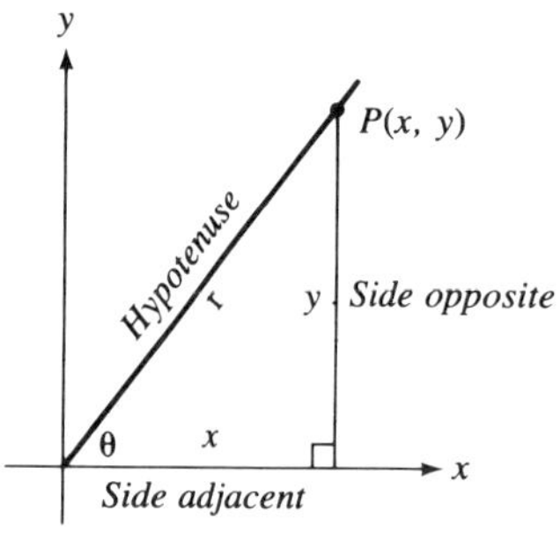

Figure 6.11

Example 5 For the triangle shown, side adjacent β is a, side opposite β is b, side adjacent α is b, side opposite α is a.

Using Figure 6.11 and the definition of the trigonometric functions, we obtain the following:

$$\sin \theta = \frac{y}{r} = \frac{\text{side opposite}}{\text{hypotenuse}}$$

$$\cos \theta = \frac{x}{r} = \frac{\text{side adjacent}}{\text{hypotenuse}}$$

$$\tan \theta = \frac{y}{x} = \frac{\text{side opposite}}{\text{side adjacent}}$$

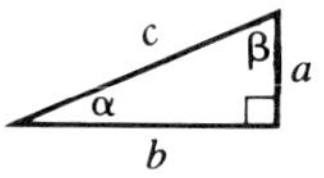

■

The reader is urged to memorize these three functions in terms of the words describing the location of the sides of the triangle relative to the angle.

Example 6 For the triangle shown, find $\sin \alpha$, $\cos \alpha$, and $\tan \alpha$.
The hypotenuse has length $\sqrt{3^2 + 6^2} = \sqrt{45} = 3\sqrt{5}$. We see then that

$$\sin \alpha = \frac{\text{side opp}}{\text{hyp}} = \frac{3}{3\sqrt{5}} = \frac{1}{\sqrt{5}}$$

$$\cos \alpha = \frac{\text{side adj}}{\text{hyp}} = \frac{6}{3\sqrt{5}} = \frac{2}{\sqrt{5}}$$

$$\tan \alpha = \frac{\text{side opp}}{\text{side adj}} = \frac{3}{6} = \frac{1}{2}$$

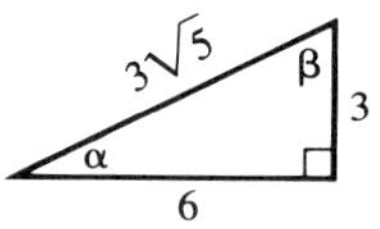

■

Example 7 For the triangle shown,

$$\sin \alpha = \frac{3}{5} \quad \cos \beta = \frac{3}{5} \quad \cot \alpha = \frac{4}{3} \quad \tan \beta = \frac{4}{3}$$

$$\cos \alpha = \frac{4}{5} \quad \sin \beta = \frac{4}{5} \quad \sec \alpha = \frac{5}{4} \quad \csc \beta = \frac{5}{4}$$

$$\tan \alpha = \frac{3}{4} \quad \cot \beta = \frac{3}{4} \quad \csc \alpha = \frac{5}{3} \quad \sec \beta = \frac{5}{3}$$

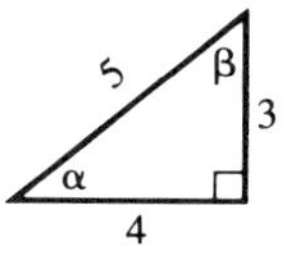

■

Complementary

Cofunction

Two acute angles α, β are **complementary** if $\alpha + \beta = 90°$. In this case, as indicated in Example 7, $\cos\alpha = \sin\beta$, $\sin\alpha = \cos\beta$, and $\tan\alpha = \cot\beta$. It is for this reason that sine and cosine are called **cofunctions.** Tangent and cotangent are likewise cofunctions, as are secant and cosecant. In general, if θ is acute, the value of a trigonometric function at θ is equal to the value of the cofunction at $(90° - \theta)$.

Example 8 $\sin 15° = \cos 75°$, $\tan\dfrac{\pi}{8} = \cot\dfrac{3\pi}{8}$, and $\sec 1 = \csc\left(\dfrac{\pi}{2} - 1\right)$. ■

Example 9 $\sin 30° = \cos 60° = \dfrac{1}{2}$, $\sin 60° = \cos 30° = \dfrac{\sqrt{3}}{2}$, and $\tan 30° = \cot 60° = \dfrac{1}{\sqrt{3}}$. ■

Exercises 6.3

The angles and sides given in these exercises refer to the standard lettering of the right triangle shown.

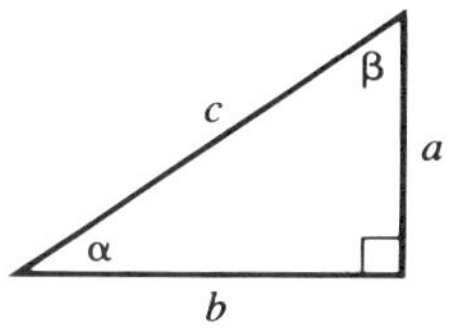

In Exercises 1–8 find the remaining sides of the triangle if $\alpha = 30°$.

1. $c = 4$ **2.** $c = 3$ **3.** $a = \frac{1}{2}$ **4.** $a = \frac{1}{4}$

5. $b = 3\sqrt{3}$ **6.** $b = \dfrac{\sqrt{3}}{2}$ **7.** $b = 5$ **8.** $b = 7$

In Exercises 9–14 find the remaining sides of the triangle if $\alpha = 45°$.

9. $b = 2$ **10.** $a = 6$ **11.** $c = 5$ **12.** $c = 12$

13. $a = \frac{1}{2}$ **14.** $b = \frac{2}{3}$

Use special angles to find the values of the trigonometric functions at the given angles in Exercises 15–28.

15. $\dfrac{3\pi}{4}$ **16.** $-\dfrac{\pi}{3}$ **17.** $-210°$ **18.** $225°$

19. $315°$ **20.** $-\dfrac{11\pi}{6}$ **21.** $\dfrac{5\pi}{3}$ **22.** $-\dfrac{7\pi}{4}$

23. $-\dfrac{5\pi}{6}$ **24.** $\dfrac{4\pi}{3}$ **25.** $\dfrac{7\pi}{6}$ **26.** $240°$

27. $-\dfrac{23\pi}{4}$ **28.** $-\dfrac{15\pi}{4}$

In Exercises 29–38 find the values of the trigonometric functions at α *and at* β *for the given sides of the triangle.*

29. $a = 5;\ b = 12$
30. $a = 4;\ b = 5$
31. $a = 7;\ b = 24$
32. $a = 3;\ b = 2$
33. $a = \sqrt{2};\ c = \sqrt{11}$
34. $a = 1;\ b = \sqrt{15}$
35. $a = 2;\ c = \sqrt{7}$
36. $a = 2;\ c = \sqrt{20}$
37. $b = 3;\ c = 4$
38. $a = 4;\ c = 7$

In Exercises 39–44 if θ *is an acute angle, express the trigonometric function values as functions of* θ *only.*

39. $\sin(90^\circ - \theta)$
40. $\cos(90^\circ - \theta)$
41. $\tan\left(\frac{\pi}{2} - \theta\right)$
42. $\sec\left(\frac{\pi}{2} - \theta\right)$
43. $\csc\left(\frac{\pi}{2} - \theta\right)$
44. $\cot(90^\circ - \theta)$

In Exercises 45–54 express the trigonometric function values as functions of an angle θ *such that* $0 \le \theta < \frac{\pi}{4}$ *or* $0^\circ \le \theta < 45^\circ$.

45. $\cos\frac{3\pi}{7}$
46. $\sin\frac{8\pi}{17}$
47. $\sin\frac{3\pi}{8}$
48. $\sec 65^\circ$
49. $\tan 80^\circ$
50. $\tan\frac{5\pi}{12}$
51. $\cot 50^\circ$
52. $\csc 70^\circ$
53. $\sec\frac{11\pi}{24}$
54. $\cos\frac{4\pi}{9}$

In Exercises 55–62 use the given information to find the remaining sides of the triangle.

55. $\sin\alpha = \frac{1}{3};\ b = 3$
56. $\tan\alpha = 4;\ b = 2$
57. $\cot\beta = 5;\ a = 8$
58. $\sin\beta = \frac{2}{\sqrt{5}};\ c = \sqrt{5}$
59. $\sec\alpha = 2;\ a = 2$
60. $\sec\beta = 3;\ c = 5$
61. $\cos\beta = \frac{1}{4};\ c = \frac{1}{2}$
62. $\cot\alpha = 8;\ a = \frac{1}{2}$

6.4 Trigonometric Functions of Real Numbers

For many applications it is essential that the trigonometric functions be viewed as functions of real numbers. Radian measure of angles allows this change of domain to be accomplished in a natural way.

For each real number t there corresponds a unique angle in standard position whose measure is t radians. Trigonometric functions of this angle are used to define the trigonometric functions of t. Thus, if $f(\theta)$ represents one of the six trigonometric functions, $f(t)$ is defined to be $f(t$ radians).

$f(t)$, t Real

Example 1 **a.** $2 \varepsilon R$. Since $\pi/2 < 2 < \pi$, the terminal side of the angle of radian measure 2 falls in the second quadrant, so that $\sin 2 > 0$ and $\cos 2 < 0$.

b. $\sqrt{2} \varepsilon R$. The values of all trigonometric functions of $\sqrt{2}$ are positive, because an angle of radian measure $\sqrt{2}$ has terminal side in the first quadrant.

c. $\frac{5\pi}{6} \varepsilon R$. To determine trigonometric function values for this number, draw a diagram with $\frac{5\pi}{6}$ in standard position as shown. Then

$$\sin\frac{5\pi}{6} = \frac{1}{2}$$
$$\cos\frac{5\pi}{6} = -\frac{\sqrt{3}}{2}$$
$$\tan\frac{5\pi}{6} = -\frac{1}{\sqrt{3}}$$

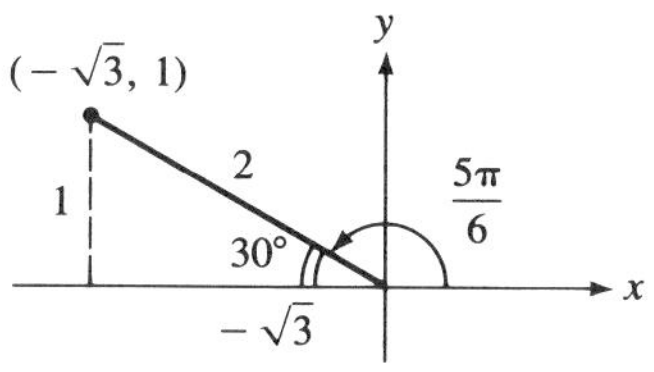

■

Recall from the discussion in Section 6.1 that radian measure of an angle is the same as "directed arc length" on the unit circle U. By this we mean that if $t \varepsilon R$, then, geometrically, one thinks of an arc length $|t|$ on U, measured from (1, 0) clockwise if $t < 0$ and counterclockwise if $t > 0$. This point on U is then used to locate the terminal side of the angle of measure t radians.

Example 2 In the illustration, $r = 1$.

a. $\sin 1 = \frac{y_1}{r} = y_1$ and $\cos 1 = \frac{x_1}{r} = x_1$

b. $\sin(-\pi) = \frac{0}{1} = 0$ and $\cos(-\pi) = \frac{-1}{1} = -1$.

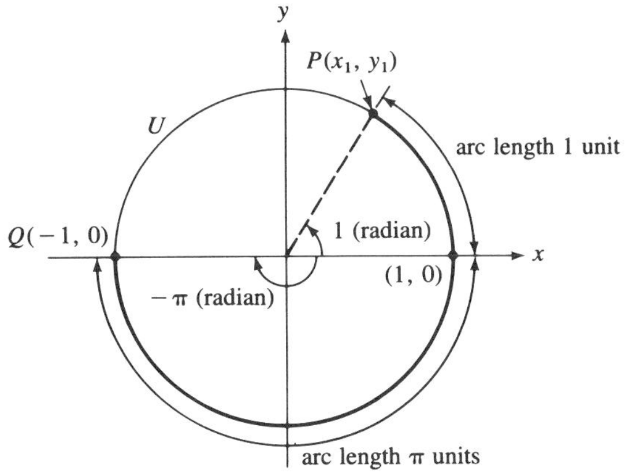

■

Notice that if $t \varepsilon R$ and (x, y) is the point on U determined by the angle of measure t radians, then $\sin t = y$, $\cos t = x$, and $\tan t = y/x$.

The above observations and examples demonstrate that the trigonometric functions can be viewed as functions whose domains are angles, or as functions

Circular functions

whose domains are real numbers. In the latter case the trigonometric functions are frequently called **circular functions,** because their values are determined by directed arc length on a circle.

Example 3 In the illustration shown we consider the numbers $\frac{\pi}{6}$, $\frac{3\pi}{4}$, and $-\frac{\pi}{3}$, and the corresponding angles.

a. For $t = \frac{\pi}{6}$ we use a 30°-60°-90° triangle to obtain the rectangular coordinates for P_1. Then

$$\sin\frac{\pi}{6} = \frac{1}{2} \quad \text{and} \quad \cos\frac{\pi}{6} = \frac{\sqrt{3}}{2}$$

b. For $t = \frac{3\pi}{4}$ use a 45°-45°-90° triangle to obtain rectangular coordinates for P_2. Then

$$\sin\frac{3\pi}{4} = \frac{1}{\sqrt{2}} \quad \text{and} \quad \cos\frac{3\pi}{4} = -\frac{1}{\sqrt{2}}$$

c. Again using a 30°-60°-90° triangle,

$$\sin\left(-\frac{\pi}{3}\right) = -\frac{\sqrt{3}}{2} \quad \text{and} \quad \cos\left(-\frac{\pi}{3}\right) = \frac{1}{2}$$

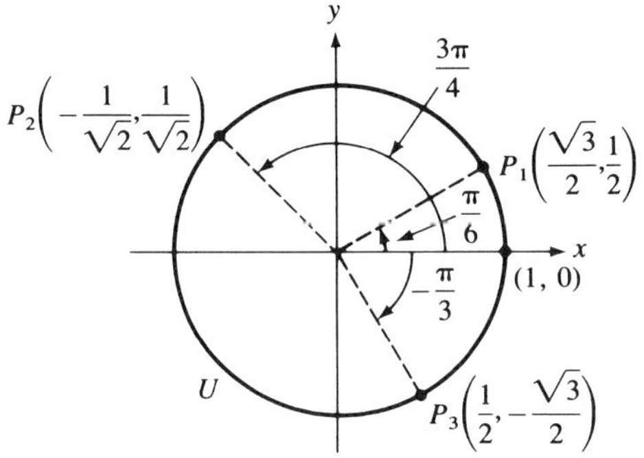

■

Example 4 In the figure shown let t be the directed arc length on U from (1, 0) to $\left(\frac{3}{4}, -\frac{\sqrt{7}}{4}\right)$. Then

$$\sin t = -\frac{\sqrt{7}}{4} \qquad \cos t = \frac{3}{4} \quad \text{and} \quad \tan t = -\frac{\sqrt{7}}{3}$$

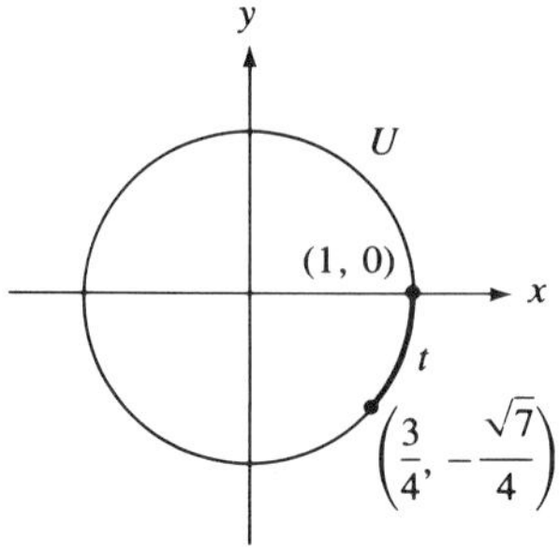

■

It must be emphasized that the trigonometric functions of real numbers are not precisely the same functions as defined in Section 6.2, because the elements in the domains there were angles rather than real numbers. It is customary to use the same abbreviations for the two sets of trigonometric functions and not to distinguish in notation when radian measure is used (or when measure is not specified). In future discussions, necessary distinctions will be clear from the context.

Because of the approach we have used for trigonometric functions of real numbers it is immediately clear that all relationships among trigonometric functions that are valid for angles are also valid for real numbers (and vice versa). For example, if $t \varepsilon R$,

$$\sin^2 t + \cos^2 t = 1 \quad \text{and} \quad \tan t = \frac{1}{\cot t}$$

Example 5 **a.** Let $t \varepsilon R$. If $\sin t < 0$ and $\tan t > 0$, what is the sign of $\cos t$?
Cos $t < 0$ since the terminal side of the angle of measure t radians is located in quadrant III.

b. Suppose $\cos t = -\frac{4}{5}$ and $\sin t > 0$. Compute $\sin t$.
Since $\sin^2 t = 1 - \cos^2 t = 1 - \frac{16}{25} = \frac{9}{25}$ and $\sin t > 0$, we see that $\sin t = \frac{3}{5}$.

c. Find $\sin t$ if $\cos t = \frac{3}{5}$ and $\tan t = -\frac{4}{3}$.

$$\sin t = \left(\frac{\sin t}{\cos t}\right) \cdot (\cos t) = (\tan t) \cdot (\cos t) = \left(-\frac{4}{3}\right) \cdot \left(\frac{3}{5}\right) = -\frac{4}{5}$$ ■

Example 6 Find the values of the remaining trigonometric functions at t if $|\cos t| = \frac{5}{13}$ and $2 < t < 3$.
Since $\pi/2 < 2 < t < 3 < \pi$, the terminal side of the angle of measure t radians is located in quadrant II, so that $\cos t = -\frac{5}{13}$. Since $\sin t > 0$,

$$\sin t = (1 - \cos^2 t)^{1/2} = \left(1 - \frac{25}{169}\right)^{1/2} = \left(\frac{144}{169}\right)^{1/2} = \frac{12}{13}$$

Then

$$\tan t = -\frac{12}{5} \qquad \cot t = -\frac{5}{12} \qquad \sec t = -\frac{13}{5} \qquad \csc t = \frac{13}{12}$$ ■

The results in Section 6.2 pertaining to domain and range for the trigonometric functions are also applicable to trigonometric functions of real numbers. Specifically:

Domain: **a.** $\sin t$ and $\cos t$ are defined for all $t \varepsilon R$.
b. $\tan t$ and $\sec t$ are not defined if $\cos t = 0$, i.e., if t is an odd multiple of $\frac{\pi}{2}$.
c. $\cot t$ and $\csc t$ are not defined if $\sin t = 0$, i.e., if t is an even multiple of $\frac{\pi}{2}$.

> Range: **a.** $|\sin t| \le 1$ and $|\cos t| \le 1$ for all $t \in R$.
> **b.** $|\sec t| \ge 1$ and $|\csc t| \ge 1$ for all $t \in R$.
> **c.** $\tan t$ and $\cot t$ assume all real values.

Example 7 Show a point P on U such that the directed arc length t from $(1, 0)$ to P is such that $\sin t = -\frac{1}{3}$.

Locate the point $-\frac{1}{3}$ on the y axis. Extend the line $y = -\frac{1}{3}$ to its intersection with U. If t is the directed arc length shown, then $\sin t = -\frac{1}{3}$. In this case, $\cos^2 t = 1 - \sin^2 t = 1 - \frac{1}{9} = \frac{8}{9}$, so that

$$\cos t = \frac{2\sqrt{2}}{3} \quad \text{and} \quad P = \left(\frac{2\sqrt{2}}{3}, -\frac{1}{3}\right)$$

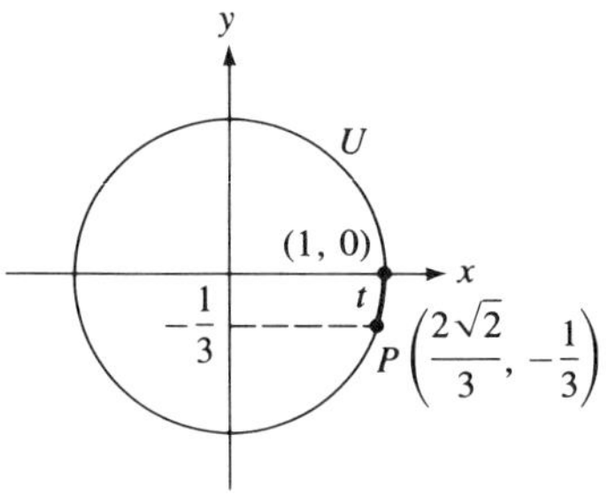

■

This section began with the observation that for many applications it is necessary to view the trigonometric functions as functions of real numbers. One such application will now be discussed.

If a mass is suspended from the end of a spring as shown in Figure 6.12 and is displaced from its equilibrium position, it oscillates according to certain physical principles. If x measures the distance from the mass to the equilibrium position, the function

$$x = A \sin(\omega t + \theta)$$

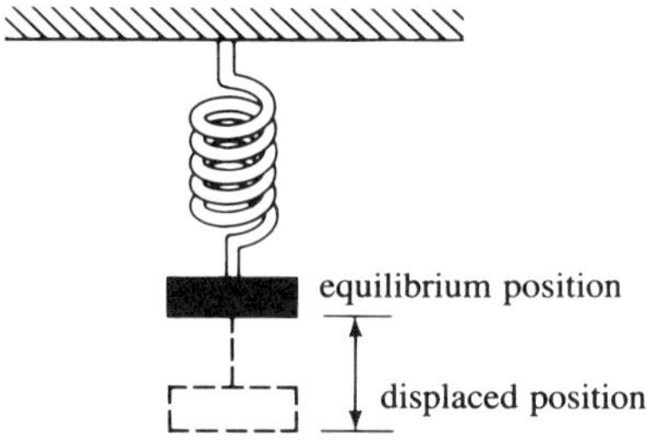

Figure 6.12

describes the motion of the mass, where A, ω, and θ are constants determined by the mass, the spring, the initial displacement, and the initial velocity. The *period* T is the time for one complete oscillation of the mass and is given by the formula $T = 2\pi/\omega$. The frequency f is the number of complete oscillations per unit time and is related to the period by the formula $f = 1/T$.

Example 8 A vibrating mass at the end of a spring oscillates according to the equation

$$x = 2 \sin\left(\frac{5\pi}{3} t + \frac{\pi}{2}\right)$$

where t is measured in seconds. The period is

$$\frac{2\pi}{5\pi/3} = \frac{6}{5} \text{ sec}$$

which is the time necessary for one complete oscillation. The frequency is $\frac{5}{6}$, which means that the mass makes $\frac{5}{6}$ of an oscillation in 1 sec. ■

Exercises 6.4

In Exercises 1–10 find the values of the trigonometric functions at the specified real number.

1. $\frac{\pi}{6}$
2. $\frac{5\pi}{6}$
3. $-\frac{3\pi}{4}$
4. $-\pi$
5. π
6. $\frac{2\pi}{3}$
7. $\frac{5\pi}{3}$
8. $-\frac{5\pi}{4}$
9. $-\frac{\pi}{2}$
10. $-\frac{5\pi}{2}$

In Exercises 11–20 find the algebraic signs of the trigonometric functions at the specified real number.

11. $\frac{5}{2}$
12. $\sqrt{5}$
13. -7
14. -3
15. $\sqrt{3}$
16. $\frac{11}{3}$
17. -10
18. $-\frac{43\pi}{4}$
19. $\frac{71\pi}{6}$
20. $\frac{34\pi}{3}$

In Exercises 21–32 find the values of the trigonometric functions at the real number in the specified interval that satisfies the given condition.

21. $\left(\pi, \frac{3\pi}{2}\right)$; $\sin t = -\frac{2}{5}$
22. $(-\pi, 0)$; $\cos t = -\frac{4}{9}$
23. $(-\pi, 0)$; $\cos t = \frac{1}{\sqrt{2}}$
24. $(0, \pi)$; $\cot t = -1$
25. $(-\pi, 0)$; $\tan t = 1$
26. $\left[-\frac{\pi}{2}, \frac{\pi}{2}\right]$; $\sin t = -1$
27. $(0, 2\pi)$; $\cos t = -1$
28. $(0, \pi)$; $\sec t = \sqrt{2}$
29. $(0, \pi)$; $\cot t = \sqrt{3}$
30. $(-\pi, 0)$; $\tan t = \sqrt{3}$
31. $\left(-\frac{\pi}{2}, \frac{\pi}{2}\right)$; $\csc t = -2$
32. $(0, \pi)$; $\csc t = 1$

In Exercises 33–42, given the two trigonometric function values at the real number t, find the values of the remaining trigonometric functions at t.

33. $\cos t = 1$; $\sin t = 0$

34. $\cot t = 1$; $\sec t = -\sqrt{2}$

35. $\cos t = -\frac{1}{2}$; $\tan t = -\sqrt{3}$

36. $\cos t = 0$; $\sin t = -1$

37. $\sin t = -\frac{\sqrt{2}}{2}$; $\tan t = -1$

38. $\tan t = \frac{\sqrt{3}}{3}$; $\csc t = -2$

39. $\tan t = \frac{3}{2}$; $\csc t = \frac{\sqrt{13}}{3}$

40. $\sin t = \frac{\sqrt{3}}{3}$; $\sec t = -\frac{3}{\sqrt{6}}$

41. $\cot t = -\frac{1}{\sqrt{15}}$; $\sec t = 4$

42. $\cot t = -\frac{24}{7}$; $\cos t = -\frac{24}{25}$

In Exercises 43–50 find the real number t satisfying the specified condition and show t on U.

43. $0 \le t \le 3$; $\cos t = -\frac{1}{2}$

44. $-\frac{3}{2} \le t \le \frac{3}{2}$; $\sin t = \frac{1}{2}\sqrt{3}$

45. $-\frac{3}{2} \le t \le \frac{3}{2}$; $\sin t = -\frac{1}{\sqrt{2}}$

46. $-\frac{\pi}{2} \le t \le \frac{\pi}{2}$; $\cot t = -\sqrt{3}$

47. $0 \le t \le \frac{\pi}{2}$; $\tan^2 t = \frac{1}{3}$

48. $-\frac{1}{5} \le t \le \frac{1}{5}$; $\cos t = 1$

49. $\pi \le t \le \frac{3\pi}{2}$; $\cot^2 t = 1$

50. $\frac{\pi}{2} \le t \le \pi$; $\tan^3 t = -1$

In Exercises 51–52 a mass at the end of a spring oscillates according to the equation $x = 3 \sin\left(\frac{5\pi}{6} t + \frac{\pi}{4}\right)$, *where t is measured in seconds.*

51. What are the values for the period and the frequency of this mass?

52. How many oscillations does this mass make in 1 min?

For Your Calculator

In Exercises 53–58 use your calculator to approximate the values of the trigonometric functions at the real number t, $0 \le t < 2\pi$, *that satisfies the given conditions. Use the procedure of Examples 5 and 6 of this section. Do not use inverse functions.*

53. $\cos t = 0.2783$; $\tan t > 0$

54. $\sin t = 0.6223$; $\cos t < 0$

55. $\sin t = -0.2345$; $\tan t > 0$

56. $\cos t = -0.5928$; $\cot t < 0$

57. $\tan t = 2.3695$; $\cos t < 0$

58. $\sec t = 10.2243$; $\sin t < 0$

6.5 Use of Reference Angle and Reference Number

Due to their periodicity, the values of the trigonometric functions are known for all angles once the values are known for those angles θ such that

$0° \leq \theta < 360°$ (or $0 \leq \theta < 2\pi$).

Likewise, the values are known for all real numbers once the values for the real numbers in the interval $[0, 2\pi)$ are determined. The reader has doubtless observed that symmetries of the rectangular coordinate system assure, for example, that $\sin 60° = \sin 120°$, as illustrated in Figure 6.13. These symmetries enable us to determine the values of trigonometric functions for all nonquadrantal angles from the values for acute angles.

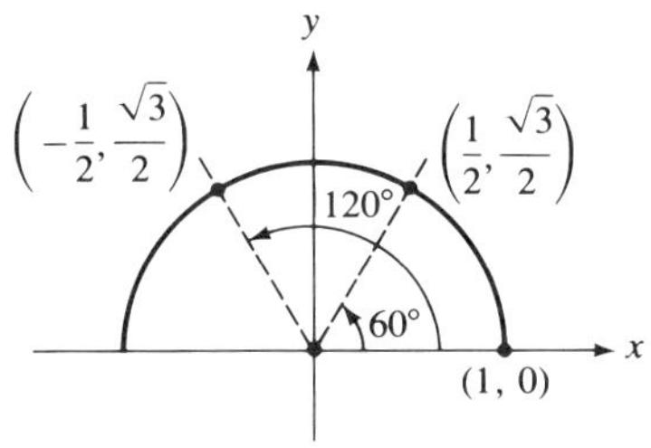

Figure 6.13

Example 1 If $\sin 50° = 0.766$, determine the values of $\sin 130°$, $\sin 230°$, and $\sin 310°$. The sine of each angle is the y coordinate of the point on U, as shown in the illustration. In each case the absolute value of the coordinate is 0.766. The sine of each angle is determined by attaching the proper algebraic sign for the quadrant to $\sin 50°$, obtaining $\sin 130° = 0.766$, $\sin 230° = -0.766$, and $\sin 310° = -0.766$.

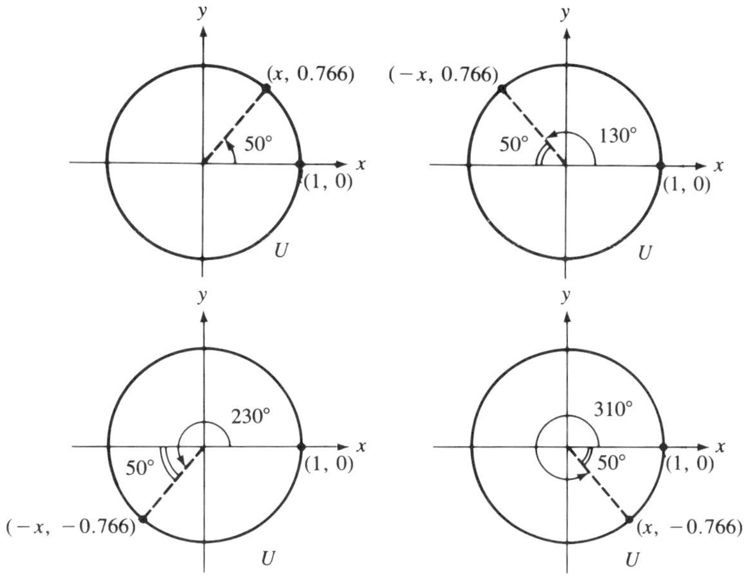

■

From Example 1 we note that in determining the absolute value of $\sin \theta$ the important consideration is the acute angle that the terminal side of θ makes with the x axis. For an angle θ in standard position, this acute angle is called the **reference angle** $\tilde{\theta}$. In general,

Reference angle

$$\tilde{\theta} = |n\pi - \theta|$$

where $n \in Z$ and n is chosen so that $\tilde{\theta}$ is an acute angle.

Reference number

If $t \in R$, the **reference number** $\tilde{t}$ is the radian measure of the reference angle for the angle of measure t radians.

Example 2 **a.** The reference angles for $\theta = 30°, 160°, -75°, -250°, \frac{7\pi}{4}, 1500°$, and $\frac{22\pi}{3}$ are, respectively, $\tilde{\theta} = 30°, 20°, 75°, 70°, \frac{\pi}{4}, 60°$, and $\frac{\pi}{3}$.

b. The reference numbers for $t = \frac{3\pi}{4}, -\frac{5\pi}{4}, \frac{11\pi}{6}, \sqrt{2}, e$, and 12 are, respectively, $\tilde{t} = \frac{\pi}{4}, \frac{\pi}{4}, \frac{\pi}{6}, \sqrt{2}, \pi - e$, and $(4\pi - 12)$. ■

The rule for finding the values of a trigonometric function for any angle θ if the value for the reference angle $\tilde{\theta}$ is known follows.

> For each trigonometric function, compute the trigonometric function of the reference angle $\tilde{\theta}$ and attach the proper algebraic sign for that trigonometric function, as determined by the quadrant in which the terminal side of θ is located.

In other words, if $f(\theta)$ is any one of the trigonometric functions, $f(\theta) = \pm f(\tilde{\theta})$. We need only determine $f(\tilde{\theta})$ and the algebraic sign.

Example 3 If $\sin 15° = 0.2588$ and $\cos 15° = 0.9659$, find $\sin 165°$, $\cos 165°$, and $\tan 165°$. Notice that 15° is the reference angle for 165° and that $\tan 15° = 0.2588/0.9659 = 0.2679$. Since the sine function is positive while both cosine and tangent are negative in quadrant II, we have $\sin 165° = \sin 15° = 0.2588$,
$\cos 165° = -\cos 15° = -0.9659$, and $\tan 165° = -\tan 15° = -0.2679$.

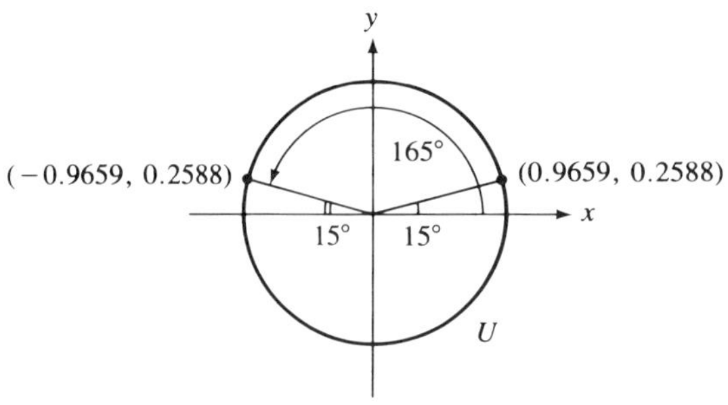

■

Example 4 Find $\sin \frac{13\pi}{12}$, $\cos \frac{13\pi}{12}$, and $\tan \frac{13\pi}{12}$.

The reference number for $t = 13\pi/12$ is $\pi/12$. An angle of radian measure $\pi/12$ is an angle of measure 15°. The terminal side of the angle $13\pi/12$ is located in quadrant III and, for such an angle θ, $\sin\theta < 0$, $\cos\theta < 0$, and $\tan\theta > 0$. Making use of Example 3, we see that

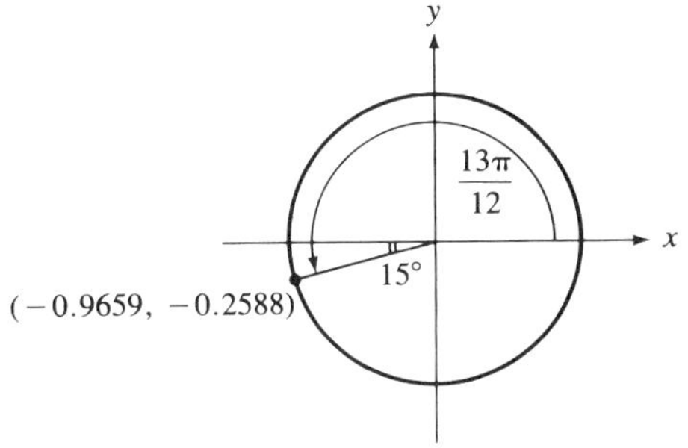

$$\sin\frac{13\pi}{12} = -\sin\frac{\pi}{12} = -0.2588 \qquad \cos\frac{13\pi}{12} = -\cos\frac{\pi}{12} = -0.9659$$

$$\tan\frac{13\pi}{12} = \tan\frac{\pi}{12} = 0.2679$$

■

Example 5 Find $\sin \frac{11\pi}{6}$, $\cos \frac{11\pi}{6}$, and $\tan \frac{11\pi}{6}$.

The terminal side of an angle of measure $11\pi/6$ falls in quadrant IV. The reference angle is $\pi/6$. Using the 30°-60°-90° triangle we obtain

$$\sin\frac{11\pi}{6} = -\sin\frac{\pi}{6} = -\frac{1}{2} \qquad \cos\frac{11\pi}{6} = \cos\frac{\pi}{6} = \frac{\sqrt{3}}{2}$$

$$\tan\frac{11\pi}{6} = -\tan\frac{\pi}{6} = -\frac{1}{\sqrt{3}}$$

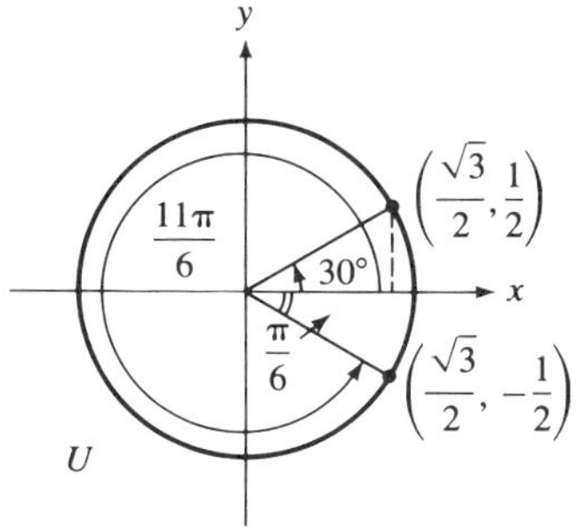

The values of the trigonometric functions at $\pi/6$ can also be read directly from Table 6.2 in Section 6.3. ■

Example 6 Find all $t \varepsilon [0, 2\pi)$ such that $\cos t = -\frac{1}{2}$.

The acute angle whose cosine is 1/2 measures $\pi/3$. Then $\tilde{t} = \pi/3$ is the reference number for t. The cosine function is negative in quadrants II and III. If $0 \leq t < 2\pi$, then t must be $2\pi/3$ or $4\pi/3$, as illustrated.

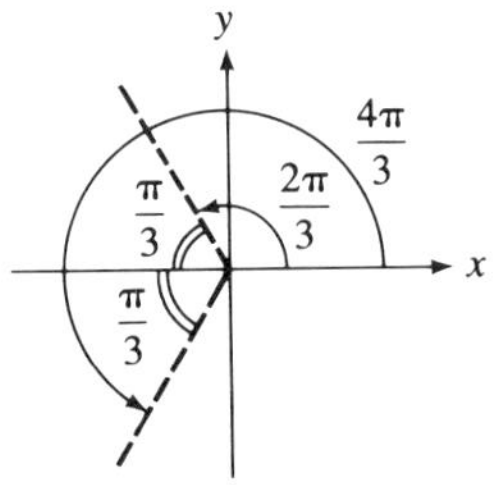

■

Using advanced techniques, tables of trigonometric values for acute angles have been constructed. The use of these tables and the concept of reference angle enable us to determine approximate trigonometric function values for any angle.

Table 6.3 is a portion of a typical table of trigonometric functions. Acute angles are shown in both degree and radian measure. We read from the top of the table down to find

$$\sin 33°20' = \sin 0.5818 = 0.5495$$

Since sine and cosine are cofunctions, we also have

$$\cos 56°40' = \sin 33°20' = 0.5495$$

This latter value can be read directly from the table by reading from the bottom of the table up. Caution: reading up to 56°40′ is a little tricky. Look carefully to see how it works.

Use of the cofunction concept and the top-down, bottom-up display permits approximate determination of all trigonometric function values from values for those angles α such that $0 \leq \alpha \leq 45°$.

Example 7 From Table 6.3 we see that

$$\sin(0.5963) = 0.5616$$

and

$$\tan 55°30' = 1.455$$

■

Example 8 Find approximate values for $\sin \sqrt{17}$ and $\cos \sqrt{17}$.

$\sqrt{17}$ is approximately 4.123. The reference angle for an angle of measure 4.123 is $4.123 - \pi$, which is about 0.9814. Taking 0.9803 as the closest value in Table 6.3 and reading bottom-up we see that

$$\sin\sqrt{17} = -\sin 0.9803 = -0.8307$$
$$\cos\sqrt{17} = -\cos 0.9803 = -0.5568$$

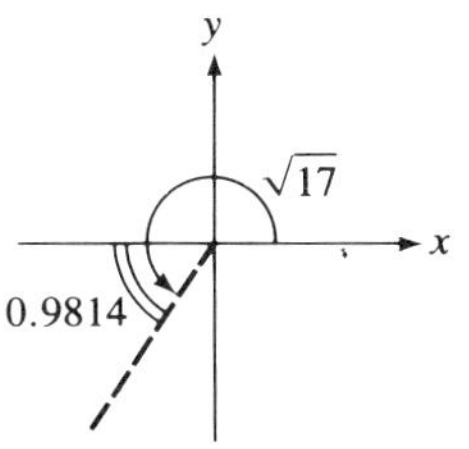

In degree measure the reference angle being used is 56°10′. Notice that we attached the proper algebraic sign for quadrant III. ■

Table 6.3

t	t degrees	$\sin t$	$\cos t$	$\tan t$	$\cot t$	$\sec t$	$\csc t$		
.5585	**32°00′**	.5299	.8480	.6249	1.600	1.179	1.887	**58°00′**	1.0123
.5614	10	.5324	.8465	.6289	1.590	1.181	1.878	50	1.0094
.5643	20	.5348	.8450	.6330	1.580	1.184	1.870	40	1.0065
.5672	30	.5373	.8434	.6371	1.570	1.186	1.861	30	1.0036
.5701	40	.5398	.8418	.6412	1.560	1.188	1.853	20	1.0007
.5730	50	.5422	.8403	.6453	1.550	1.190	1.844	10	.9977
.5760	**33°00′**	.5446	.8387	.6494	1.540	1.192	1.836	**57°00′**	.9948
.5789	10	.5471	.8371	.6536	1.530	1.195	1.828	50	.9919
.5818	20	.5495	.8355	.6577	1.520	1.197	1.820	40	.9890 ←
.5847	30	.5519	.8339	.6619	1.511	1.199	1.812	30	.9861
.5876	40	.5544	.8323	.6661	1.501	1.202	1.804	20	.9832
.5905	50	.5568	.8307	.6703	1.492	1.204	1.796	10	.9803
.5934	**34°00′**	.5592	.8290	.6745	1.483	1.206	1.788	**56°00′**	.9774
.5963	10	.5616	.8274	.6787	1.473	1.209	1.781	50	.9745
.5992	20	.5640	.8258	.6830	1.464	1.211	1.773	40	.9716
.6021	30	.5664	.8241	.6873	1.455	1.213	1.766	30	.9687
.6050	40	.5688	.8225	.6916	1.446	1.216	1.758	20	.9657
.6080	50	.5712	.8208	.6959	1.437	1.218	1.751	10	.9628
.6109	**35°00′**	.5736	.8192	.7002	1.428	1.221	1.743	**55°00′**	.9599
		$\cos t$	$\sin t$	$\cot t$	$\tan t$	$\csc t$	$\sec t$	t degrees	t

WARNING
Be sure to check twice when reading from bottom to top

Example 9 Find all angles θ such that $\tan\theta = 1.540$ and $0° \le \theta < 360°$.
From Table 6.3 the reference angle $\tilde{\theta}$ is 57°. Since $\tan\theta > 0$, the terminal side of θ is in quadrant I or III, and hence θ is 57° or θ is 180° + 57°. Thus $\theta = 57°$ or $\theta = 237°$.

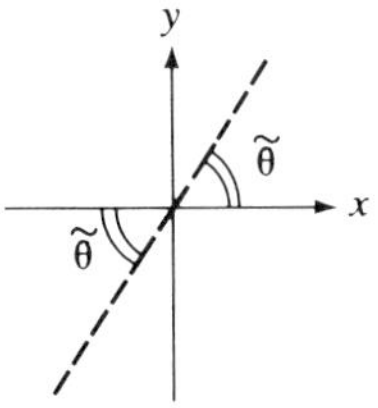

■

Example 10 The length of the hypotenuse of a right triangle is 8 meters. One angle of the triangle measures 34°40′. Find the approximate lengths of the legs of the triangle and the approximate area of the triangle.

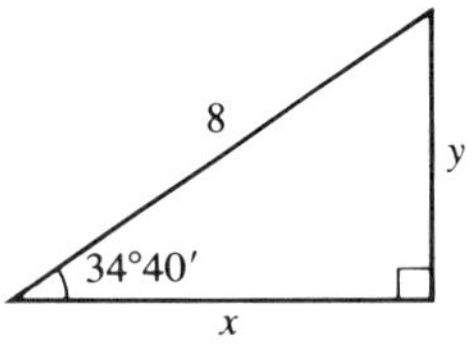

Using the sketch shown,

$$\cos 34°40' = \frac{x}{8} \quad \text{and} \quad \sin 34°40' = \frac{y}{8}$$

From Table 6.3

$$x = 8 \cos 34°40' = 8(0.8225) = 6.58 \text{ meters}$$

and

$$y = 8 \sin 34°40' = 8(0.5688) = 4.55 \text{ m}$$

The area is

$$\frac{1}{2}xy = \frac{1}{2}(6.58)(4.55) = 14.97 \text{ m}^2$$

■

Interpolation In Example 8 we wanted to look up trigonometric functions for 0.9814, but this number was not listed, so we used the closest listed number, 0.9803. For some applications more accuracy may be necessary. The process of linear interpolation used in Chapter 5 for logarithm functions can be used to obtain better approximations.

Example 11 Find $S = \sin 0.9814$ by interpolation.

$$0.0029\left\{\begin{array}{l} 0.0011\left\{\begin{array}{l}0.9803 \\ 0.9814\end{array}\right. \\ \quad\quad\quad\;\; 0.9832\end{array}\right. \quad \begin{array}{c} \textit{Angle} \\ \end{array} \quad \begin{array}{c}\textit{Sine} \\ \left.\begin{array}{c}0.8307 \\ S\end{array}\right\} x \\ 0.8323\end{array}\left.\right\} 0.0016$$

We approximate x by equating ratios:

$$\frac{x}{0.0016} = \frac{0.0011}{0.0029}$$

to obtain $x \simeq 0.0006$. Thus $S \simeq 0.8307 + 0.0006 = 0.8313$. ■

Example 12 If $\cos \theta = 0.8410$, find θ by interpolation.

$$\begin{array}{c c c c c}
 & \textit{Angle} & \textit{Cosine} & & \\
10' \left\{ \begin{array}{l} x \left\{ \begin{array}{l} 32°40' \\ \theta \end{array} \right. \\ \quad 32°50' \end{array} \right. & & \begin{array}{l} \left. \begin{array}{l} 0.8418 \\ 0.8410 \end{array} \right\} 0.0008 \\ 0.8403 \end{array} & \left. \vphantom{\begin{array}{l} a \\ b \\ c \end{array}} \right\} 0.0015 &
\end{array}$$

Equating ratios,

$$\frac{x}{10'} = \frac{0.0008}{0.0015}$$

to obtain $x \simeq 5.3'$. Thus $\theta \simeq 32°40' + 5.3' = 32°45.3' = 32°45'18''$. ■

In general, tables of trigonometric functions give approximate values. We know how to find exact values for special angles. It is also possible to find exact values for some other angles using trigonometric relationships (15° is an example). This problem is investigated in Chapter 7.

Today most college students have use of a handheld calculator, and many of these include the trigonometric functions. Calculators give more decimal-place accuracy than tables and eliminate the necessity of interpolation. As a consequence, discussion of trigonometric tables and interpolation as applied to those tables will surely become obsolete within the next few years. In the meantime, however, the student should be warned that some teachers and some mathematics departments do not permit use of the handheld calculator on examinations.

Exercises 6.5

In Exercises 1–10 find the reference angle for the given angle.

1. 75° **2.** 100° **3.** 280° **4.** $\frac{9\pi}{8}$ **5.** $\frac{13\pi}{8}$

6. $\frac{5\pi}{11}$ **7.** −820° **8.** −950° **9.** $-\frac{15\pi}{12}$ **10.** $-\frac{17\pi}{30}$

In Exercises 11–20 find the reference number for the given real number.

11. 2 **12.** 3 **13.** $\frac{13\pi}{12}$ **14.** $\frac{15\pi}{8}$ **15.** π^2

16. e^2 **17.** -4 **18.** -7 **19.** $-\frac{7\pi}{3}$ **20.** $-\frac{3\pi}{4}$

21. If $\sin 40° = 0.6428$, find $\sin 140°$, $\sin 220°$, and $\sin 320°$.
22. If $\cos 10° = 0.9848$, find $\cos 170°$, $\cos 190°$, and $\cos 350°$.
23. If $\tan \frac{\pi}{15} = 0.2125$, find $\tan \frac{14\pi}{15}$, $\tan \frac{16\pi}{15}$, and $\tan \frac{29\pi}{15}$.
24. If $\tan \frac{3\pi}{7} = 4.3813$, find $\tan \frac{4\pi}{7}$, $\tan \frac{10\pi}{7}$, and $\tan \frac{11\pi}{7}$.

In Exercises 25–30 find all values of θ, $-2\pi \le \theta \le 2\pi$, that satisfy the specified condition.

25. $\tan \theta = -1$ **26.** $\sin \theta = \frac{\sqrt{2}}{2}$ **27.** $\sin \theta = \frac{1}{2}$

28. $\cot \theta = -\sqrt{3}$ **29.** $\cos \theta = -\frac{\sqrt{2}}{2}$ **30.** $\cos \theta = -\frac{1}{2}$

Use Table V to find approximate values for the six trigonometric functions at the specified angles in Exercises 31–36.

31. 15°20′ **32.** 18°40′ **33.** 50°10′
34. 72°30′ **35.** 0.9192 **36.** 0.8407

In Exercises 37–42 use Table V to find an acute angle satisfying the given condition.

37. $\cos \theta = 0.8601$ **38.** $\cot \theta = 12.25$ **39.** $\tan \theta = 4.219$
40. $\sin \theta = 0.9250$ **41.** $\sec \theta = 1.695$ **42.** $\csc \theta = 1.071$

In Exercises 43–48 use Table V to find approximate values of the trigonometric functions at the following angles or numbers.

43. 290°20′ **44.** 143°40′ **45.** 3.9823

46. 2.4900 **47.** $\frac{7\pi}{12}$ **48.** $\frac{23\pi}{15}$

49. Find cos 1.121 by interpolation.
50. Find tan 1.5 by interpolation.

In Exercises 51–52 use interpolation to find an acute angle θ satisfying the given condition. Answer in degree measure.

51. $\sin \theta = 0.3920$ **52.** $\cos \theta = 0.4215$

In Exercises 53–58, for the right triangle illustrated and the given information, find the length of the remaining sides and the area of the triangle.

53. $\alpha = 48°20'$; $a = 5$ **54.** $\beta = 56°30'$; $b = 2$
55. $\beta = 77°10'$; $c = 10$ **56.** $\alpha = 81°40'$; $c = 10$
57. $\beta = 28°50'$; $b = 3$ **58.** $\alpha = 17°50'$; $a = 4$

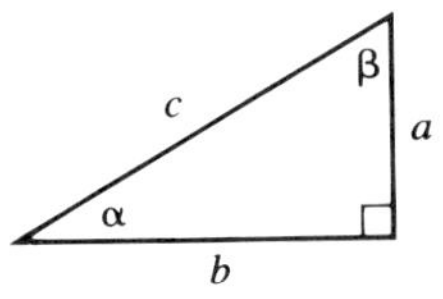

For Your Calculator

Multiples of π are needed in determining reference numbers. On most calculators one simply presses a key $\boxed{\pi}$ to display π.

In Exercises 59–60 use your calculator to approximate the reference number for the given real number t.

59. $t = e^5 + 10^{7/5}$

60. $t = \tan 1.55 - (5.1)^5$

In Exercises 61–64 use Table V and interpolation to approximate the trigonometric value. Then compare this approximation with the one obtained using your calculator.

61. $\sin 61°17'$

62. $\cos 23°53'$

63. $\tan 5$

64. $\csc 7$

65. For the right triangle illustrated in Exercise 53, approximate the length of the remaining sides and the area of the triangle if $c = \sqrt[3]{1{,}500}$ and $\beta = 47°23'37''$.

6.6 Graphs of the Trigonometric Functions

As with most functions, the graphs of the trigonometric functions are an essential ingredient in the understanding of the behavior of the function. In this section we investigate the graphs of all the trigonometric functions, with special emphasis on techniques that will enable the reader to sketch quickly a portion of the graphs of these functions.

We begin with the development of the graph of $y = f(x) = \sin x$. For the graph we take the set of real numbers R as domain for the sine function. The range is the interval $[-1, 1]$. In previous sections we have observed that the sine function is periodic with period 2π. The easiest way to sketch a periodic function is to graph 1 period and then extend the graph to as large an interval of R as desired.

Let us analyze the behavior of $\sin t$ on the interval $[0, 2\pi]$. To do this, recall that $\sin t$ is the y coordinate of the point on the unit circle U with arc length measure t.

Using Figure 6.14 we observe that as t increases from 0 to $\pi/2$, the y coordinate increases from 0 to 1. On the interval $[\pi/2, \pi]$, the y coordinate decreases from 1 to 0, then continues to decrease to -1 on $[\pi, 3\pi/2]$. On the interval $[3\pi/2, 2\pi]$ the y coordinate then increases from -1 to 0 to complete 1 period. We summarize this information in the chart shown. To illustrate the use of the arrows, row 1 is read "as t goes from 0 to $\pi/2$, y goes from 0 to 1."

t	$y = \sin t$
$0 \to \frac{\pi}{2}$	$0 \to 1$
$\frac{\pi}{2} \to \pi$	$1 \to 0$
$\pi \to \frac{3\pi}{2}$	$0 \to -1$
$\frac{3\pi}{2} \to 2\pi$	$-1 \to 0$

It is desirable to use the standard (x, y) cartesian coordinate system rather than (t, y). This should pose no special problems if we keep in mind that in the expression $\sin x$, x is the measure of arc length of U (or an angle of measure x radians) rather than the first coordinate of the point on U with arc length x.

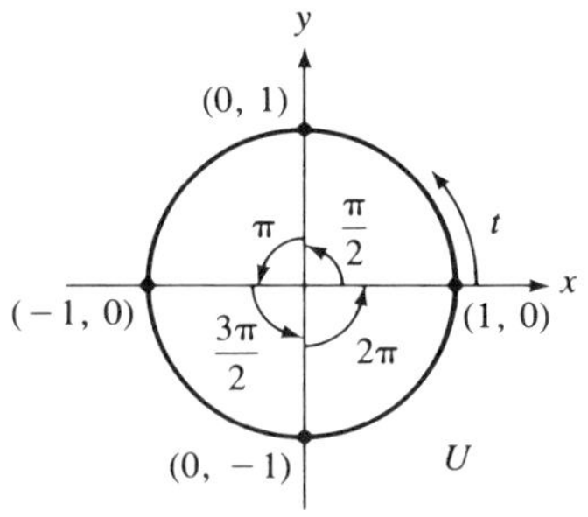

Figure 6.14

It is clear that in graphing one period of $f(x) = \sin x$ we should concentrate on $x \in \{0, \pi/2, \pi, 3\pi/2, 2\pi\}$. These five points correspond to quadrantal angles and are called the **graph-points** of $f(x) = \sin x$. To graph by the graph-points we use the following procedure.

Graph-points

> Begin at 0. Then move to the right 1/4 of the period to locate the second of the graph-points. Another move to the right of 1/4 of the period locates the third graph-point. Continue in this fashion to locate all five graph-points.

To obtain the graph of $f(x) = \sin x$ on $[0, 2\pi]$ it is easiest if we use the graph-points to set the scale. For rough graphs, we can use the approximation $\pi/2 \simeq 1.5$. Figure 6.15 illustrates the graph of $\sin x$ on $[0, 2\pi]$ obtained by using the graph-points. Of course, the graph of $\sin x$ continues indefinitely. Using periodicity, we can extend the graph to any desired interval.

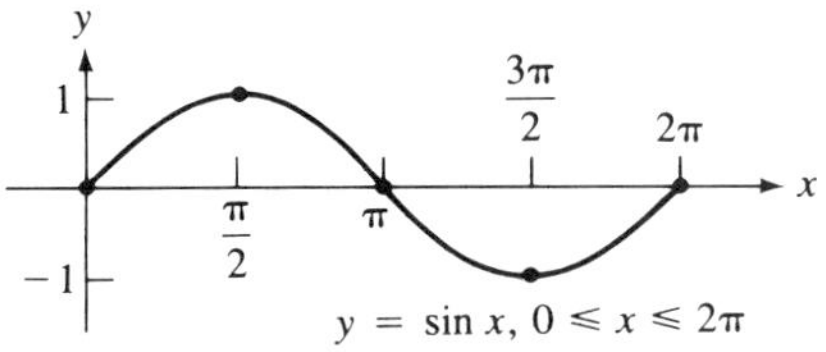

Figure 6.15

Example 1 Graph $y = \sin x$ for $-\dfrac{5\pi}{2} \leq x \leq 3\pi$.

We begin by graphing $\sin x$ on $[0, 2\pi]$, using graph-points. Then repeat the graph by moving $\frac{1}{4}$ of the period each time, continuing to retain the form of the graph.

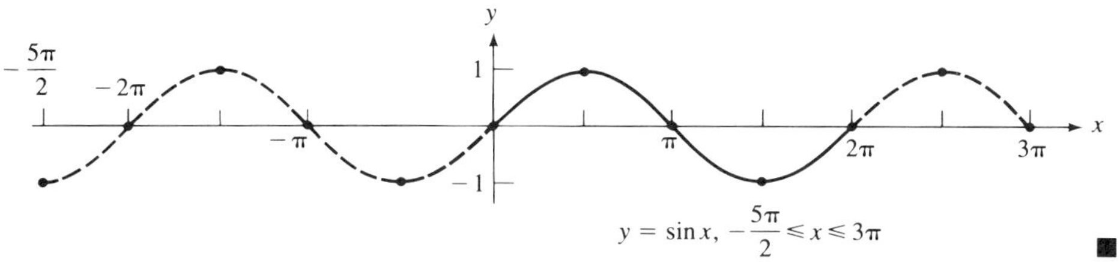

■

For $f(x) = \cos x$ we obtain the table of values shown. Then, using the graph-points, we obtain 1 period of the cosine function (see Figure 6.16).

x	$y = \cos x$
$0 \to \frac{\pi}{2}$	$1 \to 0$
$\frac{\pi}{2} \to \pi$	$0 \to -1$
$\pi \to \frac{3\pi}{2}$	$-1 \to 0$
$\frac{3\pi}{2} \to 2\pi$	$0 \to 1$

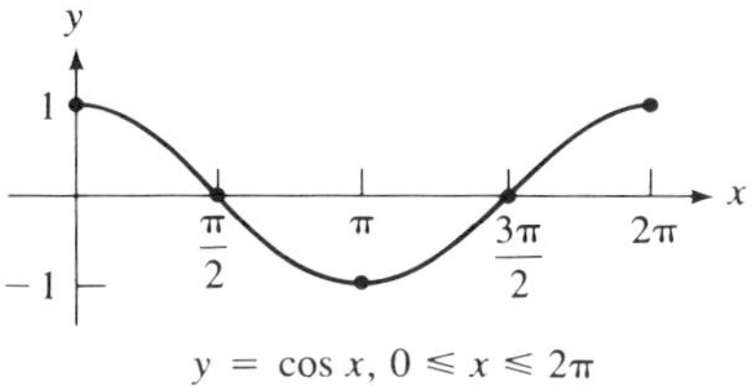

Figure 6.16

Example 2 Sketch $y = \cos x$ on the interval $\left[-\frac{5\pi}{2}, 3\pi\right]$.

We use Figure 6.16 and the procedure developed in Example 1. The result is shown in the sketch.

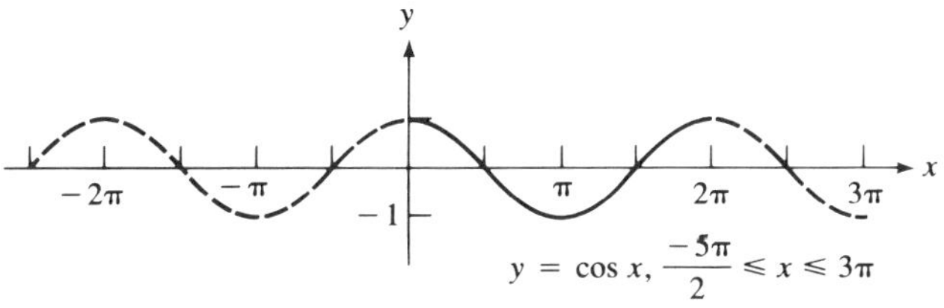

■

The reader is urged to fix the graphs of $y = \sin x$ and $y = \cos x$ on $[0, 2\pi]$ (Figures 6.15 and 6.16) firmly in mind. Reference to these graphs will yield quick and easy sketches of much more complex functions, for example $f(x) = -2 \sin(3x - \pi/6)$.

By obtaining the graphs of the tangent and cotangent functions we can easily see that both have period π. Recall that these functions are not defined for all real numbers. It is still possible to graph these functions by considering special values for x.

Example 3 Graph $y = \tan x$ on $\left[-\frac{3\pi}{2}, 2\pi\right]$.

Begin by graphing on $[0, \pi]$. Notice that if $0 < x < \pi/2$, $\tan x$ is positive and gets very large as $x \to \pi/2$. For $x > \pi/2$, $\tan x$ is negative but still large in absolute value. The line $x = \pi/2$ is an asymptote for the graph of $y = \tan x$ (see Chapter 4). A quick check will show that the behavior on the interval $(\pi, 3\pi/2)$ is a repeat of the behavior on the interval $(0, \pi/2)$. Similarly the behavior on $(3\pi/2, 2\pi)$ repeats the behavior on $(\pi/2, \pi)$. The behavior of the function is repeated on every interval of length π. The graph is shown in the sketch.

x	$y = \tan x$
0	0
$\frac{\pi}{4}$	1
$\frac{\pi}{2}$	*
$\frac{3\pi}{4}$	-1
π	0

Use of a calculator can clearly demonstrate the behavior of $\tan x$ near $\pi/2$. This will be outlined as a calculator exercise.

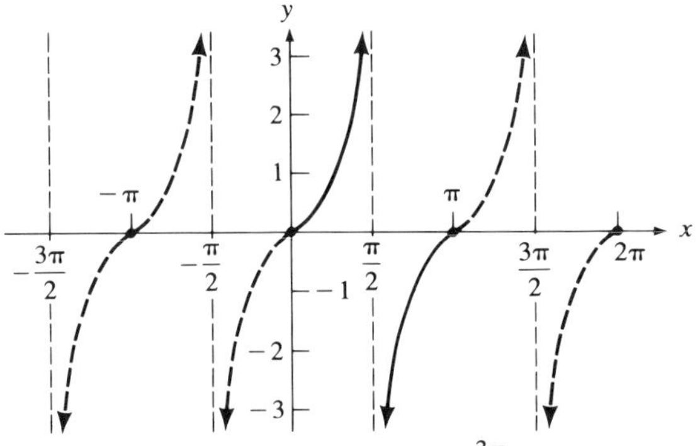

$y = \tan x, \ -\frac{3\pi}{2} < x \leq 2\pi$

■

Example 4 Graph $y = \cot x$ on $\left[-\frac{3\pi}{2}, 2\pi\right]$.

The graph for the interval $(0, \pi)$, except for $x = \pi/2$, can be obtained from the graph of $y = \tan x$ by taking reciprocals of function values. There are vertical asymptotes at each multiple of π. The graph is shown in the sketch, with the graph of $y = \tan x$ on $\left(-\frac{\pi}{2}, \pi\right)$ indicated in dashes for reference.

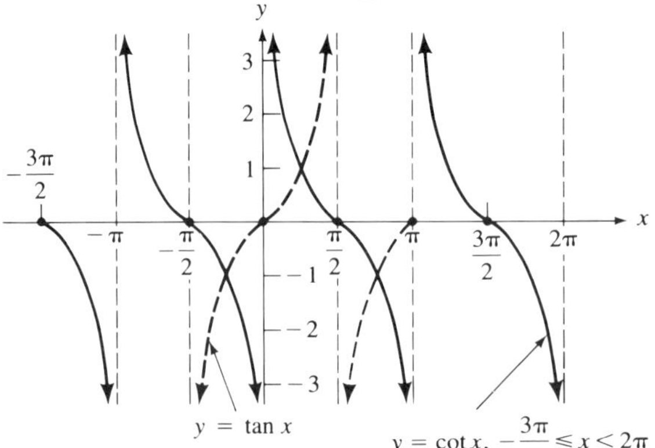

■

Example 5 Graph $y = \sec x$ on $\left[-2\pi, \frac{5\pi}{2}\right]$.

Except when $\cos x = 0$, the graph can be obtained from $y = \cos x$ by taking reciprocals of function values. There are vertical asymptotes through each odd multiple of $\pi/2$. The graph is shown in the sketch, with the graph of $y = \cos x$ on $[-\pi/2, 3\pi/2]$ indicated in dashes for reference.

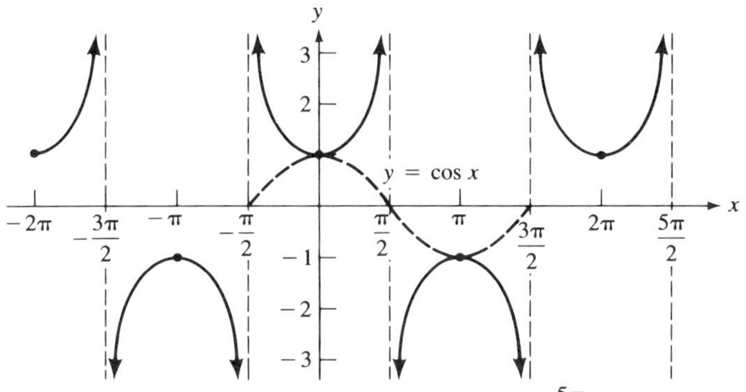

$y = \sec x,\ -2\pi \leq x \leq \frac{5\pi}{2}$ ■

Example 6 Graph $y = \csc x$ on $\left[-2\pi, \frac{5\pi}{2}\right]$.

As in Example 5, if $\sin x \neq 0$ the graph can be obtained from $y = \sin x$ by taking reciprocals. In this case there are vertical asymptotes through every integer multiple of π. The graph is shown in the sketch, with the graph of $y = \sin x$ on $[0, 2\pi]$ indicated in dashes for reference.

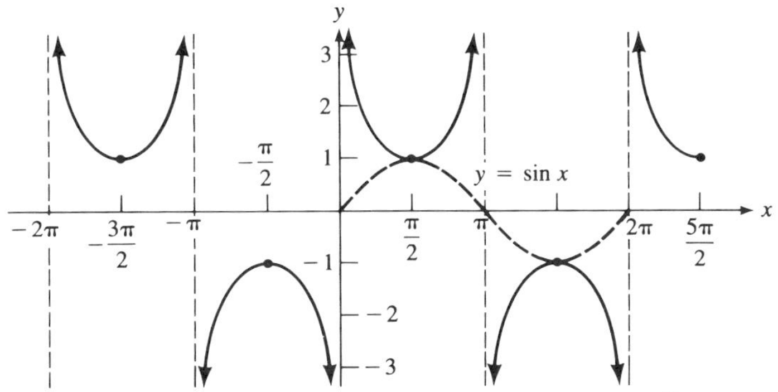

$y = \csc x,\ -2\pi \leq x \leq \frac{5\pi}{2}$ ■

It is easy to see that the graphs of the trigonometric functions sine, tangent, cotangent, and cosecant are symmetric with respect to the origin, i.e., these are odd functions. The graphs of the cosine and secant functions are symmetric with respect to the y axis, i.e., these are even functions. Thus, for all values of x we have the following relationships.

$$
\begin{aligned}
\sin(-x) &= -\sin x \\
\cos(-x) &= \cos x \\
\tan(-x) &= -\tan x \\
\cot(-x) &= -\cot x \\
\sec(-x) &= \sec x \\
\csc(-x) &= -\csc x
\end{aligned}
$$

Figure 6.17 illustrates the graphs of $y = \sin x$ and $y = \cos x$ on the same coordinate plane for the interval $[-2\pi, 3\pi]$. If the graph of $\sin x$ is moved $\pi/2$ units to the left or the graph of $\cos x$ is moved $\pi/2$ units to the right, the two graphs coincide. This is equivalent to the cofunction relation

$$\sin x = \cos(x - \pi/2).$$

We discuss this type of translation in more detail in the next section.

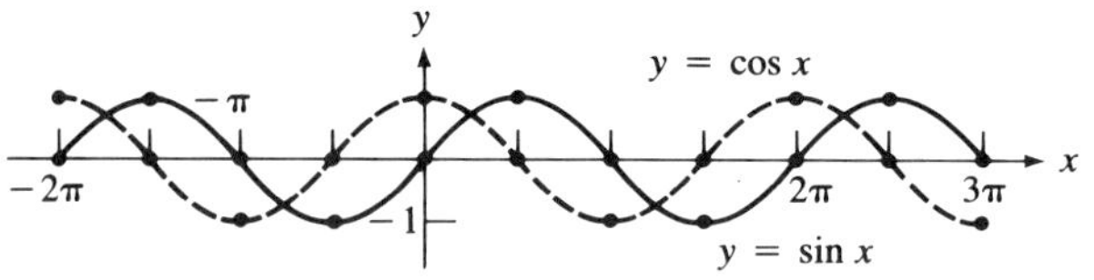

Figure 6.17

Exercises 6.6

In Exercises 1–6 sketch the graph on the interval $[2\pi, 6\pi]$.

1. $y = \sin x$
2. $y = \cos x$
3. $y = \tan x$
4. $y = \cot x$
5. $y = \sec x$
6. $y = \csc x$

In Exercises 7–16 use techniques developed in this section and previous chapters to sketch the graphs on the interval $[-2\pi, 2\pi]$. *Be creative.*

7. $y = -\sin x$
8. $y = -\cos x$
9. $y = |\sin x|$
10. $y = |\cos x|$
11. $y = \sin^2 x$
12. $y = \cos^2 x$
13. $y = \cos(x + \pi)$
14. $y = \sin(x + \pi)$
15. $y = 3 + \cos x$
16. $y = \sin x - 2$

17. Sketch the graph of $y = \sin x + \cos x$ on $[-\pi, 2\pi]$ by plotting several points.

18. Sketch the graph of $y = \sin x - \cos x$ on $[-\pi, 2\pi]$ by plotting several points.

In Exercises 19–24 analyze the graphs sketched in this section and determine the intervals in $[-\pi, 2\pi]$ for which:

19. $\sin x > \cos x$

20. $\sin x < \cos x$

21. $\tan x \geq 1$

22. $\cot x \geq 1$

23. $|\tan x| \leq 1$

24. $|\cot x| \leq 1$

In Exercises 25–30 analyze the graphs sketched in the text material for this section and determine the intervals in $[-2\pi, 2\pi]$ for which:

25. $\sec x \leq 1$

26. $\csc x \geq 1$

27. $\sin x$ is increasing

28. $\cos x$ is increasing

29. $\sec x$ is decreasing

30. $\csc x$ is decreasing

In Exercises 31–34 express $f(-\theta)$ in terms of $f(\theta)$ for the given trigonometric functions and angles.

31. $\sin(-175)$

32. $\cos\left(-\frac{121\pi}{12}\right)$

33. $\cot\left(-\frac{78\pi}{3}\right)$

34. $\tan(-1{,}000°)$

For Your Calculator

35. a. Compute $\tan t$ for $t = 1.5600, 1.5650, 1.5700, 1.5705, 1.5707, 1.57075$, and 1.57079. This illustrates that as t tends to $\frac{\pi}{2}$ from the left, $\tan t \to \infty$.

b. Compute $\tan t$ for $t = 1.5900, 1.5850, 1.5800, 1.5710, 1.5709, 1.57087$, and 1.57081. This illustrates that as t tends to $\frac{\pi}{2}$ from the right, $\tan t \to -\infty$.

36. Use $\pi = 3.1416$ to check several values for $\cot t = \frac{1}{\tan t}$ near π, on both sides, to illustrate that $x = \pi$ is a vertical asymptote for the graph of $y = \cot t$.

37. Choose 10 values for x, spaced throughout the interval $(0, 2\pi]$, and compute $\tan x$, $\tan(x + \pi)$, and $\tan(x - \pi)$ for each value. This illustrates that the tangent function has period π.

38. a. Compute $f(x) = \frac{\sin x}{x}$ in both degree and radian measure for several values of x in the interval $\left[-\frac{\pi}{2}, \frac{\pi}{2}\right]$. Include many that are small in absolute value.

b. Sketch the graph of

$$f(x) = \begin{cases} \frac{\sin x}{x}, & x \neq 0 \\ 1, & x = 0 \end{cases}$$

on the interval $\left[-\frac{\pi}{2}, \frac{\pi}{2}\right]$.

6.7 Amplitude, Period, and Phase Shift in Graphing

Variations of the trigonometric functions are frequently encountered in applications. It is not difficult to understand the behavior of these variations if we make use of the concepts of amplitude, period, and phase shift.

If $x \varepsilon R$, then $-1 \leq \sin x \leq 1$, and therefore $-2 \leq 2 \sin x \leq 2$. The graph of $f(x) = 2 \sin x$ has the same basic shape as the graph of the sine function, but this function assumes values in the interval $[-2, 2]$ rather than in $[-1, 1]$. Using the five graph-points for the sine function it is easy to obtain Figure 6.18, which illustrates the graphs of $y = \sin x$, $y = 2 \sin x$, and $y = -2 \sin x$ on $[0, 2\pi]$.

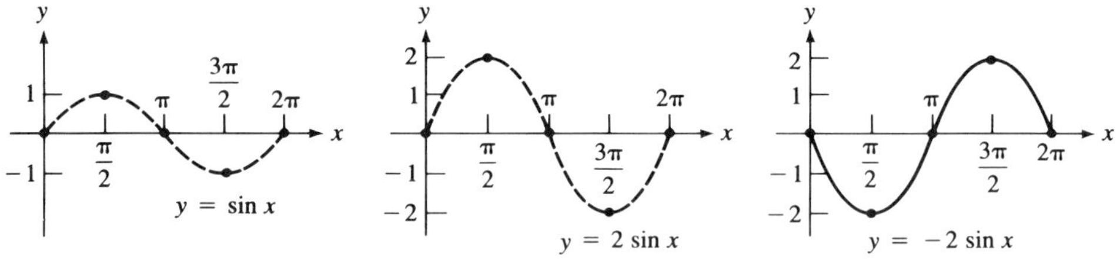

Figure 6.18

Amplitude

For the function $f(x) = a(\sin x)$, $a \varepsilon R$, we define the **amplitude** of f to be $|a|$. In this case the range of f is the interval $[-|a|, |a|]$.

We know that the period of the sine function is 2π. Consider the function $f(x) = \sin 2x$. From the chart shown we observe that $\sin 2x$ assumes all values in the interval $[-1, 1]$ for $x \varepsilon [0, \pi]$. The period of $\sin 2x$ is π. In general, for $b > 0$ we obtain 1 cycle of $\sin bx$ as bx ranges from 0 to 2π. Since x ranges from 0 to $2\pi/b$ as bx ranges from 0 to 2π, we see that the period of $\sin bx$ is $2\pi/b$.

x	$2x$	$\sin 2x$
0	0	0
$\frac{\pi}{4}$	$\frac{\pi}{2}$	1
$\frac{\pi}{2}$	π	0
$\frac{3\pi}{4}$	$\frac{3\pi}{2}$	-1
π	2π	0

To graph 1 period of $f(x) = \sin 2x$ we use knowledge of the graph of $\sin x$ and the observation that the graph-points of $\sin 2x$ are obtained by repeatedly moving to the right $\frac{1}{4}$ of the period from 0. The graph-points are

$$\{0, \pi/4, \pi/2, 3\pi/4, \pi\}$$

The sketch is shown in Figure 6.19.

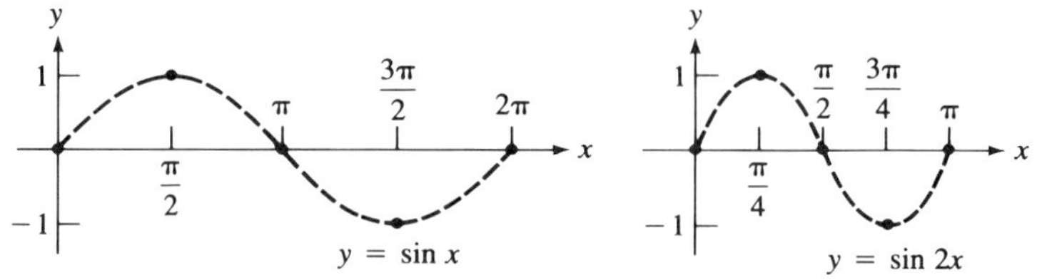

Figure 6.19

Example 1 Sketch $f(x) = -2 \cos \frac{1}{2} x$ for $-\pi \leq x \leq 4\pi$.

The amplitude of f is 2 and the period is $\frac{2\pi}{1/2} = 4\pi$. The graph-points for one period are $\{0, \pi, 2\pi, 3\pi, 4\pi\}$. We recall the graph of $y = \cos x$ shown in sketch (a). The graph of $y = 2 \cos \frac{1}{2}x$ is obtained from this as sketch (b). Sketch (c) is the desired graph after modification of (b) and extension to the desired interval.

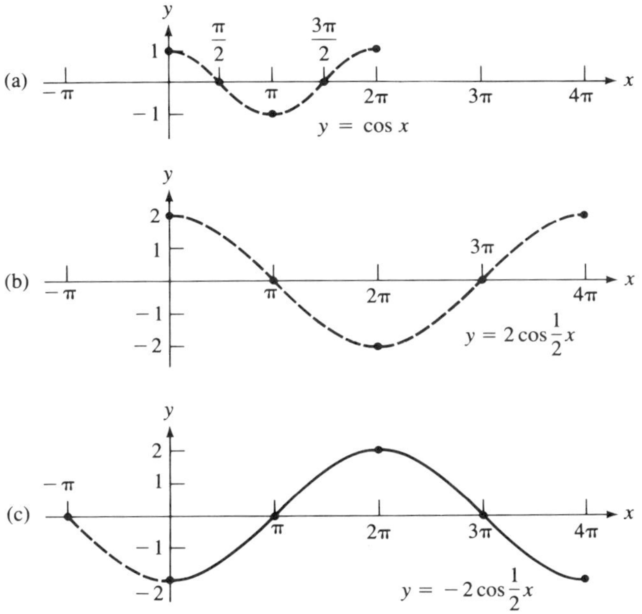

■

Sometimes it is useful to translate the graph of a trigonometric function to the left or to the right. For the functions $g(x) = \cos(x - \pi/4)$ and $f(x) = \cos x$, $g(\pi/4)$ corresponds to $f(0)$. To sketch g we simply start the graph of the cosine function at $x = \pi/4$ rather than at $x = 0$. The **phase shift** for g is $\pi/4$ units to the right. In general, the magnitude of the phase shift for $\cos(bx + c)$ is $|c/b|$. The shift is to the right if $-c/b > 0$ and to the left if $-c/b < 0$. Rather than memorize this last statement it is suggested that the point corresponding to cos 0 be located.

Phase shift

Example 2 Sketch one period of $g(x) = \cos\left(x - \frac{\pi}{4}\right)$.

The graph points are $\{\pi/4, 3\pi/4, 5\pi/4, 7\pi/4, 9\pi/4\}$. Notice that we repeatedly add $\frac{1}{4}$ of the period, starting at $\pi/4$. The sketch is easily obtained from knowledge of $y = \cos x$ on $[0, 2\pi]$.

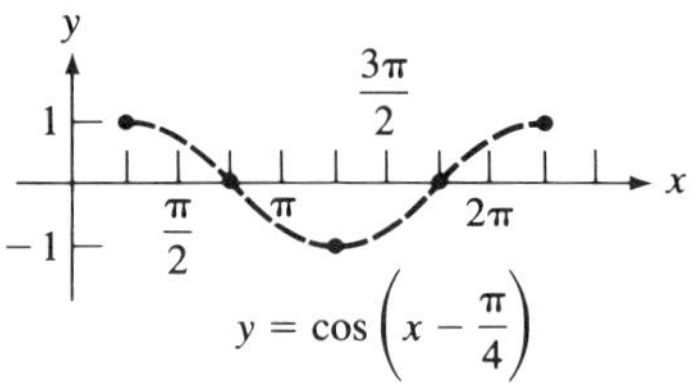

■

The general situation for sine and cosine can now be stated. For $b > 0$ and the functions $a[\sin(bx + c)]$ and $a[\cos(bx + c)]$, the amplitude of each is $|a|$, the period of each is $2\pi/b$, and the phase shift of each is $|c/b|$ units. For the case $b < 0$ we can use the relations $\sin(-\alpha) = -\sin\alpha$ and $\cos(-\alpha) = \cos\alpha$ to rewrite the expression with $b > 0$.

Example 3 Determine amplitude, period, phase shift, and direction of shift for $f(x) = 5\sin\left(-\frac{1}{3}x + 2\right)$.

We write

$$f(x) = 5\sin\left[-\left(\frac{1}{3}x - 2\right)\right] = -5\sin\left(\frac{1}{3}x - 2\right)$$

The amplitude is 5, the period is $\frac{2\pi}{1/3} = 6\pi$, and the phase shift is $\frac{2}{1/3} = 6$ units to the right. ■

Example 4 Sketch one period of $f(x) = 3\cos\left(2x + \frac{\pi}{3}\right)$.

The amplitude is 3, the period is π, and the shift is $\pi/6$ units to the left. We locate $-\pi/6$, then repeatedly move $\frac{1}{4}$ of the period ($\pi/4$ units) to the right to obtain the graph-points $\{-\pi/6, \pi/12, \pi/3, 7\pi/12, 5\pi/6\}$. The sketch is quickly obtained and is shown.

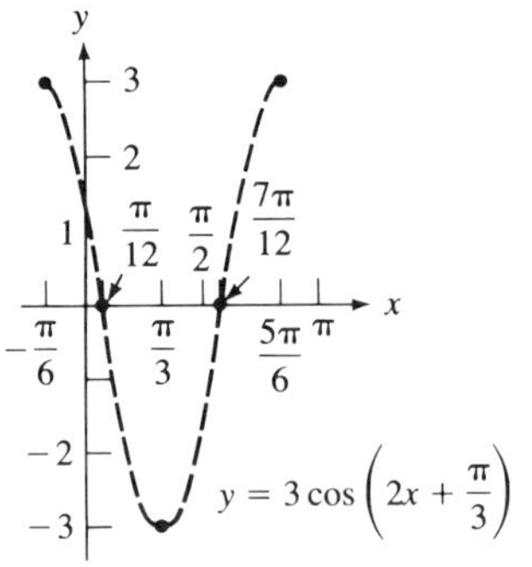

■

A similar analysis can be performed for the other trigonometric functions. The amplitude concept is of no use for these, but the concepts of sketching by period and phase shift are applicable to all the trigonometric functions.

Example 5 Sketch 1 period of $f(x) = \tan\left(\frac{1}{2}x - \frac{\pi}{4}\right)$.

The period for $\tan x$ is π, so the period of $f(x)$ is $\frac{\pi}{1/2} = 2\pi$. The phase shift is $\pi/2$ units to the right. The graphs for both $\tan x$ and $f(x)$ are shown.

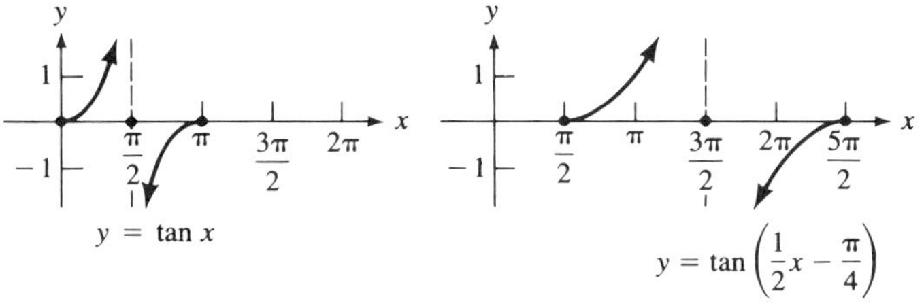

■

Addition of ordinates It is sometimes convenient to graph sums of functions by the method of addition of function values, or ordinates.

Example 6 Sketch the graph of $y = \cos 2x + \sin x$ on $[0, 2\pi]$.

We sketch the graphs of $y = \cos 2x$ and $y = \sin x$ as dashes. The graph of $y = \cos 2x + \sin x$ is then obtained by adding the y coordinates of points on the two curves having the same x coordinate. The result is the solid curve.

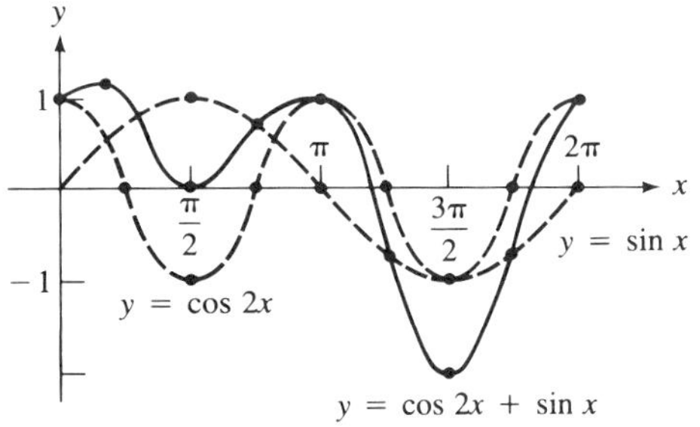

■

The next examples illustrate the significance of amplitude, period, and phase shift in two of many practical applications.

Example 7 Electromagnetic waves are described by an equation of the form $W = A \sin(Bt + C)$, where t is time, and phase shift is related to distance from the source of the wave. Radio waves are forms of electromagnetic waves, as are x rays, television waves, and microwaves. Radio stations broadcast in either AM (amplitude modulation) or FM (frequency modulation). In somewhat simplified terms, the loudness of a sound depends on the amplitude of its carrier wave, whereas the pitch of a sound depends on its frequency (recall from Section 6.4 that frequency is the reciprocal of the period). We see then that a higher pitch corresponds to a shorter period. In radio broadcast, sounds are attached to the carrier waves as sums of sine curves. Musicians agree that it is the relative amplitude of these sine curves that gives quality to sound.

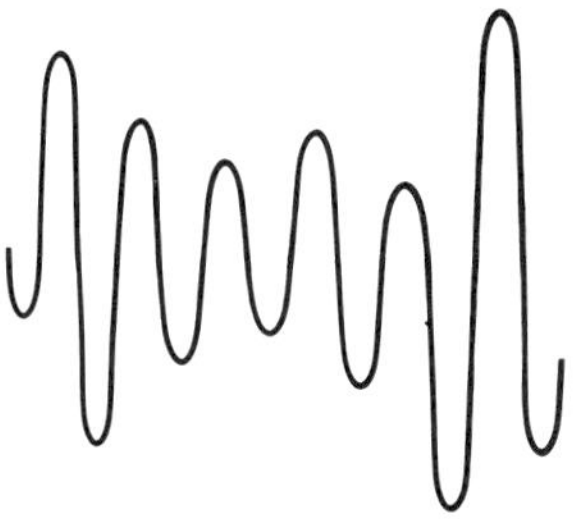

AM: The carrier wave has constant period and variable amplitude.

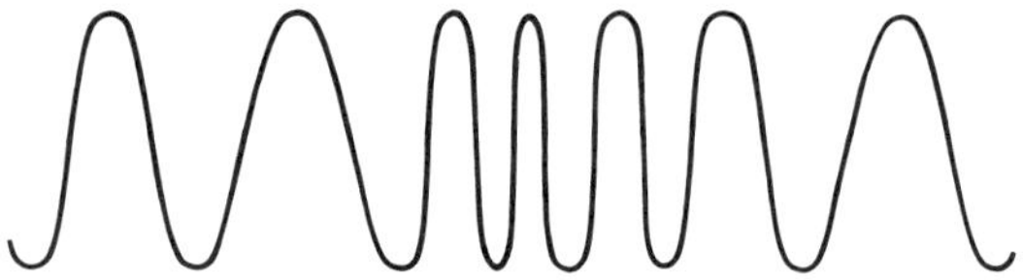

FM: The carrier wave has constant amplitude and variable period. ■

Example 8 An alternating current generator produces a current of I amperes in an electrical circuit whose value at time t seconds is given by

$$I(t) = 10e^{-t} \cos \pi t$$

Sketch the graph of $I(t)$ for $0 \leq t \leq 6$.

The function $\cos \pi t$ has period $2\pi/\pi = 2$. The amplitude of $I(t)$ is damped by the factor $10e^{-t}$ as time increases. Making use of the graph-points of $\cos \pi t$ and sketches of the graphs of the curves $y = \pm 10e^{-t}$, we obtain the sketch of $I(t)$ shown.

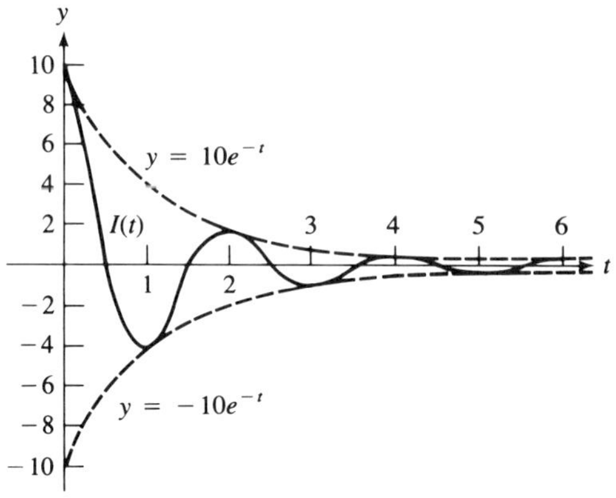

■

Exercises 6.7

In Exercises 1–8 determine the amplitude, period, phase shift, and direction of shift for the given functions.

1. $y = \sin\left(x - \frac{\pi}{2}\right)$

2. $y = \cos\left(2x - \frac{\pi}{2}\right)$

3. $y = -\frac{1}{2}\cos\left(\frac{\pi}{6} - 3x\right)$

4. $y = 4\sin\left(2 - \frac{\pi}{2}x\right)$

5. $y = \pi \sin(2\pi x + \pi)$

6. $y = -2\sin\left(\frac{1}{2}x - \pi\right)$

7. $y = -\frac{1}{3}\cos\left(\frac{1}{3}\pi x + 3\right)$

8. $y = \sqrt{2}\cos\left(\frac{1}{2}\pi x + 6\pi\right)$

In Exercises 9–36 sketch the graph of the function, including at least 1.5 periods.

9. $y = 2\cos 2x$

10. $y = 3\sin 3x$

11. $y = -3\sin\frac{1}{3}x$

12. $y = -\cos\left(-\frac{1}{2}x\right)$

13. $y = \sin 2\pi x$

14. $y = \cos\frac{\pi x}{3}$

15. $y = |-2\sin 3x|$

16. $y = \left|-2\sin\frac{1}{2}x\right|$

17. $y = \cos\left(x - \frac{\pi}{2}\right)$

18. $y = \sin\left(x - \frac{\pi}{2}\right)$

19. $y = \sin(x + 2)$

20. $y = \cos(x - 1)$

21. $y = \sin\pi(x + 2)$

22. $y = \cos\pi(x - 1)$

23. $y = 2\cos\left(2x - \frac{\pi}{3}\right)$

24. $y = \sin(3x + \pi)$

25. $y = \frac{1}{2}\sin\left(\frac{1}{2}x - \frac{\pi}{4}\right)$

26. $y = 2\cos\left(\frac{1}{2}x - \frac{\pi}{6}\right)$

27. $y = 3\cos\left(\frac{1}{2}\pi x + \frac{\pi}{6}\right)$

28. $y = 3\sin\left(\frac{1}{2}\pi x - \frac{\pi}{4}\right)$

29. $y = 4 + 2\sin\left(\pi x - \frac{\pi}{2}\right)$

30. $y = -2\cos(\pi x + \pi) - 2$

31. $y = -\cot\left(x + \frac{\pi}{4}\right)$

32. $y = -\tan\left(\frac{\pi}{4} - x\right)$

33. $y = 3\sec\left(2x - \frac{\pi}{2}\right)$

34. $y = \cot\left(\frac{1}{2}x + \frac{\pi}{4}\right)$

35. $y = -2\tan(2x - \pi)$

36. $y = 2\csc(2x + \pi)$

In Exercises 37–40 sketch the graph of the function on the interval $[-\pi, 2\pi]$.

37. $y = \sin\frac{1}{2}x + \cos 2x$

38. $y = \sin\frac{1}{2}\pi x - 2\cos\pi x$

39. $y = e^x - \cos x$

40. $y = e^{-x}\cos x$

For Your Calculator

In Exercises 41–42 sketch the graph of the function by plotting values of the function at 13 equally spaced points in the specified interval. Include the interval endpoints. This will verify the correctness of Examples 4 and 5 of this section, in which the graphs were obtained by modification of other graphs.

41. $f(x) = 3\cos\left(2x + \frac{\pi}{3}\right), \left[-\frac{\pi}{6}, \frac{5\pi}{6}\right]$

42. $f(x) = \tan\left(\frac{1}{2}x - \frac{\pi}{4}\right), \left[\frac{\pi}{2}, \frac{5\pi}{2}\right]$

6.8 Inverse Trigonometric Functions

Example 1 Find x if $\sin x = \frac{1}{2}$.

From our knowledge of the 30°-60°-90° triangle we know that $\sin \pi/6 = \frac{1}{2}$. But also, $\sin 5\pi/6 = \frac{1}{2}$. In fact, if t has reference number $\pi/6$ and the terminal side of the corresponding angle is in quadrants I or II, then $\sin t = \frac{1}{2}$. We conclude that the solution set for $\sin x = \frac{1}{2}$ is

$$\{t\colon t = \frac{\pi}{6} \pm 2k\pi \quad \text{or} \quad t = \frac{5\pi}{6} \pm 2k\pi, \qquad k \in Z\}$$ ■

Example 1 emphasizes the fact that periodic functions in general and trigonometric functions in particular are not 1-1. Thus the trigonometric functions do not have inverses. We may, however, use the procedure of Section 3.6 and restrict the domains of the trigonometric functions to make them 1-1; the resulting functions will then have inverses.

An inspection of the graph of $y = \sin x$ (Figure 6.20a) provides many choices for the restricted domain. The sine function is 1-1 on the intervals $[-3\pi/2, -\pi/2]$, $[-\pi/2, \pi/2]$, $[\pi/2, 3\pi/2]$, and infinitely many other intervals.

It is customary to choose the interval containing 0, namely $[-\pi/2, \pi/2]$. On this interval $y = \sin x$ is 1-1 and has the interval $[-1, 1]$ as range. We denote the inverse for this restricted sine function by $\arcsin x$ and read "the arc sine of x" or "the inverse sine of x." It is also standard to use the inverse function notation developed in Chapter 3, so that $\sin^{-1} x = \arcsin x$. The reader must note that the superscript -1 is used to indicate an inverse function and is *not* an exponent. This is the exception mentioned in Section 6.2. We observe that the domain of $\arcsin x$ is $[-1, 1]$ and the range of $\arcsin x$ is $[-\pi/2, \pi/2]$.

> For $-1 \le x \le 1$,
>
> $$-\frac{\pi}{2} \le \arcsin x \le \frac{\pi}{2}$$
>
> and
>
> $$\sin(\arcsin x) = x$$

The graphs of $y = \sin x$ (restricted) and $y = \arcsin x$ are symmetric with respect to the line $y = x$, as discussed in Chapter 3 (see Figure 6.20 for both graphs).

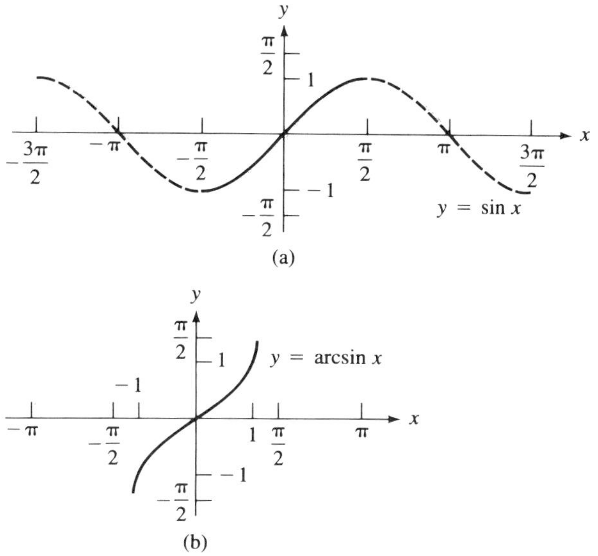

Figure 6.20

Example 2 Find **(a)** $\arcsin \frac{1}{2}$; **(b)** $\arcsin\left(-\frac{1}{\sqrt{2}}\right)$; **(c)** arcsin(sin 3); **(d)** $\sin\left(\arcsin \frac{1}{e}\right)$.

a. Refer to Example 1. Many people read $\arcsin \frac{1}{2}$ as "the number (or angle) whose sine is $\frac{1}{2}$" and keep in mind that the number must be chosen from the interval $[-\pi/2, \pi/2]$. We see that $\arcsin \frac{1}{2} = \pi/6$.

b. The reference angle for the angle whose sine is $-1/\sqrt{2}$ is $\pi/4$. Thus $\arcsin(-1/\sqrt{2}) = -\pi/4$. Notice that $7\pi/4$ is the wrong answer even though $\sin 7\pi/4 = -1/\sqrt{2}$, because $7\pi/4 \notin [-\pi/2, \pi/2]$.

c. The answer is not 3. The number 3 is not in the domain of the restricted sine function. However, sin 3 is in the domain of the arcsine function. Observe that $\sin 3 = \sin(\pi - 3)$ and $(\pi - 3) \in [-\pi/2, \pi/2]$.
Thus $\arcsin(\sin 3) = \arcsin[\sin(\pi - 3)] = \pi - 3$.

d. $\sin(\arcsin 1/e) = 1/e$. ■

It is important to give special emphasis to the statement in the solution of Example 2**a**. The number whose sine is $\frac{1}{2}$ is not 30°. The range of the inverse sine function consists of those real numbers in the interval $[-\pi/2, \pi/2]$. Use of the angle terminology is convenient, but radian measure must be used for the inverse trigonometric functions. Thus, for example, the angle whose tangent is 1 is $\pi/4$. It is not 45°.

Inverse functions for the remaining trigonometric functions are developed in an analogous manner. We will use the restrictions indicated in the following display. The reader is cautioned that some users and some authors use different intervals for the arcsecant, arccosecant, and arccotangent functions.

WARNING
Be sure to choose values of the inverse trigonometric functions that are in the appropriate interval.

Function	Domain	Range
$y = \arcsin x$	$-1 \le x \le 1$	$-\pi/2 \le y \le \pi/2$
$y = \arccos x$	$-1 \le x \le 1$	$0 \le y \le \pi$
$y = \arctan x$	R	$-\pi/2 < y < \pi/2$
$y = \operatorname{arccot} x$	R	$0 < y < \pi$
$y = \operatorname{arcsec} x$	$\lvert x\rvert \ge 1$	$0 \le y \le \pi,\ y \ne \pi/2$
$y = \operatorname{arccsc} x$	$\lvert x\rvert \ge 1$	$-\pi/2 \le y \le \pi/2$ $y \ne 0$

The graphs for four of the inverse trigonometric functions on the appropriate intervals are shown in Figure 6.21, with the graph of the corresponding restricted trigonometric function and the line $y = x$ illustrated with dashed curves.

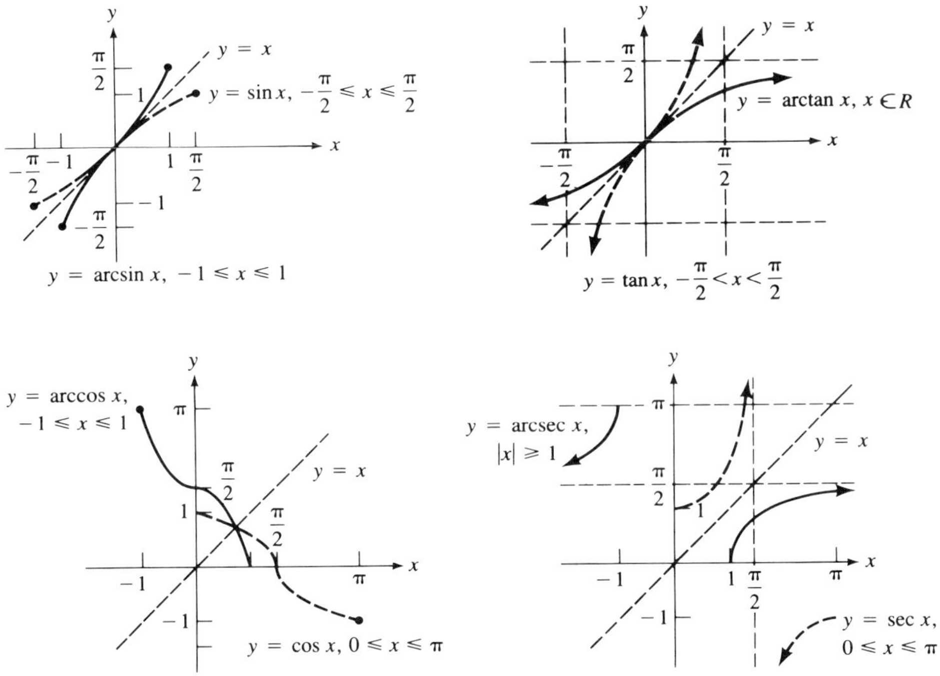

Figure 6.21

Example 3 Find **(a)** $\arccos\left(-\frac{1}{2}\right)$; **(b)** $\tan^{-1}(-1)$; **(c)** $\sec^{-1}\left(-\frac{2}{\sqrt{3}}\right)$; **(d)** $\csc^{-1}\left(\frac{1}{2}\right)$; **(e)** $\arctan(-1.501)$.

a. Let $t = \arccos\left(-\frac{1}{2}\right)$. The acute angle whose cosine is $\frac{1}{2}$ is $\pi/3$. Thus the reference angle $\tilde{t}$ is $\pi/3$. Since $\cos t < 0$ and $0 \le \cos^{-1} x \le \pi$, the terminal side of t is in quadrant II. We see that $t = \pi - \pi/3 = 2\pi/3$.

b. $\tan \pi/4 = 1$ and $-\pi/2 < \tan^{-1} x < \pi/2$. Thus $\tan^{-1}(-1) = -\pi/4$.
c. $\sec \pi/6 = 2/\sqrt{3}$ and $0 \leq \sec^{-1} x \leq \pi$. Thus $\sec^{-1}(-2/\sqrt{3}) = 5\pi/6$.
d. $\frac{1}{2}$ is not in the domain of arccsc x. This means that there is no angle whose cosecant is $\frac{1}{2}$.
e. From Table 6.3 we see that $\tan 0.9832 = 1.501$. Thus $\arctan(-1.501) = -0.9832$. ■

Example 4 Find **(a)** $\cos\left(\arcsin \frac{\sqrt{3}}{2}\right)$; **(b)** $\sec[\arctan(-1)]$; **(c)** $\tan(\tan^{-1} 1)$; **(d)** $\tan^{-1}\left(\tan \frac{3\pi}{4}\right)$.
a. $\cos(\arcsin \sqrt{3}/2) = \cos \pi/3 = \frac{1}{2}$.
b. $\sec[\arctan(-1)] = \sec(-\pi/4) = \sqrt{2}$.
c. $\tan(\tan^{-1} 1) = \tan \pi/4 = 1$. In general, the identity analogous to $\tan(\tan^{-1} x) = x$ holds for all the inverse trigonometric functions.
d. $\tan^{-1}(\tan 3\pi/4) = \tan^{-1}(-1) = -\pi/4$. The result here is a warning associated with the identity mentioned in part **(c)**. ■

In the general situation it is also possible to write expressions such as parts **(a)**, **(b)**, and **(c)** of Example 4 without using inverse function notation. We do this by using the right triangle terminology developed in Section 6.3.

Example 5 Find **(a)** $\cos(\arcsin x)$; **(b)** $\sec\left[\arctan\left(\frac{x-3}{2}\right)\right]$.
a. If $\alpha = \arcsin x$, then the reference angle $\tilde{\alpha}$ can be viewed as an angle contained in the right triangle shown.

$$\sin(\tilde{\alpha}) = \frac{x}{1} = \frac{\text{side opp}}{\text{hyp}}$$

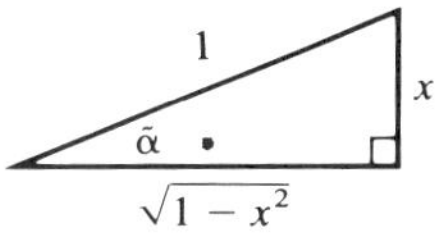

Hence the sides have the proportions indicated. We compute the side adjacent to $\tilde{\alpha}$ by using the Pythagorean theorem. Then

$$\cos \tilde{\alpha} = \frac{\text{side adj}}{\text{hyp}} = \sqrt{1 - x^2}$$

and hence

$$\cos(\arcsin x) = \sqrt{1 - x^2}$$

b. Let $\alpha = \arctan\left(\frac{x-3}{2}\right)$. Form the triangle using

$$\tan \tilde{\alpha} = \frac{\text{side opp}}{\text{side adj}} = \frac{x-3}{2}$$

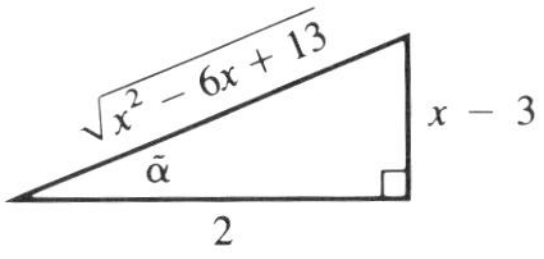

We compute the hypotenuse:

$$\sqrt{(x-3)^2+2^2}=\sqrt{x^2-6x+13}$$

and hence

$$\sec\tilde{\alpha}=\frac{\text{hyp}}{\text{side adj}}=\frac{1}{2}\sqrt{x^2-6x+13}$$

We see that

$$\sec\left[\arctan\left(\frac{x-3}{2}\right)\right]=\frac{1}{2}\sqrt{x^2-6x+13}$$ ■

The thoughtful reader might wonder about the sudden jump from $\tilde{\alpha}$ to α at the end of both parts of Example 5 and the possible loss of the correct algebraic sign. It can be verified by careful analysis of the quadrants involved that the procedure we followed will automatically include the correct sign. For example, $\cos t \geq 0$ if $-\pi/2 \leq t \leq \pi/2$, and in part **(a)** we have $-\pi/2 \leq \arcsin x \leq \pi/2$ for all x. Thus $\cos(\arcsin x) \geq 0$.

Exercises 6.8

In Exercises 1–18 find exact values of the inverse trigonometric functions.

1. $\arcsin\frac{\sqrt{3}}{2}$

2. $\arctan(-1)$

3. $\arccos(-1)$

4. $\arcsin(-1)$

5. $\arctan(-\sqrt{3})$

6. $\operatorname{arccot}\left(-\frac{1}{\sqrt{3}}\right)$

7. $\operatorname{arcsec}(-2)$

8. $\arccos\frac{1}{2}$

9. $\operatorname{arcsec} 1$

10. $\sin^{-1} 0$

11. $\tan^{-1}\left(-\frac{1}{\sqrt{3}}\right)$

12. $\cot^{-1}(-1)$

13. $\cos^{-1} 0$

14. $\sec^{-1}\left(\frac{2\sqrt{3}}{3}\right)$

15. $\sin^{-1}\left(-\frac{1}{2}\right)$

16. $\csc^{-1}(-1)$

17. $\csc^{-1}(-\sqrt{2})$

18. $\operatorname{arccsc}\sqrt{2}$

In Exercises 19–26 use Table V to approximate the values.

19. $\arcsin 0.4797$

20. $\operatorname{arccot} 3.412$

21. $\arccos(-0.8225)$

22. $\arccos(-0.6583)$

23. $\tan^{-1}(-7.429)$

24. $\sin^{-1}(-0.8760)$

25. $\sec^{-1}(-1.382)$

26. $\csc^{-1}(-1.175)$

In Exercises 27–38 find exact values.

27. $\sin^{-1}\left[\cos\left(-\frac{\pi}{4}\right)\right]$

28. $\cos^{-1}\left[\sin\left(-\frac{\pi}{2}\right)\right]$

29. $\cos^{-1}\left[\cos\left(-\frac{\pi}{6}\right)\right]$

30. $\tan^{-1}\left[\tan\left(-\frac{2\pi}{3}\right)\right]$

31. sec(arcsec 3)

32. $\tan\left(\arccos\frac{3}{\sqrt{13}}\right)$

33. $\cot\left(\operatorname{arccsc}\frac{13}{5}\right)$

34. cot[arccot(−4)]

35. arcsin(sin 6)

36. arccos[cos(−4)]

37. $\sin^{-1}\left(-\frac{\sqrt{3}}{2}\right) + \tan^{-1}(-1)$

38. $\cos^{-1}\left(-\frac{1}{2}\right) + \sec^{-1}(-\sqrt{2})$

In Exercises 39–44 write the expressions without using inverse function notation (i.e., as algebraic expressions in x).

39. $\sin(\arccos x)$

40. $\tan(\arcsin x)$

41. $\sec\left[\operatorname{arccot}\left(\frac{x+2}{4}\right)\right]$

42. $\csc\left[\arctan\left(\frac{x-2}{5}\right)\right]$

43. $\tan(\arcsin\sqrt{1-x^2})$

44. $\sin\left(\arccos\frac{\sqrt{9-4x^2}}{3}\right)$

45. On the same coordinate plane sketch the graphs of the functions $y = \operatorname{arccot} x$, $y = x$, and $y = \cot x$ for the appropriate intervals, as illustrated in Figure 6.21.

46. On the same coordinate plane sketch the graphs of the functions $y = \operatorname{arccsc} x$, $y = x$, and $y = \csc x$ for the appropriate intervals, as illustrated in Figure 6.21.

For Your Calculator

The means for computing the inverse trigonometric functions on your calculator will depend on the model you have. Usually one presses a key, [inv], [arc], or [−1] on some models, then the key for the desired trigonometric function. Also, the output can be in either the degree mode or the radian mode.

[deg] [arc] [sin] .5 answer 30 *(30 degrees)*
[rad] [arc] [sin] .5 answer .524 *($\frac{\pi}{6}$ radians)*

In Exercises 47–54 compute the first expression in both degree and radian modes and compare your answer with the second expression. Explain any difference.

47. sin(arcsin 0.5687); 0.5687

48. $\sin^{-1}(\sin 1)$; 1

49. $\sin^{-1}(\sin 3)$; 3

50. arccos(cos 200); 200

51. arctan(tan 2); 2

52. cos[arccos(−0.1234)]; −0.1234

53. arccos[cos(−100)]; 100

54. arctan(tan 300); 300

55. Sketch the graph of the function $f(x) = \arccos x$ by plotting several points in the interval $[-1, 1]$ and thereby verify the correctness of that part of Figure 6.21.

56. **a.** Compute $\arctan x$ for $x = 5, 10, 10^2, 10^3, 10^{10}, 10^{45}, 10^{98}$ in both degree and radian modes.
b. Repeat part **(a)** for the negatives of the values of x given.

Chapter Review

In Exercises 1–2 sketch the angles in standard position.

1. $110°; -\frac{7\pi}{6}; -435°$

2. $-250°; -\frac{2\pi}{3}; \frac{13\pi}{6}$

In Exercises 3–10 write the given angle using the specified measure.

3. 24°; radians

4. −105°; radians

5. $-\frac{11\pi}{6}$; degrees

6. $\frac{5\pi}{4}$; degrees

7. 31.17°; degrees, minutes, seconds

8. 83.53°; degrees, minutes, seconds

9. 41°23′17″; radians

10. 11°38′22″; radians

In Exercises 11–18 find the values of the trigonometric functions at θ if:

11. $\left(-\frac{1}{3}, 3\right)$ is on the terminal side of θ.

12. $\left(-2, -\frac{2}{3}\right)$ is on the terminal side of θ.

13. θ is in quadrant IV and the terminal side of θ lies on the line $y = -\frac{4}{3}x$.

14. θ is in quadrant III and the terminal side of θ is parallel to the line $2x - 3y = 6$.

15. $\sin\theta = -\frac{5}{\sqrt{61}}$ and $\cos\theta = -\frac{6}{\sqrt{61}}$.

16. $\tan\theta = -\frac{2}{\sqrt{3}}$ and $\cos\theta = \frac{\sqrt{3}}{\sqrt{7}}$.

17. θ is in quadrant II and $\tan\theta = -\frac{10}{\sqrt{21}}$.

18. θ is in quadrant IV and $\cos\theta = \frac{7}{25}$.

19. Find (x, y) if $\sin\theta = -\frac{6}{\sqrt{61}}$, $\cos\theta = \frac{5}{\sqrt{61}}$, and $r = 3$.

20. Find (x, y) if $\tan\theta = \frac{40}{9}$, $\cos\theta = -\frac{9}{41}$, and $r = 2$.

In Exercises 21–26 find the remaining sides for the right triangle illustrated if:

21. $\alpha = 30°; c = 4$

22. $\beta = 30°; a = 6$

23. $\alpha = 45°; b = 4$

24. $\beta = 45°; c = 2$

25. $a = 3; \cot\alpha = 10$

26. $c = 5; \sec\beta = 2$

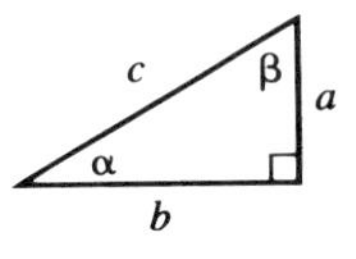

In Exercises 27–28 find the values for the trigonometric functions at α and β for the triangle illustrated above, if:

27. $a = 40;\ b = 9$

28. $b = 6;\ c = \sqrt{61}$

In Exercises 29–30 find the exact values of the trigonometric functions at the specified angles.

29. $-135°;\ \frac{5\pi}{3};\ -3\pi$

30. $-150°;\ \frac{7\pi}{4};\ -\frac{5\pi}{2}$

31. Express $\tan \frac{9\pi}{20}$ as a function of an angle θ such that $0 \le \theta \le \frac{\pi}{4}$.

32. Express $\cos(70°18'35'')$ as a function of an angle θ such that $0 \le \theta < 45°$.

In Exercises 33–42 find the exact values of the trigonometric functions at the real number t satisfying the specified conditions.

33. $t = \frac{4\pi}{3}$

34. $t = -\frac{5\pi}{4}$

35. $t = \frac{3\pi}{2}$

36. $t = \pi$

37. $t \in (0, \pi)$ and $\cos t = -\frac{4}{9}$

38. $t \in (0, \pi)$ and $\sec t = -\frac{5}{3}$

39. $t \in (-\pi, 0)$ and $\tan t = \sqrt{3}$

40. $t \in \left(-\frac{\pi}{2}, \frac{\pi}{2}\right)$ and $\sin t = -\frac{3}{\sqrt{13}}$

41. $\cos t = \frac{1}{4}$ and $\sin t = -\frac{\sqrt{15}}{4}$

42. $\cot t = -\frac{\sqrt{3}}{3}$ and $\csc t = \frac{2\sqrt{3}}{3}$

43. Find the real number t if $\sin t = -\frac{1}{2},\ -\frac{\pi}{2} \le t \le \pi$.

44. Find the real number t if $\cot t = -1,\ 0 \le t \le \pi$.

In Exercises 45–46 find the reference angles or reference numbers.

45. $\frac{11\pi}{8};\ 190°;\ 6$

46. $-\frac{11\pi}{6};\ 640°;\ \pi + 1$

In Exercises 47–52 find approximate values for the sine, cosine, and tangent functions at the angles or real numbers.

47. $51°20';\ -125°30'$

48. $170°40';\ -190°10'$

49. $\frac{11\pi}{12};\ -\frac{19\pi}{15}$

50. $\frac{5\pi}{9};\ -\frac{4\pi}{5}$

51. 0.5451 (interpolate)

52. 0.9172 (interpolate)

In Exercises 53–56 use Table V to find an acute angle θ satisfying the given condition.

53. $\sin\theta = 0.7392$

54. $\csc\theta = 1.202$

55. $\sec\theta = 1.777$ (interpolate)

56. $\cot\theta = 1.618$ (interpolate)

In Exercises 57–60 sketch the graphs of the functions on the interval $[-2\pi, 2\pi]$.

57. $y = \cos x - 3$

58. $y = 2 + \sin x$

59. $y = -3\tan x$

60. $y = -2\csc x$

In Exercises 61–62 determine the intervals in $[0, 4\pi]$ *for which the specified conditions are satisfied.*

61. $\sin x$ is increasing and $\sin x \geq \cos x$.

62. $\cos x$ is decreasing and $\cos x < \sin x$.

In Exercises 63–66 determine amplitude, period, and phase shift, and sketch the graph of the specified function for at least 1.5 periods.

63. $y = 2\cos\left(\frac{1}{2}x - \frac{\pi}{3}\right)$

64. $y = -2\sin(2x - \pi)$

65. $y = 2\tan\left(\frac{1}{2}\pi x - \pi\right)$

66. $y = 2\sec\left(\frac{1}{2}\pi x - \frac{\pi}{2}\right)$

In Exercises 67–68 sketch the graph of the function on $[-\pi, 2\pi]$.

67. $y = \cos\frac{1}{2}x + \sin(-2x)$

68. $y = \sin\pi x - \cos(-2x)$

In Exercises 69–72 find the exact value of the expression.

69. $\sin^{-1}0 + \tan^{-1}\left(-\frac{1}{\sqrt{3}}\right)$

70. $\arccos\left(-\frac{1}{2}\right) - \operatorname{arccsc}(-\sqrt{2})$

71. $\cos^{-1}\left[\sin\left(-\frac{\pi}{6}\right)\right]$

72. $\tan\left[\arccos\left(-\frac{1}{\sqrt{2}}\right)\right]$

In Exercises 73–74 use Table V to approximate the values.

73. $\arcsin(-0.7679) - \operatorname{arcsec} 1.050$

74. $\arctan(-3.018) + \operatorname{arccsc} 2.971$

In Exercises 75–76 write the trigonometric expression as an algebraic expression in x.

75. $\sin\left[\arctan\left(\frac{x-2}{4}\right)\right]$

76. $\tan\left[\arccos\left(\frac{2}{\sqrt{9x^2+4}}\right)\right]$

77. If an angle θ subtends an arc of 65 cm on a circle of radius 15 cm, what is the measure of θ and the area of the circular sector determined by θ?

78. The minute hand of a clock measures 8 cm. How far does the tip of the hand move in 12 min?

Chapter Test

In Problems 1–2 write the given angle using the specified measure.

1. 192°; radians

2. $-\frac{8\pi}{3}$; degrees

In Problems 3–4 find the values of the trigonometric functions at θ if:

3. $(-12, -5)$ is on the terminal side of θ.

4. $\cos\theta = \frac{3}{5}$ and $\cot\theta = -\frac{3}{4}$.

5. If $c = \sqrt{6}$ and $\cot\alpha = \frac{\sqrt{2}}{2}$, find a and b for the right triangle illustrated. Also find $\sin\alpha$, $\tan\beta$, and $\sec\beta$.

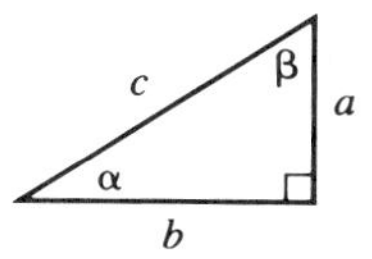

In Problems 6–9 find the exact values of the trigonometric functions at the specified angles or numbers.

6. $-225°$

7. $\frac{7\pi}{6}$

8. $\frac{5\pi}{3}$

9. t if $-\pi < t < 0$ and $\cot t = \sqrt{3}$

In Problems 10–11 use Table V to find approximate values for the sine, cosine, and tangent functions at the specified angle.

10. 258°40′

11. $-\frac{13\pi}{12}$ $\left(\text{Hint: } \frac{\pi}{12} = 15°\right)$

12. Use Table V to find an angle θ such that $0 \le \theta \le 180°$ and $\cos\theta = -0.1851$.

13. Use Table V to approximate $\arcsin(-0.8056) - \arctan 0.7002$.

In Problems 14–15 find the exact value of the expression.

14. $\sin^{-1}(-1) + \cos^{-1}(-1)$

15. $\cos^{-1}\left(\sin\frac{5\pi}{6}\right)$

16. Write $\sin\left(\text{arccot}\,\frac{2x}{5}\right)$ as an algebraic expression in x.

17. What is the measure of the arc and the area of the sector subtended by an angle of measure 12° on a circle of radius 10 cm?

In Problems 18–19 sketch the graph of the function on the interval $[0, 2\pi]$.

18. $y = 2\sin x$

19. $y = \cos(2x - \pi)$

7 Analytic Trigonometry

As a result of the discussion in Section 6.7 we know that sounds in radio broadcast can be attached to carrier waves as sums of sine or cosine functions. In some cases the sum of sine and cosine functions can be replaced by a single sine function, thereby reducing the number of functions required. For example, $\sin 2t + \sqrt{3} \cos 2t = 2 \sin(2t + \pi/3)$.

Many applications require transformations of the type discussed in the previous paragraph. In this chapter we examine trigonometric identities that provide means for transformation of equations and the solution of equations.

7.1 Equations and Identities

An equation is a statement that two expressions are the same. If the statement is true for all elements in the common domain of the left and right sides of the equation, then the equation is said to be an **identity.** If the statement is false for at least one element in the common domain, the equation is a **conditional equation.**

Identity

Conditional equation

The purpose of this section is to analyze various types of equations and identities to help clarify these concepts and to illustrate the techniques used to solve equations and verify identities. More sophisticated examples appear in later sections of this chapter as additional machinery becomes available.

Example 1 Verify the identity $(2x + y)^2 = 4x^2 + 4xy + y^2$.

This is a special case of the square of a binomial as discussed in Section 1.5. ■

Example 2 Determine whether $\tan x = \dfrac{\sin x}{\cos x}$ is an identity.

It was observed in Section 6.2 that $\tan x = \sin x/\cos x$ if $\cos x \neq 0$. If $\cos x = 0$ then x is not in the common domain, because x is not in the domain of the right-hand side. We see that $\tan x = \sin x/\cos x$ for all x in the common domain of the two sides and so this is an identity. For the record, if $\cos x = 0$, x is also not in the domain of the tangent function. ■

Example 3 Solve the following equations: **(a)** $\sin^2 x + \cos^2 x = 1$; **(b)** $\sin x = \sqrt{1 - \cos^2 x}$.

a. This is true for all real numbers (or angles), as observed in Section 6.2. This equation is an identity.

b. This is not an identity since, if $x = 3\pi/2$, $\sin x = -1$ while $\sqrt{1 - \cos^2 x} = 1$. Using part **(a)** it is easy to see that $\sin x = \sqrt{1 - \cos^2 x}$ if $\sin x \geq 0$. This is the case if the terminal side of the corresponding angle is located in quadrant I or II. The solution set for this conditional equation is $\{x: x = \alpha + 2k\pi, 0 \leq \alpha \leq \pi, k \varepsilon Z\}$. ■

Example 4 Verify the identity

$$(\sin \theta + \cos \theta)^2 = 2 \cos \theta \sin \theta + 1$$

We see that

$$\begin{aligned}(\sin \theta + \cos \theta)^2 &= \sin^2 \theta + 2 \cos \theta \sin \theta + \cos^2 \theta \\ &= 2 \cos \theta \sin \theta + (\sin^2 \theta + \cos^2 \theta) \\ &= 2 \cos \theta \sin \theta + 1\end{aligned}$$ ■

Example 4 illustrates the preferred procedure for verifying a trigonometric identity. Specifically, we try to transform one side of the equation into the other side using only reversible operations. Alternatively, one can show that the two sides of the equation are both equal to a third expression. It is not that this method is less rigorous; it is just that in most applications it is necessary to transform a given expression to a workable form. Verification of identities by the preferred method gives practice that will yield rich dividends in the future.

Example 5 Verify the identity

$$\frac{\cos^3 x - \sin^3 x}{\cos x - \sin x} = \cos^2 x + \cos x \sin x + \sin^2 x$$

$$\begin{aligned}\frac{\cos^3 x - \sin^3 x}{\cos x - \sin x} &= \frac{(\cos x - \sin x)(\cos^2 x + \cos x \sin x + \sin^2 x)}{\cos x - \sin x} \\ &= \cos^2 x + \cos x \sin x + \sin^2 x \quad \textit{(See Section 1.6.)}\end{aligned}$$

We observe that even though the right-hand side is defined for all values of x, the left-hand side is not defined when $\cos x = \sin x$. The common domain excludes those points for which $\cos x = \sin x$, and hence we are not cancelling zero when we cancel $(\cos x - \sin x)$ from numerator and denominator. ■

Example 6 Verify the identity

$$\frac{x^2 + 5x + 6}{x + 2} = x + 3$$

$$\frac{x^2 + 5x + 6}{x + 2} = \frac{(x + 2)(x + 3)}{x + 2} = x + 3 \quad \text{if} \quad x \neq -2$$

$x = -2$ is not in the common domain. ■

Example 7 Verify the identity

$$\frac{\tan^2 x}{\sin^2 x} = \sec^2 x$$

$$\frac{\tan^2 x}{\sin^2 x} = \left(\frac{\tan x}{\sin x}\right)^2 = \left[(\tan x)\left(\frac{1}{\sin x}\right)\right]^2 = \left[\left(\frac{\sin x}{\cos x}\right)\left(\frac{1}{\sin x}\right)\right]^2$$

$$= \left(\frac{1}{\cos x}\right)^2 = \frac{1}{\cos^2 x} = \sec^2 x$$ ■

Example 8 Verify the identity

$$\arcsin x + \arccos x = \frac{\pi}{2} \quad \text{for} \quad 0 < \arcsin x < \frac{\pi}{2}$$

This one requires a different approach. Let $y = \arcsin x$. Then $\sin y = x$ and

$$\arcsin x + \arccos x = y + \arccos(\sin y)$$

$$\stackrel{\star}{=} y + \left(\frac{\pi}{2} - y\right) = \frac{\pi}{2}$$

To see *, construct the triangle associated with $\sin y = x =$ side opp/hyp, as shown. Then $\arccos(\sin y)$ is the angle whose cosine is the sine of y, i.e., the angle whose cosine is x. This is the angle α illustrated; we see that $\alpha + y = \pi/2$. Thus $\arccos(\sin y) = \pi/2 - y$.

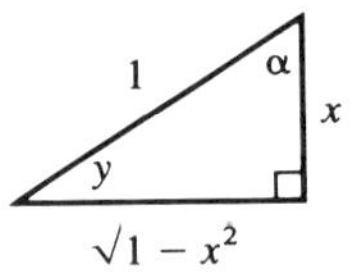

■

Example 9 Solve the equation $\tan x = -1$, $x \in R$.
If $\tan x = -1$, the reference number $\bar{x}$ is $\pi/4$, and the terminal side of the corresponding angle is located in quadrant II or IV as shown. The solution set is $\{x\colon x = 3\pi/4 + k\pi,\ k \in Z\}$. There are infinitely many solutions, but the equation is not an identity.

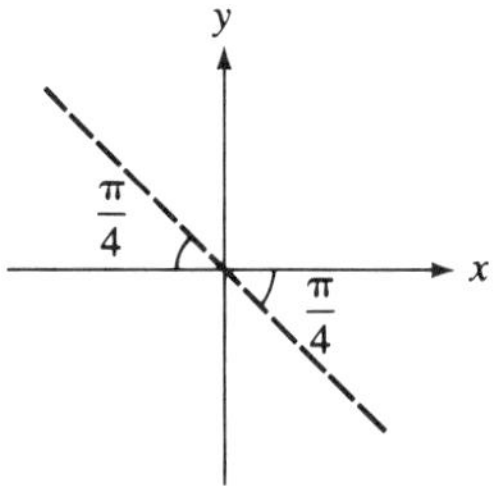

■

Example 10 Solve the equation $\cos \alpha = 3$.
Recall that $|\cos \alpha| \leq 1$ for all α. There are no α such that $\cos \alpha = 3$, and therefore the solution set is the empty set $\emptyset$. ■

Example 11 Solve the equation $\sin 2x = \frac{1}{2}$, $0 \le x \le 2\pi$.

Sin $t = \frac{1}{2}$ if $\tilde{t} = \pi/6$ and t is located in quadrant I or II. We note that if $0 \le x \le 2\pi$, then $0 \le 2x \le 4\pi$. Thus for $2x$ we need all angles $t \in [0, 4\pi]$ such that $\tilde{t} = \pi/6$ and t is in quadrant I or II. We see that

$$t = 2x \in \left\{\frac{\pi}{6}, \frac{5\pi}{6}, \frac{13\pi}{6}, \frac{17\pi}{6}\right\}$$

and hence

$$x \in \left\{\frac{\pi}{12}, \frac{5\pi}{12}, \frac{13\pi}{12}, \frac{17\pi}{12}\right\}$$

As an illustration of $t = 2x$ see Figure 7.1, in which the scale on the y axis has been enlarged to show detail.

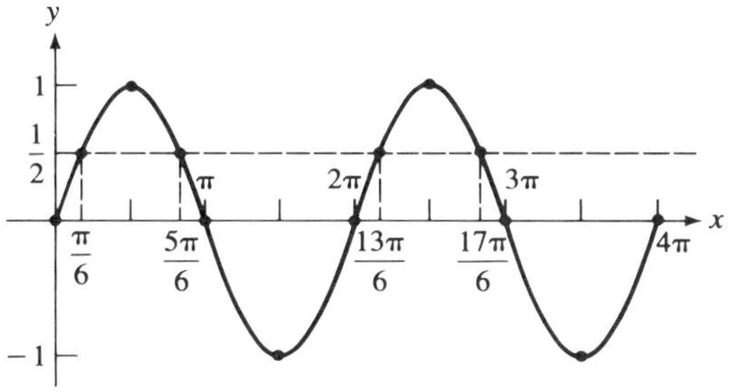

Figure 7.1 ■

Example 12 Solve the equation $\sin^{-1}(x + 2) = \frac{\pi}{3}$.

If $\sin^{-1}(x + 2) = \pi/3$ then $\sin \pi/3 = x + 2$. Also $\sin \pi/3 = \sqrt{3}/2$, so $\sqrt{3}/2 = x + 2$ and thus $x = (\sqrt{3}/2 - 2)$ is the only solution. ■

Example 13 Solve the equation

$$2 \sin^2 t + 3 \sin t + 4 = 3, \qquad 0° \le t \le 360°$$

The equation is equivalent to

$$2 \sin^2 t + 3 \sin t + 1 = 0$$

or

$$(2 \sin t + 1)(\sin t + 1) = 0$$

Setting the factors equal to 0 we obtain

$$\sin t = -\frac{1}{2} \quad \text{or} \quad \sin t = -1$$

For $\sin t = -\frac{1}{2}$ we need $\tilde{t} = 30°$, $0° \le t \le 360°$, and t in quadrant III or IV. Thus $t = 210°$ or $t = 330°$.

For $\sin t = -1$, $0° \le t \le 360°$ we must have $t = 270°$.

We see that the solution set for the equation is {210°, 270°, 330°}. In radian measure the angles are $7\pi/6$, $3\pi/2$, and $11\pi/6$. ■

Example 14 Solve the equation

$$4\tan^2 u - 4\tan u - 1 = 0, \qquad -\frac{\pi}{2} < u < \frac{\pi}{2}$$

The quadratic equation $4x^2 - 4x - 1 = 0$ has irrational roots. Using the quadratic formula we find

$$\begin{aligned}\tan u &= \frac{-(-4) \pm \sqrt{(-4)^2 - 4(4)(-1)}}{2(4)} \\ &= \frac{4 \pm \sqrt{16 + 16}}{8} \\ &= \frac{4 \pm 4\sqrt{2}}{8} = \frac{1 \pm \sqrt{2}}{2}\end{aligned}$$

The solution set is then

$$\left\{\tan^{-1}\left(\frac{1+\sqrt{2}}{2}\right), \quad \tan^{-1}\left(\frac{1-\sqrt{2}}{2}\right)\right\}$$

If an approximation is desired, we note that

$$\frac{1+\sqrt{2}}{2} \simeq 1.2071 \quad \text{and} \quad \frac{1-\sqrt{2}}{2} \simeq -0.2071$$

and using Table V we see that

$$\tan^{-1} 1.2071 \simeq 0.8785 \quad \text{and} \quad \tan^{-1}(-0.2071) \simeq -0.2036$$

In degree measure, the angles are approximately 50°20′ and −11°40′. ■

Exercises 7.1

In Exercises 1–8 show that the given equation is not an identity.

1. $\sqrt{A + B} = \sqrt{A} + \sqrt{B}$

2. $(a + b)^2 = a^2 + b^2$

3. $\cos 2x = 2\cos x$

4. $\sin^2 x = \sin 2x$

5. $\sin x = 1 + \cos x$

6. $\cos(x + y) = \cos x + \cos y$

7. $\sec x = \sqrt{1 + \tan^2 x}$

8. $\sin 2x = \sin x + \cos x$

In Exercises 9–26 verify the identity.

9. $\dfrac{1}{x - 1} - \dfrac{1}{x + 1} = \dfrac{2}{x^2 - 1}$

10. $xy + y - xz - z = (x + 1)(y - z)$

11. $\sin x \sec x = \tan x$

12. $\tan x \cos x = \sin x$

13. $\sin x = \dfrac{\tan x}{\sec x}$

14. $\tan x = \dfrac{\sec x}{\csc x}$

15. $\dfrac{\cos^2 x}{\cot^2 x} = \sin^2 x$

16. $\csc^2 x = \cot^2 x \sec^2 x$

17. $\tan x + \cot x = \sec x \csc x$

18. $\cos x + \sin x \tan x = \sec x$

19. $\cos^2 x = (1 - \sin x)(1 + \sin x)$

20. $\sec x - \sec x \sin^2 x = \cos x$

21. $\dfrac{1 + 2\cos x}{\sin x} = \csc x + 2\cot x$

22. $\cot x(\tan x + \cot x) = \csc^2 x$

23. $\dfrac{(\sin x + \cos x)^2}{\sin x} = \csc x + 2\cos x$

24. $\sin^4 x - \cos^4 x = (\sin x - \cos x)(\sin x + \cos x)$

25. $\arctan x + \operatorname{arccot} x = \dfrac{\pi}{2}, 0 < \arctan x < \dfrac{\pi}{2}$

26. $\operatorname{arcsec} x + \operatorname{arccsc} x = \dfrac{\pi}{2}, 0 < \operatorname{arcsec} x < \dfrac{\pi}{2}$

In Exercises 27–34 find the general solution to the equation.

27. $\sin x + 1 = 0$

28. $\cos^2 x = 0$

29. $\tan^2 x = 1$

30. $2\sin x + \sqrt{2} = 0$

31. $2\sin x = 1$

32. $\sin x \cot x = 1$

33. $\tan^{-1}(x - 2) = -\dfrac{\pi}{3}$

34. $\cos^{-1}(4 - x^2) = \dfrac{2\pi}{3}$

In Exercises 35–48 solve the equation for $x \in [0, 2\pi]$.

35. $\cos x - \sin x \cos x = 0$

36. $2\sin x \cos x = \sin x$

37. $2\cos^2 x - 1 = 0$

38. $1 = 2\sin^2 x$

39. $\tan 2x = 1$

40. $3\sec 2x = 2\sqrt{3}$

41. $2\cos^2 x - 3\cos x + 1 = 0$

42. $\sin^2 x - 2\sin x + 1 = 0$

43. $\tan^2 x = \tan x$

44. $2\sin^2 x = \sin x$

45. $4\sin^2 x + 4\sin x + 5 = 4$

46. $\cos^2 x = -\dfrac{1}{2}(3\cos x + 1)$

47. $2\sin x \cos x - \sin x - 2\cos x + 1 = 0$

48. $2\sin x \cos x + 2\sin x + \cos x + 1 = 0$

In Exercises 49–50 solve the equation in the specified interval.

49. $\sin^2 x + \sin x - 1 = 0, x \in \left[-\dfrac{\pi}{2}, \dfrac{\pi}{2}\right]$

50. $2\cos^2 x - 4\cos x + 1 = 0, x \in [0, \pi]$

For Your Calculator

In Exercises 51–52 approximate the solutions of the equation that are in the specified interval.

51. $\tan^2 u - 6\tan u + 1 = 0, u \in \left(-\dfrac{\pi}{2}, \dfrac{\pi}{2}\right)$

52. $2\cos u + 3\cos u - 1 = 0, u \in [0, \pi]$

In Exercises 53–58 compute the value of each side of the equation to verify that the identity is true for the specified value(s).

53. $\arcsin x + \arccos x = \dfrac{\pi}{2}$, $x = \ln 2$

54. $\sin 2x = 2 \sin x \cos x$, $x = \tan 10$

55. $\sin^2 x + \cos^2 x = 1$, $x = 5$

56. $\cos 5x = 16 \cos^5 x - 20 \cos^3 x + 5 \cos x$, $x = \arctan 3$

57. $\tan(x + y) = \dfrac{\tan x + \tan y}{1 - \tan x \tan y}$; $x = 23°$, $y = 34°$

58. $\cos(x + y) = \cos x \cos y - \sin x \sin y$; $x = 5$, $y = -2$

In Exercises 59–60 show that the equation is not an identity by testing the specified value(s).

59. $\sin(x - y) = \sin x - \sin y$; $x = 318°$, $y = -65°$

60. $\cos\left(\dfrac{1}{2}x\right) = \dfrac{1}{2}\cos x$, $x = \pi^2$

7.2 The Fundamental Identities

There are eight trigonometric identities that are basic to the analysis of trigonometric expressions and functions. These are called **fundamental identities.** The fundamental identities must be memorized by the reader before continuing in Chapter 7.

The fundamental identities

1. $\sec x = \dfrac{1}{\cos x}$ 2. $\csc x = \dfrac{1}{\sin x}$ 3. $\cot x = \dfrac{1}{\tan x}$
4. $\tan x = \dfrac{\sin x}{\cos x}$ 5. $\cot x = \dfrac{\cos x}{\sin x}$
6. $\sin^2 x + \cos^2 x = 1$
7. $\tan^2 x + 1 = \sec^2 x$
8. $1 + \cot^2 x = \csc^2 x$

These identities need not be memorized by the numbers. They are numbered to help in communication between reader and author in the examples and exercises. We will also number other identities in later sections of this chapter when appropriate.

The first six of these identities are relations inherent in the definitions of the trigonometric functions and were used in Chapter 6. The last two are easily obtained using the first six as follows.

Example 1 Verify the following identities: **(a)** $\tan^2 x + 1 = \sec^2 x$; **(b)** $1 + \cot^2 x = \csc^2 x$.

a.
$$\begin{aligned}\tan^2 x + 1 &= (\tan x)^2 + 1\\ &\overset{4}{=} \left(\frac{\sin x}{\cos x}\right)^2 + 1\\ &= \frac{\sin^2 x}{\cos^2 x} + 1\\ &= \frac{\sin^2 x + \cos^2 x}{\cos^2 x}\\ &\overset{6}{=} \frac{1}{\cos^2 x} \overset{1}{=} \sec^2 x\end{aligned}$$

We use the number above the equal sign to indicate the specific identity being used for that step. Steps without numbers are algebraic in nature.

b.
$$\begin{aligned}1 + \cot^2 x &\overset{5}{=} 1 + \frac{\cos^2 x}{\sin^2 x}\\ &= \frac{\sin^2 x + \cos^2 x}{\sin^2 x}\\ &\overset{6}{=} \frac{1}{\sin^2 x} \overset{2}{=} \csc^2 x\end{aligned}$$ ■

The statement of an identity in a particular form does not preclude the use of the identity in an equivalent algebraic form. For example, the identity $\cos^2 x = 1 - \sin^2 x$ is still Identity 6 and $\sin x = \dfrac{1}{\csc x}$ is still Identity 2. In verification of an identity by the preferred method it is usually (but not always) best to start with the "most complex looking" side.

Example 2 Verify the identity

$$(\sin^2 u - 1)\sec^2 u = -1$$

$$\begin{aligned}(\sin^2 u - 1)\sec^2 u &\overset{1}{=} (\sin^2 u - 1)\frac{1}{\cos^2 u}\\ &\overset{6}{=} (-\cos^2 u)\frac{1}{\cos^2 u} = -1\end{aligned}$$ ■

Example 3 Verify the identity

$$\frac{\cos x + \sin x}{\cos x - \sin x} = \frac{1 + \tan x}{1 - \tan x}$$

$$\begin{aligned}\frac{\cos x + \sin x}{\cos x - \sin x} &= \frac{\dfrac{1}{\cos x}(\cos x + \sin x)}{\dfrac{1}{\cos x}(\cos x - \sin x)}\\ &= \frac{1 + \dfrac{\sin x}{\cos x}}{1 - \dfrac{\sin x}{\cos x}}\\ &\overset{4}{=} \frac{1 + \tan x}{1 - \tan x}\end{aligned}$$ ■

Most of the work in Example 3 was algebraic manipulation. The multiplication of both numerator and denominator by the same expression is a standard "trick" and is frequently described as multiplication by a "well-chosen 1." We note that the common domain for the two sides of the equation excludes only those x for which $\tan x = 1$, and for these values $\cos x \neq 0$. We see then that multiplication by $1/\cos x$ presented no problems. This sort of thing must always be checked, sometimes mentally, even though it is not explicitly mentioned.

Example 4 Verify the identity

$$\frac{\sec^2\theta - 1}{\sin^2\theta} = \frac{1}{1 - \sin^2\theta}$$

$$\frac{\sec^2\theta - 1}{\sin^2\theta} \stackrel{7}{=} \frac{\tan^2\theta}{\sin^2\theta} = \tan^2\theta \, \frac{1}{\sin^2\theta}$$

$$\stackrel{4}{=} \frac{\sin^2\theta}{\cos^2\theta}\,\frac{1}{\sin^2\theta} = \frac{1}{\cos^2\theta}$$

$$\stackrel{6}{=} \frac{1}{1 - \sin^2\theta}$$

■

Example 5 Suppose $\sec x = -4$ and $\sin x > 0$. Find the values of the remaining trigonometric functions at x by using the fundamental identities.

Since $\sec x < 0$ and $\sin x > 0$, the terminal side of x is located in quadrant II. Using Identity 1,

$$\cos x \stackrel{1}{=} \frac{1}{\sec x} = \frac{1}{-4} = -\frac{1}{4}$$

Using Identity 6,

$$\sin^2 x \stackrel{6}{=} 1 - \cos^2 x = 1 - \frac{1}{16} = \frac{15}{16}$$

so that $\sin x = \frac{1}{4}\sqrt{15}$.

Using Identity 2,

$$\csc x \stackrel{2}{=} \frac{1}{\sin x} = \frac{4}{\sqrt{15}}$$

Using Identity 7,

$$\tan^2 x \stackrel{7}{=} \sec^2 x - 1 = 16 - 1 = 15$$

so that, noting that $\tan x < 0$ in quadrant II, we see that $\tan x = -\sqrt{15}$. Using Identity 3,

$$\cot x \stackrel{3}{=} \frac{1}{\tan x} = \frac{-1}{\sqrt{15}}$$

We note that an alternative method would have been to obtain these trigonometric function values from the triangle associated with the statement $\sec \tilde{x} = 4$, as shown.

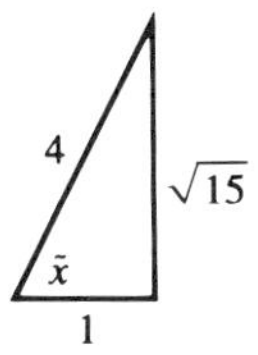

■

Example 6 Find all solutions in $[0, 2\pi]$ for the equation

$$\sin x = \cos x$$

Note that $\cos x = 0$ only if x is an odd multiple of $\pi/2$, and for these values $\sin x = \pm 1$. Thus $\sin x = \cos x$ is equivalent to

$$1 = \frac{\sin x}{\cos x} = \tan x$$

The only solutions to $\tan x = 1$ are such that $\bar{x} = \pi/4$. Since $\tan x = 1 > 0$, the solution set in $[0, 2\pi]$ is $\{\pi/4, 5\pi/4\}$. ■

Example 7 Find all solutions in $[0, 2\pi]$ for the equation

$$\sin x + \cos x = 1$$

Square both sides of the equation to obtain

$$\begin{aligned} \sin^2 x + 2 \sin x \cos x + \cos^2 x &= 1 \\ (\sin^2 x + \cos^2 x) + 2 \sin x \cos x &= 1 \\ 1 \qquad\qquad + 2 \sin x \cos x &= 1 \end{aligned}$$

Thus $2 \sin x \cos x = 0$. This is true if $\sin x = 0$ or $\cos x = 0$. The solution set for $2 \sin x \cos x = 0$ in $[0, 2\pi]$ is $S = \{0, \pi/2, \pi, 3\pi/2, 2\pi\}$.

WARNING
Be sure to check for extraneous solutions when solving an equation by squaring both sides.

Since the equation $2 \sin x \cos x = 0$ was obtained from the original by squaring both sides, all the solutions of the original equation $\sin x + \cos x = 1$ are contained in S. However, some elements of S may be extraneous solutions of the original equation. In fact, π and $3\pi/2$ do not satisfy $\sin x + \cos x = 1$.

The solution set for $\sin x + \cos x = 1$ in $[0, 2\pi]$ is seen to be $\{0, \pi/2, 2\pi\}$. ■

Example 8 Find all solutions in $[-2\pi, 2\pi]$ for

$$\sin^2 x - \cos^2 x - 3 \sin x - 1 = 0$$

We replace $\cos^2 x$ by $1 - \sin^2 x$ to obtain an equivalent equation involving only one trigonometric function:

$$\begin{aligned} \sin^2 x - \cos^2 x - 3 \sin x - 1 &= 0 \\ \sin^2 x - (1 - \sin^2 x) - 3 \sin x - 1 &= 0 \\ 2 \sin^2 x - 3 \sin x - 2 &= 0 \\ (2 \sin x + 1)(\sin x - 2) &= 0 \end{aligned}$$

There are no solutions corresponding to the factor $(\sin x - 2) = 0$ because $|\sin x| \leq 1$ for all x. If $2 \sin x + 1 = 0$ we see that $\sin x = -\dfrac{1}{2}$ and $\tilde{x} = \pi/6$. The solution set in $[-2\pi, 2\pi]$ is then $\{-5\pi/6, -\pi/6, 7\pi/6, 11\pi/6\}$. ■

In calculus it is sometimes necessary to alter the form of an algebraic expression by performing a trigonometric substitution. The resulting trigonometric expression is frequently easier to use. For example, the substitution $x = 5 \tan \theta$ will yield the trigonometric expression $5 \sec \theta$ for the expression $\sqrt{x^2 + 25}$. To see this, observe that

$$\begin{aligned}\sqrt{x^2 + 25} &= \sqrt{(5 \tan \theta)^2 + 25} = \sqrt{25 \tan^2 \theta + 25} \\ &= \sqrt{25(\tan^2 \theta + 1)} = \sqrt{25 \sec^2 \theta} = 5 \sec \theta\end{aligned}$$

Of course, depending on the application, some restrictions on domain must be noted. Since $\sqrt{x^2 + 25} \geq 0$ for all x, this result is valid only for $\sec \theta \geq 0$.

Example 9 Use the substitution $x = 2 \sec \theta$ to find a trigonometric expression associated with $\dfrac{(x^2 - 4)^{3/2}}{x}$.

$$\begin{aligned}\frac{(x^2 - 4)^{3/2}}{x} &= \frac{(4 \sec^2 \theta - 4)^{3/2}}{2 \sec \theta} = \frac{[4(\sec^2 \theta - 1)]^{3/2}}{2 \sec \theta} \\ &= \frac{(4 \tan^2 \theta)^{3/2}}{2 \sec \theta} = \frac{8 \tan^3 \theta}{2 \sec \theta} \\ &= 4 \tan^3 \theta \cos \theta\end{aligned}$$

■

Exercises 7.2

In Exercises 1–26 verify the identity.

1. $\dfrac{\sin^2 x}{1 - \cos x} = 1 + \cos x$

2. $1 + \tan^2 x = \dfrac{1}{1 - \sin^2 x}$

3. $\dfrac{\cos x}{1 + \sin x} = \dfrac{1 - \sin x}{\cos x}$

4. $\dfrac{1}{1 + \sin x} + \dfrac{1}{1 - \sin x} = 2 \sec^2 x$

5. $1 + \sin x = \dfrac{\cos x}{\sec x - \tan x}$

6. $\dfrac{1 - \cos x}{1 + \cos x} = \dfrac{\sec x - 1}{\sec x + 1}$

7. $\sin^2 x (1 + \cot^2 x) = 1$

8. $\cos^2 x \sin x + \sin^3 x = \sin x$

9. $2 \cos^2 x - 1 = 1 - 2 \sin^2 x$

10. $\cot^2 x - \cos^2 x = \cos^2 x \cot^2 x$

11. $(1 - \sin x)(\sec x + \tan x) = \cos x$

12. $\cos(-x) \csc(-x) = -\cot x$

13. $\cos(-x) \tan(-x) = -\sin x$

14. $\sec(-x) \sin(-x) = -\tan x$

15. $\cot(-x) \sin(-x) = \cos x$

16. $\dfrac{\cot x}{\tan x} = \dfrac{1 + \cot^2 x}{1 + \tan^2 x}$

17. $\dfrac{\csc x - \cot x}{\tan x + \cot x} = \cos x - \cos^2 x$

18. $\tan x + \dfrac{2}{\tan x} = \dfrac{2 - \sin^2 x}{\sin x \cos x}$

19. $\dfrac{1 + \sin x}{\cos x} + \dfrac{\cos x}{1 + \sin x} = 2 \sec x$

20. $\dfrac{\cos x}{1 + \sin x} + \tan x = \sec x$

21. $\dfrac{1 - \cot x}{1 + \cot x} = \dfrac{\tan x - 1}{\tan x + 1}$

22. $\dfrac{\cos^3 x + \sin^3 x}{\cos x + \sin x} = 1 - \cos x \sin x$

23. $\dfrac{\tan x - \sin x}{\sec x} = \dfrac{\sin^3 x}{1 + \cos x}$

24. $1 + \cos x = \dfrac{\sin x}{\csc x - \cot x}$

25. $\sin^2 x \tan^2 x + \cos^2 x \tan^2 x + \sin^2 x + \cos^2 x = \sec^2 x$

26. $\csc^2 x \sec^2 x - \cot^2 x \sec^2 x - \csc^2 x \tan^2 x + \cot^2 x \tan^2 x = 1$

In Exercises 27–28 use the fundamental identities to find the values of the six trigonometric functions at x, given that:

27. $\tan x = 3$; $\cos x < 0$

28. $\csc x = -5$; $\tan x > 0$

In Exercises 29–52 find the solutions of the equation that are in the interval $[0, 2\pi]$.

29. $\cos x - \sec x = 0$

30. $\sec^2 x = 2$

31. $\cos x \cot x = \cos x$

32. $\sin x \tan x = \sin x$

33. $\sin x - \cos x = 1$

34. $\cos x - \sin x = 1$

35. $\sin x + \cos x = \sqrt{2}$

36. $3 \sec x = \tan^2 x + 3$

37. $\tan^2 x + \sec^2 x = 3$

38. $\sin^2 x + 4 \sin x + 3 = 0$

39. $2 \cos^2 2x - 3 \cos 2x + 1 = 0$

40. $2 \sin^2 2x - 3 \sin 2x + 1 = 0$

41. $\sec x - \tan x = -1$

42. $\tan x - \sec x = -1$

43. $\sin^2 x - \cos x + 1 = 0$

44. $\cos^2 x - \sin x + 1 = 0$

45. $\cos^2 x - \sin^2 x + 1 = 0$

46. $\sin^2 2x - \cos^2 2x + 1 = 0$

47. $2 \sin x = \tan x$

48. $2 \cos x = \cot x$

49. $\sin^2 x - \cos^2 x - 5 \sin x + 3 = 0$

50. $\cos^2 x - \sin^2 x + \cos x = 0$

51. $\cos x - \sin^2 x \cos x - 1 + \sin^2 x = 0$

52. $2 \tan^2 x \sin x + 2 \sin x - \tan^2 x - 1 = 0$

In Exercises 53–54 find the solutions of the equation in the indicated interval.

53. $\cos^2 x - \sin x = 0, \left[-\dfrac{\pi}{2}, \dfrac{\pi}{2}\right]$

54. $2 \sin^2 x + 4 \cos x - 3 = 0, [0, \pi]$

In Exercises 55–58 use the given substitution to find a trigonometric expression associated with the given algebraic expression for $0 < \theta < \dfrac{\pi}{2}$.

55. $\sqrt{9 - x^2}$; $x = 3 \sin \theta$

56. $\sqrt{x^2 - 16}, x = 4 \sec \theta$

57. $\dfrac{(36 + x^2)^{3/2}}{x^2}, x = 6 \tan \theta$

58. $\dfrac{32x}{(64 - x^2)^{3/2}}, x = 8 \sin \theta$

7.3 The Addition Formulas

Additive function

An **additive function** f is one for which $f(x + y) = f(x) + f(y)$ for all $x, y \in R$. Very few functions are additive, and it would be surprising indeed if the trigonometric functions satisfied such an identity. That they are not additive is easy to show. For example, $\sin(60° + 30°) = \sin 90° = 1$, but $\sin 60° + \sin 30° = \sqrt{3}/2 + \dfrac{1}{2} \simeq 1.366$.

It is, however, possible and helpful to express a trigonometric function of a sum in terms of trigonometric functions of the individual angles (or numbers). We did this for some special angles in Section 6.3, where, for example, we observed that $\sin(\pi/2 - x) = \cos x$.

To obtain the general addition formulas we start with the identity $\cos(x - y) = \cos x \cos y + \sin x \sin y$. The others will then be easily obtained from this one and the fundamental identities.

Recall from Section 6.4 that if $t \in R$ and (a, b) is the point on the unit circle U determined by the angle of measure t radians, then $a = \cos t$ and $b = \sin t$.

Consider $x, y \in R$ and refer to Figure 7.2.

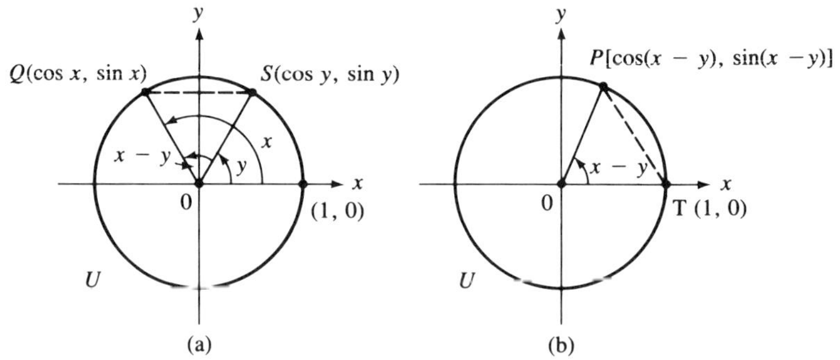

Figure 7.2

Triangle OSQ of Figure 7.2a is congruent to triangle OTP of Figure 7.2b. Then $(\overline{SQ})^2 = (\overline{TP})^2$, and using the distance formula (Section 3.2) we see that

$$\begin{aligned}(\overline{SQ})^2 &= (\cos x - \cos y)^2 + (\sin x - \sin y)^2 \\ &= \cos^2 x - 2\cos x \cos y + \cos^2 y + \sin^2 x - 2\sin x \sin y + \sin^2 y \\ &= (\cos^2 x + \sin^2 x) + (\cos^2 y + \sin^2 y) - 2(\cos x \cos y + \sin x \sin y) \\ &= 1 + 1 - 2(\cos x \cos y + \sin x \sin y)\end{aligned}$$

and that

$$\begin{aligned}(\overline{TP})^2 &= [\cos(x - y) - 1]^2 + [\sin(x - y) - 0]^2 \\ &= \cos^2(x - y) - 2\cos(x - y) + 1 + \sin^2(x - y) \\ &= 1 + [\cos^2(x - y) + \sin^2(x - y)] - 2\cos(x - y) \\ &= 1 + 1 - 2\cos(x - y)\end{aligned}$$

Since the expressions for $(\overline{SQ})^2$ and $(\overline{TP})^2$ are equal, we see that

$$\cos(x - y) = \cos x \cos y + \sin x \sin y$$

If we replace y by $-y$ in this last expression we obtain

$$\begin{aligned}\cos(x + y) &= \cos[x - (-y)]\\ &= \cos x \cos(-y) + \sin x \sin(-y)\\ &= \cos x \cos y - \sin x \sin y\end{aligned}$$

since $\cos(-\alpha) = \cos \alpha$ and $\sin(-\alpha) = -\sin \alpha$ for all α.

We have now obtained identities that express the cosine of the sum or difference of two angles (numbers) in terms of the cosine and sine of the individual angles (numbers). This is the first addition formula.

The addition formulas

9. $\cos(x \pm y) = \cos x \cos y \mp \sin x \sin y$
10. $\sin(x \pm y) = \sin x \cos y \pm \cos x \sin y$
11. $\tan(x \pm y) = \dfrac{\tan x \pm \tan y}{1 \mp \tan x \tan y}$

Recall that we called Identities 1–8 the fundamental identities. The use of the $\pm$ or $\mp$ notation makes it possible to combine the statement of two identities. If the top sign is used on the left side then the top sign must also be used throughout the right side.

These identities must also be memorized. Many people find that learning these in terms of position is most helpful. To do this we would read the bottom half of Identity 10 as "the sine of the difference is the sine of the first cosine of the second minus the cosine of the first sine of the second."

We will verify two more of the addition formulas as Examples 1 and 2. The remaining two are left as exercises.

Example 1 Verify the formula

$$\sin(x + y) = \sin x \cos y + \cos x \sin y$$

$$\begin{aligned}\sin(x + y) &= \cos\left[\frac{\pi}{2} - (x + y)\right] = \cos\left[\left(\frac{\pi}{2} - x\right) - y\right]\\ &\overset{9}{=} \cos\left(\frac{\pi}{2} - x\right)\cos y + \sin\left(\frac{\pi}{2} - x\right)\sin y\\ &= \sin x \cos y + \cos x \sin y\end{aligned}$$

■

Example 2 Verify the formula

$$\tan(x - y) = \frac{\tan x - \tan y}{1 + \tan x \tan y}$$

$$\tan(x - y) \overset{4}{=} \frac{\sin(x - y)}{\cos(x - y)} \overset{9,10}{=} \frac{\sin x \cos y - \cos x \sin y}{\cos x \cos y + \sin x \sin y}$$

$$= \frac{\dfrac{1}{\cos x \cos y}(\sin x \cos y - \cos x \sin y)}{\dfrac{1}{\cos x \cos y}(\cos x \cos y + \sin x \sin y)}$$

$$= \frac{\dfrac{\sin x \cos y}{\cos x \cos y} - \dfrac{\cos x \sin y}{\cos x \cos y}}{\dfrac{\cos x \cos y}{\cos x \cos y} + \dfrac{\sin x \sin y}{\cos x \cos y}}$$

$$\overset{4}{=} \frac{\tan x - \tan y}{1 + \tan x \tan y}$$ ■

Example 3 Verify the identity

$$\cos\left(\frac{3\pi}{2} + \theta\right) = \sin \theta$$

$$\cos\left(\frac{3\pi}{2} + \theta\right) \overset{9}{=} \cos\frac{3\pi}{2}\cos\theta - \sin\frac{3\pi}{2}\sin\theta$$
$$= (0)\cos\theta - (-1)\sin\theta$$
$$= \sin\theta$$ ■

Example 4 Verify the identity

$$\sin(x + y)\sin(x - y) = \sin^2 x - \sin^2 y$$

$$\sin(x + y)\sin(x - y)$$
$$\overset{10}{=} (\sin x \cos y + \cos x \sin y)(\sin x \cos y - \cos x \sin y)$$
$$= \sin^2 x \cos^2 y - \cos^2 x \sin^2 y$$
$$\overset{6}{=} \sin^2 x(1 - \sin^2 y) - (1 - \sin^2 x)\sin^2 y$$
$$= \sin^2 x - \sin^2 x \sin^2 y - \sin^2 y + \sin^2 x \sin^2 y$$
$$= \sin^2 x - \sin^2 y$$ ■

Example 5 Find $\cos(\alpha + \beta)$ if α is in quadrant III, $\sin\alpha = -\frac{3}{5}$, and β is in quadrant IV, $\cos\beta = \frac{4}{7}$.

We draw the triangles associated with the reference angles as shown. Noting that, due to quadrant location, $\cos\alpha < 0$ and $\sin\beta < 0$, we have

$$\cos(\alpha + \beta) \overset{9}{=} \cos\alpha\cos\beta - \sin\alpha\sin\beta$$
$$= \left(-\frac{4}{5}\right)\left(\frac{4}{7}\right) - \left(-\frac{3}{5}\right)\left(-\frac{\sqrt{33}}{7}\right)$$
$$= -\left(\frac{16}{35}\right) - \left(\frac{3\sqrt{33}}{35}\right)$$
$$= -\frac{16 + 3\sqrt{33}}{35}$$

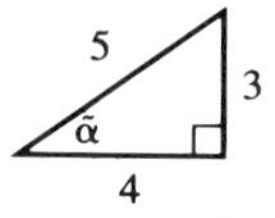

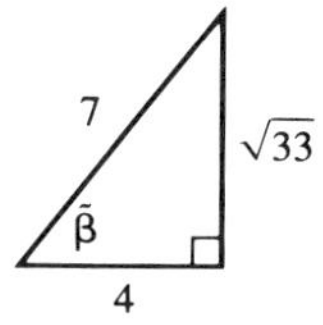

■

Example 6 Find exact values for **(a)** sin 285°; **(b)** $\csc\left(\dfrac{5\pi}{12}\right)$.

WARNING
Be sure to use the appropriate addition formula to compute sin(225° + 60°). This is not the same as sin 225° + sin 60°.

a.
$$\begin{aligned}\sin 285° &= \sin(225° + 60°)\\ &\overset{10}{=} \sin 225° \cos 60° + \cos 225° \sin 60°\\ &= \left(-\frac{1}{\sqrt{2}}\right)\left(\frac{1}{2}\right) + \left(-\frac{1}{\sqrt{2}}\right)\left(\frac{\sqrt{3}}{2}\right)\\ &= -\frac{1+\sqrt{3}}{2\sqrt{2}}\end{aligned}$$

We used the facts that the reference angle for 225° is 45° and that 225° is located in quadrant III, where cosine and sine are both negative.

b.
$$\begin{aligned}\sin\frac{5\pi}{12} &= \sin\left(\frac{\pi}{4} + \frac{\pi}{6}\right)\\ &\overset{10}{=} \sin\frac{\pi}{4}\cos\frac{\pi}{6} + \cos\frac{\pi}{4}\sin\frac{\pi}{6}\\ &= \left(\frac{1}{\sqrt{2}}\right)\left(\frac{\sqrt{3}}{2}\right) + \left(\frac{1}{\sqrt{2}}\right)\left(\frac{1}{2}\right)\\ &= \frac{\sqrt{3}+1}{2\sqrt{2}}\end{aligned}$$

Thus, using Identity 2,

$$\csc\frac{5\pi}{12} \overset{2}{=} \frac{1}{\sin\dfrac{5\pi}{12}} = \frac{2\sqrt{2}}{\sqrt{3}+1}$$ ■

Example 7 Find all solutions in $[0, 2\pi]$ for the equation

$$\sin 2x - \cos x = 0$$

If $\sin 2x - \cos x = 0$ then

$$\begin{aligned}\sin(x + x) - \cos x &= 0\\ (\sin x \cos x + \cos x \sin x) - \cos x &\overset{10}{=} 0\\ 2\sin x \cos x - \cos x &= 0\\ \cos x\,(2\sin x - 1) &= 0\end{aligned}$$

If $x \varepsilon [0, 2\pi]$ then $\cos x = 0$ for $x = \pi/2$ or $x = 3\pi/2$, and $\sin x = \dfrac{1}{2}$ for $x = \pi/6$ or $x = 5\pi/6$. We see that the solution set is $\{\pi/6, \pi/2, 5\pi/6, 3\pi/2\}$. ■

Example 8 Evaluate $\cos\left(\arccos\dfrac{1}{2} - \arcsin\dfrac{3}{4}\right)$.

Let $x = \arccos\dfrac{1}{2}$ and $y = \arcsin\dfrac{3}{4}$. By definition of the inverse trigonometric functions, x and y are acute angles. Draw the triangles associated with these angles as shown. Then we see that

$$\cos\left(\arccos\frac{1}{2} - \arcsin\frac{3}{4}\right) = \cos(x - y)$$
$$\overset{9}{=} \cos x \cos y + \sin x \sin y$$
$$= \left(\frac{1}{2}\right)\left(\frac{\sqrt{7}}{4}\right) + \left(\frac{\sqrt{3}}{2}\right)\left(\frac{3}{4}\right)$$
$$= \frac{\sqrt{7} + 3\sqrt{3}}{8}$$

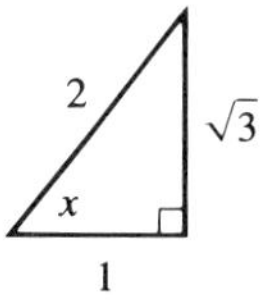

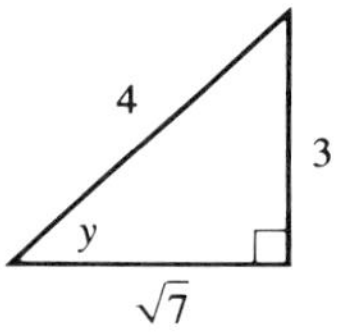

■

Exercises 7.3

In Exercises 1–4 determine whether the equation is an identity and justify your answer.

1. $\sin(x - y) = \sin x - \sin y$

2. $\sin(x + x) = 2 \sin x$

3. $\cos(x + x) = 2 \cos x - 1$

4. $\arccos\left[\sin\left(x + \frac{\pi}{2}\right)\right] = x$

Note that $15° = 45° - 30°$, $75° = 30° + 45°$, $105° = 60° + 45°$, $195° = 225° - 30°$, $\pi/12 = \pi/3 - \pi/4$, $7\pi/12 = 3\pi/4 - \pi/6$, $11\pi/12 = 3\pi/4 + \pi/6$. *In Exercises 5–12 find exact values of the specified function at the given angle.*

5. $\sin 15°$

6. $\cos 105°$

7. $\tan \frac{\pi}{12}$

8. $\tan 195°$

9. $\tan \frac{11\pi}{12}$

10. $\cos \frac{\pi}{12}$

11. $\cos 75°$

12. $\tan \frac{7\pi}{12}$

In Exercises 13–15 find the requested information if α *is in quadrant I,* β *is in quadrant II,* $\sin \alpha = \frac{3}{5}$ *and* $\cos \beta = -\frac{\sqrt{3}}{3}$.

13. $\cos(\alpha - \beta)$

14. $\tan(\alpha + \beta)$

15. The quadrant of $\alpha + \beta$

In Exercises 16–18 find the requested information if α is in quadrant II, β is in quadrant II, $\tan \alpha = -\frac{7}{24}$, *and* $\cos \beta = -\frac{5}{13}$.

16. $\sin(\alpha + \beta)$

17. $\tan(\alpha + \beta)$

18. The quadrant of $\alpha + \beta$

In Exercises 19–20 verify the formulas.

19. $\sin(x - y) = \sin x \cos y - \cos x \sin y$

20. $\tan(x + y) = \dfrac{\tan x + \tan y}{1 - \tan x \tan y}$

In Exercises 21–36 verify the identity.

21. $\sin(x - \pi) = -\sin x$

22. $\sin\left(x + \frac{\pi}{2}\right) = \cos x$

23. $\tan\left(\frac{3\pi}{2} + x\right) = -\cot x$

24. $\tan(\pi + x) = \tan x$

25. $\tan\left(x + \frac{\pi}{4}\right) = \dfrac{1 + \tan x}{1 - \tan x}$

26. $\tan\left(x - \frac{\pi}{4}\right) = \dfrac{\tan x - 1}{\tan x + 1}$

27. $\sin\left(x + \frac{\pi}{3}\right) - \cos\left(x + \frac{\pi}{6}\right) = \sin x$

28. $\cos\left(x - \frac{\pi}{6}\right) - \cos\left(x + \frac{\pi}{6}\right) = \sin x$

29. $\tan\left(x + \frac{\pi}{4}\right) - \tan\left(x - \frac{3\pi}{4}\right) = 0$

30. $\cos(x + y) + \cos(x - y) = 2 \cos x \cos y$

31. $\sin(x + y) - \sin(x - y) = 2 \cos x \sin y$

32. $(\sin x \cos y + \sin y \cos x)^2 + (\cos x \cos y - \sin x \sin y)^2 = 1$

33. $\tan x + \tan y = \dfrac{\sin(x + y)}{\cos x \cos y}$

34. $\sec(x - y) = \dfrac{\sec x \sec y}{1 + \tan x \tan y}$

35. $\cot(x + y) = \dfrac{\cot x \cot y - 1}{\cot x + \cot y}$

36. $\tan\left(x + \frac{\pi}{3}\right) + \tan\left(x - \frac{\pi}{3}\right) = \dfrac{8 \tan x}{1 - 3 \tan^2 x}$

In Exercises 37–40 express as a function of x alone.

37. $\cos(180° + x)$

38. $\cot\left(x - \frac{\pi}{4}\right)$

39. $\cos\left(x + \frac{\pi}{2}\right)$

40. $\cos(\pi - x)$

In Exercises 41–42 find the value of x if $0 \leq x \leq \pi/2$.

41. $\sin \frac{2\pi}{7} \sin \frac{5\pi}{7} - \cos \frac{2\pi}{7} \cos \frac{5\pi}{7} = \cos x$

42. $\sin x = 2 \sin \frac{\pi}{20} \cos \frac{\pi}{20}$

In Exercises 43–48 evaluate the expression.

43. $\sin\left(\arcsin \frac{1}{2} + \arccos 0\right)$

44. $\sin\left(\arccos \frac{1}{2} - \arcsin \frac{1}{4}\right)$

45. $\tan\left(\arccos \frac{3}{5} + \arccos \frac{4}{5}\right)$

46. $\tan[\arccos(-1) - \arctan(-1)]$

47. $\cos\left[\arccos\left(-\frac{1}{2}\right) + \arcsin\left(\frac{1}{\sqrt{2}}\right)\right]$

48. $\cos\left[\arctan(-1) + \arccos\left(-\frac{\sqrt{15}}{4}\right)\right]$

49. Show that a linear function $f(x) = mx + b$ is additive if, and only if, $b = 0$.

50. Let $0 < x < \frac{\pi}{2}$ and $0 < y < \frac{\pi}{2}$. Find $\sin x$ if $\cos(x + y) = -\frac{12}{13}$ and $\sin y = \frac{2}{3}$.

51. Show that $\cos(x + 5\pi) = -\cos x$ and $\cos(x + 6\pi) = \cos x$. Express $\cos(x + k\pi)$, $k \varepsilon Z$, in terms of $\cos x$.

52. Show that $\sin(x + 5\pi) = -\sin x$ and $\sin(x + 6\pi) = \sin x$. Express $\sin(x + k\pi)$, $k \varepsilon Z$, in terms of $\sin x$.

7.4 Double-Angle and Half-Angle Formulas

Formulas for trigonometric functions of integer multiples of a real number can be obtained from the addition formulas.

Example 1 Find a formula for $\cos 2x$ in terms of $\sin x$ and $\cos x$.

$$\cos 2x = \cos(x + x) \stackrel{9}{=} \cos x \cos x - \sin x \sin x$$
$$= \cos^2 x - \sin^2 x$$ ■

The procedure of Example 1 can be used to obtain formulas for $\sin 2x$ and $\tan 2x$ also. These formulas are widely used and are called the **double-angle formulas.**

Double-angle formulas

12. $\sin 2x = 2 \sin x \cos x$

13. $\cos 2x \stackrel{a}{=} \cos^2 x - \sin^2 x$
$\stackrel{b}{=} 2\cos^2 x - 1$
$\stackrel{c}{=} 1 - 2\sin^2 x$

14. $\tan 2x = \dfrac{2 \tan x}{1 - \tan^2 x}$

Notice that 13b is easily obtained from 13a by replacing $\sin^2 x$ by $1 - \cos^2 x$. Similarly, 13c is obtained from 13a.

Example 2 Verify the identities displayed below.

$$\sin^2 x = \frac{1}{2}(1 - \cos 2x)$$

$$\cos^2 x = \frac{1}{2}(1 + \cos 2x)$$

Using 13c we see that $2 \sin^2 x = 1 - \cos 2x$, so that the first identity displayed is just an algebraic variation of 13c. The second identity is an algebraic variation of 13b. ■

The versions of Identity 13 displayed in Example 2 are extensively used in applications and so were singled out for special notice. These forms will also be described as 13c and 13b, respectively. To emphasize their importance these identities are included in Section 7.6 on the basic list of formulas as separate lines.

The following example illustrates a technique that is indispensable in calculus.

Example 3 Write $\sin^4 x$ as an expression involving only first powers of trigonometric functions.

$$\begin{aligned}\sin^4 x = (\sin^2 x)^2 &\overset{13c}{=} \left[\frac{1}{2}(1 - \cos 2x)\right]^2 \\ &= \frac{1}{4}(1 - 2\cos 2x + \cos^2 2x) \\ &\overset{13b}{=} \frac{1}{4}\left[1 - 2\cos 2x + \frac{1}{2}(1 + \cos 4x)\right] \\ &= \frac{1}{8}(3 - 4\cos 2x + \cos 4x)\end{aligned}$$ ■

Example 4 Verify the identity

$$\csc 2y - \cot 2y = \tan y$$

$$\begin{aligned}\csc 2y - \cot 2y &\overset{2,5}{=} \frac{1}{\sin 2y} - \frac{\cos 2y}{\sin 2y} \\ &= \frac{1 - \cos 2y}{\sin 2y} \\ &\overset{13c}{\underset{12}{=}} \frac{1 - (1 - 2\sin^2 y)}{2 \sin y \cos y} \\ &= \frac{2\sin^2 y}{2 \sin y \cos y} \\ &= \frac{\sin y}{\cos y} \overset{4}{=} \tan y\end{aligned}$$ ■

Example 5 If α is in quadrant II and $\cos \alpha = -\frac{4}{9}$, find **(a)** $\sin 2\alpha$; **(b)** $\cos 2\alpha$; **(c)** $\tan 2\alpha$.

Draw the triangle associated with $\tilde{\alpha}$ as shown. Noting that $\sin \alpha > 0$, $\cos \alpha < 0$, and $\tan \alpha < 0$, we have

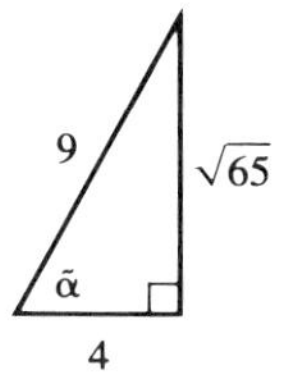

a. $\sin 2\alpha \overset{12}{=} 2 \sin \alpha \cos \alpha$

$$= 2\left(\frac{\sqrt{65}}{9}\right)\left(-\frac{4}{9}\right) = -\frac{8\sqrt{65}}{81}$$

b. $\cos 2\alpha \overset{13a}{=} \cos^2 \alpha - \sin^2 \alpha$

$$= \left(-\frac{4}{9}\right)^2 - \left(\frac{\sqrt{65}}{9}\right)^2 = -\frac{49}{81}$$

c. $\tan 2\alpha \overset{4}{=} \dfrac{\sin 2\alpha}{\cos 2\alpha} = \dfrac{8\sqrt{65}}{49}$

Note that we could have used Identity 14 to obtain $\tan 2\alpha$, but knowing $\sin 2\alpha$ and $\cos 2\alpha$ it is easier just to use Identity 4. ■

Example 6 Find all solutions of $2 \cos 2x = \sin 4x$ for $x \in [0, 2\pi]$.
If $2 \cos 2x = \sin 4x$ then

$$2 \cos 2x \overset{12}{=} 2 \sin 2x \cos 2x$$

so that

$$2 \cos 2x(\sin 2x - 1) = 0$$

Then we have $\cos 2x = 0$ or $\sin 2x = 1$. Since $x \in [0, 2\pi]$ we must consider $2x \in [0, 4\pi]$:

$$\text{if} \quad \cos 2x = 0, \qquad 2x = \frac{\pi}{2}, \frac{3\pi}{2}, \frac{5\pi}{2}, \frac{7\pi}{2}$$

and

$$\text{if} \quad \sin 2x = 1, \qquad 2x = \frac{\pi}{2}, \frac{5\pi}{2}$$

The solutions of the second equation are included in the solutions of the first and we see that the solution set for $x \in [0, 2\pi]$ is $\{\pi/4, 3\pi/4, 5\pi/4, 7\pi/4\}$. ■

Notice that if we replace x by $\frac{1}{2}x$ in Identity 13c (Example 2) we obtain $\sin^2\left(\frac{1}{2}x\right) = \frac{1}{2}(1 - \cos x)$ so that

$$\sin \frac{1}{2}x = \pm\sqrt{\frac{1 - \cos x}{2}}$$

Similarly, using Identity 13b we obtain

$$\cos\frac{1}{2}x = \pm\sqrt{\frac{1+\cos x}{2}}$$

From these two and Identity 4 we obtain

$$\tan\frac{1}{2}x = \pm\sqrt{\frac{1-\cos x}{1+\cos x}}$$

These formulas are known as the **half-angle formulas.**

Half-angle formulas

15. $\sin\frac{1}{2}x = \pm\sqrt{\frac{1-\cos x}{2}}$

16. $\cos\frac{1}{2}x = \pm\sqrt{\frac{1+\cos x}{2}}$

17. $\tan\frac{1}{2}x \overset{a}{=} \pm\sqrt{\frac{1-\cos x}{1+\cos x}}$

$$\overset{b}{=} \frac{1-\cos x}{\sin x}$$

$$\overset{c}{=} \frac{\sin x}{1+\cos x}$$

In using Identities 15, 16, and 17a the correct sign must be determined by analysis of the quadrant involved. Equations 17b and 17c for $\tan\frac{1}{2}x$ are easier to use and include the correct sign. These are obtained using a slightly different approach.

Example 7 Verify that

a. $\tan\frac{1}{2}x = \frac{\sin x}{1+\cos x}$

b. $\tan\frac{1}{2}x = \frac{1-\cos x}{\sin x}$

a. $$\tan\frac{1}{2}x \overset{4}{=} \frac{\sin\frac{1}{2}x}{\cos\frac{1}{2}x} = \frac{2\sin\frac{1}{2}x\cos\frac{1}{2}x}{2\cos\frac{1}{2}x\cos\frac{1}{2}x}$$

$$\underset{13b}{\overset{12}{=}} \frac{\sin 2\left(\frac{1}{2}x\right)}{1+\cos 2\left(\frac{1}{2}x\right)} = \frac{\sin x}{1+\cos x}$$

b. $\tan \frac{1}{2}x \overset{17c}{=} \frac{\sin x}{1 + \cos x} = \frac{\sin x(1 - \cos x)}{(1 + \cos x)(1 - \cos x)}$

$$= \frac{\sin x(1 - \cos x)}{1 - \cos^2 x}$$

$$\overset{6}{=} \frac{\sin x(1 - \cos x)}{\sin^2 x}$$

$$= \frac{1 - \cos x}{\sin x}$$

■

Example 8 Find $\sin \frac{1}{2}\alpha$ and $\cos \frac{1}{2}\alpha$ if α is in quadrant III and $\tan \alpha = \frac{4}{3}$.

If $\pi \le \alpha \le 3\pi/2$ then $\pi/2 \le \frac{1}{2}\alpha \le 3\pi/4$, so that the terminal side of $\frac{1}{2}\alpha$ is in quadrant II. Thus $\sin \frac{1}{2}\alpha > 0$, $\cos \frac{1}{2}\alpha < 0$, $\cos \alpha < 0$, and

$$\sin \frac{1}{2}\alpha \overset{15}{=} \sqrt{\frac{1 - \cos \alpha}{2}}$$

$$= \sqrt{\frac{1 - \left(-\frac{3}{5}\right)}{2}}$$

$$= \frac{2}{\sqrt{5}}$$

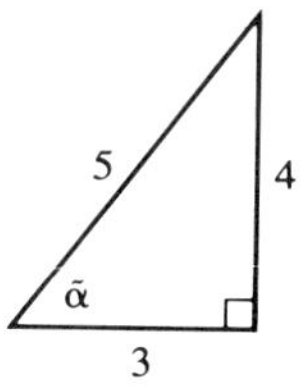

and

$$\cos \frac{1}{2}\alpha = -\sqrt{\frac{1 + \cos \alpha}{2}}$$

$$= -\sqrt{\frac{1 + \left(-\frac{3}{5}\right)}{2}} = -\frac{1}{\sqrt{5}}$$

■

Example 9 Find the exact values of **(a)** $\tan \frac{\pi}{8}$; **(b)** $\cos \frac{\pi}{12}$.

a. $\tan \frac{\pi}{8} = \tan\left(\frac{1}{2} \cdot \frac{\pi}{4}\right) \overset{17b}{=} \frac{1 - \cos \frac{\pi}{4}}{\sin \frac{\pi}{4}}$

$$= \frac{1 - \left(\frac{1}{\sqrt{2}}\right)}{\frac{1}{\sqrt{2}}} = \frac{\sqrt{2} - 1}{1} = \sqrt{2} - 1$$

In degree measure, $\tan 22.5^\circ = \sqrt{2} - 1$.

b. $\cos \frac{\pi}{12} = \cos\left(\frac{1}{2} \cdot \frac{\pi}{6}\right) \overset{16}{=} \sqrt{\frac{1 + \cos \pi/6}{2}}$

$$= \sqrt{\frac{1 + \left(\frac{\sqrt{3}}{2}\right)}{2}} = \sqrt{\frac{2 + \sqrt{3}}{4}}$$
$$= \frac{1}{2}\sqrt{2 + \sqrt{3}}$$

In degree measure,

$$\cos 15° = \frac{1}{2}\sqrt{2 + \sqrt{3}}$$

Notice that $\pi/12 = \pi/4 - \pi/6$. Using Identity 9,

$$\cos\frac{\pi}{12} = \cos\left(\frac{\pi}{4} - \frac{\pi}{6}\right) \overset{9}{=} \cos\frac{\pi}{4}\cos\frac{\pi}{6} + \sin\frac{\pi}{4}\sin\frac{\pi}{6}$$
$$= \left(\frac{1}{\sqrt{2}}\right)\left(\frac{\sqrt{3}}{2}\right) + \left(\frac{1}{\sqrt{2}}\right)\left(\frac{1}{2}\right)$$
$$= \left(\frac{1}{2\sqrt{2}}\right)(\sqrt{3} + 1)$$

It is reasonable to wonder if these results are the same. The reader should verify that

$$\frac{1}{2}\sqrt{2 + \sqrt{3}} = \left(\frac{1}{2\sqrt{2}}\right)(\sqrt{3} + 1)$$ ■

Example 10 Verify the identity

$$\left(\sin\frac{1}{2}x - \cos\frac{1}{2}x\right)^2 = 1 - \sin x$$

$$\left(\sin\frac{1}{2}x - \cos\frac{1}{2}x\right)^2 = \sin^2\frac{1}{2}x - 2\sin\frac{1}{2}x\cos\frac{1}{2}x + \cos^2\frac{1}{2}x$$
$$= \left(\sin^2\frac{1}{2}x + \cos^2\frac{1}{2}x\right) - 2\sin\frac{1}{2}x\cos\frac{1}{2}x$$
$$\overset{6}{=} 1 - 2\sin\frac{1}{2}x\cos\frac{1}{2}x$$
$$\overset{12}{=} 1 - \sin 2\left(\frac{1}{2}x\right) = 1 - \sin x$$ ■

The following example illustrates calculations necessary in analytic geometry for rotation of rectangular coordinate systems.

Example 11 Find $\sin\theta$ and $\cos\theta$ if $0 \le \theta < \frac{\pi}{2}$ and $\cot 2\theta = -\frac{5}{12}$.

Since $\cot 2\theta < 0$, θ acute, we must have 2θ located in quadrant II, so that $\cos 2\theta < 0$ but $\sin\theta > 0$, $\cos\theta > 0$. Then

$$\sin\theta \overset{15}{=} \sqrt{\frac{1-\cos 2\theta}{2}}$$

$$= \sqrt{\frac{1-\left(-\frac{5}{13}\right)}{2}}$$

$$= \sqrt{\frac{1+\frac{5}{13}}{2}} = \sqrt{\frac{18}{26}} = \sqrt{\frac{9}{13}} = \frac{3}{\sqrt{13}}$$

$$\cos\theta \overset{16}{=} \sqrt{\frac{1+\cos 2\theta}{2}} = \sqrt{\frac{1+\left(-\frac{5}{13}\right)}{2}}$$

$$= \sqrt{\frac{8}{26}} = \frac{2}{\sqrt{13}}$$

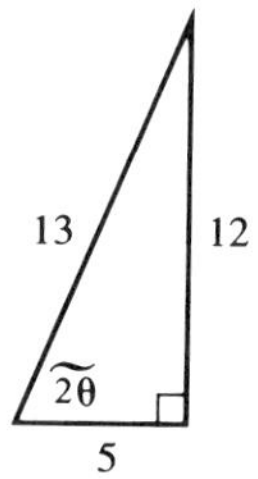

■

Exercises 7.4

In Exercises 1–6 if $\sin x = -\frac{4}{5}$ *and* $\pi < x < \frac{3\pi}{2}$, *find:*

1. $\sin 2x$ **2.** $\cos 2x$ **3.** $\tan 2x$

4. $\sin \frac{1}{2}x$ **5.** $\cos \frac{1}{2}x$ **6.** $\tan \frac{1}{2}x$

In Exercises 7–12 if $\cos x = -\frac{5}{13}$ *and* $0 < x < \pi$, *find:*

7. $\sin \frac{1}{2}x$ **8.** $\cos \frac{1}{2}x$ **9.** $\tan \frac{1}{2}x$

10. $\sin 2x$ **11.** $\cos 2x$ **12.** $\tan 2x$

In Exercises 13–18 find exact values.

13. $\cos 22.5°$ **14.** $\tan 15°$ **15.** $\sin \frac{9\pi}{8}$

16. $\cos \frac{\pi}{24}$ **17.** $\sin \frac{23\pi}{12}$ **18.** $\tan \frac{3\pi}{8}$

In Exercises 19–22 find $\sin\theta$ *and* $\cos\theta$ *if* θ *is acute and* $\cot 2\theta$ *equals:*

19. $\frac{4}{3}$ **20.** $\sqrt{15}$ **21.** $-\frac{7}{24}$ **22.** $-\frac{3}{2}$

In Exercises 23–28 find exact values.

23. $\sin(2 \arctan 1)$ **24.** $\cos\left[2 \arcsin\left(-\frac{1}{2}\right)\right]$

25. $\tan\left(2 \sin^{-1} \frac{3}{5}\right)$ **26.** $\sin\left[2 \sin^{-1}\left(-\frac{1}{4}\right)\right]$

27. $\cos\left(\frac{1}{2} \tan^{-1} \frac{4}{3}\right)$ **28.** $\tan\left[\frac{1}{2} \cos^{-1}\left(-\frac{5}{13}\right)\right]$

In Exercises 29–42 verify the identity.

29. $\sin 2x = \dfrac{2 \tan x}{1 + \tan^2 x}$

30. $2 \sin^2 x + \cos 2x = 1$

31. $\dfrac{1 + \cos 2x}{\sin 2x} = \tan x$

32. $\dfrac{2}{1 + \cos 2x} = \sec^2 x$

33. $\cos 2x = \dfrac{1 - \tan^2 x}{1 + \tan^2 x}$

34. $\dfrac{\sin^3 x + \cos^3 x}{\sin x + \cos x} = 1 - \dfrac{1}{2} \sin 2x$

35. $\dfrac{\sin 3x}{\sin x} - \dfrac{\cos 3x}{\cos x} = 2$

36. $\csc 2x + \cot 2x = \cot x$

37. $2 \cot 2x = \cot x - \tan x$

38. $\sin \frac{1}{2}x \cos \frac{1}{2}x = \frac{1}{2} \sin x$

39. $\tan \frac{1}{2}x \sin x = 2 \sin^2 \frac{1}{2}x$

40. $2 \tan \frac{1}{2}x \csc x = \sec^2 \frac{1}{2}x$

41. $4 \sin x \cos^2 x - \sin 3x = \sin x$

42. $\tan \frac{1}{2}x = \dfrac{\sin x - \cos x + 1}{\sin x + \cos x + 1}$

In Exercises 43–56 find the solutions of the equation for $0 \le x < 2\pi$.

43. $\cos 2x = \frac{1}{2}$

44. $\cos 2x = \sin 4x$

45. $\sin 2x \cos x = 0$

46. $\sin x(\tan 2x - 1) = 0$

47. $\sin 2x - \cos x = 0$

48. $\cos 2x - \sin x = 0$

49. $\sin 2x - \cos 2x = 0$

50. $\cos 2x + \sec 2x = -2$

51. $\sin 2x + \sin 4x = 0$

52. $\tan 2x + 2 \sin x = 0$

53. $\sec \frac{1}{2}x - \cos \frac{1}{2}x = 0$

54. $\csc \frac{1}{2}x - \sin \frac{1}{2}x = 0$

55. $\cos^2 \frac{1}{2}x = \cos^2 x$

56. $\sin^2 \frac{1}{2}x = \cos^2 x$

57. Write $\cos^4 x$ as an expression involving only first powers of trigonometric functions.

58. Write $\sin^6 x$ as a polynomial in $\cos 2x$.

59. Verify that $\frac{1}{2}\sqrt{2 + \sqrt{3}} = \dfrac{1}{2\sqrt{2}}(\sqrt{3} + 1)$ (see Example 9, this section).

7.5 Miscellaneous Formulas

In integral calculus it is helpful to be able to express products of the sine and cosine functions as sums or differences of these trigonometric functions.

Notice that adding the two identities incorporated in Identity 10 yields the result

$$\sin(x + y) + \sin(x - y) = 2 \sin x \cos y$$

so that

$$\sin x \cos y = \frac{1}{2}[\sin(x+y) + \sin(x-y)]$$

This is the first of the product formulas.

Product formulas

$$18.\ \sin x \cos y = \frac{1}{2}[\sin(x+y) + \sin(x-y)]$$

$$19.\ \cos x \cos y = \frac{1}{2}[\cos(x+y) + \cos(x-y)]$$

$$20.\ \sin x \sin y = \frac{1}{2}[\cos(x-y) - \cos(x+y)]$$

The reader can obtain Identities 19 and 20 from Identity 9 by addition and subtraction, respectively.

Example 1 Write the product $\sin 5x \cos 7x$ as a difference of sine functions.

$$\begin{aligned}\sin 5x \cos 7x &\overset{18}{=} \frac{1}{2}[\sin(5x+7x) + \sin(5x-7x)] \\ &= \frac{1}{2}[\sin 12x + \sin(-2x)] \\ &= \frac{1}{2}[\sin 12x - \sin 2x]\end{aligned}$$

■

In this example we used the relation $\sin(-\alpha) = -\sin\alpha$ for all α.

Example 2 Verify the identity

$$\sin 6x \sin 4x \sin 2x = \frac{1}{4}(\sin 4x + \sin 8x - \sin 12x)$$

$$\begin{aligned}\sin 6x \sin 4x \sin 2x &= (\sin 6x \sin 4x)\sin 2x \\ &\overset{20}{=} \frac{1}{2}(\cos 2x - \cos 10x)\sin 2x \\ &= \frac{1}{2}(\sin 2x \cos 2x) - \frac{1}{2}(\sin 2x \cos 10x) \\ &\overset{18}{=} \frac{1}{2}\left[\frac{1}{2}(\sin 4x + \sin 0)\right] - \frac{1}{2}\left\{\frac{1}{2}[\sin 12x + \sin(-8x)]\right\} \\ &= \frac{1}{4}\sin 4x - \frac{1}{2}\left(\frac{1}{2}\sin 12x - \frac{1}{2}\sin 8x\right) \quad \textit{(since } \sin 0 = 0 \textit{ and } \sin(-8x) = -\sin 8x\textit{)} \\ &= \frac{1}{4}\sin 4x - \frac{1}{4}\sin 12x + \frac{1}{4}\sin 8x \\ &= \frac{1}{4}(\sin 4x + \sin 8x - \sin 12x)\end{aligned}$$

■

When solving conditional equations involving sums or differences of sine and cosine functions it is frequently necessary to reverse the above process and write sums or differences as products. This technique is usually called factoring.

If we let $x + y = u$ and $x - y = v$, then $u + v = 2x$ and $u - v = 2y$, so that $x = \frac{1}{2}(u + v)$ and $y = \frac{1}{2}(u - v)$. We may then write Identity 18 as

$$\sin(x + y) + \sin(x - y) = 2 \sin x \cos y$$

or

$$\sin u + \sin v = 2 \sin\left(\frac{u + v}{2}\right) \cos\left(\frac{u - v}{2}\right)$$

This last equation is the first of the factoring formulas. Since this is an identity for all values of u and v, we revert to our usual x, y notation in the statement of the formulas.

Factoring formulas

21. $\sin x + \sin y = 2 \sin\left(\frac{x + y}{2}\right) \cos\left(\frac{x - y}{2}\right)$

22. $\cos x + \cos y = 2 \cos\left(\frac{x + y}{2}\right) \cos\left(\frac{x - y}{2}\right)$

23. $\cos x - \cos y = 2 \sin\left(\frac{x + y}{2}\right) \sin\left(\frac{y - x}{2}\right)$

The reader may easily obtain 22 from 19 and 23 from 20. Be sure to note the order change in the second factor of 23.

It is natural to wonder about the absence of a formula for $\sin x - \sin y$. This case is most easily handled in practice by using Identity 21 and the relation $\sin(-y) = -\sin y$.

Example 3 Verify the identity

$$\sin x - \sin y = 2 \cos\left(\frac{x + y}{2}\right) \sin\left(\frac{x - y}{2}\right)$$

$$\begin{aligned}\sin x - \sin y &= \sin x + \sin(-y)\\ &\overset{21}{=} 2 \sin\left[\frac{x + (-y)}{2}\right] \cos\left[\frac{x - (-y)}{2}\right]\\ &= 2 \sin\left(\frac{x - y}{2}\right) \cos\left(\frac{x + y}{2}\right)\end{aligned}$$

■

Example 4 Factor the expression $\cos 4x - \cos 8x$.

$$\cos 4x - \cos 8x \overset{23}{=} 2 \sin\left(\frac{4x + 8x}{2}\right) \sin\left(\frac{8x - 4x}{2}\right)$$

$$= 2 \sin 6x \sin 2x$$

■

Example 5 Find all solutions of

$$\sin 3x + \sin 2x = 0 \quad \text{for} \quad x \in [0, 2\pi)$$

$$\sin 3x + \sin 2x \overset{21}{=} 2 \sin\left(\frac{5}{2}x\right) \cos\left(\frac{1}{2}x\right) = 0$$

if $\sin\left(\frac{5}{2}x\right) = 0$ or $\cos\left(\frac{1}{2}x\right) = 0$. We note that if $0 \le x < 2\pi$, then $0 \le \frac{5}{2}x < 5\pi$ and $0 \le \frac{1}{2}x < \pi$. For $x \in [0, 2\pi)$ we see that

$$\sin\left(\frac{5}{2}x\right) = 0 \quad \text{if} \quad \frac{5}{2}x = 0, \pi, 2\pi, 3\pi, 4\pi$$

i.e., if $x = 0, 2\pi/5, 4\pi/5, 6\pi/5, 8\pi/5$, and

$$\cos\left(\frac{1}{2}x\right) = 0 \quad \text{if} \quad \frac{1}{2}x = \frac{\pi}{2}$$

i.e., if $x = \pi$. The solution set is $\{0, 2\pi/5, 4\pi/5, \pi, 6\pi/5, 8\pi/5\}$. ■

Example 6 Verify the identity

$$\sin 6x + \sin 4x + \sin 2x = 4 \sin 3x \cos 2x \cos x$$

$$\begin{aligned}
\sin 6x + \sin 4x + \sin 2x &= (\sin 6x + \sin 4x) + \sin 2x \\
&\overset{21}{=} 2 \sin 5x \cos x + \sin 2x \\
&\overset{12}{=} 2 \sin 5x \cos x + 2 \sin x \cos x \\
&= 2 \cos x(\sin 5x + \sin x) \\
&\overset{21}{=} 2 \cos x(2 \sin 3x \cos 2x) \\
&= 4 \sin 3x \cos 2x \cos x
\end{aligned}$$

■

Exercises 7.5

In Exercises 1–4 factor the expression.

1. $\sin 6x - \sin 2x$

2. $\sin 6x + \sin 2x$

3. $\cos \frac{1}{2} x - \cos \frac{3}{2} x$

4. $\cos \frac{1}{2} x + \cos \frac{3}{2} x$

In Exercises 5–8 express as a sum or difference of trigonometric functions.

5. $\sin \frac{3}{2} x \cos \frac{1}{2} x$

6. $\cos \frac{1}{4} x \sin \frac{3}{4} x$

7. $\sin \frac{\pi}{5} \sin \frac{\pi}{10}$

8. $\cos \frac{7\pi}{12} \cos \frac{11\pi}{12}$

In Exercises 9–14 find the exact value of the expression.

9. $\sin 75° + \sin 15°$

10. $\cos 165° + \cos 105°$

11. $\cos 105° - \cos 15°$

12. $\cos 15° \cos 45°$

13. $\sin 165° \cos 105°$

14. $\sin 105° \sin 45°$

In Exercises 15–18 verify the formulas.

15. $\cos x \cos y = \frac{1}{2}[\cos(x + y) + \cos(x - y)]$

16. $\sin x \sin y = \frac{1}{2}[\cos(x - y) - \cos(x + y)]$

17. $\cos x + \cos y = 2 \cos\left(\frac{x + y}{2}\right) \cos\left(\frac{x - y}{2}\right)$

18. $\cos x - \cos y = 2 \sin\left(\frac{x + y}{2}\right) \sin\left(\frac{y - x}{2}\right)$

In Exercises 19–28 verify the identity.

19. $\dfrac{\cos x + \cos 5x}{\cos x - \cos 5x} = \dfrac{\cot 3x}{\tan 2x}$

20. $\dfrac{\cos 3x - \cos x}{\cos x + \cos 3x} = -\tan 2x \tan x$

21. $\dfrac{\sin x + \sin y}{\cos x - \cos y} = -\cot \frac{1}{2}(x - y)$

22. $\dfrac{\sin x + \sin y}{\sin x - \sin y} = \dfrac{\tan \frac{1}{2}(x + y)}{\tan \frac{1}{2}(x - y)}$

23. $\dfrac{\cos x + \cos y}{\cos y - \cos x} = \cot \frac{1}{2}(x + y) \cot \frac{1}{2}(x - y)$

24. $\dfrac{\sin x + \sin 2x}{\cos x + \cos 2x} = \tan \frac{3}{2}x$

25. $2 \sin 3x \cos x = \sin 4x + \sin 2x$

26. $\sin 5x \cos 3x = \frac{1}{2}(\sin 8x + \sin 2x)$

27. $\cos(x + y) \cos(x - y) = \cos^2 y - \sin^2 x$

28. $\sin(x + y) \cos y - \cos(x + y) \sin y = \sin x$

In Exercises 29–38 solve the equation for $0 \leq x < \pi$.

29. $\cos 4x + \cos 2x = 0$

30. $\cos 2x + \cos 3x = 0$

31. $\cos x - \cos 3x = 0$

32. $\sin x + \sin 3x = 0$

33. $\cos x + \cos 3x = 2 \cos 2x$

34. $\cos x - \cos 3x = 2 \sin x$

35. $\sin x - \cos 2x = \sin 3x$

36. $2 \sin 3x + \cos x - \cos 5x = 0$

37. $\sin 5x - \sin 3x - \sin x = 0$

38. $\cos 5x + \cos x = 2 \cos 2x$

39. Verify the identity

$$\sin 5x + 2 \sin 3x + \sin x = 4 \cos^2 x \sin 3x$$

40. Write $\sin(x + h) - \sin x$ as a product.

7.6 Identities and Equations Revisited

In this chapter we have developed a number of important formulas. No doubt the reader has developed some skill in the application of these formulas to identities and equations. It is inevitable, however, that in the development of any subject the exercises that appear in a section require primarily some mastery of the formulas and techniques of that particular section. Much more difficult in practice is the decision as to the formula or technique required for a specific problem. The nature of present secondary- and college-level mathematics instruction is such that the student receives little practice in making this

Basic Formulas

1. $\sec x = \dfrac{1}{\cos x}$
2. $\csc x = \dfrac{1}{\sin x}$
3. $\cot x = \dfrac{1}{\tan x}$
4. $\tan x = \dfrac{\sin x}{\cos x}$
5. $\cot x = \dfrac{\cos x}{\sin x}$
6. $\sin^2 x + \cos^2 x = 1$
7. $\tan^2 x + 1 = \sec^2 x$
8. $1 + \cot^2 x = \csc^2 x$
9. $\cos(x \pm y) = \cos x \cos y \mp \sin x \sin y$
10. $\sin(x \pm y) = \sin x \cos y \pm \cos x \sin y$
11. $\tan(x \pm y) = \dfrac{\tan x \pm \tan y}{1 \mp \tan x \tan y}$
12. $\sin 2x = 2 \sin x \cos x$
13. $\cos 2x \overset{a}{=} \cos^2 x - \sin^2 x \overset{b}{=} 2\cos^2 x - 1 \overset{c}{=} 1 - 2\sin^2 x$
14. $\tan 2x = \dfrac{2 \tan x}{1 - \tan^2 x}$
15. $\sin \dfrac{1}{2} x = \pm\sqrt{\dfrac{1 - \cos x}{2}}$
16. $\cos \dfrac{1}{2} x = \pm\sqrt{\dfrac{1 + \cos x}{2}}$
17. $\tan \dfrac{1}{2} x \overset{a}{=} \pm\sqrt{\dfrac{1 - \cos x}{1 + \cos x}} \overset{b}{=} \dfrac{1 - \cos x}{\sin x} \overset{c}{=} \dfrac{\sin x}{1 + \cos x}$
18. $\sin x \cos y = \dfrac{1}{2}[\sin(x + y) + \sin(x - y)]$
19. $\cos x \cos y = \dfrac{1}{2}[\cos(x + y) + \cos(x - y)]$
20. $\sin x \sin y = \dfrac{1}{2}[\cos(x - y) - \cos(x + y)]$
21. $\sin x + \sin y = 2 \sin\left(\dfrac{x + y}{2}\right) \cos\left(\dfrac{x - y}{2}\right)$
22. $\cos x + \cos y = 2 \cos\left(\dfrac{x + y}{2}\right) \cos\left(\dfrac{x - y}{2}\right)$
23. $\cos x - \cos y = 2 \sin\left(\dfrac{x + y}{2}\right) \sin\left(\dfrac{y - x}{2}\right)$
24. $\sin^2 x = \dfrac{1 - \cos 2x}{2}$
25. $\cos^2 x = \dfrac{1 + \cos 2x}{2}$

type of decision. It is the purpose of this section to encourage and develop such practice.

We begin with a list of the basic formulas of analytic trigonometry. There is some disagreement as to which of these should be memorized by the student, and the instructor will usually specify which should be learned. In any case, the authors are not aware of any instructor who would object to the student knowing all of these identities.

Example 1 Verify the identity

$$\frac{\sin 5x}{\sin x} - \frac{\cos 5x}{\cos x} = 4 \cos 2x$$

$$\begin{aligned}\frac{\sin 5x}{\sin x} - \frac{\cos 5x}{\cos x} &= \frac{\sin 5x \cos x - \sin x \cos 5x}{\sin x \cos x}\\ &\overset{18}{\underset{\star}{=}} \frac{\frac{1}{2}[\sin 6x + \sin 4x] - \frac{1}{2}[\sin 6x + \sin(-4x)]}{\sin x \cos x}\\ &= \frac{\frac{1}{2}\sin 6x + \frac{1}{2}\sin 4x - \frac{1}{2}\sin 6x + \frac{1}{2}\sin 4x}{\sin x \cos x}\\ &\underset{\star\star}{=} \frac{\sin 4x}{\sin x \cos x} \overset{12}{=} \frac{2 \sin 2x \cos 2x}{\sin x \cos x}\\ &\overset{12}{=} \frac{2(2 \sin x \cos x) \cos 2x}{\sin x \cos x} = 4 \cos 2x\end{aligned}$$

■

There frequently are several tracks one can follow in the verification of an identity. Notice in Example 1 that it is possible to step from * to ** by use of Identity 10. We emphasize that the possibility of different approaches to a specific problem does not necessarily mean that one is invalid.

Example 2 Verify the identity

$$\sin x \sin 3x = \sin^2 2x - \sin^2 x$$

$$\begin{aligned}\sin x \sin 3x &\overset{20}{=} \frac{1}{2}[\cos(-2x) - \cos 4x]\\ &= \frac{1}{2}(\cos 2x - \cos 4x) \quad \textit{(since } \cos(-\alpha) = \cos\alpha \textit{ for all } \alpha)\\ &\overset{13}{=} \frac{1}{2}[\cos 2x - (\cos^2 2x - \sin^2 2x)]\\ &\overset{6}{=} \frac{1}{2}[\cos 2x - (1 - \sin^2 2x) + \sin^2 2x]\\ &= \frac{1}{2}(2 \sin^2 2x + \cos 2x - 1)\\ &\overset{13c}{=} \frac{1}{2}[2 \sin^2 2x + (1 - 2 \sin^2 x) - 1]\\ &= \frac{1}{2}(2 \sin^2 2x - 2 \sin^2 x) = \sin^2 2x - \sin^2 x\end{aligned}$$

■

Example 3 Verify the identity

$$\cos 5x = 16 \cos^5 x - 20 \cos^3 x + 5 \cos x$$

$$\begin{aligned}
\cos 5x &= \cos(2x + 3x) \\
&\overset{9}{=} \cos 2x \cos 3x - \sin 2x \sin 3x \\
&= \cos 2x \cos(2x + x) - \sin 2x \sin(2x + x) \\
&\overset{9,10}{=} \cos 2x(\cos 2x \cos x - \sin 2x \sin x) \\
&\quad - \sin 2x(\sin 2x \cos x + \cos 2x \sin x) \\
&= \cos^2 2x \cos x - \sin 2x \sin x \cos 2x \\
&\quad - \sin^2 2x \cos x - \sin 2x \sin x \cos 2x \\
&= \cos^2 2x \cos x - 2 \sin x \sin 2x \cos 2x - \sin^2 2x \cos x \\
&= (2 \overset{13b}{\cos^2} x - 1)^2 \cos x - 2 \sin x[2 \overset{12}{\sin} x \cos x(2 \cos^2 x \overset{13b}{-} 1)] \\
&\quad - (2 \sin x \overset{12}{\cos} x)^2 \cos x \\
&= (4 \cos^4 x - 4 \cos^2 x + 1) \cos x - 4 \sin^2 x \cos x(2 \cos^2 x - 1) \\
&\quad - 4 \sin^2 x \cos^2 x \cos x \\
&= 4 \cos^5 x - 4 \cos^3 x + \cos x - 4 \cos x(1 \overset{6}{-} \cos^2 x)(2 \cos^2 x - 1) \\
&\quad - 4(1 \overset{6}{-} \cos^2 x) \cos^3 x \\
&= 4 \cos^5 x - 4 \cos^3 x + \cos x - 4 \cos x(3 \cos^2 x - 2 \cos^4 x - 1) \\
&\quad - 4 \cos^3 x + 4 \cos^5 x \\
&= 8 \cos^5 x - 8 \cos^3 x + \cos x - 12 \cos^3 x + 8 \cos^5 x + 4 \cos x \\
&= 16 \cos^5 x - 20 \cos^3 x + 5 \cos x
\end{aligned}$$

■

Example 4 Verify the identity

$$\sin 6x = \sin 2x(2 \cos 4x + 1)$$

$$\begin{aligned}
\sin 6x &= \sin(2x + 4x) \\
&\overset{10}{=} \sin 2x \cos 4x + \cos 2x \sin 4x \\
&\overset{12}{=} \sin 2x \cos 4x + \cos 2x(2 \sin 2x \cos 2x) \\
&= \sin 2x(\cos 4x + 2 \cos^2 2x) \\
&\overset{13b}{=} \sin 2x\left[\cos 4x + 2 \left(\frac{1}{2}\right)(1 + \cos 4x)\right] \qquad \textit{(See Example 2, Section 7.4.)} \\
&= \sin 2x(\cos 4x + 1 + \cos 4x) \\
&= \sin 2x(2 \cos 4x + 1)
\end{aligned}$$

■

Example 5 Solve the equation

$$2 \sin x - \cos x = 1 - \sin 2x$$

for $x \in [-\pi, \pi]$. Since $\sin 2x \overset{12}{=} 2 \sin x \cos x$ the equation is equivalent to

$$2 \sin x \cos x + 2 \sin x - \cos x - 1 = 0$$

We factor by grouping:

$$2 \sin x(\cos x + 1) - (\cos x + 1) = 0$$
$$(2 \sin x - 1)(\cos x + 1) = 0$$

We see that $\sin x = \frac{1}{2}$ or $\cos x = -1$.

$$\sin x = \frac{1}{2} \quad \text{for} \quad \tilde{x} = \frac{\pi}{6}; \qquad \text{for } -\pi \leq x \leq \pi,\ x = \frac{\pi}{6}, \frac{5\pi}{6}$$

$$\cos x = -1 \quad \text{for} \quad -\pi \leq x \leq \pi \quad \text{when} \quad x = \pi, -\pi$$

The solution set is then $\{-\pi, \pi/6, 5\pi/6, \pi\}$. ■

Example 6 Solve the equation

$$\cos x - \sin x = \sqrt{2} \quad \text{for} \quad x \in [0, 2\pi]$$

None of the factoring formulas are helpful, so we square both sides of the equation to obtain

$$\cos^2 x - 2 \sin x \cos x + \sin^2 x = 2$$
$$(\sin^2 x + \cos^2 x) - 2 \sin x \cos x = 2$$
$$1 - 2 \sin x \cos x = 2$$
$$\sin x \cos x = -\frac{1}{2}$$
$$\frac{1}{2} \sin 2x \overset{12}{=} -\frac{1}{2}$$
$$\sin 2x = -1$$

For $0 \leq 2x \leq 4\pi$ the solutions of this equation are

$$2x = \frac{3\pi}{2}, \frac{7\pi}{2} \quad \text{or} \quad x = \frac{3\pi}{4}, \frac{7\pi}{4}$$

$x = 3\pi/4$ does not satisfy the original equation and hence must be discarded as an extraneous solution. The solution set is $\{7\pi/4\}$. ■

Example 7 Find all solutions of the equation

WARNING
Be sure to remember that if $0 \leq x \leq 2\pi$ then $0 \leq 2x \leq 4\pi$.

$$\cos 4x = 3 \sin 2x - 4 \quad \text{for} \quad x \in [0, 2\pi]$$
$$3 \sin 2x - 4 = \cos 4x = \cos 2(2x) \overset{13c}{=} 1 - 2 \sin^2 2x$$

if

$$2 \sin^2 2x + 3 \sin 2x - 5 = 0$$
$$(2 \sin 2x + 5)(\sin 2x - 1) = 0$$

The solutions require $\sin 2x = -\frac{5}{2}$ or $\sin 2x = 1$. Sin $2x = -\frac{5}{2}$ has no solutions since $|\sin \alpha| \leq 1$ for all α. Sin $2x = 1$ for $x \in [0, 2\pi]$ if $2x = \pi/2$ or $2x = 5\pi/2$. The solution set is $\{\pi/4, 5\pi/4\}$. ■

Exercises 7.6

In Exercises 1–14 verify the identity.

1. $(\sin \frac{1}{2}x + \cos \frac{1}{2}x)^2 = 1 + \sin x$

2. $\dfrac{\cos x}{\sec 2x} - \dfrac{\sin 2x}{\csc x} = \cos 3x$

3. $\dfrac{1 - \tan x}{1 + \tan x} = \dfrac{1 - \sin 2x}{\cos 2x}$

4. $\sin 3x = 4 \sin x \cos^2 x - \sin x$

5. $\cos 4x = 8 \cos^4 x - 8 \cos^2 x + 1$

6. $\sin 4x = 4 \cos x(\sin x - 2 \sin^3 x)$

7. $\dfrac{\sin 4x - \sin 2x}{\cos 4x + \cos 2x} = \tan x$

8. $\dfrac{\sin x}{\sin 3x} - \dfrac{\cos x}{\cos 3x} = -2 \sin 2x \csc 6x$

9. $\dfrac{\sin x + \sin 3x + \sin 5x}{\cos x + \cos 3x + \cos 5x} = \tan 3x$

10. $\cos^4 x = \frac{1}{8}(\cos 4x + 4 \cos 2x + 3)$

11. $\dfrac{\sin 4x - \sin 6x}{\sin 3x \sin 2x - \cos 3x \cos 2x} = 2 \sin x$

12. $\sin 4x - \sin 2x = 2 \cos 3x \sin x$

13. $\cos x \cos 2x \cos 4x = \dfrac{\sin 8x}{8 \sin x}$

14. $2 \sin x - \sin 3x = 4 \sin^3 x - \sin x$

15. Write $\sin 3x$ as a polynomial in $\sin x$.

16. Write $\cos 3x$ as a polynomial in $\cos x$.

In Exercises 17–30 find the solutions of the equation for $0 \leq x < 2\pi$.

17. $\cos x - 1 = \sin x$

18. $\cos 2x \cos 3x + \sin 2x \sin 3x = -\frac{1}{2}$

19. $\sin x = -\sqrt{3} \cos x$

20. $\sin 2x - \cos 2x = 0$

21. $\cos x = \sin \frac{1}{2}x$

22. $\sin 4x = 2 \sin 2x$

23. $\sin 2x + \sin x + 2 \cos x = -1$

24. $\sin x + \sin 3x + \sin 5x + \sin 7x = 0$

25. $\sin 3x - \cos 3x = \cos x - \sin x$

26. $\sin 3x = \sin 7x$

27. $\dfrac{\sin x + \sin 2x}{\cos x + \cos 2x} = -1$

28. $\tan \frac{1}{2}x \sin x = -1$

29. $\sin 4x + \cos 3x = \cos x$

30. $2 \sin x + \tan 2x = 0$

Chapter Review

In Exercises 1–2 use fundamental identities to find the values of the six trigonometric functions at x if:

1. $\sec x = -3$; $\sin x < 0$

2. $\tan x = -3$; $\sin x > 0$

In Exercises 3–14 find exact values.

3. $\cos 15°$

4. $\sin 285°$

5. $\tan \dfrac{13\pi}{12}$

6. $\sin \dfrac{\pi}{8}$

7. $\cos\left(-\dfrac{5\pi}{12}\right)$

8. $\sin 165° + \sin 105°$

9. $\cos 45° \sin 105°$

10. $\sin\left[2 \arcsin\left(-\frac{\sqrt{3}}{2}\right)\right]$

11. $\cos\left[2 \tan^{-1} \frac{3}{4}\right]$

12. $\tan\left[\frac{1}{2}\cos^{-1}\left(-\frac{7}{25}\right)\right]$

13. $\sin\left[\arccos \frac{3}{5} + \arctan(-1)\right]$

14. $\tan\left[\arcsin\left(-\frac{1}{2}\right) - \arccos(-1)\right]$

In Exercises 15–18 evaluate the given expression if α is in quadrant II, β is in quadrant III, $\sin \alpha = \frac{7}{25}$ *and* $\cos \beta = -\frac{4}{5}$.

15. $\sin(\alpha - \beta)$

16. $\cos(\alpha + \beta)$

17. $\sin \frac{1}{2}\alpha$

18. $\tan 2\beta$

19. Find $\sin \theta$ and $\cos \theta$ if θ is acute and $\cot 2\theta = -\frac{3}{4}$.

20. Factor $\sin 5x - \sin 3x$.

21. Express $\sin \frac{3}{2} x \cos \frac{5}{2} x$ as a sum of trigonometric functions.

In Exercises 22–23 use the given substitution to find a trigonometric expression associated with the given algebraic expression for $0 < \theta < \frac{\pi}{2}$.

22. $\sqrt{16 - 9x^2}$, $x = \frac{4}{3} \sin \theta$

23. $x(49 + x^2)^{3/2}$, $x = 7 \tan \theta$

In Exercises 24–30 prove the equation is an identity or provide a counterexample.

24. $\cos x = 1 - \sin x$

25. $\tan^2 x + \cot^2 x = 1$

26. $\arcsin(\sin x) = x$

27. $\tan\left(\frac{3\pi}{2} + x\right) = -\cot x$

28. $\sin^2 x = \frac{1}{2}(1 - \cos 2x)$

29. $2 \cos \frac{1}{2}x = 1 + \cos x$

30. $\sin x \cos y = \frac{1}{2}[\sin(x - y) - \sin(x + y)]$

In Exercises 31–48 verify the identity.

31. $\cot x + 1 = \csc x(\cos x + \sin x)$

32. $\cos\left(x - \frac{3\pi}{2}\right) = -\sin x$

33. $\cos^2 x \tan^2 x + \sin^2 x \cot^2 x = 1$

34. $\dfrac{1 + \cos x}{\sin x} + \dfrac{\sin x}{1 + \cos x} = 2 \csc x$

35. $\sin^2 x - \cos^2 x = \dfrac{\tan x - \cot x}{\tan x + \cot x}$

36. $1 - \dfrac{\cos^2 x}{1 + \sin x} = \sin x$

37. $\sin(-x) \sec(-x) \cot(-x) = 1$

38. $\cos(x + y) - \cos(x - y) = -2 \sin x \sin y$

39. $\sin^2 x \csc^2 x + \cos^2 x \csc^2 x - \sin^2 x - \cos^2 x = \cot^2 x$

40. $\dfrac{\cos^3 x + \sin^3 x}{\cos x + \sin x} = \cos x(\sec x - \sin x)$

41. $\tan\left(x + \frac{\pi}{4}\right) - \tan\left(x - \frac{\pi}{4}\right) = \frac{2}{1 - \tan^2 x}$

42. $\sin 2x = 2 \sin^2 x \cot x$

43. $\dfrac{1 + \tan \frac{1}{2}x}{1 - \tan \frac{1}{2}x} = \dfrac{\cos x}{1 - \sin x}$

44. $\cot x + \tan x = 2 \csc 2x$

45. $\dfrac{\sin 3x}{\sin x} + \dfrac{\cos 3x}{\cos x} = 4 \cos 2x$

46. $\sin 3x + \sin 2x + \sin x = \sin 2x(1 + 2 \cos x)$

47. $\dfrac{\cos 3x - \cos x}{\sin x - \sin 3x} = \tan 2x$

48. $\dfrac{\sin 3x - \sin x}{\cos 3x + \cos x} = \tan x$

In Exercises 49–51 find the general solution of the equation.

49. $\arcsin(x^2 - 9) = -\frac{\pi}{6}$

50. $\cos^2 x = \cos x$

51. $\cot 2x = 1$

In Exercises 52–66 find the solutions for $0 \le x < 2\pi$.

52. $2 \sin^2 x + \sin x - 1 = 0$

53. $\cos^2 x + \cos x - 1 = 0$

54. $\sec^2 x + 2 \tan^2 x = 2$

55. $\csc x + \cot x = \sqrt{3}$

56. $2 \sin x = 1 - \cos x$

57. $\sin^2 x - 3 \sin x + 2 = 0$

58. $\sin 4x = \sin 2x$

59. $\cos 2x + \cos x + 1 = 0$

60. $\cos x + \sin \frac{1}{2}x = 1$

61. $\cos x + \cos 3x = 0$

62. $\sin x + \sin 3x = 2 \sin 2x$

63. $\sin 3x + \sin 2x + \sin x = 0$

64. $\cos x + \cos 3x + \cos 5x = 0$

65. $4 \sin^2 x + 2(1 - \sqrt{2}) \sin x = \sqrt{2}$

66. $2 \cos x \sin x + 2 \cos x + \sin x + 1 = 0$

Chapter Test

1. Use the fundamental identities to find $\tan x$ if $\sec x = \frac{13}{12}$ and $\csc x < 0$.

2. Find the exact value of $\cos \frac{\pi}{12}$ by using:

a. $\frac{\pi}{12} = \frac{\pi}{4} - \frac{\pi}{6}$ **b.** $\frac{\pi}{12} = \left(\frac{1}{2}\right)\left(\frac{\pi}{6}\right)$

In Problems 3–6 find the exact value.

3. $\sin 165° \sin 105°$

4. $\cos \frac{\pi}{12} - \cos \frac{5\pi}{12}$

5. $\tan\left[\frac{1}{2} \arcsin\left(-\frac{3}{5}\right)\right]$

6. $\cos\left[\arccos\left(-\frac{1}{3}\right) + \arcsin \frac{1}{2}\right]$

7. Find $\sin \theta$ if θ is acute and $\cot 2\theta = \frac{7}{24}$.

In Problems 8–9 show that the equation is not an identity.

8. $\sin x = \sin x \cos x$

9. $\cos(x - y) = \cos x - \cos y$

In Problems 10–12 verify the identity.

10. $\sin\left(x - \frac{3\pi}{2}\right) = \cos(-x)$

11. $\frac{\sin 2x}{1 + \cos 2x} = \tan x$

12. $(\sin \frac{1}{2}x - \cos \frac{1}{2}x)^2 = 1 - \sin x$

In Problems 13–15 find the solution of the equation for $0 \leq x < 2\pi$.

13. $4 \cos^2 x - 3 = 0$

14. $\cos 3x = \cos x$

15. $2 \sin 2x + 4 \cos x + \sin x + 1 = 0$

8 Solution of Triangles

Because of the Franklin Mountains, it is not possible to drive directly from Coronado Tower to Hondo Pass in El Paso, Texas. Instead one can drive southeasterly to the Plaza, then somewhat northeast to Hondo Pass, as in Figure 8.1. What is the air distance from Hondo Pass to Coronado Tower? What is the heading for an aircraft to fly from Hondo Pass to Coronado Tower?

Questions of this type require solution of triangles and are quite common in applications. This chapter is devoted primarily to the solution of triangles.

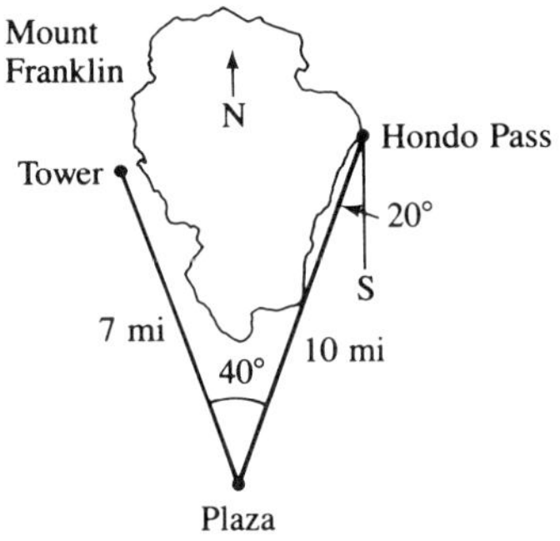

Figure 8.1

8.1 Solving Right Triangles

Solve a triangle

To **solve a triangle** means to find the measures of all three sides and all three angles.

Standard lettering for triangles

It will be helpful in this chapter to utilize a **standard lettering for triangles.** We label the vertices A, B, and C, the corresponding angles with the Greek letters α, β, and γ, respectively, and the sides opposite each vertex with the lowercase letters a, b, and c, respectively, as indicated in Figure 8.2.

By using the trigonometric functions and the techniques developed in Section 6.5, it is possible to solve a right triangle if two sides or an acute angle and one side are known.

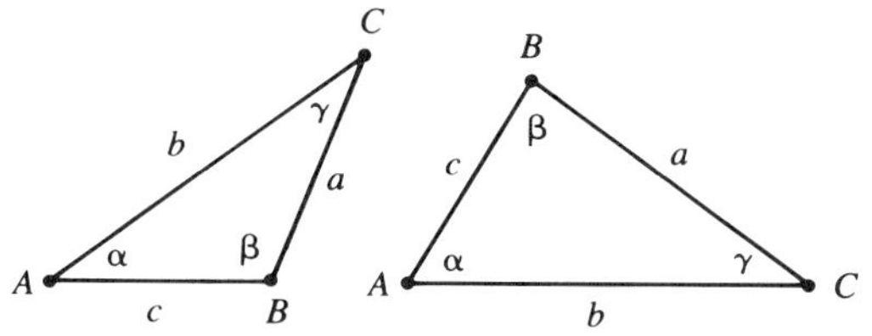

Figure 8.2

Example 1 Solve the triangle ABC if $b = 5$, $c = 3$, and $\beta = 90°$.

The given information is shown in the illustration. Since $\cos \alpha = \frac{3}{5} = 0.6$, we see that $\alpha \simeq 53°$. Since $\gamma + \alpha = 90°$, $\gamma \simeq 37°$. Side a can be computed using the Pythagorean theorem: $a = \sqrt{25 - 9} = 4$. Side a can also be computed using one of the trigonometric functions. For example, since $\tan \alpha = a/3$, we see that $a = 3 \tan \alpha \simeq 3(1.327) \simeq 3.98 \simeq 4$.

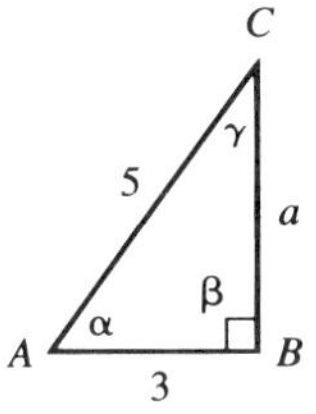

■

Example 2 Solve the triangle ABC if $\gamma = 90°$, $\alpha = 23°$, and $c = 10$.

The given information is shown in the illustration. We see that $\beta = 90° - \alpha$, so that $\beta = 67°$. Since $\cos \alpha = b/10$, $b = 10 \cos 23° \simeq 10(0.9205) \simeq 9.2$. Also, $\sin \alpha = a/10$, hence $a = 10 \sin \alpha \simeq 10(0.3907) \simeq 3.9$. Of course, other trigonometric functions and parts of the triangle could have been used.

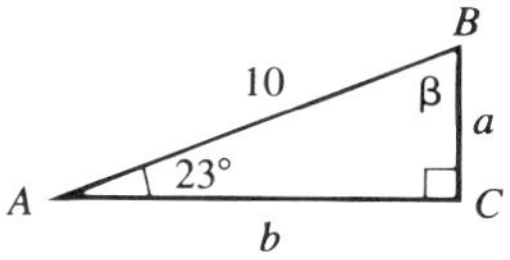

■

Results obtained in the solution of triangles cannot be more accurate than the given or known information. It is evident from Examples 1 and 2 that some agreement on significant digits is necessary before proceeding to physical applications. We adopt the rules stated in Table 8.1 to determine the accuracy to be included in answers, and use the usual rounding conventions. Justification of these rules is beyond the scope of this text.

Angle of elevation

Some technical terms occur frequently in applications. By the **angle of elevation** we mean the angle between the horizontal and the line of sight from observer to object when the object is above the observer, as illustrated in

Table 8.1

Significant Digits for Sides	Angles to Nearest
1 or 2	Degree
3	10 minutes
4	Minute

Angle of depression Figure 8.3. The **angle of depression** is identical to the angle of elevation, except that the object observed is below the observer.

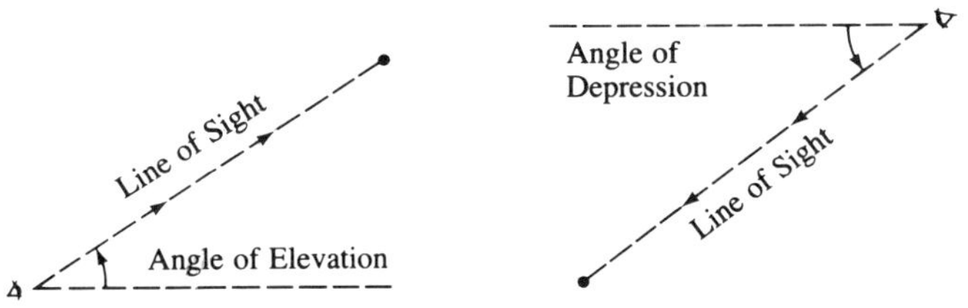

Figure 8.3

Example 3 From the top of a lighthouse 30 meters high, the angle of depression of a ship at sea is 19°. How far is the ship from the base of the lighthouse?
We first draw a figure to illustrate the situation. Note that $\alpha = 90° - 19° = 71°$. If d denotes the requested distance, then $\tan 71° = d/30$, so that

$$d = 30 \tan 71° \simeq 30(2.904) \simeq 87.12 \simeq 87 \text{ m}$$

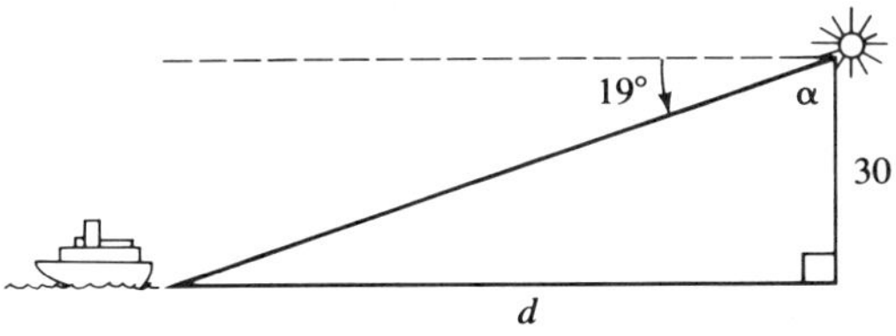

■

Example 4 A small aircraft at a constant altitude of 10,000 feet flies directly toward an observer on the ground. At first glance, the angle of elevation is 22°. A few moments later the angle of elevation is 55°. Approximate the distance traveled by the aircraft.
The situation is depicted in the illustration. We seek the distance d.

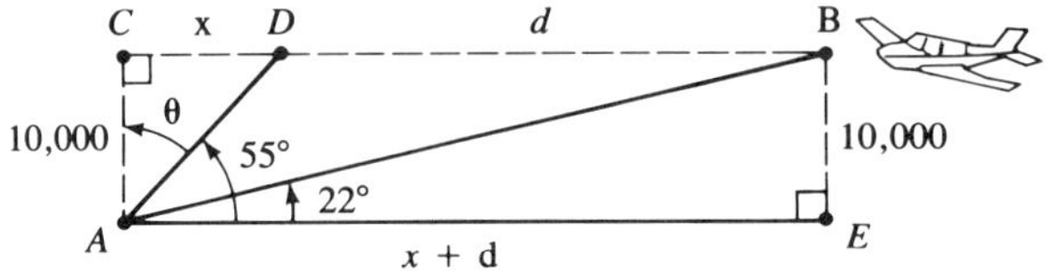

From triangle AEB we see that

$$\cot 22^\circ = \frac{x + d}{10{,}000}$$

and hence

$$* \; x + d = 10{,}000 \cot 22^\circ$$

In triangle ADC, $\theta = 35^\circ$, so that

$$\tan 35^\circ = \frac{x}{10{,}000} \quad \text{and} \quad x = 10{,}000 \tan 35^\circ$$

Therefore, from *,

$$\begin{aligned} d &= 10{,}000 \cot 22^\circ - x = 10{,}000 \cot 22^\circ - 10{,}000 \tan 35^\circ \\ &\simeq 10{,}000(2.475 - 0.7002) = 10{,}000(1.7748) = 17{,}748 \\ &\simeq 18{,}000 \text{ ft} \end{aligned}$$

■

Bearings

In some sea navigation and surveying problems, directions are given by **bearings,** which specify an initial north or south direction, followed by an acute angle and its direction from the initial direction. This is illustrated in Figure 8.4. N35°E can be read "north 35° east" or "35° east of north."

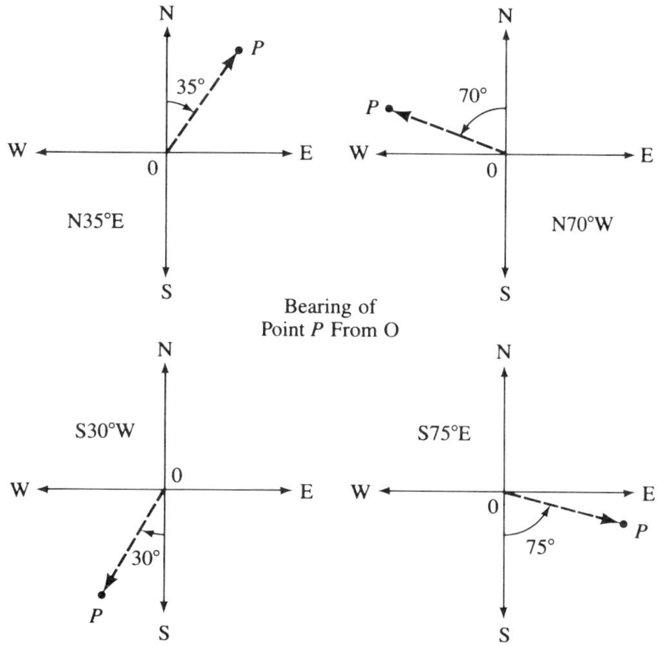

Figure 8.4

Example 5 A ship moves S50°E for 4 hours at 22 mi/hr. Approximately how far east and how far south does it travel?

The ship travels 88 mi as shown. Notice that $\sin 50° = x/88$ and $\cos 50° = y/88$, so that

$$x = 88 \sin 50° \simeq 67.41$$
$$y = 88 \cos 50° \simeq 56.57$$

The ship has traveled approximately 67 mi east and 57 mi south.

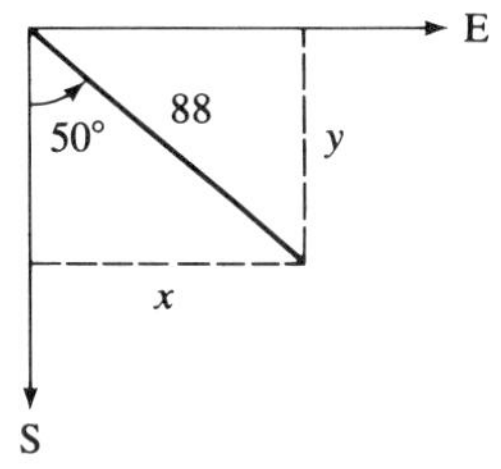

■

Heading In aeronautics, the **heading** of an airplane is the direction in which the airplane is pointed, measured clockwise from the north as determined by a compass reading.

Example 6 In the illustration shown, the heading of OP is 40° and the heading of OQ is 320°.

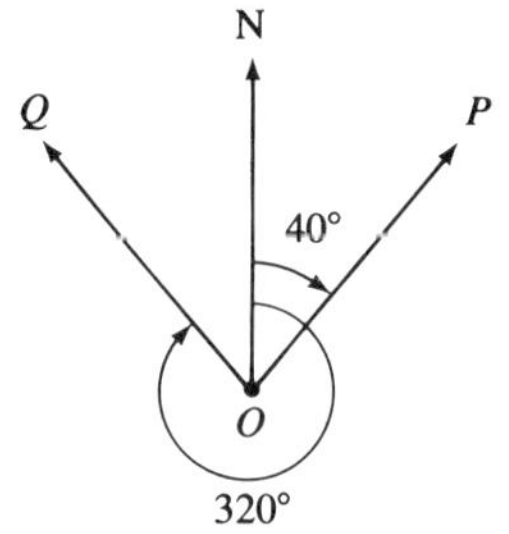

■

Exercises 8.1

In Exercises 1–10 solve the triangle using the given information. In each case, $\gamma = 90°$.

1. $\alpha = 66°$; $b = 17$

2. $\alpha = 58°$; $c = 45$

3. $\beta = 20°$; $c = 20$

4. $\beta = 18°$; $a = 10$

5. $a = 5$; $b = 13$

6. $b = 24$; $c = 25$

7. $\alpha = 32°40'$; $a = 67.3$

8. $a = 40.18$; $b = 117.4$

9. $a = 17.72$; $c = 43.08$

10. $\beta = 23°20'$; $b = 11.7$

11. Suppose that the angle of elevation of the sun is 35°. How long a shadow will a man 6 feet 1 inch tall cast on level ground?

12. From an observation balloon 3,970 feet high the angle of depression of a lake is 33°. How far is the lake from the point on the ground directly beneath the balloon?

13. From a point 4,987 feet above sea level on

Mount Franklin the angle of depression of the Rio Grande river is $11°18'$. The Rio Grande is 3,650 feet above sea level. What is the line of sight distance of the observer from the river?

14. The diagonal of a rectangle is 22 centimeters, and one side is 15 centimeters. What is the length of the other side? What is the measure of the angles between the diagonal and the sides?

15. The height of a tower is 125 feet. What is the angle of elevation of the top of the tower as seen from a point on the ground 100 yards from its base?

16. At a certain distance from a building the angle of elevation, from ground level, of the top of the building is 30°. Moving 10 yards towards the building makes the angle of elevation 45°. How tall is the building?

17. A flagpole is 82.12 feet high and stands on top of a building. From a point on level ground the angles of elevation of the bottom and top of the flagpole are $37°24'$ and $46°47'$, respectively. How tall is the building?

18. A man having a shoulder height of 5 feet shoots at a clay pigeon. If the disk is 40 feet from the shooter (line of sight) and the shotgun makes an angle of elevation of 37° at the moment the shooter fires, how high above the ground is the clay pigeon?

19. A rockwall is 4.3 feet high and stands 5.5 feet from a house 25 feet high. A ladder is laid across the wall and just touches the top of the house. What is the length of the ladder?

20. A boat leaves a dock and travels N21°E for 20 minutes, then turns and travels S70°W for 30 minutes. If the boat's speed is a constant 22 miles per hour, how far is the boat from the dock?

21. A pirate's notes state that in order to find the treasure you begin at a certain tree, walk S52°W for 183 feet, and then walk N19°W for 295 feet. How far and in what direction is the treasure from the tree?

22. A drug smuggler flies through Anthony Gap at midnight on a heading of 125°. At that moment, the border patrol gives chase from a point 20 miles due south of Anthony Gap at the same speed on a heading of 55°. How far will the patrol fly before intercepting the smuggler?

23. A crop duster leaves Fabens and flies at a heading of 110° for 4 miles, then flies at a heading of 250° for 2 miles. How far is the plane from Fabens? What heading should the pilot use to return to Fabens?

8.2 Law of Sines

In general one does not expect the triangles encountered in applications to be right triangles. It is easy, however, to decompose an arbitrary triangle into right triangles, so the methods of Section 8.1 can be used to solve the triangle, provided that sufficient data is available. When this is done one finds that there is considerable repetition in the techniques utilized. Thus for purposes of efficiency and simplicity it is desirable to develop relationships for solving triangles that are not right triangles. Such triangles are called **oblique** triangles.

Oblique

We begin with the theorem known as the **Law of Sines.**

Law of Sines

If ABC is any triangle, lettered in the standard way, then

$$\frac{a}{\sin \alpha} = \frac{b}{\sin \beta} = \frac{c}{\sin \gamma}$$

In words, the ratio of each side to the sine of the opposite angle is the same for all sides. Since ABC is a triangle, none of the six numbers appearing in the Law of Sines can be zero, so the theorem may also be written

$$\frac{\sin \alpha}{a} = \frac{\sin \beta}{b} = \frac{\sin \gamma}{c}$$

To verify the Law of Sines we consider the three possible cases, depending on whether the triangle has three acute angles, one obtuse angle, or a right angle, as in Figure 8.5.

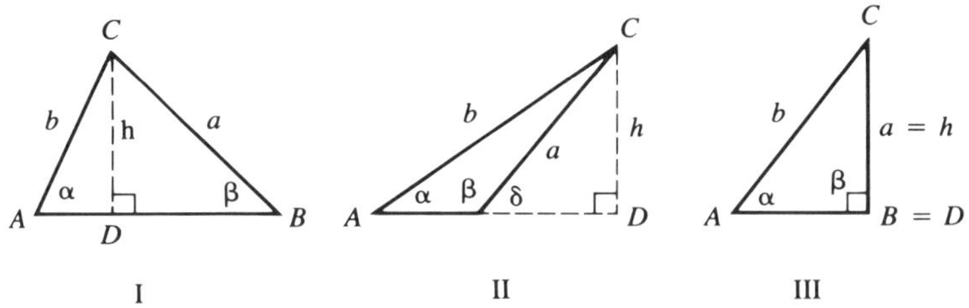

Figure 8.5

In each case, using triangle ACD, $\sin \alpha = h/b$, so that $h = b \sin \alpha$.

For I, using triangle BDC we see that $\sin \beta = h/a$, so that $h = a \sin \beta$.

For II, again using triangle BDC, we have $\sin \delta = h/a$, so that $h = a \sin \delta$. But $a \sin \delta = a \sin(180° - \delta) = a \sin \beta$, so again $h = a \sin \beta$.

For III, $\sin \beta = 1 = h/a$, so once again $h = a \sin \beta$. Thus in every case $a \sin \beta = h = b \sin \alpha$, or

$$\frac{a}{\sin \alpha} = \frac{b}{\sin \beta}$$

Similarly, by drawing the perpendicular from B, we obtain

$$\frac{a}{\sin \alpha} = \frac{c}{\sin \gamma}$$

This completes the verification.

One side, two angles (SAA)

If **one side and two angles** of a triangle are given, the triangle is uniquely determined and the Law of Sines can be used to solve it.

Example 1 Solve the triangle ABC if $\alpha = 35°10'$, $\beta = 57°40'$, and $c = 21.4$.

First draw a figure representing the given information.

Since $\alpha + \beta + \gamma = 180°$, we see that

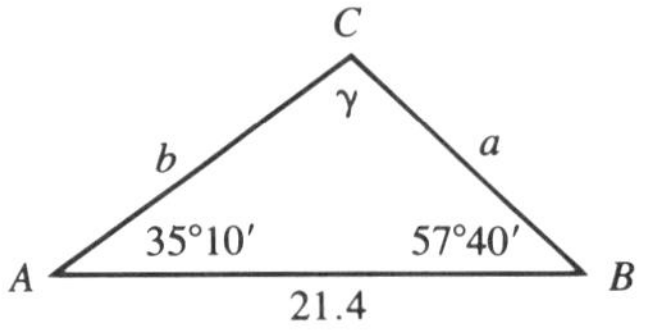

$$\gamma = 180° - (35°10' + 57°40') = 87°10'$$

To find a:

$$\frac{a}{\sin \alpha} = \frac{c}{\sin \gamma}$$

so that

$$a = \frac{c \sin \alpha}{\sin \gamma} = \frac{21.4 \sin 35°10'}{\sin 87°10'} \simeq 12.3$$

To find b:

$$b = \frac{c \sin \beta}{\sin \gamma} = \frac{21.4 \sin 57°40'}{\sin 87°10'} \simeq 18.1$$ ■

Trigonometric function values needed in Example 1 can be obtained using a table or a handheld calculator. Computations can be done using the calculator or logarithm tables.

Two sides, angle opposite one (SSA)

If only one angle of a triangle is given, then the Law of Sines can be used to "solve" the triangle, provided that two sides are given and the angle is opposite one of the given sides. The reason for the quotation marks is that in this situation there may be 0, 1, or 2 triangles satisfying the given data. This is illustrated in Figure 8.6 for an acute angle α and in Figure 8.7 for an obtuse angle α. In both figures α and b are fixed. For these reasons, this case, SSA, is

Ambiguous

frequently called the **ambiguous** case.

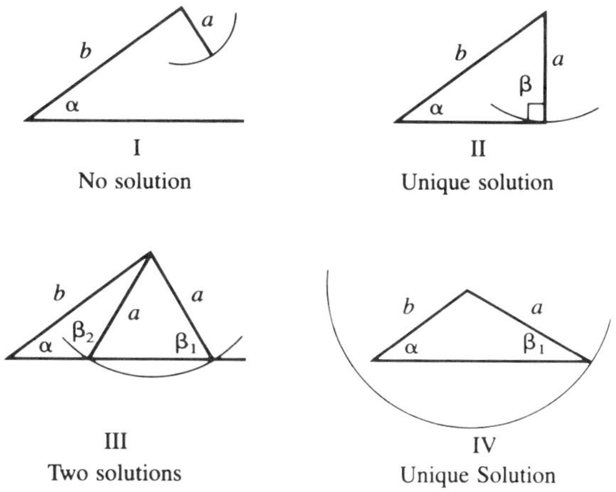

Figure 8.6

It is possible to develop inequalities that depend on the measure of α and the relative sizes of a and b and permit the determination of the exact situation, but in practice it is easier to attempt the solution. The results obtained will point to the specific case involved. For the lettering a, b, α as in Figure 8.6, we would first determine $\sin \beta$ from the Law of Sines:

$$\sin \beta = \frac{b \sin \alpha}{a}$$

1. $\sin \beta > 1$. Since $-1 \leq \sin \theta \leq 1$ for all θ, this is not possible. The situation is as in Figure 8.6,I. No triangle exists.
2. $\sin \beta = 1$. In this case $\beta = 90°$. There is a unique (right) triangle, as in Figure 8.6,II.
3. $0 < \sin \beta < 1$. There are two angles β_1, β_2, which have the same sine. One of these, say β_1, is acute; the other, β_2, is obtuse. If $\alpha + \beta_2 < 180°$, the situation is as in Figure 8.6,III. Both β_1 and β_2 satisfy the data. There are two triangles. This is the only case that is ambiguous. If $\alpha + \beta_2 > 180°$, the situation is as in Figure 8.6,IV. The angle β_2 will not work. There is a unique triangle.

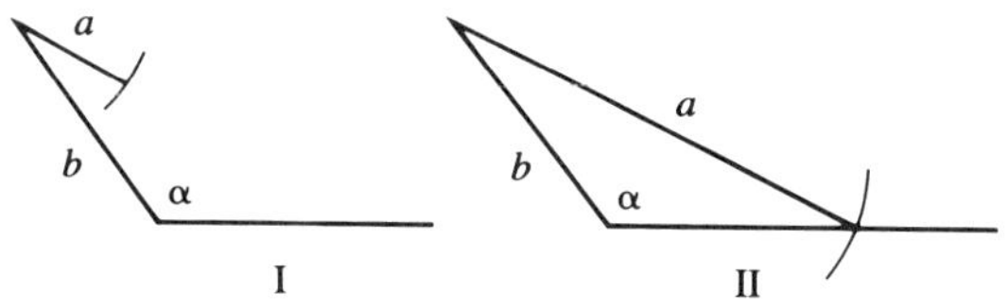

Figure 8.7

For α obtuse there are only two possibilities. The situation Figure 8.7,I will occur if

$$\sin \beta = \frac{b \sin \alpha}{a} > 1$$

and that in Figure 8.7,II will occur if

$$0 < \sin \beta < 1$$

Example 2 Solve the triangle ABC if $\beta = 70°$, $a = 30$, and $b = 20$.

From the Law of Sines,

$$\sin \alpha = \frac{a \sin \beta}{b} = \frac{30 \sin 70°}{20} \simeq (1.5)(0.9397) \simeq 1.4096 > 1$$

There is no triangle that has the given parts. ■

Example 3 Solve the triangle ABC if $\alpha = 45°40'$, $a = 811$, and $b = 553$.

From the Law of Sines,

$$\sin \beta = \frac{b \sin \alpha}{a} = \frac{553 \sin 45°40'}{811} \simeq (0.6819)(0.7153) \simeq 0.4878$$

Thus $\beta = \beta_1 \simeq 29°10'$ or $\beta = \beta_2 \simeq 180° - 29°10' = 150°50'$. Since $\alpha + \beta_2 > 180°$, there is no triangle with the given parts and β_2. We therefore have a unique triangle. Continuing, we see that $\gamma \simeq 180° - 45°40' - 29°10' = 105°10'$.

We use the Law of Sines again to obtain

$$c = \frac{a \sin \gamma}{\sin \alpha} = \frac{811 \sin 105°10'}{\sin 45°40'} \simeq \frac{(811)(0.9652)}{0.7153} \simeq 1{,}090$$

The triangle is as shown in the illustration.

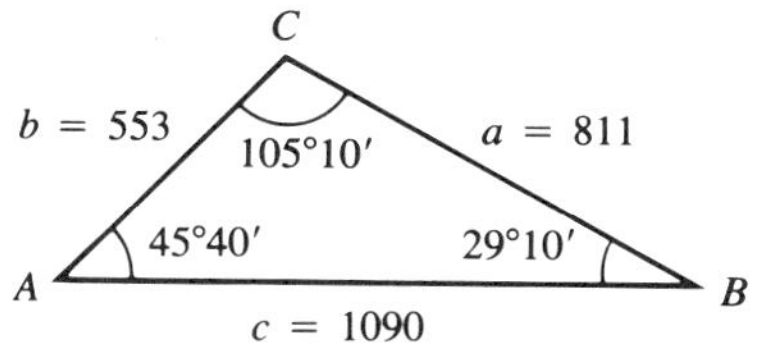

■

Example 4 Solve the triangle $\gamma = 30°$, $a = 7.1$, and $c = 4.9$.

From the Law of Sines,

$$\sin \alpha = \frac{a \sin \gamma}{c} = \frac{(7.1)(0.5)}{4.9} \simeq 0.7245$$

Thus $\alpha = \alpha_1 \simeq 46°$ or $\alpha = \alpha_2 \simeq 180° - 46° = 134°$. Since $134° + \gamma = 134° + 30° = 164° < 180°$, there are two triangles.

Using α_1, $\beta_1 \approx 180° - 30° - 46° = 104°$ and

$$b = \frac{c \sin \beta_1}{\sin \gamma} = \frac{4.9 \sin 104°}{\sin 30°} \simeq \frac{(4.9)(0.9703)}{0.5} \simeq 9.5$$

Using α_2, $\beta_2 \simeq 180° - 30° - 134° = 16°$ and

$$b = \frac{c \sin \beta_2}{\sin \gamma} = \frac{4.9 \sin 16°}{\sin 30°} \simeq \frac{(4.9)(0.2756)}{0.5} \simeq 2.7$$

The triangles are shown in the illustration. For α_1 the triangle is A_1CB and for α_2 the triangle is A_2CB.

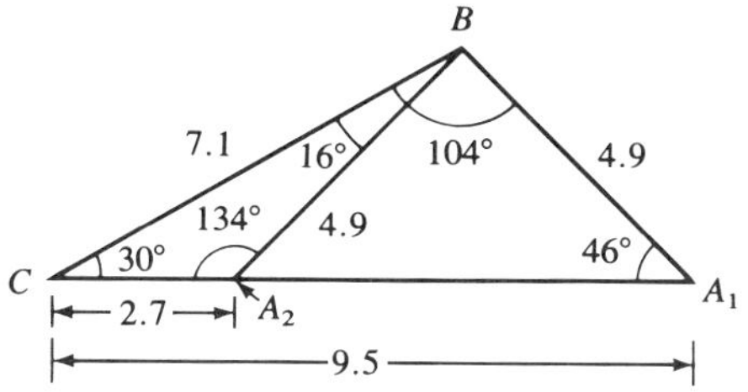

■

Exercises 8.2

In Exercises 1–14 solve the triangle ABC.

1. $\alpha = 47°$; $\beta = 65°$; $a = 12$

2. $\beta = 23°$; $\gamma = 43°$; $b = 23$

3. $\beta = 73°20'$; $b = 20.4$; $c = 27.1$

4. $\alpha = 36°50'$; $a = 3.2$; $b = 5.8$

5. $\alpha = 27°40'$; $\gamma = 68°50'$; $b = 6.31$

6. $\alpha = 37°10'$; $\beta = 74°40'$; $a = 37.6$

7. $\gamma = 28°20'$; $b = 2.38$; $c = 4.81$

8. $\alpha = 31°$; $a = 7$; $b = 5$

9. $\beta = 21°20'$; $b = 128$; $c = 146$

10. $\gamma = 30°$; $b = 19$; $c = 15$

11. $\beta = 30°$; $b = 16$; $c = 32$

12. $\alpha = 39°30'$; $a = 131$; $c = 104$

13. $\alpha = 127°20'$; $a = 427$; $c = 358$

14. $\beta = 110°40'$; $a = 23.7$; $b = 53.2$

15. Hueco Tanks is 20 mi due east of El Paso. A plane leaves El Paso and flies at a heading of 35°. Moments later the plane is spotted from Hueco Tanks at a heading of 307°. How far is the plane from Hueco Tanks at the moment it is sighted?

16. From a point on level ground the angle of elevation of the top of a building is 40°. From a point 15 meters nearer to the building, the angle of elevation is 64°. How tall is the building?

17. A parallelogram has sides of 186 and 123 meters. A diagonal makes an angle of 37° with the longer side. How long is the diagonal?

18. One diagonal of a parallelogram measures 40 cm and determines angles of 28° and 39° with the two sides. How long are the sides of the parallelogram?

19. A hill is inclined at an angle of 20° from the horizontal. A vertical tower stands on the hill. From a point 100 ft down the hill from the tower, the angle of elevation of the top of the tower is 38°. How tall is the tower?

20. Two buildings 25 feet apart are on one side of a street. A bus stop across the street is located between the two so that it makes an angle of 71°10′ with one building and the street and an angle of 79°40′ with the other building and the street. How wide is the street?

21. Passengers on a ship traveling due north observe an iceberg at a bearing of N20°E. After another 10 miles the iceberg is at a bearing of N37°E. How close will the ship come to the iceberg?

22. A ship leaves harbor at noon and travels N45°E at 15 miles per hour. When the ship leaves the harbor, the harbormaster headquarters is located at a bearing of N12°W, and at 2 P.M. the headquarters is at a bearing of S58°W. How far was the ship from the headquarters at noon? At 2 P.M.?

23. It is proposed that a tunnel be dug through Mount Franklin from Coronado Country Club to Mountain Park. A point is chosen on Transmountain Road that is 7,000 yards from Coronado Country Club and 6,500 yards from Mountain Park. Find the length of the tunnel if the angle from Coronado Country Club to Mountain Park to the point on Transmountain Road is 78°.

24. A cross 25 feet high stands on Mount Cristo Rey. From the top of the cross the angle of depression of a point on the Rio Grande is 20°, and from the base of the cross the angle of depression of the same point is 19°30′. What is the height (above the river) of Mount Cristo Rey at the base of the cross?

25. The equation

$$\frac{a-b}{c} = \frac{\sin \frac{1}{2}(\alpha - \beta)}{\cos \frac{1}{2}\gamma}$$

is valid for any triangle ABC and is known as **Mollweide's formula.** This formula is frequently used to check a solution of a triangle, because it contains all six parts of the triangle. Check Problem 1 of this exercise set using Mollweide's formula.

26. The equation

$$\frac{a+b}{c} = \frac{\cos \frac{1}{2}(\alpha - \beta)}{\sin \frac{1}{2}\gamma}$$

is valid for any triangle ABC and is known as **Newton's formula.** Check Problem 1 using Newton's formula.

8.3 Law of Cosines

As discussed in Section 8.2, the Law of Sines can be used to solve oblique triangles if triangular parts of the form SAA or SSA are available. This requires that both parts be known for one of the ratios of the Law of Sines. If all three sides or two sides and the included angle of a triangle are the only known parts, the Law of Sines cannot be used directly. In these cases the **Law of Cosines** can be used to begin solving the triangle. The solution would then normally be completed using the Law of Sines.

Law of Cosines

> If ABC is any triangle, lettered in the standard way, then
>
> $$a^2 = b^2 + c^2 - 2bc \cos \alpha$$
> $$b^2 = a^2 + c^2 - 2ac \cos \beta$$
> $$c^2 = a^2 + b^2 - 2ab \cos \gamma$$

For simplicity of use it is recommended that the Law of Cosines be learned by the nature of the relationship of the parts of the triangle involved rather than by specific letters. In words, the Law of Cosines is: *the square of any side of a triangle is equal to the sum of the squares of the other two sides minus twice the product of these two sides with the cosine of their included angle.*

To verify the Law of Cosines, as in Section 8.2, we consider the three possible cases, depending on whether the triangle has three acute angles, one obtuse angle, or a right angle, as in Figure 8.8. We use the middle lettering of the statement of the Law of Cosines. The remaining two equations can be obtained by cyclic changes of the letters.

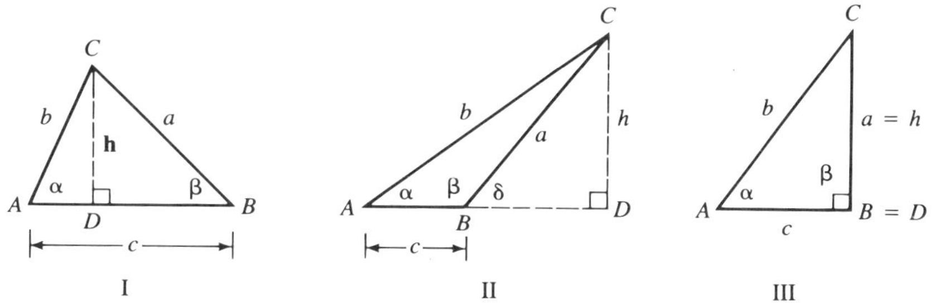

Figure 8.8

For III, since $\cos \beta = \cos 90° = 0$, the Law of Cosines is simply the Pythagorean theorem.

For I and II, in both cases, using triangle ACD and the Pythagorean theorem, $b^2 = h^2 + (\overline{AD})^2$.

For I, using triangle BDC, $h = a \sin \beta$ and $\overline{AD} = c - \overline{DB} = c - a \cos \beta$.

For II, again using triangle BDC, we have

$$h = a \sin \delta = a \sin(180° - \delta) = a \sin \beta$$

and

$$\overline{AD} = c + \overline{BD} = c + a \cos \delta = c + a[-\cos(180° - \delta)] = c - a \cos \beta$$

Thus for both I and II,

$$\begin{aligned} b^2 &= h^2 + (\overline{AD})^2 = (a \sin \beta)^2 + (c - a \cos \beta)^2 \\ &= a^2 \sin^2 \beta + c^2 - 2ac \cos \beta + a^2 \cos^2 \beta \\ &= a^2(\sin^2 \beta + \cos^2 \beta) + c^2 - 2ac \cos \beta \\ &= a^2 + c^2 - 2ac \cos \beta \end{aligned}$$

since $\sin^2 \beta + \cos^2 \beta = 1$. This completes the verification.

Since the Law of Cosines reduces to the Pythagorean theorem if the triangle is a right triangle, the Law of Cosines is frequently referred to as the **generalized Pythagorean theorem.**

Three sides (SSS) If three sides of a triangle are given, the triangle is uniquely determined, and the Law of Cosines is used to determine one angle of the triangle. *Usually it is easiest to find the largest angle (the one opposite the longest side), since this will guarantee that the remaining two angles are acute. One of the remaining angles can then be found using the Law of Sines without fear of ambiguity.* Of course, the Law of Cosines could also be used to find a second angle, but this would be a more difficult computation.

Example 1 Solve the triangle ABC if $a = 41.9$, $b = 31.2$, and $c = 19.1$.

Since a is the longest side, we first find α. From the Law of Cosines, $a^2 = b^2 + c^2 - 2bc \cos \alpha$, so that

$$\cos \alpha = \frac{b^2 + c^2 - a^2}{2bc} = \frac{(31.2)^2 + (19.1)^2 - (41.9)^2}{2(31.2)(19.1)}$$
$$\simeq \frac{-417.4}{1{,}191.8} \simeq -0.3502$$

Since $0 < \alpha < 180°$ and $\cos \alpha < 0$, we note that α must be obtuse. The reference angle $\tilde{\alpha}$ is approximately $69°30'$, so that $\alpha \simeq \arccos(-0.3502) \simeq 110°30'$.

To find β:

$$\sin \beta = \frac{b \sin \alpha}{a} \simeq \frac{31.2 \sin 110°30'}{41.9}$$

$$\simeq (0.7446)(0.9367) \simeq 0.6975$$

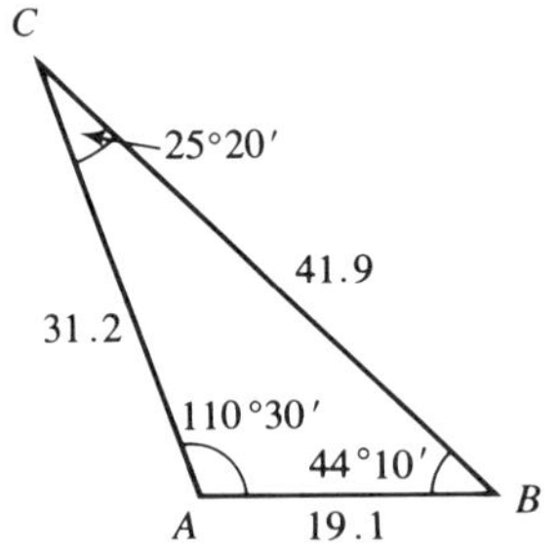

Thus $\beta \simeq 44°10'$.

To find γ:

$$\gamma = 180° - (\alpha + \beta) \simeq 180° - (154°40')$$
$$= 25°20'$$

The triangle is shown in the illustration. ■

Two sides, included angle (SAS)

If two sides and the included angle of a triangle are given, the triangle is also uniquely determined and the Law of Cosines is used to find the third side. A second angle can then be found using the Law of Sines. *When this approach is used, one wants to find the angle opposite the shorter of the two given sides. This will avoid ambiguity, because this angle must be acute.*

Example 2 Solve the triangle ABC if $a = 2.7$, $b = 4.3$, and $\gamma = 23°$.

We first find the third side:

$$c^2 = a^2 + b^2 - 2ab\cos\gamma = (2.7)^2 + (4.3)^2 - 2(2.7)(4.3)\cos 23°$$
$$\simeq 7.29 + 18.49 - (23.22)(0.9205) \simeq 25.78 - 21.37 \simeq 4.41$$

so that $c \simeq 2.1$.

Since side a is shorter than side b, we find the (necessarily) acute angle α from the Law of Sines:

$$\sin\alpha = \frac{a\sin\gamma}{c} \simeq \frac{2.7\sin 23°}{2.1} \simeq (1.286)(0.3907)$$
$$\simeq 0.5024$$

Thus $\alpha \simeq 30°$.

To find β:

$$\beta = 180° - (\alpha + \gamma) \simeq 180° - 53° \simeq 127°$$

The triangle is shown in the illustration.

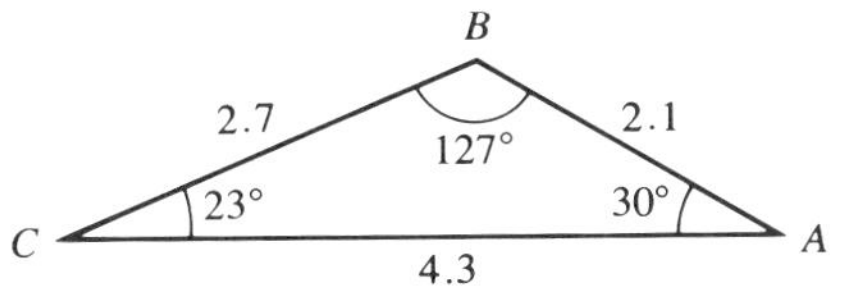

■

When using the Law of Cosines to find a third side, as in Example 2, a common error is to forget that the cosine of an obtuse angle is negative.

Example 3 For the triangle ABC, find a if $b = 7$, $c = 11$, and $\alpha = 130°$.

From the Law of Cosines,

WARNING Be sure to note that the cosine of an obtuse angle is negative.

$$a^2 = b^2 + c^2 - 2bc\cos\alpha = (7)^2 + (11)^2 - 2(7)(11)\cos 130°$$
$$\simeq 49 + 121 - 154(-0.6428) \simeq 170 + 99 = 269$$

Thus $a \simeq 16$. ■

We are now ready to solve the problem posed at the beginning of this chapter.

Example 4 What is the air distance from Hondo Pass to Coronado Tower? What is the heading for an aircraft to fly from Hondo Pass to Coronado Tower?

The situation is shown in the illustration, in which Figure 8.1 is reproduced as part I.

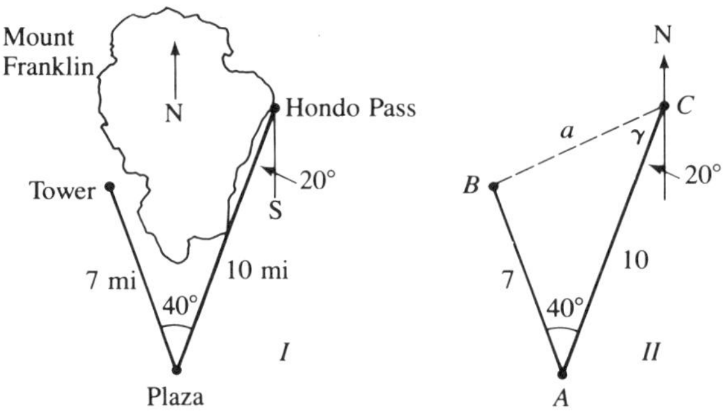

From the Law of Cosines,

$$a^2 = b^2 + c^2 - 2bc \cos \alpha = 100 + 49 - 2(10)(7) \cos 40^\circ \simeq 42$$

so that $a \simeq 6.5$.

From the Law of Sines,

$$\sin \gamma = \frac{c \sin \alpha}{a} \simeq \frac{7 \sin 40^\circ}{6.5} \simeq 0.6922$$

so that $\gamma \simeq 44^\circ$.

Thus, the air distance from Hondo Pass to Coronado Tower is approximately 6.5 mi on a heading of approximately 244°. ■

Exercises 8.3

In Exercises 1–12 solve the triangle ABC.

1. $\beta = 47^\circ$; $a = 23$; $c = 31$

2. $\gamma = 92^\circ$; $a = 49$; $b = 37$

3. $a = 15.8$; $b = 21.7$; $c = 27.2$

4. $a = 2.36$; $b = 1.76$; $c = 3.49$

5. $\alpha = 34^\circ$; $\beta = 72^\circ$; $a = 21$

6. $\alpha = 54^\circ 40'$; $a = 590$; $b = 432$

7. $\gamma = 105^\circ$; $a = 16$; $b = 23$

8. $\alpha = 117^\circ$; $b = 3.5$; $c = 2.1$

9. $\alpha = 40^\circ 50'$; $b = 545$; $c = 480$

10. $\beta = 66^\circ 20'$; $a = 34.4$; $c = 22.5$

11. $a = 122$; $b = 131$; $c = 115$

12. $a = 584$; $b = 377$; $c = 282$

13. The sides of a triangle measure 12, 15, and 19 centimeters. What is the approximate measure of the largest angle of this triangle?

14. What is the measure of the smallest angle of the triangle that has vertices at the points (2, 2), (−1, −3), and (−2, 8)?

15. An isosceles triangle has a base of 2.7 inches and a perimeter of 10.3 inches. What is the measure of the vertex angle of the triangle?

16. The radius of a circle is 16.8 centimeters. If a chord of the circle has a central angle of 58°40′, what is the length of the chord?

17. A straight canal 100 mi long is cut through the earth so that both ends are at sea level. Assuming that the earth is a sphere of radius 4,000 miles, what is the maximum distance below sea level of the canal?

18. Two sides of a triangular pasture touch the corner of a farmhouse. One side measures 1.33 miles at a bearing of S84°20′E and the other measures 1.12 miles at a bearing of S12°50′W. What is the bearing of the third side from each of its endpoints? (Two answers.)

19. A plane leaves El Paso at noon and flies on a heading of 130° at a constant speed of 270 miles per hour. Twenty minutes later a second plane leaves El Paso on a heading of 250° at a speed of 300 miles per hour. How far apart are the planes at 1:00 P.M.?

20. The sides of a parallelogram are 35 meters and 48 meters. One angle of the parallelogram measures 115°. What are the lengths of the two diagonals?

21. Juarez Avenue is 4.3 miles due north of Cordova Island. Fort Bliss is located to the east of both locations and is 5.1 miles from Cordova Island, 6.2 miles from Juarez Avenue. What is the bearing of Cordova Island from Fort Bliss?

22. A ship starts out for a destination 600 mi from port on a course of N84°E. However, the ship must first stop at an island 100 mi due east of the port. What bearing should the ship use on leaving the island?

23. The equation

$$\frac{a-b}{a+b} = \frac{\tan \frac{1}{2}(\alpha - \beta)}{\tan \frac{1}{2}(\alpha + \beta)}$$

is valid for any triangle ABC and is known as the **Law of Tangents.** This formula is sometimes used in conjunction with the Law of Sines to solve triangles of the type SAS. This is done because the Law of Tangents lends itself to logarithmic computation much better than does the Law of Cosines. Solve Problem 1 of this exercise set using the Law of Tangents instead of the Law of Cosines.

24. Solve Problem 2 using the Law of Tangents.

25. Check Problem 1 using Mollweide's formula (see page 296, Problem 25).

26. Check Problem 2 using Newton's formula (see page 296, Problem 26).

8.4 Vector Algebra

Length and volume are examples of physical quantities that can be described by size alone. However, in the applications of mathematics to many areas, we are concerned with the description of quantities such as force, velocity, and acceleration that possess both direction and magnitude.

Vector By a **vector v** we shall mean a directed line segment. We use boldface type to denote vectors. Such a quantity can be represented geometrically by a line segment in the form of an arrow with length equal to the magnitude and pointing in the appropriate direction, as in Figure 8.9.

Initial point *Terminal point* $|\mathbf{v}|$ If the endpoints of a vector **v** are labeled as in Figure 8.9, we also designate **v** as **AB,** where the point A is the **initial point** and B the **terminal point** of **AB.** The magnitude of **v** is represented by $|\mathbf{v}|$.

Vectors **AB** and **CD** are **equal** if they have the same magnitude and direction. In Figure 8.9 notice that $\mathbf{AB} = \mathbf{CD} = \mathbf{RS} \neq \mathbf{QP} \neq \mathbf{XY}$. Thus, a vector can be freely translated throughout space as long as its magnitude and direction remain unchanged. If a cartesian coordinate system is introduced, a vector with initial point at the origin is called a **position vector.**

Position vector

Since every vector **v** in the plane can be represented by a position vector, we see that there is a one-to-one correspondence between vectors and points in the

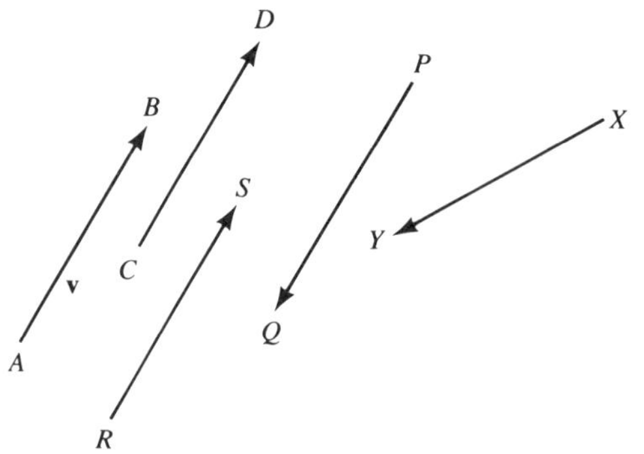

Figure 8.9

plane, as illustrated in Figure 8.10. The point in the plane is the terminal point of the position vector representative of **v.**

Vectors as ordered pairs $\langle a, b\rangle$

Consequently, a vector can also be described as an ordered pair of real numbers. It is customary to use the notation $\langle a, b\rangle$ to represent the vector $\mathbf{v} = \mathbf{OP}$, as in Figure 8.10, to avoid confusion with the point (a, b) or the interval (a, b).

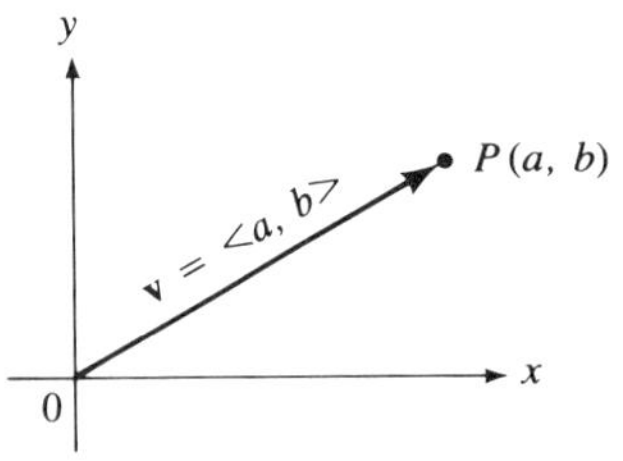

Figure 8.10

Example 1 If $A = (-2, 4)$ and $B = (3, 1)$, describe the vector **AB** as an ordered pair. What is $|\mathbf{AB}|$?

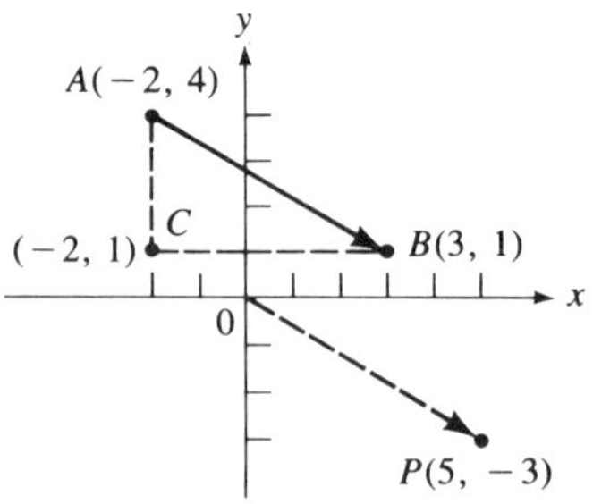

In the triangle shown, $\overline{AC} = 4 - 1 = 3$, and $\overline{CB} = 3 - (-2) = 5$. Thus

$$|\mathbf{AB}| = \sqrt{3^2 + 5^2} = \sqrt{34}$$

The directed line segment **OP** has the same magnitude and direction as **AB.** Thus, **AB** = $\langle 5, -3 \rangle$. ■

In general, if $P_1(x_1, y_1)$ and $P_2(x_2, y_2)$ are points in the plane, we note that the position vector representing the vector $\mathbf{P_1P_2}$ is

$$\mathbf{P_1P_2} = \langle x_2 - x_1, y_2 - y_1 \rangle$$

and

$$|\mathbf{P_1P_2}| = \sqrt{(x_2 - x_1)^2 + (y_2 - y_1)^2}$$

Example 2 Find the magnitude and direction of the vectors $\mathbf{v} = \langle -2, 3 \rangle$ and $\mathbf{u} = \langle -3, 0 \rangle$. We describe the direction of a vector by use of an angle in standard position. Thus $|\mathbf{u}| = 3$, $\theta_1 = \pi$, and $|\mathbf{v}| = \sqrt{13}$, $\theta_2 = \arccos(-2/\sqrt{13})$. The angle θ_2 can also be expressed as $\pi - \arctan 1.5$.

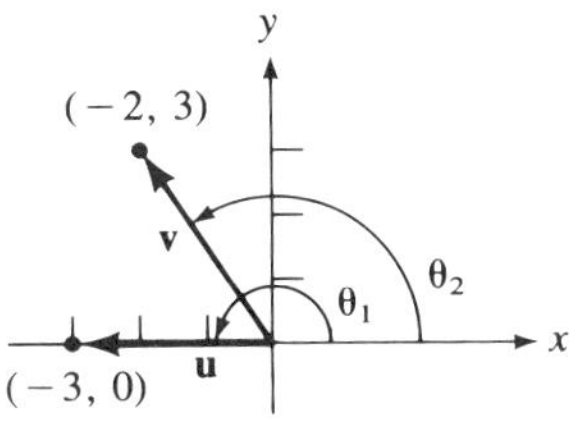

■

Scalar *rv* If $\mathbf{v} = \langle a, b \rangle$ is a vector and r is a real number, then $r\mathbf{v} = r\langle a, b \rangle$ is defined to be $\langle ra, rb \rangle$. In this context, real numbers are called **scalars,** and this operation is called **multiplication of a vector by a scalar.** Geometrically, $r\mathbf{v}$ is a vector whose magnitude is $|r|\,|\mathbf{v}|$ and whose direction is the same as $\mathbf{v}$ if $r > 0$ and opposite that of $\mathbf{v}$ if $r < 0$, as in Figure 8.11.

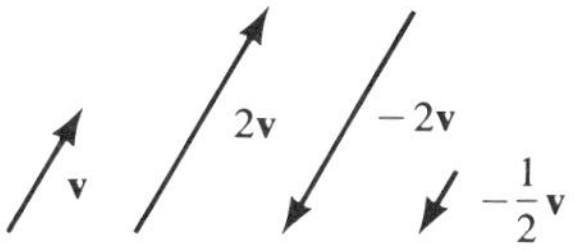

Figure 8.11

Zero vector For operations on vectors it is convenient to consider the origin as a vector whose magnitude is zero. The vector $\langle 0, 0 \rangle$ is the **zero vector 0.** If $\langle a, b \rangle$ is any vector, notice that $0\langle a, b \rangle = \langle 0a, 0b \rangle = \langle 0, 0 \rangle = \mathbf{0}$.

Vector sum For vectors $\mathbf{u} = \langle a, b \rangle$ and $\mathbf{v} = \langle c, d \rangle$, we define the **sum** of **u** and **v, u + v,** to be

$$\langle a, b \rangle + \langle c, d \rangle = \langle a + c, b + d \rangle$$

Geometrically, if $\mathbf{u} = \mathbf{OP}$, $\mathbf{v} = \mathbf{OQ}$, and $\mathbf{u} + \mathbf{v} = \mathbf{OR}$, it can be shown that O, P, R, and Q are the vertices of a parallelogram, as illustrated in Figure 8.12. To see this, note that the slope of the line segment PR is

$$\frac{(b + d) - b}{(a + c) - a} = \frac{d}{c}$$

and the slope of the line segment OQ is also d/c. Thus PR and OQ are parallel. Similarly, OP and QR are parallel. It is for this reason that it is said that addition of vectors is done using the **parallelogram law.**

Parallelogram law

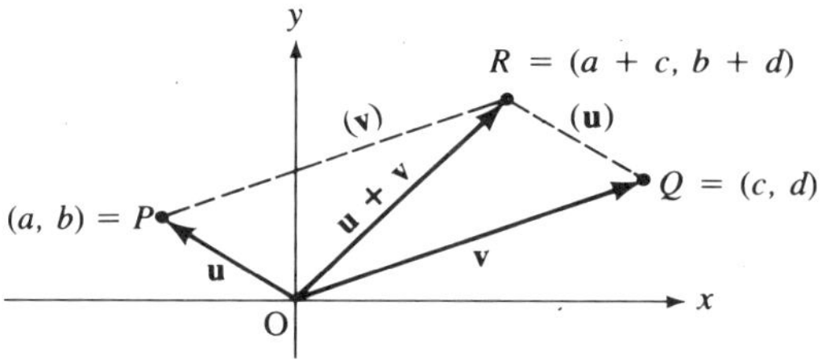

Figure 8.12

Notice in Figure 8.12 that **QR** = **OP,** so that **u** + **v** = **OR** can be obtained by attaching the initial point of **u** to the terminal point of **v.** Then **u** + **v** is the vector from the initial point of **v** to the terminal point of **u.** Similarly, the initial point of **v** can be attached to the terminal point of **u** to obtain **u** + **v.**

Example 3 Find the sum $\langle -2, 3\rangle + \langle 3, 4\rangle$. Illustrate geometrically.

By the definition,

$$\langle -2, 3\rangle + \langle 3, 4\rangle = \langle -2 + 3, 3 + 4\rangle = \langle 1, 7\rangle$$

The sum is the vector **OP** in the figure shown.

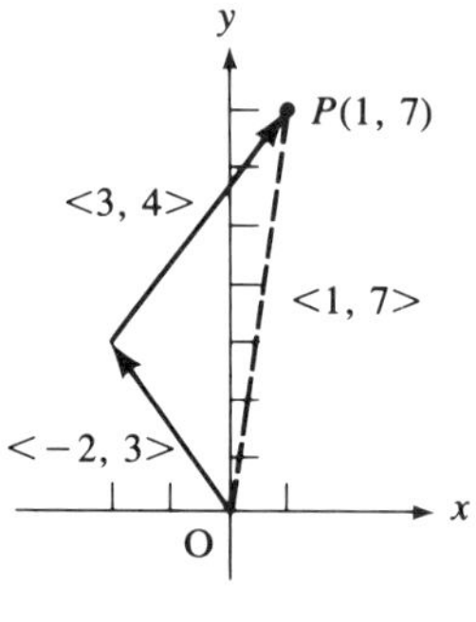

■

−v The vector $-\mathbf{v}$ is defined by the relationship $\mathbf{v} + (-\mathbf{v}) = \mathbf{0}$. If $\mathbf{v} = \langle a, b\rangle$, notice that $\langle a, b\rangle + \langle -a, -b\rangle = \langle 0, 0\rangle$. Thus $-\mathbf{v} = \langle -a, -b\rangle = (-1)\langle a, b\rangle$.

Vector difference For vectors **u** and **v** the **vector difference** is defined by $\mathbf{u} - \mathbf{v} = \mathbf{u} + (-\mathbf{v})$. Notice that

$$\langle a, b\rangle - \langle c, d\rangle = \langle a - c, b - d\rangle$$

Example 4 For $\mathbf{u} = \langle 1, -1\rangle$ and $\mathbf{v} = \langle 2, 3\rangle$, illustrate **u, v, −v,** and **u − v** geometrically.

The situation is shown in the figure. Note that there are two possible representations of $\mathbf{u} - \mathbf{v}$. Thus geometrically, as well as algebraically, $\mathbf{u} - \mathbf{v}$ is that vector that must be added to $\mathbf{v}$ to get $\mathbf{u}$.

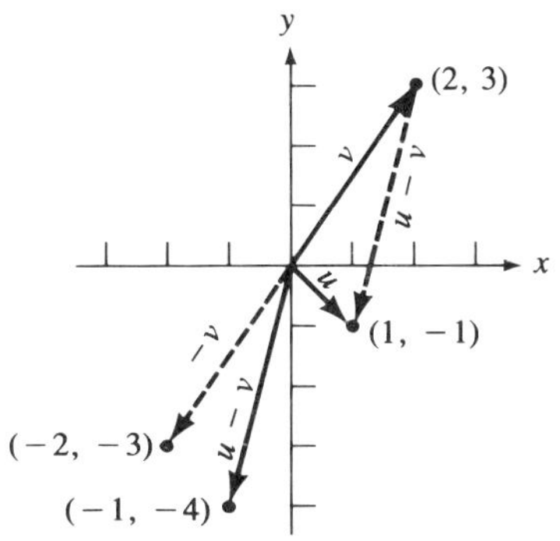

■

The following display summarizes the basic properties of vector addition and scalar multiplication.

Let $\mathbf{u}, \mathbf{v}, \mathbf{w}$ be vectors and r, s be real numbers. Then

$\mathbf{u} + \mathbf{v} = \mathbf{v} + \mathbf{u}$	$(\mathbf{u} + \mathbf{v}) + \mathbf{w} = \mathbf{u} + (\mathbf{v} + \mathbf{w})$
$\mathbf{u} + \mathbf{0} = \mathbf{u}$	$\mathbf{u} + (-\mathbf{u}) = \mathbf{0}$
$r(s\mathbf{u}) = (rs)\mathbf{u}$	$(r + s)\mathbf{u} = r\mathbf{u} + s\mathbf{u}$
$r(\mathbf{u} + \mathbf{v}) = r\mathbf{u} + r\mathbf{v}$	$1\mathbf{u} = \mathbf{u}$
$(-1)\mathbf{u} = -\mathbf{u}$	$0\mathbf{u} = \mathbf{0}$
$r\mathbf{0} = \mathbf{0}$	

The proofs of the above results are included in the exercises since they are all straightforward. For example, if $\mathbf{u} = \langle a, b\rangle$, $\mathbf{v} = \langle c, d\rangle$, we see that

$$\begin{aligned} r(\mathbf{u} + \mathbf{v}) &= r(\langle a, b\rangle + \langle c, d\rangle) = r\langle a + c, b + d\rangle \\ &= \langle r(a + c), r(b + d)\rangle = \langle ra + rc, rb + rd\rangle \\ &= \langle ra, rb\rangle + \langle rc, rd\rangle = r\langle a, b\rangle + r\langle c, d\rangle \\ &= r\mathbf{u} + r\mathbf{v} \end{aligned}$$

i, j
Unit vector

There is another standard notation for the representation of vectors in a given cartesian coordinate system. The symbols $\mathbf{i}$ and $\mathbf{j}$ are defined by $\mathbf{i} = \langle 1, 0\rangle$ and $\mathbf{j} = \langle 0, 1\rangle$. The vectors $\mathbf{i}$ and $\mathbf{j}$ are examples of **unit vectors,** because each has magnitude equal to 1 unit. These vectors are useful because every vector can be easily represented as a **linear combination** of $\mathbf{i}$ and $\mathbf{j}$. By this we mean that for every vector $\mathbf{v}$ there are scalars a and b such that $\mathbf{v} = a\mathbf{i} + b\mathbf{j}$, as in Figure 8.13.

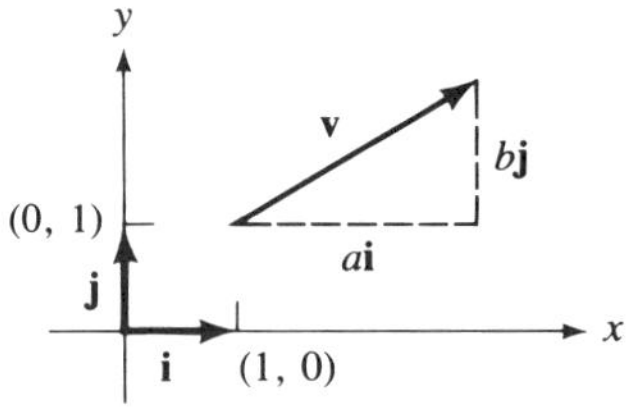

Figure 8.13

Horizontal component
Vertical component

We say that a is the **horizontal component** of **v** and b is the **vertical component** of **v**. Notice that if $\mathbf{v} = \langle a, b\rangle$, then $a\mathbf{i} + b\mathbf{j} = a\langle 1, 0\rangle + b\langle 0, 1\rangle = \langle a, 0\rangle + \langle 0, b\rangle = \langle a, b\rangle = \mathbf{v}$. Because of the basic properties of vector addition and scalar multiplication, we can treat linear combinations of **i** and **j** using the usual algebraic techniques.

Example 5 Let $\mathbf{u} = 3\mathbf{i} - 2\mathbf{j}$ and $\mathbf{v} = -2\mathbf{i} + 5\mathbf{j}$. Write the vector $2\mathbf{u} - 3\mathbf{v}$ in the **i, j** notation.

$$\begin{aligned} 2\mathbf{u} - 3\mathbf{v} &= 2(3\mathbf{i} - 2\mathbf{j}) - 3(-2\mathbf{i} + 5\mathbf{j}) = (6\mathbf{i} - 4\mathbf{j}) + (6\mathbf{i} - 15\mathbf{j}) \\ &= 12\mathbf{i} - 19\mathbf{j} \end{aligned}$$

■

Example 6 Find the horizontal and vertical components of the vector illustrated.

We note that

$$a = 4 \cos 340° = 4 \cos 20° \simeq 3.8$$

and

$$b = 4 \sin 340° = -4 \sin 20° \simeq -1.4$$

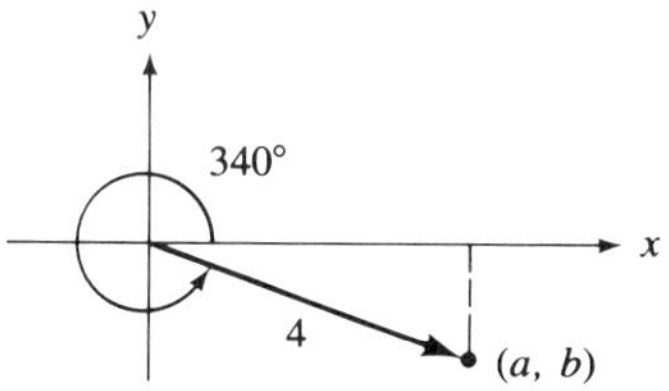

■

We conclude this section with a statement of the general result illustrated in Example 6.

> If $\mathbf{v} = \langle a, b\rangle$ makes an angle θ with the positive x axis, then
>
> $$a = |\mathbf{v}| \cos \theta$$
>
> and
>
> $$b = |\mathbf{v}| \sin \theta$$

Exercises 8.4

In Exercises 1–8 give the magnitude of the vector **v** *and describe the direction by stating the angle* θ, $0 \le \theta < 2\pi$, *from the positive x axis to the position vector* **v**. *Illustrate geometrically.*

1. $\mathbf{v} = \langle -3, 0 \rangle$

2. $\mathbf{v} = \langle 2, 2\sqrt{3} \rangle$

3. $\mathbf{v} = \langle -5, -5 \rangle$

4. $\mathbf{v} = \langle 0, -4 \rangle$

5. $\mathbf{v} = \langle -2, 7 \rangle$

6. $\mathbf{v} = \langle 3, 9 \rangle$

7. $\mathbf{v} = \mathbf{PQ}$; $P = (1, 4)$; $Q = (2, 5)$

8. $\mathbf{v} = \mathbf{PQ}$; $P = (-1, -7)$; $Q = (1, -3)$

In Exercises 9–10 illustrate **u, v, u + v,** *and* **u − v** *geometrically.*

9. $\mathbf{u} = \langle -1, 5 \rangle$; $\mathbf{v} = \langle 1, -3 \rangle$

10. $\mathbf{u} = \langle 2, 3 \rangle$; $\mathbf{v} = \langle -2, 1 \rangle$

Let $\mathbf{u} = \langle 4, 7 \rangle$, $\mathbf{v} = \langle 2, -5 \rangle$, $\mathbf{w} = \langle -2, -2 \rangle$. *In Exercises 11–20 find the indicated vectors or scalars.*

11. $\mathbf{u} + 4\mathbf{v}$

12. $-\mathbf{u} + 2\mathbf{w}$

13. $\frac{1}{2}(\mathbf{u} + \mathbf{v})$

14. $\mathbf{w} - 2\mathbf{u}$

15. $-\mathbf{u} + \mathbf{0}$

16. $2\mathbf{u} - 3\mathbf{v}$

17. $\mathbf{u} + 2\mathbf{v} - 3\mathbf{w}$

18. $-3\mathbf{u} - 2\mathbf{v} + \mathbf{w}$

19. $|\mathbf{u} - \mathbf{w}|$

20. $\left| -\frac{1}{3}\mathbf{v} \right|$

Let $\mathbf{u} = \langle -3, 4 \rangle$, $\mathbf{v} = \langle -6, -1 \rangle$. *In Exercises 21–24 write the specified vector using* **i, j** *notation.*

21. $\mathbf{v} - \mathbf{u}$

22. $\mathbf{u} + \mathbf{v} + \mathbf{u}$

23. $\frac{1}{2}\mathbf{u} - 2\mathbf{v}$

24. $-2\mathbf{u} + \frac{1}{2}\mathbf{v}$

In Exercises 25–28 find x and y if:

25. $\langle -2, 3 \rangle + \langle x, y \rangle = 2\langle 4, 5 \rangle$

26. $\langle x, y \rangle - 2\langle 7, -3 \rangle = -\langle 1, -6 \rangle$

27. $2\langle x, -2 \rangle + \langle -1, y \rangle = \langle 0, -3 \rangle$

28. $\langle 8, -3 \rangle - \langle x, 10 \rangle = \langle -4, 2y \rangle$

In Exercises 29–32 find the horizontal and vertical components of the vector **v** *if* θ *is the angle between the positive x axis and the position vector* **v**.

29. $|\mathbf{v}| = 3$; $\theta = -\frac{\pi}{3}$

30. $|\mathbf{v}| = 4$; $\theta = \frac{\pi}{6}$

31. $|\mathbf{v}| = \sqrt{2}$; $\theta = -\frac{5\pi}{4}$

32. $|\mathbf{v}| = 1$; $\theta = \pi$

In Exercises 33–34 find all angles θ, $0 \le \theta < 2\pi$, *from the positive x axis to the position vector* **u**.

33. $|\mathbf{u}| = 13$ and **u** has horizontal component -12.

34. $|\mathbf{u}| = 5$ and **u** has vertical component -4.

35. Find the coordinates of Q if $\mathbf{PQ} = \mathbf{RS}$, $P = (-1, 3)$, $R = (-7, -4)$, and $S = (6, 3)$.

36. A parallelogram has vertices at $(-1, -2)$, $(4, 0)$, and $(0, 3)$. What are the coordinates of the fourth vertex?

Let **u, v, w** *be vectors and r, s be real numbers. In Exercises 37–46 prove the basic properties of vector addition and scalar multiplication indicated.*

37. $\mathbf{u} + \mathbf{v} = \mathbf{v} + \mathbf{u}$

38. $(\mathbf{u} + \mathbf{v}) + \mathbf{w} = \mathbf{u} + (\mathbf{v} + \mathbf{w})$

39. $\mathbf{u} + \mathbf{0} = \mathbf{u}$

40. $\mathbf{u} + (-\mathbf{u}) = \mathbf{0}$

41. $r(s\mathbf{u}) = (rs)\mathbf{u}$

42. $(r + s)\mathbf{u} = r\mathbf{u} + s\mathbf{u}$

43. $(-1)\mathbf{u} = -\mathbf{u}$

44. $1\mathbf{u} = \mathbf{u}$

45. $r\mathbf{0} = \mathbf{0}$

46. $0\mathbf{u} = \mathbf{0}$

47. Show that $\mathbf{u} = (\cos \alpha)\mathbf{i} + (\sin \alpha)\mathbf{j}$ is a unit vector for each angle α.

48. Show that if $\mathbf{v}$ is a vector, $\mathbf{v} \neq \mathbf{0}$, then $\frac{1}{|\mathbf{v}|}(\mathbf{v})$ is a unit vector.

8.5 Vector Applications

There are basically two types of multiplication defined for vectors, and both products have numerous mathematical and physical applications. In two dimensions the useful product is the dot product, and we therefore restrict our attention to this type of vector multiplication. It is easiest to begin with an algebraic definition, after which we discuss the geometric interpretation.

Dot product

Let $\mathbf{u} = \langle a, b\rangle$ and $\mathbf{v} = \langle c, d\rangle$ be vectors. By the **dot product** of **u** and **v**, $\mathbf{u} \cdot \mathbf{v}$, we mean the scalar (real number) given by the formula

$$\mathbf{u} \cdot \mathbf{v} = \langle a, b\rangle \cdot \langle c, d\rangle = ac + bd$$

We emphasize that *the dot product of two vectors is not a vector.* The dot product is also called the **inner product** or the **scalar product.**

Example 1 For $\mathbf{v} = \langle -4, 1\rangle$, $\mathbf{w} = \langle 3, 2\rangle$, and $\mathbf{u} = \langle 2, -3\rangle$, find $\mathbf{v} \cdot \mathbf{u}$ and $\mathbf{w} \cdot \mathbf{u}$. By definition,

$$\mathbf{v} \cdot \mathbf{u} = (-4)(2) + (1)(-3) = -8 - 3 = -11$$

and

$$\mathbf{w} \cdot \mathbf{u} = (3)(2) + (2)(-3) = 6 - 6 = 0$$

■

The following display summarizes the important properties of the dot product.

Let **u, v, w** be vectors and r be a scalar. Then

$$\mathbf{u} \cdot \mathbf{v} = \mathbf{v} \cdot \mathbf{u} \qquad \mathbf{u} \cdot (\mathbf{v} + \mathbf{w}) = \mathbf{u} \cdot \mathbf{v} + \mathbf{u} \cdot \mathbf{w}$$

$$r(\mathbf{u} \cdot \mathbf{v}) = (r\mathbf{u}) \cdot \mathbf{v} = \mathbf{u} \cdot (r\mathbf{v}) \qquad \mathbf{0} \cdot \mathbf{u} = 0$$

$$|\mathbf{u}|^2 = \mathbf{u} \cdot \mathbf{u}$$

Proofs for the above properties are straightforward. For example, if $\mathbf{u} = \langle a, b \rangle$, $\mathbf{v} = \langle c, d \rangle$, $r \in R$, then

$$\begin{aligned} r(\mathbf{u} \cdot \mathbf{v}) &= r(ac + bd) = (ra)c + (rb)d \\ &= \langle ra, rb \rangle \cdot \langle c, d \rangle = (r\mathbf{u}) \cdot \mathbf{v} \end{aligned}$$

Angle between vectors

By the **angle θ between two nonzero vectors u** and **v** we mean the smaller of the angles between the corresponding position vectors, measured in the positive direction. Thus $0 \le \theta \le \pi$. This is illustrated in Figure 8.14.

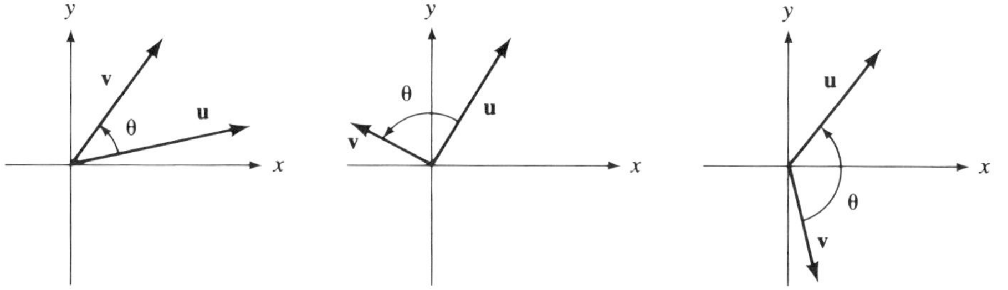

Figure 8.14

We now give the geometric interpretation of the dot product.
If θ is the angle between the nonzero vectors **u** and **v**, then

$$\mathbf{u} \cdot \mathbf{v} = |\mathbf{u}|\,|\mathbf{v}| \cos \theta$$

To see this, assume that $\mathbf{u} = \langle a, b \rangle$ and $\mathbf{v} = \langle c, d \rangle$ are not parallel $(0 < \theta < \pi)$ and refer to the triangle illustrated in Figure 8.15.

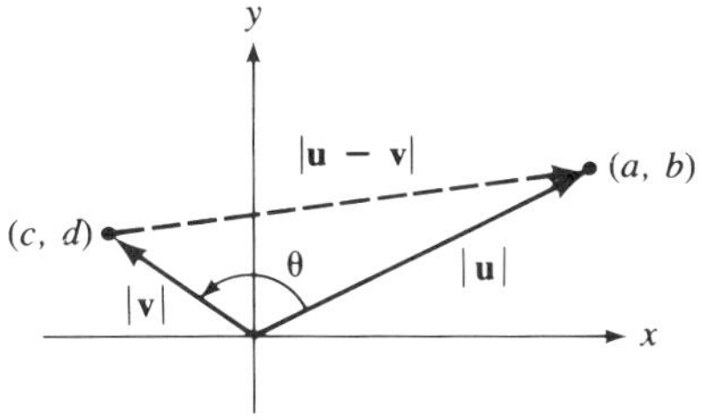

Figure 8.15

Notice that $\mathbf{u} - \mathbf{v} = \langle a - c, b - d \rangle$, so that

$$\begin{aligned} |\mathbf{u} - \mathbf{v}|^2 &= (\mathbf{u} - \mathbf{v}) \cdot (\mathbf{u} - \mathbf{v}) = (a - c)^2 + (b - d)^2 \\ &= (a^2 + b^2) + (c^2 + d^2) - 2(ac + bd) \\ &= \mathbf{u} \cdot \mathbf{u} + \mathbf{v} \cdot \mathbf{v} - 2(\mathbf{u} \cdot \mathbf{v}) \\ &= |\mathbf{u}|^2 + |\mathbf{v}|^2 - 2(\mathbf{u} \cdot \mathbf{v}) \end{aligned}$$

We also have, from the Law of Cosines,

$$|\mathbf{u} - \mathbf{v}|^2 = |\mathbf{u}|^2 + |\mathbf{v}|^2 - 2|\mathbf{u}|\,|\mathbf{v}| \cos \theta$$

Equating the two expressions for $|\mathbf{u} - \mathbf{v}|^2$, we obtain $\mathbf{u} \cdot \mathbf{v} = |\mathbf{u}|\,|\mathbf{v}| \cos \theta$. The case for **u** and **v** parallel is included as an exercise.

The above result is useful in finding the angle between two vectors. Indeed,

$$\theta = \arccos \frac{\mathbf{u} \cdot \mathbf{v}}{|\mathbf{u}|\,|\mathbf{v}|}$$

Orthogonal

> The vectors **u** and **v** are **orthogonal** (perpendicular) if, and only if, $\theta = \frac{\pi}{2}$, i.e., $\cos \theta = 0$. This is true if, and only if, $\mathbf{u} \cdot \mathbf{v} = 0$.

Example 2 Find the angle θ between $\langle 3, 1\rangle$ and $\langle -2, 4\rangle$.

$$\cos \theta = \frac{\langle 3, 1\rangle \cdot \langle -2, 4\rangle}{|\langle 3, 1\rangle|\,|\langle -2, 4\rangle|} = \frac{(3)(-2) + (1)(4)}{\sqrt{10}\sqrt{20}}$$
$$= \frac{-2}{\sqrt{200}} = \frac{-1}{\sqrt{50}} \simeq -0.1414$$

Thus $\theta \simeq 1.7$ radians or, in degree measure, $\theta \simeq 98°$. ■

Example 3 Find all nonzero vectors orthogonal to $\langle 1, -3\rangle$.

If $\mathbf{u} = \langle a, b\rangle$ is orthogonal to $\langle 1, -3\rangle$, then $0 = \langle a, b\rangle \cdot \langle 1, -3\rangle = a - 3b$, so that $a = 3b$. Thus the set of nonzero vectors orthogonal to $\langle 1, -3\rangle$ is

$$\{\langle 3r, r\rangle : r \in R, r \neq 0\}$$ ■

Work The dot product has an immediate application to the computation of work done by a force. Refer to Figure 8.16, and suppose a constant force **F** acts on a body b, resulting in a displacement **d.** The **work** W done by **F** in the displacement is defined as the product of **d** and the component of **F** in the direction of **d.** Thus,

$$W = |\mathbf{d}|\,(|\mathbf{F}| \cos \theta) = \mathbf{F} \cdot \mathbf{d}$$

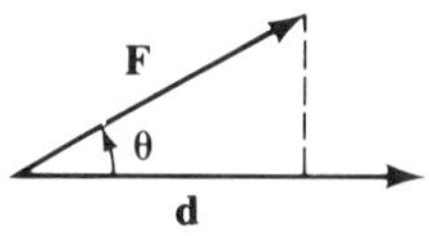

Figure 8.16

We will use units of pounds to describe magnitude of force and feet to describe magnitude of displacement.

Example 4 Find the work done by the force $\mathbf{F} = 3\mathbf{i} + 2\mathbf{j}$ acting on a body, if the body is displaced from the point $P(1, 3)$ to the point $Q(8, 2)$.
Since $\mathbf{PQ} = \langle 7, -1\rangle$, $W = \langle 3, 2\rangle \cdot \langle 7, -1\rangle = 21 - 2 = 19$ ft-lbs. ■

Many of the earlier problems of this chapter can be stated and solved in vector terminology.

Example 5 A motorboat travels at a speed of 30 mi/hr in still water. Suppose the boat starts upriver at an angle of 40° with the riverbank against a 5 mi/hr current. At what angle with the riverbank is the boat traveling and at what effective speed?

The situation is shown in the diagram. The solution is obtained from the vector **OQ.** Since $|\mathbf{OP}| = 30$,

$$\mathbf{OP} = \langle 30 \cos 50°, 30 \sin 50°\rangle \simeq \langle 19.3, 23.0\rangle$$

and since $|\mathbf{PQ}| = 5$, $\mathbf{PQ} = \langle 0, -5\rangle$. Thus

$$\mathbf{OQ} = \mathbf{OP} + \mathbf{PQ} = \langle 19.3, 18.0\rangle$$

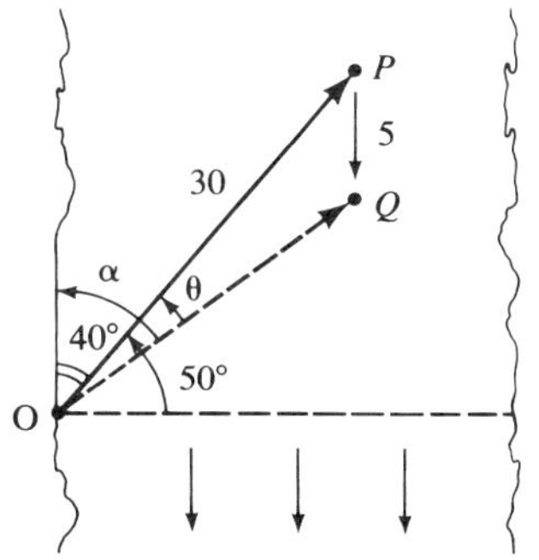

The effective speed is

$$|\mathbf{OQ}| = \sqrt{(19.3)^2 + (18.0)^2} \simeq \sqrt{696.5} \simeq 26.4 \text{ mi/hr}$$

Since

$$\cos\theta = \frac{\mathbf{OP} \cdot \mathbf{OQ}}{|\mathbf{OP}|\,|\mathbf{OQ}|} \simeq \frac{(19.3)^2 + (23.0)(18.0)}{(30)(26.4)}$$
$$\simeq \frac{786.5}{792} \simeq 0.9931$$

we have $\theta \simeq 6°46'$. The boat is traveling at an angle $\alpha = (40° + 6°46') = 46°46'$ with the riverbank. ■

Resultant The **resultant** of forces is the sum of the vectors that represent the forces. It is the single force that produces the same effect as the combined individual forces.

Example 6 A boat is towed through a canal by two ropes so that the resultant force is parallel to the canal, as in the figure shown. If the angle between the ropes is 60° and the forces applied by the ropes are 350 and 500 lb, find the magnitude $|\mathbf{F}|$ of the resultant force $\mathbf{F}$ on the boat. Also find the angles that the ropes make with the sides of the canal.

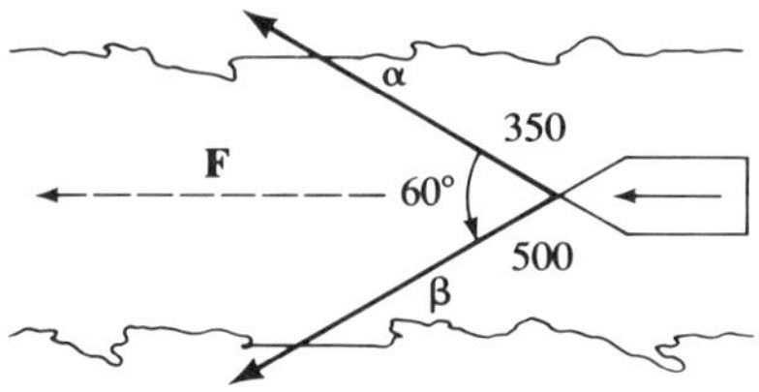

We draw a triangular diagram as shown. We seek $|\mathbf{F}|$, α, and β. Using the Law of Cosines,

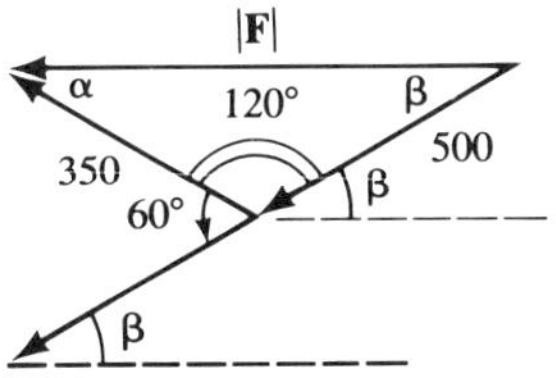

$$
\begin{aligned}
|\mathbf{F}|^2 &= (350)^2 + (500)^2 - 2(350)(500)\cos 120^\circ \\
&= 122{,}500 + 250{,}000 - (350{,}000)(-0.5) \\
&= 547{,}500
\end{aligned}
$$

Thus $|\mathbf{F}| \simeq 740$ lb. Using the Law of Sines,

$$\frac{500}{\sin \alpha} = \frac{740}{\sin 120^\circ}$$

so that

$$
\begin{aligned}
\sin \alpha &= \frac{(\sin 120^\circ)(500)}{740} \\
&\simeq \frac{(0.8660)(500)}{740} \simeq 0.5851
\end{aligned}
$$

Thus $\alpha \simeq 35^\circ 50'$, and

$$\beta \simeq 180^\circ - 120^\circ - 35^\circ 50' = 24^\circ 10'$$ ■

Exercises 8.5

In Exercises 1–8 find the value of the expressions if $\mathbf{u} = \langle -3, 2\rangle$, $\mathbf{v} = \langle -1, -4\rangle$ *and* $\mathbf{w} = \langle 3, -3\rangle$.

1. $\mathbf{u} \cdot \mathbf{v}$

2. $\mathbf{v} \cdot \mathbf{w}$

3. $(\mathbf{u} - \mathbf{v}) \cdot \mathbf{w}$

4. $(\mathbf{v} + 3\mathbf{u}) \cdot \mathbf{u}$

5. $\mathbf{u} \cdot \mathbf{w} - \mathbf{v} \cdot \mathbf{w}$

6. $\mathbf{v} \cdot \mathbf{u} + 3\mathbf{u} \cdot \mathbf{u}$

7. $(3\mathbf{v} - \mathbf{w}) \cdot (2\mathbf{u} + 3\mathbf{v})$

8. $(2\mathbf{w} - 3\mathbf{u}) \cdot (\mathbf{v} + 4\mathbf{u})$

In Exercises 9–14 find the angles between the vectors and illustrate.

9. $\langle -2, 0\rangle$; $\langle 0, -1\rangle$

10. $\langle 2, 0\rangle$; $\langle -2, 0\rangle$

11. $\langle \sqrt{3}, -1\rangle$; $\langle -1, \sqrt{3}\rangle$

12. $\langle -1, \sqrt{3}\rangle$; $\langle -1, -\sqrt{3}\rangle$

13. $2\mathbf{i} - 3\mathbf{j}$; $\mathbf{i} + 4\mathbf{j}$

14. $-3\mathbf{i} + \mathbf{j}$; $2\mathbf{i} - \mathbf{j}$

15. Find all nonzero vectors orthogonal to $\langle -2, -3\rangle$.

16. Find all nonzero vectors orthogonal to $\langle -4, 2\rangle$.

17. Find x so that $\mathbf{v} = x\mathbf{i} + 2\mathbf{j}$ is orthogonal to $\mathbf{u} = 2\mathbf{i} - 3\mathbf{j}$.

18. Find y if $\langle -2, 7\rangle$ is orthogonal to $\langle 3, y\rangle$.

Let **u, v, w** *be vectors and r a real number. In Exercises 19–24 prove the given basic properties of the dot product.*

19. $\mathbf{u} \cdot (\mathbf{v} + \mathbf{w}) = \mathbf{u} \cdot \mathbf{v} + \mathbf{u} \cdot \mathbf{w}$

20. $\mathbf{u} \cdot \mathbf{v} = \mathbf{v} \cdot \mathbf{u}$

21. $r(\mathbf{u} \cdot \mathbf{v}) = \mathbf{u} \cdot (r\mathbf{v})$

22. $\mathbf{0} \cdot \mathbf{u} = 0$

23. $|\mathbf{u}|^2 = \mathbf{u} \cdot \mathbf{u}$

24. If $\mathbf{u} \cdot \mathbf{u} = 0$ then $\mathbf{u} = \mathbf{0}$.

In Exercises 25–26 find the angles of the triangles with the vertices indicated.

25. $(0, 0)$; $(3, 4)$; $(2, -3)$

26. $(0, 0)$; $(-1, 3)$; $(4, 2)$

27. Find two unit vectors orthogonal to $\mathbf{i} + 2\mathbf{j}$.

28. Find two unit vectors orthogonal to $\mathbf{i} - \sqrt{3}\,\mathbf{j}$.

In Exercises 29–30 find the work done by the force **F** *acting on a body if the body is displaced from P to Q.*

29. $\mathbf{F} = 4\mathbf{i} - 3\mathbf{j}$; $P = (-1, 4)$; $Q = (7, -3)$

30. $\mathbf{F} = -2\mathbf{i} + 5\mathbf{j}$; $P = (3, 4)$; $Q = (1, 10)$

31. Verify $\mathbf{u} \cdot \mathbf{v} = |\mathbf{u}|\,|\mathbf{v}| \cos\theta$ in the cases in which **u** and **v** are parallel ($\theta = 0$ or $\theta = \pi$).

32. The handle of a lawn mower is inclined at an angle of 45° with the horizontal. The horizontal force needed to move the mower across level ground is 50 pounds. What force must be applied to the handle?

33. A river flows south with a current of 5 mi/hr. A motorboat travels at a speed of 25 mi/hr in still water. In what direction must the boat be headed in order to move directly east across the river? What is its speed in the easterly direction?

34. A river flows south with a current of 4 mi/hr. A swimmer who can swim at 1.5 mi/hr in still water heads due east across the river. At what angle with the riverbank is the swimmer actually traveling and at what effective speed?

35. What is the resultant of forces of 10 pounds toward the south and 18 pounds toward S60°W?

36. What is the resultant of a force of 25 pounds toward N45°E and a force of 20 pounds toward S30°E?

37. A jeep in the desert is driven 5 mi east, then 3 mi northeast, then 7 mi southeast, and finally 4 mi in the direction of $-\mathbf{i} + \sqrt{3}\,\mathbf{j}$. How far and in what direction is it from the starting point?

38. A force of 120 pounds is applied to the origin toward the east. A second force is applied in a direction N30°E and the magnitude of the resultant force is 300 pounds. What is the magnitude of the second force? What are the angles that the resultant makes with the two forces?

Chapter Review

In Exercises 1–10 solve the triangle ABC.

1. $\alpha = 28°30'$; $\beta = 67°$; $c = 20$

2. $\gamma = 40°$; $b = 12$; $c = 9$

3. $a = 7$; $b = 10$; $\gamma = 90°$

4. $\alpha = 22°20'$; $\beta = 90°$; $b = 10$

5. $a = 40$; $b = 30$; $c = 20$

6. $a = 3.21$; $b = 2.12$; $c = 1.75$

7. $b = 12$; $c = 22$; $\alpha = 140°$

8. $a = 5$; $b = 7$; $\gamma = 20°$

9. $\alpha = 50°$; $c = 140$; $a = 100$

10. $\beta = 47°$; $a = 30$; $b = 12$

In Exercises 11–19 let $\mathbf{u} = \langle -2, 4\rangle$, $\mathbf{v} = \langle 1, -3\rangle$, *and* $\mathbf{w} = \langle -5, -4\rangle$. *Find:*

11. $\mathbf{u} - 3\mathbf{v}$

12. $\mathbf{w} - \frac{1}{2}\mathbf{v}$

13. $2\mathbf{u} - 3\mathbf{v} + \mathbf{w}$

14. $\left|2\mathbf{u} - \frac{1}{3}\mathbf{w}\right|$

15. $\mathbf{u} \cdot \mathbf{w} - 2\mathbf{v} \cdot \mathbf{u}$

16. $(2\mathbf{u} + 3\mathbf{v}) \cdot \mathbf{w}$

17. The angle between **u** and **w**.

18. The angle between $\mathbf{u} - 4\mathbf{v}$ and **v**.

19. A unit vector with the same direction as **w**.

In Exercises 20–23 find x and/or y such that the given condition holds.

20. $\frac{1}{2}\langle x, -1\rangle - \langle -1, y\rangle = \langle -3, -3\rangle$

21. $\langle 6, -4\rangle + 2\langle x, -2\rangle = \langle 0, -y\rangle$

22. $x\mathbf{i} - 3\mathbf{j}$ is orthogonal to $-\mathbf{i} - 2\mathbf{j}$.

23. $\mathbf{u} = x\mathbf{i} + y\mathbf{j}$ is orthogonal to $2\mathbf{i} - 3\mathbf{j}$, $x > 0$, and $|\mathbf{u}| = 1$.

24. Find the horizontal and vertical components of **v** if $|\mathbf{v}| = 2$ and the angle between **v** and $\langle 1, 0\rangle$ is $2\pi/3$.

25. If $|\mathbf{u}| = 5$ and **u** has vertical component -3, what are the possible directions of **u**?

26. A parallelogram has vertices at $(-3, 0)$, $(2, 3)$, and $(5, -2)$. What are the coordinates of the fourth vertex?

27. What are the angles of the triangle with vertices at $(-3, 1)$, $(3, 2)$, and $(4, -5)$?

28. A force $\mathbf{F} = 7\mathbf{i} - 3\mathbf{j}$ displaces a body from $(-2, 5)$ to $(2, -4)$. What is the work done by **F**?

29. Horizon City is 20 miles due south of El Paso. A plane leaves El Paso and flies at a heading of 100°. Moments later the plane is spotted from Horizon City at a heading of 60°. How far is the plane from Horizon City at the moment it is sighted?

30. A boat leaves the dock and travels N75°W for 15 minutes, then turns and travels S15°W for 30 minutes. If the boat's speed is a constant 30 mi/hr, how far is the boat from the dock?

31. The sides of a triangle measure 10, 14, and 20 meters. What is the approximate measure of the largest angle of this triangle?

32. An aircraft is flying at a heading of 40° with an airspeed of 600 mi/hr. The wind is blowing from the south at 25 mi/hr. Approximate the actual course of the aircraft. What is the approximate ground speed?

33. A tower is 200 feet high. What is the angle of elevation of the top of the tower as seen from a point on the ground 75 yards from its base?

34. A ship starts out for an island 1,000 miles from port on a course of S20°W. However, the ship must first stop at a lighthouse located 400 miles due south of the port. What bearing should the ship use on leaving the lighthouse? How far is it from the lighthouse to the island?

35. A force of 12 pounds is applied to the origin in the direction of $\mathbf{i} + \sqrt{3}\mathbf{j}$, and a second force of 20 pounds is applied in the direction of $\mathbf{i} + \mathbf{j}$. Find the magnitude and direction of the resultant. Find the magnitude and direction of the force that will balance these two.

36. One diagonal of a parallelogram measures 50 inches and makes angles of 32° and 36° with the two sides. How long are the sides of the parallelogram?

Chapter Test

Calculators and/or tables are not needed.

1. Let $\mathbf{u} = \langle 2, -3\rangle$, $\mathbf{v} = \langle -4, -1\rangle$. Find:

a. $3\mathbf{u} - 2\mathbf{v}$

b. $\mathbf{u} \cdot \mathbf{v}$

c. $|\mathbf{u}|$

d. A unit vector orthogonal to $\mathbf{u}$

e. The cosine of the angle between $\mathbf{u}$ and $\mathbf{v}$

f. Illustrate $\mathbf{u}$, $\mathbf{v}$, and $\mathbf{u} - \mathbf{v}$ on the same cartesian coordinate system.

2. Solve the triangle ABC for c if $\gamma = 120°$, $a = 4$, $b = 2$.

3. Solve the triangle ABC for angle α if $a = 8$, $c = 5$, $\gamma = 30°$. Express the result(s) in terms of inverse trigonometric functions.

4. A ship moves S50°W for 3 hr at 25 mi/hr. Express the distance traveled west and the distance traveled south in terms of the sine and cosine functions.

5. Use an inverse trigonometric function to express the measure of the smallest angle of the triangle that has vertices at (0, 0), (2, 2), and (4, −3).

6. From the top of a radio tower 100 meters high, the angle of depression of the station is 10°. Use trigonometric functions to give the distance from the base of the tower to the station.

7. What is the resultant of forces of 10 pounds toward the east and 20 pounds toward N30°E?

9 Complex Numbers and Polar Coordinates

In Chapter 5 we saw how computations involving powers and roots of real numbers can be greatly simplified by the use of the logarithmic function. For example, the cube root of 13 can be approximated by the number whose logarithm is $\frac{1}{3}\log 13$. In this chapter we show how computation of powers and roots of complex numbers can be simplified by use of the trigonometric form of a complex number.

9.1 Geometric Representation and Trigonometric Form

The set C of complex numbers was defined in Section 2.4 as

$$C = \{a + bi\colon a, b \in R\}$$

At that time the usual terminology associated with complex numbers was discussed, along with the basic operations of addition, multiplication, subtraction, and division on C.

Each element $a + bi$ in C can be associated with the point in the plane having rectangular coordinates (a, b). Also, each point (x, y) in the plane can be viewed as the geometric representation of the complex number $x + yi$. In this way a one-to-one correspondence is established between the set of complex numbers and the set of points in the plane. When the cartesian coordinate system is used to represent complex numbers it is called the **complex plane.**

Complex plane

Example 1 Illustrate the numbers $2 + 3i$, $-2 + 3i$, $3 - i$, -3, $2i$, and $-\frac{3}{2} - 2i$ in the complex plane.

The numbers are illustrated in the figure shown.

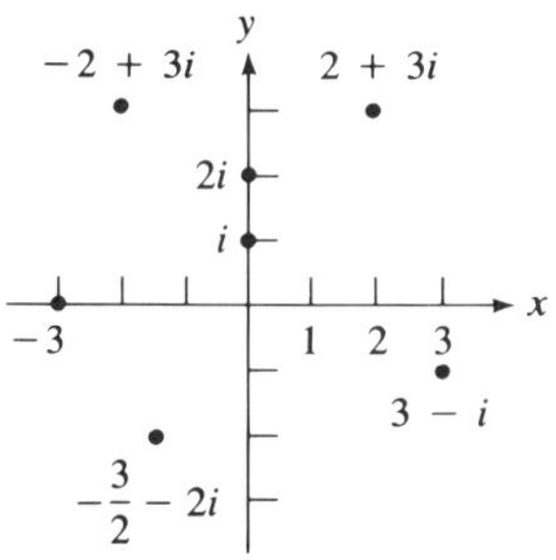

■

Let $z = (a, b)$ be a point in the complex plane. If $b = 0$, then z is a real number and is a point on the x axis. Similarly, if $a = 0$, z is purely imaginary and is a point on the y axis. For these reasons it is customary in the complex plane to refer to the x axis as the **real axis** and to the y axis as the **imaginary axis.**

Real axis
Imaginary axis

The complex number $z = a + bi$ may also be represented geometrically as the position vector $\mathbf{z} = \langle a, b \rangle$ with terminal point located at (a, b) and initial point at the origin. With this representation, addition of complex numbers corresponds to addition of vectors.

Complex numbers as vectors

Example 2 Let $z = -3 - 2i$ and $w = 1 + 5i$. Represent the position vectors $\mathbf{z} = \langle -3, -2 \rangle$, $\mathbf{w} = \langle 1, 5 \rangle$, and $\mathbf{z} + \mathbf{w}$ in the complex plane.
Notice that,

$$\begin{aligned} z + w &= (-3 - 2i) + (1 + 5i) \\ &= (-3 + 1) + (-2 + 5)i = -2 + 3i \end{aligned}$$

and

$$\mathbf{z} + \mathbf{w} = \langle -3, -2 \rangle + \langle 1, 5 \rangle = \langle -2, 3 \rangle$$

so that the vector corresponding to the complex number $z + w$ is the same as the vector corresponding to $\mathbf{z} + \mathbf{w}$.

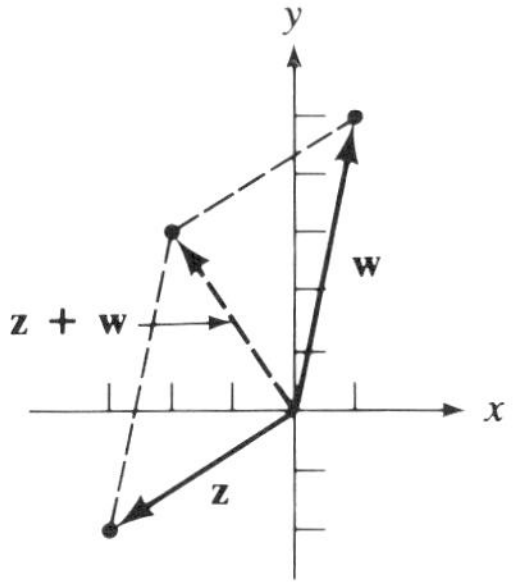

■

Although addition and subtraction of complex numbers are easy using any of the representations $a + bi$, (a, b), or $\langle a, b \rangle$, multiplication and division pose special problems in the complex plane. These will be resolved in the next section.

There is a natural extension of the concept of absolute value to complex numbers. For $z = a + bi$ we define the **absolute value** of z, $|z|$, to be

Absolute value

$$|z| = |a + bi| = \sqrt{a^2 + b^2}$$

Notice that $|z|$ is the magnitude of the vector $\mathbf{z} = \langle a, b \rangle$. Just as $|x|$ is the distance of the real number x from 0 on the number line, $|z|$ is likewise the distance from the origin to the point (a, b) in the complex plane. In the complex plane the absolute value of z is called the **modulus** or **magnitude** of z.

Modulus

Example 3 Find the moduli of the complex numbers $3 - 5i$, $\overline{3 - 5i}$, $-3i$, and 4.

From the definition of absolute value,

$$|3 - 5i| = \sqrt{(3)^2 + (-5)^2} = \sqrt{9 + 25} = \sqrt{34}$$
$$|\overline{3 - 5i}| = |3 + 5i| = \sqrt{9 + 25} = \sqrt{34}$$
$$|-3i| = |0 - 3i| = \sqrt{0^2 + (-3)^2} = \sqrt{9} = 3$$
$$|4| = |4 + 0i| = \sqrt{4^2 + 0^2} = \sqrt{16} = 4$$

Notice that the last two are easily obtained by inspection, using the concept of distance from the origin. ■

In Figure 9.1 the complex number $z = a + bi$ is represented by the point $P(a, b)$ in the complex plane. This point can also be located by specifying the angle θ and the distance r. In fact, using the triangle illustrated, $a = r\cos\theta$ and $b = r\sin\theta$, so that in terms of r and θ,

$$(a, b) = (r\cos\theta, r\sin\theta)$$

We see then that

$$a + bi = (r\cos\theta) + (r\sin\theta)i$$

or

$$z = r(\cos\theta + i\sin\theta)$$

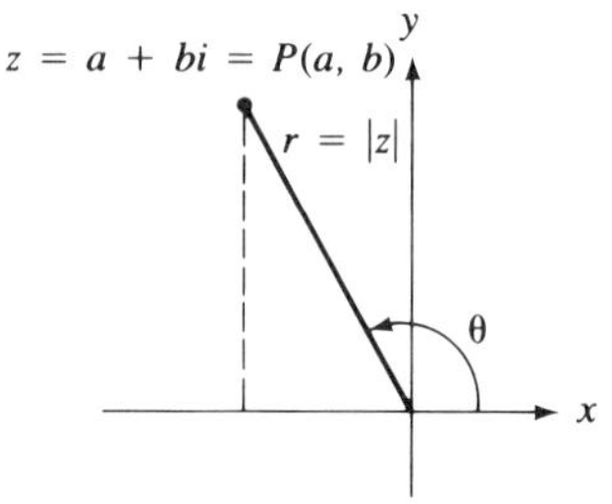

Figure 9.1

Trigonometric form
Polar form

The form $r(\cos\theta + i\sin\theta)$ is called the **trigonometric form** (or **polar form**) of z. It is useful in multiplication and division of complex numbers and in computation of powers and roots of complex numbers.

Argument

Principal argument

In the trigonometric form $z = r(\cos\theta + i\sin\theta)$ the number r is the modulus of z and is uniquely determined. However, θ is any angle in standard position that has terminal side on the line segment OP. Any such angle is called an **argument** of z. The unique argument θ such that $0 \le \theta < 2\pi$ (or $0° \le \theta < 360°$) is called the **principal argument** of z. Arguments can be expressed in any measure. Radian and degree measure are both used frequently.

Example 4 Find the principal argument of $z = 1 - i$. Give two additional arguments of z. In the triangle illustrated, $\alpha = 45°$. The principal argument of z is $\theta = 7\pi/4$. The angles $-45°$ and $15\pi/4$ are also arguments of z.

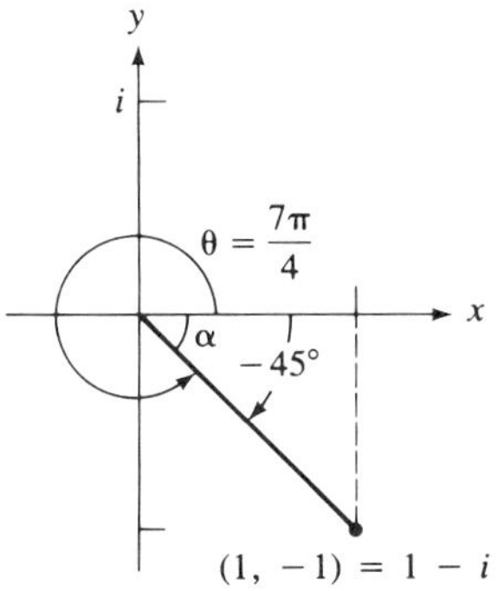

■

Any complex number $z = a + bi$ can be written in trigonometric form simply by factoring the number $r = |z| = \sqrt{a^2 + b^2}$ from $a + bi$. Thus,

$$\begin{aligned} z &= \sqrt{a^2 + b^2}\left(\frac{a}{\sqrt{a^2 + b^2}} + i\,\frac{b}{\sqrt{a^2 + b^2}}\right) \\ &= r(\cos\theta + i\sin\theta) \end{aligned}$$

While it is true that $\tan\theta = b/a$, the statement $\theta = \arctan b/a$ is not sufficient to describe θ. This is because $-\pi/2 < \arctan u < \pi/2$ by definition. Usually an argument of z is obtained from the quadrant of z and the reference angle $\tilde{\theta}$.

Example 5 Write the complex number $z = -3 - 5i$ in trigonometric form.
The modulus is $r = |z| = \sqrt{34}$. Thus,

$$\begin{aligned} z &= \sqrt{34}\left(\frac{-3}{\sqrt{34}} + i\,\frac{-5}{\sqrt{34}}\right) \\ &= \sqrt{34}(\cos\theta + i\sin\theta) \end{aligned}$$

where $\tan\tilde{\theta} = \dfrac{-5}{-3} = \dfrac{5}{3}$, and the terminal side of θ is in quadrant III.

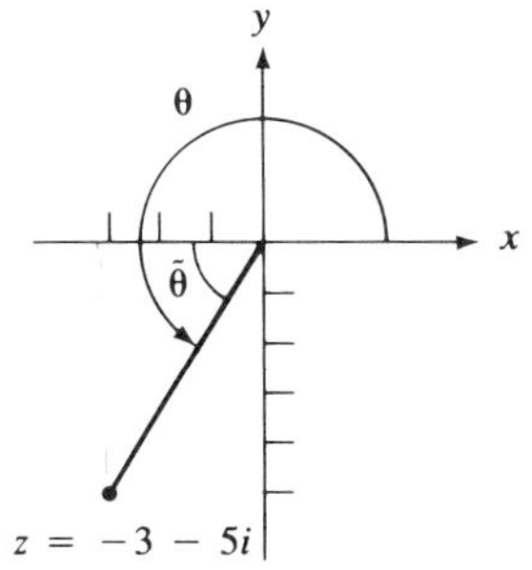

■

Example 6 Write the complex number $z = -\sqrt{3} + i$ in trigonometric form.
The modulus is $r = |z| = \sqrt{4} = 2$. For the argument we need a quadrant II angle θ such that $\tan\theta = -1/\sqrt{3}$.
The reference angle $\tilde{\theta}$ is 30°. The principal argument of z is $5\pi/6$, as shown, and in trigonometric form

$$z = 2\left(\cos\frac{5\pi}{6} + i\sin\frac{5\pi}{6}\right)$$

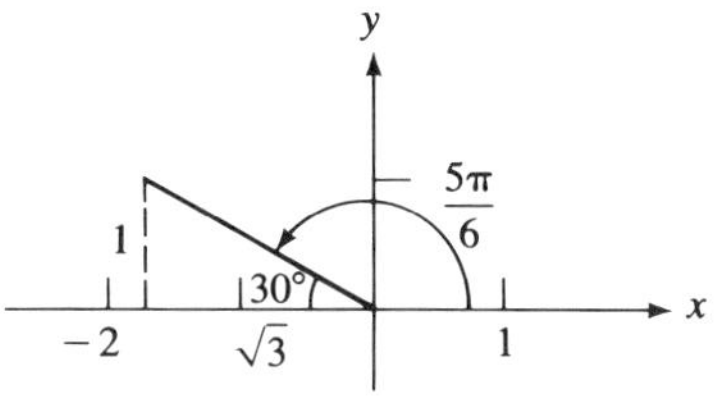

■

Example 7 Write the complex number $z = 2\left(\cos\frac{11\pi}{6} + i\sin\frac{11\pi}{6}\right)$ in the form $a + bi$.

$$\cos\frac{11\pi}{6} = \cos\frac{\pi}{6} = \frac{\sqrt{3}}{2} \quad \text{and} \quad \sin\frac{11\pi}{6} = -\sin\frac{\pi}{6} = -\frac{1}{2}$$

Thus,

$$z = 2\left[\frac{\sqrt{3}}{2} + i\left(-\frac{1}{2}\right)\right] = \sqrt{3} - i$$

■

Exercises 9.1

In Exercises 1–8 illustrate the point that represents the complex number and the point that represents its conjugate in the complex plane.

1. 0 **2.** -3 **3.** $-3i$ **4.** $2i$
5. $3 - 4i$ **6.** $-1 + 5i$ **7.** $\sqrt{3} - 2i$ **8.** $1 - \sqrt{2}\,i$

In Exercises 9–10 represent the position vectors for u, v, and $u - v$ in the complex plane.

9. $u = 2 - 3i$; $v = 3 + i$ **10.** $u = -2 + i$; $v = 1 + 3i$

In Exercises 11–18 find the modulus of each complex number.

11. $1 - i$ **12.** $-5i$ **13.** $-2 + 2\sqrt{3}\,i$
14. $-\frac{1}{2} - \frac{\sqrt{3}}{2}i$ **15.** 0 **16.** $3 - 2i$
17. $a - 2i$ **18.** $-3 - bi$

In Exercises 19–24 find an argument of each complex number.

19. 1
20. $-\pi i$
21. $-1 - i$
22. $\frac{1}{2} - \frac{1}{2}i$
23. $3\sqrt{3} - 3i$
24. $-1 - \sqrt{3}\,i$

In Exercises 25–30 write the complex number in standard form, given the specified modulus and argument.

25. $2, \frac{\pi}{4}$
26. $5, 2\pi$
27. $\frac{1}{2}, \frac{3\pi}{2}$
28. $\sqrt{2}, \frac{7\pi}{6}$
29. $\sqrt{3}, -\frac{7\pi}{6}$
30. $\frac{1}{3}, -\frac{2\pi}{3}$

In Exercises 31–44 write the complex number in trigonometric form.

31. $-1 + i$
32. $-1 - i$
33. 3
34. $-2i$
35. $-1 + \sqrt{3}\,i$
36. $3\sqrt{3} - 3i$
37. πi
38. -8
39. $7 - 7i$
40. $\sqrt{2} + \sqrt{2}\,i$
41. $\frac{3}{2} + \frac{3\sqrt{3}}{2}i$
42. $-\frac{5\sqrt{3}}{2} + \frac{5}{2}i$
43. $-4 + 7i$
44. $2 - 5i$
45. Show that for $u \in C$, $|\bar{u}| = |u|$.
46. If $u, v \in C$, show that $|uv| = |u|\,|v|$.
47. For $u, v \in C$, verify that $\frac{u}{v} = \frac{u\bar{v}}{|v|^2}$.
48. If $u = r(\cos\theta + i\sin\theta)$, show that $\bar{u} = r[\cos(-\theta) + i\sin(-\theta)]$.

9.2 Multiplication and Division in Trigonometric Form

One of the advantages of the trigonometric form representation for complex numbers is the simplicity of product and quotient computations.

Let $u = r(\cos\alpha + i\sin\alpha)$ and $v = s(\cos\beta + i\sin\beta)$ be any complex numbers written in trigonometric form. Recall the addition formulas:

$$\cos(\alpha + \beta) = \cos\alpha\cos\beta - \sin\alpha\sin\beta$$
$$\sin(\alpha + \beta) = \sin\alpha\cos\beta + \cos\alpha\sin\beta$$

Then

$$\begin{aligned} uv &= [r(\cos\alpha + i\sin\alpha)][s(\cos\beta + i\sin\beta)] \\ &= rs[(\cos\alpha\cos\beta - \sin\alpha\sin\beta) + i(\sin\alpha\cos\beta + \cos\alpha\sin\beta)] \\ &= rs[\cos(\alpha + \beta) + i\sin(\alpha + \beta)] \end{aligned}$$

Thus, the product of any two complex numbers is the complex number whose magnitude is the product of the two magnitudes and whose argument is the sum of the two arguments.

$$[r(\cos\alpha + i\sin\alpha)][s(\cos\beta + i\sin\beta)] = rs[\cos(\alpha+\beta) + i\sin(\alpha+\beta)]$$

Example 1 Let $u = -\sqrt{3} + i$ and $v = -\frac{3}{2} - \frac{3\sqrt{3}}{2}i$. Write u and v in trigonometric form and **(a)** find the product in trigonometric form; **(b)** Check by computation of the product using standard form; **(c)** Illustrate u, v, and uv geometrically.

In trigonometric form,

$$u = 2\left(\cos\frac{5\pi}{6} + i\sin\frac{5\pi}{6}\right) \quad \text{and} \quad v = 3\left(\cos\frac{4\pi}{3} + i\sin\frac{4\pi}{3}\right)$$

a. $$uv = (2)(3)\left[\cos\left(\frac{5\pi}{6} + \frac{4\pi}{3}\right) + i\sin\left(\frac{5\pi}{6} + \frac{4\pi}{3}\right)\right]$$
$$= 6\left(\cos\frac{13\pi}{6} + i\sin\frac{13\pi}{6}\right) = 6\left(\cos\frac{\pi}{6} + i\sin\frac{\pi}{6}\right)$$

b. $$(-\sqrt{3} + i)\left(-\frac{3}{2} - \frac{3\sqrt{3}}{2}i\right) = \left(\frac{3\sqrt{3}}{2} + \frac{3\sqrt{3}}{2}\right) + i\left(-\frac{3}{2} + \frac{9}{2}\right)$$
$$= 3\sqrt{3} + 3i = 6\left(\frac{\sqrt{3}}{2} + \frac{1}{2}i\right) = 6\left(\cos\frac{\pi}{6} + i\sin\frac{\pi}{6}\right)$$

c.

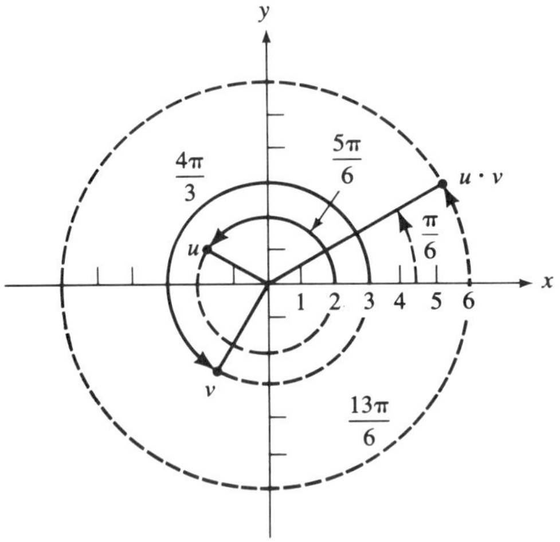

■

Example 2 Use trigonometric form to compute $(1 - i)^2$.

$$1 - i = \sqrt{2}(\cos 315° + i\sin 315°)$$

so that

$$\begin{aligned}(1 - i)^2 = (1 - i)(1 - i) &= \sqrt{2}\,\sqrt{2}[\cos(315^\circ + 315^\circ) + i \sin(315^\circ + 315^\circ)] \\ &= 2(\cos 630^\circ + i \sin 630^\circ) = 2(\cos 270^\circ + i \sin 270^\circ) \\ &= 2[0 + i(-1)] = -2i\end{aligned}$$

■

Recall that division of complex numbers was accomplished using multiplication and the conjugate of the divisor as follows:

$$\frac{u}{v} = \frac{u}{v}\,\frac{\bar{v}}{\bar{v}} = \frac{u\bar{v}}{v\bar{v}}$$

This works because $v\bar{v}$ is a real number. In fact, $v\bar{v} = |v|^2$, so that

$$\frac{u}{v} = \frac{u\bar{v}}{|v|^2}$$

Recall also the identities

$$\sin(-\theta) = -\sin\theta \quad \text{and} \quad \cos(-\theta) = \cos\theta$$

Again let

$$u = r(\cos\alpha + i \sin\alpha) \quad \text{and} \quad v = s(\cos\beta + i\sin\beta)$$

Notice that

$$\bar{v} = s(\cos\beta - i\sin\beta) = s[\cos(-\beta) + i\sin(-\beta)]$$

Then

$$\begin{aligned}\frac{u}{v} = \frac{u\bar{v}}{|v|^2} &= \frac{[r(\cos\alpha + i\sin\alpha)]\{s[\cos(-\beta) + i\sin(-\beta)]\}}{|s(\cos\beta + i\sin\beta)|^2} \\ &= \frac{rs[\cos(\alpha - \beta) + i\sin(\alpha - \beta)]}{s^2} \\ &= \frac{r}{s}[\cos(\alpha - \beta) + i\sin(\alpha - \beta)]\end{aligned}$$

Thus, the quotient of any two complex numbers is the complex number whose magnitude is the quotient of the two magnitudes and whose argument is the difference (in the correct order) of the two arguments.

$$\frac{r(\cos\alpha + i\sin\alpha)}{s(\cos\beta + i\sin\beta)} = \frac{r}{s}[\cos(\alpha - \beta) + i\sin(\alpha - \beta)]$$

Example 3 Use trigonometric form to compute $(2i) \div (3 + 3i)$.
Since

$$2i = 2\left(\cos\frac{\pi}{2} + i\sin\frac{\pi}{2}\right) \quad \text{and} \quad 3 + 3i = 3\sqrt{2}\left(\cos\frac{\pi}{4} + i\sin\frac{\pi}{4}\right)$$

we see that

$$\frac{2i}{3+3i} = \frac{2}{3\sqrt{2}}\left[\cos\left(\frac{\pi}{2}-\frac{\pi}{4}\right) + i\sin\left(\frac{\pi}{2}-\frac{\pi}{4}\right)\right]$$
$$= \frac{\sqrt{2}}{3}\left(\cos\frac{\pi}{4} + i\sin\frac{\pi}{4}\right) = \frac{\sqrt{2}}{3}\left(\frac{1}{\sqrt{2}} + \frac{1}{\sqrt{2}}i\right)$$
$$= \frac{1}{3} + \frac{1}{3}i$$

■

Example 4 Let $u = -\sqrt{3} + i$, $v = -\frac{3}{2} - \frac{3\sqrt{3}}{2}i$, as in Example 1. **(a)** Compute $u \div v$; **(b)** write the quotient in standard form; **(c)** illustrate geometrically.

a. $$\frac{u}{v} = \frac{2}{3}\left[\cos\left(\frac{5\pi}{6}-\frac{4\pi}{3}\right) + i\sin\left(\frac{5\pi}{6}-\frac{4\pi}{3}\right)\right]$$
$$= \frac{2}{3}\left[\cos\left(-\frac{3\pi}{6}\right) + i\sin\left(-\frac{3\pi}{6}\right)\right] = \frac{2}{3}\left[\cos\left(-\frac{\pi}{2}\right) + i\sin\left(-\frac{\pi}{2}\right)\right]$$

b. $$\frac{u}{v} = \frac{2}{3}\left[\cos\left(-\frac{\pi}{2}\right) + i\sin\left(-\frac{\pi}{2}\right)\right] = \frac{2}{3}[0 + i(-1)]$$
$$= -\frac{2}{3}i$$

c.

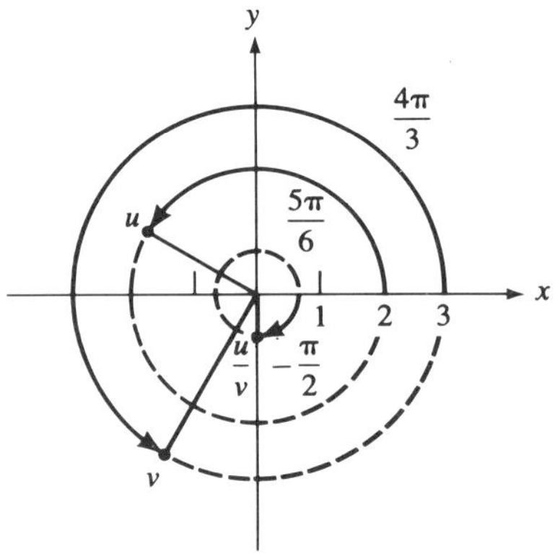

■

Example 5 Write the quotient $\frac{u}{v}$ in trigonometric form if $u = 3 + 4i$ and $v = -2 + i$.

$u = 5(\cos\alpha + i\sin\alpha)$

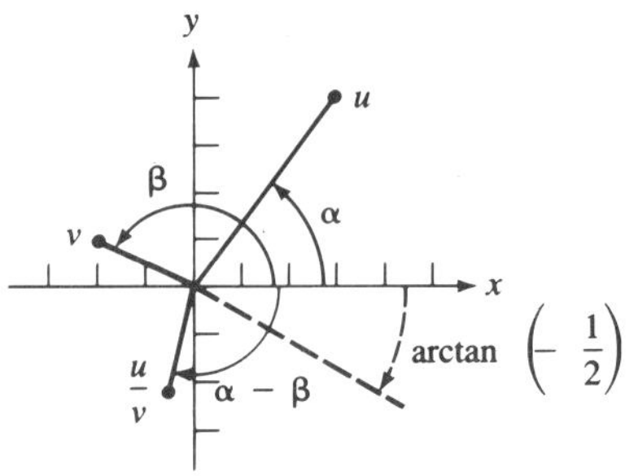

where

$$\tan \alpha = \frac{4}{3} \quad \text{and} \quad 0 < \alpha < \frac{\pi}{2}$$

$$v = \sqrt{5}(\cos \beta + i \sin \beta)$$

where

$$\tan \beta = -\frac{1}{2} \quad \text{and} \quad \frac{\pi}{2} < \beta < \pi$$

Thus

$$\frac{u}{v} = \frac{5}{\sqrt{5}}[\cos(\alpha - \beta) + i \sin(\alpha - \beta)]$$
$$= \sqrt{5}[\cos \theta + i \sin \theta]$$

where

$$\theta = \alpha - \beta = \left(\arctan \frac{4}{3}\right) - \left[\arctan\left(-\frac{1}{2}\right) + \pi\right]$$

If an approximation is desired, we note that $\alpha \simeq 53.1°$, $\beta \simeq 153.4°$, and $\theta \simeq -100.3°$. ■

Example 6 If $u = -1 + i$, compute u^{-2}.
In trigonometric form,

$$u = \sqrt{2}\left(\cos \frac{3\pi}{4} + i \sin \frac{3\pi}{4}\right)$$

Then

$$u^{-2} = \frac{1}{u^2} = \frac{1}{uu} = \frac{\cos 0 + i \sin 0}{2\left(\cos \frac{3\pi}{2} + i \sin \frac{3\pi}{2}\right)}$$
$$= \frac{1}{2}\left[\cos\left(0 - \frac{3\pi}{2}\right) + i \sin\left(0 - \frac{3\pi}{2}\right)\right]$$
$$= \frac{1}{2}(0 + i) = \frac{1}{2}i$$

■

Example 7 If $u = 1 + \sqrt{3}\, i$, compute u^3.
In trigonometric form,

$$u = 2(\cos 60° + i \sin 60°)$$

Then

$$u^3 = uu^2 = u\{[2(\cos 60° + i \sin 60°)][2(\cos 60° + i \sin 60°)]\}$$
$$= u[4(\cos 120° + i \sin 120°)]$$
$$= [2(\cos 60° + i \sin 60°)][4(\cos 120° + i \sin 120°)]$$
$$= 8(\cos 180° + i \sin 180°) = 8(-1 + 0)$$
$$= -8$$

■

Exercises 9.2

In Exercises 1–4 perform the indicated computation in trigonometric form and illustrate u, v, and the computation geometrically.

1. $u = -2i;\ v = -3 - 3i;\ uv$

2. $u = 1 - i;\ v = \sqrt{3} + i;\ uv$

3. $u = -1 + \sqrt{3}\,i;\ v = \sqrt{3} + i;\ \dfrac{u}{v}$

4. $u = 6i;\ v = -3 - 3i;\ \dfrac{u}{v}$

In Exercises 5–14 compute uv and $u \div v$ by using trigonometric form and the results of this section. Also write each result in standard form.

5. $u = 3i;\ v = -2i$

6. $u = -4;\ v = -2i$

7. $u = 1 - i;\ v = 1 + i$

8. $u = -1 + i;\ v = -1 - i$

9. $u = -8;\ v = -\sqrt{2} - \sqrt{2}\,i$

10. $u = 1 + \sqrt{3}\,i;\ v = \sqrt{3} + i$

11. $u = 1 - \sqrt{3}\,i;\ v = -2\sqrt{3} - 2i$

12. $u = \dfrac{3}{2} + \dfrac{3\sqrt{3}}{2}\,i;\ v = -\dfrac{5}{2}\sqrt{3} + \dfrac{5}{2}\,i$

13. $u = 1 + \sqrt{3}\,i;\ v = 1 - i$

14. $u = 1 + i;\ v = \sqrt{3} - i$

In Exercises 15–18 perform the indicated operation and write the result in trigonometric form.

15. $(2 + 3i)(-3 + 2i)$

16. $\dfrac{1 + 3i}{2 - 5i}$

17. $\dfrac{(1 + i)(-2 + 3i)}{-3 + i}$

18. $\dfrac{4 - i}{(1 - i)(-2 + 4i)}$

In Exercises 19–26 perform the computation in trigonometric form and write the result in standard form.

19. $(-1 + i)^3$

20. $(-2 + 2\sqrt{3}\,i)^2$

21. $\left(-\dfrac{1}{2} - \dfrac{\sqrt{3}}{2}\,i\right)^3$

22. $\left(-\dfrac{1}{2} + \dfrac{\sqrt{3}}{2}\,i\right)^3$

23. $(\sqrt{3} + i)^{-2}$

24. $(1 - i)^{-2}$

25. $[3(\cos 15° + i \sin 15°)][5(\cos 75° + i \sin 75°)]$

26. $6\left(\cos \dfrac{5\pi}{12} + i \sin \dfrac{5\pi}{12}\right) \div 3\left(\cos \dfrac{\pi}{12} + i \sin \dfrac{\pi}{12}\right)$

27. Show that $[r(\cos \theta + i \sin \theta)]^{-1} = \dfrac{1}{r}(\cos \theta - i \sin \theta)$.

28. Let $u = r(\cos \alpha + i \sin \alpha)$, $v = s(\cos \beta + i \sin \beta)$. Show that

$$|u - v| = \sqrt{r^2 + s^2 - 2rs \cos(\alpha - \beta)}$$

29. Let $u = r(\cos \theta + i \sin \theta)$. Show that

$$u^2 = r^2(\cos 2\theta + i \sin 2\theta)$$

30. Let $u = r(\cos \alpha + i \sin \alpha)$. Show that $v = \sqrt{r}(\cos \frac{1}{2}\alpha + i \sin \frac{1}{2}\alpha)$ is a square root of u.

31. Show that $(-1 + \sqrt{3}\,i)$ is a cube root of 8.

9.3 DeMoivre's Theorem and *n*th Roots of Complex Numbers

As seen in Section 9.2, multiplication of complex numbers in trigonometric form is easily accomplished. Since positive integer exponentiation is merely repeated multiplication, it is natural to expect that such powers of complex numbers in trigonometric form are also easy to compute. This is indeed the case. Let $u = r(\cos\theta + i\sin\theta)$.
Then

$$\begin{aligned} u^2 &= [r(\cos\theta + i\sin\theta)][r(\cos\theta + i\sin\theta)] \\ &= r^2[\cos(\theta + \theta) + i\sin(\theta + \theta)] \\ &= r^2(\cos 2\theta + i\sin 2\theta) \end{aligned}$$

and

$$\begin{aligned} u^3 = uu^2 &= [r(\cos\theta + i\sin\theta)][r^2(\cos 2\theta + i\sin 2\theta)] \\ &= r^3[\cos(\theta + 2\theta) + i\sin(\theta + 2\theta)] \\ &= r^3(\cos 3\theta + i\sin 3\theta) \end{aligned}$$

These computations make plausible the following famous result, credited to the 18th-century French mathematician Abraham DeMoivre.

DeMoivre's theorem

> For each natural number n,
>
> $$[r(\cos\theta + i\sin\theta)]^n = r^n(\cos n\theta + i\sin n\theta)$$

A proof of DeMoivre's theorem requires a method called mathematical induction and hence will be postponed until Chapter 12.

Example 1 Compute u^{10} if $u = 2(\cos 15° + i\sin 15°)$.
We use DeMoivre's theorem to obtain

$$\begin{aligned} u^{10} &= 2^{10}[\cos(10)(15°) + i\sin(10)(15°)] \\ &= 1{,}024(\cos 150° + i\sin 150°) \end{aligned}$$

In standard form,

$$u^{10} = 1{,}024\left(-\frac{\sqrt{3}}{2} + \frac{1}{2}i\right) = -512\sqrt{3} + 512i$$

■

Example 2 Compute u^2 and u^3 for $u = -1 + \sqrt{3}\,i$.
In trigonometric form,

$$u = 2\left(\cos\frac{2\pi}{3} + i\sin\frac{2\pi}{3}\right)$$

Thus,

$$\begin{aligned} u^2 &= 2^2\left[\cos(2)\left(\frac{2\pi}{3}\right) + i\sin(2)\left(\frac{2\pi}{3}\right)\right] \\ &= 4\left(\cos\frac{4\pi}{3} + i\sin\frac{4\pi}{3}\right) \\ &= 4\left(-\frac{1}{2} - \frac{\sqrt{3}}{2}i\right) = -2 - 2\sqrt{3}\,i \end{aligned}$$

and

$$\begin{aligned} u^3 &= 2^3\left[\cos(3)\left(\frac{2\pi}{3}\right) + i\sin(3)\left(\frac{2\pi}{3}\right)\right] = 8(\cos 2\pi + i\sin 2\pi) \\ &= 8 \end{aligned}$$

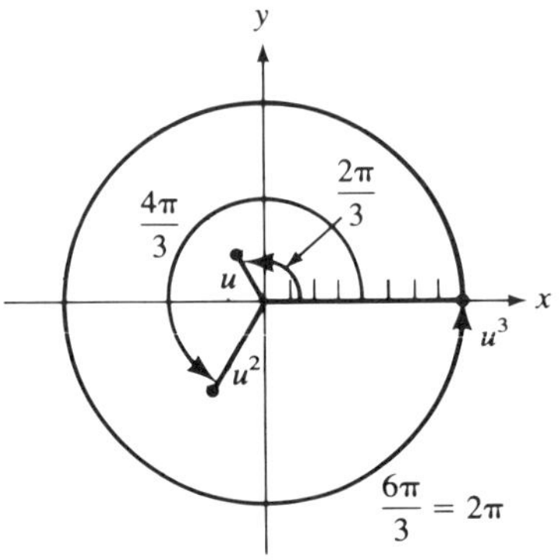

■

Example 3 Use DeMoivre's theorem to compute $(\sqrt{2} - \sqrt{2}\,i)^3$.
In trigonometric form,

$$\sqrt{2} - \sqrt{2}\,i = 2\left[\cos\left(-\frac{\pi}{4}\right) + i\sin\left(-\frac{\pi}{4}\right)\right]$$

Thus,

$$\begin{aligned} (\sqrt{2} - \sqrt{2}\,i)^3 &= 2^3\left[\cos\left(-\frac{3\pi}{4}\right) + i\sin\left(-\frac{3\pi}{4}\right)\right] \\ &= 8\left(-\frac{\sqrt{2}}{2} - \frac{\sqrt{2}}{2}i\right) = -4\sqrt{2} - 4\sqrt{2}\,i \end{aligned}$$

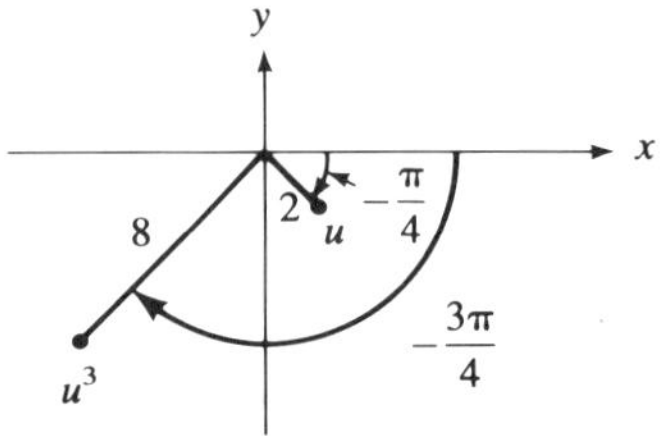

■

Example 4 Compute $(1 + i)^{15}$.
Since

$$1 + i = \sqrt{2}\left(\cos \frac{\pi}{4} + i \sin \frac{\pi}{4}\right)$$

we have

$$\begin{aligned}(1 + i)^{15} &= (2^{1/2})^{15}\left(\cos \frac{15\pi}{4} + i \sin \frac{15\pi}{4}\right)\\ &= 128\sqrt{2}\left(\cos \frac{7\pi}{4} + i \sin \frac{7\pi}{4}\right) = 128(1 - i)\end{aligned}$$

■

nth root If $z \varepsilon C$, by an **nth root** of z we mean a number w such that $w^n = z$. DeMoivre's theorem has the following consequence, which is used to find nth roots of complex numbers.

> If $n \varepsilon Z$, $n > 0$, the nth roots of $z = r(\cos \theta + i \sin \theta)$ are given by
>
> $$\sqrt[n]{r}\left[\cos\left(\frac{\theta}{n} + \frac{2k\pi}{n}\right) + i \sin\left(\frac{\theta}{n} + \frac{2k\pi}{n}\right)\right]$$
>
> for $k = 0, 1, \ldots, n - 1$.

Using DeMoivre's theorem it is easy to verify that these are nth roots of $z = r(\cos \theta + i \sin \theta)$.

$$\begin{aligned}&\left\{\sqrt[n]{r}\left[\cos\left(\frac{\theta}{n} + \frac{2k\pi}{n}\right) + i \sin\left(\frac{\theta}{n} + \frac{2k\pi}{n}\right)\right]\right\}^n\\ &= (\sqrt[n]{r})^n\left[\cos n\left(\frac{\theta}{n} + \frac{2k\pi}{n}\right) + i \sin n\left(\frac{\theta}{n} + \frac{2k\pi}{n}\right)\right]\\ &= r[\cos(\theta + 2k\pi) + i \sin(\theta + 2k\pi)]\\ &= r(\cos \theta + i \sin \theta) = z\end{aligned}$$

We now argue that there are no other nth roots. If $w = s(\cos \alpha + i \sin \alpha)$ is an nth root of z, then

$$r(\cos \theta + i \sin \theta) = z = w^n = s^n(\cos n\alpha + i \sin n\alpha)$$

Equating modulus, real part, and imaginary part of these two complex numbers we obtain

$$r = s^n \qquad \cos \theta = \cos n\alpha \qquad \sin \theta = \sin n\alpha$$

Simultaneous solution of the last two equations requires that θ and $n\alpha$ be coterminal, $n\alpha = \theta + k(2\pi)$ for some integer k. Therefore $\alpha = (\theta + 2k\pi)/n$ for some $k \varepsilon Z$. If we substitute $k = 0, 1, \ldots, n - 1$ we obtain equally spaced points on a circle of radius $\sqrt[n]{r}$ and thus distinct roots of z. Substitution of $k = n$ repeats the root for $k = 0$, and $k = n + 1$ repeats the root obtained from $k = 1$.

In general, additional values of k (positive or negative) simply repeat roots already obtained. We have demonstrated then that each complex number has n distinct nth roots and that these are specified by the formula displayed above.

Part of the above argument deserves special emphasis.

> The nth roots of z are n points equally spaced around a circle with center at the origin and radius $\sqrt[n]{r}$.

Example 5 Find the fourth roots of -16.

In trigonometric form, $-16 = 16(\cos \pi + i \sin \pi)$; hence the four fourth roots are given by

$$2\left[\cos\left(\frac{\pi}{4} + \frac{2k\pi}{4}\right) + i \sin\left(\frac{\pi}{4} + \frac{2k\pi}{4}\right)\right] \quad \text{for} \quad k = 0, 1, 2, 3$$

We obtain

$$u_0 = 2\left(\cos\frac{\pi}{4} + i \sin\frac{\pi}{4}\right) = \sqrt{2} + \sqrt{2}\, i$$

$$u_1 = 2\left[\cos\left(\frac{\pi}{4} + \frac{2\pi}{4}\right) + i \sin\left(\frac{\pi}{4} + \frac{2\pi}{4}\right)\right] = -\sqrt{2} + \sqrt{2}\, i$$

$$u_2 = 2\left[\cos\left(\frac{\pi}{4} + \frac{4\pi}{4}\right) + i \sin\left(\frac{\pi}{4} + \frac{4\pi}{4}\right)\right] = -\sqrt{2} - \sqrt{2}\, i$$

$$u_3 = 2\left[\cos\left(\frac{\pi}{4} + \frac{6\pi}{4}\right) + i \sin\left(\frac{\pi}{4} + \frac{6\pi}{4}\right)\right] = \sqrt{2} - \sqrt{2}\, i$$

The roots are shown in the diagram. Note how the angle between the vectors from the origin to the roots is always $\pi/2$, which is $2\pi/4$.

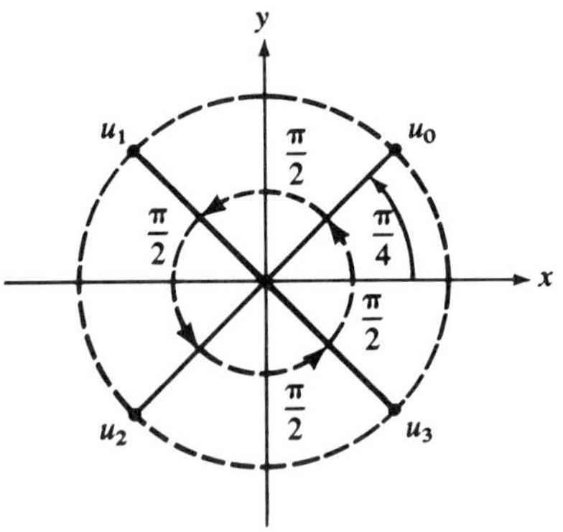

■

Example 6 Find the three cube roots of $1 + i$. Leave the results in trigonometric form.

In trigonometric form,

$$1 + i = \sqrt{2}\left(\cos\frac{\pi}{4} + i \sin\frac{\pi}{4}\right)$$

Thus the cube roots are

$$u_0 = \sqrt[6]{2}\left(\cos\frac{\pi}{12} + i\sin\frac{\pi}{12}\right)$$

$$u_1 = \sqrt[6]{2}\left[\cos\left(\frac{\pi}{12} + \frac{2\pi}{3}\right) + i\sin\left(\frac{\pi}{12} + \frac{2\pi}{3}\right)\right] = \sqrt[6]{2}\left(\cos\frac{9\pi}{12} + i\sin\frac{9\pi}{12}\right)$$

$$u_2 = \sqrt[6]{2}\left[\cos\left(\frac{\pi}{12} + \frac{4\pi}{3}\right) + i\sin\left(\frac{\pi}{12} + \frac{4\pi}{3}\right)\right] = \sqrt[6]{2}\left(\cos\frac{17\pi}{12} + i\sin\frac{17\pi}{12}\right)$$

The roots are illustrated in the sketch. Note that the angle between the vectors from the origin to the roots is always $2\pi/3$ in this case.

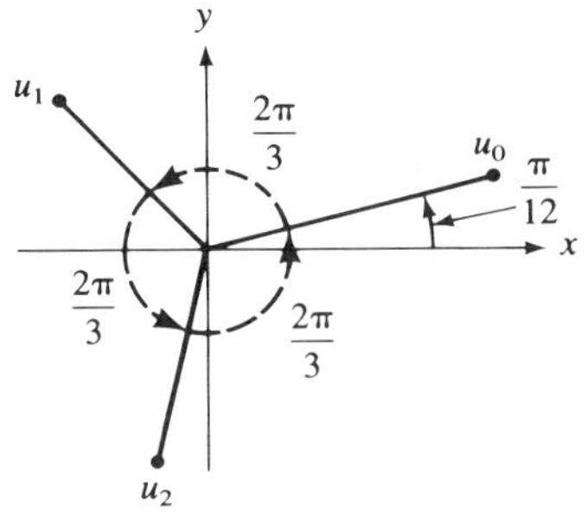

■

Roots of unity When $z = 1$, the nth roots of z are called the ***n*th roots of unity.**

Example 7 Find the six sixth roots of unity.
In trigonometric form, $1 = \cos 0 + i\sin 0$. The sixth roots then are given by

$$\cos\left(\frac{0}{6} + \frac{2k\pi}{6}\right) + i\sin\left(\frac{0}{6} + \frac{2k\pi}{6}\right)$$
$$= \cos\frac{k\pi}{3} + i\sin\frac{k\pi}{3} \quad \text{for} \quad k = 0, 1, 2, 3, 4, 5$$

The roots are

$$u_0 = \cos 0 + i\sin 0 = 1$$
$$u_1 = \cos\frac{\pi}{3} + i\sin\frac{\pi}{3} = \frac{1}{2} + \frac{\sqrt{3}}{2}i$$
$$u_2 = \cos\frac{2\pi}{3} + i\sin\frac{2\pi}{3} = -\frac{1}{2} + \frac{\sqrt{3}}{2}i$$
$$u_3 = \cos\pi + i\sin\pi = -1$$
$$u_4 = \cos\frac{4\pi}{3} + i\sin\frac{4\pi}{3} = -\frac{1}{2} - \frac{\sqrt{3}}{2}i$$
$$u_5 = \cos\frac{5\pi}{3} + i\sin\frac{5\pi}{3} = \frac{1}{2} - \frac{\sqrt{3}}{2}i$$

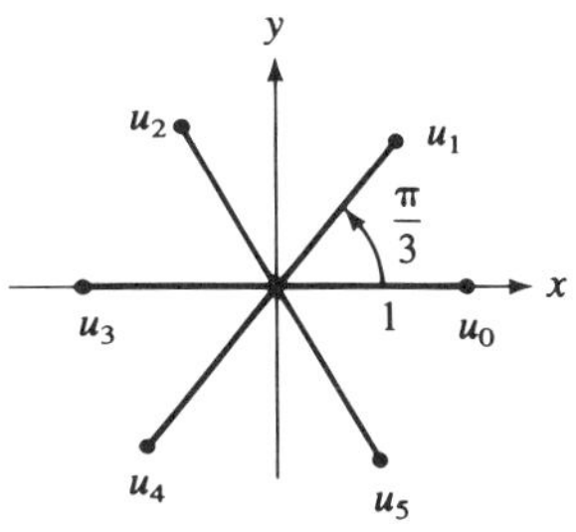

■

Example 8 Use DeMoivre's theorem to express $\cos 3\theta$ in terms of $\cos\theta$ and $\sin 3\theta$ in terms of $\sin\theta$.

By DeMoivre's theorem,

$$(\cos\theta + i\sin\theta)^3 = \cos 3\theta + i\sin 3\theta$$

By direct multiplication as a binomial we obtain

$$\begin{aligned}(\cos\theta + i\sin\theta)^3 &= \cos^3\theta + 3i\cos^2\theta\sin\theta - 3\cos\theta\sin^2\theta - i\sin^3\theta\\ &= (\cos^3\theta - 3\sin^2\theta\cos\theta) + i(3\sin\theta\cos^2\theta - \sin^3\theta)\end{aligned}$$

Equating real and imaginary parts of these two expansions, we have

$$\begin{aligned}\cos 3\theta &= \cos^3\theta - 3\sin^2\theta\cos\theta = \cos^3\theta - 3(1-\cos^2\theta)\cos\theta\\ &= 4\cos^3\theta - 3\cos\theta\end{aligned}$$

$$\begin{aligned}\sin 3\theta &= 3\sin\theta\cos^2\theta - \sin^3\theta = 3\sin\theta(1-\sin^2\theta) - \sin^3\theta\\ &= 3\sin\theta - 4\sin^3\theta\end{aligned}$$

■

Exercises 9.3

In Exercises 1–12 compute the powers and express the result in standard form.

1. $(\cos 12° + i\sin 12°)^{10}$

2. $[2(\cos 7.5° + i\sin 7.5°)]^{12}$

3. $\left(\cos\dfrac{\pi}{180} + i\sin\dfrac{\pi}{180}\right)^{90}$

4. $\left[3\left(\cos\dfrac{\pi}{24} + i\sin\dfrac{\pi}{24}\right)\right]^4$

5. $(1+i)^8$

6. $(2-2i)^{10}$

7. $(-2+2\sqrt{3}\,i)^3$

8. $(-\sqrt{2}+\sqrt{2}\,i)^6$

9. $(\sqrt{3}-i)^4$

10. $\left(-\dfrac{1}{2}-\dfrac{1}{2}i\right)^5$

11. $\left(-\dfrac{\sqrt{3}}{2}+\dfrac{1}{2}i\right)^{20}$

12. $\left(\dfrac{\sqrt{2}}{2}-\dfrac{\sqrt{2}}{2}i\right)^{50}$

In Exercises 13–18 compute the n distinct nth *roots, express the result in standard form, and sketch.*

13. Cube roots of i

14. Cube roots of unity

15. Eighth roots of unity

16. Square roots of $1-\sqrt{3}\,i$

17. Fourth roots of $-8-8\sqrt{3}\,i$

18. Fourth roots of $-\dfrac{1}{2}-\dfrac{\sqrt{3}}{2}i$

In Exercises 19–26 compute the n distinct nth *roots, express the result in trigonometric form, and sketch.*

19. Fifth roots of $4-4i$

20. Fourth roots of $1+\sqrt{3}\,i$

21. Cube roots of $4\sqrt{3}-4i$

22. Fifth roots of $16-16\sqrt{3}\,i$

23. Fourth roots of $2+2\sqrt{3}\,i$

24. Cube roots of $\sqrt{3}-i$

25. Fourth roots of $81\left(\dfrac{1}{2}+\dfrac{\sqrt{3}}{2}i\right)$

26. Sixth roots of $64\left(-\dfrac{\sqrt{3}}{2}+\dfrac{1}{2}i\right)$

27. Find the three solutions to the equation $x^3 + i = 0$.

28. Find the five solutions to the equation $x^5 + 32 = 0$.

29. Find four solutions to the equation $2x^4 - (1 - \sqrt{3}\,i) = 0$.

30. Find four solutions to the equation $x^4 - (2 + 2\sqrt{3}\,i) = 0$.

31. Perform the computations: $(1 + i)^5 \div (1 - i)^6$.

32. Perform the computations: $(-2i) \div (1 - \sqrt{3}\,i)^5$.

33. Perform the computations: $(1 - i)^7 \div (1 + \sqrt{3}\,i)^4$.

34. Use DeMoivre's theorem to express $\cos 2\theta$ in terms of $\cos \theta$.

35. Use DeMoivre's theorem to express $\sin 4\theta$ in terms of $\sin \theta$ and/or $\cos \theta$.

36. Use the relationship $z^{-n} = (z^{-1})^n$ to show that

$$z^{-n} = \frac{1}{r^n}(\cos n\theta - i \sin n\theta)$$

if $z = r(\cos \theta + i \sin \theta)$

9.4 Polar Coordinates

There are many coordinate systems used in the applications of mathematics. Prior to this chapter we located points in the plane by use of rectangular cartesian coordinates. Now we use the trigonometric form of the complex number to form a basis for another commonly used coordinate system called the **polar coordinate system.**

Polar coordinate system

We begin with a point O, called the **pole,** and a horizontal ray (or half-line) extending indefinitely to the right, called the **polar axis.**

Pole

Polar axis

To determine the polar coordinates for a point P in the plane, rotate the polar axis through an angle θ until the point P lies on the ray and measure the distance r from O to P. The pair (r, θ) form **polar coordinates** for P (see Figure 9.2).

Polar coordinates

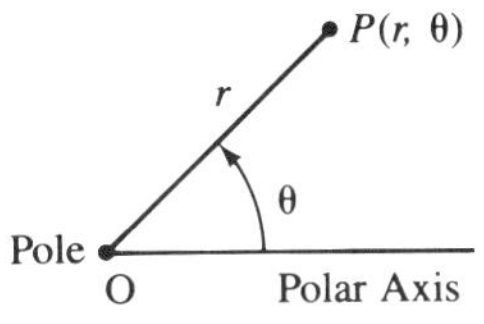

Figure 9.2

As is customary, positive angles are measured counterclockwise and negative angles are measured clockwise. Usually the angle θ is measured in radians, although degree measure is permitted if clearly indicated. Thus the angle θ in the point represented by (2, 1) in polar coordinates is 1 radian. The coordinates (2, 1°) and (2, $\pi/180$) represent the same point, which is not the same as the point represented by (2, 1).

The polar coordinates (r, θ) for P are not unique, because infinitely many angles have the terminal side OP. Figure 9.3 illustrates three different polar coordinates for the same point.

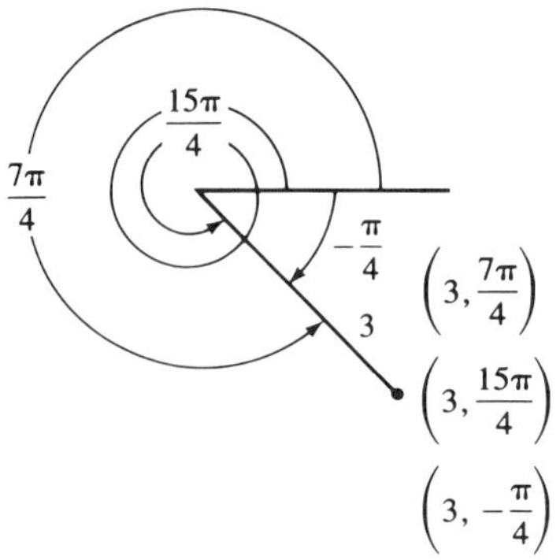

Figure 9.3

It is convenient to be able to use negative values for r as well as θ in the pair (r, θ). To do this, for negative r we extend the terminal side of θ through the pole, as indicated by the dashed line in Figure 9.4, and mark off $|r|$ on the extension.

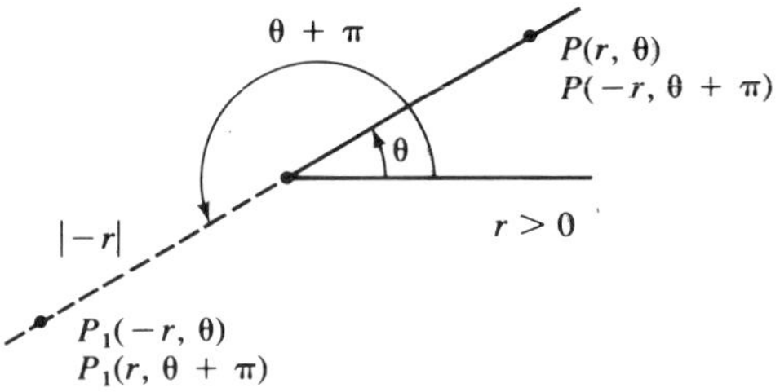

Figure 9.4

It is clear from Figures 9.3 and 9.4 that if (r, θ) are polar coordinates for P, then $(r, \theta + 2n\pi)$ and $(-r, \theta + [2n + 1]\pi)$ are also polar coordinates for P for each integer n.

Polar coordinates for the pole are $(0, \theta)$, where θ is any angle.

Example 1 Plot the point with polar coordinates $\left(1, \dfrac{\pi}{4}\right)$. Give polar coordinates for the same point such that **(a)** $0 < \theta < 2\pi$ and $r < 0$; **(b)** $-2\pi < \theta < 0$ and $r > 0$.

The solutions are shown in the illustration.

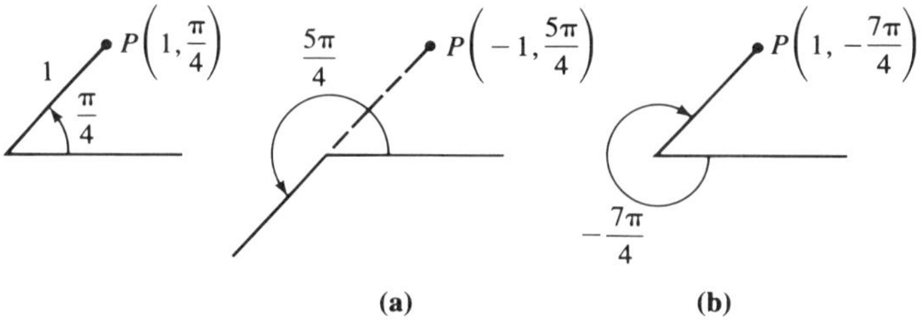

■

Example 2 Plot the point with polar coordinates $\left(-2, \frac{\pi}{3}\right)$. Give polar coordinates for the same point such that **(a)** $0 < \theta < 2\pi$ and $r > 0$; **(b)** $-2\pi < \theta < 0$ and $r < 0$. The solutions are shown in the illustration.

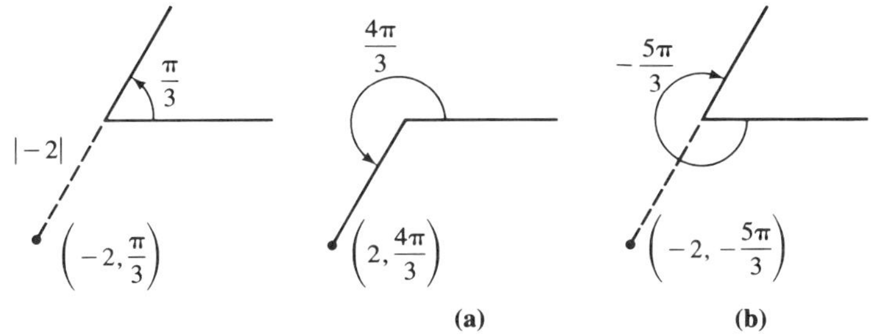

■

To establish the relationships between rectangular and polar coordinates, we superimpose a rectangular coordinate system on a polar coordinate system in such a way that the origin is also the pole and the positive x axis coincides with the polar axis, as in Figure 9.5.

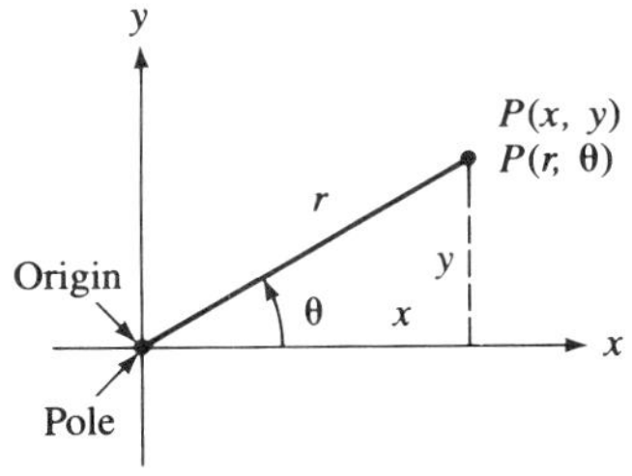

Figure 9.5

For $r > 0$, the following relations are immediate. It can be shown that they are also valid for $r < 0$.

$$x = r\cos\theta \qquad y = r\sin\theta$$

$$r^2 = x^2 + y^2 \qquad \tan\theta = \frac{y}{x} \quad \text{if } x \neq 0$$

We emphasize that given polar coordinates (r, θ), the rectangular coordinates (x, y) are uniquely determined. However, if the rectangular coordinates are given, there are infinitely many polar coordinates for the point.

Example 3 Find the rectangular coordinates for the following points given in polar coordinates: **(a)** $\left(3, \frac{\pi}{4}\right)$; **(b)** $(-2, \pi)$; **(c)** $\left(\frac{1}{2}, -\frac{\pi}{3}\right)$; **(d)** $(4, -2\pi)$.

a. $x = 3\cos\dfrac{\pi}{4} = \dfrac{3}{\sqrt{2}} \qquad y = 3\sin\dfrac{\pi}{4} = \dfrac{3}{\sqrt{2}}$

b. $x = -2\cos\pi = 2 \qquad y = -2\sin\pi = 0$

c. $x = \dfrac{1}{2}\cos\left(-\dfrac{\pi}{3}\right) = \dfrac{1}{4} \qquad y = \dfrac{1}{2}\sin\left(-\dfrac{\pi}{3}\right) = -\dfrac{\sqrt{3}}{4}$

d. $x = 4\cos(-2\pi) = 4 \qquad y = 4\sin(-2\pi) = 0$

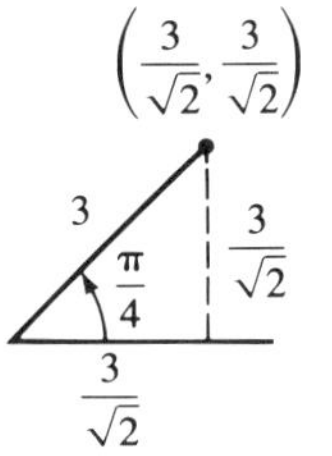

(a)

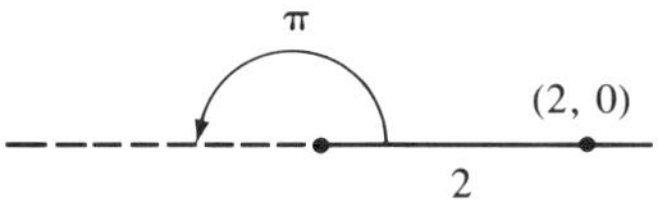

(b)

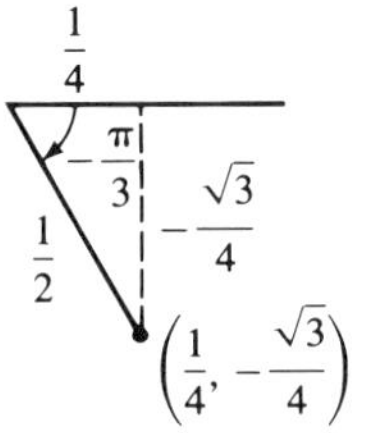

(c)

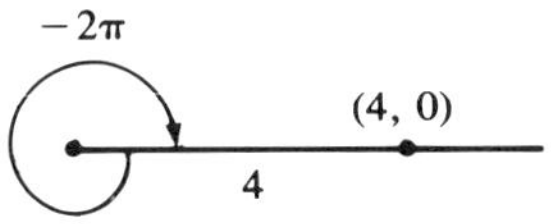

(d) ■

Example 4 Find polar coordinates for the following points given in rectangular coordinates: **(a)** (0, 4); **(b)** (−3, −3); **(c)** $(2\sqrt{3}, -2)$.

It is usually a good idea to make a sketch. It is especially helpful in this situation because there are many choices for the polar coordinates.

a. When $x = 0$, either $\pi/2$ or $-\pi/2$ will serve for θ. In this case we choose $\theta = \pi/2$ and $r = 4$.

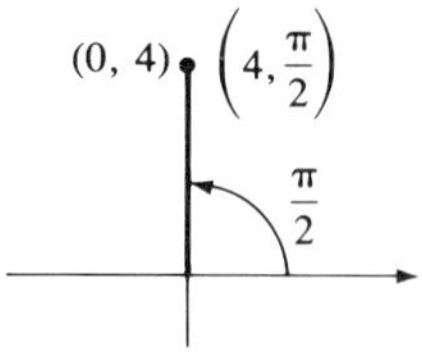

b. Here $\tan\theta = -3/-3 = 1$, and since the point is in the third quadrant, we choose $\theta = 5\pi/4$ and $r = \sqrt{18}$.

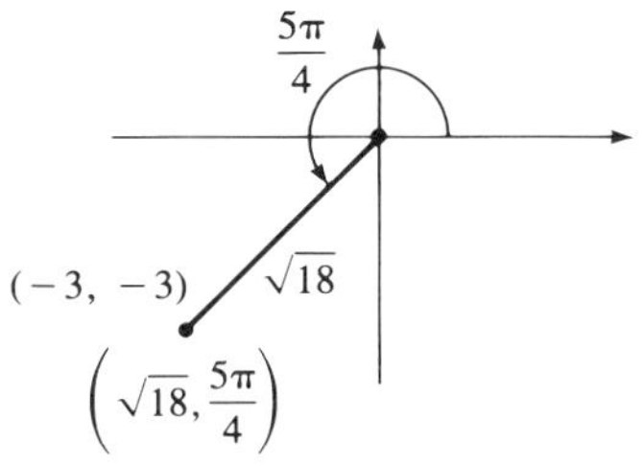

c. $\tan\theta = -2/2\sqrt{3} = -1/\sqrt{3}$, and since the point is in the fourth quadrant, choose $\theta = -\pi/6$. Then

$$r^2 = (2\sqrt{3})^2 + 2^2 = 16$$
$$r = 4$$

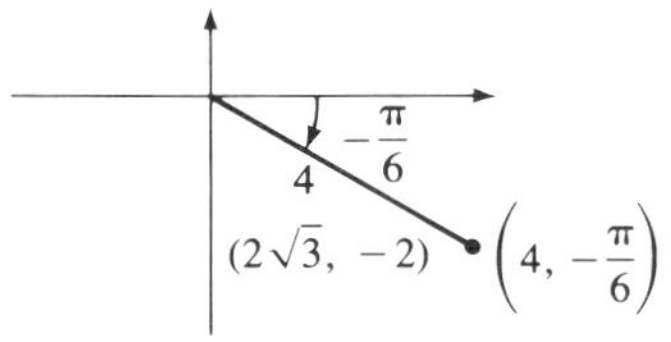

■

Polar equations

Using the relationships displayed earlier, it is easy to convert rectangular equations in x and y to **polar equations,** that is, equations in r and θ.

Example 5 Convert $x^2 + y^2 - 2x = 9$ to polar form.

We can always replace x by $r\cos\theta$ and y by $r\sin\theta$ and simplify. In this case, however, the relation $r^2 = x^2 + y^2$ can be used to save some work. The polar form is immediate:

$$r^2 - 2r\cos\theta = 9$$ ■

Converting a polar equation to rectangular form can require more ingenuity. It gets easier with practice.

Example 6 Convert the equation $r = 3\sin\theta$ to rectangular form.

It is helpful to study and perhaps modify the equation before converting to rectangular form. In this case, we multiply both sides by r to obtain

$$r^2 = 3r\sin\theta$$

Thus $x^2 + y^2 = 3y$ is the rectangular form. ■

Example 7 Convert the equation $r^3 = \sec\theta$ to rectangular form.

We note that $r^2r = 1/\cos\theta$, so that $r^2r\cos\theta = 1$. The rectangular form is then $(x^2 + y^2)x = 1$. ■

Exercises 9.4

In Exercises 1–12 plot the points with the given polar coordinates and give two other sets of polar coordinates for each point.

1. $\left(2, -\frac{\pi}{2}\right)$ **2.** $\left(3, -\frac{\pi}{2}\right)$ **3.** $\left(-1, \frac{\pi}{3}\right)$ **4.** $\left(-4, \frac{\pi}{3}\right)$

5. $(2, \pi)$ **6.** $(7, \pi)$ **7.** $\left(1, \frac{7\pi}{6}\right)$ **8.** $\left(2, -\frac{7\pi}{6}\right)$

9. $\left(-2, -\frac{3\pi}{4}\right)$ **10.** $\left(-6, \frac{3\pi}{4}\right)$ **11.** $\left(1, \frac{5\pi}{4}\right)$ **12.** $\left(-1, -\frac{5\pi}{4}\right)$

In Exercises 13–20 find rectangular coordinates for the points given in polar coordinates.

13. $\left(3, \frac{\pi}{6}\right)$ **14.** $\left(4, \frac{\pi}{4}\right)$ **15.** $\left(-4, \frac{\pi}{4}\right)$ **16.** $\left(-2, \frac{\pi}{3}\right)$

17. $(2, -\pi)$ **18.** $(-4, -2\pi)$ **19.** $\left(-1, \frac{3\pi}{2}\right)$ **20.** $\left(4, -\frac{3\pi}{2}\right)$

In Exercises 21–28 find polar coordinates for the points given in rectangular coordinates.

21. $(0, -2)$ **22.** $(1, 0)$ **23.** $(-2, 0)$ **24.** $(0, 3)$

25. $(-2, 2)$ **26.** $(3, -3)$ **27.** $(\sqrt{3}, -1)$ **28.** $(-1, -\sqrt{3})$

In Exercises 29–36 convert the equation to polar form.

29. $x^2 + y^2 + 2x - 2y = 0$ **30.** $y^2 = 4x$

31. $x^2 - y^2 = 1$ **32.** $2x - 3y = 1$

33. $y = 5x$ **34.** $x^2 + y^2 - 4x = 0$

35. $(x + 1)^2 + (y + 2)^2 = 1$ **36.** $x^2 + y - 4x = 0$

In Exercises 37–51 convert the equations to rectangular form.

37. $r \sin \theta = 1$ **38.** $r = 2 \sec \theta$

39. $r = 5$ **40.** $r = 1 - \cos \theta$

41. $\cot \theta = -3$ **42.** $r = 4 \sin \theta$

43. $r = 3 + 2 \cos \theta$ **44.** $r(1 - \cos \theta) = 3$

45. $r^2 = 2r \cos \theta + r \sin \theta$ **46.** $r \cos \left(\theta - \frac{\pi}{4}\right) = 1$

47. $r = 2 \cos \left(\theta - \frac{\pi}{3}\right)$ **48.** $r = \sin 2\theta$

49. $r \sin 2\theta = 0$ **50.** $r^2 = \cos 2\theta$

51. $r^2 = \sin 2\theta$

9.5 Graphs in Polar Coordinates

Polar graphs

Just as the graph of an equation in x and y represents the set of all points (x, y) in the cartesian plane whose coordinates are solutions of the equation, the graph of an equation in r and θ represents the set of all points (r, θ) in the $r\theta$ plane that satisfy the equation. Graphs of polar equations are called **polar graphs.**

It is possible of course, using the techniques of Section 9.4, to sketch polar graphs by converting the polar equation to rectangular form and sketching the rectangular form. However, the conversion and simplification can be time consuming, and the resultant equation may itself be difficult to sketch. The purpose of this section is to discuss the basic ideas involved in sketching polar graphs. These techniques frequently permit rapid sketching of polar equations.

Example 1 Sketch the polar graph of $r = 1$.

Since r represents the directed distance from the pole to the point along the ray, the set of all points that have $r = 1$ is the circle of radius 1 with center at the pole.

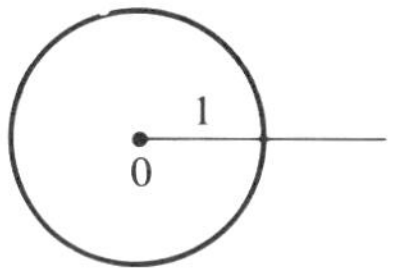

■

Example 2 Sketch the polar graph of $\theta = \dfrac{\pi}{3}$.

The graph consists of all points lying along the line that makes an angle of $\pi/3$ with the polar axis.

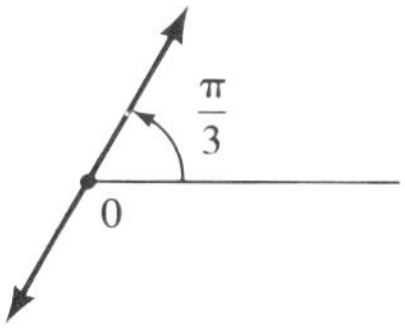

■

Example 3 Sketch the polar graph of $r = 2 \cos \theta$.

We make a table of values.

θ	0	$\frac{\pi}{6}$	$\frac{\pi}{3}$	$\frac{\pi}{2}$	$\frac{2\pi}{3}$	$\frac{5\pi}{6}$	π
r	2	$\sqrt{3}$	1	0	-1	$-\sqrt{3}$	-2

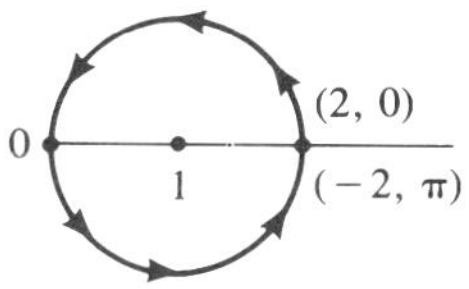

The arrowheads on the curve indicate the path of a point moving along the curve as θ increases from 0 to π. Notice also that as θ increases from π to 2π the point traces out the curve again. ■

The graph in Example 3 appears to be a circle. To verify that this is the case, multiply both sides of the equation by r to obtain the rectangular form:

$$r^2 = 2r\cos\theta$$
$$x^2 + y^2 = 2x$$

By completing the square in x this equation can be written

$$(x - 1)^2 + y^2 = 1$$

which is the equation of a circle with center at (1, 0) and radius 1.

Example 4 Sketch the polar graph of $r = 2 - 4\sin\theta$.

We make a table of values, rounded off to one decimal place.

θ	0	$\frac{\pi}{6}$	$\frac{\pi}{3}$	$\frac{\pi}{2}$	$\frac{2\pi}{3}$	$\frac{5\pi}{6}$	π	$\frac{7\pi}{6}$	$\frac{4\pi}{3}$	$\frac{3\pi}{2}$	$\frac{5\pi}{3}$	$\frac{11\pi}{6}$	2π
r	2	0	−1.5	−2	−1.5	0	2	4	5.5	6	5.5	4	2

Limacon The graph is shown in the illustration. The arrowhead shows the path traced by a point as θ increases from 0 to 2π. This curve is called a **limacon.**

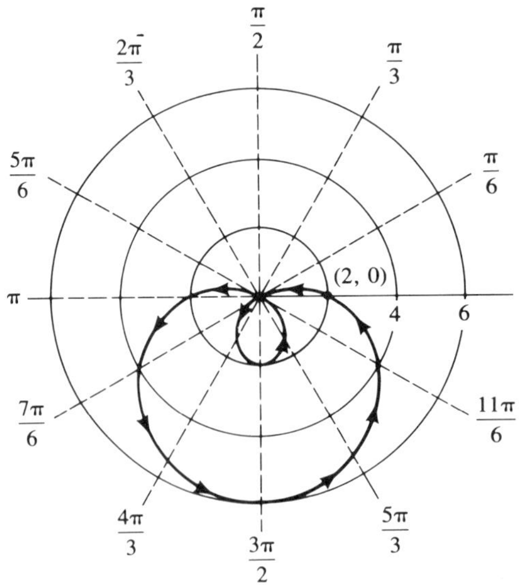

■

Frequently it is only necessary to know the general shape of a polar graph, for which a few accurately placed points may be sufficient. The graph points of the sine and cosine functions developed in Section 6.6 are useful for this purpose.

Example 5 Sketch the graph of the equation $r = 2 + 2\cos\theta$.

The values for $\cos\theta$ are immediate for $\theta = 0, \pi/2, \pi, 3\pi/2, 2\pi$. We analyze the behavior of the equation as shown in the table.

θ	r
$0 \to \frac{\pi}{2}$	$4 \to 2$
$\frac{\pi}{2} \to \pi$	$2 \to 0$
$\pi \to \frac{3\pi}{2}$	$0 \to 2$
$\frac{3\pi}{2} \to 2\pi$	$2 \to 4$

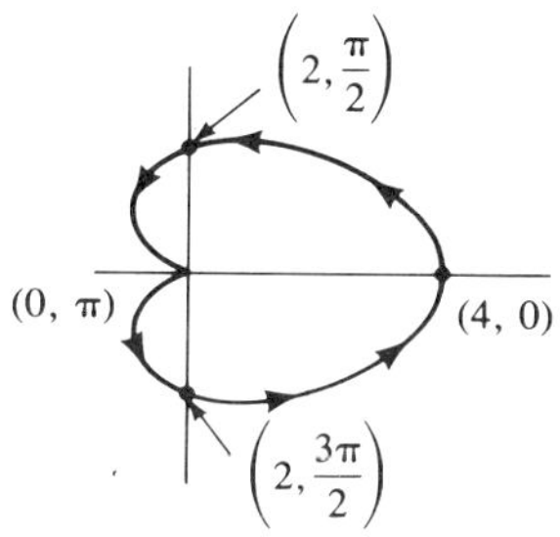

Recall that the first row of the display reads, "as θ goes from 0 to $\pi/2$, r goes from 4 to 2." This information gives the sketch shown. This curve is called a

Cardioid **cardioid.** ■

Example 6 Sketch the graph of the equation $r = 2\sin 2\theta$.

The values of $\sin\alpha$ are immediate if α is a multiple of $\pi/2$. Thus for the graph of $r = 2\sin 2\theta$ we want to plot the points (r, θ) for which 2θ is a multiple of $\pi/2$. To do this construct the display as follows, using the fourth column to note the corresponding portion of the curve in the figure.

θ	2θ	r	
$0 \to \frac{\pi}{4}$	$0 \to \frac{\pi}{2}$	$0 \to 2$	a
$\frac{\pi}{4} \to \frac{\pi}{2}$	$\frac{\pi}{2} \to \pi$	$2 \to 0$	b
$\frac{\pi}{2} \to \frac{3\pi}{4}$	$\pi \to \frac{3\pi}{2}$	$0 \to -2$	c
$\frac{3\pi}{4} \to \pi$	$\frac{3\pi}{2} \to 2\pi$	$-2 \to 0$	d
$\pi \to \frac{5\pi}{4}$	$2\pi \to \frac{5\pi}{2}$	$0 \to 2$	e
$\frac{5\pi}{4} \to \frac{3\pi}{2}$	$\frac{5\pi}{2} \to 3\pi$	$2 \to 0$	f
$\frac{3\pi}{2} \to \frac{7\pi}{4}$	$3\pi \to \frac{7\pi}{2}$	$0 \to -2$	g
$\frac{7\pi}{4} \to 2\pi$	$\frac{7\pi}{2} \to 4\pi$	$-2 \to 0$	h

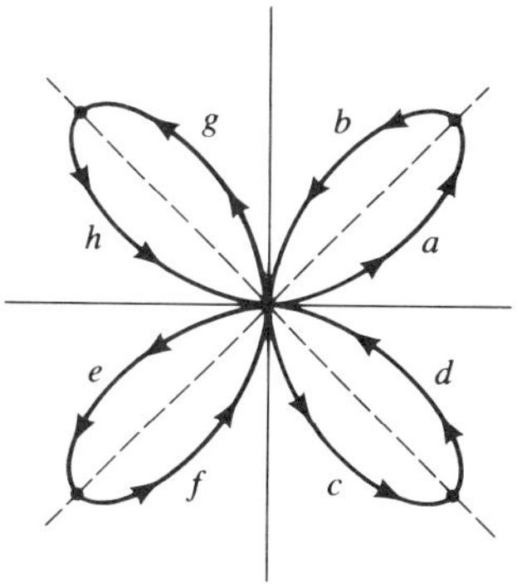

Notice that the last four rows of the display repeat the first four rows in the r column. This is because 2θ is simply making another complete revolution. The pattern is such that the sketch could have been completed without explicitly displaying these last four rows. ■

Four-leafed rose
Rose curves

The curve in Example 6 is called a **four-leafed rose.** In general, the **rose curves** are the graphs of

$$r = a \sin n\theta \quad \text{and} \quad r = a \cos n\theta$$

for $n \varepsilon Z$ and $a \varepsilon R$. The number of leaves is n if n is odd and $2n$ if n is even.

Example 7 Sketch the graph of $r^2 = 16 \cos 2\theta$.

Notice that if $\cos 2\theta$ is negative there are no values for r, while if $\cos 2\theta$ is positive there are two values for r.

θ	2θ	r	
$0 \to \frac{\pi}{4}$	$0 \to \frac{\pi}{2}$	$4 \to 0$	a
		$-4 \to 0$	b
$\frac{\pi}{4} \to \frac{\pi}{2}$	$\frac{\pi}{2} \to \pi$	No	
$\frac{\pi}{2} \to \frac{3\pi}{4}$	$\pi \to \frac{3\pi}{2}$	No	
$\frac{3\pi}{4} \to \pi$	$\frac{3\pi}{2} \to 2\pi$	$0 \to 4$	c
		$0 \to -4$	d

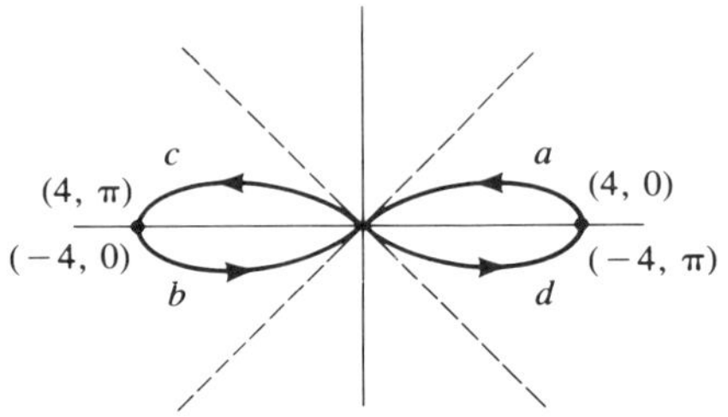

Lemniscate

As θ increases from π to 2π, the curve is retraced. This propeller-shaped curve is called a **lemniscate.** ■

Polar coordinate graph paper is commercially available. This paper is similar to that shown in the sketch of Example 4, but includes rays at other angles. Such paper makes it easier to sketch polar graphs.

Exercises 9.5

In Exercises 1–30 sketch the polar graph of the equations.

1. $r = \frac{1}{2}$

2. $r = -\frac{3}{4}$

3. $\theta = -\frac{\pi}{6}$

4. $\theta = -\frac{3\pi}{4}$

5. $\sin \theta = -\frac{1}{2}$

6. $\cos \theta = \frac{\sqrt{2}}{2}$

7. $r \cos \theta = -3$

8. $r \sin \theta = \pi$

9. $r = -3 \cos \theta$

10. $r = 2 \sin \theta$

11. $r = 6 \sin \theta$

12. $r = -2 \cos \theta$

13. $r = \cos\left(\theta + \frac{\pi}{4}\right)$

14. $r = \sin\left(\theta + \frac{\pi}{3}\right)$

15. $r = \cos 2\theta$ (four-leafed rose)

16. $r = 4 \sin 2\theta$ (four-leafed rose)

17. $r = \sin 5\theta$ (five-leafed rose)

18. $r = 3 \sin 3\theta$ (three-leafed rose)

19. $r = 3 \cos 3\theta$ (three-leafed rose)

20. $r = 2 \sin 5\theta$ (five-leafed rose)

21. $r = 1 - \sin \theta$ (cardioid)

22. $r = 2 + 2 \sin \theta$ (cardioid)

23. $r = 1 - 2 \cos \theta$ (limacon)

24. $r = 2 - \cos \theta$ (limacon)

25. $r = 3 + 2 \cos \theta$ (limacon)

26. $r = 1 + 2 \sin \theta$ (limacon)

27. $r^2 = 25 \cos 2\theta$ (lemniscate)

28. $r^2 = 4 \sin 2\theta$ (lemniscate)

29. $r = \theta, \theta \geq 0$ (spiral)

30. $r = 2^\theta$ (spiral)

31. Find the center and radius of the circle $r = a \sin \theta$ for $a \in R$, $a \neq 0$.

32. Find the center and radius of the circle $r = a \cos \theta$ for $a \in R$, $a \neq 0$.

33. Write in polar form the equation of the parabola with focus at the pole and directrix $r \cos \theta = -2$ (the line $x = -2$).

Chapter Review

In Exercises 1–4 locate the complex number z in the complex plane, find the modulus of z, and specify one argument for z.

1. $z = -3$

2. $z = 5i$

3. $z = -1 + \sqrt{3}\, i$

4. $z = -\sqrt{2} - \sqrt{2}\, i$

In Exercises 5–6 write the complex number in standard form, given the specified modulus and argument.

5. $8, -\dfrac{5\pi}{4}$

6. $1, \dfrac{7\pi}{6}$

In Exercises 7–10 write the complex number in trigonometric form.

7. $-7i$

8. $6\sqrt{3} - 6i$

9. $\pi - \pi i$

10. $-5 + 2i$

In Exercises 11–14 compute uv and u ÷ v using trigonometric form and write the results in standard form.

11. $u = -1 + i;\ v = 1 - i$

12. $u = -5i;\ v = 4$

13. $u = 1 - \sqrt{3}\, i;\ v = \sqrt{3} + i$

14. $u = -8i;\ v = 4 - 4i$

In Exercises 15–16 perform the computations and write the results in trigonometric form.

15. $(5 - 2i)(-2 + 7i)$

16. $\dfrac{1 - 3i}{3 - 4i}$

In Exercises 17–18 perform the computations in trigonometric form and write the results in standard form.

17. $[4(\cos 75° + i \sin 75°)][3(\cos 195° + i \sin 195°)]$

18. $10\left(\cos \frac{\pi}{12} + i \sin \frac{\pi}{12}\right) \div 5\left(\cos \frac{5\pi}{12} + i \sin \frac{5\pi}{12}\right)$

In Exercises 19–22 compute the powers and express the results in standard form.

19. $(\sqrt{3} + i)^6$

20. $\left(-\frac{1}{\sqrt{2}} + \frac{1}{\sqrt{2}}i\right)^{25}$

21. $\left(\frac{1}{2} - \frac{\sqrt{3}}{2}i\right)^{30}$

22. $(-3 + 3\sqrt{3}\,i)^4$

In Exercises 23–24 compute the n distinct nth *roots, express the results in standard form, and sketch.*

23. Cube roots of -64

24. Fourth roots of unity

In Exercises 25–26 compute the n distinct nth *roots, express the results in trigonometric form, and sketch.*

25. Fifth roots of $-\sqrt{3} + i$

26. Cube roots of $-1 - i$

27. Find the three solutions to the equation $x^3 - i = 0$.

28. Find the five solutions to the equation $x^5 - 1{,}024 = 0$.

In Exercises 29–32 plot the points with the given polar coordinates, give two other sets of polar coordinates for each point, and find the rectangular coordinates.

29. $(-2, -\pi)$

30. $\left(-5, \frac{3\pi}{2}\right)$

31. $\left(2, \frac{4\pi}{3}\right)$

32. $\left(4, \frac{7\pi}{4}\right)$

In Exercises 33–34 find polar coordinates for the points given in rectangular coordinates.

33. $(-4, 4)$

34. $(3\sqrt{3}, -3)$

In Exercises 35–36 convert to polar form.

35. $x^2 = 4y + 1$

36. $x^2 + y^2 - 4y = 0$

In Exercises 37–40 convert to rectangular form.

37. $r = -4 \csc \theta$

38. $r = 2 \cos \theta$

39. $r^3 = 3 \csc \theta$

40. $r = 3 - \sin \theta$

In Exercises 41–48 sketch the polar graph of the equations.

41. $r = 3 \sin \theta$ (circle)

42. $\tan \theta = -1$ (line)

43. $r = 3 \sin 2\theta$ (four-leafed rose)

44. $r = 4 \cos 3\theta$ (three-leafed rose)

45. $r = 1 - \cos \theta$ (cardioid)

46. $r^2 = 9 \cos 2\theta$ (lemniscate)

47. $r = 2 - 3 \cos \theta$ (limacon)

48. $r = \theta^{-1}, \theta > 0$ (spiral)

49. Find the center and radius of the circle $r = a \sin \theta + b \cos \theta$ for $a, b \in R, a \neq 0, b \neq 0$.

Chapter Test

1. Let $z = \sqrt{3} - i$.
   a. Plot z in the complex plane.
   b. Find the modulus of z.
   c. Find an argument of z.
   d. Write z in trigonometric form.
   e. Compute z^3 using trigonometric form.
   f. Write z^3 in standard form.
2. Use trigonometric form to compute $(\sqrt{3} - i) \div (1 - \sqrt{3}\, i)$. Write the result in standard form.
3. Compute $\left(\frac{\sqrt{2}}{2} + \frac{\sqrt{2}}{2} i\right)^{20}$. Write the result in standard form.
4. Find the three cube roots of $-i$.
5. a. Write $(-2, 2)$ in polar coordinates (r, θ) with $r < 0$.
   b. Give the rectangular coordinates for the point $\left(-1, \frac{4\pi}{3}\right)$.
6. Convert the equation $2x^2 + 2y^2 + 3x = 0$ to polar form.
7. Convert the equation $2r = \sin\theta$ to rectangular form.
8. Give a polar equation for the line passing through the pole and the point $\left(1, \frac{5\pi}{6}\right)$.
9. Sketch the polar graph of the equation $r = 2\sin 3\theta$.

10 Systems of Equations, Inequalities, and Matrices

The activities of grazing animals can be classified in three categories: grazing, moving to new grazing areas, and resting. The energy gain from grazing is 200 calories per hour (cal/hr) and losses from moving and resting are 150 cal/hr and 50 cal/hr, respectively. How many hours should be spent in each activity so that the energy gain exactly compensates for the energy losses?
Let

x = number of hours grazing
y = number of hours moving
z = number of hours resting

To meet the requirement that energy gain equal energy loss we write the equation

$$200x = 150y + 50z$$

and since there are but 24 hours in a day,

$$x + y + z = 24$$

To solve this problem we must seek solutions to both equations or solutions to the *system* of equations

$$\begin{aligned} 200x &= 150y + 50z \\ x + y + z &= 24 \end{aligned}$$

In this chapter we examine methods for solving systems of equations and inequalities.

10.1 Systems of Equations in Two Unknowns

A **solution** for an equation in two unknowns, such as $3x + 2y = 6$, is a pair (a, b) such that the substitution of a for x and b for y makes the equation a true statement. A solution to a system of equations in two unknowns is a pair (a, b) that is a solution to *each* of the equations in the system. One method for finding the solution to a system is the method of **substitution.**

Method of substitution

Example 1 Solve the system

$$2x + 3y = 5$$
$$x + y = 2$$

by substitution.
Solve the second equation for y and substitute that expression in the first equation for y:

$$y = 2 - x$$
$$2x + 3(2 - x) = 5$$
$$x = 1$$

From $y = 2 - x$ we obtain $y = 1$. ■

In Example 1 we found that (1, 1) is a solution to the system. Is it the only solution? A graphic interpretation of the situation answers the question. Each of the equations $x + y = 2$ and $2x + 3y = 5$ is the equation of a line, and the solution we seek is the point of intersection of the two lines (see Figure 10.1). Since the lines intersect in exactly one point, the solution $x = 1$, $y = 1$ is the only solution.

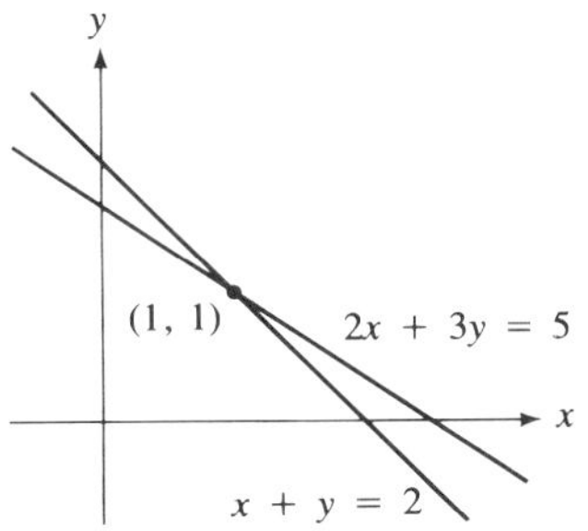

Figure 10.1

Given a system of two linear equations in two unknowns, there are exactly three possibilities for the solution set:

1. The lines are not parallel; they intersect in exactly one point and the system has exactly one solution.
2. The lines coincide; that is, the equations are equivalent and any point on the line is a solution.
3. The lines are parallel and distinct; the lines do not intersect and there are no solutions to the system. The system is said to be **inconsistent.**

Example 2 Solve the system

$$x - y = 1$$
$$2y - 2x = -1$$

Solve the first equation for x and substitute in the second equation:

$$\begin{aligned} x &= y + 1 \\ 2y - 2(y + 1) &= -1 \\ -2 &= -1 \end{aligned}$$

This last equation is always false, and the system therefore has no solutions. ■

Method of elimination Another method of solution is called the method of **elimination.** The basic idea here is to combine the equations in a way that eliminates one of the unknowns.

Example 3 Solve the system

$$\begin{aligned} 4x - 2y &= 9 \\ x + 2y &= 6 \end{aligned}$$

by the method of elimination.

We add the second equation to the first to eliminate y, obtaining

$$\begin{aligned} 5x &= 15 \\ x &= 3 \end{aligned}$$

Substituting $x = 3$ into the first equation, we obtain $y = \frac{3}{2}$.

It's not a bad idea to check your solutions:

$$4(3) - 2\left(\frac{3}{2}\right) = 12 - 3 = 9 \qquad \text{OK}$$

$$3 + 2\left(\frac{3}{2}\right) = 3 + 3 = 6 \qquad \text{OK}$$

■

Example 4 Solve the system

$$\begin{aligned} 5x - 3y &= -5 \\ 3x + 2y &= 16 \end{aligned}$$

Multiply the first equation by 2 and the second equation by 3 and add:

$$\begin{aligned} 10x - 6y &= -10 \\ 9x + 6y &= 48 \\ \hline 19x \qquad &= 38 \\ x &= 2 \end{aligned}$$

Substituting $x = 2$ in the first equation yields $y = 5$. ■

Example 5 Solve the system

$$\begin{aligned} 4x + 10y &= 6 \\ 6x + 15y &= 9 \end{aligned}$$

Multiply the first equation by 3 and the second equation by -2 and add:

$$\begin{array}{r} 12x + 30y = 18 \\ -12x - 30y = -18 \\ \hline 0 = 0 \end{array}$$

The lines are coincident; that is, the equations are equivalent. ■

Nonlinear systems may sometimes be solved by the same methods as those used for linear systems.

Example 6 Solve the system

$$x^2 + y^2 = 9$$
$$x + y = 3$$

The first equation is the equation of a circle and the second the equation of a line. We know that the circle and line have either no points of intersection, exactly one point of intersection (if the line is tangent to the circle), or two points of intersection. From the graph it appears that there are two points of intersection; that is, two solutions to the system.

Solve the second equation for y and substitute in the first equation:

$$\begin{aligned} y &= 3 - x \\ x^2 + (3 - x)^2 &= 9 \\ x^2 + 9 - 6x + x^2 &= 9 \\ 2x^2 - 6x &= 0 \\ x(x - 3) &= 0 \end{aligned}$$

The possible values for x are 0 and 3. Substituting in the second equation we obtain:

$$x = 0 \qquad y = 3 - 0 = 3$$
$$x = 3 \qquad y = 3 - 3 = 0$$

The solutions are (0, 3) and (3, 0).

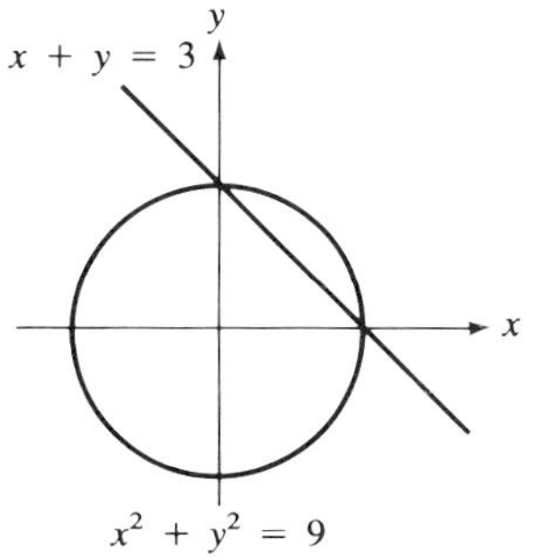

■

Example 7 Solve the system

$$y = x^2 + 2x + 1$$
$$y - x = 3$$

Looking at the graph, we expect to find two solutions.

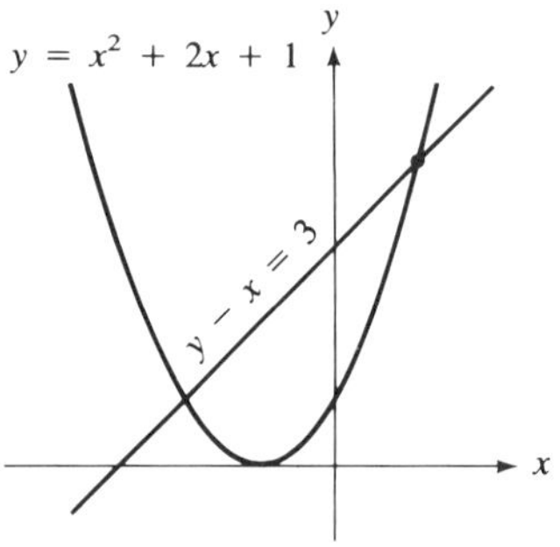

Solve the second equation for y and substitute in the first equation:

$$\begin{aligned} y &= x + 3 \\ x + 3 &= x^2 + 2x + 1 \\ 0 &= x^2 + x - 2 \\ 0 &= (x + 2)(x - 1) \end{aligned}$$

The possible values for x are -2 and 1. Substituting into the second equation, we obtain

$$x = -2 \qquad y = 3 + (-2) = 1$$
$$x = 1 \qquad y = 3 + 1 = 4$$

The solutions are $(-2, 1)$ and $(1, 4)$. It should be checked that these are solutions to *both* equations. ■

Example 8 Solve the system

$$\begin{aligned} 2x^2 - 3xy + y^2 &= 8 \\ x^2 - xy \qquad &= -4 \end{aligned}$$

Multiply the second equation by 2 and add:

$$4x^2 - 5xy + y^2 = 0$$

Factor the left member:

$$(4x - y)(x - y) = 0$$

For (x, y) to be a solution we must have either

$$4x - y = 0 \quad \text{or} \quad x - y = 0$$

If $4x - y = 0$, then $y = 4x$ and, substituting in the second equation,

$$x^2 - x(4x) = -4$$
$$x^2 = \frac{4}{3}$$
$$x = \pm\frac{2\sqrt{3}}{3}$$
$$y = \pm\frac{8\sqrt{3}}{3}$$

If $x - y = 0$, then $y = x$ and the resulting equation is

$$x^2 - x(x) = -4$$
$$0 = -4$$

which has no solutions. The only solutions are $(2\sqrt{3}/3, 8\sqrt{3}/3)$ and $(-2\sqrt{3}/3, -8\sqrt{3}/3)$. ■

WARNING Check your solutions in both equations.

A word of advice is in order for nonlinear systems: it is usually worthwhile to check your solutions in both equations before accepting them.

Example 9 The perimeter of a rectangle is 30 in. and the length is $\frac{3}{2}$ of the width. What are the dimensions?

Let x = width of the rectangle and y = length of the rectangle. Then

$$2x + 2y = 30 \quad (\textit{The perimeter})$$
$$y = \frac{3}{2}x$$

Substitute $\frac{3}{2}x$ for y in the first equation:

$$2x + 2\left(\frac{3}{2}x\right) = 30$$
$$5x = 30$$
$$x = 6$$
$$y = 9$$

The rectangle is 9 in. long and 6 in. wide. ■

Example 10 It is desired to add 120 lb of nitrogen and 90 lb of phosphoric acid to a field using two types of fertilizer with the contents shown in the table. How much of each should be used?

Fertilizer	Nitrogen	Phosphoric Acid
Fast Green	20%	10%
All Crop	6%	6%

Let x = number of pounds of Fast Green used and y = number of pounds of All Crop used.
Then

$$0.2x + 0.06y = 120 \quad (\textit{Nitrogen needed})$$
$$0.1x + 0.06y = 90 \quad (\textit{Phosphoric acid needed})$$

Multiply the second equation by -1 and add to the first equation:

$$0.1x = 30$$
$$x = 300$$

From the first equation, with x replaced by 300,

$$(0.2)(300) + 0.06y = 120$$
$$0.06y = 60$$
$$y = 1{,}000$$

The fertilizer needs can be met with 300 lb of Fast Green and 1,000 lb of All Crop. ■

Example 11 The perimeter of a 4-acre rectangular lot is 112 rods. If 1 acre = 160 square rods, what are the dimensions of the lot?
Let x = width of the lot and y = length of the lot. Then

$$xy = 640 \quad (\textit{Area} = \textit{4 acres} = \textit{640 square rods})$$
$$2x + 2y = 112 \quad (\textit{Perimeter} = \textit{112 rods})$$

Solve the second equation for y and substitute in the first equation:

$$(56 - x)x = 640$$
$$x^2 - 56x + 640 = 0$$
$$(x - 40)(x - 16) = 0$$

The possible solutions are $x = 40$ and $x = 16$. Notice that $x = 40$ yields $y = 16$ and $x = 16$ yields $y = 40$. Consequently, the lot is 40 rods long and 16 rods wide. ■

Example 12 A publisher calculates that it costs \$6,000 to prepare a book for publication and \$0.75 to print each copy. If the book sells for \$2.25, how many copies must be sold before the company breaks even?

The cost to publish x copies is given by the function

$$C(x) = 6{,}000 + 0.75x$$

The revenue from selling x copies is given by the function

$$R(x) = 2.25x$$

If we graph both functions, the point where the two graphs intersect is called the *break-even point,* that is, the point where revenue equals cost of production (see the graph.) If we solve the system

$$y = 6{,}000 + 0.75x$$
$$y = 2.25x$$

we find that the coordinates of the break-even point are $x = 4{,}000$, $y = 9{,}000$. If 4,000 copies are printed and sold, the company will spend 9,000 and receive 9,000; that is, it will break even. Notice that for $x > 4{,}000$, the graph of the revenue function is above the graph of the cost function and the company makes a profit. Likewise, if $x < 4{,}000$, the cost exceeds the revenue and the company loses money.

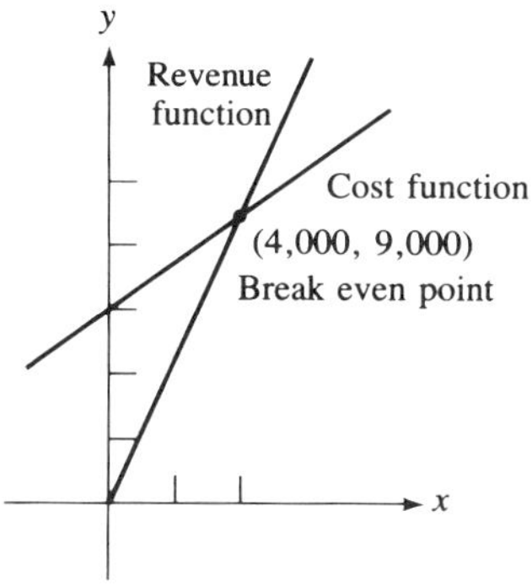

■

Exercises 10.1

Use either substitution or elimination to find the solution sets for those systems that are not inconsistent.

1. $2x - 2y = 0$, $x - y = 4$

2. $4x + 2y = -8$, $2x + y = 4$

3. $2x - 3y = 5$, $3x + y = 2$

4. $2x - 3y = 5$, $5x - 4y = 9$

5. $4x + 3y = 5$, $3x - 2y = 8$

6. $3x - 2y = 8$, $5x + 7y = 3$

7. $3x - 2y = 6$, $6x - 4y = 12$

8. $x + 2y = 4$, $3x + 6y = 12$

9. $3x + 4y = 5$, $-2x + 3y = 8$

10. $3x + 4y = 5$, $7x + 5y = 3$

11. $4x + 3y = 10$, $x = 0.5y$

12. $3x + 2y = 8$, $0.5x = y$

13. $0.2x = 0.3y$, $0.4x - 0.2y = 0.2$

14. $0.4x = 0.3y$, $0.8x + 0.2y = 0.4$

15. $\frac{1}{3}x + \frac{1}{4}y = -\frac{5}{12}$, $\frac{1}{10}x - \frac{1}{5}y = \frac{1}{2}$

16. $5x = 2y + 1$, $10x - 4y = 2$

17. $3x + 6y = 9$, $x = 3 - 2y$

18. $\frac{1}{4}x - \frac{1}{3}y = \frac{5}{12}$, $\frac{1}{10}x + \frac{1}{5}y = -\frac{1}{2}$

19. $\frac{x}{3} - \frac{y}{2} = 6$, $\frac{2x}{3} - \frac{y}{6} = 1$

20. $\frac{x}{4} + \frac{y}{5} = 2$, $\frac{3x}{2} + \frac{y}{3} = -2$

21. $x^2 - xy + y^2 = 7$, $x - y = 1$

22. $x^2 - 2xy + y^2 = 0$
$x - 2y = 1$

23. $x - y = 5$
$xy = 6$

24. $x + y = 7$
$xy = 12$

25. $x^2 + y^2 = 9$
$x^2 = 9 - 4y$

26. $16x^2 + 9y^2 = 144$
$x^2 = 8 - 2y$

27. $xy = 4$
$x^2 + 4y^2 = 20$

28. $3x^2 + y^2 = 13$
$xy = 4$

29. $5x^2 - 14xy + 8y^2 = 0$
$2xy = 5$

30. $3x^2 - 11xy + 6y^2 = 0$
$2xy = 9$

31. $4x^2 - 5xy + 3y^2 = 24$
$2x^2 - 3xy + 2y^2 = 16$

32. $2x^2 - 3xy + 2y^2 = 16$
$2x^2 - 2xy + y^2 = 8$

33. Julia wishes to plant 75 acres in onions and lettuce. If it costs \$550 per acre to raise lettuce and \$300 per acre to raise onions and Julia plans to invest \$35,000 in the crop, how many acres of each should she plant?

34. The sum of the digits of a two-digit number is 12 and the difference is 4. What is the number? Is there more than one answer?

35. If 3 pounds of salt and 5 pounds of sugar cost 94¢ and 6 pounds of salt and 2 pounds of sugar cost 76¢, what is the cost per pound for sugar?

36. Ed went to the store to buy 3 dozen oranges and 4 dozen apples for \$3.20 but discovered he didn't have enough money. He did, however, buy 4 dozen oranges and 3 dozen apples for \$3.10. What is the price of a dozen oranges?

37. Jack and Bob together had 30 marbles. Jack gave half his marbles to Vena, and Bob then had 4 times as many marbles as did Jack. They discovered later that Bob had lost his marbles. How many marbles did Vena, Jack, and Bob then have?

38. Patricia wants to apply 124 pounds of nitrogen and 44 pounds of phosphorus per acre to her farm. Type I fertilizer has 30 pounds of nitrogen and 6 pounds of phosphorus per 100 pounds, and type II fertilizer has 16 pounds of nitrogen and 8 pounds of phosphorus per 100 pounds. How much of each type of fertilizer should she use?

39. How many liters of nonfat dry milk that is 21% protein should be mixed with whole milk that is 9% protein to produce 2 liters of milk that is 12% protein?

40. The sum of the digits of a two-digit number is 12. If the digits are interchanged, the number is increased by 54. What is the number?

41. How much water should be mixed with a 12% acid solution to produce 16 liters of 9% acid solution?

42. The area of a rectangle is 300 square feet and the length of the diagonal is 25 feet. What are the dimensions of the rectangle?

43. Carl invests a sum of money at a certain rate of interest for a year and earns \$75 interest. Fred invests \$250 more than Carl at 1% less interest and also earns \$75 interest. How much did Fred invest and at what rate?

44. Susan has a rectangular flower garden 504 square feet in area surrounded by a path 3 feet wide. The area of the path is 312 square feet. Find the length and the width of Susan's flower garden.

45. The cost of manufacturing a game is \$3,000 plus 75¢ for each game produced. If the game sells for \$2.75, how many games must be sold to break even?

46. A hall rents for \$150 plus \$5 for each person attending the event. How many tickets must be sold at \$7.50 each to break even?

47. A club spends \$40 on advertising and plans to sell candy at \$2.50 per box. If the candy costs \$1.70 per box, how many boxes must the club sell to break even?

10.2 Systems of Inequalities

A department store has 21 strings of Christmas lights and 30 packages of ornaments left over after Christmas. The company has two plans for packaging and selling leftover lights and ornaments.

Plan I: 2 strings of lights and 5 packages of ornaments for $4.99
Plan II: 3 strings of lights and 3 packages of ornaments for $5.99

How should the merchandise be packaged to maximize the income?
Let x = number of packages under plan I and y = number of packages under plan II. Then x and y are subject to the following system of inequalities:

$$\begin{aligned} x &\geq 0 \\ y &\geq 0 \\ 2x + 3y &\leq 21 \\ 5x + 3y &\leq 30 \end{aligned}$$

Our first task is to find the solution set for the system of inequalities. You may want to review Section 4.8 quickly before reading further.

Example 1 Solve the system

$$\begin{aligned} x &\geq 0 \\ y &\geq 0 \\ 2x + 3y &\leq 21 \\ 5x + 3y &\leq 30 \end{aligned}$$

The solution set for each of the inequalities is shown in Figure 10.2.

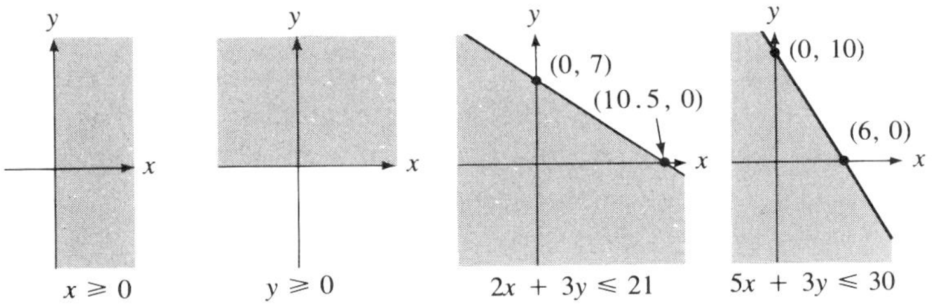

Figure 10.2

The solution set for the system is the intersection of the four sets shown (see Figure 10.3).

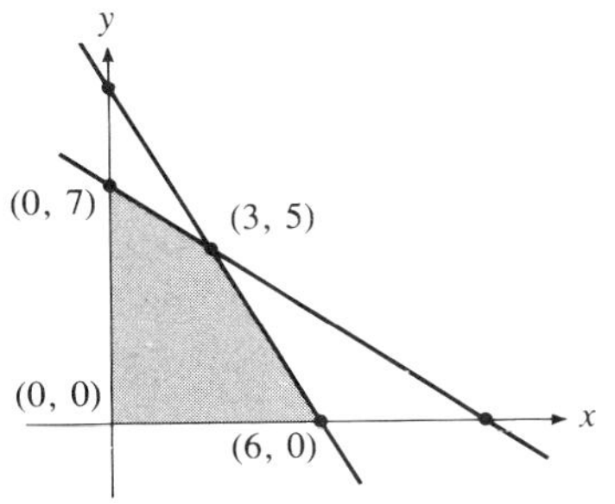

Figure 10.3

■

The graph in Figure 10.3 is the solution set for the system of inequalities in Example 1. To find the packaging scheme that will maximize the income, one could simply check the income for each plan corresponding to each point in the solution set with integer coordinates and find the one that gives the largest income. However, there is a better way, which is discussed in the next section.

Example 2 Graph the solution set of the system

$$x \geq 0$$
$$y \geq 0$$
$$x - y \leq 1$$

The solution set of each of these inequalities is shown in Figure 10.4. The solution set is the intersection of these three sets, as shown in Figure 10.5.

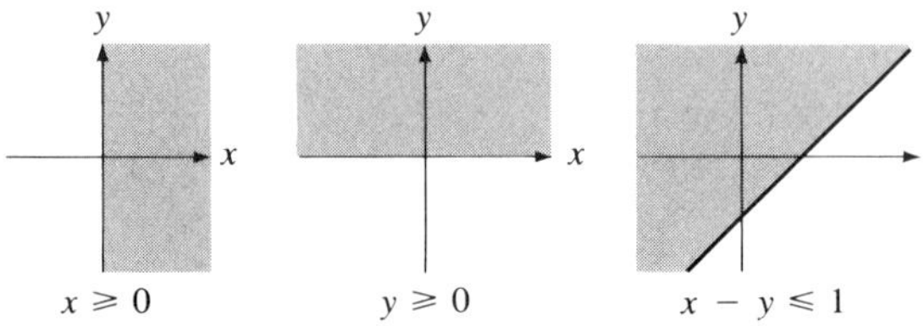

Figure 10.4

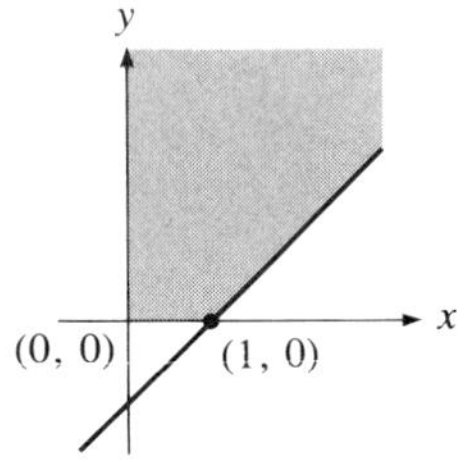

Figure 10.5 ■

A polygonal region in the plane is said to be **convex** if the line segment connecting any two points in the region is entirely within the region. In Figure 10.6, regions A, B, and C are convex, whereas region D is not convex.

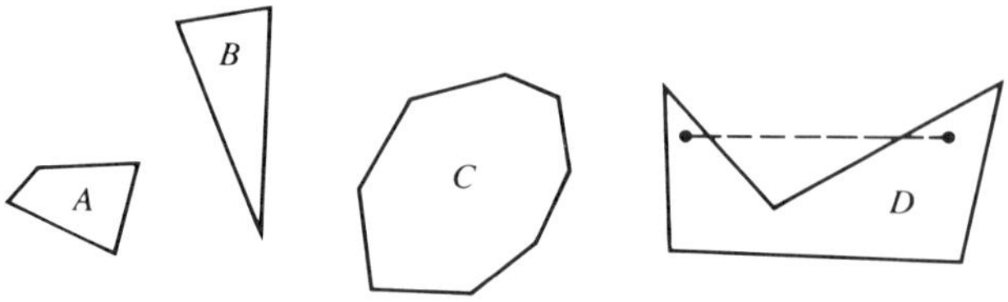

Figure 10.6

The solution set for the systems of inequalities in Examples 1 and 2 turned out to be convex polygonal regions, which is as it should be. Although we will not prove it here, it is true that the solution set of a system of linear inequalities is either the empty set or a convex polygonal region in the plane. Conversely, every convex polygonal region is the solution set of a system of linear inequalities. In the example below, we show how to find the system of inequalities, given the convex polygonal set that is the solution set of the system.

Example 3 Find the system of linear inequalities that has the set shown in the illustration as solution set.

First we need to find the equations of the lines that form the edges of the region. Segment AB has slope -2, so from the point-slope formula, the line through A and B has the equation

$$2x + y = 5 \qquad AB$$

In similar fashion, the other three lines have the equations

$$6y - x = 17 \qquad BC$$
$$2x - y = 10 \qquad CD$$
$$4y - x = 2 \qquad AD$$

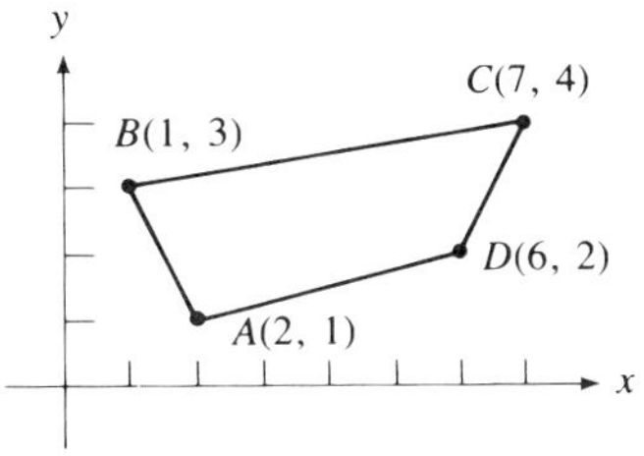

The system of inequalities is then

$$2x + y \; ? \; 5$$
$$6y - x \; ? \; 17$$
$$2x - y \; ? \; 10$$
$$4y - x \; ? \; 2$$

where the ? in each case must be replaced by either $\leq$ or $\geq$. This is most easily done by choosing a point inside the region and substituting its coordinates into each inequality. In this case, (3, 2) is in the region. So, for example, the first inequality

$$2x + y \; ? \; 5$$

must have (3, 2) as a solution. In other words, after substituting 3 for x and 2 for y,

$$2(3) + 2 \; ? \; 5 \quad \text{or} \quad 8 \; ? \; 5$$

The proper inequality is

$$2x + y \geq 5$$

In the same manner, we discover that the system of inequalities desired is

$$2x + y \geq 5$$
$$6y - x \leq 17$$
$$2x - y \leq 10$$
$$4y - x \geq 2$$

■

Systems of inequalities, some of which are nonlinear, are graphed in a manner similar to that used in graphing linear inequalities.

Example 4 Graph the solution set of the system

$$\begin{aligned} x^2 + y^2 &\le 1 \\ y^2 &\le x \end{aligned}$$

The graphs of the two inequalities are shown in Figure 10.7 and the graph of the solution set is shown in Figure 10.8.

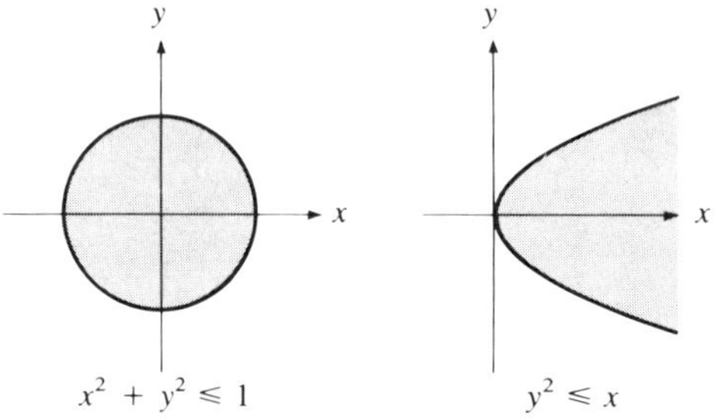

Figure 10.7

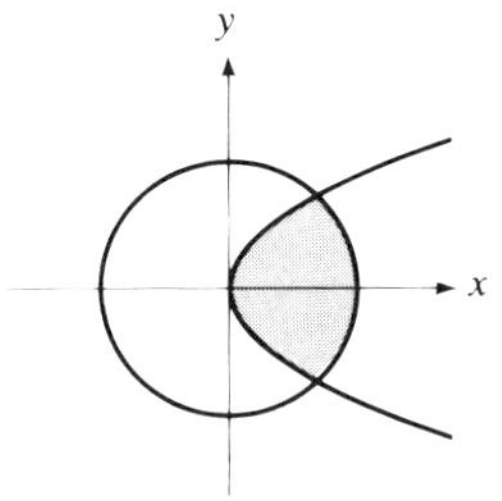

Figure 10.8 ■

Exercises 10.2

Graph the solution set of each of the systems of inequalities.

1. $\begin{aligned} 2y + x &\le 1 \\ y - x &\le 2 \\ y &\ge 0 \end{aligned}$

2. $\begin{aligned} 2x + y &\ge 1 \\ x - y &\le 2 \\ y &\le 1 \end{aligned}$

3. $\begin{aligned} 2y - x &\le 2 \\ y + x &\le 4 \\ x &\ge 0 \\ y &\ge 0 \end{aligned}$

4. $\begin{aligned} 2x - y &\le 2 \\ x + y &\ge 4 \\ x &\ge 0 \\ y &\le 6 \end{aligned}$

5. $\begin{aligned} 2y - 3x &\ge 6 \\ x &\ge -2 \\ y &\le 3 \end{aligned}$

6. $\begin{aligned} 2y - 3x &\le 6 \\ x &\le 2 \\ y &\ge 3 \end{aligned}$

7. $\begin{aligned} x + y &\le 4 \\ x + y &\ge -1 \\ y &\ge -1 \\ y &\le 4 \end{aligned}$

8. $\begin{aligned} x - y &\le 4 \\ x - y &\ge -1 \\ x &\le 4 \\ x &\ge -1 \end{aligned}$

9. $\begin{aligned} 2x + 4y &\ge 6 \\ 4x + 2y &\ge 6 \\ y &\le 3 \\ x &\le 3 \end{aligned}$

10. $\begin{aligned} 2x - 4y &\ge 6 \\ 4x - 2y &\ge 6 \\ x &\le 3 \\ y &\ge -3 \end{aligned}$

11. $\begin{aligned} x^2 + y^2 &\le 36 \\ y^2 - x^2 &\le 1 \end{aligned}$

12. $\begin{aligned} x^2 + y^2 &\le 25 \\ y^2 - x^2 &\ge 1 \end{aligned}$

13. $\begin{aligned} y &\le x + 2 \\ y &\ge x^2 \end{aligned}$

14. $\begin{aligned} y &\le x - 1 \\ y^2 &\ge x \end{aligned}$

15. $\begin{aligned} 9 \le x^2 + y^2 &\le 36 \\ x^2 &\ge y \end{aligned}$

16. $\begin{aligned} 4 \le x^2 + y^2 &\le 16 \\ y^2 &\ge x \end{aligned}$

In Exercises 17–24 each of the convex polygonal sets is the solution set for a system of inequalities. Find the systems.

17.

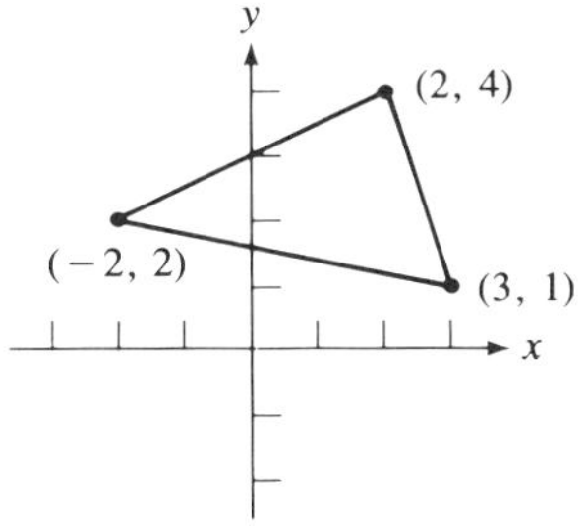

18.

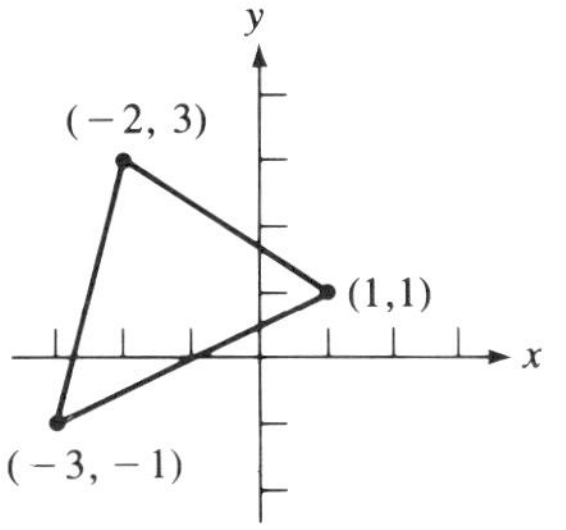

19.

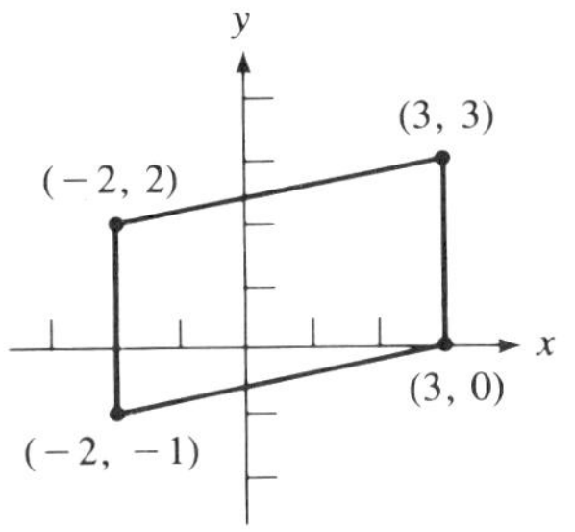

20.

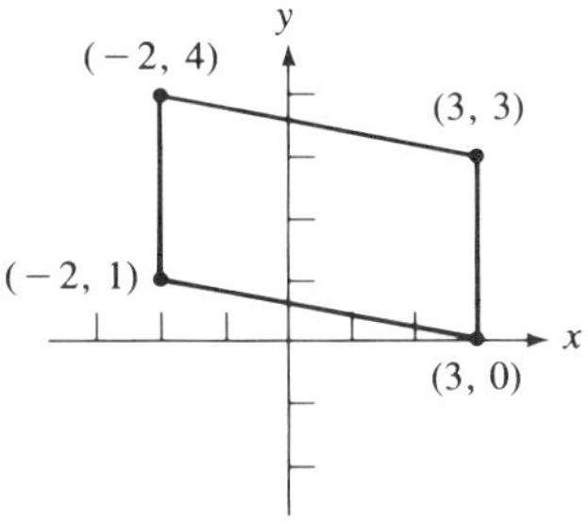

21.

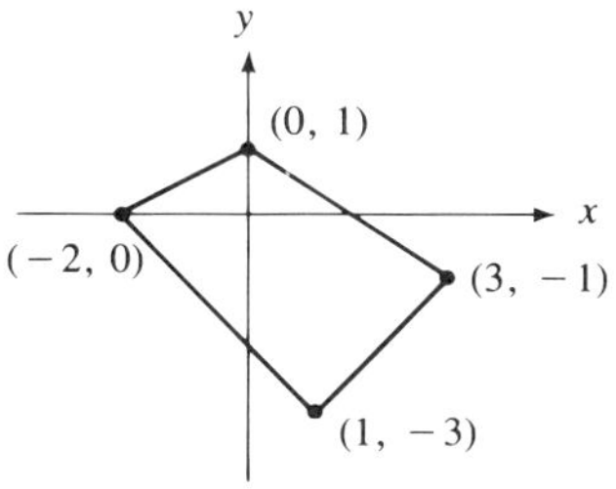

22.

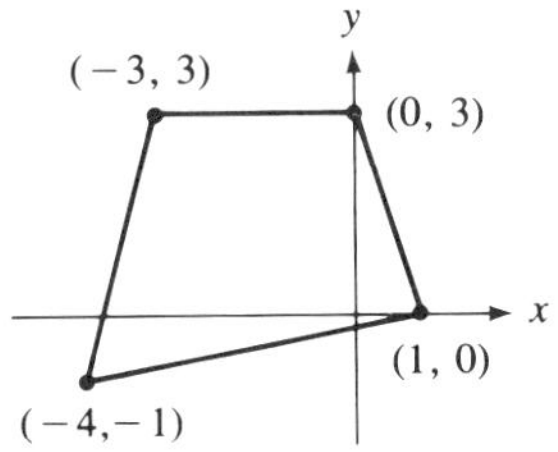

23.

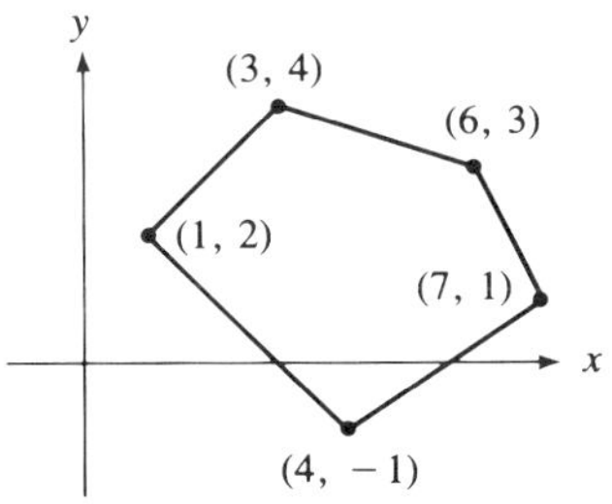

24.

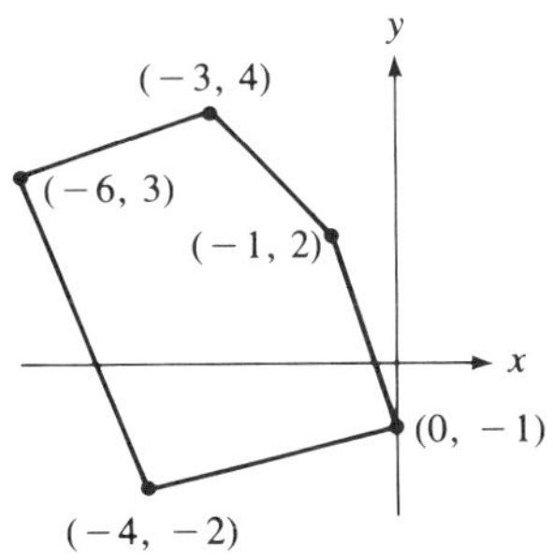

25. A farmer has 100 acres to plant in onions and chile. The cost of raising a crop of onions is \$500 per acre and the cost of raising a crop of chile is \$300 per acre. If he can invest at most \$40,000, how many acres of each crop can he afford?

26. The owner of a motorcycle shop stocks two types of motorcycle helmets. Type I costs \$50 each and type II costs \$65 each. The owner can display at most 50 helmets and has at most \$3,055 to invest. How many helmets of each type can be ordered?

27. A farmer has 10 days in which to plant cotton and milo. He can plant milo at the rate of 12 acres per day and cotton at the rate of 8 acres per day. He has 100 acres to plant. How many acres of each crop can he plant?

28. A queen-size mattress costs \$60 and takes 20 cubic feet of storage space; a king-size mattress costs \$75 and takes 24 cubic feet of storage. If 980 cubic feet of storage space is available and \$3,000 is available to buy mattresses, how many can be purchased and stored?

10.3 Linear Programming

Let us restate the problem posed at the beginning of Section 10.2. A store has 21 strings of lights and 30 packages of ornaments that may be packaged and sold in two different ways:

Plan I: 2 strings of lights and 5 packages of ornaments for \$4.99

Plan II: 3 strings of lights and 3 packages of ornaments for \$5.99

How should the merchandise be packaged to maximize the income?

Let x = number of packages under plan I and y = number of packages under plan II. Then x and y are subject to the following system of inequalities:

$$\begin{aligned} x &\geq 0 \\ y &\geq 0 \\ 2x + 3y &\leq 21 \\ 5x + 3y &\leq 30 \end{aligned}$$

The solution set of this system is the convex polygonal set shown in Figure 10.9.

This region is called the **feasible** region, because only feasible solutions to the problem must be points in this set. We wish to find the point or points in the feasible region at which the income function $I = 4.99x + 5.99y$ has the largest value. To answer this question, ask a slightly different one. Can we package the merchandise to obtain an income of \$50? This means, are there points (x, y) that give the income function a value of \$50? Such points satisfy the equation

$$4.99x + 5.99y = 50$$

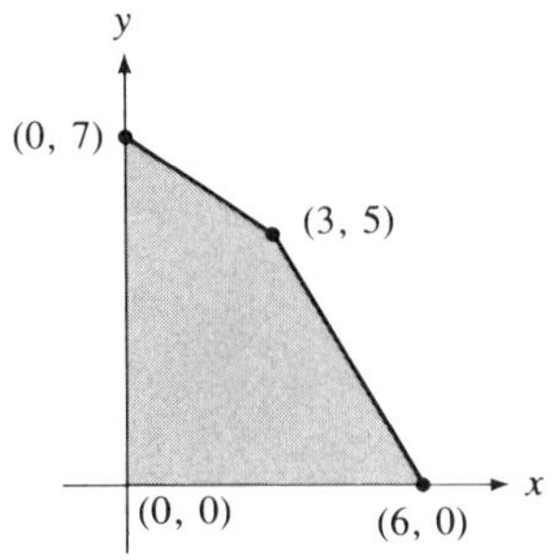

Figure 10.9

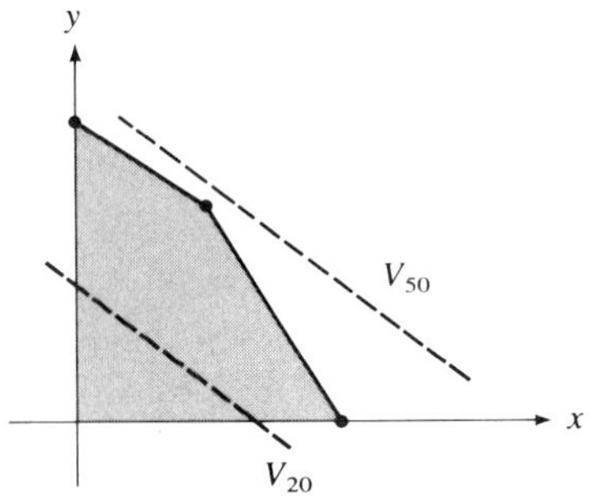

Figure 10.10

which is the equation of a line, call is V_{50}. Can we package the merchandise to obtain an income of \$20? Are there points that give the income function a value of \$20? Such points satisfy the equation

$$4.99x + 5.99y = 20$$

(See Figure 10.10.)

The lines V_{50} and V_{20} are what are called lines of **constant value** for the income function because the function $I = 4.99x + 5.99y$ has the same value at all points on those lines. The line V_{50} doesn't intersect the feasible region, but some line parallel to V_{50} will intersect the feasible region. As the lines of constant value are moved to the left, the value of the income function decreases, so we expect the largest value of I to occur when the line of constant value first touches the feasible region, in this case at (3, 5). For a convex polygonal region this will always occur at a corner, or, as it is often called, an **extreme** point; alternatively, the lines of constant value may be parallel to an edge and have the maximum value at every point of that edge.

The maximum income is realized by making 3 packages under plan I and 5 under plan II, for an income of

$$I = 4.99(3) + 5.99(5) = \$44.92$$

One might think that this is the obvious solution, since it is the only way to use up all the merchandise. However, if the packages under plan II sold for \$9.99, then the maximum income would occur when only plan II packages were used.

What is being used here is a method of decision-making called **linear programming,** developed in the 1940s by a group of mathematicians. What we will discuss here is the graphical method, which only applies when there are only two unknowns. If there are more than two variables, there are other ways of solving the problem and computer programs designed to do the computations.

Bounded set

A set of points in the plane is said to be **bounded** if for some number r, all points in the set are inside the circle $x^2 + y^2 = r^2$.

Linear programming

Linear programming is based on the following principle: If one wishes to maximize (or minimize) a linear function subject to a system of linear inequalities, and the solution set is bounded, then the maximum (or minimum) value of the function will occur at one of the corners (extreme points) or along one edge of the feasible region. If the set is unbounded, the linear function may not have either a maximum or a minimum value, but if it does have one or both, they will occur at one of the corners or along an edge of the feasible region.

Another statement of this principle is that on a convex polygonal region a linear function always has its maximum and minimum values (if they exist) at a corner or along one edge. This is illustrated in the next example.

Example 1 Find the largest and smallest value of the function $I = 3x + 4y$ on the region shown in the illustration.

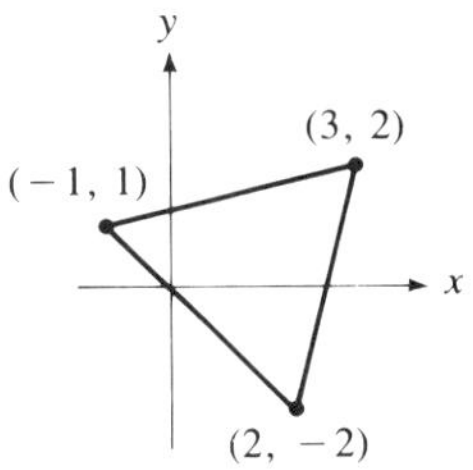

The extreme points are $(-1, 1)$, $(2, -2)$ and $(3, 2)$. Check the value of I at the extreme points:

$(-1, 1)$ $I = 3(-1) + 4(1) = 1$
$(2, -2)$ $I = 3(2) + 4(-2) = -2$ *(Minimum)*
$(3, 2)$ $I = 3(3) + 4(2) = 17$ *(Maximum)* ■

Example 2 The Unicycle Shop makes and sells two types of unicycles, a Whizzer and a Buzzer. The shop makes \$150 on each Whizzer it makes and sells and \$200 on each Buzzer it makes and sells. The labor requirements for each are shown in the table.

	Assembly	Painting	Testing
Whizzer	150 hr	50 hr	10 hr
Buzzer	60 hr	40 hr	20 hr

There are 30,000 hr available for assembly, 13,000 hr available for painting, and 5,000 hr available for testing. How many of each model will give the Unicycle Shop the maximum profit?

Let x = number of Whizzers to be made and sold and y = number of Buzzers to be made and sold. Then

$$x \geq 0$$
$$y \geq 0$$
$$150x + 60y \leq 30{,}000 \quad \textit{(Assembly time available)}$$
$$50x + 40y \leq 13{,}000 \quad \textit{(Painting time available)}$$
$$10x + 20y \leq 5{,}000 \quad \textit{(Testing time available)}$$

The graph is shown in the illustration.

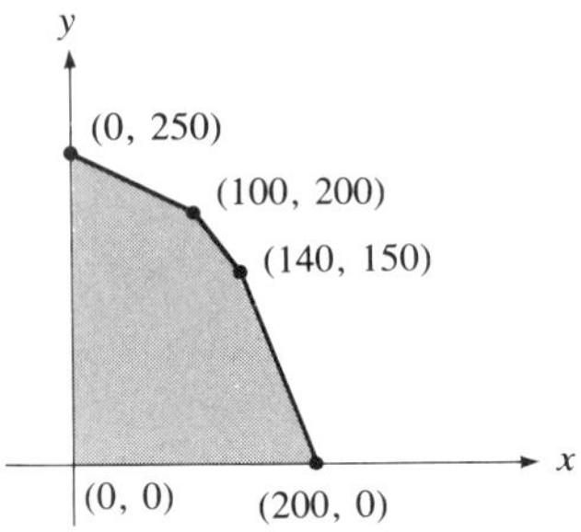

The extreme points are (0, 0), (0, 250), (100, 200), (140, 150), and (200, 0). The profit function is $P = 150x + 200y$. Checking at the extreme points:

	(0, 0)	(0, 250)	(100, 200)	(140, 150)	(200, 0)
Profit	0	50,000	55,000	51,000	30,000

A maximum profit of \$55,000 is obtained by making 100 Whizzers and 200 Buzzers. ■

Example 3 A diet calls for 15 units of Vitamin C and 12 units of Vitamin E. If two vitamin supplements are available with contents as shown in the table, and if type I costs 3¢ per gram and type II costs 4¢ per gram, how much of each should be used to minimize the cost?

Supplement	Vitamin C	Vitamin E
Type I	3 units/g	2 units/g
Type II	3 units/g	4 units/g

Let x = number of grams of type I to be used and y = number of grams of type II to be used. Then

$$x \geq 0$$

$$y \geq 0$$

$$3x + 3y \geq 15 \quad (\textit{Vitamin C needs})$$

$$2x + 4y \geq 12 \quad (\textit{Vitamin E needs})$$

The cost function is

$$C = 0.03x + 0.04y.$$

The solution set is graphed in the illustration.

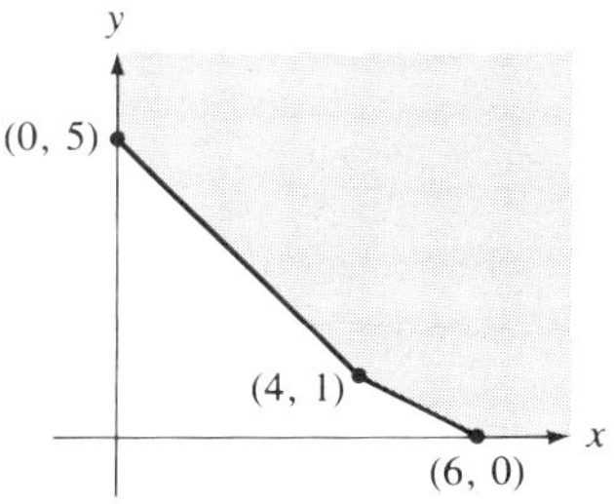

In this case the feasible region is not bounded, but the cost function does have a minimum value and it must be at one of the corners.

	(0, 5)	(4, 1)	(6, 0)
Cost	0.20	0.16	0.18

The cost is minimized by using 4 grams of type I and 1 gram of type II. ■

Exercises 10.3

Find the maximum and minimum values of the function on the solution set of the system of inequalities. The number in parentheses refers to a similar problem in the exercise set in Section 10.2.

1. $C = 3x + 3y$ (1)
$2y + x \leq 1$
$y - x \leq 2$
$y \geq 0$

2. $C = 3x + 3y$ (2)
$2x + y \geq 1$
$x - y \leq 2$
$y \leq 1$

3. $C = 2x - 5y$ (3)
$2y - x \leq 2$
$y + x \leq 4$
$x \geq 0$
$y \geq 0$

4. $C = 5x - 2y$ (4)
$2x - y \leq 2$
$x + y \geq 4$
$x \geq 0$
$y \leq 6$

5. $R = 2y + 3x$ (5)
$2y - 3x \geq 6$
$x \geq -2$
$y \leq 3$

6. $R = 3x + 2y$ (6)
$2y - 3x \leq 6$
$x \leq 2$
$y \geq 3$

7. $F = 2x - 3y - 5$ (7)
$x + y \leq 4$
$x + y \geq -1$
$y \geq -1$
$y \leq 4$

8. $F = 3x - 4y + 7$ (8)
$x - y \leq 4$
$x - y \geq -1$
$x \leq 4$
$x \geq -1$

9. $P = 5 - 4x - 3y$ (9)
$2x + 4y \geq 6$
$4x + 2y \geq 6$
$y \leq 3$
$x \leq 3$

10. $P = 12 + x - 2y$ (10)
$2x - 4y \geq 6$
$4x - 2y \geq 6$
$x \leq 3$
$y \geq -3$

11. A farmer has 100 acres to plant in onions and chile. The cost of raising a crop of onions is \$500/acre and the cost of raising a crop of chile is \$300/acre. If the farmer can invest at most \$40,000, how many acres of each crop should he plant if he expects a return (after expenses) of \$800 per acre for onions and \$450 per acre for chile?

12. The owner of a motorcycle shop stocks two types of motorcycle helmets. Type I costs \$50 each and type II costs \$65 each. The owner can display 50 helmets and has at most \$3,055 to invest. If the profit on type I helmets is \$8.00 and the profit on type II helmets is \$16.40, how many helmets of each type should he buy to maximize the profit?

13. A diet calls for 8 units of protein and 12 units of carbohydrates. Food A contains 1 unit of protein and 1 unit of carbohydrates per pound and food B contains 2 units of protein and 6 units of carbohydrates per pound. If food A costs 90¢/pound and food B costs \$1.20/pound, how many pounds of each should be used to meet the requirements and minimize the cost?

14. An oil company has contracts to supply at least 30,000 barrels of low-grade heating oil and 20,000 barrels of high-grade heating oil. Refinery A can produce 3,000 barrels of low-grade oil and 5,000 barrels of high-grade oil per day. Refinery B can produce 6,000 barrels of low-grade oil and 2,500 barrels of high-grade oil per day. If it costs \$15,000 per day to operate refinery A and \$20,000 per day to operate refinery B, how many days should each operate to meet the contracted requirements and minimize the cost?

15. Carla has two duplicating machines. Machine I can make 3 copies per minute and copies cost 5¢ each. Machine II makes 5 copies per minute and copies cost 6¢ each. If Carla must make 60 copies in 10 minutes, how can she utilize both machines to minimize the cost?

16. Corina's Cocina makes two types of burritos. The super burrito contains 3 oz of beef and 1 oz of pork. The poquito burrito contains 2 oz of pork and 1 oz of beef. Corina has 4 lbs of beef and 3 lbs of pork to use in making the burritos. Her profit on a super burrito is 40¢ and her profit on a poquito burrito is 15¢. How many of each type of burrito should she make?

17. A publisher produces a book in both hardbound and paperback editions. The profit on a paperback is \$3 and the profit on a hardbound is \$4. The hardbound takes 5 hours to print and 3 hours to bind. The paperback takes 4 hours to print and 2 hours to bind. In a given week there are 200 hours of printing time available and 108 hours of binding time available. How many books of each type should be produced to maximize the profits?

18. Carl's Manufacturing produces two types of tire sealant, super and regular. The profit per bottle for super is 9¢ and for regular 7¢. The company has 60,000 empty bottles available and uses a bottling machine that can produce 1,000 bottles of super in 4 hours and 1,000 bottles of regular in 3 hours. If 200 hours of bottling time are available, how many of each type should be bottled to maximize the profit?

19. In 1 hour machine A can produce 240 bolts and 100 nuts, and machine B can produce 160 bolts and 150 nuts. The cost of operating machine A is \$2.50/hour and the cost of operating machine B is \$3.50/hour. The combined operating time for both machines is 40 hours. If at least 4,000 bolts and at least 3,000 nuts must be produced, how can this be done with minimum cost?

10.4 Systems of Linear Equations in Three Unknowns

The problem posed at the beginning of this chapter led to a system of equations in *three* unknowns:

$$\begin{aligned} x + y + z &= 24 \\ 200x - 150y - 50z &= 0 \end{aligned}$$

A solution to this system will be a **triple** (x, y, z) of numbers satisfying both equations. Just as a pair of numbers may be used to represent points in the plane, triples of numbers may be used to represent points in space. In space, the solution set for a linear equation in three unknowns is a plane. This idea may be used to illustrate geometrically the types of solutions to expect when solving systems of equations in three unknowns.

A solution to a system of two equations in three unknowns will correspond to a point common to both planes. The intersection of two planes is either a line or, if the planes are parallel, the empty set. In the first case there are infinitely many solutions, and in the second case there are no solutions (see Figure 10.11).

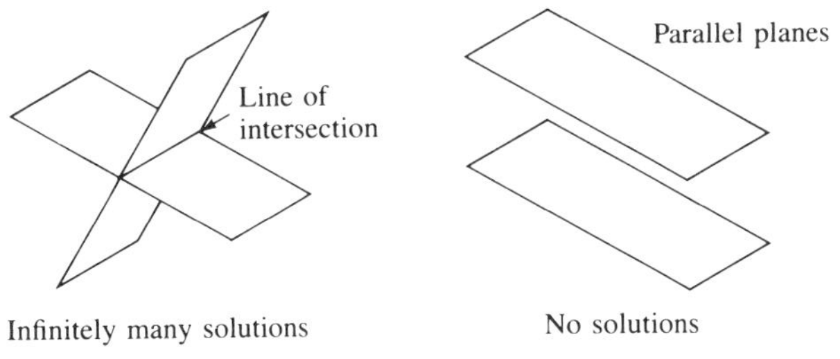

Figure 10.11

For a system of three equations in three unknowns, the planes may intersect in a single point, in a line, or have no point in common (see Figure 10.12).

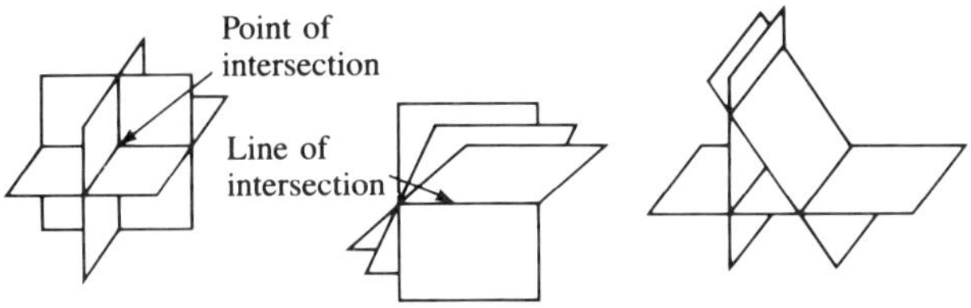

Unique solution | Infinitely many solutions | No point in common, no solution

Figure 10.12

Example 1 Solve the system

$$\begin{aligned} x + y + z &= 24 \\ 200x - 150y - 50z &= 0 \end{aligned}$$

We will try to eliminate one of the variables and solve the resulting equation for one of the remaining variables in terms of the other. We first eliminate z. Divide the second equation by 50 and add:

$$\begin{aligned} x + y + z &= 24 \\ 4x - 3y - z &= 0 \\ \hline 5x - 2y &= 24 \end{aligned}$$

Solve for y:

$$y = 2.5x - 12$$

Substitute for y in the first equation:

$$x + (2.5x - 12) + z = 24$$

Solve for z:

$$z = 36 - 3.5x$$

Since the system consists of only two equations, there will be infinitely many solutions, namely the triples (x, y, z) such that

$$\begin{aligned} y &= 2.5x - 12 \\ z &= 36 - 3.5x \end{aligned}$$

■

General and particular solutions

The solution found in Example 1 is what is called a **general solution,** in that the equations describe every solution. A **particular solution** is a triple of numbers which is a solution to the system. In the case of Example 1, a particular solution can be found by choosing a value for x and using the general solution to find y and z.

Example 2 Find the particular solution to the system in Example 1 when $x = 7$.
Using the general solution:

$$\begin{aligned} y &= 2.5(7) - 12 = 5.5 \\ z &= 36 - 3.5(7) = 11.5 \end{aligned}$$

The desired particular solution is (7, 5.5, 11.5). ■

Example 3 Solve the system

$$\begin{aligned} x + y + z &= 1 \\ x - y + 2z &= 7 \\ x + y - z &= -3 \end{aligned}$$

Subtract the first equation from the second to eliminate x:

$$-2y + z = 6$$

Subtract the first equation from the third to obtain

$$-2z = -4$$

The system has now been reduced to the system

$$\begin{aligned} x + y + z &= 1 \\ -2y + z &= 6 \\ -2z &= -4 \end{aligned}$$

From the last equation we get $z = 2$, and substituting in the second equation,

$$\begin{aligned} -2y + 2 &= 6 \\ y &= -2 \end{aligned}$$

Substituting $z = 2$ and $y = -2$ in the first equation,

$$\begin{aligned} x - 2 + 2 &= 1 \\ x &= 1 \end{aligned}$$

The solution is $x = 1$, $y = -2$, $z = 2$. ■

Example 4 The Olde Cabinet Shoppe has three patterns for doors, sides, and drawers. The patterns are laid out on sheets of plywood, and each pattern yields doors, sides, and drawers as shown in the table.

	Pattern 1	Pattern 2	Pattern 3
Doors	2	1	10
Sides	1	6	2
Drawers	7	3	4

How many sheets of each pattern does the shop need in order to obtain 75 doors, 50 sides, and 105 drawers?

Let x = number of sheets with pattern 1, y = number of sheets with pattern 2, and z = number of sheets with pattern 3. Then

$$\begin{aligned} 2x + y + 10z &= 75 \quad \textit{(Number of doors)} \\ x + 6y + 2z &= 50 \quad \textit{(Number of sides)} \\ 7x + 3y + 4z &= 105 \quad \textit{(Number of drawers)} \end{aligned}$$

is the system to solve. Multiply the second equation by -2 and add to the first equation:

$$-11y + 6z = -25$$

Multiply the second equation by -7 and add to the third equation:

$$-39y - 10z = -245$$

Now solve the two equations in y and z:

$$\begin{aligned} -11y + 6z &= -25 \\ -39y - 10z &= -245 \end{aligned}$$

Multiply the first equation by 5 and the second by 3 and add:

$$\begin{aligned} -55y + 30z &= -125 \\ -117y - 30z &= -735 \\ \hline -172y &= -860 \\ y &= 5 \end{aligned}$$

From the equation $-11y + 6z = -25$,

$$z = 5$$

and from the equation $x + 6y + 2z = 50$,

$$x = 10$$

The Olde Cabinet Shoppe should use 10 sheets with pattern 1, 5 sheets with pattern 2, and 5 sheets with pattern 3. ■

Example 5 Solve the system

$$\begin{aligned} 3x + 5y + 8z &= 21 \\ 2x - 3y - z &= -5 \\ x + 8y + 9z &= 26 \end{aligned}$$

Multiply the third equation by -2 and add to the second equation:

$$-19y - 19z = -57$$

Multiply the third equation by -3 and add to the first equation:

$$-19y - 19z = -57$$

We have reduced the original system to the equivalent system

$$\begin{aligned} x + 8y + 9z &= 26 \\ -19y - 19z &= -57 \\ -19y - 19z &= -57 \end{aligned}$$

Since the last two equations are the same, we look for a general solution yielding infinitely many particular solutions. The equation $-19y - 19z = -57$ yields the equation

$$y = 3 - z$$

and from the first equation

$$x = 2 - z.$$

This is the general solution. ■

Example 6 Solve the system

$$\begin{aligned} x + 2y + z &= 3 \\ x + 4y - z &= 1 \\ x + y + 2z &= 5 \end{aligned}$$

Subtract the first equation from the second:

$$2y - 2z = -2$$

Subtract the third equation from the second:

$$3y - 3z = -4$$

The system has been reduced to the equivalent system

$$\begin{aligned} x + 4y - z &= 1 \\ 2y - 2z &= -2 \\ 3y - 3z &= -4 \end{aligned}$$

Divide the second equation by 2 and the third equation by 3:

$$\begin{aligned} y - z &= -1 \\ y - z &= -\frac{4}{3} \end{aligned}$$

These two equations are inconsistent; thus the system is inconsistent and hence has no solutions. ■

Exercises 10.4

In Exercises 1–12 find the general solution in terms of the variable listed and the particular solution for the value given.
Find the general solution in terms of x and the particular solution for $x = -1$.

1. $\begin{aligned} 3x + y + 2z &= 2 \\ -x + 7y + 3z &= 3 \end{aligned}$

2. $\begin{aligned} x + 3y - 2z &= 2 \\ -7x + y + 3z &= -3 \end{aligned}$

3. $\begin{aligned} x + y - z &= -2 \\ 3x + 2y - z &= -2 \end{aligned}$

4. $\begin{aligned} 5x - y + z &= 4 \\ x + y + z &= 4 \end{aligned}$

Find the general solution in terms of y and the particular solution for $y = 2$.

5. $\begin{aligned} 2x + 4y - 3z &= 0 \\ -x + 3y - z &= 15 \end{aligned}$

6. $\begin{aligned} 2x + 4y - 3z &= 4 \\ -x + 5y - 2z &= 12 \end{aligned}$

7. $\begin{aligned} 2x + 8y - 3z &= 4 \\ x - y + z &= 7 \end{aligned}$

8. $\begin{aligned} x + 2y + z &= 0 \\ 2x - 5y + 3z &= 2 \end{aligned}$

Find the general solution in terms of z and the particular solution for $z = -4$.

9. $\begin{aligned} x - y - z &= 0 \\ 3x + y - 2z &= 0 \end{aligned}$

10. $\begin{aligned} x - y - 2z &= 0 \\ x + y - z &= 0 \end{aligned}$

11. $\begin{aligned} x - 2y + z &= 0 \\ 2x - 5y + 3z &= 4 \end{aligned}$

12. $\begin{aligned} x + y - 3z &= 0 \\ 2x + 3y - 8z &= 1 \end{aligned}$

In Exercises 13–30 find the solution if the system is not inconsistent.

13. $\begin{aligned} x + y + z &= 0 \\ 2y + z &= 1 \\ -3z &= -9 \end{aligned}$

14. $\begin{aligned} 2x - y + 4z &= -7 \\ y + z &= 3 \\ 2z &= -2 \end{aligned}$

15. $\begin{aligned} x + 2y + 3z &= 5 \\ -x + 3y + 3z &= -1 \\ -x + 3y + 2z &= 0 \end{aligned}$

16. $\begin{aligned} x + 2y + z &= 4 \\ -2x - 3y + z &= -3 \\ -2x - y + z &= -5 \end{aligned}$

17. $\begin{aligned} a - b + 2c &= -1 \\ a - 3b - 2c &= -17 \\ a + 2b + c &= 9 \end{aligned}$

18. $\begin{aligned} -3s + t + r &= -10 \\ 2s + t + 3r &= 1 \\ s - t + 4r &= -6 \end{aligned}$

19. $\begin{aligned} -3x + 2y + 4z &= 5 \\ 6x - y + 6z &= 0 \\ 3x + y + 2z &= 1 \end{aligned}$

20. $\begin{aligned} 2m + 3n - 2p &= 4 \\ m + 6n - 14p &= -3 \\ -5m - 9n + 6p &= -10 \end{aligned}$

21. $\begin{aligned} x - 3y + 2z &= -17 \\ 2y - 3z &= 12 \\ y + 4z &= 6 \end{aligned}$

22. $\begin{aligned} 2a - b + c &= 8 \\ 2a + 3b &= 1 \\ a + 4b &= -2 \end{aligned}$

23. $\begin{aligned} 2s + r - 4t &= 5 \\ s + 2r - 5t &= -2 \\ 3s - 3t &= 12 \end{aligned}$

24. $\begin{aligned} m + n - p &= 1 \\ 2m - n - 5p &= -1 \\ 3m - 6p &= 0 \end{aligned}$

25. $\begin{aligned} x - 3y + z &= -15 \\ 5x - y - 2z &= -5 \\ 8x - 10y + z &= -50 \end{aligned}$

26. $\begin{aligned} 2a - b + c &= 0 \\ a - 2b - c &= -6 \\ a + b + 2c &= 6 \end{aligned}$

27. $\begin{aligned} \frac{5}{x} + \frac{6}{y} - \frac{1}{z} &= 6 \\ -\frac{1}{x} + \frac{1}{y} + \frac{4}{z} &= -1 \\ \frac{3}{x} + \frac{4}{y} + \frac{1}{z} &= 0 \end{aligned}$

28. $\begin{aligned} \frac{1}{x} - \frac{4}{y} + \frac{3}{z} &= -5 \\ -\frac{3}{x} + \frac{5}{y} + \frac{2}{z} &= -7 \\ \frac{1}{x} - \frac{1}{y} + \frac{2}{z} &= -5 \end{aligned}$

29. $\begin{aligned} t + r + 3s &= -1 \\ 2r + 6s - t &= -8 \\ -16 - 3r + 3t &= 6s \end{aligned}$

30. $\begin{aligned} 8x + 1 + 2z &= -y \\ 4x + 2 &= y + z \\ 2y + 3z &= 4x + 1 \end{aligned}$

31. Carla's Nut Shoppe wants to make a mixture containing 45 oz of peanuts, 10 oz of pecans, and 15 oz of Brazil nuts. They use three mixes of nuts in bags marked A, B, and C as shown in the table.

	A	B	C
Peanuts	3 oz	5 oz	3 oz
Pecans	1 oz	3 oz	0 oz
Brazil nuts	1 oz	2 oz	1 oz

How many bags of each should be used to fill the order?

32. A collector collected 77 coins—nickels, dimes, and quarters—from parking meters for a total of \$6.40. There were 11 more dimes than quarters. How many of each kind of coin did the collector collect?

33. The equation of a parabola whose axis of symmetry is parallel to the y axis is

$$y = ax^2 + bx + c.$$

Find the equation of such a parabola that passes through the points (1, 0), (2, 5), and (−1, 2).

34. The equation of a circle is

$$x^2 + y^2 + ax + by + c = 0.$$

Find the equation of the circle that passes through (4, 8), (−2, 8), and (5, 1). (Hint: if you use some facts about circles, you will not need three equations in three unknowns.)

35. Sean is furnishing a restaurant with 3 kinds of tables, some of which seat 4 persons, some of which seat 6 persons, and some of which seat 8 persons. Fire regulations limit the capacity of the room to 220 persons, and Sean thinks the room can accommodate 45 tables. How can Sean furnish his restaurant?

36. From Exercise 35, if Sean thinks that 6 of the 8-person tables are proper for his needs, how many of each of the other kinds of tables should he use?

37. A turkey costs \$9, a duck \$7, and a chicken \$2. Alexandra buys 30 birds, spends \$100, and buys at least 1 duck, 1 chicken, and 1 turkey. How many chickens does she buy? (Hint: the number of each kind of bird bought must be a *positive integer*.)

38. Carl counts his cows and chickens by counting the number of heads and legs. As he finishes counting, his wife Mary Alice laughingly points out that he has counted the legs on at least one of their three-legged milking stools. Carl says, "That's OK, I counted 35 heads and 78 legs, so I know how many cows and chickens I have anyway." How many cows does Carl have?

10.5 Matrices

In solving a system of equations such as

$$\begin{aligned} 2x + y + 10z &= 75 \\ 3x + 6y + 2z &= 4 \\ -5x - 3z &= 0 \end{aligned}$$

we normally are careful to keep all the x terms in a column, all the y terms in a column, and all the z terms in a column. This makes it easier to add and subtract equations in the process of elimination. We could, in fact, omit the unknowns and let the position of the coefficient indicate the unknown involved. The result is a rectangular array of numbers called a matrix.

Matrix

i, j Entry

Equality

A **matrix** is a rectangular array of numbers. If A is a matrix with m rows and n columns, we say A is an $m \times n$ matrix. We use the notation $a_{i,j}$ to denote the ***i, j* entry,** that is, the number in the ith row and jth column. Two matrices A and B are **equal** if and only if (1) A and B have the same number of rows and columns and (2) $a_{i,j} = b_{i,j}$ for all i and j. In other words, $A = B$ if and only if A and B are identical—the same size with corresponding entries equal.

Example 1 $\begin{bmatrix} 6 & -5 \\ 4 & 1 \end{bmatrix}$ is a 2×2 matrix; $a_{1,1} = 6$, $a_{1,2} = -5$
$a_{2,1} = 4$, $a_{2,2} = 1$ ■

Example 2 $[1 \quad -3]$ is a 1×2 matrix, sometimes called a **row matrix** or **row vector.** ■

Row matrix (vector)

Example 3 $\begin{bmatrix} 2 & 3 & -5 \\ 1 & 0 & 6 \end{bmatrix}$ is a 2×3 matrix. ■

Example 4 *Column matrix (vector)*

$\begin{bmatrix} 4 \\ 5 \\ -9 \\ 0 \end{bmatrix}$ is a 4×1 matrix, sometimes called a **column matrix** or a **column vector.** ■

Example 5 The matrices $\begin{bmatrix} 14 & -2 & 3 \\ 0 & x & 5 \end{bmatrix}$ and $\begin{bmatrix} 14 & y & 3 \\ 0 & 6 & 5 \end{bmatrix}$ are equal if and only if $y = -2$ and $x = 6$. ■

Suppose you place an order at a lumber yard for 10 pounds of nails, 20 2×4s, 15 2×6s, and 40 sheets of plywood. This order can be written as a row vector:

$$[10 \quad 20 \quad 15 \quad 40]$$

If the cost of each of these items is

nails \$0.35 per pound
2×4s \$2.25 each
2×6s \$3.50 each
plywood \$12.00 per sheet

then the prices can be written as a column vector:

$$\begin{bmatrix} 0.35 \\ 2.25 \\ 3.50 \\ 12.00 \end{bmatrix}$$

The total cost of the order is then

$$[10 \quad 20 \quad 15 \quad 40]\begin{bmatrix} 0.35 \\ 2.25 \\ 3.50 \\ 12.00 \end{bmatrix} = (10)(0.35) + (20)(2.25) + (15)(3.50) + (40)(12.00) = \$581$$

This method of combining a row vector and a column vector to obtain a number is called the *scalar product* of the row and column vectors.

Scalar product

If A is a row vector with m entries and B is a column vector with m entries (note that the number of entries is the same for both), then the **scalar product** of A and B is defined as follows.

$$AB = [a_1 \quad a_2 \cdots a_m]\begin{bmatrix} b_1 \\ b_2 \\ \cdot \\ \cdot \\ \cdot \\ b_m \end{bmatrix} = a_1b_1 + a_2b_2 + \cdots + a_mb_m$$

Example 6 $[1 \quad -2 \quad 9]\begin{bmatrix} 4 \\ 1 \\ -2 \end{bmatrix} = (1)(4) + (-2)(1) + (9)(-2) = -16$ ■

Example 7 $[0 \quad -2 \quad 3 \quad 4 \quad 5]\begin{bmatrix} -2 \\ -3 \\ 4 \\ 0 \\ 3 \end{bmatrix} = (0)(-2) + (-2)(-3) + (3)(4) + (4)(0) + (5)(3) = 33$ ■

Example 8 $[2 \quad -3 \quad 4 \quad 1]\begin{bmatrix} x \\ y \\ z \\ w \end{bmatrix} = 2x - 3y + 4z + w$ ■

We can define the product of two matrices (subject to some restrictions on the number of rows and columns) by using the scalar product of the rows of one matrix and the columns of the second matrix.

Product of matrices

> If A is an $m \times n$ matrix and B is an $n \times p$ matrix, we define the **product** AB to be the $m \times p$ matrix such that the entry in the ith row and jth column is the scalar product of the ith row of A and the jth column of B. (Note that the number of columns of A must be the same as the number of rows of B.)

Example 9 Compute the product

$$\begin{bmatrix} 2 & 3 \\ -1 & 5 \end{bmatrix}\begin{bmatrix} 6 & -1 \\ 4 & -2 \end{bmatrix}$$

The entry in the first row and first column is the scalar product of the first row of the first matrix and the first column of the second matrix.

$$\begin{bmatrix} 2 & 3 \\ & \end{bmatrix}\begin{bmatrix} 6 & \\ 4 & \end{bmatrix} = \begin{bmatrix} (2)(6) + (3)(4) & \\ & \end{bmatrix}$$

In the first row and second column is the scalar product of first row and second column, as shown.

$$\begin{bmatrix} 2 & 3 \\ & \end{bmatrix}\begin{bmatrix} & -1 \\ & -2 \end{bmatrix} = \begin{bmatrix} & (2)(-1) + (3)(-2) \\ & \end{bmatrix}$$

In the second row and first column is the scalar product of the second row and first column, as shown.

$$\begin{bmatrix} & \\ -1 & 5 \end{bmatrix}\begin{bmatrix} 6 & \\ 4 & \end{bmatrix} = \begin{bmatrix} & \\ (-1)(6) + (5)(4) & \end{bmatrix}$$

Finally,

$$\begin{bmatrix} & \\ -1 & 5 \end{bmatrix}\begin{bmatrix} & -1 \\ & -2 \end{bmatrix} = \begin{bmatrix} & \\ & (-1)(-1) + (5)(-2) \end{bmatrix}$$

The product is

$$\begin{bmatrix} 2 & 3 \\ -1 & 5 \end{bmatrix}\begin{bmatrix} 6 & -1 \\ 4 & -2 \end{bmatrix} = \begin{bmatrix} 24 & -8 \\ 14 & -9 \end{bmatrix}$$ ■

Example 10 Compute the product

$$\begin{bmatrix} 2 & 3 \\ 4 & 1 \\ 1 & 0 \end{bmatrix}\begin{bmatrix} 1 & 1 \\ 2 & -1 \end{bmatrix}$$

The computation is as follows:

$$\begin{bmatrix} 2 & 3 \\ 4 & 1 \\ 1 & 0 \end{bmatrix}\begin{bmatrix} 1 & 1 \\ 2 & -1 \end{bmatrix} = \begin{bmatrix} (2)(1) + (3)(2) & (2)(1) + (3)(-1) \\ (4)(1) + (1)(2) & (4)(1) + (1)(-1) \\ (1)(1) + (0)(2) & (1)(1) + (0)(-1) \end{bmatrix} = \begin{bmatrix} 8 & -1 \\ 6 & 3 \\ 1 & 1 \end{bmatrix}$$

Note that the product

$$\begin{bmatrix} 1 & 1 \\ 2 & -1 \end{bmatrix}\begin{bmatrix} 2 & 3 \\ 4 & 1 \\ 1 & 0 \end{bmatrix}$$

is not defined because the first matrix has *two* columns, whereas the second matrix has *three* rows. ■

Example 11

$$\begin{bmatrix} 2 & 3 \\ 1 & 6 \\ 4 & -7 \end{bmatrix}\begin{bmatrix} 1 \\ 4 \end{bmatrix} = \begin{bmatrix} (2)(1) + (3)(4) \\ (1)(1) + (6)(4) \\ (4)(1) + (-7)(4) \end{bmatrix} = \begin{bmatrix} 14 \\ 25 \\ -24 \end{bmatrix}$$ ■

Example 12

$$\begin{bmatrix} a & b \\ c & d \end{bmatrix}\begin{bmatrix} 1 & 0 \\ 0 & 1 \end{bmatrix} = \begin{bmatrix} a & b \\ c & d \end{bmatrix}$$ ■

Example 13

$$\begin{bmatrix} 1 & 0 \\ 0 & 1 \end{bmatrix}\begin{bmatrix} a & b \\ c & d \end{bmatrix} = \begin{bmatrix} a & b \\ c & d \end{bmatrix}$$ ■

Examples 12 and 13 show that the matrix

$$I = \begin{bmatrix} 1 & 0 \\ 0 & 1 \end{bmatrix}$$

functions as the multiplicative identity, just as the number 1 does for multiplication of real numbers. It is not hard to see that

$$\begin{bmatrix} 1 & 0 & 0 \\ 0 & 1 & 0 \\ 0 & 0 & 1 \end{bmatrix}$$

works the same way for 3×3 matrices, and

$$I_n = \begin{bmatrix} 1 & 0 & 0 \ldots 0 \\ 0 & 1 & 0 \ldots 0 \\ \cdot & & \\ \cdot & & \\ \cdot & & \\ 0 & 0 & 0 \ldots 1 \end{bmatrix} \qquad (n \textit{ rows and } n \textit{ columns})$$

is the multiplicative identity for $n \times n$ matrices.

We can also use matrix multiplication to write systems of linear equations in matrix notation. Consider the system

$$\begin{aligned} 4x - 3y + 2z &= 6 \\ -5x + 2y - z &= 0 \\ x \qquad + 4z &= 6 \end{aligned}$$

The matrix made up of the coefficients of the unknowns is called the **coefficient matrix.** In this case it is

$$A = \begin{bmatrix} 4 & -3 & 2 \\ -5 & 2 & -1 \\ 1 & 0 & 4 \end{bmatrix}$$

If we write the unknowns as a column vector

$$X = \begin{bmatrix} x \\ y \\ z \end{bmatrix}$$

then

$$AX = \begin{bmatrix} 4 & -3 & 2 \\ -5 & 2 & -1 \\ 1 & 0 & 4 \end{bmatrix} \begin{bmatrix} x \\ y \\ z \end{bmatrix} = \begin{bmatrix} 4x - 3y + 2z \\ -5x + 2y - z \\ x \qquad + 4z \end{bmatrix}$$

If we write the constant terms in a column vector as

$$B = \begin{bmatrix} 6 \\ 0 \\ 6 \end{bmatrix}$$

then the matrix equation $AX = B$ is simply a restatement of the system

$$\begin{aligned} 4x - 3y + 2z &= 6 \\ -5x + 2y - z &= 0 \\ x \qquad + 4z &= 6 \end{aligned}$$

Example 14 Write the following system as a matrix equation:

$$\begin{aligned} 5a - 6b + 4c &= 9 \\ -2a + 9b - c &= 10 \end{aligned}$$

The coefficient matrix is

$$\begin{bmatrix} 5 & -6 & 4 \\ -2 & 9 & -1 \end{bmatrix}$$

so the matrix equation is

$$\begin{bmatrix} 5 & -6 & 4 \\ -2 & 9 & -1 \end{bmatrix} \begin{bmatrix} a \\ b \\ c \end{bmatrix} = \begin{bmatrix} 9 \\ 10 \end{bmatrix}$$

■

Example 15 Write the following matrix equation as a system of equations:

$$\begin{bmatrix} 0 & -2 & 4 & -5 \\ -9 & 0 & 1 & 0 \\ 4 & 4 & 5 & 1 \end{bmatrix} \begin{bmatrix} u \\ v \\ w \\ x \end{bmatrix} = \begin{bmatrix} 9 \\ -7 \\ 8 \end{bmatrix}$$

The system is

$$\begin{aligned} -2v + 4w - 5x &= 9 \\ -9u \qquad + w \qquad &= -7 \\ 4u + 4v + 5w + x &= 8 \end{aligned}$$

■

Exercises 10.5

In Exercises 1–6, determine, if possible, the values of x and y that make the matrices equal.

1. $\begin{bmatrix} 1 & 3 \\ x & y \end{bmatrix}; \begin{bmatrix} 1 & 3 \\ 0 & 4 \end{bmatrix}$

2. $\begin{bmatrix} -4 & x \\ 0 & y \end{bmatrix}; \begin{bmatrix} -4 & 4 \\ 0 & -3 \end{bmatrix}$

3. $\begin{bmatrix} x + 2y & 0 \\ 1 & 4 \end{bmatrix}; \begin{bmatrix} -4 & 0 \\ 1 & 3x - 2y \end{bmatrix}$

4. $\begin{bmatrix} 5 & x + y \\ 1 & 5 \end{bmatrix}; \begin{bmatrix} 5 & 4 \\ x & x + 2y \end{bmatrix}$

5. $\begin{bmatrix} x - 2y & 1 \\ x & -6 \end{bmatrix}; \begin{bmatrix} 6 & 1 \\ 3 & x + 2y \end{bmatrix}$

6. $\begin{bmatrix} y - 2x & 9 \\ 3 & 0 \end{bmatrix}; \begin{bmatrix} 12 & 9 \\ 3 & y + 2x \end{bmatrix}$

In Exercises 7–30 compute each product, if defined.

7. $[2 \quad -1] \begin{bmatrix} 4 \\ 8 \end{bmatrix}$

8. $[4 \quad -2] \begin{bmatrix} 2 \\ 3 \end{bmatrix}$

9. $[-2 \quad 3 \quad 1] \begin{bmatrix} 9 \\ 5 \\ -2 \end{bmatrix}$

10. $[4 \quad -2 \quad 8] \begin{bmatrix} 4 \\ 5 \\ -1 \end{bmatrix}$

11. $\begin{bmatrix} 2 & -1 \\ 3 & 6 \end{bmatrix}\begin{bmatrix} 4 & -2 \\ 8 & -4 \end{bmatrix}$

12. $\begin{bmatrix} 4 & -2 \\ 8 & -4 \end{bmatrix}\begin{bmatrix} 2 & -1 \\ 3 & 6 \end{bmatrix}$

13. $\begin{bmatrix} -3 & 3 \\ 2 & -2 \end{bmatrix}\begin{bmatrix} 2 & 3 \\ 4 & 6 \end{bmatrix}$

14. $\begin{bmatrix} 2 & 3 \\ 4 & 6 \end{bmatrix}\begin{bmatrix} -3 & 3 \\ 2 & -2 \end{bmatrix}$

15. $\begin{bmatrix} 1 & -1 \\ -1 & 3 \end{bmatrix}\begin{bmatrix} -4 & 6 \\ 1 & -2 \end{bmatrix}$

16. $\begin{bmatrix} -4 & 6 \\ 1 & -2 \end{bmatrix}\begin{bmatrix} 1 & -1 \\ -1 & 3 \end{bmatrix}$

17. $\begin{bmatrix} 1 & 0 & 3 \\ 1 & 1 & 6 \end{bmatrix}\begin{bmatrix} 5 & 1 \\ 0 & 2 \\ 0 & -9 \end{bmatrix}$

18. $\begin{bmatrix} 1 & 0 & -1 \\ 2 & 1 & 3 \end{bmatrix}\begin{bmatrix} -1 & 2 \\ 5 & 10 \\ -1 & 2 \end{bmatrix}$

19. $\begin{bmatrix} -1 & 2 & 1 \\ 4 & 0 & 2 \end{bmatrix}\begin{bmatrix} 0 & -1 \\ 1 & 3 \end{bmatrix}$

20. $\begin{bmatrix} 0 & -1 \\ 4 & 0 \end{bmatrix}\begin{bmatrix} -1 & 2 & 1 \\ 2 & 1 & 3 \end{bmatrix}$

21. $\begin{bmatrix} 1 & -2 \\ 0 & 3 \\ 1 & -4 \end{bmatrix}\begin{bmatrix} 1 & 1 \\ 3 & -5 \end{bmatrix}$

22. $\begin{bmatrix} 2 & -1 \\ 3 & 0 \\ -1 & 2 \end{bmatrix}\begin{bmatrix} 4 & -5 \\ 3 & -5 \end{bmatrix}$

23. $\begin{bmatrix} -2 & 3 \\ 4 & 4 \end{bmatrix}\begin{bmatrix} -2 & 5 \\ 1 & 0 \\ -3 & 1 \end{bmatrix}$

24. $\begin{bmatrix} -2 & 5 \\ 1 & 0 \\ -3 & 1 \end{bmatrix}\begin{bmatrix} -2 & 3 \\ 4 & 4 \end{bmatrix}$

25. $\begin{bmatrix} 5 & 0 & -4 \\ 1 & -4 & -4 \end{bmatrix}\begin{bmatrix} 8 \\ 1 \\ 6 \end{bmatrix}$

26. $\begin{bmatrix} -5 & 6 & 0 \\ -2 & 3 & 5 \end{bmatrix}\begin{bmatrix} 6 \\ 8 \\ 1 \end{bmatrix}$

27. $\begin{bmatrix} -8 \\ 2 \\ 3 \end{bmatrix}\begin{bmatrix} 2 & -3 & 1 \end{bmatrix}$

28. $\begin{bmatrix} 7 \\ 3 \\ -1 \end{bmatrix}\begin{bmatrix} 4 & -2 & 3 \end{bmatrix}$

29. $\begin{bmatrix} 13 & -6 \\ -2 & 1 \end{bmatrix}\begin{bmatrix} 1 & 6 \\ 2 & 13 \end{bmatrix}$

30. $\begin{bmatrix} -2 & 5 \\ 1 & -2 \end{bmatrix}\begin{bmatrix} 2 & 5 \\ 1 & 2 \end{bmatrix}$

In Exercises 31–38 write each system of equations as a matrix equation.

31. $2x - 3y = 7$
$4y = x$

32. $3u + 4v = 10$
$v - 6 = 4u$

33. $2t + 5r = 9s$
$r + 14s = 2t$

34. $4a - 5b + 7 = 19c$
$3b - 4c + 2 = 5b$

35. $4r + 5s - 9t = 0$
$-r + 3s + 8t = 0$
$-3r + t = 9$

36. $4x - 3y + 8z = 7$
$3x - 4y + 8z = 1$
$9y - z = 0$

37. $x - 7y + 4z = 9$
$2y - z = 1$
$2y + 4x - 9 = 6z$

38. $5m - 2n + 4p = 18$
$n - 5p = 12$
$4p - 2n = m$

In Exercises 39–44 write each matrix equation as a system of equations.

39. $\begin{bmatrix} 2 & -5 & 6 \\ 0 & -6 & 4 \end{bmatrix}\begin{bmatrix} a \\ b \\ c \end{bmatrix} = \begin{bmatrix} 7 \\ 9 \end{bmatrix}$

40. $\begin{bmatrix} 5 & -2 & 0 \\ 0 & 4 & -1 \end{bmatrix}\begin{bmatrix} m \\ n \\ p \end{bmatrix} = \begin{bmatrix} 6 \\ 2 \end{bmatrix}$

41. $\begin{bmatrix} 1 & 0 \\ 0 & 1 \end{bmatrix}\begin{bmatrix} s \\ t \end{bmatrix} = \begin{bmatrix} 8 \\ 7 \end{bmatrix}$

42. $\begin{bmatrix} 1 & 0 \\ 0 & 1 \end{bmatrix}\begin{bmatrix} x \\ y \end{bmatrix} = \begin{bmatrix} 9 \\ 0 \end{bmatrix}$

43. $\begin{bmatrix} 7 & -6 \\ 4 & -5 \\ -2 & 1 \end{bmatrix}\begin{bmatrix} a \\ b \end{bmatrix} = \begin{bmatrix} 9 \\ 4 \\ 3 \end{bmatrix}$

44. $\begin{bmatrix} -7 & 8 \\ 6 & -7 \\ -8 & 1 \end{bmatrix}\begin{bmatrix} a \\ b \end{bmatrix} = \begin{bmatrix} 10 \\ 11 \\ 14 \end{bmatrix}$

In each of Exercises 45–50 write the order vector as a row vector and the price vector as a column vector and compute the scalar product to find the total cost of the order.

45. Fifteen pounds of nails, 24 2×4s, 16 2×6s, and 20 sheets of plywood. Nails are 36¢/pound, 2×4s are $2.50 each, 2×6s are $3.50 each, and plywood is $14 per sheet.

46. Eighteen pounds of nails, 18 2×4s 12 2×6s, and 15 sheets of plywood. Costs are the same as in Exercise 45.

47. Eleven pounds of nails, 25 2×6s, and 8 sheets of plywood. Same prices as in Exercise 45.

48. Fourteen 2×4s, 11 2×6s, and 32 sheets of plywood. Same prices as in Exercise 45.

49. Twenty pounds of potatoes at 99¢/pound, 15 pounds of apples at 36¢/pound and 20 pounds of chile at 45¢/pound.

50. Eighteen pounds of chile, 16 pounds of potatoes, and 14 pounds of apples. Same prices as in Exercise 49.

10.6 Gaussian Elimination

Consider the system

$$\begin{aligned} 3x - 2y + 5z &= 4 \\ x - y + z &= 16 \\ 2x + y - 3z &= -1 \end{aligned}$$

We will solve this system by the method of Gaussian elimination and show you how to use the *augmented* matrix to simplify this method of solution.

Augmented matrix of a system

The **augmented matrix of a system** is obtained by adjoining (augmenting) the column of constants to the coefficient matrix. For the system above, the augmented matrix is

$$\begin{bmatrix} 3 & -2 & 5 & 4 \\ 1 & -1 & 1 & 16 \\ 2 & 1 & -3 & -1 \end{bmatrix}$$

↑ (*Column of coefficients*)

To solve the system of equations by elimination, certain operations are carried out with the equations. Such operations may be represented by similar operations on the rows of the augmented matrix. Follow the example below and see how it works.

Equations		Augmented matrix
$3x - 2y + 5z = 4$	(E-1)	$\begin{bmatrix} 3 & -2 & 5 & 4 \\ 1 & -1 & 1 & 16 \\ 2 & 1 & -3 & -1 \end{bmatrix}$
$x - y + z = 16$	(E-2)	
$2x + y - 3z = -1$	(E-3)	

Switch equations (E-1) and (E-2).

$$\begin{aligned} x - y + z &= 16 & \text{(E-1}'\text{)} \\ 3x - 2y + 5z &= 4 & \text{(E-2}'\text{)} \\ 2x + y - 3z &= -1 & \text{(E-3)} \end{aligned}$$

Switch rows 1 and 2.

$$\begin{bmatrix} 1 & -1 & 1 & 16 \\ 3 & -2 & 5 & 4 \\ 2 & 1 & -3 & -1 \end{bmatrix}$$

Multiply equation (E-1′) by −3 and add to (E-2′).

$$\begin{aligned} x - y + z &= 16 & \text{(E-1}'\text{)} \\ y + 2z &= -44 & \text{(E-2}''\text{)} \\ 2x + y - 3z &= -1 & \text{(E-3)} \end{aligned}$$

Multiply row 1 by −3 and add to row 2.

$$\begin{bmatrix} 1 & -1 & 1 & 16 \\ 0 & 1 & 2 & -44 \\ 2 & 1 & -3 & -1 \end{bmatrix}$$

Multiply (E-1′) by −2 and add to (E-3).

$$\begin{aligned} x - y + z &= 16 & \text{(E-1}'\text{)} \\ y + 2z &= -44 & \text{(E-2}''\text{)} \\ 3y - 5z &= -33 & \text{(E-3}'\text{)} \end{aligned}$$

Multiply row 1 by −2 and add to row 3.

$$\begin{bmatrix} 1 & -1 & 1 & 16 \\ 0 & 1 & 2 & -44 \\ 0 & 3 & -5 & -33 \end{bmatrix}$$

Multiply (E-2″) by −3 and add to (E-3′).

$$\begin{aligned} x - y + z &= 16 & \text{(E-1}'\text{)} \\ y + 2z &= -44 & \text{(E-2}''\text{)} \\ -11z &= 99 & \text{(E-3}''\text{)} \end{aligned}$$

Multiply row 2 by −3 and add to row 3.

$$\begin{bmatrix} 1 & -1 & 1 & 16 \\ 0 & 1 & 2 & -44 \\ 0 & 0 & -11 & 99 \end{bmatrix}$$

Divide (E-3″) by −11.

$$\begin{aligned} x - y + z &= 16 \\ y + 2z &= -44 \\ z &= -9 \end{aligned}$$

Divide row 3 by −11.

$$\begin{bmatrix} 1 & -1 & 1 & 16 \\ 0 & 1 & 2 & -44 \\ 0 & 0 & 1 & -9 \end{bmatrix}$$

The system is now reduced to a form in which the solution is easy to find. Substituting $z = -9$ in the second equation:

$$\begin{aligned} y + 2(-9) &= -44 \\ y &= -26 \end{aligned}$$

Substituting $z = -9$ and $y = -26$ in the first equation:

$$\begin{aligned} x - (-26) + (-9) &= 16 \\ x &= -1 \end{aligned}$$

The solution is $x = -1$, $y = -26$, and $z = -9$.

To solve a system of linear equations our strategy has been to convert the system to an equivalent system that is easier to solve. An equivalent system is obtained if we

1. interchange two equations,
2. multiply (or divide) one of the equations by a nonzero number, or
3. add a multiple of one equation to another.

Elementary row operations

These three operations correspond to the **elementary row operations** on the augmented matrix:

1. interchange two rows;
2. multiply (or divide) one row by a nonzero number;
3. add a multiple of one row to another row.

In the examples that follow, we solve systems by performing elementary row operations on the augmented matrix to reduce the system to one that is easy to solve.

Example 1 Solve the system

$$\begin{aligned} 3x + 7y - z &= -1 \\ 2x - y + 6z &= 8 \\ x + 4y + z &= 7 \end{aligned}$$

The augmented matrix is

$$\begin{bmatrix} 3 & 7 & -1 & -1 \\ 2 & -1 & 6 & 8 \\ 1 & 4 & 1 & 7 \end{bmatrix}$$

Interchange rows 1 and 3 (it makes calculations easier).

$$\begin{bmatrix} 1 & 4 & 1 & 7 \\ 2 & -1 & 6 & 8 \\ 3 & 7 & -1 & -1 \end{bmatrix}$$

Multiply row 1 by -2 and add to row 2.

$$\begin{bmatrix} 1 & 4 & 1 & 7 \\ 0 & -9 & 4 & -6 \\ 3 & 7 & -1 & -1 \end{bmatrix}$$

Multiply row 1 by -3 and add to row 3.

$$\begin{bmatrix} 1 & 4 & 1 & 7 \\ 0 & -9 & 4 & -6 \\ 0 & -5 & -4 & -22 \end{bmatrix}$$

Multiply row 2 by $-\dfrac{1}{9}$.

$$\begin{bmatrix} 1 & 4 & 1 & 7 \\ 0 & 1 & -\dfrac{4}{9} & \dfrac{2}{3} \\ 0 & -5 & -4 & -22 \end{bmatrix}$$

Notice how this sequence of calculations was designed to obtain zeros in the first column, which corresponds to eliminating x from the second and third equations.

Multiply row 2 by 5 and add to row 3.

$$\begin{bmatrix} 1 & 4 & 1 & 7 \\ 0 & 1 & -\dfrac{4}{9} & \dfrac{2}{3} \\ 0 & 0 & -\dfrac{56}{9} & -\dfrac{56}{3} \end{bmatrix}$$

Multiply row 3 by $-\dfrac{9}{56}$.

$$\begin{bmatrix} 1 & 4 & 1 & 7 \\ 0 & 1 & -\dfrac{4}{9} & \dfrac{2}{3} \\ 0 & 0 & 1 & 3 \end{bmatrix}$$

We can stop at this stage and solve the system

$$\begin{aligned} x + 4y + z &= 7 \\ y - \frac{4}{9}z &= \frac{2}{3} \\ z &= 3 \end{aligned}$$

or continue to work with the augmented matrix. To illustrate, we continue.

Multiply row 3 by $\frac{4}{9}$ and add to row 2. $\begin{bmatrix} 1 & 4 & 1 & 7 \\ 0 & 1 & 0 & 2 \\ 0 & 0 & 1 & 3 \end{bmatrix}$

Multiply row 3 by -1 and add to row 1. Multiply row 2 by -4 and add to row 1. $\begin{bmatrix} 1 & 0 & 0 & -4 \\ 0 & 1 & 0 & 2 \\ 0 & 0 & 1 & 3 \end{bmatrix}$

This is the augmented matrix of the system

$$\begin{aligned} x &= -4 \\ y &= 2 \\ z &= 3 \end{aligned}$$

■

Use your own judgment as to how far to carry the row operations on the augmented matrix. The object is to solve the system, so stop at the stage at which it seems easiest to you to finish the solution.

Example 2 Solve the system

$$\begin{aligned} x + 2y + z &= 2 \\ y + z &= 3 \\ x + 3y + 2z &= 5 \end{aligned}$$

The augmented matrix is $\begin{bmatrix} 1 & 2 & 1 & 2 \\ 0 & 1 & 1 & 3 \\ 1 & 3 & 2 & 5 \end{bmatrix}$

Multiply row 1 by -1 and add to row 3. $\begin{bmatrix} 1 & 2 & 1 & 2 \\ 0 & 1 & 1 & 3 \\ 0 & 1 & 1 & 3 \end{bmatrix}$

Multiply row 2 by -1 and add to row 3. $\begin{bmatrix} 1 & 2 & 1 & 2 \\ 0 & 1 & 1 & 3 \\ 0 & 0 & 0 & 0 \end{bmatrix}$

Multiply row 2 by -2 and add to row 1. $\begin{bmatrix} 1 & 0 & -1 & -4 \\ 0 & 1 & 1 & 3 \\ 0 & 0 & 0 & 0 \end{bmatrix}$

This is the augmented matrix for the system

$$\begin{aligned} x - z &= -4 \\ y + z &= 3 \end{aligned}$$

The general solution is

$$x = z - 4$$
$$y = 3 - z$$

■

Example 3 Solve the system

$$\begin{aligned} x - 2y + 3z &= 4 \\ x - 3y - 2z &= 6 \\ x - 5y - 12z &= 0 \end{aligned}$$

The augmented matrix is $\begin{bmatrix} 1 & -2 & 3 & 4 \\ 1 & -3 & -2 & 6 \\ 1 & -5 & -12 & 0 \end{bmatrix}$

Subtract row 1 from rows 2 and 3. $\begin{bmatrix} 1 & -2 & 3 & 4 \\ 0 & -1 & -5 & 2 \\ 0 & -3 & -15 & -4 \end{bmatrix}$

Multiply row 2 by -3 and add to row 3. $\begin{bmatrix} 1 & -2 & 3 & 4 \\ 0 & -1 & -5 & 2 \\ 0 & 0 & 0 & -10 \end{bmatrix}$

This is the augmented matrix for the system

$$\begin{aligned} x - 2y + 3z &= 4 \\ -y - 5z &= 2 \\ 0 &= -10 \end{aligned}$$

which is inconsistent. The system has no solution. ■

Exercises 10.6

In Exercises 1–28 use the augmented matrix to solve the system.

1. $3x - 4y = 6$
$2x - 3y = -7$

2. $5x + 2y = 9$
$3x + y = -7$

3. $15a + 10b = 8$
$9a + 6b = 9$

4. $8x - 12y = 9$
$6x - 9y = 12$

5. $7m + 5n = 15$
$10m + 7n = 12$

6. $-8u + 9v = 12$
$7u - 8v = 10$

7. $2a + 4b - 4c = 16$
$a + 2c = 12$

8. $2x - 4y + 4z = 8$
$x - 2z = 10$

9. $2x + 3y - 4z = 6$
$x + y - z = 1$

10. $2x + 3y + 5z = -6$
$x + y - 12z = 4$

11. $3x + 4y - 7z = 9$
$2x - 3y + 8z = -9$

12. $2x - 5y - 6z = 8$
$3x - 7y + 8z = -4$

13. $\begin{aligned} 2x + 10y - 3z &= -6 \\ 3x + 4y + z &= 13 \\ x + z &= 5 \end{aligned}$

14. $\begin{aligned} 4a + b - 3c &= 13 \\ b - c &= 7 \\ 10a - 3b - 2c &= -6 \end{aligned}$

15. $\begin{aligned} 3a + 4b + c &= 12 \\ 2a + 10b - 3c &= -3 \\ a + c &= 6 \end{aligned}$

16. $\begin{aligned} a - 4b - 3c &= 9 \\ 3a + 10b - 3c &= -3 \\ a - c &= 6 \end{aligned}$

17. $\begin{aligned} x - 3y - z &= 3 \\ 3x + y + 2z &= 19 \\ -2x + 3y + z &= -5 \end{aligned}$

18. $\begin{aligned} x - 3y - 2z &= 2 \\ 6x + y + 4z &= 19 \\ 4x - 3y - 2z &= 5 \end{aligned}$

19. $\begin{aligned} x + 2y - z &= 3 \\ 3z + 2x - y &= -4 \\ 2y - 3z - 3x &= 1 \end{aligned}$

20. $\begin{aligned} x - 2y - z &= 3 \\ 2z - 3x - y &= 4 \\ 2y + 3z + 3x &= -1 \end{aligned}$

21. $\begin{aligned} x + z &= 1 \\ z + y &= 2 \\ x + y &= 1 \end{aligned}$

22. $\begin{aligned} x + y &= 0 \\ y + z &= 1 \\ x + z &= 3 \end{aligned}$

23. $\begin{aligned} \frac{1}{2}x + y + \frac{1}{2}z &= 1 \\ x - 2y + \frac{2}{3}z &= \frac{2}{3} \\ \frac{2}{5}x - \frac{1}{5}z &= \frac{8}{5} \end{aligned}$

24. $\begin{aligned} \frac{1}{2}x - 2y + \frac{1}{2}z &= \frac{1}{2} \\ x - 2y + \frac{1}{3}z &= \frac{4}{3} \\ \frac{1}{3}x + \frac{1}{3}z &= 1 \end{aligned}$

25. $\begin{aligned} \frac{1}{x} - \frac{1}{y} + \frac{2}{z} &= 7 \\ \frac{1}{x} + \frac{1}{y} - \frac{3}{2z} &= -1 \\ -\frac{3}{x} + \frac{1}{y} + \frac{1}{z} &= 1 \end{aligned}$

26. $\begin{aligned} -\frac{7}{x} + \frac{1}{y} + \frac{4}{z} &= -16 \\ \frac{1}{x} + \frac{1}{y} - \frac{4}{z} &= -8 \\ -\frac{1}{x} + \frac{7}{y} - \frac{4}{z} &= 16 \end{aligned}$

27. $\begin{aligned} \frac{1}{x} + \frac{2}{y} + \frac{1}{z} &= 2 \\ \frac{1}{y} + \frac{1}{z} &= 3 \\ \frac{1}{x} + \frac{3}{y} + \frac{2}{z} &= 5 \end{aligned}$

28. $\begin{aligned} \frac{1}{x} - \frac{2}{y} - \frac{3}{z} &= -6 \\ \frac{2}{x} + \frac{1}{y} - \frac{1}{z} &= -2 \\ \frac{3}{x} + \frac{2}{y} - \frac{1}{z} &= -2 \end{aligned}$

29. Three circles whose centers lie on a line are tangent, as shown. If the centers are at A, B, and C, and $\overline{AC} = 1$ cm, $\overline{MN} = 1.75$ cm, and $\overline{BN} = 0.675$ cm, find the radius of each circle.

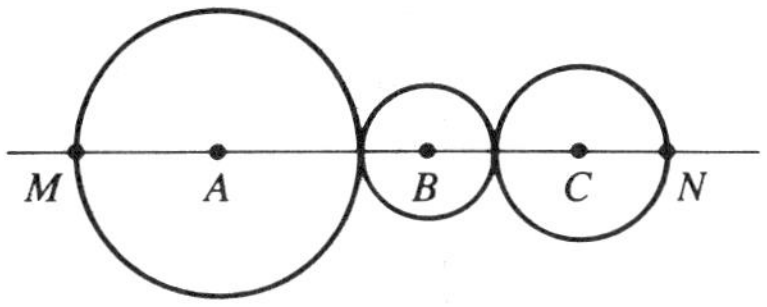

30. A scale drawing is to be made of the three circles shown. The circles are tangent at the points of intersection. Find the radii of the circles.

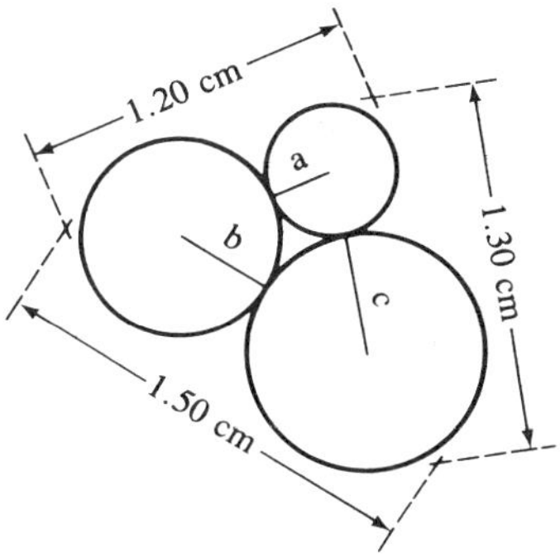

31. Three circles are tangent, with centers as shown. Find the radii of the circles.

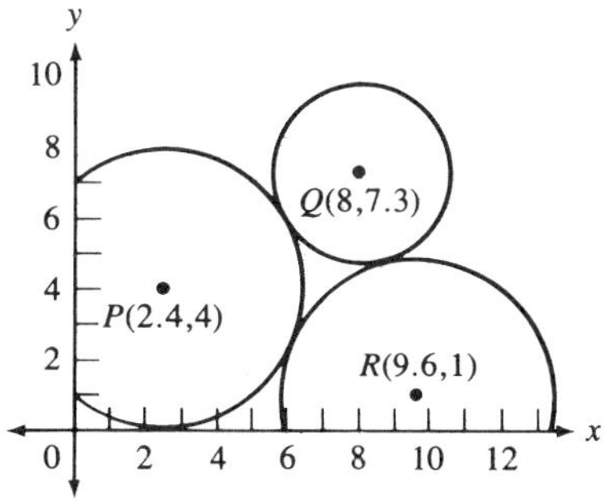

10.7 Matrices and Matrix Inversion

In Example 10 in Section 10.1 we solved the problem of adding 120 pounds of nitrogen and 90 pounds of phosphoric acid to a field, using two types of fertilizer. This led to the system

$$\begin{aligned} 0.2x + 0.06y &= 120 \\ 0.1x + 0.06y &= 90 \end{aligned}$$

In that example we found the solution $x = 300$, $y = 1{,}000$. Suppose we now move to a different field with different fertilizer requirements. Then we change the constant terms in the system and obtain a new system to be solved. The manner of solution by eliminating variables might be exactly the same in each case, but because the constant terms have changed, the entire process would have to be repeated. In this section, we shall see how this repetition can be avoided.

In Section 10.5 we saw how systems of equations can be written as matrix equations in the form

$$AX = B$$

where A is the coefficient matrix. If A, X, and B were real numbers rather than matrices, we would solve the equation by dividing by A or, equivalently, by multiplying by A^{-1}. What follows is an adaptation of that method of solution to matrix equations.

Inverse of a matrix

If A is an $n \times n$ matrix, we say that B is the **inverse** of A if and only if $AB = BA = I$, where I is the $n \times n$ identity matrix. The inverse, if indeed one exists, is unique and is written A^{-1}.

The strategy we pursue is the following: if we were to solve this system by using the augmented matrix, we would begin with the matrix

$$\begin{bmatrix} 0.2 & 0.06 & 120 \\ 0.1 & 0.06 & 90 \end{bmatrix}$$

and by means of elementary row operations transform the matrix into the form

$$\begin{bmatrix} 1 & 0 & 300 \\ 0 & 1 & 1{,}000 \end{bmatrix}$$

What we want to do is record these elementary row operations in such a way that they can be easily used again if the constant terms are changed. This is accomplished in the following way.

Method for finding the inverse

1. Attach a 2×2 identity matrix to the coefficient matrix:

$$\begin{bmatrix} 0.2 & 0.06 & 1 & 0 \\ 0.1 & 0.06 & 0 & 1 \end{bmatrix}$$

2. By elementary row operations, transform the left half of the matrix into a 2×2 identity matrix:

$$\begin{bmatrix} 0.2 & 0.06 & 1 & 0 \\ 0.1 & 0.06 & 0 & 1 \end{bmatrix}$$

Multiply the first row by 5 and the second row by 10.
$$\begin{bmatrix} 1 & 0.3 & 5 & 0 \\ 1 & 0.6 & 0 & 10 \end{bmatrix}$$

Subtract row 1 from row 2.
$$\begin{bmatrix} 1 & 0.3 & 5 & 0 \\ 0 & 0.3 & -5 & 10 \end{bmatrix}$$

Subtract row 2 from row 1.
$$\begin{bmatrix} 1 & 0 & 10 & -10 \\ 0 & 0.3 & -5 & 10 \end{bmatrix}$$

Multiply row 2 by $\frac{10}{3}$.
$$\begin{bmatrix} 1 & 0 & 10 & -10 \\ 0 & 1 & -\frac{50}{3} & \frac{100}{3} \end{bmatrix}$$

3. The right half of the resulting matrix is the inverse of the coefficient matrix:
$$A = \begin{bmatrix} 0.2 & 0.06 \\ 0.1 & 0.06 \end{bmatrix} \qquad A^{-1} = \begin{bmatrix} 10 & -10 \\ -\frac{50}{3} & \frac{100}{3} \end{bmatrix}$$
Indeed,
$$\begin{bmatrix} 0.2 & 0.06 \\ 0.1 & 0.06 \end{bmatrix} \begin{bmatrix} 10 & -10 \\ -\frac{50}{3} & \frac{100}{3} \end{bmatrix} = \begin{bmatrix} 1 & 0 \\ 0 & 1 \end{bmatrix}$$

How do we use A^{-1} to solve the system? Write the system as a matrix equation:

$$AX = B$$

and multiply by A^{-1} to obtain

$$A^{-1}AX = A^{-1}B$$

or

$$IX = A^{-1}B \quad (\textit{I is the } n \times n \textit{ identity matrix})$$
$$X = A^{-1}B$$

In this case

$$X = \begin{bmatrix} x \\ y \end{bmatrix} \quad \text{and} \quad B = \begin{bmatrix} 120 \\ 90 \end{bmatrix}$$

so the solution is

$$\begin{bmatrix} x \\ y \end{bmatrix} = \begin{bmatrix} 10 & -10 \\ -\frac{50}{3} & \frac{100}{3} \end{bmatrix} \begin{bmatrix} 120 \\ 90 \end{bmatrix} = \begin{bmatrix} 300 \\ 1{,}000 \end{bmatrix}$$

Notice that once we have found A^{-1} we can solve the system very easily for any choice of constant terms—that is, for any choice of B. For any square matrix we can use the same process to seek the inverse, if there is one; not all square matrices have inverses.

We summarize as follows.

General procedure for finding the inverse of a matrix

To find the inverse (if there is one) of an $n \times n$ matrix A:

1. Attach an $n \times n$ identity matrix to the right of A.
2. By elementary row operations, transform this new matrix so that the left half is the $n \times n$ identity matrix.
3. The right half of the resulting matrix is the inverse of A.
4. If you obtain a row of zeros in the left half of the matrix at some stage, stop; the matrix A has no inverse.

Example 1 Find the inverse of

$$A = \begin{bmatrix} 2 & 3 & 1 \\ 1 & 2 & 1 \\ -1 & 1 & 1 \end{bmatrix}$$

Attach the 3×3 identity matrix.
$$\begin{bmatrix} 2 & 3 & 1 & 1 & 0 & 0 \\ 1 & 2 & 1 & 0 & 1 & 0 \\ -1 & 1 & 1 & 0 & 0 & 1 \end{bmatrix}$$

Interchange the first and second rows.
$$\begin{bmatrix} 1 & 2 & 1 & 0 & 1 & 0 \\ 2 & 3 & 1 & 1 & 0 & 0 \\ -1 & 1 & 1 & 0 & 0 & 1 \end{bmatrix}$$

Add the first row to the third row and multiply the first row by -2 and add to the second row.
$$\begin{bmatrix} 1 & 2 & 1 & 0 & 1 & 0 \\ 0 & -1 & -1 & 1 & -2 & 0 \\ 0 & 3 & 2 & 0 & 1 & 1 \end{bmatrix}$$

Multiply the second row by -1.
$$\begin{bmatrix} 1 & 2 & 1 & 0 & 1 & 0 \\ 0 & 1 & 1 & -1 & 2 & 0 \\ 0 & 3 & 2 & 0 & 1 & 1 \end{bmatrix}$$

Multiply row 2 by -3 and add to row 3; multiply row 2 by -2 and add to row 1.
$$\begin{bmatrix} 1 & 0 & -1 & 2 & -3 & 0 \\ 0 & 1 & 1 & -1 & 2 & 0 \\ 0 & 0 & -1 & 3 & -5 & 1 \end{bmatrix}$$

Multiply row 3 by -1.
$$\begin{bmatrix} 1 & 0 & -1 & 2 & -3 & 0 \\ 0 & 1 & 1 & -1 & 2 & 0 \\ 0 & 0 & 1 & -3 & 5 & -1 \end{bmatrix}$$

Add row 3 to row 1; subtract row 3 from row 2.
$$\begin{bmatrix} 1 & 0 & 0 & -1 & 2 & -1 \\ 0 & 1 & 0 & 2 & -3 & 1 \\ 0 & 0 & 1 & -3 & 5 & -1 \end{bmatrix}$$

The inverse is the right half of this matrix:

$$A^{-1} = \begin{bmatrix} -1 & 2 & -1 \\ 2 & -3 & 1 \\ -3 & 5 & -1 \end{bmatrix}$$

■

Example 2 Solve the system

$$\begin{aligned} 2x + 3y + z &= 1 \\ x + 2y + z &= -1 \\ -x + y + z &= 3 \end{aligned}$$

The matrix A in Example 1 is the coefficient matrix, so we may use the inverse of A found in Example 1 to obtain

$$X = A^{-1}B$$

where

$$B = \begin{bmatrix} 1 \\ -1 \\ 3 \end{bmatrix}$$

The solution is

$$\begin{bmatrix} x \\ y \\ z \end{bmatrix} = \begin{bmatrix} -1 & 2 & -1 \\ 2 & -3 & 1 \\ -3 & 5 & -1 \end{bmatrix} \begin{bmatrix} 1 \\ -1 \\ 3 \end{bmatrix} = \begin{bmatrix} -6 \\ 8 \\ -11 \end{bmatrix}$$

■

Example 3 Solve the system

$$\begin{aligned} 2x + 3y + z &= 4 \\ x + 2y + z &= 9 \\ -x + y + z &= 0 \end{aligned}$$

The coefficient matrix is again the matrix A of Example 1, so the solution is

$$\begin{bmatrix} x \\ y \\ z \end{bmatrix} = \begin{bmatrix} -1 & 2 & -1 \\ 2 & -3 & 1 \\ -3 & 5 & -1 \end{bmatrix} \begin{bmatrix} 4 \\ 9 \\ 0 \end{bmatrix} = \begin{bmatrix} 14 \\ -19 \\ 33 \end{bmatrix}$$

■

Example 4 Find the inverse of

$$\begin{bmatrix} 1 & 2 & -1 \\ 3 & 0 & 1 \\ 4 & 2 & 0 \end{bmatrix}$$

Attach the 3×3 identity.

$$\begin{bmatrix} 1 & 2 & -1 & 1 & 0 & 0 \\ 3 & 0 & 1 & 0 & 1 & 0 \\ 4 & 2 & 0 & 0 & 0 & 1 \end{bmatrix}$$

Multiply row 1 by −3 and add to row 2; multiply row 1 by −4 and add to row 3.

$$\left[\begin{array}{ccc|ccc} 1 & 2 & -1 & 1 & 0 & 0 \\ 0 & -6 & 4 & -3 & 1 & 0 \\ 0 & -6 & 4 & -4 & 0 & 1 \end{array}\right]$$

Subtract row 2 from row 3.

$$\left[\begin{array}{ccc|ccc} 1 & 2 & -1 & 1 & 0 & 0 \\ 0 & -6 & 4 & -3 & 1 & 0 \\ 0 & 0 & 0 & -1 & -1 & 1 \end{array}\right]$$

Since we have a row of zeros in the left half of the matrix, the process stops and the matrix does *not* have an inverse. ■

Exercises 10.7

In Exercises 1–12 find the inverse matrix, if it exists.

1. $\begin{bmatrix} 9 & -4 \\ -2 & 1 \end{bmatrix}$

2. $\begin{bmatrix} 7 & -3 \\ -5 & 2 \end{bmatrix}$

3. $\begin{bmatrix} 6 & 4 \\ -6 & -5 \end{bmatrix}$

4. $\begin{bmatrix} -4 & -6 \\ 10 & 15 \end{bmatrix}$

5. $\begin{bmatrix} -6 & 14 \\ 9 & -21 \end{bmatrix}$

6. $\begin{bmatrix} -5 & 9 \\ -7 & 12 \end{bmatrix}$

7. $\begin{bmatrix} 3 & -1 & 1 \\ 1 & 0 & 1 \\ -9 & 3 & -4 \end{bmatrix}$

8. $\begin{bmatrix} -1 & 0 & 3 \\ 3 & 1 & -9 \\ 1 & 1 & -4 \end{bmatrix}$

9. $\begin{bmatrix} -1 & -3 & 3 \\ 0 & 4 & 5 \\ 4 & 16 & -7 \end{bmatrix}$

10. $\begin{bmatrix} 1 & -2 & 6 \\ 2 & -1 & -3 \\ 12 & -9 & -3 \end{bmatrix}$

11. $\begin{bmatrix} 4 & -5 & 2 \\ 1 & -2 & 3 \\ 2 & -1 & -4 \end{bmatrix}$

12. $\begin{bmatrix} 4 & 1 & 2 \\ 5 & 2 & 1 \\ 2 & 3 & -4 \end{bmatrix}$

13. Find the inverse of

$$\begin{bmatrix} 2 & 1 & 1 \\ -5 & -1 & 1 \\ -8 & -2 & 1 \end{bmatrix}$$

14. Solve the system

$$\begin{aligned} 2x + y + z &= 18 \\ -5x - y + z &= -2 \\ -8x - 2y + z &= 7 \end{aligned}$$

15. Solve the system

$$\begin{aligned} 2a + b + c &= 1 \\ -5a - b + c &= 12 \\ -8a - 2b + c &= 0 \end{aligned}$$

16. Solve the system

$$\begin{aligned} -8s - 2r + t &= -9 \\ 2s + r + t &= -10 \\ -5s - r + t &= 1 \end{aligned}$$

17. Find the inverse of

$$\begin{bmatrix} 3 & 5 & 3 \\ 3 & 5 & 1 \\ 2 & 4 & 1 \end{bmatrix}$$

18. Solve the system

$$\begin{aligned} 3x + 5y + 3z &= 8 \\ 3x + 5y + z &= 6 \\ 2x + 4y + z &= 4 \end{aligned}$$

19. Solve the system

$$\begin{aligned} 3a + 5b + 3c &= 12 \\ 3a + 5b + c &= -10 \\ 2a + 4b + c &= -8 \end{aligned}$$

20. Solve the system

$$\begin{aligned} 2m + 4n + p &= 16 \\ 3m + 5n + p &= -2 \\ 3m + 5n + 3p &= 0 \end{aligned}$$

21. Find the inverse of

$$\begin{bmatrix} 2 & -7 & 3 \\ 5 & -3 & 2 \\ 2 & -2 & 1 \end{bmatrix}$$

22. Solve the system

$$\begin{aligned} 2x - 7y + 3z &= 15 \\ 5x - 3y + 2z &= -3 \\ 2x - 2y + z &= 0 \end{aligned}$$

23. Solve the system

$$\begin{aligned} 2a + 3b - 7c &= 1 \\ 5a + 2b - 3c &= -3 \\ 2a + b - 2c &= -6 \end{aligned}$$

24. Solve the system

$$\begin{aligned} 3m + 2n - 7p &= 9 \\ 2m + 5n - 3p &= -2 \\ m + 2n - 2p &= -1 \end{aligned}$$

In Exercises 25–32 find the value(s) of the unknowns that make the matrices inverses of each other.

25. $\begin{bmatrix} 7 & -4 \\ 10 & -6 \end{bmatrix}$ and $\begin{bmatrix} 3 & -2 \\ 5 & x \end{bmatrix}$

26. $\begin{bmatrix} 5 & 9 \\ 3 & 6 \end{bmatrix}$ and $\begin{bmatrix} 2 & -3 \\ -1 & a \end{bmatrix}$

27. $\begin{bmatrix} 8 & 5 \\ -4 & -3 \end{bmatrix}$ and $\begin{bmatrix} a & b \\ -1 & -2 \end{bmatrix}$

28. $\begin{bmatrix} 2 & -7 \\ -3 & 9 \end{bmatrix}$ and $\begin{bmatrix} -3 & m \\ -1 & p \end{bmatrix}$

29. $\begin{bmatrix} 6 & -19 \\ 2 & -6 \end{bmatrix}$ and $\begin{bmatrix} -3 & a \\ b & 3 \end{bmatrix}$

30. $\begin{bmatrix} -2 & 8 \\ -6 & 25 \end{bmatrix}$ and $\begin{bmatrix} s & 4 \\ -3 & t \end{bmatrix}$

31. $\begin{bmatrix} 0 & 2 & 3 \\ 2 & -1 & -2 \\ 1 & 1 & 1 \end{bmatrix}$ and $\begin{bmatrix} 1 & 1 & -1 \\ -4 & a & 6 \\ 3 & b & -4 \end{bmatrix}$

32. $\begin{bmatrix} -1 & -1 & 1 \\ 0 & 5 & -4 \\ -8 & -2 & 3 \end{bmatrix}$ and $\begin{bmatrix} 7 & x & -1 \\ 32 & 5 & -4 \\ 40 & y & -5 \end{bmatrix}$

Carla's Nut Shoppe has three mixes of nuts in bags marked A, B, and C with contents as shown in the table.

	A	B	C
Peanuts	3 oz	5 oz	3 oz
Pecans	3 oz	5 oz	1 oz
Brazil nuts	2 oz	4 oz	1 oz

33. Carla has an order for a mixture containing 25 oz of peanuts, 19 oz of pecans and 15 oz of Brazil nuts. Write the system of equations as a matrix equation and use the inverse matrix to solve the system and tell Carla how to fill the order.

34. Carla has another order for a mixture containing 36 oz of peanuts, 28 oz of pecans, and 22 oz of Brazil nuts. Use your inverse from Exercise 33 to solve this problem and tell Carla how many bags of each mix to use to fill the order.

35. Carla has another order for a mixture containing 21 oz of peanuts, 19 oz of pecans, and 15 oz of Brazil nuts. Use the inverse from Exercise 33 to tell Carla how to fill the order.

36. Bryan manages one of Carla's Nut Shoppes and receives an order for a mixture containing 40 oz of peanuts, 16 oz of pecans, and 16 oz of Brazil nuts. Can he fill the order? Explain your answer.

10.8 Determinants

There is another aspect of matrices related to systems of equations that we should not overlook. Associated with each square matrix A is a number called the *determinant* of A, denoted det A. In this section we will see how to compute det A, and in the next section how to use determinants to solve systems of equations.

Determinant of a 2×2 matrix

For a 2×2 matrix, the **determinant** is the product of the diagonal entries from upper left to lower right minus the product of the diagonal entries from upper right to lower left as shown below:

$$\det \begin{bmatrix} a & b \\ c & d \end{bmatrix} = ad - bc$$

Example 1

$$\det \begin{bmatrix} 4 & -3 \\ 1 & 2 \end{bmatrix} = (4)(2) - (1)(-3) = 11$$

■

For the 3×3 matrix the determinant can be computed in several ways. The first method presented is similar to that used for the 2×2 matrix.

To compute the determinant of the matrix

$$\begin{bmatrix} 2 & 3 & -1 \\ 1 & 0 & 7 \\ 2 & -1 & -3 \end{bmatrix}$$

first, repeat the first two columns to the right of the matrix:

$$\begin{bmatrix} 2 & 3 & -1 \\ 1 & 0 & 7 \\ 2 & -1 & -3 \end{bmatrix} \begin{matrix} 2 & 3 \\ 1 & 0 \\ 2 & -1 \end{matrix}$$

Now form the sum of the products along the diagonals from upper left to lower right and subtract the sum of the products along the diagonals from upper right to lower left:

Determinant of a 3×3 matrix

$$\begin{bmatrix} 2 & 3 & -1 \\ 1 & 0 & 7 \\ 2 & -1 & -3 \end{bmatrix} \begin{matrix} 2 & 3 \\ 1 & 0 \\ 2 & -1 \end{matrix}$$

$$-(0 - 14 - 9) \quad + \quad (0 + 42 + 1) = 66$$

Thus, det $A = 66$.

Example 2 Compute

$$\det \begin{bmatrix} -6 & 0 & 6 \\ 3 & -4 & 1 \\ 2 & -1 & -2 \end{bmatrix}$$

Repeat the first two columns and compute the diagonal products:

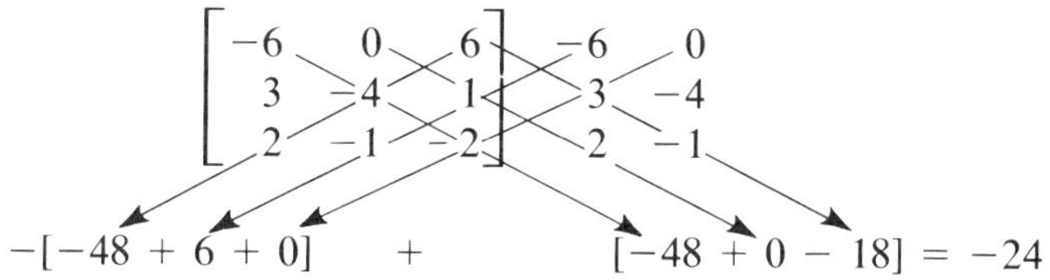

The determinant is −24. ■

This type of calculation cannot be modified to work with 4 × 4 or larger matrices, but we now give a second way of finding the determinant, which can be used for all square matrices.

Minor
Cofactor

> If A is an $n \times n$ matrix, we define the ***i, j* minor** of A, denoted M_{ij}, to be the $(n-1) \times (n-1)$ matrix obtained by deleting the ith row and jth column of A. The ***i, j* cofactor** of A, written A_{ij}, is defined by
>
> $$A_{ij} = (-1)^{i+j} \det M_{ij}$$

Example 3 If

$$A = \begin{bmatrix} -6 & 0 & 6 \\ 3 & -4 & 1 \\ 2 & -1 & -2 \end{bmatrix}$$

then

$$M_{11} = \begin{bmatrix} \not{-6} & \not{0} & \not{6} \\ \not{3} & -4 & 1 \\ \not{2} & -1 & -2 \end{bmatrix} = \begin{bmatrix} -4 & 1 \\ -1 & -2 \end{bmatrix}$$

$$M_{21} = \begin{bmatrix} \not{-6} & 0 & 6 \\ \not{3} & \not{-4} & \not{1} \\ \not{2} & -1 & -2 \end{bmatrix} = \begin{bmatrix} 0 & 6 \\ -1 & -2 \end{bmatrix}$$

$$M_{31} = \begin{bmatrix} \not{-6} & 0 & 6 \\ \not{3} & -4 & 1 \\ \not{2} & \not{-1} & \not{-2} \end{bmatrix} = \begin{bmatrix} 0 & 6 \\ -4 & 1 \end{bmatrix}$$

and

$$A_{11} = (-1)^2 \det M_{11} = 9$$
$$A_{21} = (-1)^3 \det M_{21} = -6$$
$$A_{31} = (-1)^4 \det M_{31} = 24$$ ■

Expansion by cofactors

> If A is an $n \times n$ matrix with a_{ij} denoting the entry in the ith row and jth column and A_{ij} denoting the i, j cofactor, then
>
> for each i, $\det A = a_{i1}A_{i1} + a_{i2}A_{i2} + \ldots + a_{in}A_{in}$ (1)
>
> for each j, $\det A = a_{1j}A_{1j} + a_{2j}A_{2j} + \ldots + a_{nj}A_{nj}$ (2)
>
> Equation (1) is called the expansion of det A by cofactors of the ith row, and Equation (2) is the expansion of det A by cofactors of the jth column.

Since the minors of a 4×4 matrix are 3×3 matrices, etc., this gives a method for computing the determinant of larger matrices.

Example 4 Compute

$$\det \begin{bmatrix} 2 & 3 & -1 \\ 1 & 0 & 7 \\ 2 & -1 & -3 \end{bmatrix}$$

by using cofactors of the second column.

$$a_{12} = 3 \quad \text{and} \quad A_{12} = (-1)^3 \det \begin{bmatrix} 1 & 7 \\ 2 & -3 \end{bmatrix} = 17$$

$$a_{22} = 0 \quad \text{and} \quad A_{22} = (-1)^4 \det \begin{bmatrix} 2 & -1 \\ 2 & -3 \end{bmatrix} = -4$$

$$a_{32} = -1 \quad \text{and} \quad A_{32} = (-1)^5 \det \begin{bmatrix} 2 & -1 \\ 1 & 7 \end{bmatrix} = -15$$

Thus

$$\det A = (3)(17) + (0)(-4) + (-1)(-15) = 66$$

which agrees with the previous computation at the beginning of the section. ■

You have no doubt noticed in Example 4 that it was unnecessary to compute A_{22}, because $a_{22} = 0$. In fact, in using cofactors to compute the determinant, it is often easiest to use a row or column that contains zeros. In Section 10.6 we used elementary row operations on a matrix to obtain zeros in certain positions. With certain restrictions, we may use the same operations on a matrix to make the computation of the determinant easier.

The **determinant function** has the following properties:

Properties of the determinant function

1. If every entry in a row or column of a matrix A is zero, then $\det A = 0$.
2. If two rows or columns of A are interchanged to obtain the matrix B, then $\det B = -\det A$.
3. If every entry of a row or column of A is multiplied by k to obtain the matrix B, then $\det B = k \det A$.

4. If a multiple of one row (or column) is added to another row (or column) of A to obtain a matrix B, then $\det B = \det A$.
5. If one row (or column) of A is a multiple of another row (or column) of A, then $\det A = 0$.

We illustrate each of the properties with an example.

Example 5 $\det \begin{bmatrix} 0 & 0 \\ 9 & 5 \end{bmatrix} = 0$ because all entries in the first row are 0. ■

Example 6

$$\det \begin{bmatrix} 2 & 3 & 0 \\ 1 & 1 & -5 \\ 0 & 4 & 9 \end{bmatrix} = -\det \begin{bmatrix} 1 & 1 & -5 \\ 2 & 3 & 0 \\ 0 & 4 & 9 \end{bmatrix}$$

because the first and second rows were interchanged. ■

Example 7

$$\det \begin{bmatrix} 7 & -7 & 14 \\ 4 & 1 & 0 \\ 5 & 6 & 9 \end{bmatrix} = 7 \det \begin{bmatrix} 1 & -1 & 2 \\ 4 & 1 & 0 \\ 5 & 6 & 9 \end{bmatrix} \quad \text{by Property 3.}$$

■

Notice how, in Example 7, Property 3 is used to remove common factors from a row (or column) and make the numbers in the matrix smaller, thereby reducing the computation necessary to find the determinant.

Example 8

$$\det \begin{bmatrix} 2 & -8 & 9 \\ 1 & -2 & 9 \\ 4 & -1 & 3 \end{bmatrix} \overset{(-2)}{=} \det \begin{bmatrix} 0 & -4 & -9 \\ 1 & -2 & 9 \\ 4 & -1 & 3 \end{bmatrix} {\scriptstyle(-4)}$$

$$= \det \begin{bmatrix} 0 & -4 & -9 \\ 1 & -2 & 9 \\ 0 & 7 & -33 \end{bmatrix}$$

The second matrix is obtained by multiplying row 2 by -2 and adding to row 1. The third matrix is obtained from the second matrix by multiplying row 2 by -4 and adding to row 3. By Property 4, all the matrices have the same determinant. ■

Example 9

$$\det \begin{bmatrix} 6 & 7 & 9 \\ 0 & 4 & 0 \\ 2 & 4 & 3 \end{bmatrix} = 0$$

since the first column is $\frac{2}{3}$ of the third column. ■

Example 10 Compute the determinant of

$$\begin{bmatrix} 2 & -8 & 9 \\ 1 & -2 & 9 \\ 4 & -1 & 3 \end{bmatrix}$$

Looking back at Example 8, we see that

$$\det\begin{bmatrix} 2 & -8 & 9 \\ 1 & -2 & 9 \\ 4 & -1 & 3 \end{bmatrix} = \det\begin{bmatrix} 0 & -4 & -9 \\ 1 & -2 & 9 \\ 4 & -1 & 3 \end{bmatrix} = \det\begin{bmatrix} 0 & -4 & -9 \\ 1 & -2 & 9 \\ 0 & 7 & -33 \end{bmatrix}$$

The first column in the last matrix has only one nonzero entry, so expand by cofactors of that column:

$$\det\begin{bmatrix} 0 & -4 & -9 \\ 1 & -2 & 9 \\ 0 & 7 & -33 \end{bmatrix} = (1)(-1)^{1+2}\det\begin{bmatrix} -4 & -9 \\ 7 & -33 \end{bmatrix} = -195$$

■

Example 11 Compute the determinant of

$$\begin{bmatrix} 4 & 3 & 1 & -1 \\ 1 & -5 & 2 & 1 \\ 6 & -2 & 4 & 3 \\ -4 & 1 & -3 & 2 \end{bmatrix}$$

To avoid fractions, we choose a row or column that contains a 1 or -1 and use Property 4 to obtain a row or column with only one nonzero entry. The operations are shown with the arrows.

$$\det\begin{bmatrix} 4 & 3 & 1 & -1 \\ 1 & -5 & 2 & 1 \\ 6 & -2 & 4 & 3 \\ -4 & 1 & -3 & 2 \end{bmatrix} = \det\begin{bmatrix} 4 & 3 & 0 & -1 \\ 1 & -5 & 3 & 1 \\ 6 & -2 & 7 & 3 \\ -4 & 1 & -1 & 2 \end{bmatrix}$$

$$= \det\begin{bmatrix} 4 & 0 & 0 & -1 \\ 1 & -2 & 3 & 1 \\ 6 & 7 & 7 & 3 \\ -4 & 7 & -1 & 2 \end{bmatrix}$$

$$= \det\begin{bmatrix} 0 & 0 & 0 & -1 \\ 5 & -2 & 3 & 1 \\ 18 & 7 & 7 & 3 \\ 4 & 7 & -1 & 2 \end{bmatrix} = (-1)(-1)^{1+4}\det\begin{bmatrix} 5 & -2 & 3 \\ 18 & 7 & 7 \\ 4 & 7 & -1 \end{bmatrix}$$

This last equality is the result of expansion by cofactors of the first row and consists of one term,since there is only one nonzero entry in that row. To continue,

$$\det\begin{bmatrix} 5 & -2 & 3 \\ 18 & 7 & 7 \\ 4 & 7 & -1 \end{bmatrix} \overset{(3)}{=} \det\begin{bmatrix} 17 & 19 & 0 \\ 18 & 7 & 7 \\ 4 & 7 & -1 \end{bmatrix} \overset{(7)}{=} \det\begin{bmatrix} 17 & 19 & 0 \\ 46 & 56 & 0 \\ 4 & 7 & -1 \end{bmatrix}$$

$$= (-1)(-1)^{3+3}\det\begin{bmatrix} 17 & 19 \\ 46 & 56 \end{bmatrix} = -78$$

The determinant of the original matrix is -78. ■

Exercises 10.8

Compute the determinants of the given matrices.

1. $\begin{bmatrix} 1 & -6 \\ -2 & 3 \end{bmatrix}$

2. $\begin{bmatrix} 4 & -8 \\ 2 & 6 \end{bmatrix}$

3. $\begin{bmatrix} 5 & 0 \\ -4 & 2 \end{bmatrix}$

4. $\begin{bmatrix} 2 & -4 \\ 1 & -2 \end{bmatrix}$

5. $\begin{bmatrix} 1 & -3 \\ 2 & -6 \end{bmatrix}$

6. $\begin{bmatrix} -6 & 5 \\ 1 & 8 \end{bmatrix}$

7. $\begin{bmatrix} 3 & 1 & 2 \\ -2 & -1 & 1 \\ 5 & 1 & -3 \end{bmatrix}$

8. $\begin{bmatrix} 5 & 1 & -3 \\ -2 & -1 & 1 \\ 4 & 16 & -1 \end{bmatrix}$

9. $\begin{bmatrix} 3 & 4 & 5 \\ 1 & 16 & 1 \\ 2 & -1 & -3 \end{bmatrix}$

10. $\begin{bmatrix} 3 & -2 & 4 \\ 1 & -1 & 16 \\ 2 & 1 & -1 \end{bmatrix}$

11. $\begin{bmatrix} 3 & -1 & 1 \\ 1 & -1 & 16 \\ 5 & -2 & 4 \end{bmatrix}$

12. $\begin{bmatrix} 3 & -1 & -5 \\ -1 & 16 & 4 \\ -2 & -1 & -3 \end{bmatrix}$

13. $\begin{bmatrix} 0 & 2 & 3 \\ 2 & -1 & -2 \\ 1 & 1 & 1 \end{bmatrix}$

14. $\begin{bmatrix} -1 & -1 & 1 \\ 0 & 5 & -4 \\ -8 & -2 & 3 \end{bmatrix}$

15. $\begin{bmatrix} 7 & 1 & 1 \\ 32 & 5 & 4 \\ 40 & 6 & -5 \end{bmatrix}$

16. $\begin{bmatrix} 1 & 1 & -1 \\ -4 & -3 & 6 \\ 3 & 2 & -4 \end{bmatrix}$

17. $\begin{bmatrix} 1 & -2 & 2 \\ 2 & 1 & -3 \\ 3 & 2 & -1 \end{bmatrix}$

18. $\begin{bmatrix} 1 & 4 & 2 \\ 2 & 1 & -3 \\ 3 & 5 & -1 \end{bmatrix}$

19. $\begin{bmatrix} 2 & -7 & 3 \\ 5 & -3 & 2 \\ 2 & -2 & 1 \end{bmatrix}$

20. $\begin{bmatrix} 2 & -7 & 1 \\ 5 & -3 & 4 \\ 2 & -2 & 1 \end{bmatrix}$

21. $\begin{bmatrix} 1 & -2 & 3 & 4 \\ -1 & 0 & 2 & 2 \\ 3 & 4 & -3 & 4 \\ 0 & -2 & -1 & 3 \end{bmatrix}$

22. $\begin{bmatrix} 2 & 3 & -4 & 1 \\ -1 & 2 & -1 & 0 \\ 3 & 5 & 2 & -4 \\ 2 & -3 & 4 & -1 \end{bmatrix}$

10.9 Cramer's Rule

Consider the system

$$ax + by = h$$
$$cx + dy = k$$

To solve the system, multiply the first equation by $-c$ and the second equation by a to obtain

$$(ad - bc)\, y = ak - hc$$

If $ad - bc \neq 0$, then

$$y = \frac{ak - hc}{ad - bc}$$

In a similar fashion, we find

$$x = \frac{dh - bk}{ad - bc}$$

Now notice that

$$\det \begin{bmatrix} a & b \\ c & d \end{bmatrix} = ad - bc$$

$$\det \begin{bmatrix} a & h \\ c & k \end{bmatrix} = ak - hc$$

$$\det \begin{bmatrix} h & b \\ k & d \end{bmatrix} = dh - bk$$

If we write

$$A = \begin{bmatrix} a & b \\ c & d \end{bmatrix}, \qquad A_x = \begin{bmatrix} h & b \\ k & d \end{bmatrix}, \quad \text{and} \quad A_y = \begin{bmatrix} a & h \\ c & k \end{bmatrix}$$

then we have discovered that if $\det A \neq 0$, the solution to the system

$$\begin{aligned} ax + by &= h \\ cx + dy &= k \end{aligned}$$

is

$$x = \frac{\det A_x}{\det A} \qquad y = \frac{\det A_y}{\det A}$$

Cramer's Rule for 2×2 systems

where A is the matrix of coefficients, A_x is obtained from A by replacing the column of coefficients of x by the column of constant terms, and A_y is obtained from A by replacing the column of coefficients of y by the column of constant terms. This result is called **Cramer's Rule** for systems of two equations in two unknowns.

Example 1 Use Cramer's Rule to solve the system

$$\begin{aligned} 2x - 3y &= 6 \\ x + y &= 4 \end{aligned}$$

$$A = \begin{bmatrix} 2 & -3 \\ 1 & 1 \end{bmatrix} \qquad A_x = \begin{bmatrix} 6 & -3 \\ 4 & 1 \end{bmatrix} \qquad A_y = \begin{bmatrix} 2 & 6 \\ 1 & 4 \end{bmatrix}$$

and

$$\det A = 5 \qquad \det A_x = 18 \qquad \det A_y = 2$$

Thus

$$x = \frac{\det A_x}{\det A} = \frac{18}{5}$$

$$y = \frac{\det A_y}{\det A} = \frac{2}{5}$$ ■

The general form of Cramer's Rule we state without proof.

Cramer's Rule

> Let A be the coefficient matrix of a system of n linear equations in n unknowns $x_1, x_2, \ldots, x_n$. If $\det A \neq 0$, then the system has a unique solution given by
>
> $$x_i = \frac{\det A_{x_i}}{\det A} \quad \text{for} \quad i = 1, 2, \ldots, n$$
>
> where A_{x_i} is the matrix obtained from A by replacing the column of coefficients of x_i by the column of constant terms.

Example 2 Use Cramer's Rule to solve the system

$$\begin{aligned} 3x - 2y + 5z &= 4 \\ x - y + z &= 16 \\ 2x + y - 3z &= -1 \end{aligned}$$

$$A = \begin{bmatrix} 3 & -2 & 5 \\ 1 & -1 & 1 \\ 2 & 1 & -3 \end{bmatrix} \quad \text{so} \quad \det A = 11$$

$$A_x = \begin{bmatrix} 4 & -2 & 5 \\ 16 & -1 & 1 \\ -1 & 1 & -3 \end{bmatrix} \quad \text{so} \quad \det A_x = -11$$

$$A_y = \begin{bmatrix} 3 & 4 & 5 \\ 1 & 16 & 1 \\ 2 & -1 & -3 \end{bmatrix} \quad \text{so} \quad \det A_y = -286$$

$$A_z = \begin{bmatrix} 3 & -2 & 4 \\ 1 & -1 & 16 \\ 2 & 1 & -1 \end{bmatrix} \quad \text{so} \quad \det A_z = -99$$

By Cramer's Rule

$$x = \frac{\det A_x}{\det A} = \frac{-11}{11} = -1$$

$$y = \frac{\det A_y}{\det A} = \frac{-286}{11} = -26$$

$$z = \frac{\det A_z}{\det A} = \frac{-99}{11} = -9$$ ■

Exercises 10.9

Solve the systems by Cramer's Rule.

1. $4x + 3y = 7$
$3x + 2y = 1$

2. $3x - 5y = 1$
$2x + 7y = 0$

3. $3x - y = 9$
$5x - y = 3$

4. $x - 2y = 8$
$3x + 2y = 1$

5. $\frac{1}{2}x - \frac{3}{4}y = -2$
$\frac{1}{4}x + \frac{1}{4}y = -1$

6. $\frac{2}{3}x - \frac{1}{3}y = 4$
$-\frac{1}{3}x + \frac{1}{2}y = -1$

7. $3x + y + 3z = 8$
$x - 2y - z = 1$
$2x + 5y + 2z = 1$

8. $3x - 2y + z = 5$
$x + y - 2z = -2$
$2x - 2y + 4z = 8$

9. $2a + 5b + 2c = -9$
$-7a - 3b - 2c = 2$
$3a + 2b + c = 0$

10. $3m - 5n + 3p = 12$
$-m + 3n - 2p = -1$
$2m + n - p = -4$

11. $x + y = 6$
$y + w = 0$
$x + y + z = -4$
$2z + w = -2$

12. $x + w = 15$
$2x + y = -8$
$3x + y + z = 0$
$y + w = -5$

13. If $\det \begin{bmatrix} a & b \\ c & d \end{bmatrix} = 0$, prove that the lines $ax + by = h$ and $cx + dy = k$ are either parallel or coincident. Conclude that the system

$$ax + by = h$$
$$cx + dy = k$$

has either no solutions or infinitely many solutions if

$$\det \begin{bmatrix} a & b \\ c & d \end{bmatrix} = 0$$

14. If $\det \begin{bmatrix} a & b \\ c & d \end{bmatrix} \neq 0$, show that

$$\begin{bmatrix} \dfrac{d}{\det A} & \dfrac{-b}{\det A} \\ \dfrac{-c}{\det A} & \dfrac{a}{\det A} \end{bmatrix}$$

is the inverse of $\begin{bmatrix} a & b \\ c & d \end{bmatrix}$.

15. Use the method of Problem 14 to find the inverses of

$$\begin{bmatrix} 7 & -2 \\ 10 & -3 \end{bmatrix} \quad \text{and} \quad \begin{bmatrix} 2 & 3 \\ 1 & 5 \end{bmatrix}$$

Chapter Review

Solve the systems in Exercises 1–7.

1. $x + 3y = 8$
$y = x + 2$

2. $4x - 3y = 6$
$3x - 4y = 1$

3. $0.2x + 0.3y = 0.9$
$0.4x + 0.3y = 0.8$

4. $2x - 3y = 5$
$3x - 5y = 9$

5. $2x - y = 6$
$x^2 - 6x + y^2 + 4 = 0$

6. $x^2 + y^2 = 25$
$12x^2 + 7xy - 12y^2 = 0$

7. $8x^2 - y^2 + 9 = 0$
$3x^2 + y^2 = 31$

8. Craig invested \$800 at one rate and \$1,000 at another rate, and his interest for 1 year was \$136. His sister invested the same amounts with the rates interchanged, and her interest check was \$134. Find the rates.

9. Jean arranged a banquet for a group of people at a cost of \$140. The next year the cost per person went up 50¢ and two more people attended the banquet for a cost of \$165. How many people attended the first banquet?

Sketch the solution sets for the systems in Exercises 10–13.

10. $\begin{aligned} 2x + 3y &\geq 5 \\ 4x + 2y &\geq 5 \\ y &\leq 3 \\ x &\leq 3 \end{aligned}$

11. $\begin{aligned} y &\geq x - 1 \\ y &\leq 1 - x \\ x &\geq 0 \end{aligned}$

12. $\begin{aligned} 3x^2 + 4y^2 &\leq 192 \\ 3x^2 - y^2 &\geq 12 \end{aligned}$

13. $\begin{aligned} x^2 + y^2 &\leq 34 \\ y - x &\geq 2 \end{aligned}$

14. Machine A produces 30 units/hr and costs \$20/hr to operate. Machine B produces 50 units/hr and costs \$30/hr to operate. Production demands that 650 units be produced in a 10-hr day. How can this be done at a minimum cost?

For the systems in Exercises 15–18 find the general solution in terms of z and find a particular solution for $z = -2$.

15. $\begin{aligned} 2x - 3y + z &= 6 \\ 3x + 5y - z &= 8 \end{aligned}$

16. $\begin{aligned} x + y - 4z &= 1 \\ 2x - 3y - z &= 4 \end{aligned}$

17. $\begin{aligned} 4x &= y + 4z - 2 \\ y &= x + 2z \end{aligned}$

18. $\begin{aligned} z &= 5 + 2x \\ y &= 2x - 3z + 1 \end{aligned}$

Solve the system in Exercises 19–22 if the system is not inconsistent.

19. $\begin{aligned} x + 3y &= -5 \\ 2x - 3z &= -4 \\ 2y + 5z &= 6 \end{aligned}$

20. $\begin{aligned} -10x - 6y + 4z &= 2 \\ 2x + 3y + z &= 2 \\ -7x + 2y + 9z &= 1 \end{aligned}$

21. $\begin{aligned} 2x - 7y - 3z &= -11 \\ x - 3y - z &= -4 \\ x - 2y &= -1 \end{aligned}$

22. $\begin{aligned} x + y + 3z &= 0 \\ x - 3y - z &= -4 \\ 2x - 7y - 3z &= -11 \end{aligned}$

23. The sum of the angles of a triangle is 180°. If the largest angle exceeds the smallest angle by 60° and if the remaining angle is 3 times the size of the smallest angle, what is the size of each angle?

24. The equation of a circle is $x^2 + y^2 + ax + by + c = 0$. Find the equation if the circle passes through the points (4, 8), (−2, 8), and (5, 1).

Compute each product if defined.

25. $\begin{bmatrix} 1 & 3 \\ -2 & 7 \end{bmatrix} \begin{bmatrix} 5 \\ 4 \end{bmatrix}$

26. $\begin{bmatrix} 1 & 3 \\ 7 & 9 \end{bmatrix} \begin{bmatrix} 1 & 6 \\ 3 & 2 \end{bmatrix}$

27. $\begin{bmatrix} 4 & -1 \\ 1 & 0 \end{bmatrix} \begin{bmatrix} 9 & 4 \\ 1 & -11 \end{bmatrix}$

28. $\begin{bmatrix} 1 & 4 & 0 \\ 2 & 2 & -1 \\ 6 & 1 & 1 \end{bmatrix} \begin{bmatrix} 0 \\ 1 \\ 1 \end{bmatrix}$

29. $\begin{bmatrix} 1 & 1 & 7 \\ 2 & 0 & 4 \\ 1 & 3 & -2 \end{bmatrix} \begin{bmatrix} 2 & -3 & 4 \\ 6 & 1 & 0 \\ -2 & 5 & -7 \end{bmatrix}$

30. $\begin{bmatrix} 1 & 3 \\ -2 & 5 \end{bmatrix} \begin{bmatrix} 4 & 1 \\ 6 & -2 \end{bmatrix} \begin{bmatrix} 0 & 9 \\ 3 & -6 \end{bmatrix}$

Write each of the following systems in matrix form.

31. $\begin{aligned} 3x + 7y &= 16 \\ x - 5y &= 17 \end{aligned}$

32. $\begin{aligned} 4x - 17y &= 11 \\ -x + 6y &= 0 \end{aligned}$

33. $\begin{aligned} 3x - 2y + z &= 4 \\ y - z &= 9 \end{aligned}$

34. $\begin{aligned} 4x + 6y - 11z &= 19 \\ 2x - y \qquad &= 8 \end{aligned}$

Write the matrix equation as a system of equations.

35. $\begin{bmatrix} 5 & 4 \\ 2 & 3 \\ -1 & 0 \end{bmatrix} \begin{bmatrix} x \\ y \end{bmatrix} = \begin{bmatrix} 1 \\ -1 \\ 6 \end{bmatrix}$

36. $\begin{bmatrix} 1 & -2 & 3 \\ 4 & 0 & 2 \\ 2 & -3 & 0 \end{bmatrix} \begin{bmatrix} x \\ y \\ z \end{bmatrix} = \begin{bmatrix} 0 \\ 1 \\ 8 \end{bmatrix}$

Solve the systems by using Gaussian elimination and the augmented matrix.

37. $\begin{aligned} x + 2y + 3z &= 4 \\ x + y + z &= 1 \\ x + 3y + 7z &= 13 \end{aligned}$

38. $\begin{aligned} 2z - y + 3x &= 4 \\ y + x \qquad &= 4 \\ z + x - 1 &= 0 \end{aligned}$

Solve the systems by inverting the coefficient matrix.

39. $\begin{aligned} x + 3y + 3z &= 9 \\ x + 4y + 3z &= 6 \\ x + 3y + 4z &= 6 \end{aligned}$

40. $\begin{aligned} x + 2y + 3z &= -7 \\ 2x + 4y + 5z &= 3 \\ 3x + 5y + 6z &= 4 \end{aligned}$

Solve by Cramer's Rule.

41. $\begin{aligned} 3x + 7y &= 16 \\ x - 5y &= 17 \end{aligned}$

42. $\begin{aligned} x - 5y + 3z &= -2 \\ x - 2y + z &= 14 \\ x - 7y + 4z &= -9 \end{aligned}$

Chapter Test

Solve the systems.

1. $\begin{aligned} 2x + y &= 2 \\ x - y &= 4 \end{aligned}$

2. $\begin{aligned} 2x + 7y &= 1 \\ 3x - 5y &= 7 \end{aligned}$

3. Clarence and Luis own 50 pigs in partnership, with Luis owning 4 times as many as Clarence. How many does each own?

4. Find the general solution in terms of z:

$$\begin{aligned} 10x + 9y - 10z &= -10 \\ 4x + y - 8z &= -2 \end{aligned}$$

5. Find the particular solution to the system in Problem 4 when $z = 1$.

6. Solve the system

$$\begin{aligned} x^2 + 2y^2 &= 22 \\ 2x - 3y &= -5 \end{aligned}$$

7. Graph the solution set for the system

$$y \le 3x + 4$$
$$y \le 4 - x$$
$$y \ge -1$$

8. Find the minimum and maximum values of $C = 2x - 5y$ on the solution set for the system in Problem 7.

9. Use the augmented matrix to solve the system

$$\begin{aligned} 2x + 3y - z &= 0 \\ x + 4y + 2z &= -5 \\ 2x - y - z &= 4 \end{aligned}$$

10. Solve by finding the inverse of the coefficient matrix:

$$\begin{aligned} 3x - 5y + 3z &= 2 \\ x - 3y + 2z &= -1 \\ 2x + y - z &= 5 \end{aligned}$$

11. Compute the determinant of the matrix:

$$\begin{bmatrix} -1 & 2 & 3 \\ 0 & 5 & 4 \\ 3 & 9 & 1 \end{bmatrix}$$

12. Solve by Cramer's Rule:

$$\begin{aligned} 2x - 3y &= 6 \\ 3x - 4y &= 1 \end{aligned}$$

11 Polynomial Equations

A storage tank can be emptied with any of three pumps, A, B, and C. With all pumps operating, the tank can be emptied in 1 hr. It takes 1 hr longer to drain the tank with pump B than with pump A and 4 hr longer with pump C than with pump A. How long does it take each pump to empty the tank by itself?

Let x be the number of hours it takes for pump A to empty the tank. Then $x + 1$ is the number of hours it takes pump B to drain the tank and $x + 4$ is the number of hours it takes pump C to drain the tank. Thus,

$$\frac{1}{x} = \text{fraction of the tank emptied by pump A in 1 hr}$$

$$\frac{1}{x+1} = \text{fraction of the tank emptied by pump B in 1 hr}$$

$$\frac{1}{x+4} = \text{fraction of the tank emptied by pump C in 1 hr}$$

With all pumps running, the tank empties in 1 hr, so

$$\frac{1}{x} + \frac{1}{x+1} + \frac{1}{x+4} = 1$$

Simplifying this equation, we obtain

$$x^3 + 2x^2 - 6x - 4 = 0$$

This is a new type of equation, and in this chapter we develop some techniques for solving such equations.

11.1 Remainder and Factor Theorems

Before discussing the Remainder and Factor Theorems we need to review long division with polynomials. The process is very similar to long division with real numbers.

Example 1 Divide $x^2 + 2x - 15$ by $x - 3$.

Division of polynomials

Step 1 Write the division problem in a format similar to that for real numbers:

$$x - 3\overline{)\,x^2 + 2x - 15}$$

Step 2 Divide x^2 (the term of highest degree of the dividend) by x (the term of highest degree in the divisor). Multiply the result by $x - 3$ and subtract from the dividend:

$$\begin{array}{r|l} & x \\ \hline x - 3 & x^2 + 2x - 15 \\ & \underline{x^2 - 3x} \\ & \quad\quad 5x - 15 \end{array}$$

Step 3 Divide $5x$ by x and multiply the result, 5, by $x - 3$ and subtract:

$$\begin{array}{r|l} & x + 5 \\ \hline x - 3 & x^2 + 2x - 15 \\ & \underline{x^2 - 3x} \\ & \quad\quad 5x - 15 \\ & \quad\quad \underline{5x - 15} \\ & \quad\quad\quad\quad\quad 0 \end{array}$$

Since the remainder is 0,

$$\frac{x^2 + 2x - 15}{x - 3} = x + 5$$ ■

Example 2 Divide $5x^4 - 3x^2 + 2x + 1$ by $x - 3$.

$$\begin{array}{r|l} & 5x^3 + 15x^2 + 42x + 128 \\ \hline x - 3 & 5x^4 \qquad\quad - 3x^2 + 2x + 1 \\ & \underline{5x^4 - 15x^3} \\ & \quad 15x^3 - 3x^2 \\ & \quad \underline{15x^3 - 45x^2} \\ & \qquad\quad 42x^2 + 2x \\ & \qquad\quad \underline{42x^2 - 126x} \\ & \qquad\qquad\quad 128x + 1 \\ & \qquad\qquad\quad \underline{128x - 384} \\ & \qquad\qquad\qquad\quad 385 \end{array}$$ ■

There are several things to note about this example. First of all, since there was no x^3 term in the dividend, a space was left so that the computations could be carried out more systematically. Second, we only brought down from the dividend the terms that were necessary at each stage. From Example 2 we see that there was a remainder of 385, which means that

$$5x^4 - 3x^2 + 2x + 1 = (x - 3)(5x^3 + 15x^2 + 42x + 128) + 385$$

If $P(x)$ is any polynomial in x of degree n and r is any number, then we can carry out the long division as shown above and find a polynomial $Q(x)$ of degree $n - 1$ and a number R such that for all x,

$$P(x) = (x - r)Q(x) + R$$

In the example above,

$$P(x) = 5x^4 - 3x^2 + 2x + 1$$
$$x - r = x - 3$$
$$Q(x) = 5x^3 + 15x^2 + 42x + 128$$

and

$$R = 385$$

Example 3 Divide $2x^4 - 3x + 7$ by $x + 2$.

$$\begin{array}{r|l} & 2x^3 - 4x^2 + 8x - 19 \\ \hline x + 2 & 2x^4 \qquad\qquad\qquad - 3x + 7 \\ & \underline{2x^4 + 4x^3} \\ & -4x^3 \\ & \underline{-4x^3 - 8x^2} \\ & 8x^2 - 3x \\ & \underline{8x^2 + 16x} \\ & -19x + 7 \\ & \underline{-19x - 38} \\ & 45 \end{array}$$

Thus,

$$2x^4 - 3x + 7 = (x + 2)(2x^3 - 4x^2 + 8x - 19) + 45$$ ■

What is the point of this discussion? If

$$P(x) = (x - r)Q(x) + R$$

then

$$P(r) = (r - r)Q(r) + R = R$$

This result is called the Remainder Theorem.

Remainder Theorem

> Let $P(x)$ be a polynomial of degree $n \geq 1$ and r be any number. Then there is a polynomial $Q(x)$ of degree $n - 1$ and a number R such that for all x,
>
> $$P(x) = (x - r)Q(x) + R$$
>
> In particular,
>
> $$R = P(r)$$

To check what this means, look at Example 3. The remainder after dividing $P(x) = 2x^4 - 3x + 7$ by $x + 2 = x - (-2)$ is 45; hence by the Remainder Theorem, $P(-2) = 45$. Let's check.

$$P(-2) = 2(-2)^4 - 3(-2) + 7 = 32 + 6 + 7 = 45$$

Example 4 Use the Remainder Theorem to compute $P(-4)$, where $P(x) = x^3 + x^2 - 13x - 4$.

We want to divide $P(x)$ by $x - (-4) = x + 4$:

$$\begin{array}{r}
x^2 - 3x - 1 \\
x + 4 \overline{) \, x^3 + x^2 - 13x - 4} \\
\underline{x^3 + 4x^2} \\
-3x^2 - 13x \\
\underline{-3x^2 - 12x} \\
-x - 4 \\
\underline{-x - 4} \\
0 \text{ remainder}
\end{array}$$

So, by the Remainder Theorem, $P(-4) = 0$. ■

Notice something about Example 4: because the remainder is zero,

$$P(x) = (x + 4)(x^2 - 3x - 1)$$

that is, $x + 4$ is a *factor* of $P(x)$.

Let $P(x)$ be any polynomial of degree $n \geq 1$ and r any number. Then there is a polynomial $Q(x)$ such that

$$P(x) = (x - r)Q(x) + P(r)$$

In particular, $x - r$ is a factor of $P(x)$ if and only if $P(r) = 0$.

Factor Theorem

> Let $P(x)$ be any polynomial of degree $n \geq 1$ and r any number. Then $x - r$ is a factor of $P(x)$ if and only if $P(r) = 0$.

Example 5 Use the Factor Theorem to determine whether $x - 2$ is a factor of $P(x) = x^5 - 3x^4 + 4x^2 - 2x + 4$.

By the Factor Theorem, $x - 2$ is a factor of $P(x)$ if and only if $P(2) = 0$. Therefore,

$$\begin{aligned}
P(2) &= (2)^5 - 3(2)^4 + 4(2)^2 - 2(2) + 4 \\
&= 32 - 48 + 16 - 4 + 4 \\
&= 0
\end{aligned}$$

Thus, $x - 2$ *is* a factor of $x^5 - 3x^4 + 4x^2 - 2x + 4$; in fact,

$$x^5 - 3x^4 + 4x^2 - 2x + 4 = (x - 2)(x^4 - x^3 - 2x^2 - 2)$$ ■

Exercises 11.1

In Exercises 1–12 write $P(x) = (x - r)Q(x) + R$, where $P(x)$ and r are as indicated and R is a real number.

1. $P(x) = x^4 - x^3 - x^2 + x - 1$; $r = 3$

2. $P(x) = x^5 + 4x^3 - 3x^2 + 6x - 1$; $r = 2$

3. $P(x) = x^3 - 3x^2 + 3x - 1$; $r = 1$

4. $P(x) = x^3 - 6x^2 + 12x - 8$; $r = 2$

5. $P(x) = 4x^5 - 3x^2 + 2x - 9$; $r = 0$

6. $P(x) = 6x^3 - 3x^2 + 32x - 8$; $r = 0$

7. $P(x) = x^4 - 6x^3 + 5x + 2$; $r = -1$

8. $P(x) = x^4 - 6x^3 + 5x + 2$; $r = -2$

9. $P(x) = \frac{1}{2}x^2 - \frac{1}{3}x + \frac{1}{6}$; $r = \frac{1}{4}$

10. $P(x) = \frac{1}{6}x^2 - \frac{1}{12}x + \frac{1}{4}$; $r = \frac{1}{4}$

11. $P(x) = 2x^4 - 3x^3 - 12x^2 + 7x + 6$; $r = 3$

12. $P(x) = x^4 + 2x^3 - 15x^2 - 32x - 16$; $r = 4$

Both the Factor Theorem and the Remainder Theorem hold for complex numbers. In Exercises 13–16 write $P(x) = (x - r)Q(x) + R$, where $P(x)$ and r are as given and R is a complex number.

13. $P(x) = 4x^4 - 3x^2 + 2x$; $r = i$

14. $P(x) = 4x^4 - 3x^2 + 2x$; $r = -i$

15. $P(x) = x^3 - 5ix^2 + 4x + 4i$; $r = 2i$

16. $P(x) = x^3 + 5ix^2 - 9x - 9i$; $r = -3i$

In Exercises 17–24 use the Remainder Theorem to compute the indicated numbers.

17. $P(x) = 2x^2 - 6x - 1$; find $P(1)$.

18. $P(x) = 2x^2 - 6x - 1$; find $P(-1)$.

19. $P(x) = 6x^3 - 6x^2 + 2x + 7$; find $P(3)$.

20. $P(x) = 4x^5 - 2x^4 - 6x^3 - 2x - 3$; find $P(2)$.

21. $P(x) = 5x^3 + 2x^2 - 6x - 8$; find $P(-3)$.

22. $P(x) = 5x^3 + 2x^2 - 6x - 8$; find $P(-2)$.

23. $P(x) = 18x^4 - 3x^2 + 7$; find $P(i)$.

24. $P(x) = 14x^4 + 4x^2 + 1$; find $P(-i)$.

In Exercises 25–32 use the Factor Theorem to determine whether the indicated binomial is a factor of the indicated polynomial.

25. $x - 2$; $P(x) = x^3 - 2x^2 + 4x - 8$

26. $x + 3$; $P(x) = x^3 - 3x^2 - 2x + 6$

27. $x + 2$; $P(x) = x^3 - 5x^2 - 13x + 2$

28. $x + 3$; $P(x) = x^3 + 2x^2 - 2x - 3$

29. $x + 4$; $P(x) = 2x^3 + 4x^2 - 8x - 12$

30. $x - 6$; $P(x) = x^3 - 8x^2 + 9x - 18$

31. $2x - 1$; $P(x) = 2x^3 - 3x^2 - 3x + 2$

32. $2x - 3$; $P(x) = 2x^3 - x^2 - 7x + 6$

33. Show that $x + a$ is a factor of $x^n + a^n$ if n is an odd positive integer.

34. Show that $x - a$ is a factor of $x^n - a^n$ if n is a positive integer.

35. Under what circumstances is $x + a$ a factor of $x^n - a^n$?

11.2 Synthetic Division

In carrying out long division with polynomials there is a certain amount of wasted effort. *Synthetic division* is a method of conserving some of that wasted effort. We illustrate with an example.

Divide $3x^2 - 2x + 5$ by $x - 3$.

$$\begin{array}{r}
3x + 7 \\
x - 3\,\overline{)\,3x^2 - 2x + 5} \\
\underline{3x^2 - 9x} \\
7x + 5 \\
\underline{7x - 21} \\
26
\end{array}$$

First omit all the x's and allow the positions of the numbers to indicate the powers of x involved:

$$\begin{array}{r}
3 + \textcircled{7} \\
\textcircled{1} - 3\,\overline{)\,3 \quad -2 \quad +5} \\
\underline{\textcircled{3} \quad -9} \\
7 \quad +\textcircled{5} \\
\underline{\textcircled{7} \; -21} \\
26
\end{array}$$

The numbers circled in this display are either unnecessary or repetitive, so they may be omitted and the work shortened:

$$\begin{array}{rrrr}
 & 3 & & \\
-3) & 3 & -2 & 5 \\
 & & -9 & -21 \\
\hline
 & & 7 & 26
\end{array}$$

Two more simplifications and we are there. This is division by $x - r$, and so in the process we multiply by $-r$ and subtract. But, it is simpler to multiply by r and add. Also, move the number above the line down to the bottom row of numbers:

$$\begin{array}{rrrr}
3) & 3 & -2 & 5 \\
 & & 9 & 21 \\
\hline
 & 3 & 7 & 26
\end{array}$$

quotient $3x + 7$ remainder

To summarize the process, follow the next example.

Example 1 Divide $2x^4 - 5x^3 + 2x - 1$ by $x - 3$.

Synthetic division

Step 1 Write the coefficients of the dividend in order of decreasing powers of x, with a 0 to denote missing powers of x.

$$2 \quad -5 \quad 0 \quad 2 \quad -1$$

Step 2 To divide by $x - r$, place r to the left.

$$3)\; 2 \quad -5 \quad 0 \quad 2 \quad -1$$

Step 3 Bring down the first number, multiply by r, and write the product below the next number.

$$\begin{array}{r|rrrrr} 3) & 2 & -5 & 0 & 2 & -1 \\ & & 6 & & & \\ \hline & 2 & & & & \end{array}$$

Step 4 Add these two numbers and write the sum below the line.

$$\begin{array}{r|rrrrr} 3) & 2 & -5 & 0 & 2 & -1 \\ & & 6 & & & \\ \hline & 2 & 1 & & & \end{array}$$

Step 5 Multiply this number by r, write the number below the next entry on the top line, and add. Write the sum below the line.

$$\begin{array}{r|rrrrr} 3) & 2 & -5 & 0 & 2 & -1 \\ & & 6 & 3 & & \\ \hline & 2 & 1 & 3 & & \end{array}$$

Step 6 Continue.

$$\begin{array}{r|rrrrr} 3) & 2 & -5 & 0 & 2 & -1 \\ & & 6 & 3 & 9 & 33 \\ \hline & 2 & 1 & 3 & 11 & 32 \end{array}$$

Step 7 The bottom line gives the quotient and remainder.

$2x^3 + x^2 + 3x + 11$ (quotient) $\qquad 32$ (remainder)

Thus,

$$2x^4 - 5x^3 + 2x - 1 = (x - 3)(2x^3 + x^2 + 3x + 11) + 32$$

This is the process of **synthetic division.** ■

Example 2 Use synthetic division to divide $6x^5 - 3x^4 + 2x^2 + 5$ by $x + 1$.
In this example we have $x + 1 = x - (-1)$, so $r = -1$. Remember that r is used in synthetic division.

$$\begin{array}{r|rrrrrr} -1) & 6 & -3 & 0 & 2 & 0 & 5 \\ & & -6 & 9 & -9 & 7 & -7 \\ \hline & 6 & -9 & 9 & -7 & 7 & -2 \end{array}$$

$6x^4 - 9x^3 + 9x^2 - 7x + 7$ (quotient) $\qquad -2$ (remainder)

Thus,

$$6x^5 - 3x^4 + 2x^2 + 5 = (x + 1)(6x^4 - 9x^3 + 9x^2 - 7x + 7) - 2$$ ■

Example 3 If $P(x) = 2x^4 - x^3 + 6x^2 - 7x + 4$, find $P(2)$ by synthetic division and the Remainder Theorem.
Remember, $P(2)$ is the remainder when $P(x)$ is divided by $x - 2$.

$$\begin{array}{r|rrrrr} 2) & 2 & -1 & 6 & -7 & 4 \\ & & 4 & 6 & 24 & 34 \\ \hline & 2 & 3 & 12 & 17 & 38 \leftarrow \text{remainder} \end{array}$$

Thus, $P(2) = 38$. ■

Example 4 Is $x - 5$ a factor of $x^4 - 3x^3 - 11x^2 + 3x + 10$?

$$\begin{array}{r|rrrrr} 5) & 1 & -3 & -11 & 3 & 10 \\ & & 5 & 10 & -5 & -10 \\ \hline & 1 & 2 & -1 & -2 & 0 \end{array}$$

Therefore, $x - 5$ is a factor of $x^4 - 3x^3 - 11x^2 + 3x + 10$; in fact,

$$x^4 - 3x^3 - 11x^2 + 3x + 10 = (x - 5)(x^3 + 2x^2 - x - 2)$$ ■

Exercises 11.2

In Exercises 1–12 use synthetic division to perform the indicated division.

1. $x^4 + 6x^3 - 7x^2 + 6x - 1$ by $x + 3$

2. $x^4 - 6x^3 + 7x^2 - 6x + 1$ by $x + 3$

3. $2x^5 \quad 3x^4 + x + 1$ by $x - 2$

4. $3x^5 - 2x^3 + x - 1$ by $x - 3$

5. $x^4 - 1$ by $x - 1$

6. $x^4 - 1$ by $x + 1$

7. $x^3 + 1$ by $x + 1$

8. $x^3 - 1$ by $x - 1$

9. $x^4 + x^3 + x^2 + x + 1$ by $x - 1$

10. $x^4 - x^3 + x^2 - x + 1$ by $x + 1$

11. $x^3 + 3x^2 - x + 6$ by $x - \frac{1}{2}$

12. $x^4 + 4x^3 + 7x - 6$ by $x + \frac{1}{3}$

In Exercises 13–24 use synthetic division to compute the indicated values.

13. $P(x) = 5x^4 - 6x^3 + 2x + 9$; $P(-2)$

14. $P(x) = 5x^4 - 6x^3 + 2x + 9$; $P(-3)$

15. $P(x) = x^4 - 8x^3 - x^2 + 3x - 19$; $P(3)$

16. $P(x) = x^4 + 5x^3 - 9x^2 - 3x - 8$; $P(2)$

17. $P(x) = 14x^3 - 13x^2 - 2x + 1$; $P(1)$

18. $P(x) = 32x^3 + 18x^2 - 16x - 2$; $P(-1)$

19. $P(x) = -x^4 + x^3 - 3x - 5$; $P(2i)$

20. $P(x) = -x^5 + x^2 + 6x + 8$; $P(-2i)$

21. $P(x) = x^4 - 6x^3 + 10x^2 - 6x + 1$; $P\left(\frac{1}{2}\right)$

22. $P(x) = x^4 - 2x^3 + 10x - 9$; $P\left(-\frac{1}{4}\right)$

23. $P(x) = x^4 + x^3 + x^2 + x + 1$; $P(1 + i)$

24. $P(x) = x^4 - x^3 + x^2 - x + 1$; $P(1 - i)$

For Your Calculator

Synthetic division is also referred to as "synthetic substitution" and is related to the nested form for polynomials and the Remainder Theorem. This may be seen by comparing the key sequence for computing values of a polynomial by both methods. The example below is for a calculator with algebraic logic and hierarchy of operations.

If $P(x) = 3x^4 - 4x^3 + 7x^2 + 6x - 8$, compute $P(2)$.

By synthetic division

$$\begin{array}{r|rrrrr} 2) & 3 & -4 & 7 & 6 & -8 \\ & & 6 & 4 & 22 & 56 \\ \hline & 3 & 2 & 11 & 28 & 48 \end{array}$$

To carry out this computation on the calculator, the key sequence is

$3\ \boxed{\times}\ 2\ \boxed{-}\ 4\ \boxed{=}\ \boxed{\times}\ 2\ \boxed{+}\ 7\ \boxed{=}\ \boxed{\times}\ 2\ \boxed{+}\ 6\ \boxed{=}\ \boxed{\times}\ 2\ \boxed{-}\ 8\ \boxed{=}$ answer 48

In nested form, $P(x) = \{[(3x - 4)x + 7]x + 6\}x - 8$, and the key sequence for computing $P(2)$ is

3 [×] 2 [−] 4 [=] [×] 2 [+] 7 [=] [×] 2 [+] 6 [=] [×] 2 [−] 8 [=]   answer 48

It is no surprise that the final display is 48 in both cases, because the first method gives the result by the Remainder Theorem and the second by computation using the nested form. But it is interesting to note that the key sequence is *identical* for the two methods. Synthetic division (or synthetic substitution if you choose) is not only a short way to carry out long division, but also provides a quick way to compute values of polynomials without writing them in nested form.

Use your calculator to compute the indicated values.

25. $P(x) = x^4 - 6x^3 + 5x + 2$; $P(-1)$

26. $P(x) = x^3 - 6x^2 + 12x - 8$; $P(2)$

27. $S(a) = a^4 - a^3 - a^2 + a - 1$; $S(3)$

28. $R(t) = t^5 + 4t^3 - 3t^2 + 6t - 1$; $R(2)$

29. $Y(u) = 18u^4 - 3u^2 + 7$; $Y(-2)$

30. $U(x) = 5x^3 + 2x^2 - 6x - 4$; $U(-3)$

11.3 Roots of Polynomial Equations

A polynomial equation of degree n is an equation of the form

$$a_nx^n + a_{n-1}x^{n-1} + \ldots + a_1x + a_0 = 0, \qquad a_n \neq 0$$

Methods for solving polynomial equations of degree one and two are quite familiar, but for degree three or higher the situation is more complicated. A polynomial equation of degree three is called a **cubic** equation and a polynomial equation of degree four is called a **quartic** or **biquadratic** equation. Although we will not present them here, there are methods for solving any cubic or quartic equation. Those methods are significantly more complicated than the quadratic formula. It is a rather startling result that there is *no* general method for solving a polynomial equation of degree n by a finite number of algebraic operations if $n \geq 5$. By algebraic operations we mean addition, subtraction, multiplication, division, and taking nth roots. However, there are some techniques that can be helpful in finding the roots of polynomial equations of degree greater than 2, and those are the subject of our study.

The main theorem we shall use is called the **Fundamental Theorem of Algebra.** Its statement is deceptively simple, but the proof (which we shall omit) involves concepts beyond our experience at this time.

Fundamental Theorem of Algebra

> If $P(x)$ is a polynomial of degree greater than zero, then there is at least one number r (possibly complex) such that $P(r) = 0$.

If $P(x)$ is a polynomial of degree $n \geq 1$, then the theorem above guarantees that there is a number r_1 such that $P(r_1) = 0$. By the factor theorem, $x - r_1$ is a factor of $P(x)$; hence there is a polynomial $Q(x)$ of degree $n - 1$ such that

$$P(x) = (x - r_1)Q(x)$$

If $n - 1 > 0$, then by the Fundamental Theorem of Algebra there is a number r_2 such that $Q(r_2) = 0$. Again by the Factor Theorem, $x - r_2$ is a factor of $Q(x)$. This means that there is a polynomial $T(x)$ of degree $n - 2$ such that

$$P(x) = (x - r_1)Q(x) = (x - r_1)(x - r_2)T(x)$$

We continue the process to obtain

$$P(x) = a(x - r_1)(x - r_2) \ldots (x - r_n)$$

Notice that r_i is a solution to the equation $P(x) = 0$ for each i among $1, 2, \ldots, n$. Moreover, if $P(r) = 0$, then from the factorization, r must be one of the r_i. We summarize as follows.

Multiplicity of a root

If $x - r$ occurs k times in the factorization of $P(x)$, then we say that r is a root of $P(x) = 0$ of **multiplicity** k.

Roots of a polynomial equation of degree n

> *Every polynomial equation $P(x) = 0$ of degree n has exactly n roots, counting multiplicities. If the roots are $r_1, r_2, \ldots, r_n$, then*
>
> $$P(x) = a(x - r_1)(x - r_2) \ldots (x - r_n)$$

Example 1 Find the solutions for the equation

$$x^4 - x^3 - 7x^2 + 13x - 6 = 0$$

For the time being, we use trial and error to seek the roots.

$$\begin{array}{r|rrrrr} 1) & 1 & -1 & -7 & 13 & -6 \\ & & 1 & 0 & -7 & 6 \\ \hline & 1 & 0 & -7 & 6 & 0 \end{array}$$

The number 1 is a root, and $x - 1$ is therefore a factor of the polynomial:

$$x^4 - x^3 - 7x^2 + 13x - 6 = (x - 1)(x^3 - 7x + 6)$$

We continue by examining the polynomial $x^3 - 7x + 6$:

$$\begin{array}{r|rrrr} 1) & 1 & 0 & -7 & 6 \\ & & 1 & 1 & -6 \\ \hline & 1 & 1 & -6 & 0 \end{array}$$

The number 1 is again a root (a double root of the original equation), and

$$x^4 - x^3 - 7x^2 + 13x - 6 = (x - 1)(x - 1)(x^2 + x - 6)$$

The polynomial $x^2 + x - 6$ is easily factored by the methods given in Chapter 1:

$$x^4 - x^3 - 7x^2 + 13x - 6 = (x - 1)(x - 1)(x - 2)(x + 3)$$

The solutions for the equation are 1, 2, and -3, with 1 a double root. ■

Example 2 Solve the equation $x^4 - 3x^3 - 3x^2 + 12x - 4 = 0$.
We again proceed by trial and error; but, remembering the last example, when we find a root the bottom line in the synthetic division process represents the next polynomial to use.

$$\begin{array}{r|rrrrr l}
1) & 1 & -3 & -3 & 12 & -4 & \\
 & & 1 & -2 & -5 & 7 & \\
\hline
 & 1 & -2 & -5 & 7 & 3 & \text{1 is not a root} \\
2) & 1 & -3 & -3 & 12 & -4 & \\
 & & 2 & -2 & -10 & 4 & \\
\hline
-2) & 1 & -1 & -5 & 2 & 0 & \text{2 is a root; we continue with the bottom line} \\
 & & -2 & 6 & -2 & & \\
\hline
 & 1 & -3 & 1 & 0 & & -2 \text{ is a root}
\end{array}$$

So far we know that

$$x^4 - 3x^3 - 3x^2 + 12x - 4 = (x - 2)(x + 2)(x^2 - 3x + 1)$$

We can use the quadratic formula to find the roots of the equation

$$x^2 - 3x + 1 = 0$$

They are

$$r_1 = \frac{3 - \sqrt{5}}{2} \quad \text{and} \quad r_2 = \frac{3 + \sqrt{5}}{2}$$

The factorization is

$$x^4 - 3x^3 - 3x^2 + 12x - 4 = (x - 2)(x + 2)\left(x - \frac{3 - \sqrt{5}}{2}\right)\left(x - \frac{3 + \sqrt{5}}{2}\right)$$

The roots are 2, −2, $(3 - \sqrt{5})/2$, and $(3 + \sqrt{5})/2$. ■

In Chapter 2 we saw that complex roots of quadratic equations with real coefficients came in conjugate pairs; that is, if $a + bi$ is a root of a quadratic equation, then $a - bi$ is also a root ($b \neq 0$ of course). The same is true of polynomial equations of higher degree. You are referred to the exercises for a proof of this result.

Roots in conjugate pairs

> *If $a + bi$, $b \neq 0$, is a root of a polynomial equation with real coefficients, then $a - bi$ is also a root.*

Example 3 Solve the equation $x^4 - 7x^3 + 19x^2 - 23x + 10 = 0$, given that $2 - i$ is a root.
Since $2 - i$ is a root and all the coefficients are real numbers, $2 + i$ is also a root. Hence $x - 2 - i$ and $x - 2 + i$ are factors, and this means that

$$(x - 2 - i)(x - 2 + i) = x^2 - 4x + 5$$

is a factor of $x^4 - 7x^3 + 19x^2 - 23x + 10$. To find the remaining factors we use long division:

$$
\begin{array}{r}
x^2 - 3x + 2 \\
x^2 - 4x + 5 \overline{\smash{)}\, x^4 - 7x^3 + 19x^2 - 23x + 10} \\
\underline{x^4 - 4x^3 + 5x^2} \\
- 3x^3 + 14x^2 - 23x \\
\underline{- 3x^3 + 12x^2 - 15x} \\
2x^2 - 8x + 10 \\
\underline{2x^2 - 8x + 10} \\
0
\end{array}
$$

The polynomial $x^2 - 3x + 2$ factors as $(x - 1)(x - 2)$.
The factorization of the polynomial is

$$x^4 - 7x^3 + 19x^2 - 23x + 10 = (x - 2 - i)(x - 2 + i)(x - 2)(x - 1)$$

The roots of the equation are $2 - i$, $2 + i$, 2, and 1. ■

Exercises 11.3

In Exercises 1–24 find the roots of the polynomial equations under the given conditions. Give the factorization of the polynomial.

1. $x^3 - 8x^2 + 5x + 14 = 0$; 2 is a root.

2. $x^3 - x^2 - 16x - 20 = 0$; 5 is a root.

3. $4x^3 + 15x + 19 = 0$; -1 is a root.

4. $3x^3 - 12x - 9 = 0$; -1 is a root.

5. $x^4 - 2x^2 - 3x - 2 = 0$; -1 and 2 are roots.

6. $x^4 - 4x^3 - 2x^2 + 7x + 4 = 0$; -1 and 4 are roots.

7. $2x^4 - 5x^3 - 10x^2 + 15x + 18 = 0$; 2 and 3 are roots.

8. $2x^4 + 11x^3 + 4x^2 - 29x + 12 = 0$; 1 and -4 are roots.

9. $x^3 - 11x^2 + 40x - 50 = 0$; $3 - i$ is a root.

10. $x^3 + x^2 - 4x + 6 = 0$; $1 + i$ is a root.

11. $2x^3 - 3x^2 + 18x - 27 = 0$; $-3i$ is a root.

12. $3x^3 - 2x^2 + 48x - 32 = 0$; $4i$ is a root.

13. $y^4 + 3y^3 + 12y - 16 = 0$; $2i$ is a root.

14. $y^4 - 6y^3 + 33y^2 - 150y + 200 = 0$; $-5i$ is a root.

15. $x^3 - 9x^2 + 23x - 12 = 0$; the equation has an integer root between -5 and 5.

16. $x^3 + 5x^2 + 4x - 4 = 0$; the equation has an integer root between -4 and 4.

17. $x^4 - 7x^3 + 12x^2 + 4x - 16 = 0$; the equation has two integer roots between -5 and 5.

18. $x^4 + x^3 - 2x^2 + 4x - 24 = 0$; the equation has two integer roots between -5 and 5.

19. $2x^3 + 5x^2 - 11x + 4 = 0$; $\frac{1}{2}$ is a root.

20. $2x^3 + 11x^2 + 20x + 12 = 0$; $-\frac{3}{2}$ is a root.

21. $x^4 - 2x^2 + 1 = 0$; 1 is a double root.

22. $y^4 - 8y^2 + 16 = 0$; 2 is a double root.

23. $x^4 + 7x^3 - 20x^2 - 132x + 144 = 0$; -6 is a double root.

24. $x^4 + 2x^3 - 20x^2 - 66x - 45 = 0$; -3 is a double root.

25. Factor $6x^2 + x - 2$ into linear factors.

26. Factor $4x^2 - 4x - 3$ into linear factors.

27. Factor $P(x) = 36x^4 + 48x^3 - 23x^2 - 17x + 6$ into linear factors, given that $P\left(\frac{1}{2}\right) = 0$ and $P\left(\frac{1}{3}\right) = 0$.

28. Factor the polynomial

$Q(x) = 2x^4 + 9x^3 + 13x^2 + 81x - 45$

into linear factors, given that $Q(3i) = 0$.

If a and b are real numbers, recall that the conjugate of $a + bi$, written $\overline{a + bi}$, is the complex number $a - bi$.

29. If v and u are complex numbers, prove that $\overline{uv} = \overline{u}\,\overline{v}$.

30. If v and u are complex numbers, prove that $\overline{u + v} = \overline{u} + \overline{v}$.

31. If u is a complex number, prove that u is a real number if and only if $\overline{u} = u$.

32. If $P(x)$ is a polynomial with real coefficients, prove that $P(\overline{u}) = \overline{P(u)}$ for any complex number u. (Hint: use Exercises 29, 30, and 31.)

33. If $P(x)$ is a polynomial with real coefficients and a is a root of the equation $P(x) = 0$, prove that $\overline{a}$ is also a root.

11.4 Locating Roots

In this section we develop some means of locating roots of a polynomial equation without prior information.

The roots of the polynomial equation $P(x) = 0$ can be found by locating the x intercepts of the graph of $y = P(x)$. For example, the graph of the function $y = x^3 - 2x^2 + x$ is shown in Figure 11.1.

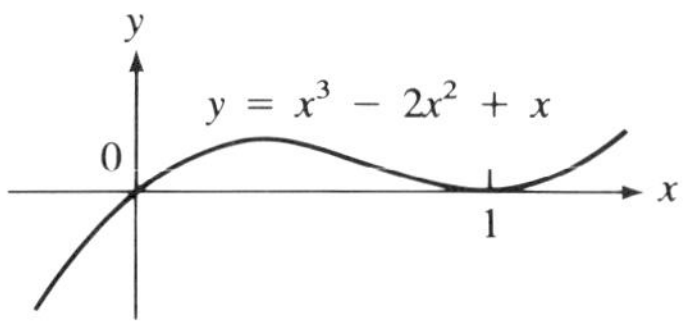

Figure 11.1

The graph crosses the x axis at $x = 0$ and touches the x axis at $x = 1$; and, indeed, the roots of the equation $x^3 - 2x^2 + x = 0$ are 0 and 1, with 1 as a double root.

The observation concerning the graph of the polynomial function and the roots of the corresponding polynomial equation is useful, however, in another way. Consider the polynomial $P(x) = \frac{1}{2}x^3 - 3x^2 + 3x + 2$. From the fact that $P(1) = \frac{5}{2}$ and $P(3) = -\frac{5}{2}$, we see that the graph of $y = P(x)$ must pass through the points $\left(1, \frac{5}{2}\right)$ and $\left(3, -\frac{5}{2}\right)$ (see Figure 11.2). Since the graph of a polynomial function must be an unbroken curve, the curve in this case must cross the x axis somewhere between 1 and 3. Thus the equation $P(x) = 0$ has a root somewhere between 1 and 3. This argument is based on intuition, but in more advanced courses a rigorous proof can be given. We summarize in the form of a theorem.

> **Theorem** If $P(x)$ is a polynomial and $P(a)$ and $P(b)$ have opposite signs, then the equation $P(x) = 0$ has at least one root between a and b.

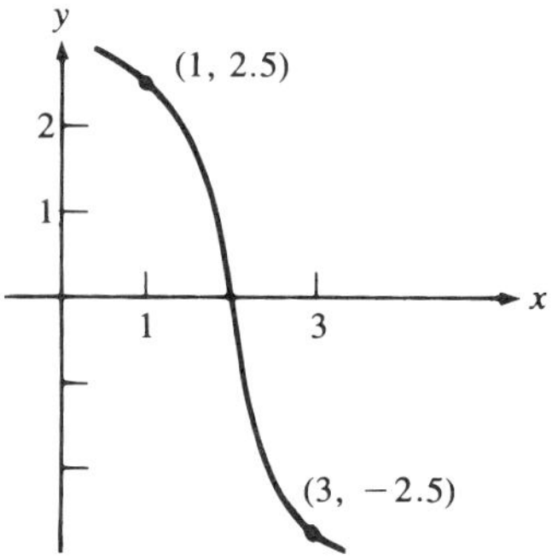

Figure 11.2

Example 1 Isolate the roots of $x^3 - 3x^2 - 6x + 8 = 0$.
Make a table of values for $P(x) = x^3 - 3x^2 - 6x + 8$.

x	$P(x)$
-3	-28
root →	
0	8
root →	
2	-8
root →	
5	28

Each sign change in the value of $P(x)$ indicates a root of the equation $P(x) = 0$. There is a root between -3 and 0, between 0 and 2, and between 2 and 5. For your information, the roots are -2, 1, and 4. ■

Suppose all the coefficients of $P(x)$ are nonnegative. Then for any positive real number r, $P(r)$ is positive. Consequently, the equation $P(x) = 0$ has no positive roots. Similarly, if all the coefficients are nonpositive, the equation $P(x) = 0$ will have no positive roots.

Example 2 The equation $2x^4 + 3x^2 + 4x + 7 = 0$ has no positive roots. ■

Example 3 The equation $-x^3 - 4x^2 - 3x - 7 = 0$ has no positive roots. ■

Consider now the equation $P(x) = x^3 - 4x^2 + 3x - 7 = 0$. Since some of the signs are negative and some are positive, we can say nothing about the positive roots. What about the negative roots? The negative roots of $P(x) = 0$ are the positive roots of $P(-x) = 0$. We find that $P(-x) = -x^3 - 4x^2 - 3x - 7$, and this means that $P(-x) = 0$ has no positive roots; that is, $P(x) = 0$ has no negative roots. Notice that replacing x by $-x$ in a polynomial changes the sign of the terms of odd degree and leaves unchanged the signs of the terms of even degree.

We summarize as follows.

> If all coefficients of $P(x)$ have the same sign, $P(x) = 0$ has no positive roots. If all coefficients of $P(-x)$ have the same sign, the equation $P(x) = 0$ has no negative roots. The signs of the coefficients of $P(-x)$ will be the same if the signs of the coefficients of the terms of odd degree are opposite to those of even degree.

Example 4 The equation $P(x) = x^4 - 3x^3 - 4x + 3 = 0$ has no negative roots. Notice that the terms of odd degree, $-3x^3$ and $-4x$, have negative coefficients, and the terms of even degree, x^4 and 3, have positive coefficients. Also note that

$$P(-x) = x^4 + 3x^3 + 4x + 3$$

■

If we divide $P(x)$ by $x - r$, the bottom row obtained by synthetic division gives the coefficients of the quotient, $Q(x)$. Since

$$P(x) = (x - r)Q(x) + R$$

the coefficients of Q and the remainder tell us something about the roots of the equation $P(x) = 0$.

If $r > 0$ and all the coefficients of $Q(x)$ and the remainder R are nonnegative, then $P(x)$ will be positive for all $x > r$. In particular, the equation $P(x) = 0$ will have no solution greater than r. Similarly if all the coefficients of $Q(x)$ and the remainder R are nonpositive, then $P(x)$ will be negative for all $x > r$, and as before, $P(x) = 0$ will have no roots greater than r.

If $r < 0$, a similar argument applies with $P(-x)$. We summarize in the following theorem.

> **Theorem** If $P(x)$ is divided by $x - r$, then
>
> 1. if $r > 0$ and the numbers on the bottom row are all either nonnegative or nonpositive, then the equation $P(x) = 0$ has no roots larger than r;
> 2. if $r < 0$ and the numbers on the bottom row alternate in sign (counting 0 as positive or negative as necessary) the equation $P(x) = 0$ has no roots less than r.

Example 5 Find upper and lower bounds for the roots of the equation $x^4 - 6x^3 + 4x^2 + 3x + 6 = 0$.

Some trial and error yields

$$\begin{array}{r|rrrrr} 5) & 1 & -6 & 4 & 3 & 6 \\ & & 5 & -5 & -5 & -10 \\ \hline & 1 & -1 & -1 & -2 & -4 \end{array}$$

$$\begin{array}{r|rrrrr} 6) & 1 & -6 & 4 & 3 & 6 \\ & & 6 & 0 & 24 & 162 \\ \hline & 1 & 0 & 4 & 27 & 168 \end{array}$$

The equation has no roots larger than 6 and, in fact, has a root between 5 and 6:

$$\begin{array}{r|rrrrr} -1) & 1 & -6 & 4 & 3 & 6 \\ & & -1 & 7 & -11 & 8 \\ \hline & 1 & -7 & 11 & -8 & 14 \end{array}$$

The signs alternate on the bottom row; the equation has no roots less than -1. ■

Another well-known theorem concerning the roots of a polynomial equation is Descartes's rule of signs. We present it without proof.

If the terms of a polynomial are arranged in order of descending powers of x and two successive coefficients differ in sign, we say that the polynomial has a **variation** of sign. For example, the polynomial

$$x^4 \overset{①}{\frown} 3x^3 \overset{②}{\frown} x \overset{③}{\frown} 1 \quad\text{i.e.}\quad x^4 - 3x^3 + x - 1$$

has three variations of sign as indicated.

Descartes's rule of signs

> Suppose $P(x)$ is a polynomial of degree greater than or equal to 1 and $P(x)$ has k variations in sign. Then the number of positive roots of $P(x) = 0$ is either k or $k - 2m$, where m is a positive integer. If $S(x) = P(-x)$ has j variations in sign, the equation $P(x) = 0$ has either j negative roots or $j - 2s$ negative roots, where s is a positive integer.

Example 6 By Descartes's rule of signs, $x^3 - 3x + 2 = 0$ has either two positive roots or none, since the polynomial $P(x) = x^3 - 3x + 2$ has two variations in sign. The polynomial $P(-x) = -x^3 + 3x + 2$ has one variation in sign, and the equation $x^3 - 3x + 2 = 0$ therefore has one negative root. ■

Let us apply all the results of this section to isolate the roots of a polynomial equation.

Example 7 Isolate the roots of $8x^3 - 20x^2 - 18x + 45 = 0$.

The polynomial $P(x) = 8x^3 - 20x^2 - 18x + 45$ has two variations in sign, and the equation $P(x) = 0$ therefore has either two positive roots or none. The polynomial $P(-x) = -8x^3 - 20x^2 + 18x + 45$ has one variation in sign, and $P(x) = 0$ therefore has one negative root. We can make a table to indicate the possible combinations of roots, recalling that for polynomial equations with real coefficients, the complex roots come in conjugate pairs.

Positive Roots	Negative Roots	Complex Roots
2	1	0
0	1	2

Now we seek upper and lower bounds for the roots:

$$\begin{array}{r|rrrr} 3) & 8 & -20 & -18 & 45 \\ & & 24 & 12 & -18 \\ \hline & 8 & 4 & -6 & 27 \end{array}$$

$$\begin{array}{r|rrrr} 4) & 8 & -20 & -18 & 45 \\ & & 32 & 48 & 120 \\ \hline & 8 & 12 & 30 & 165 \end{array}$$

The equation has no roots larger than 4:

$$\begin{array}{r|rrrr} -2) & 8 & -20 & -18 & 45 \\ & & -16 & 72 & -108 \\ \hline & 8 & -36 & 54 & -63 \end{array}$$

The equation has no roots smaller than -2.

So far, we know that all real roots are between -2 and 4 and that there is one negative root. To look further, we see that $P(-2) = -63$ and $P(-1) = 35$. The negative root is between -2 and -1. More calculations show that $P(1) = 15$, $P(2) = -7$, and $P(3) = 37$. The positive roots are located between 1 and 2 and between 2 and 3. This concludes the search. Just for your information, the roots are $-\frac{3}{2}, \frac{3}{2}$, and $\frac{5}{2}$. ■

Exercises 11.4

In Exercises 1–12 find an integer n such that the equation has a root between n and n + 1, inclusive; that is, such that either n or n + 1 or both could be roots.

1. $x^3 - 4x^2 + x + 3 = 0$

2. $x^3 + 4x^2 - x - 3 = 0$

3. $3x^4 - 4x^3 + 6x^2 - 2x - 5 = 0$

4. $6x^4 - 7x^3 - 4x^2 + 7x - 2 = 0$

5. $4x^3 - 4x^2 + 6x + 6 = 0$

6. $6x^3 - 6x^2 + 4x + 4 = 0$

7. $x^3 + 3x - 20 = 0$

8. $x^3 + 2x^2 - 5 = 0$

9. $9x^3 + 4x^2 - 22 = 0$

10. $x^3 - 5x - 45 = 0$

11. $x^4 + 10x - 100 = 0$

12. $x^4 + 8x - 65 = 0$

In Exercises 13–20 find integers A and B such that all real roots of the equation lie between A and B.

13. $3x^4 + 4x^3 + 5x^2 + 2x + 1 = 0$

14. $5x^4 + x^3 + x^2 + 62x + 4 = 0$

15. $x^4 - 5x^3 + 8x^2 - 9x = 0$

16. $x^6 - 9x^5 + 8x^2 - 9x = 0$

17. $x^5 - 4x^4 + 2x^3 - x + 1 = 0$

18. $x^5 + 4x^4 - 2x^3 + x + 1 = 0$

19. $x^4 - x^3 + x^2 - x - 1 = 0$

20. $x^5 - x^4 + x^3 - x + 1 = 0$

In Exercises 21–30 use Descartes's rule of signs to determine the possible number of positive roots, the possible number of negative roots, and the possible number of complex roots.

21. $4x^5 - 3x^2 - 3x + 16 = 0$

22. $3x^5 - 4x^4 - 3x^3 + 12 = 0$

23. $-x^4 - x^3 + 6x^2 + 6x \quad 12 = 0$

24. $-x^4 - 4x^3 - 6x^2 + 7x + 12 = 0$

25. $-x^3 - x^2 - x + 1 = 0$

26. $-x^3 + x^2 + x + 1 = 0$

27. $x^4 = x^2 + 9$

28. $x^6 = x^4 + 8$

29. $x^4 - x = 69$

30. $x^4 + x = 42$

11.5 Rational Roots of Polynomial Equations

Let

$$P(x) = a_n x^n + a_{n-1}x^{n-1} + \ldots + a_1 x + a_0$$

where $a_n, a_{n-1}, \ldots, a_0$ are integers, and suppose that p/q is a rational root of the equation $P(x) = 0$. We may suppose that p and q are relatively prime, that is, p/q is reduced to lowest terms. Then

$$a_n \left(\frac{p}{q}\right)^n + a_{n-1}\left(\frac{p}{q}\right)^{n-1} + \ldots + a_1\left(\frac{p}{q}\right) + a_0 = 0$$

or

$$a_n p^n + a_{n-1}p^{n-1}q + \ldots + a_1 p q^{n-1} + a_0 q^n = 0$$

Consequently,

$$a_{n-1}p^{n-1}q + \ldots + a_1 p q^{n-1} + a_0 q^n = -a_n p^n$$

Each term on the left is divisible by q, so $-a_n p^n$ must be divisible by q. But p and q are relatively prime, so q must be a divisor of a_n. Similarly, we can show that p is a divisor of a_0. We summarize as follows.

Requirements for rational roots

> If $a_n, a_{n-1}, \ldots, a_1, a_0$ are integers and p/q is a rational root of the equation
>
> $$a_n x^n + a_{n-1}x^{n-1} + \ldots + a_1 x + a_0 = 0$$
>
> then p is a divisor of a_0 and q is a divisor of a_n.

Example 1 Find the rational roots of $x^3 - 3x^2 + 15x - 26 = 0$. Any rational root must be of the form p/q, where p is a divisor of -26 and q is a divisor of 1. The possible candidates are -26, -13, -2, -1, 1, 2, 13, and 26. The results from Section 11.4 will be helpful in shortening the search:

$$\begin{array}{r|rrrr} 3) & 1 & -3 & 15 & -26 \\ & & 3 & 0 & 45 \\ \hline & 1 & 0 & 15 & 19 \end{array}$$

There are no roots larger than 3; the remaining positive candidates are 1 and 2:

$$\begin{array}{r|rrrr} 2) & 1 & -3 & 15 & -26 \\ & & 2 & -2 & 26 \\ \hline & 1 & -1 & 13 & 0 \end{array}$$

Since 2 is a solution,

$$x^3 - 3x^2 + 15x - 26 = (x - 2)(x^2 - x + 13)$$

The roots of $x^2 - x + 13 = 0$ are

$$\frac{1 + \sqrt{51}\, i}{2} \quad \text{and} \quad \frac{1 - \sqrt{51}\, i}{2}$$ ■

Example 2 Find the rational roots of $2x^3 + 3x^2 - 2x - 3 = 0$.
Any rational root must be of the form p/q, where p is a divisor of -3 and q is a divisor of 2. The possible choices are -3, $-\frac{3}{2}$, -1, $-\frac{1}{2}$, $\frac{1}{2}$, 1, $\frac{3}{2}$, and 3. The polynomial $P(x) = 2x^3 + 3x^2 - 2x - 3$ has only one variation in sign; the equation therefore has exactly one positive real root. Let's look for that one:

$$\begin{array}{r|rrrr} \frac{1}{2}) & 2 & 3 & -2 & -3 \\ & & 1 & 2 & 0 \\ \hline & 2 & 4 & 0 & -3 \end{array}$$

$$\begin{array}{r|rrrr} 1) & 2 & 3 & -2 & -3 \\ & & 2 & 5 & 3 \\ \hline & 2 & 5 & 3 & 0 \end{array}$$

Thus,

$$2x^3 + 3x^2 - 2x - 3 = (x - 1)(2x^2 + 5x + 3)$$

The roots of $2x^2 + 5x + 3 = 0$ are -1 and $-\frac{3}{2}$, so the roots of the original equation are 1, -1, and $-\frac{3}{2}$. ■

Example 3 Solve the equation $8x^5 - 12x^4 + 6x^3 - 5x^2 + 3 = 0$.
Descartes's rule of signs tells us that there are four, two, or zero positive roots and exactly one negative root. In seeking upper and lower bounds for the roots

we discover that 1 is a root and

$$8x^5 - 12x^4 + 6x^3 - 5x^2 + 3 = (x - 1)(8x^4 - 4x^3 + 2x^2 - 3x - 3)$$

Continuing the search for the roots of $8x^4 - 4x^3 + 2x^2 - 3x - 3 = 0$, we discover that 1 is again a root and

$$8x^4 - 4x^3 + 2x^2 - 3x - 3 = (x - 1)(8x^3 + 4x^2 + 6x + 3)$$

The equation $8x^3 + 4x^2 + 6x + 3 = 0$ has no positive roots. We find that

$$\begin{array}{r|rrrr} -1) & 8 & 4 & 6 & 3 \\ & & -8 & 4 & -10 \\ \hline & 8 & -4 & 10 & -7 \end{array}$$

so the equation has no roots less than -1. The possible rational roots that are between -1 and 0 are $-\frac{3}{4}$, $-\frac{1}{2}$, $-\frac{3}{8}$, $-\frac{1}{4}$, and $-\frac{1}{8}$. Trial and error shows that $-\frac{1}{2}$ is the only negative root (Descartes's rule of signs indicated only one negative root). Synthetic division shows that

$$8x^5 - 12x^4 + 6x^3 - 5x^2 + 3 = (x - 1)(x - 1)\left(x + \frac{1}{2}\right)(8x^2 + 6)$$

The roots of $8x^2 + 6 = 0$ are $-(\sqrt{3}/2)i$ and $(\sqrt{3}/2)i$. The roots of the original equation are 1, $-1/2$, $-(\sqrt{3}/2)i$, and $(\sqrt{3}/2)i$, with 1 as a double root. ■

The methods of this section are only useful in locating rational roots or in reducing the polynomial equation by finding rational roots to the point where the resulting polynomial equation can be solved by other means. If the roots are all irrational or complex, other means are necessary. Moreover, if the coefficients have an abundance of divisors, the task can be wearisome.

By the way, we haven't solved the problem posed at the beginning of the chapter. Recall that we had a storage tank with three pumps, A, B, and C. With all pumps running, the tank can be drained in 1 hr; pump B takes 1 hr longer to empty the tank than does pump A; and pump C takes 4 hr longer than does pump A to drain the tank. This led to the equation

$$\frac{1}{x} + \frac{1}{x + 1} + \frac{1}{x + 4} = 1$$

which is equivalent to the equation

$$x^3 + 2x^2 - 6x - 4 = 0$$

We are interested only in positive roots of the equation, and Descartes's rule of signs tells us that there is exactly one positive root. The only possible rational candidates are 1, 2, and 4. A simple calculation shows that 2 is the only positive root. Thus it takes 2 hr to empty the tank with pump A, 3 hr with pump B, and 6 hr with pump C.

Exercises 11.5

In Exercises 1–20 examine the equations for rational roots. Where possible, find all solutions.

1. $x^3 - x^2 - 8x + 12 = 0$

2. $x^3 - 4x^2 - 3x + 18 = 0$

3. $x^3 - 2x^2 - 5x + 6 = 0$

4. $x^3 - 4x^2 + x + 6 = 0$

5. $6x^3 - 25x^2 - 29x + 20 = 0$

6. $8x^3 - 26x^2 + 3x + 9 = 0$

7. $6x^4 - 31x^3 + 56x^2 - 41x = -10$

8. $10x^4 + 19x^3 - 35x^2 + 9 = 3x$

9. $x^3 - 3x^2 - 8x = 10$

10. $x^3 - 7x^2 + 12x = 10$

11. $x^3 - x^2 - 18x - 10 = 0$

12. $x^3 + 2x^2 - 22x - 8 = 0$

13. $2x^3 - 9x^2 - 2x + 3 = 0$

14. $x^3 - 11x^2 + 28x + 12 = 0$

15. $2x^3 - x^2 + 9x - 18 = 0$

16. $3x^3 + 2x^2 + 23x - 8 = 0$

17. $2x^4 + 3x^3 - 12x^2 - 18x = 5$

18. $3x^4 + 25x^3 + 45x^2 - 30x = 8$

19. $2x^4 - 5x^3 - 3x^2 + 4x = -2$

20. $3x^4 + 4x^3 + 21x^2 + 8x = -12$

21. The dimensions of a box are $8 \times 10 \times 12$ inches. If each dimension is increased by the same amount, the volume is increased by 720 cubic inches. How much is each dimension increased?

22. The dimensions of a box are $8 \times 12 \times 16$ inches. If each dimension is increased by an amount proportional to that dimension, the volume is increased by 1,464 cubic inches. What is the amount of change in each dimension?

23. The integers n, $n - 1$, and $2n$ have the property that the sum of their reciprocals is 1. What are the integers?

24. An open box is made from a rectangular piece of tin, which measures 20 inches $\times$ 10 inches, by cutting equal squares from each corner and folding up the sides. If the resulting volume is 168 cubic inches, what are the dimensions of the box?

25. Mark can do a job in n hours, John can do the same job in $n + 1$ hours, and Jack can do the same job in $n + 3$ hours. Working together, they can do the job in 1 hour and 20 minutes. How long would it take Mark to do the job?

26. The integers n, $n - 1$, and $2n - 4$ have the property that the sum of their reciprocals is $\frac{5}{6}$. Find the integers.

11.6 Irrational Roots of Polynomial Equations

It would be foolish to assume that every polynomial equation has at least one rational root, and in fact in many situations one would expect the roots to be irrational. In this section we examine a way of approximating the irrational roots of an equation. The calculations can be done by hand, but since the method is the important idea here, we urge you to use a calculator for the more tedious calculations. To evaluate the polynomials, you may wish to write them in nested form.

A hollow hemisphere of radius 2 feet is weighted with a total of 185 lb (including the weight of the hemisphere itself). Will it float in water?

Archimedes's principle says that if a body floats in water, it will displace its own weight. The question is how deep the hemisphere will sink to displace 185 pounds of water.

Figure 11.3 shows a sketch of the situation, with x denoting the depth to which the hemisphere sinks.

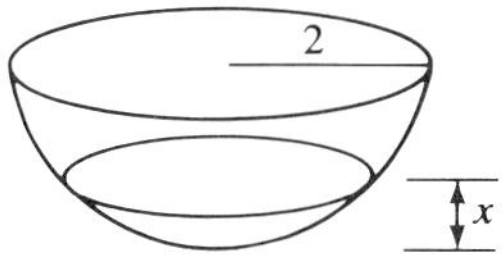

Figure 11.3

The volume of the water displaced is given by the formula

$$V = \frac{\pi}{3}(6x^2 - x^3) \qquad 0 \le x \le 2$$

Now we want to find the value of x such that the weight of the water displaced is 185 lb. Water weighs approximately 62.4 lb/ft^3, so the hemisphere displaces

$$\frac{185}{62.4} = 2.96 \text{ ft}^3$$

if it floats. We seek to solve the cubic equation

$$\frac{\pi}{3}(6x^2 - x^3) = 2.96$$

With a little rounding off, the equation reduces to

$$x^3 - 6x^2 + 2.826 = 0$$

By Descartes's rule of signs, this equation has either two positive roots or none. If we let $P(x) = x^3 - 6x^2 + 2.826$, we discover that

$$P(0) = 2.826 \quad \text{and} \quad P(1) = -2.174$$

so there is a root between 0 and 1. This means that the hemisphere will float; the question now is how deep will it float in the water?
Now

$$P(0) = 2.826$$
$$P(1) = -2.174$$

so there is a root between 0 and 1. Let's try 0.5:

$$P(0.5) = 1.451$$

so the root is between 0.5 and 1, but $P(0.5)$ is closer to zero, so we suspect the root is closer to 0.5. Try 0.7:

$$P(0.7) = 0.229$$

The root is between 0.7 and 1 and probably closer to 0.7.
Try 0.75:

$$P(0.75) = -0.127$$

Try 0.74:

$$P(0.74) = -0.054$$

Try 0.73:

$$P(0.73) = 0.018$$

Aha! The root is between 0.73 and 0.74. Let us settle for two decimal places, so try 0.735:

$$P(0.735) = -0.018$$

The root is between 0.73 and 0.735, so to two decimal places the root is 0.73. The hemisphere will sink to a depth of 0.73 ft.

There are a number of methods for approximating the roots of a polynomial equation that you may learn later; for the time being, we content ourselves with using the theorem of Section 11.4 on smaller and smaller intervals.

Example 1 Locate to two decimal places the root of $x^3 - 5.2x^2 + 6x - 1 = 0$ that is between 0 and 1.

In nested form (for your calculator).

$$P(x) = [(x - 5.2)x + 6]x - 1$$
$$P(0) = -1$$
$$P(1) = 0.8$$

Halving the interval,

$$P(0.5) = 0.825$$

The root is between 0 and 0.5. Halve the interval again:

$$P(0.25) = 0.19$$

The root is between 0 and 0.25. Try 0.2:

$$P(0.2) = 0$$

The root is 0.2 ■

Example 2 The equation $x^3 + 8x^2 + 8x - 8 = 0$ has a root between -6 and -7. Find that root to two decimal places.

In nested form

$$P(x) = [(x + 8)x + 8]x - 8$$
$$P(-6) = 16$$
$$P(-7) = -15$$

Halve the interval:

$$P(-6.5) = 3.375$$

The root is between -6.5 and -7:

$$P(-6.75) = -5.047$$

The root is between -6.75 and -6.5:

$P(-6.6) = 0.184$

The root is between -6.6 and -6.75:

$P(-6.61) = -0.148$

The root is between -6.61 and -6.6:

$P(-6.605) = 0.18$

The root is between -6.605 and -6.61, so to two decimal places the root is -6.61. ■

Exercises 11.6

In Exercises 1–10 find the indicated root correct to two decimal places.

1. $x^3 + 3x - 20 = 0$; root between 2 and 3

2. $x^3 + 2x^2 - 5 = 0$; root between 1 and 2

3. $x^3 + 4x - 7 = 0$; root between 1 and 2

4. $8x^3 - 2x^2 - 4x + 1 = 0$; root between -1 and 0

5. $3x^3 + 14x^2 + 13x - 2 = 0$; root between 0 and 1

6. $x^3 - 9x - 5 = 0$; root between -3 and -2

7. $2x^3 - 3x^2 + 0.657 = 0$; root between 0 and 1

8. $3x^3 - x^2 + 3x - 0.75 = 0$; root between 0 and 1

9. $x^3 - 2x - 100 = 0$; root between 4 and 5

10. $x^4 - 2x - 100 = 0$; root between 3 and 4

11. The diameter of a water pipe 200 ft long which is to discharge 100 ft^3/sec under a head of 10 ft is given by the real root of the equation

$$x^5 - 38x - 101 = 0$$

Find the diameter to two decimal places.

12. A vat is in the form of a rectangular parallelepiped (box) and measures $8 \times 10 \times 12$ ft. The volume is increased 500 cubic feet by increasing each dimension by the same amount. Find the amount by which each dimension is increased to two decimal places.

13. The width of the strongest beam that can be cut from a log 12 in. in diameter is given by the positive root of the equation

$$x^3 - 144x + 665 = 0$$

Find the width to two decimal places.

Chapter Review

Write $P(x) = (x - r)Q(x) + R$, where $P(x)$ and r are as indicated and R is a real number.

1. $P(x) = 3x^3 - 6x^2 + 17x - 1$; $r = 2$
2. $P(x) = 4x^5 - 6x^3 + x - 19$; $r = -1$
3. $P(x) = x^6 - x^4 + x^3 + 3x^2 - x + 1$; $r = -2$
4. $P(x) = 3x^4 - 6x^3 + x^2 - 2x + 4$; $r = 1$

Use the Remainder Theorem to evaluate the following.

5. $P(x) = 9x^5 - 17x^4 - 3x^3 + 16x^2 - 2x + 12$; find $P(2)$.
6. $P(x) = x^3 - x^2 - x - 12$; find $P(2i)$.
7. $P(x) = x^3 - 6x^2 + 9x - 2$; find $P(2 - \sqrt{3})$.
8. $P(x) = x^4 - 6x^2 + 17$; find $P(3 - i)$.

Use the Factor Theorem to determine whether the indicated binomial is a factor of the indicated polynomial.

9. $x + 2$; $3x^4 + 11x^3 + 3x^2 - 20x - 12$
10. $x + 2$; $5x^4 + 18x^2 - 6x + 3$
11. $x - 3$; $2x^3 - 3x^2 - 4x - 15$
12. $x + 2$; $3x^4 - 2x^2 + 6x - 28$
13. Solve the equation $4x^3 - 4x^2 - 11x + 6 = 0$, given that $\frac{1}{2}$ is a root.
14. Solve the equation $x^4 - 6x^3 + 4x^2 - 6x + 3 = 0$, given that i is a root.
15. Solve the equation $x^4 - 5x^3 + 5x^2 + x - 2 = 0$, given that 1 is a double root.
16. Factor the polynomial $2x^2 - 4x - 8$ into linear factors.
17. Factor the polynomial $P(x) = x^4 - 7x^3 + 22x^2 - 32x + 16$, given that $P(2 + 2i) = 0$.

Find an integer n such that the equation has a root between n and n + 1.

18. $x^3 + 2x^2 - 5x - 5 = 0$
19. $2x^3 - 3x^2 + 8x = 12$
20. $x^4 + 4x^3 - x^2 - 16x - 14 = 0$

Find integers A and B such that all real roots of the equation lie between A and B.

21. $2x^5 + 5x^3 + x + 1 = 0$
22. $3x^4 - x^3 - x^2 + 7 = 0$

Use Descartes's rule of signs to determine the possible number of positive and negative roots of the equation.

23. $6x^3 - 5x^2 + x + 2 = 0$
24. $7x^3 - 3x^2 + x = 0$
25. $8x^4 + 3x^3 + 2x^2 - 9 = 0$

Find the rational roots of the equations.

26. $6x^3 + x^2 - 19x + 6 = 0$
27. $5x^3 + 24x^2 + 25x = 6$
28. $2x^4 - 13x^3 + 7x^2 + 37x + 15 = 0$
29. $30x^3 + 10x = 31x^2 + 1$

Find the root indicated, correct to two decimal places.

30. $x^4 + x - 3 = 0$; root between 1 and 2
31. $x^3 + x^2 - 1 = 0$; root between 0 and 1

Chapter Test

Write $P(x) = (x - r)Q(x) + R$, where $P(x)$ and r are as given and R is a number.

1. $P(x) = 5x^3 - 3x^2 + 2x + 9$; $r = 3$
2. $P(x) = 6x^3 - x - 2$; $r = -2$

Use the Remainder Theorem to evaluate.

3. $P(3)$, where $P(x) = 5x^3 - 3x^2 + 2x + 9$
4. $P(-2)$, where $P(x) = 6x^3 - x - 2$

Use the Factor Theorem to determine if the indicated binomial is a factor of the indicated polynomial.

5. $x + 4;\ x^3 - x^2 - 10x + 8$

6. $x - 5;\ x^3 + 5x^2 - x - 5$

Solve the equation, using the given information.

7. $2x^4 - 11x^3 + 11x^2 + 15x - 9 = 0$, given that 3 and -1 are roots

8. $x^3 - 4x^2 + 5x - 6 = 0$, given that 3 is a root

9. Find the rational roots of $6x^3 - 19x^2 + x + 6 = 0$.

10. Determine, to two decimal places, the root of $x^3 + 7x^2 - 2x = 14$ that lies between 1 and 2.

12 Sequences, Series, and Induction

In Section 5.1 we saw how an asset can depreciate by the method of declining balances. If the present value of an asset is \$10,000 and the annual rate of depreciation is 40%, then the values of the asset at the end of 1 year, 2 years, and so on, are as follows:

$$10{,}000(0.6),\ 10{,}000(0.6)^2,\ 10{,}000(0.6)^3,\ \ldots,\ 10{,}000(0.6)^n,\ \ldots$$

This array of numbers is an example of what is called a *sequence*. In this chapter we study some of the properties of sequences and series.

12.1 Sequences

A sequence is a collection of numbers arranged in a particular order. We have a precise mathematical way of describing a sequence.

Sequence

A **sequence** is a function whose domain is either the set of all positive integers or the set $\{1, 2, 3, \ldots, n\}$, where n is a positive integer. If the domain is the set of all positive integers, we say the sequence is **infinite;** otherwise the sequence is **finite.**

Terms of a sequence

Most often, we describe a sequence by listing the values of the function, the **terms** of the sequence, in order. For example, if the values of the function are denoted by $a(n)$, we write the sequence as

$$a(1), a(2), a(3), \ldots, a(n), \ldots$$

It is also customary to write a_1 instead of $a(1)$, a_2 instead of $a(2)$, so that we would write

$$a_1, a_2, a_3, \ldots, a_n, \ldots$$

The number a_1 is called the first *term,* a_2 is the second *term,* and a_n is the nth *term.*

Example 1 List the first five terms of the sequence given by the function $a_n = n^2 - 1$.

$$a_1 = 1^2 - 1 = 0 \qquad a_2 = 2^2 - 1 = 3$$
$$a_3 = 3^2 - 1 = 8 \qquad a_4 = 4^2 - 1 = 15$$
$$a_5 = 5^2 - 1 = 24$$

In order, the first five terms are 0, 3, 8, 15, 24. ■

Example 2 List the first five terms of the sequence given by

$$a_n = \begin{cases} 2n - 1 & \text{if } n \text{ is odd} \\ \dfrac{n}{2} & \text{if } n \text{ is even} \end{cases}$$

$$a_1 = 2(1) - 1 = 1 \qquad a_2 = \frac{2}{2} = 1$$

$$a_3 = 2(3) - 1 = 5 \qquad a_4 = \frac{4}{2} = 2$$

$$a_5 = 2(5) - 1 = 9$$

In order, the first five terms are 1, 1, 5, 2, 9. ■

The function used to define the sequence can be given in a variety of ways; the following example illustrates another method.

Example 3 List the first six terms of the sequence defined as follows:

$$a_1 = 4, \quad \text{and for} \quad n \geq 2, \quad a_n = a_{n-1} + 3$$

Notice how each term is defined in terms of the preceding term.

$$a_1 = 4 \qquad a_2 = 4 + 3 = 7$$
$$a_3 = 7 + 3 = 10 \qquad a_4 = 10 + 3 = 13$$
$$a_5 = 13 + 3 = 16 \qquad a_6 = 16 + 3 = 19$$

■

Arithmetic sequence

A sequence of the type shown in Example 3, in which the difference between consecutive terms is always the same, is called an **arithmetic sequence.** The pattern for an arithmetic sequence is such that we can write a formula for the *n*th term without reference to the preceding term.

If we have an arithmetic sequence with first term a_1 and with a difference between successive terms of d—that is, $a_n = a_{n-1} + d$—then

$$a_2 = a_1 + d$$
$$a_3 = a_2 + d = a_1 + 2d$$
$$a_4 = a_3 + d = a_1 + 3d$$

nth Term of an arithmetic sequence

and so on. The *n*th term is given by

$$a_n = a_1 + (n - 1)d$$

Example 4 Find the sixth term of an arithmetic sequence with $a_1 = -3$ and $a_n = a_{n-1} + 2$. In this case $d = 2$, so

$$a_n = a_1 + (n - 1)2$$

and

$$a_6 = -3 + (5)(2) = 7$$ ■

Example 5 Find the 10th term of the arithmetic sequence with $a_1 = 16$ and $a_n = a_{n-1} - 4$. We have $d = -4$, so

$$a_n = 16 + (n - 1)(-4)$$
$$a_{10} = 16 + (9)(-4) = -20$$ ■

An arithmetic sequence is sometimes referred to as an arithmetic **progression.**

Suppose you borrow \$5,000 from a bank at 15% interest. Every 6 months you pay the interest due and reduce the principal by \$500. The payments you make are:

First payment: $a_1 = 500 + (5{,}000)(0.075) = 875$

Second payment: $a_2 = 500 + (4{,}500)(0.075) = 837.50$

Third payment: $a_3 = 500 + (4{,}000)(0.075) = 800$

Each payment is \$37.50 less than the preceding payment, because the principal has been reduced by \$500 each time a payment is made, and the interest is therefore reduced by $500(0.075) = 37.50$. This is an example of an arithmetic sequence. The formula for the nth payment is

$$\begin{aligned} a_n &= a_1 + (n - 1)d \\ &= 875 + (n - 1)(-37.50) \end{aligned}$$

The eighth payment is

$$a_8 = 875 + (7)(-37.50) = 612.50$$

The last payment, the 10th, is

$$a_{10} = 875 + (9)(-37.50) = 537.50$$

Example 6 Suppose you borrow \$8,000 from a bank at 12% interest with payments every 6 months of \$400 plus interest due.

First payment: $a_1 = 400 + (8{,}000)(0.06) = 880$

Second payment: $a_2 = 400 + (7{,}600)(0.06) = 856$

Each successive payment is decreased by \$24. The nth payment is

$$a_n = 880 + (n - 1)(-24)$$

The last payment is

$$a_{20} = 880 + (19)(-24) = 424$$ ■

Geometric sequence

A sequence in which each term is a constant multiple of the preceding term is called a **geometric sequence,** or sometimes a geometric **progression.** The constant multiplier is called the **ratio** for the sequence. If we know the first term a_1 and the ratio r, we can develop a formula for the nth term of the sequence:

$$a_2 = a_1 r$$
$$a_3 = a_2 r = a_1 r^2$$
$$a_4 = a_3 r = a_1 r^3$$

and hence

nth Term of a geometric sequence

$$a_n = a_1 r^{n-1}$$

Example 7 Find the sixth term of a geometric sequence with $a_1 = 9$ and $r = \frac{1}{3}$.

$$a_6 = a_1 r^5 = 9\left(\frac{1}{3}\right)^5 = \frac{1}{27}$$ ■

Example 8 Find the third term of the geometric sequence with $a_1 = 125$ and $r = -\frac{1}{2}$.

$$a_3 = a_1 r^2 = 125\left(-\frac{1}{2}\right)^2 = \frac{125}{4}$$ ■

Future value

The **future value** of an investment is the value of the investment at some future date, given an interest rate and compounded over some specified time interval. For example, if \$100 is invested at 9% interest compounded annually, then the future value at the end of n years is given by the formula

Future value after n years = \$100(1.09)^n

Example 9 The future value of an investment after 5 years at 8% interest compounded semiannually is \$1,776. How much was originally invested?
The unknown is A, the amount invested originally. The future value is \$1,776 and the rate per period (half-year) is 4%, so the formula is

$$1{,}776 = A(1.04)^{10} \quad \textit{(10 interest periods)}$$
$$A = 1{,}200$$ ■

Present value

The term **present value** is used to describe the value of an investment today. Thus, after n interest periods at a rate of r (expressed as a decimal) per interest period, the present value PV and the future value FV are related by the equations

$$FV = PV(1 + r)^n$$
$$PV = FV(1 + r)^{-n}$$

Example 10 A bond is offered for sale that is redeemable for \$1,000 in 5 years, and the present interest rate is 12%, compounded quarterly. What is the present value of the bond?

In this case, $FV = 1{,}000$, $n = 20$ (interest computed quarterly for 5 years), and $r = 0.03$ (0.12 divided by 4). Therefore,

$$PV = FV(1 + r)^{-n} = 1{,}000(1.03)^{-20} = 554$$

The present value of the bond is \$554. ■

Exercises 12.1

List the first four terms of the sequences in Exercises 1–12.

1. $a_n = n^2 - n$

2. $a_n = n^3 - n^2$

3. $a_n = 3 - 2n^2$

4. $a_n = 4 - 5n^2$

5. $a_n = (-1)^n(n + 1)$

6. $a_n = (-1)^{3n}(n - 1)$

7. $a_n = (-1)^{3n-1}$

8. $a_n = (-1)^{1-3n}$

9. $a_n = \begin{cases} n^3 - 1 & \text{if } n \text{ is even} \\ n^2 & \text{if } n \text{ is odd} \end{cases}$

10. $a_n = \begin{cases} 2n - n^2 & \text{if } n \text{ is odd} \\ n^2 - n & \text{if } n \text{ is even} \end{cases}$

11. $a_1 = 6,\ a_2 = 4,\ a_n = \frac{1}{2}(a_{n-1} + a_{n-2})$, for $n \geq 3$

12. $a_1 = 8,\ a_2 = 5,\ a_n = \frac{1}{4}(a_{n-1}a_{n-2})$, for $n \geq 3$

In Exercises 13–20 write the first four terms of the arithmetic sequence and a formula for the nth term.

13. $a_1 = 5;\ d = -3$

14. $a_1 = 4;\ d = -5$

15. $a_1 = -3;\ d = 2$

16. $a_1 = -7;\ d = 3$

17. $a_2 = 6;\ d = -1$

18. $a_2 = 4;\ d = 2$

19. $a_4 = 5;\ a_2 = -11$

20. $a_3 = 9;\ a_1 = 2$

In Exercises 21–30 write the first four terms of the geometric sequence and a formula for the nth term.

21. $a_1 = 5;\ r = 0.2$

22. $a_1 = 8;\ r = 0.2$

23. $a_1 = -8;\ r = -1$

24. $a_1 = 6;\ r = -1$

25. $a_1 = \frac{1}{3};\ r = 2$

26. $a_1 = \frac{2}{5};\ r = -2$

27. $a_4 = \frac{4}{9};\ r = \frac{2}{3}$

28. $a_3 = \frac{2}{5};\ r = \frac{1}{2}$

29. $a_2 = \frac{1}{2};\ a_4 = \frac{9}{2}$

30. $a_1 = \frac{4}{7};\ a_4 = \frac{49}{2}$

31. You borrow \$6,000 from a bank and promise to pay \$300 every 3 months plus interest at an annual rate of 9%. Write a formula for the nth payment. What is the amount of the last payment?

32. You borrow \$5,000 from a bank at an interest rate of 1.5% per month and must make monthly payments of \$100 per month plus interest due. Write a formula for the amount of the nth payment. What is the amount of the last payment?

33. What is the future value of an investment of \$500 at 8% compounded annually for 5 years?

34. What is the future value of an investment of \$100 at 9% compounded annually for 4 years?

35. What is the present value of a bond that is redeemable at \$500 after 4 years, if interest is 8% compounded semiannually?

36. What is the present value of a bond that is redeemable at \$600 after 4 years at 9% compounded quarterly?

12.2 Series

We are repaying a loan of \$8,000 by making 20 quarterly payments of \$400 each plus interest due. How much interest will we pay for the loan? We could list all 20 payments and find the total to answer the question, but there is an easier way. The payments form an arithmetic sequence, so we look at the sum of the terms of a finite arithmetic sequence.

If $a_1, a_2, a_3, \ldots, a_n$ is a finite sequence, we refer to the sum

$$a_1 + a_2 + \ldots + a_n$$

as a **series** or the **sum** of the sequence. We use the notation

Sigma notation

$$\sum_{k=1}^{n} a_k = a_1 + a_2 + a_3 + \ldots + a_n$$

to represent the sum. This is often called the **sigma notation** because Σ is the Greek letter sigma, used to represent the word *sum*.

Example 1

$$\sum_{k=1}^{5} k^2 = 1^2 + 2^2 + 3^2 + 4^2 + 5^2 = 55$$

■

Example 2

$$\sum_{j=2}^{5} (3j + 1) = 7 + 10 + 13 + 16 = 46$$

■

Example 3

$$2 + 4 + 6 + 8 + 10 = \sum_{n=1}^{5} 2n$$

■

Example 4

$$1 + 3 + 5 + 7 + 9 + 11 = \sum_{j=1}^{6} (2j - 1)$$

■

Suppose that $a_1, a_2, \ldots, a_n$ is an arithmetic sequence. Then

$$\begin{aligned}\sum_{k=1}^{n} a_k &= a_1 + a_2 + \ldots + a_n \\ &= a_1 + (a_1 + d) + (a_1 + 2d) + \ldots + [a_1 + (n - 1)d] \\ &= na_1 + d[1 + 2 + 3 + \ldots + (n - 1)]\end{aligned}$$

If we knew the sum $1 + 2 + 3 + \ldots + n - 1$, we would have a formula for the sum of an arithmetic sequence.
Let

$$S = 1 + 2 + \ldots + (n - 2) + (n - 1)$$

Then also

$$S = (n - 1) + (n - 2) + \ldots + 2 + 1$$

Adding the two sums together:

$$\begin{aligned} 2S &= (n - 1) + 1 + (n - 2) + 2 + \ldots + (n - 1) + 1 \\ &= n + n + \ldots + n \qquad (n - 1 \text{ terms}) \\ &= n(n - 1) \end{aligned}$$

so

$$S = \frac{n(n - 1)}{2}$$

Sum of an arithmetic sequence

Returning to the sum of the arithmetic sequence,

$$\sum_{k=1}^{n} a_k = na_1 + d\,\frac{n(n - 1)}{2}$$

Example 5 Find the sum of the arithmetic sequence with $a_1 = 10$, $d = 2$, and $n = 6$. The sum is given by the formula

$$\begin{aligned} S &= na_1 + d\,\frac{n(n - 1)}{2} \\ &= 6(10) + 2\,\frac{6(5)}{2} \\ &= 90 \end{aligned}$$

■

Example 6 Find the sum of the arithmetic sequence with $a_1 = 880$, $d = -24$, and $n = 20$. By the formula, the sum S is

$$\begin{aligned} S &= 20(880) + (-24)\,\frac{20(19)}{2} \\ &= 13{,}040 \end{aligned}$$

■

Referring to Example 6 in Section 12.1, we see that this is the total of the payments on the loan. Since the principal was \$8,000, the interest paid is

$$13{,}040 - 8{,}000 = 5{,}040$$

Notice that the first term of the arithmetic sequence is a_1 and the last term is $l = a_1 + (n - 1)d$. We can rewrite the formula for the sum of the sequence as follows:

$$S = na_1 + d\,\frac{n(n-1)}{2} = \frac{n}{2}\,[a_1 + a_1 + (n-1)d]$$
$$= \frac{n}{2}\,(a_1 + l)$$
$$= n\left(\frac{a_1 + l}{2}\right)$$

Thus we can find the sum of an arithmetic sequence by multiplying the number of terms n by the average of the first and last terms.

Example 7 Find the sum of the arithmetic sequence with $a_1 = 45$, $n = 6$, and $d = -6$. The first term is $a_1 = 45$ and the last term is $a_6 = 45 + (5)(-6) = 15$. The sum S is

$$S = 6\left(\frac{45 + 15}{2}\right) = 180$$

■

In a similar fashion we can find a formula for the sum of the geometric sequence

$$a,\ ar,\ ar^2,\ \ldots,\ ar^{n-1}$$

Denote the sum by S:

$$\begin{aligned}
S &= a + ar + ar^2 + ar^3 + \ldots + ar^{n-1} \\
rS &= \phantom{a + {}} ar + ar^2 + ar^3 + \ldots + ar^{n-1} + ar^n \\
S - rS &= a - ar^n = a(1 - r^n) \\
S(1 - r) &= a(1 - r^n) \\
S &= \frac{a(1 - r^n)}{1 - r}
\end{aligned}$$

Sum of a geometric sequence

We summarize as follows.

$$\sum_{k=0}^{n-1} ar^k = \frac{a(1 - r^n)}{1 - r} \qquad \text{if} \quad r \neq 1$$

Example 8 Find the sum of the geometric sequence with $a = 4$, $r = \frac{1}{2}$, and $n = 5$.

$$S = \frac{4\left[1 - \left(\frac{1}{2}\right)^5\right]}{1 - \frac{1}{2}} = \frac{31}{4}$$

■

There are some variations on the sum formula for a geometric series that can be very useful.

1. $a + ar + ar^2 + ar^3 + \ldots + ar^{n-1} = \dfrac{a(1 - r^n)}{1 - r}$

2. $ar + ar^2 + ar^3 + \ldots + ar^n = r(a + ar + ar^2 + \ldots + ar^{n-1})$
$\qquad = r\dfrac{a(1 - r^n)}{1 - r}$

3. $1 + r + r^2 + \ldots + r^{n-1} = \dfrac{1 - r^n}{1 - r}$

all, of course, with the stipulation that $r \neq 1$.

As an application, consider the future value of an account to which you add \$100 each month for 60 months and for which interest is compounded monthly at an annual rate of 9%. The future value will be computed beginning with the first deposit and ending at the end of the 60th month, just before the next deposit is made.

Future value of first payment:	$100(1.0075)^{60}$
Future value of second payment:	$100(1.0075)^{59}$
. . . .	
Future value of 60th payment:	$100(1.0075)$

The future value of the account is the sum of the future values of the individual deposits; that is, the sum of a geometric sequence,

$$100(1.0075),\ 100(1.0075)^2,\ \ldots,\ 100(1.0075)^{59},\ 100(1.0075)^{60}$$

Here $r = 1.0075$, the first term is $100(1.0075)$, and there are 60 terms. The sum is (using Formula 2 above)

$$\begin{aligned}&100(1.0075) + 100(1.0075)^2 + \ldots + 100(1.0075)^{60}\\ &\quad = 1.0075\,\frac{100[1 - (1.0075)^{60}]}{1 - 1.0075}\\ &\quad = 7{,}598.98\end{aligned}$$

Example 9 If you deposit \$50 in a savings account every month for 48 months at 9% annual interest compounded monthly, find the amount in the account at the end of 4 years.

The future value is the sum of the geometric series

$$50(1.0075)^{48},\ 50(1.0075)^{47},\ \ldots,\ 50(1.0075)$$

or

$$\begin{aligned}&50(1.0075) + 50(1.0075)^2 + \ldots + 50(1.0075)^{48}\\ &\quad = 1.0075\,\frac{50[1 - (1.0075)^{48}]}{1 - 1.0075}\\ &\quad = 2{,}897.61\end{aligned}$$

■

Exercises 12.2

In Exercises 1–12 compute the sum.

1. $\sum_{t=1}^{4} 2^t$

2. $\sum_{k=1}^{4} 3^k$

3. $\sum_{k=1}^{5} (2k + 3)$

4. $\sum_{k=1}^{6} (2k - 3)$

5. $\sum_{s=1}^{4} \left(-\frac{1}{2}\right)^s$

6. $\sum_{s=1}^{4} \left(-\frac{1}{3}\right)^s$

7. $\sum_{t=1}^{6} (t + 2)^2$

8. $\sum_{t=1}^{6} (t - 1)^2$

9. $\sum_{j=2}^{6} \frac{j + 1}{j - 1}$

10. $\sum_{j=2}^{6} \frac{j - 1}{j + 1}$

11. $\sum_{i=0}^{3} \left(\frac{i - 1}{i + 1}\right)^2$

12. $\sum_{i=0}^{3} \left(\frac{i + 2}{i + 3}\right)^2$

In Exercises 13–18 write each of the sums in sigma notation.

13. $4 + 7 + 10 + 13 + 16$

14. $5 + 9 + 13 + 17 + 21$

15. $2 - 4 + 8 - 16 + 32$

16. $4 - 16 + 64 - 256$

17. $8 - 4 + 2 - 1$

18. $27 - 9 + 3 - 1$

In Exercises 19–26 find the sum.

19. Arithmetic sequence with $a_1 = 5$, $n = 12$, and $d = -3$

20. Arithmetic sequence with $a_1 = 9$, $n = 11$, and $d = -2$

21. Arithmetic sequence with $a_1 = -3$, $n = 8$, and $d = 4$

22. Arithmetic sequence with $a_1 = 5$, $n = 9$, and $d = 2$

23. Geometric sequence with $a_1 = 6$, $n = 5$, and $r = 0.5$

24. Geometric sequence with $a_1 = 9$, $n = 6$, and $r = 1.1$

25. Geometric sequence with $a_1 = 3$, $n = 5$, and $r = -1$

26. Geometric sequence with $a_1 = 6$, $n = 6$, and $r = -1$

27. You borrow \$6,000 and promise to pay \$300 every 3 months plus interest due at 9% annual interest. How much interest will you pay for the loan?

28. You borrow \$5,000 at an annual interest rate of 12% and make payments of \$100 per month plus interest due each month. How much interest will you pay for the loan?

29. You deposit \$500 each 6 months in a savings account at 9% interest compounded semiannually. How much money will be in the account at the end of 6 years?

30. You deposit \$400 quarterly (four times per year) in a savings account at 8% interest compounded quarterly. How much will you have in the account at the end of 5 years?

12.3 Mathematical Induction

In the spring of 1976, Bob Speca of the University of Pennsylvania set a world's record by toppling 22,221 dominoes at one time. This is done by setting the dominoes in a row, so that as each falls it knocks down the next domino, and so on. There are two basic ingredients in this game: someone must knock down the first domino, and then as each falls it must knock down the next. If both conditions are satisfied, the process continues until all the dominoes have fallen. In a rather simple fashion, domino toppling is an example that illustrates the principle of **mathematical induction.**

Mathematical induction

Principle of mathematical induction

> Suppose $P(n)$ is a statement about the positive integer n. If $P(1)$ is true, and if whenever $P(r)$ is true it follows that $P(r + 1)$ is true, then $P(n)$ is true for every positive integer n.

To compare mathematical induction to domino toppling: verifying that $P(1)$ is true corresponds to knocking over the first domino, and the requirement that whenever $P(r)$ is true then $P(r + 1)$ is true corresponds to the condition that each domino as it falls must hit the next domino. The conclusion that $P(n)$ is true for all n corresponds to the result that all the dominoes fall.

Example 1 Use induction to prove that

$$\sum_{k=1}^{n} k = \frac{n(n+1)}{2}$$

First, we must check that the statement is true for $n = 1$. If $n = 1$, the sum consists of just the single term 1. The expression on the right is

$$\frac{1(1+1)}{2} = 1$$

so the statement is true for $n = 1$. Now assume the statement is true for the integer r.

$$1 + 2 + 3 + \ldots + r = \frac{r(r+1)}{2}$$

Induction hypothesis

This statement is often called the **induction hypothesis.** We want to use this result to show that the statement is true for the integer $r + 1$.

$$1 + 2 + 3 + \ldots + r + (r + 1) = \frac{r(r+1)}{2} + (r + 1)$$

Notice how the sum $1 + 2 + 3 + \ldots + r$ was replaced by the expression $r(r + 1)/2$. Again, this is the induction hypothesis. To continue:

$$\begin{aligned} 1 + 2 + 3 + \ldots + r + (r + 1) &= \frac{r(r+1)}{2} + (r + 1) \\ &= \frac{r(r+1)}{2} + \frac{2(r+1)}{2} \\ &= \frac{(r+1)(r+2)}{2} \end{aligned}$$

This is exactly the statement for the integer $r + 1$. So by the principle of mathematical induction, the statement is true for all n. ■

In Chapter 9 we discussed and used DeMoivre's theorem and promised a proof in Chapter 12. So, true to our word, the next example is a proof of that theorem.

Example 2 Prove DeMoivre's theorem:

$$[r(\cos \theta + i \sin \theta)]^n = r^n(\cos n\theta + i \sin n\theta)$$

For the case $n = 1$, the right and left members of the equation are identical. Assume the statement is true for the integer k; that is,

$$[r(\cos \theta + i \sin \theta)]^k = r^k(\cos k\theta + i \sin k\theta)$$

Again, this is the induction hypothesis. We want to use this statement to prove the statement true for the integer $k + 1$.

$$\begin{aligned}[r(\cos \theta + i \sin \theta)]^{k+1} &= [r(\cos \theta + i \sin \theta)]^k\, r(\cos \theta + i \sin \theta) \\ &= r^k(\cos k\theta + i \sin k\theta)\, r(\cos \theta + i \sin \theta)\end{aligned}$$

Notice how the induction hypothesis was used to replace $[r(\cos \theta + i \sin \theta)]^k$ by $r^k(\cos k\theta + i \sin k\theta)$. To continue:

$$\begin{aligned}&[r(\cos \theta + i \sin \theta)]^{k+1} \\ &\quad = [r(\cos \theta + i \sin \theta)]^k\, r(\cos \theta + i \sin \theta) \\ &\quad = r^k(\cos k\theta + i \sin k\theta)\, r(\cos \theta + i \sin \theta) \\ &\quad = r^{k+1}[(\cos k\theta \cos \theta - \sin k\theta \sin \theta) + i(\sin k\theta \cos \theta + \cos k\theta \sin \theta)] \\ &\quad = r^{k+1}[\cos(k + 1)\theta + i \sin(k + 1)\theta]\end{aligned}$$

This is exactly DeMoivre's theorem for the integer $k + 1$. By induction, the theorem is true for all positive integers n. ■

The principle of induction may be modified as follows. We must have a place to begin, a domino to knock down first to start the process, and it need not be the first domino in the row. In the induction process, we need not start with the integer 1; we need only verify that the statement is true for some integer and then use the induction principle to show that it is true for all larger integers.

Example 3 Show that $\left(1 - \frac{1}{4}\right)\left(1 - \frac{1}{9}\right) \cdots \left(1 - \frac{1}{n^2}\right) = \frac{n + 1}{2n}$ for all positive integers $n \geq 2$.

We will verify the statement for $n = 2$ and proceed from there. For $n = 2$,

$$1 - \frac{1}{4} = \frac{3}{4} = \frac{2 + 1}{2(2)}$$

which is true. Now we assume the statement true for $n = r$:

$$\left(1 - \frac{1}{4}\right)\left(1 - \frac{1}{9}\right) \cdots \left(1 - \frac{1}{r^2}\right) = \frac{r + 1}{2r}$$

For $r + 1$,

$$\left(1 - \frac{1}{4}\right)\left(1 - \frac{1}{9}\right) \ldots \left(1 - \frac{1}{r^2}\right)\left(1 - \frac{1}{(r + 1)^2}\right)$$

$$= \frac{r + 1}{2r}\left(1 - \frac{1}{(r + 1)^2}\right) \quad \textit{(by the induction hypothesis)}$$

$$= \frac{r + 1}{2r} \frac{r^2 + 2r}{(r + 1)^2}$$

$$= \frac{r + 2}{2(r + 1)} = \frac{(r + 1) + 1}{2(r + 1)}$$

This last is the correct expression for $r + 1$. By induction, the statement is true for all $n \geq 2$. ■

Example 4 Show that $n^3 - n + 3$ is divisible by 3 for each positive integer n. For $n = 1$ the statement is that 3 is evenly divisible by 3, which is a true statement. The induction hypothesis is that the number $r^3 - r + 3$ is divisible by 3. We will use this fact to show that $(r + 1)^3 - (r + 1) + 3$ is divisible by 3.

$$\begin{aligned}(r + 1)^3 - (r + 1) + 3 &= r^3 + 3r^2 + 3r + 1 - r - 1 + 3 \\ &= (r^3 - r + 3) + 3r^2 + 3r \\ &= (r^3 - r + 3) + 3(r^2 + r)\end{aligned}$$

The expression $r^3 - r + 3$ is divisible by 3 by the induction hypothesis. The expression $3(r^2 + r)$ is divisible by 3 because r is an integer. Therefore, the sum

$$(r^3 - r + 3) + 3(r^2 + r)$$

is divisible by 3, and that is what was to be proved. By the principle of mathematical induction, $n^3 - n + 3$ is evenly divisible by 3 for each positive integer n. ■

Exercises 12.3

1. Show that

$$1^2 + 2^2 + \ldots + n^2 = \frac{n(2n + 1)(n + 1)}{6}$$

for each positive integer n.

2. Show that $13^n - 7^n$ is divisible by 6 for each positive integer n.
3. If x and y are integers, show that $x^n - y^n$ is divisible by $x - y$ for each positive integer n.
4. Show that a set with n elements has 2^n subsets.
5. If $x \neq 1$, show that

$$1 + x + x^2 + \ldots + x^n = \frac{1 - x^{n+1}}{1 - x}$$

for each positive integer n.

6. Show that $1 + 3 + \ldots + (2n - 1) = n^2$ for each positive integer n.

7. Show that

$$\frac{1}{1\cdot 2}+\frac{1}{2\cdot 3}+\frac{1}{3\cdot 4}+\cdots+\frac{1}{n(n+1)}=\frac{n}{n+1}$$

for each positive integer n.

8. For all integers $n \geq 5$, show that $2^n > n^2$.

9. If $h \geq -1$, show that $(1+h)^n \geq 1 + nh$ for all positive integers n.

10. Show that

$$\sum_{k=1}^{n}\frac{k(k+1)}{2}=\frac{n(n+1)(n+2)}{6}$$

for all positive integers n.

11. Show that $n^2 + n$ is divisible by 2 for each positive integer n. (You don't really need induction for this one, but the result is useful in Exercise 12.)

12. Prove that $n^3 + 5n$ is divisible by 6 for each positive integer n.

13. Prove that

$$\sum_{k=1}^{n}(4k-2)=2n^2$$

for all positive integers n.

14. Prove that $n^3 + 2n$ is divisible by 3 for each positive integer n.

15. Prove that

$$\sum_{k=1}^{n}(2k-1)^3=n^2(2n^2-1)$$

for each positive integer n.

16. Prove that

$$\sum_{k=1}^{n}\frac{1}{\sqrt{k}}\geq\sqrt{n}$$

12.4 Binomial Theorem

In Chapter 1 we saw some powers of binomials:

$$(x+y)^1 = x + y$$
$$(x+y)^2 = x^2 + 2xy + y^2$$
$$(x+y)^3 = x^3 + 3x^2y + 3xy^2 + y^3$$

With a bit of effort, we can extend the list:

$$(x+y)^4 = x^4 + 4x^3y + 6x^2y^2 + 4xy^3 + y^4$$
$$(x+y)^5 = x^5 + 5x^4y + 10x^3y^2 + 10x^2y^3 + 5xy^4 + y^5$$

In this section we prove the Binomial Theorem, which gives a formula for the coefficients in the expansion of $(x+y)^n$.

For each positive integer n, we define

$$n! = 1 \cdot 2 \cdot 3 \cdot 4 \cdots n \quad \textit{(pronounced ``n factorial'')}$$

Factorial notation

In other words, $n!$ is the product of the first n integers if $n \geq 2$, and $1! = 1$.

Example 1 $6! = 1 \cdot 2 \cdot 3 \cdot 4 \cdot 5 \cdot 6 = 720$ ■

Binomial coefficient

We use the factorial notation to define the **binomial coefficients.** If n and k are positive integers with $n > k$, we define

$$\binom{n}{k}=\frac{n!}{k!(n-k)!}$$

We adopt the convention that

$$\binom{n}{0} = 1 \quad \text{and} \quad \binom{n}{n} = 1$$

Example 2 $$\binom{4}{3} = \frac{4!}{3!\,1!} = \frac{24}{6} = 4$$ ■

Example 3 $$\binom{7}{4} = \frac{7!}{4!\,3!} = \frac{1 \cdot 2 \cdot 3 \cdot 4 \cdot 5 \cdot 6 \cdot 7}{1 \cdot 2 \cdot 3 \cdot 4 \cdot 1 \cdot 2 \cdot 3} = \frac{5 \cdot 6 \cdot 7}{1 \cdot 2 \cdot 3} = 35$$ ■

The cancellation that we did in Example 3 can be done in the general case:

$$\binom{n}{k} = \frac{n!}{k!(n-k)!} = \frac{1 \cdot 2 \cdot 3 \cdots (n-k)(n-k+1) \cdots n}{k![1 \cdot 2 \cdot 3 \cdot 4 \cdots (n-k)]}$$
$$= \frac{(n-k+1)(n-k+2) \cdots (n-1)n}{k!}$$

In this form, both the numerator and denominator have k factors.

Example 4 $$\binom{8}{3} = \frac{8 \cdot 7 \cdot 6}{1 \cdot 2 \cdot 3} = 56$$ ■

Example 5 $$\binom{10}{4} = \frac{10 \cdot 9 \cdot 8 \cdot 7}{1 \cdot 2 \cdot 3 \cdot 4} = 210$$ ■

At the risk of being repetitious, we note that in Examples 4 and 5 the numerator and the denominator have the same number of factors.

Binomial theorem

> For each positive integer n,
>
> $$(x + y)^n = x^n + nx^{n-1}y + \ldots + \binom{n}{k} x^{n-k}y^k + \ldots + y^n$$

Our proof of this theorem is by induction.

1. Is the theorem true for $n = 1$? Yes, for $n = 1$, the left and right members are identical.
2. Assume the statement is true for $n = r$; that is,

$$(x + y)^r = x^r + rx^{r-1}y + \ldots + \binom{r}{k} x^{r-k}y^k + \ldots + y^r$$

3. Check on the statement when $n = r + 1$.

$$\begin{aligned}(x + y)^{r+1} &= (x + y)(x + y)^r \\ &= (x + y)\left(x^r + rx^{r-1}y + \ldots + \binom{r}{k} x^{r-k}y^k + \ldots + y^r\right) \\ &= x^{r+1} + (r + 1)x^r y + \ldots + \left[\binom{r}{k} + \binom{r}{k-1}\right] x^{(r+1)-k}y^k \\ &\quad + \ldots + y^{k+1}\end{aligned}$$

If we can show that

$$\binom{r}{k} + \binom{r}{k-1} = \binom{r+1}{k}$$

then the induction will be complete.

$$\begin{aligned}\binom{r}{k} + \binom{r}{k-1} &= \frac{r!}{k!(r-k)!} + \frac{r!}{(k-1)!(r-k+1)!} \\ &= \frac{r!(r-k+1)}{k!(r-k+1)!} + \frac{r!k}{k!(r-k+1)!} \\ &= \frac{r!(r+1)}{k![(r+1)-k]!} \\ &= \binom{r+1}{k}\end{aligned}$$

This completes the proof of the Binomial Theorem.

Let us see how the Binomial Theorem fits in with the examples shown earlier. Example 6 makes use of the following values:

$$\binom{5}{0} = 1 \qquad \binom{5}{1} = 5 \qquad \binom{5}{2} = \frac{5 \cdot 4}{1 \cdot 2} = 10 \qquad \binom{5}{3} = \frac{5 \cdot 4 \cdot 3}{1 \cdot 2 \cdot 3} = 10$$

$$\binom{5}{4} = \frac{5 \cdot 4 \cdot 3 \cdot 2}{1 \cdot 2 \cdot 3 \cdot 4} = 5 \qquad \binom{5}{5} = 1$$

Example 6

$$\begin{aligned}(x + y)^5 &= x^5 + \binom{5}{1} x^4y + \binom{5}{2} x^3y^2 + \binom{5}{3} x^2y^3 + \binom{5}{4} xy^4 + y^5 \\ &= x^5 + 5x^4y + 10x^3y^2 + 10x^2y^3 + 5xy^4 + y^5\end{aligned}$$

■

Example 7 makes use of the following values:

$$\binom{8}{0} = 1 \qquad \binom{8}{1} = 8 \qquad \binom{8}{2} = \frac{8 \cdot 7}{1 \cdot 2} = 28 \qquad \binom{8}{3} = \frac{8 \cdot 7 \cdot 6}{1 \cdot 2 \cdot 3} = 56$$

$$\binom{8}{4} = \frac{8 \cdot 7 \cdot 6 \cdot 5}{1 \cdot 2 \cdot 3 \cdot 4} = 70 \qquad \binom{8}{5} = \frac{8 \cdot 7 \cdot 6 \cdot 5 \cdot 4}{1 \cdot 2 \cdot 3 \cdot 4 \cdot 5} = 56$$

$$\binom{8}{6} = \frac{8 \cdot 7 \cdot 6 \cdot 5 \cdot 4 \cdot 3}{1 \cdot 2 \cdot 3 \cdot 4 \cdot 5 \cdot 6} = 28$$

$$\binom{8}{7} = \frac{8 \cdot 7 \cdot 6 \cdot 5 \cdot 4 \cdot 3 \cdot 2}{1 \cdot 2 \cdot 3 \cdot 4 \cdot 5 \cdot 6 \cdot 7} = 8 \qquad \binom{8}{8} = 1$$

Example 7 $(x + y)^8 = x^8 + 8x^7y + 28x^6y^2 + 56x^5y^3 + 70x^4y^4 + 56x^3y^5 + 28x^2y^6 + 8xy^7 + y^8$ ■

Notice the symmetry between pairs of coefficients. We can verify an identity about binomial coefficients that shows how this occurs.

For $n > k$,

$$\binom{n}{n-k} = \frac{n!}{(n-k)![n-(n-k)]!} = \frac{n!}{(n-k)!k!} = \binom{n}{k}$$

We can use this identity to simplify the calculations in the binomial expansion. Example 8 makes use of the following values:

$$\binom{6}{0} = \binom{6}{6} = 1 \qquad \binom{6}{1} = \binom{6}{5} = 6 \qquad \binom{6}{2} = \binom{6}{4} = \frac{6 \cdot 5}{1 \cdot 2} = 15$$

$$\binom{6}{3} = \frac{6 \cdot 5 \cdot 4}{1 \cdot 2 \cdot 3} = 20$$

Example 8 $(x + y)^6 = x^6 + 6x^5y + 15x^4y^2 + 20x^3y^3 + 15x^2y^4 + 6xy^5 + y^6$ ■

Another way to compute binomial coefficients is to use what is called **Pascal's triangle.**

Pascal's triangle

1
1 1
1 2 1
1 3 3 1
1 4 6 4 1
1 5 10 10 5 1
1 6 15 20 15 6 1
1 7 21 35 35 21 7 1
1 8 28 56 70 56 28 8 1

Notice that the first row (ignoring the 1 on top) consists of the coefficients of $(x + y)^1$, the second row consists of the coefficients of $(x + y)^2$, the third row the coefficients of $(x + y)^3$, and so on. Moreover, we can generate new rows for the triangle by noticing that each number is the sum of the two numbers in the row directly above that number.

Example 9 Write the next row of Pascal's triangle.

1 8 28 56 70 56 28 8 1
1 9 36 84 126 126 84 36 9 1

■

Example 10 Expand $(x + y)^9$ using Pascal's triangle.

$$(x + y)^9 = x^9 + 9x^8y + 36x^7y^2 + 84x^6y^3 + 126x^5y^4 + 126x^4y^5 + 84x^3y^6 + 36x^2y^7 + 9xy^8 + y^9$$ ■

Example 11 $(x - 2)^5 = [x + (-2)]^5 = x^5 + 5x^4(-2) + 10x^3(-2)^2 + 10x^2(-2)^3 + 5x(-2)^4 + (-2)^5$

$$= x^5 - 10x^4 + 40x^3 - 80x^2 + 80x - 32$$ ■

Example 12 Find the x^3y^2 term in the expansion of $(x + 2y)^5$.
The term is

$$\binom{5}{2} x^3(2y)^2 = 10x^3(2y)^2 = 40x^3y^2$$ ■

Exercises 12.4

Use the Binomial Theorem to expand the power.

1. $(a - 3b)^3$
2. $(x - 2y)^3$
3. $(a - b)^6$
4. $(s - t)^5$
5. $(2x + y)^5$
6. $(3a + b)^5$
7. $(u - 0.5)^4$
8. $(p - 0.5)^5$
9. $(1 + 0.05)^4$
10. $(1 - 0.05)^4$
11. $(x^2 - y^2)^3$
12. $(a^2 + b^2)^4$
13. $(2a^2 - 1)^4$
14. $(3x^2 + 1)^4$
15. $(x^3 - 1)^3$
16. $(y^3 + 1)^3$

Use the Binomial Theorem to find the indicated term.

17. The x^3y^2 term in $(x + 2y)^5$
18. The a^5b term in $(a - 3b)^6$
19. The p^4q^2 term in $(p - q)^6$
20. The r^4t^2 term in $(t - r)^6$
21. Generate the 10th row of Pascal's triangle.
22. Generate the 11th row of Pascal's triangle.

Chapter Review

List the first five terms of the indicated sequence.

1. $a_n = 4n - 3$
2. $a_n = \dfrac{1 - n^2}{2n}$
3. $a_n = n^2 - n - 41$
4. $a_1 = 4;\ a_n = \frac{1}{3} a_{n-1}$
5. $a_1 = 4;\ a_n = a_{n-1} + 3$
6. $a_1 = 12;\ a_2 = -2;\ a_n = \frac{1}{2}(a_{n-1} + a_{n-2})$

Find the indicated term of the sequence.

7. The sixth term of the arithmetic sequence with $a_1 = 0$ and $d = -4$.
8. The fourth term of the arithmetic sequence with $a_1 = 6$ and $d = 3$.
9. The seventh term of the geometric sequence with $a_1 = \frac{1}{4}$ and $r = -2$.
10. The fifth term of the geometric sequence with $a_1 = 2$ and $r = -3$.
11. You borrow \$4,000 at 12% annual interest. You pledge to make annual payments of \$1,000 plus interest owed. Give the amount of each payment. What is the total of the four payments?
12. What is the future value of \$6,000 invested at 8% compounded annually for 4 years?
13. What is the present value of a bond that is redeemable at \$500 after 4 years if you use an annual rate of 6% compounded annually?

Compute the sum.

14. $\sum_{k=1}^{6} (k^2 - k)$

15. $\sum_{t=2}^{6} (t^2 - 1)$

16. $\sum_{k=2}^{7} (2k - 1)$

17. $\sum_{t=0}^{6} (t + 1)^{-1}$

Find the sum of each of the following.

18. An arithmetic sequence with $a_1 = -2$, $d = 3$, and $n = 6$.
19. An arithmetic sequence with $a_1 = 16$ and last term $a_8 = -12$.
20. A geometric sequence with $a_1 = 4$, $r = \frac{1}{3}$, and $n = 4$.
21. A geometric sequence with $a_1 = 9$, $r = -3$, and $n = 4$.

Write the sum using sigma notation.

22. $4 + 6 + 8 + 10$

23. $3 + 7 + 11 + 15 + 19$

24. $4 + 6 + 9 + 13.5$

25. $6 - 8 + 10 - 12$

26. $3 + 8 + 15 + 24 + 35$

27. $9 + 28 + 65 + 126$

28. You borrow \$4,000 and repay \$100 each month plus interest at an annual rate of 12%. How much interest will you pay in repaying the loan?
29. You deposit \$50 per month in a savings account that pays 6% compounded monthly. How much will you have in the account at the end of 2 years?
30. Write the power $(a - 5b)^5$ as a polynomial.
31. Find the x^5 term of $(2x - 1)^8$.
32. Write $(2x - 3y)^7$ as a polynomial.
33. Prove by induction: $1 \cdot 2 + 2 \cdot 3 + \ldots + n(n + 1) = \dfrac{n^3 + 3n^2 + 2n}{3}$.
34. Prove that $15^n - 7^n$ is divisible by 8 for every positive integer n.

Chapter Test

List the first five terms of the indicated sequence.

1. $a_n = 2n - n^2$

2. $a_1 = 6;\ a_n = \frac{1}{2} a_{n-1} + 1$

Find the indicated term.

3. The fourth term of an arithmetic sequence with $a_1 = 4$ and $d = -2$
4. The fifth term of an arithmetic sequence with $a_1 = -5$ and $d = 2$
5. The fourth term of a geometric sequence with $a_1 = 2$ and $r = -2$
6. The sixth term of a geometric sequence with $a_1 = -\frac{1}{4}$ and $r = 2$
7. You borrow \$4,000 at 10% annual interest. You pledge to repay the loan in four equal annual payments plus interest owed with each payment. Give the amount of each payment.
8. Compute the sum $\sum_{k=1}^{4} (k^2 + 2k)$.
9. Write using sigma notation: $-2 + 0 + 2 + 4 + 6$.
10. Find the sum of the arithmetic sequence with $a_1 = 3$, $d = -2$, and $n = 5$.
11. Find the sum of the geometric sequence with $a_1 = 4$, $r = 2$, and $n = 5$.
12. Prove by induction: $1^2 + 2^2 + 3^2 + \ldots + (2n - 1)^2 = \dfrac{4n^3 - n}{3}$.
13. Expand by the Binomial Theorem: $(2 - ab)^5$.
14. Find the x^5 term of the expansion of $(2x - y)^7$.

Answers

Exercises 1.1, page 5

1. -8 **3.** -7 **5.** 3 **7.** 75 **9.** $-\frac{7}{4}$ **11.** 4.9 **13.** 2 **15.** 3 **17.** 10.9
19. -12 **21.** 63 **23.** -10 **25.** 42 **27.** 20 **29.** -2 **31.** 5 **33.** $-\frac{3}{2}$
35. 18 **37.** $\frac{24}{5}$ **39.** $\frac{8}{9}$ **41.** $\frac{3}{4}$ **43.** v **45.** 0 **47.** $-11t$ **49.** $s + 6$
51. $-6w - 24$ **53.** $6s + 8t$ **55.** -5 **57.** 0 **69.** -1.83 **71.** 12.1 **73.** -2.8
75. -9.5 **77.** -4.313 **79.** 4.402 **81.** 22.5 **83.** 2.4

Exercises 1.2, page 10

1. $<$ **3.** $>$ **5.** $>$ **7.** $<$ **9.** $-1, 3, 7$ **11.** $-5, -3.4, -2.3$ **13.** $1.001, 1.01, 1.1$
15. $-2.2, -2.02, -2.002$ **17.** $-\frac{4}{5}, -\frac{3}{4}, \frac{7}{12}, \frac{2}{3}$ **19.** $<$ **21.** $>$ **23.** $>$ **25.** $<$
27. $>$ **29.** 7 **31.** 2.3 **33.** 8 **35.** 15 **37.** $\frac{3}{4}$ **39.** $\frac{5}{4}$ **41.** -1 **43.** 1
45. $-x - 9$ **47.** $|x + y| \leq |x| + |y|$ is true.

Exercises 1.3, page 17

1. $\frac{37}{8}$ **3.** -1 **5.** 11 **7.** 3 **9.** 2 **11.** 10 **13.** 0 **15.** 1 **17.** 4 **19.** 0
21. $\frac{1}{7}$ **23.** $-\frac{2}{3}$ and $\frac{16}{3}$ **25.** 36 and 24 **27.** $\frac{2x - 5}{3}$ **29.** $\frac{s - 5 + t}{t}, t \neq 0$ **31.** $x > -6$
33. $x \leq \frac{14}{3}$ **35.** $s \leq -55$ **37.** $x > 9$ **39.** $x < -3$ **41.** $m \geq -3$ **43.** $-\frac{4}{3} < x < -\frac{1}{3}$
45. $-1 < x < 5$ **47.** $2 \leq x \leq 6$ **49.** $-1 \leq x \leq 2$ **51.** $-8 < x < -\frac{8}{3}$ **53.** $x \geq \frac{3}{5}$ or $x \leq -1$
55. $x > 27$ or $x < 3$ **57.** $n > \frac{s - 6 + 2t}{t}$ **59.** $\frac{v - k}{g} < t < \frac{-v - k}{g}$

Exercises 1.4, page 21

1. 54 **3.** 20 by 90 **5.** 34 **7.** 650 **9.** 15 **11.** 2.50 **13.** 24
15. 2,000 at 6% and 4,000 at 8% **17.** 750 at 5%, 1,250 at 7% **19.** 7,600
21. Barbara's rate is \$5 and Karen's rate is \$6.50. **23.** 30 nickels and 20 dimes **25.** 4 qt
27. $20 < C < 22\frac{2}{9}$

Exercises 1.5, page 26

1. degree 2 in u **3.** degree 3 in a, degree 2 in b, degree 2 in a and b **5.** degree 7 in y **7.** 1
9. 127 **11.** $a^3 + 7a^2 + 12a + 1$ **13.** 4 **15.** -36 **17.** $-8x + 6$ **19.** $-a - 5b - 2c$
21. b **23.** $s + 8t$ **25.** $x^2 - 2xy$ **27.** $3u^2 + 3uv$ **29.** $-x^2y + xy^2$ **31.** $r^2 - 16t^2$
33. $81 - p^2$ **35.** $9s^2 - 4t^2$ **37.** $x^2 - 18x + 81$ **39.** $4x^2 + 28xy + 49y^2$

41. $25t^2 - 10st + s^2$ **43.** $16t^2 + 24rt + 9r^2$ **45.** $w^2 + 11w + 28$ **47.** $b^2 - 4b - 5$
49. $6r^2 - 11rt - 7t^2$ **51.** $8t^2 + 38st + 35s^2$ **53.** $36n^2 - 35mn + 6m^2$
55. $w^3 - 6w^2 + 12w - 8$ **57.** $x^3 + 6x^2y + 12xy^2 + 8y^3$ **59.** x^{11} **61.** $x^3y^4 - x^2y^4$
63. $3m^4n^3 - 6m^3n^3$ **65.** $a^3 - 8$ **67.** $m^3 + 64$ **69.** $x^4 - 1$ **71.** $x^2y^4 - 25$
73. $-4xy + 2y^2$ **75.** $-4ab$ **77.** $x^2 + 2xy + y^2 - 4xz - 4yz + 4z^2$ **79.** $3x^2 + 8x + 13$
81. $10h + 3h^2$

Exercises 1.6, page 31

1. $5y(y - 5)$ **3.** $7xy(z - 2x)$ **5.** $-3x(1 + 3x)$ **7.** $3xy(1 - 2y)$ **9.** $ab(a - 3 + 5ab)$
11. $3a^2b^2(5b - 4a)$ **13.** $(x - 5)(x + 5)$ **15.** $(7 - s)(7 + s)$ **17.** $5(n - 4m)(n + 4m)$
19. $(w - 9)^2$ **21.** $(u - 8v)^2$ **23.** $(p + 6)^2$ **25.** $2(p + 5)^2$ **27.** $(w - 2)(w + 8)$
29. $(w - 9)(w + 3)$ **31.** $4(r + 4)(r - 2)$ **33.** $2(b - 4)(b - 3)$ **35.** $5(m - 9)(m + 8)$
37. $(5x + y)(x + 2y)$ **39.** $(3a - 1)(a - 2)$ **41.** $(5x + y)(x - 2y)$ **43.** $(3u + 2v)(u - 6v)$
45. $(5z + 2)(2z + 3)$ **47.** $2(2r - 3)(r - 1)$ **49.** $(a + 3)(a^2 - 3a + 9)$
51. $(2z - 3t)(4z^2 + 6zt + 9t^2)$ **53.** $(a + 1)(x + y)$ **55.** $(x - 2)(x + 2)(x + 7)$ **57.** -9
59. 3.528 **61.** 0.4454 **63.** 1.28864

Chapter Review, page 32

1. -12 **2.** 4 **3.** 2 **4.** 19.5 **5.** 14 **6.** $\frac{5}{8}$ **7.** -28 **8.** 35 **9.** -72 **10.** 36
11. 5.76 **12.** -8 **13.** -3 **14.** 2 **15.** -2 **16.** -50 **17.** $\frac{9}{8}$ **18.** $\frac{3}{4}$ **19.** $8a$
20. u **21.** $-7y$ **22.** $-7s$ **23.** t **24.** $7p + 1$ **25.** $3r - 2$ **26.** $-6m + 10$
27. $-9, -6, 0, 1, 4$ **28.** $-3.8, -3.4, -2.1, 0$ **29.** $-0.01, -0.002, 1, 2, 6$ **30.** $-\frac{5}{6}, -\frac{7}{9}, -\frac{3}{4}$
31. $\frac{9}{11}, \frac{11}{13}, \frac{6}{7}, \frac{13}{15}$ **32.** $-2.5, -2, 2 = |-2|, |-2.5|$ **33.** -8 **34.** $-\frac{5}{4}$ **35.** 60 **36.** -56
37. $-\frac{3}{4}$ **38.** -100 **39.** $-\frac{8}{3}$ **40.** -1 **41.** $\frac{13}{2}$ **42.** 1 **43.** $-\frac{14}{5}$ **44.** $-\frac{1}{7}$
45. $m = \frac{2n - pq}{3}$ **46.** 5 and -2 **47.** $s = \frac{1 - 2t}{1 - t}$ **48.** $t = \frac{s - 1}{s - 2}$ **49.** 3 and $\frac{1}{3}$ **50.** $x > 4$
51. $x \le 1$ **52.** $x > \frac{11}{6}$ **53.** $x \ge -\frac{29}{3}$ **54.** $-\frac{4}{3} < x < \frac{4}{3}$ **55.** $-3 \le x \le 5$
56. $-5 < x < -1$ **57.** $x \ge 6$ or $x \le -3$ **58.** $\frac{1}{2} < w < \frac{9}{2}$ **59.** Rose has 20, Jill has 10.
60. Joe took 81 strokes and Chuck took 83. **61.** She must score between 162 and 222.
62. 2 mi/hr **63.** The lengths are 9, 10, and 17. **64.** 16 **65.** 7 **66.** 31 **67.** 52
68. $x^2 - 4y^2$ **69.** $9a^2 - 25b^2$ **70.** $s^2 - 4st + 4t^2$ **71.** $m^2 + 6mn + 9n^2$
72. $x^2 + xy - 6y^2$ **73.** $a^2 - 3ab - 10b^2$ **74.** $6a^2 + 9a - 6$ **75.** $6m^2 + 11m - 10$
76. $-4x$ **77.** $8a$ **78.** $(a - 9)(a + 9)$ **79.** $5(m - 2n)(m + 2n)$ **80.** $3(n - 4m)(n + 4m)$
81. $(a - 4)^2$ **82.** $(s - 5)^2$ **83.** $3(s + 5)(s + 2)$ **84.** $2(x - 4)(x - 3)$
85. $(5a + b)(a + 2b)$ **86.** $(2t + s)(t + 3s)$ **87.** $(7q - 2)(q + 5)$ **88.** $(3v + 2)(v - 6)$
89. $(a - 3)(a + 3)(a + 4)$ **90.** $(a - 4)(a^2 + 4a + 16)$ **91.** $3(x + 2y)(x^2 - 2xy + 4y^2)$
92. $(a + 7)(2x - 3y)$

Chapter Test, page 34

1. 0 **2.** -7 **3.** 7 **4.** 9.3 **5.** 5 **6.** $-\frac{3}{10}$ **7.** -36 **8.** -18 **9.** $-x + 16$
10. $\frac{7}{4}t$ **11.** $-2.2, -1.7, 1.2, 3.4$ **12.** $-\frac{3}{7}, -\frac{2}{5}, \frac{4}{7}, \frac{3}{5}$ **13.** 1 **14.** $\frac{1}{2}$ **15.** $x < 2$
16. $p = \frac{19}{2r}, r \neq 0$ **17.** $-\frac{15}{2} \le x$ **18.** $x \ge 3$ or $x \le -9$ **19.** 66 oz
20. $P(-2) = -29, P(3) = 61$ **21.** $4x^2 - 1$ **22.** $9x^2 - 12xy + 4y^2$ **23.** $6x^2 - 7xy - 20y^2$
24. $6x^3 + 4x^2 - 10x$ **25.** $5(x - 2y)(x + 2y)$ **26.** $(a + 11)^2$ **27.** $(x - 13)(x + 4)$
28. $(4a + b)(3a - 2b)$ **29.** $(3z - 2)(9z^2 + 6z + 4)$

Exercises 2.1, page 40

1. $2ab$ **3.** $\frac{3b^2c}{a^2}$ **5.** $-y - z$ **7.** $\frac{x+2}{2}$ **9.** $\frac{1}{x+2y}$ **11.** $x + 2y$ **13.** $\frac{-3}{7+x}$
15. $\frac{x-6}{2}$ **17.** $\frac{x-4}{x+2}$ **19.** 3 **21.** 0 **23.** $\frac{x-13}{(x+2)(x-3)}$ **25.** $\frac{y^2+6y-10}{3(y-2)}$
27. $\frac{-3x-6y}{(2x-y)(x-2y)}$ **29.** $\frac{m^2-2m+5}{m-2}$ **31.** $\frac{-x+9y}{(x-y)(x-y)(x+y)}$ **33.** $\frac{5a^2-4b^2+ab-a}{a(a-b)(a+b)}$
35. $-9z^2w$ **37.** $\frac{9}{4}$ **39.** $\frac{3}{(a-1)(3a+1)}$ **41.** $\frac{2}{15}$ **43.** $\frac{xy}{-6z}$ **45.** $\frac{xy}{2(y-x)}$ **47.** $\frac{3-y}{2y-3}$
49. $2a^2$ **51.** $r - s$ **53.** $\frac{6x-1}{6x+1}$ **55.** $\frac{x+1}{x-1}$ **57.** $\frac{3x-2y}{2x-3y}$ **59.** 8 **61.** $\frac{14}{11}$ **63.** 7
65. -1 **67.** $-\frac{8}{3}$ **69.** 90 hr **71.** 4.5 hr **73.** $\frac{2}{3}$ hr **75.** 50.025 (rounded off)

Exercises 2.2, page 46

1. a^4b^7 **3.** x **5.** $\frac{1}{m^4}$ **7.** x^8 **9.** a^4b^2 **11.** $\frac{x^4}{y^2}$ **13.** $\frac{a^2c}{b^2}$ **15.** $\frac{z^2}{x^4y^6}$ **17.** $\frac{81y^8}{16x^4z^4}$
19. $\frac{17}{72}$ **21.** $\frac{x^2}{4y^3}$ **23.** 1 **25.** $x + \frac{1}{x}$ **27.** $\frac{x}{x^2+1}$ **29.** $a - \frac{b}{a}$ **31.** $\frac{a}{b} - \frac{b}{a}$ **33.** $\frac{5}{6}$
35. $s + t$ **37.** $\frac{b}{a-b}$ **39.** $2xy$ **41.** 1.55×10^2 **43.** 1.55×10 **45.** 4.5×10^{-4}
47. 4.3×10^{-6} **49.** 1.4×10^4 **51.** 2.3×10^{-2} **53.** 1,300 **55.** 0.00025 **57.** 845
59. 2.5 **61.** 2,400 **63.** 0.000008 **65.** 10^4 **67.** 24 **69.** 1,750 **71.** 192 sec
73. 753.3 sec **75.** 9,500 sec **77.** 19,533.3 sec **79.** 4.8×10^3 sec.

Exercises 2.3, page 52

1. 5 **3.** 4 **5.** 2 **7.** 67 **9.** -3 **11.** 12 **13.** 8 **15.** -3 **17.** $6^{1/2}$ **19.** $x^{5/3}$
21. $a^{3/2}b^{5/2}$ **23.** $(x+y)^{1/2}$ **25.** $\sqrt[4]{(xy)^3}$ **27.** $\sqrt{(a+b)^5}$ **29.** $\sqrt[5]{m^{-3}}$ **31.** $\sqrt[6]{a^{15}b^{10}}$
33. 8 **35.** 125 **37.** 16 **39.** 625 **41.** 4 **43.** 4 **45.** $a^{7/10}$ **47.** $t^{7/6}$ **49.** $x^{1/6}$
51. $\frac{y^{15/2}}{x^{3/2}}$ **53.** $\frac{x^{5/3}y^{5/2}}{w^{10/3}}$ **55.** $\frac{ab^{1/2}}{c^{3/4}}$ **57.** $5\sqrt{10}$ **59.** 20 **61.** $3\sqrt[3]{3}$ **63.** $4x$ **65.** $-13a^2$
67. $4x^2y^4\sqrt{2}$ **69.** 10 **71.** 4 **73.** 0.03 **75.** 5×10^{-2} **77.** $4\sqrt{7}$ **79.** $5a\sqrt{2a}$
81. $4\sqrt{3x}$ **83.** 0 **85.** $\frac{\sqrt{3}}{3}$ **87.** $-\frac{1}{2\sqrt{2}}$ **89.** $3 + 3\sqrt{5}$ **91.** $2 + \sqrt[3]{12}$ **93.** 1
95. $-10 - 3\sqrt{6}$ **97.** $2 + x - 4\sqrt{x-2}$ **99.** $\frac{\sqrt{14}}{7}$ **101.** $\frac{\sqrt[3]{3}}{3}$ **103.** $\frac{\sqrt{30b}}{6ab^2}$
105. $\frac{\sqrt[3]{15m^2n}}{5m^2n}$ **107.** $\sqrt{2} - 1$ **109.** $\frac{42+7\sqrt{3}}{11}$ **111.** $-1 - \sqrt{2}$ **113.** $\frac{x-1+\sqrt{x-1}}{2-x}$
115. $\frac{2}{(x^2+1)^{3/2}}$ **117.** $\frac{4x+30}{(x+5)^{4/3}}$

Exercises 2.4, page 57

1. $5i$ **3.** $-2i$ **5.** $3 - 5i$ **7.** $36 - 6i$ **9.** $8 + 4i$ **11.** 8 **13.** $-7i$ **15.** $7 - 3i$
17. $3 + i$ **19.** 2 **21.** $22 + 6i$ **23.** 40 **25.** 16 **27.** $-1 + 5i$ **29.** 6 **31.** $-1 - i$
33. $-4i$ **35.** $-i$ **37.** 1 **39.** 1 **41.** $\frac{2}{5} + \frac{i}{5}$ **43.** $\frac{3+2i}{13}$ **45.** $\frac{-5-4i}{2}$ **47.** -2
49. i **51.** $2i$ **53.** $2 + 2i$ **55.** $2i + 2$ **57.** no

Exercises 2.5, page 63

1. 4, −4 **3.** $\frac{1}{3}, -\frac{1}{3}$ **5.** 5, −3 **7.** 2, 6 **9.** −6, 8 **11.** $-3 \pm \sqrt{13}$ **13.** 4 a double root **15.** $5 \pm \sqrt{20}$ **17.** $\frac{1}{2}(3 \pm \sqrt{19}\,i)$ **19.** $\frac{1}{2}, \frac{1}{3}$ **21.** 0, 1 **23.** 1, 2 **25.** $1, \frac{4}{7}$ **27.** 1 a double root **29.** −2, 7 **31.** $3 \pm \sqrt{5}$ **33.** $\frac{1}{2}(5 \pm \sqrt{43})$ **35.** $\frac{1}{8}(-3 \pm \sqrt{23}\,i)$ **37.** $\frac{1}{2}(2 \pm \sqrt{10})$ **39.** $\frac{2}{3}, -3$ **41.** 0, −2 **43.** 6, 1 **45.** 6, −2 **47.** 2, −5 **49.** $-1, -\frac{2}{5}$ **51.** 2, −2 **53.** $a = -b$ **55.** $y = \frac{1}{4}x,\ y = -\frac{2}{3}x$ **57.** $x = -\frac{1}{2}(7 \pm \sqrt{65})z$ **59.** 15 in. by 15 in. **61.** $\frac{1}{2}$ m **63.** 6 m by 18 m **65.** 6 and 7 **67.** 6 and 7 **69.** 11 and 13

Exercises 2.6, page 68

1. 64 **3.** $16, \frac{1}{9}$ **5.** $3, \frac{1}{4}$ **7.** 1, −1 **9.** $1, -1, i, -i$ **11.** 4 **13.** 6, −7 **15.** 0, −1 **17.** $\frac{1}{2}$ **19.** $-\frac{7}{12}$ **21.** no solutions **23.** 13 **25.** no solutions **27.** no solutions

29. $-1 < x < 0$ 

31. $x < -5$ or $x > 1$ 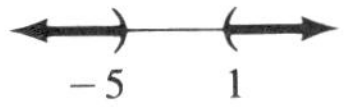

33. $-5 < x < -4$ 

35. all x

37. $x = \frac{1}{2}$ 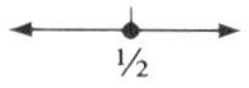

39. $x < -3$ or $x > 4$ 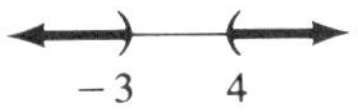

41. $0 \le x \le 2$ 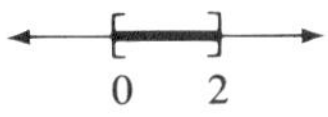

43. $x \le -5$ or $x \ge 1$ 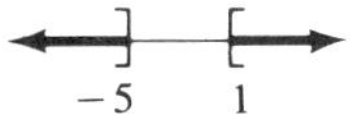

45. $-3 < x < 0$ or $x > 3$ 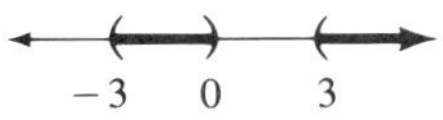

47. $x \le -3$ or $0 \le x \le 1$ 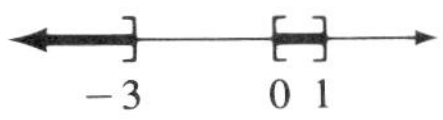

49. $x > -4$ or $x \le -9$ 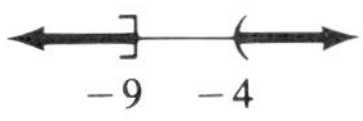

51. $x < -5$ or $x > 2$ 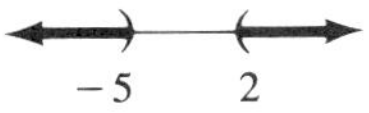

53. $x < 0$ or $2 < x \le 4$ 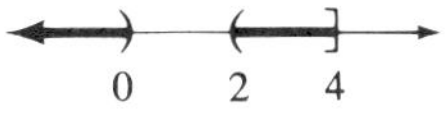

55. $3 < t < 13$

Chapter Review, page 69

1. $\frac{4y}{x}$ **2.** $\frac{4b}{a^2}$ **3.** $\frac{1}{x + 3y}$ **4.** $\frac{1}{s - 2t}$ **5.** $x + 5$ **6.** $a + 2$ **7.** $\frac{m - 8}{m + 1}$ **8.** $\frac{u - 3}{u - 1}$ **9.** $\frac{1 - x^2}{x}$ **10.** $\frac{2a^2}{a^2 - 1}$ **11.** $\frac{1 - x - y}{(x - y)(x + y)}$ **12.** 6 **13.** 6 **14.** $\frac{4acx}{by}$ **15.** $\frac{12b}{a}$ **16.** $\frac{3}{(a - 2)(a - 3)}$ **17.** $\frac{3}{(x - 1)(x + 3)}$ **18.** 9 **19.** $-\frac{16}{13}$ **20.** $-\frac{2}{41}$ **21.** 12 **22.** $\frac{1}{3}$ **23.** $-\frac{1}{2}$ **24.** $\frac{9}{2}$ **25.** 2 **26.** x^6 **27.** a^8b^{12} **28.** x **29.** $\frac{b^4}{a^3c}$ **30.** $\frac{8a^9}{27b^{18}}$ **31.** $x - \frac{1}{x^2}$ **32.** 0 **33.** $\frac{1}{a^2b^7}$ **34.** $\frac{b}{a - 2b}$ **35.** 1.4×10^{-4} **36.** 1.3×10^6 **37.** 2.34×10 **38.** 0.000044 **39.** 2,300 **40.** 0.011 **41.** $\frac{1}{x}$ **42.** $\frac{1}{m^2n^{5/2}}$ **43.** $\frac{1}{x^{2/3}y}$ **44.** $\frac{2\sqrt[3]{18x^2a^2}}{3a}$

45. $3x^2y\sqrt[3]{y}$ **46.** $4ab\sqrt{2a}$ **47.** $5x\sqrt{3y}$ **48.** $\frac{2\sqrt[3]{5}}{x^2}$ **49.** $0.02\sqrt[3]{100}$ **50.** $2 + 2\sqrt[3]{4}$
51. $\frac{-2\sqrt{3}}{9}$ **52.** 20 **53.** 5×10^{-3} **54.** $10 - 5\sqrt{3}$ **55.** $\frac{2\sqrt{x} - 2\sqrt{y}}{x - y}$ **56.** $6 + 8i$
57. $3\sqrt{2}\,i$ **58.** $5 + 2i$ **59.** $2i$ **60.** $-8 - 16i$ **61.** $1 - i$ **62.** $\frac{1}{5}(3 - i)$
63. $\frac{4}{65}(63 + 16i)$ **64.** $2 - 2i$ **65.** $4 - 7i$ **66.** 0 **67.** $0, -7, -\frac{4}{3}$ **68.** $\pm\frac{1}{3}$
69. $-7, 2$ **70.** 3, 4 **71.** $-\frac{7}{2}, 1$ **72.** $\frac{1}{4}, -\frac{2}{3}$ **73.** $-3 \pm \sqrt{7}$ **74.** $\frac{1}{2}(-5 \pm \sqrt{33})$
75. $\frac{1}{2}(-5 \pm \sqrt{3}\,i)$ **76.** $\frac{1}{8}(7 \pm \sqrt{95}\,i)$ **77.** 2, 4 **78.** $\frac{1}{2}(-1 \pm \sqrt{65})$ **79.** $-2, 5$
80. $\frac{2}{5}, -\frac{1}{3}$ **81.** 7 **82.** no solutions **83.** 16 **84.** 64 **85.** 1 **86.** $i, -i, \sqrt{2}, -\sqrt{2}$
87. $-1 < x < 6$ 

88. all x **89.** $-7 < x < 2$ 

90. $-4 \le x \le 2$ 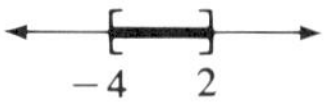

91. $-3 - \sqrt{7} \le x \le -3 + \sqrt{7}$ 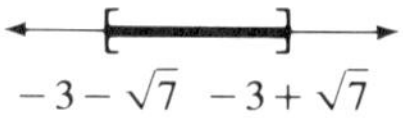

92. $x \ge 3$ or $-2 \le x \le 1$ 

93. 3.75 hr **94.** 10 min **95.** 3×4.5

Chapter Test, page 72

1. i. $-7ab$ **ii.** $-(t + 2)$ **2. i.** $\frac{-7}{3 - 6x}$ **ii.** $\frac{(6a - 5)(a + 6)}{(a - 9)(a + 4)}$ **iii.** $\frac{t^2}{6(t + 2)(5t + 2)}$
3. i. a^4b^{-8} **ii.** $\frac{4r^6}{s^2t^8}$ **iii.** $-\frac{31}{108}$ **4.** 1.2×10^{-4} **5.** 23,000 **6.** $y^{1/2}x^{-1/2}$
7. i. $(4a^2b^2c)(\sqrt{2bc})$ **ii.** $4r\sqrt{3rs}$ **iii.** $-\sqrt{7} - \sqrt{3}$ **iv.** $\frac{6(2\sqrt{a} + 3\sqrt{b})}{4a - 9b}$ **8. a.** $6 - 8i$
b. $34i$ **c.** $-i$ **d.** $2\sqrt{2}\,i + 3i$ **9. a.** $\pm\frac{5}{2}$ **b.** $\frac{7}{9}, -\frac{5}{9}$ **c.** $\frac{3}{2} \pm \frac{\sqrt{11}}{2}i$ **d.** $-2, 8$
e. $\frac{1}{2}, -3$ **f.** 1, 4 **g.** 9 **10. a.** $-2 \le x \le 11$ 

b. $x < -1$ or $x > -\frac{1}{2}$ 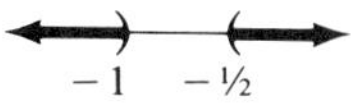

11. 3 by 12 **12.** 3.75 min

Exercises 3.1, page 77

1. $\{-2, -1, 2, 7, 14\}$ **3.** $\left\{3, 1, \frac{3}{5}, \frac{3}{7}, \frac{1}{3}\right\}$ **5.** $\{3, 1, -1, -3, -5\}$ **7.** 4 **9.** $8x^2 - 6x - 1$
11. -18 **13.** $2(a + 1)\left(a^2 + 2a + 1 - \frac{1}{a + 1}\right)$ **15.** 7 **17.** -4 **19.** -3 **21.** 32
23. 25 **25.** all real numbers **27.** $\left\{x: x \ne \frac{1}{2}\right\}$ **29.** $\left\{a: a \le \frac{2}{3}\right\}$ **31.** $\{x: x \ne 3, x \ne -3\}$
33. $\{y: y \ge 0\}$ **35.** $\{y: y \le 1\}$ **37.** 1-1 **39.** not 1-1 **41.** $V = x(14 - 2x)(18 - 2x)$
43. $V = \frac{\pi}{9}r^3$ **45.** not 1-1; $f(\sqrt{2}) = 2$ and $f\left(\frac{3}{2}\right) = 2$ **47.** $f(x) = \sqrt{x} + 1$ **49.** $f(x) = \sqrt{x + 1}$
51. domain $= \{x: x \ge 1$ or $x \le -1\}$ [x^2] [−] [1] [=] [√]
53. domain $= \{x: -2 < x < 2\}$ [x^2] [chs] [+] [4] [=] [√] [1/x]

Exercises 3.2, page 84

1–9.

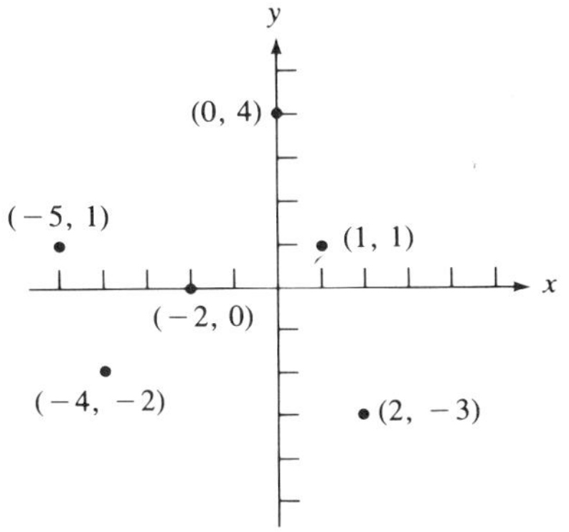

11. 1st **13.** 2nd **15.** 4th **17.** 3rd

19. $\left(-\frac{3}{2}, \frac{1}{2}\right)$, $5\sqrt{2}$ **21.** (3, 3), 2 **23.** (3, 6), 2 **25.** (−1, 1), 10 **27.** $x^2 + y^2 = 25$

29. $(x + 4)^2 + (y - 5)^2 = 4$ **31.** $(x - 2)^2 + (y + 3)^2 = 1$ **33.** is a function

35. is not a function, (4, −3) and (4, 3) belong to the set **37.** is a function

39. is not a function, (3, 0) and (−3, 0) belong to the set

41. $y = 2x - 3$

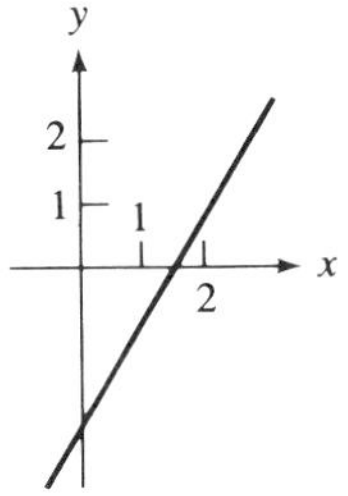

43. $f(x) = 8 - x^2$

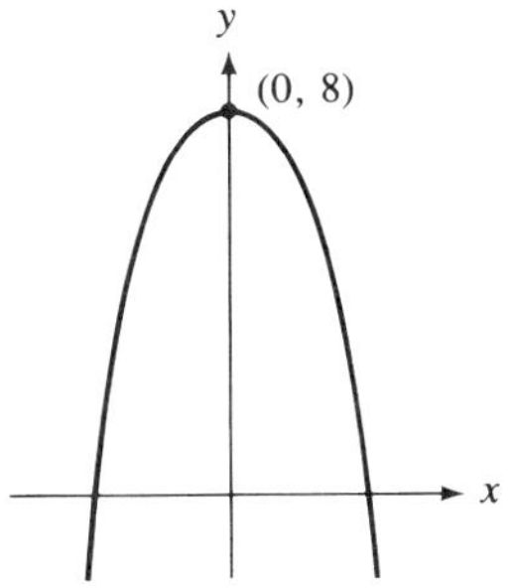

45. $y = 2x$

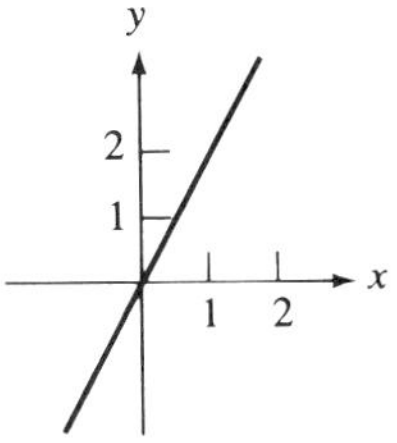

47. $y = 3x - 1$

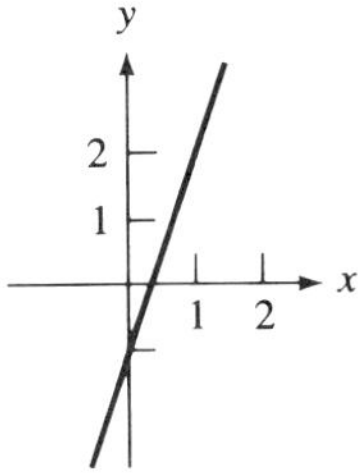

49. $f(x) = 2 - 3x$

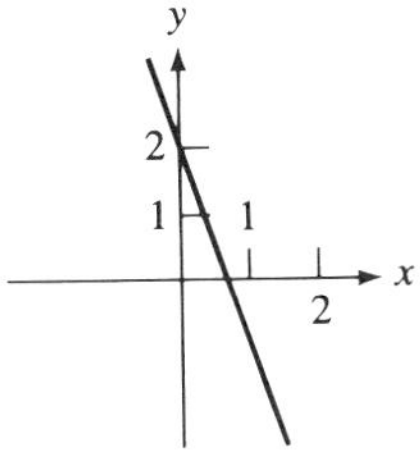

51. $y = -3$

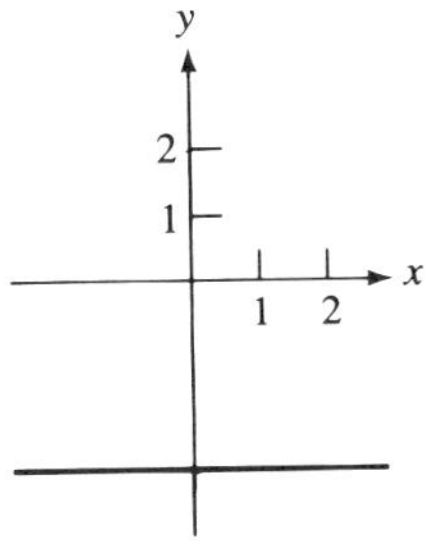

53. $h(x) = -2x$

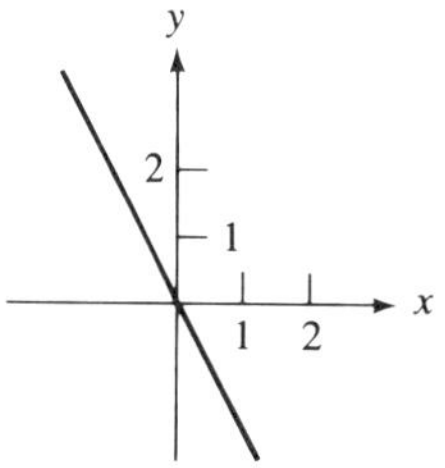

55. $L(x) = \frac{1}{2}x$

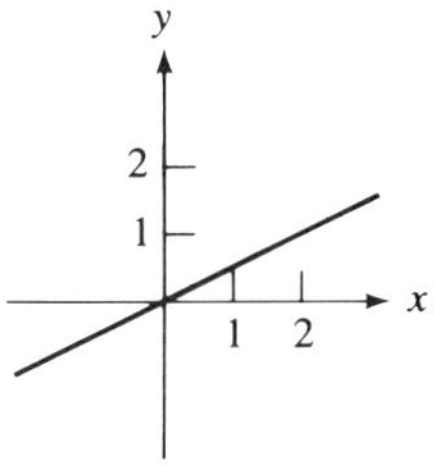

57. $y = \frac{2}{3}x + 1$

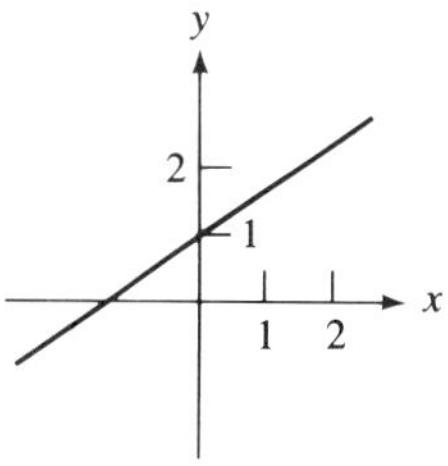

59. $y = -\sqrt{16 - x^2}$ **61.** $y = \sqrt{x^2 - 25}$

63. $y = \frac{1}{5}(2x - 14)$ **65.** 15 **67.** 13 **69.**

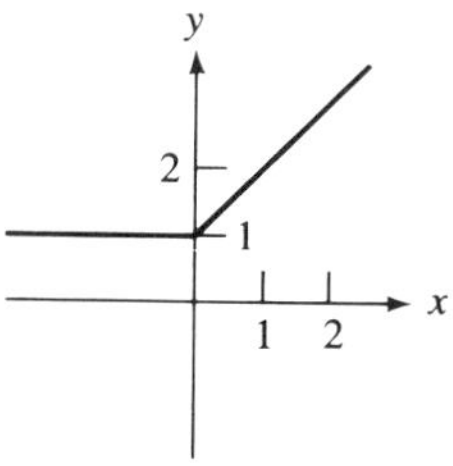

71. $y = -3x$ **73.** $y = x^{1/4}$

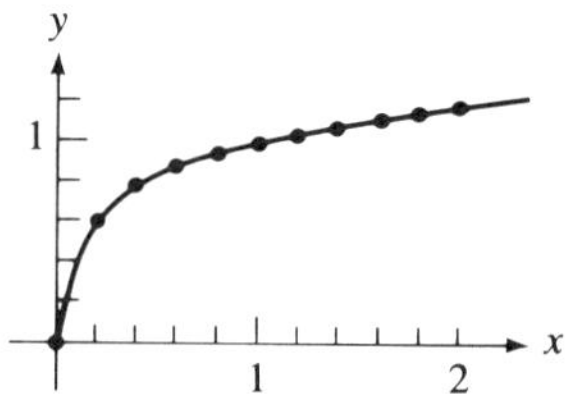

75. $f(x) = x^{-1/3}$

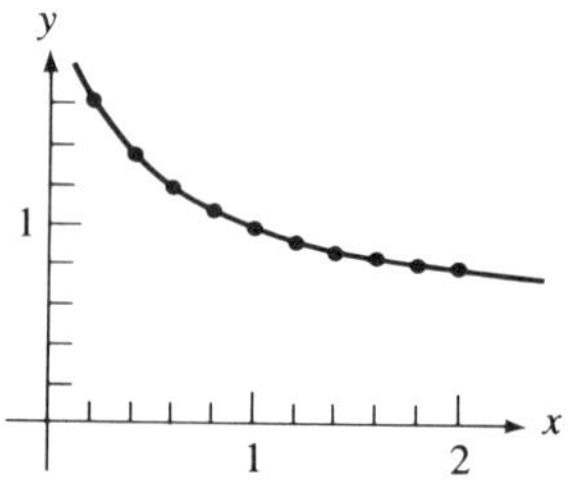

77. $y = x^{-3/2}$

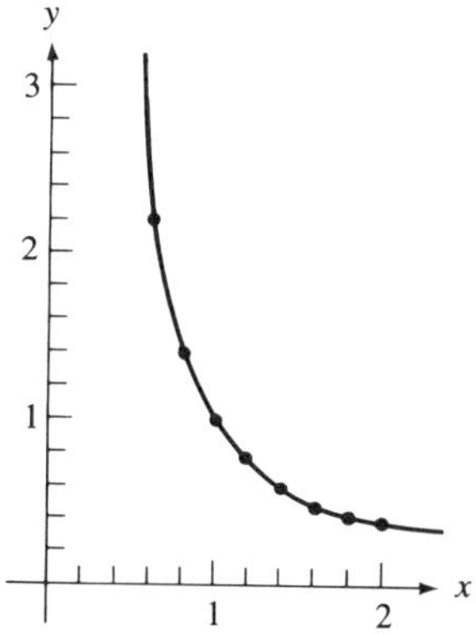

Exercises 3.3, page 92

1. 1 **3.** $\frac{1}{3}$ **5.** $-\frac{5}{3}$ **7.** $-\frac{2}{3}$ **9.** 0 **11.** not defined **13.** $y = 4x + 12$
15. $y = -2x + 10$ **17.** $y = 1$ **19.** $x = -2$ **21.** $y = \frac{1}{3}x - 3$ **23.** $3x + 4y = 12$
25. $y = 2x + 2$ **27.** $y = -\frac{2}{3}x - \frac{7}{3}$ **29.** $y = x + 5$ **31.** $x = 3$ **33.** $y = 6$

35. $m = 2$

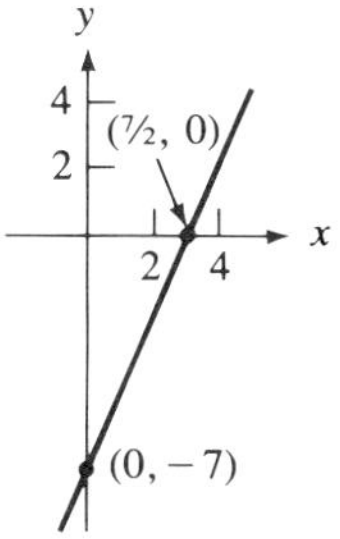

37. $m = \frac{1}{3}$

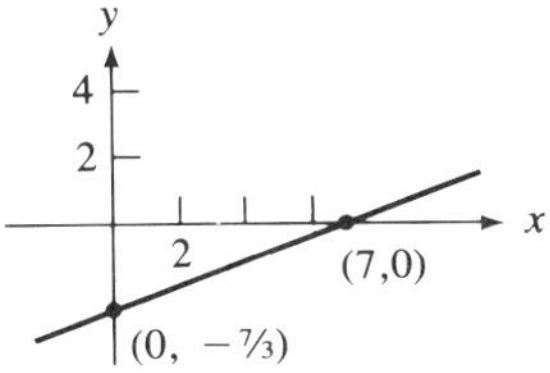

39. $m = \frac{4}{5}$

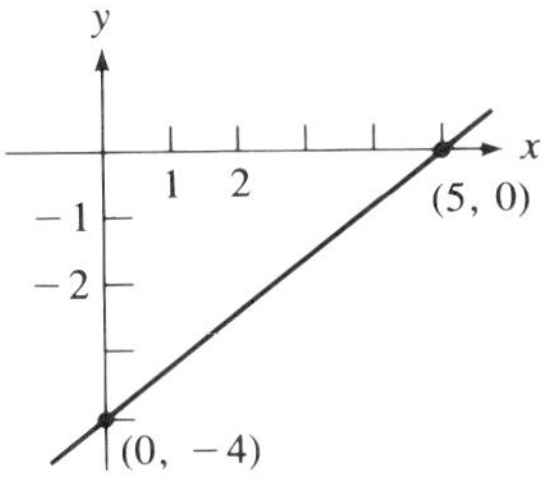

41. $m = -\frac{5}{4}$

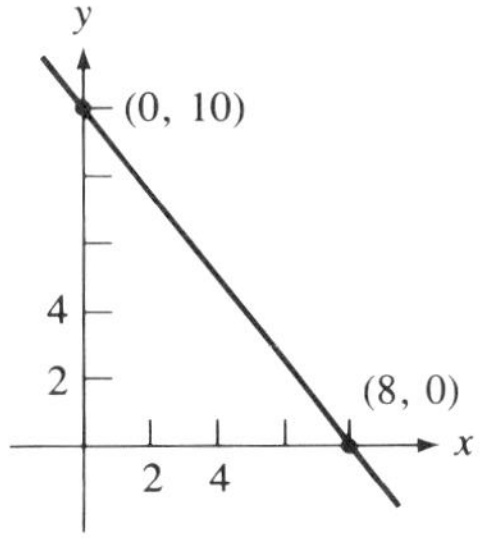

43. $m = 0$

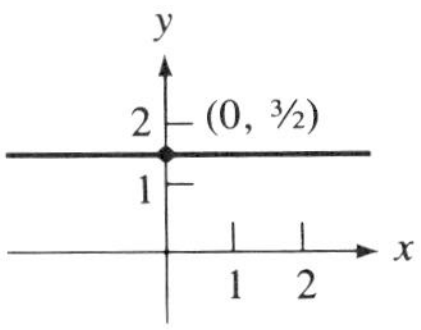

45. slope undefined

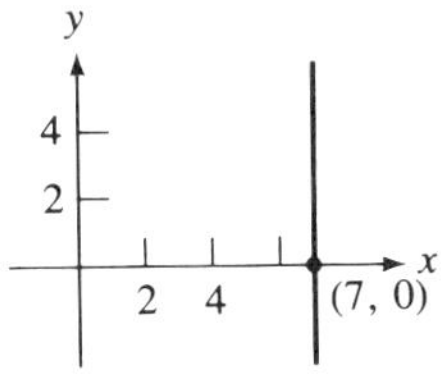

47. $5x - 3y = 11$ **49.** $x + 3y = 6$ **51.** $x = 6$ **53.** $y = 7$ **57.** (4, 2), (2, −4) or (0, 6)

Exercises 3.4, page 97

1. with respect to the y axis **3.** with respect to the y axis
5. with respect to both axes and the origin **7.** with respect to the origin
9. with respect to the origin **11.** no symmetry **13.** with respect to the x axis
15. no symmetry

17. a.

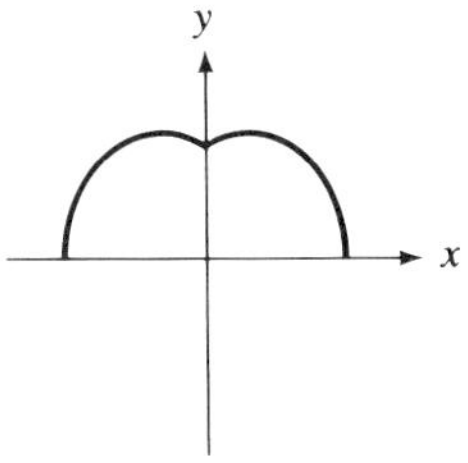

b.

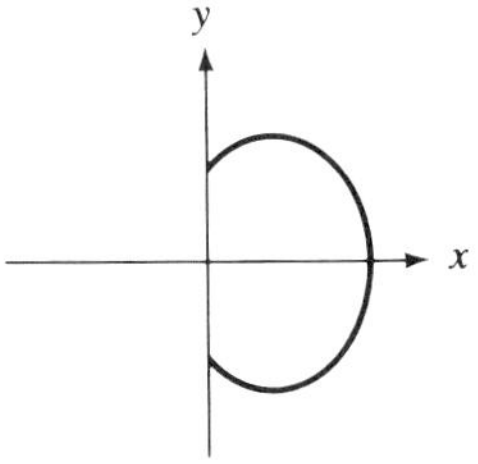

c.

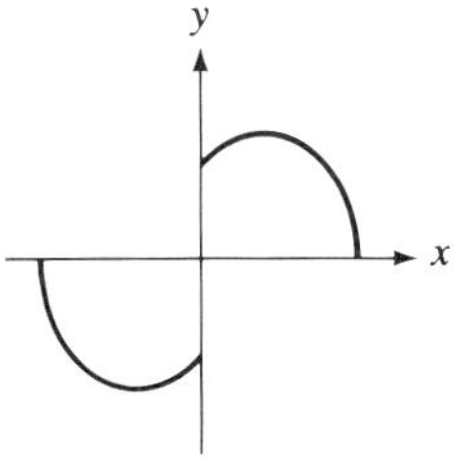

19. a.

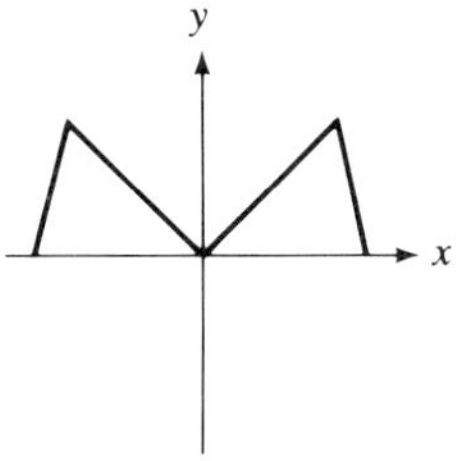

b.

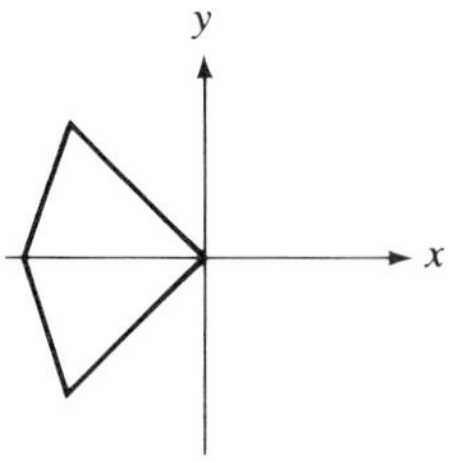

c.

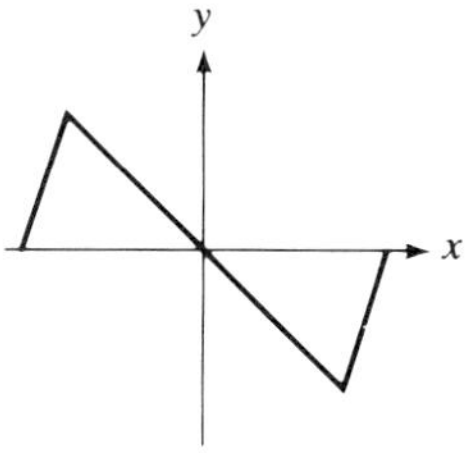

21. $y = x^2 - 3$

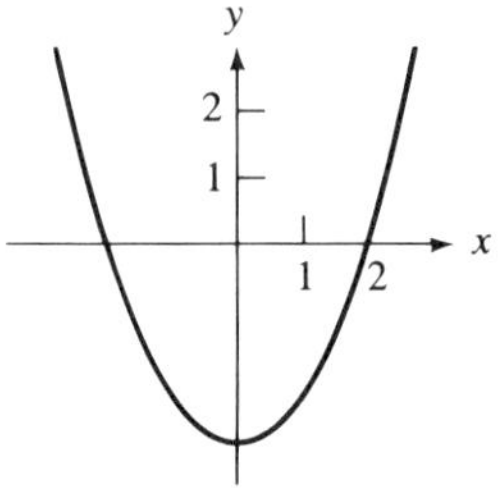

23. $y = (x - 1)^2$

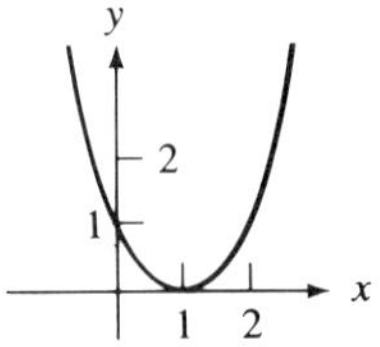

25. $y = x^3 + 2$

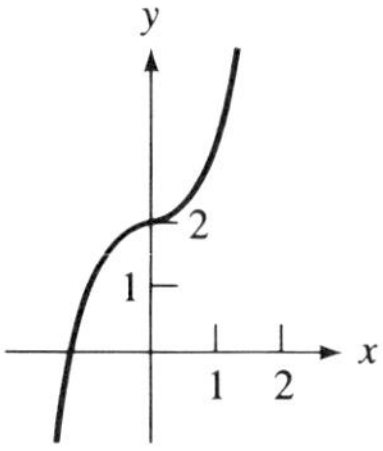

27. $y = (x + 2)^3$

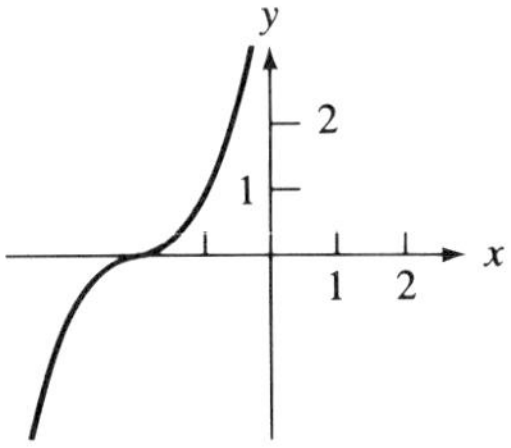

29. $x - 2 = y^2 + 1$

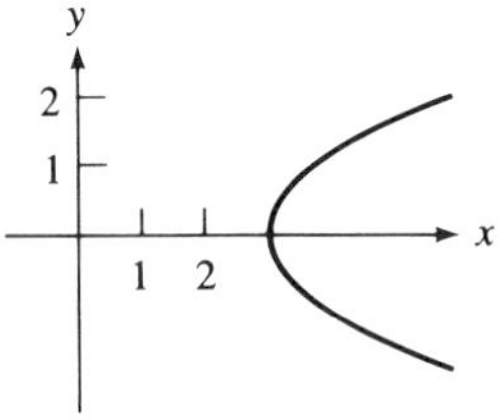

31. $y = |x - 2| + 1$

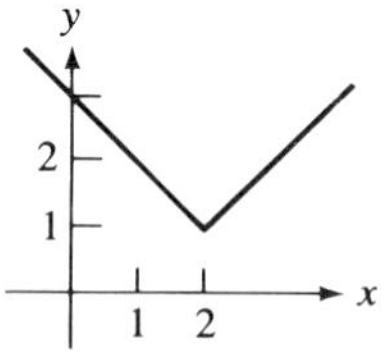

33. $y = \sqrt[3]{x - 2} + 1$

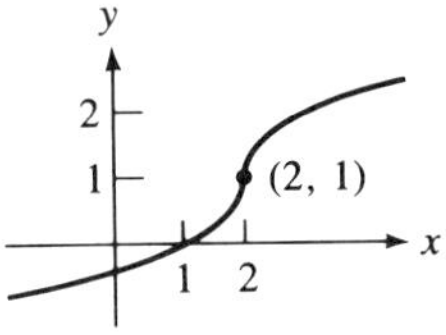

39. 29.13947 **41.** 4.75183

Exercises 3.5, page 102

1. $x^2 - x - 31,\ -x^2 + 5x + 25,\ 2x^3 - 9x^2 - 47x + 84,\ \dfrac{2x - 3}{x^2 - 3x - 28}$

3. $\sqrt{2x - 3} + \sqrt{3x + 1},\ \sqrt{2x - 3} - \sqrt{3x + 1},\ \sqrt{6x^2 - 7x - 3},\ \sqrt{\dfrac{2x - 3}{3x + 1}}$

5. $|x + 7| + |4 - 3x|,\ |x + 7| - |4 - 3x|,\ |28 - 17x - 3x^2|,\ \left|\dfrac{x + 7}{4 - 3x}\right|$

7. $x^3 - 3x^2 + 7 + \frac{1}{x}$, $x^3 - 3x^2 + 7 - \frac{1}{x}$, $x^2 - 3x + \frac{7}{x}$, $x^4 - 3x^3 + 7x$

9. $2x - 2$, $6 - 2x$, $4x - 8$, $\frac{1}{x - 2}$

11. $x^2 + x - 2$, $x - x^2$, $x^3 - x^2 - x + 1$, $\frac{1}{x + 1}$

13. $y = 2|x|$

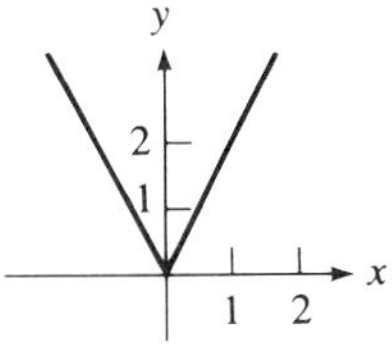

15. $y = \frac{1}{2}x^3$

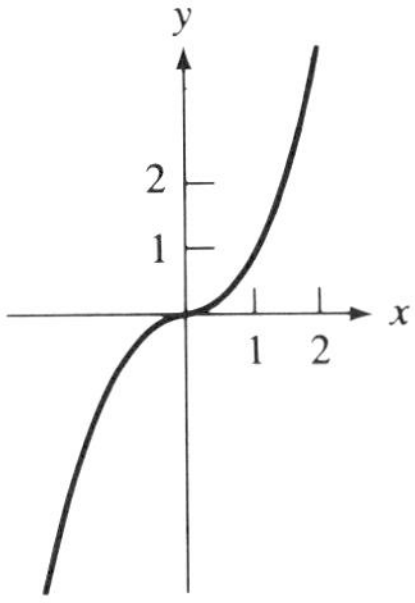

17. $y = -\sqrt[3]{x}$

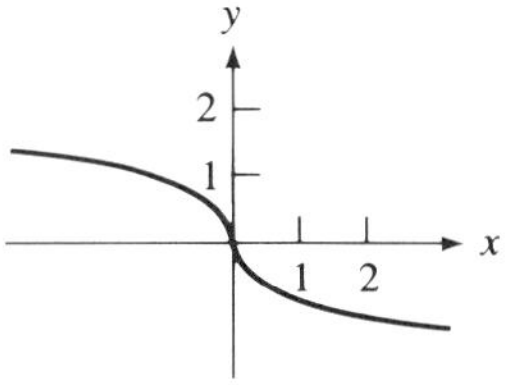

19. $y = |x| + x$

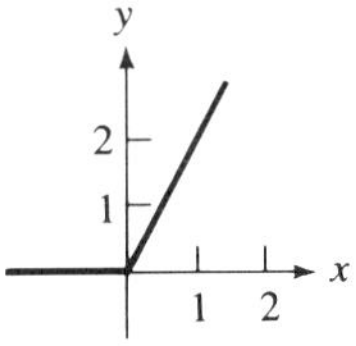

$f(x) = 0 \quad x \leq 0$
$f(x) = 2x \quad x > 0$

21. $y = x^2 - x$

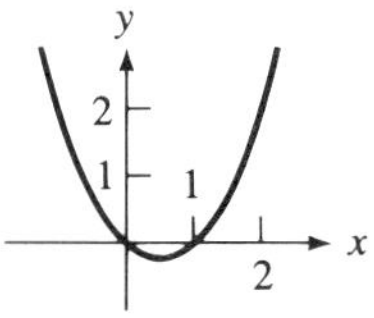

23. $y = |x + 1| + |x - 1|$

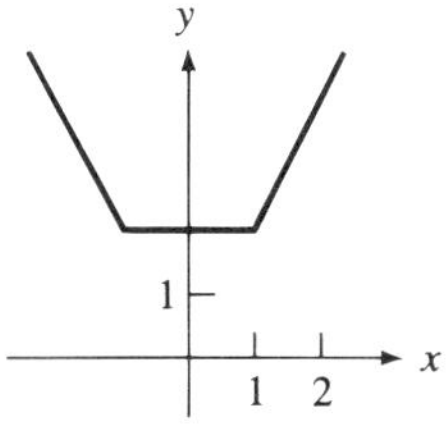

25. $y = x + \sqrt{x - 1}$

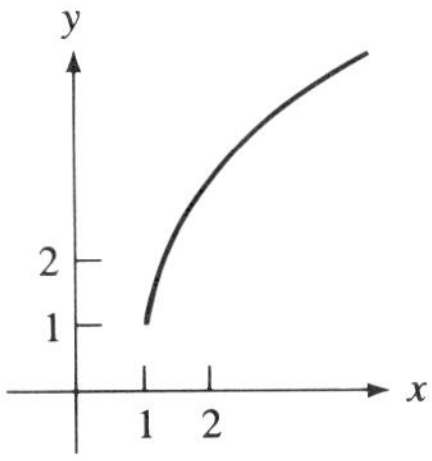

27. $F \cdot G(x) = 18x^2 - 12x - 6$
$G \cdot F(x) = 6x^2 - 25$

29. $F \cdot G(x) = \frac{1}{3x + 5}$
$G \cdot F(x) = \frac{3 + 5x}{x}$

31. $F \cdot G(x) = x - 1$
$G \cdot F(x) = \frac{x}{1 - x}$

33. $F \cdot G(x) = G \cdot F(x) = x$

35. $f \cdot g \cdot h(x) = (3x + 2)^2$

37. $f \cdot g \cdot h(x) = \frac{1}{x^2 - 4}$

39. $f \cdot g \cdot h(x) = \sqrt{x^2 - 5}$

41. $x \xrightarrow{\text{multiply by 2}} 2x \xrightarrow{\text{subtract 8}} 2x - 8$

43. $x \xrightarrow{\text{add } 9} x + 9 \xrightarrow{\text{multiply by } 1/2} \frac{1}{2}(x + 9)$

45. $x \xrightarrow{\text{subtract } 5} x - 5 \xrightarrow{\text{square}} (x - 5)^2 \xrightarrow{\text{multiply by } 2} 2(x - 5)^2$

47. $x \xrightarrow{\text{add } 7} x + 7 \xrightarrow{\text{square root}} \sqrt{x + 7} \xrightarrow{\text{invert}} \frac{1}{\sqrt{x + 7}} \xrightarrow{\text{add } 4} \frac{1}{\sqrt{x + 7}} + 4$

57. [×] 2 [−] 5 [=] [√] [−] 8 [=]

59. [−] 4 [=] [1/x] [×] 2 [chs] [+] 4 [=]

61. [×] 2 [chs] [+] 4 [=] [√] [1/x] [×] 4 [chs] [−] 6 [=]

63. [×] 5 [−] 2 [=] [x^2] [√] [+] 1 [=]

65. [+] 3 [=] [x^2] [−] 13 [=]

Exercises 3.6, page 107

1. 1-1, domain $\{x: x \geq -2\}$, range $\{y: y \geq -1\}$ **3.** not 1-1

5. 1-1, domain $\{x: x \neq 0\}$, range $\{y: y \neq 3\}$ **7.** not 1-1

9. 1-1, domain $\{x: x \geq 3\}$, range $\{y: y \geq 0\}$ **11.** not 1-1 **13.** $f^{-1}(x) = \frac{1}{4}(2 - x)$

15. $f^{-1}(x) = \frac{5}{2}\left(\frac{3}{2} - x\right)$ **17.** $f^{-1}(x) = \sqrt{\frac{x + 4}{3}}$ **19.** $h^{-1}(t) = 3 - \sqrt{\frac{t - 1}{2}}$

21. $f^{-1}(x) = (x - 1)^2 + 2$ **23.** $f^{-1}(x) = \frac{5 + 3x}{2x}$ **25.** $f^{-1}(x) = \sqrt{9 - x}$

27. $g^{-1}(t) = 4 + \sqrt{\frac{t - 5}{2}}$ **29.** $h^{-1}(r) = \frac{5r + 2}{r - 1}$ **31.** $f^{-1}(m) = \frac{2m + 2}{m + 3}$

33. $y = 4 - x^2$
$y = -\sqrt{4 - x}$

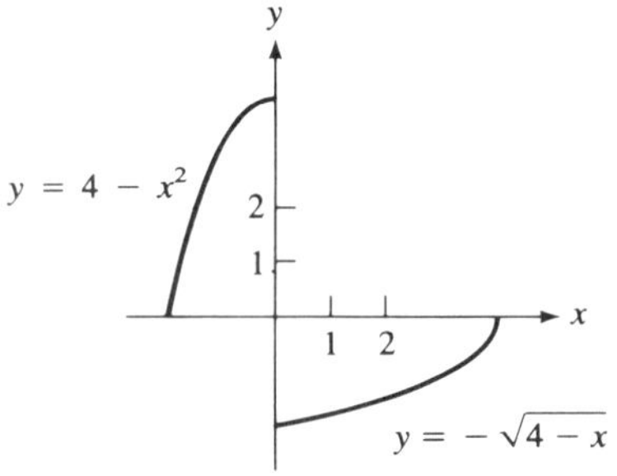

35. $a \leq -2$ **37.** $0 \leq c \leq 4$

39. $a \leq -4$ or $a \geq 4$

Exercises 3.7, page 111

1. $I = \frac{k}{d^2}$ **3.** $T = \frac{k}{s}$ **5.** $D = kd$ **7.** $W = kEI$, E for voltage, I for current **9.** $L = kI^2$

11. $A = 1.5F$ **13.** $D = 5t^2$, D = distance in meters, t = time in seconds

15. $D = 0.055P$, where D is deductions and P is pay. \$66

17. $R = \frac{1}{6}v^2$, where R is range in feet and v is velocity in feet per second. 33,750 ft

19. $KE = \frac{1}{2}mv^2$, where KE is the kinetic energy in ergs, m is the mass in grams, and v is the velocity in centimeters per second. 6,400 ergs

21. base charge is \$11, charge per kWh is 3¢, cost for 1,550 kWh is \$51.50 **23.** 19¢/mi, \$22/day

Chapter Review, page 112

1. $\{x: x \neq 0\}$ **2.** all real numbers **3.** $\{x: x \neq 3, x \neq -3\}$ **4.** $\left\{x: x > \frac{1}{2}\right\}$ **5.** $\{x: x \neq 4\}$
6. $\{y: y \geq -2\}$ **7.** $\{y: y \neq 0\}$ **8.** $\{y: y \leq 1\}$ **9.** $\{y: y \geq 0\}$ **10.** 1-1 **11.** not 1-1
12. not 1-1 **13.** not 1-1 **14.** -51 **15.** 29 **16.** $10\sqrt{2} - 15$
17. $2x^3 + 2x^2 + 4x - 3$ **18.** 2 **19.** 13 **20.** $|3t + 1|$ **21.** $A(t) = (10 - 2t)[8 - 2(t + 1)]$
22. $d = \sqrt{x^2 + x - 1}$ **23.** $\sqrt{2}$ **24.** $\sqrt{17}$ **25.** $\sqrt{20}$ **26.** $\sqrt{13}$ **27.** $\sqrt{41}$
28. $\sqrt{50}$ **29.** is a function **30.** not a function **31.** is a function **32.** not a function
33. $m = \frac{3}{2}, y = \frac{3}{2}x$ **34.** $m = \frac{1}{8}, y = \frac{1}{8}x - \frac{3}{4}$
35. $m = \frac{3}{7}, y = \frac{3}{7}x - \frac{27}{7}$ **36.** $m = -\frac{3}{2}, y = -\frac{3}{2}x + 6$ **37.** $m = 0, y = 3$
38. slope undefined, $x = -2$ **39.** $y = -x + 4$ **40.** $y = 2x + 6$ **41.** $y = 5x + 8$ **42.** $y = 5$
43. $x = 1$ **44.** $y = -\frac{3}{4}x + \frac{27}{2}$ **45.** $3x - 5y = 21$ **46.** $5x + 3y = 1$ **47.** $x = 5$
48. $y = -6$ **49.** $m = \frac{3}{4}$; intercepts $(0, -3)$, $(4, 0)$

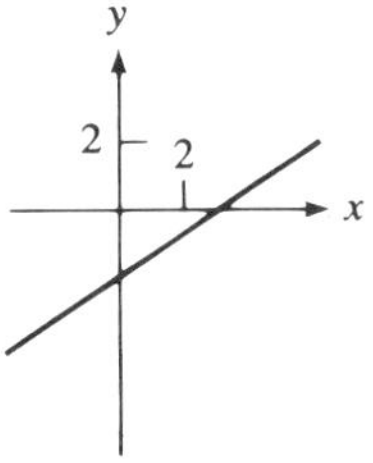

50. $m = -\frac{2}{5}$; intercepts $(0, 2)$, $(5, 0)$

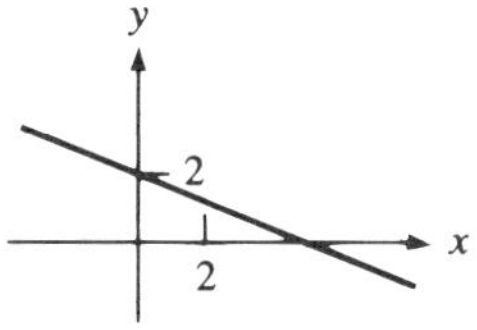

51. $m = \frac{1}{3}$; intercepts $\left(0, \frac{7}{3}\right)$, $(-7, 0)$

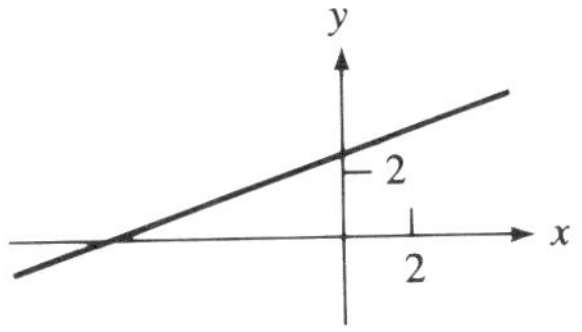

52. $m = 3$; intercepts $(0, -9)$, $(3, 0)$

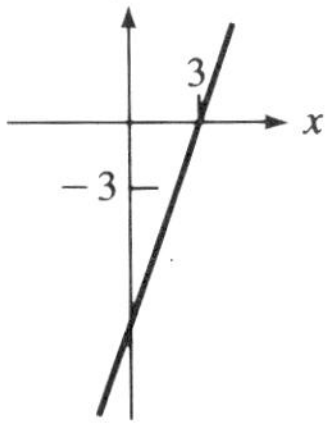

53. symmetric with respect to the x axis, y axis, and origin **54.** no symmetries
55. symmetric with respect to the x axis **56.** symmetric with respect to the y axis

57. $y = x^2 + \dfrac{1}{2}$

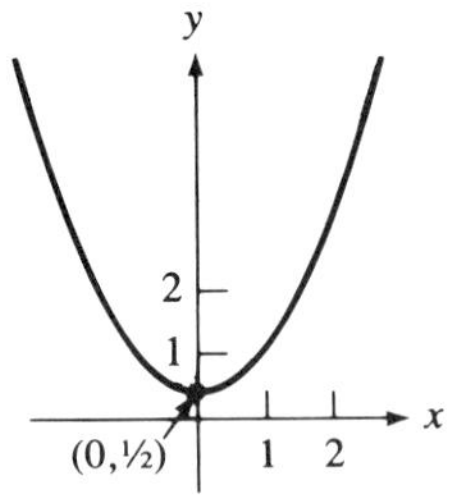

58. $y = 2 - x^3$

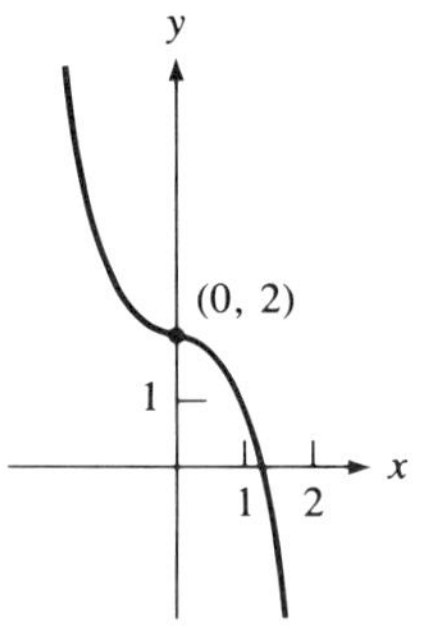

59. $y = |x + 1| - 2$

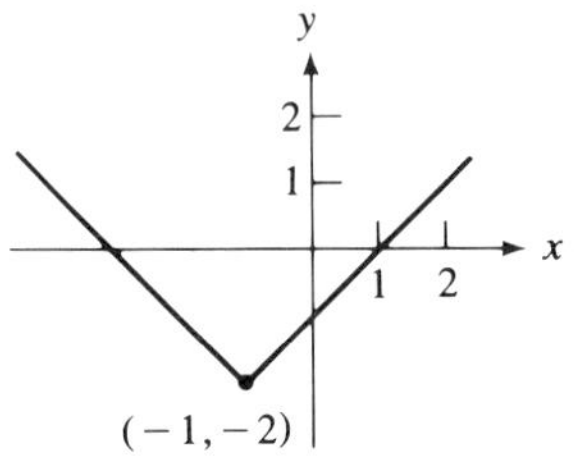

60. $y = (x - 1)^2 + 1$

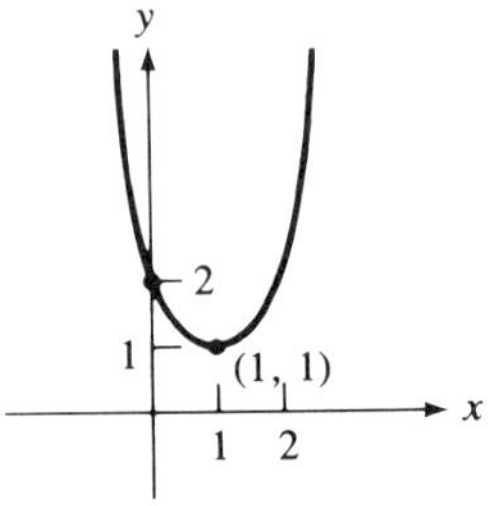

61. $y = \sqrt{x - 2} + 1$

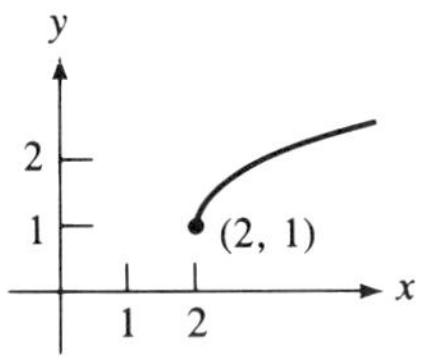

62. $y = \sqrt[3]{x + 2} - 4$

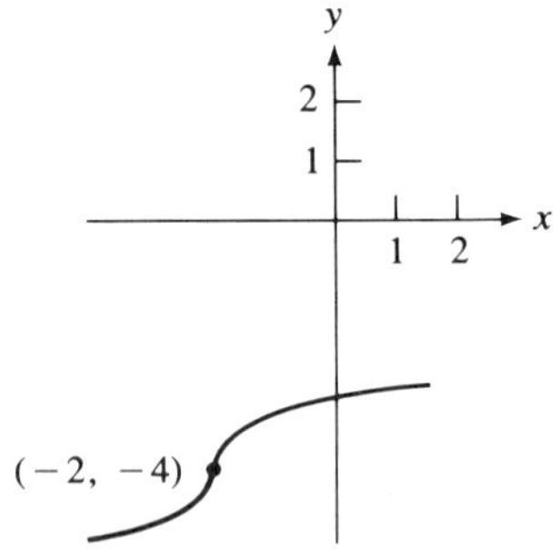

63. $y = x - |x|$

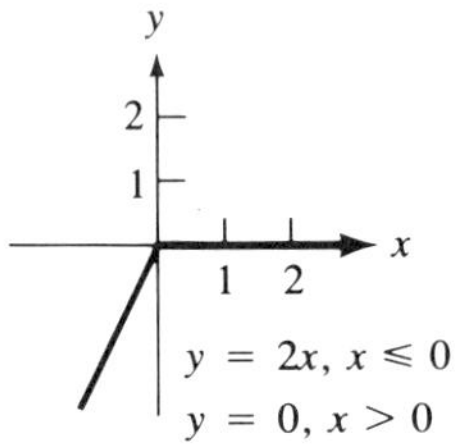

64. $y = x + \dfrac{1}{2}|x|$

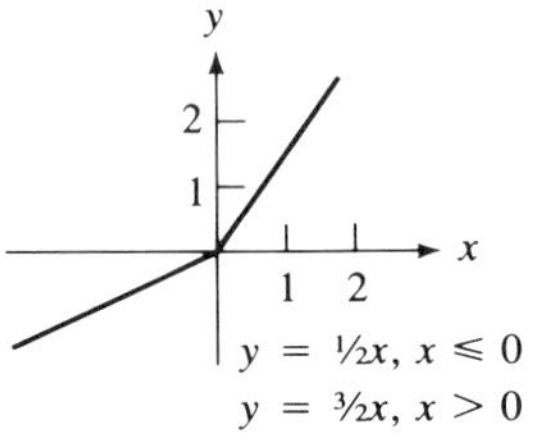

65. $f \cdot g(t) = \left(t + \dfrac{2}{t}\right)^2 - 4 = t^2 + \dfrac{4}{t^2}$ **66.** $t^2 - 4 + \dfrac{2}{t^2 - 4}$ **67.** $t^2 + t - 4 + \dfrac{2}{t}$

68. $t^2 - t - 4 - \dfrac{2}{t}$ **69.** $\dfrac{t^2 + 2}{t^3 - 4t}$ **70.** $t^3 - 2t - \dfrac{8}{t}$ **71.** $\{t\colon t \neq 0\}$ for Exercises 65, 67, 68, and 70.
For Exercise 69, $\{t\colon t \neq 2, t \neq -2, t \neq 0\}$. For Exercise 66, $\{t\colon t \neq 2, t \neq -2\}$.

72. $x \xrightarrow{\text{square}} x^2 \xrightarrow{\text{subtract 5}} x^2 - 5$ **73.** $s \xrightarrow{\text{multiply by 2/3}} \dfrac{2}{3}s \xrightarrow{\text{add 5}} \dfrac{2}{3}s + 5$

74. $x \xrightarrow{\text{subtract 3}} x - 3 \xrightarrow{\text{square}} (x - 3)^2 \xrightarrow{\text{square root}} \sqrt{(x - 3)^2} \xrightarrow{\text{add 2}} |x - 3| + 2$

75. $r \xrightarrow{\text{add 1}} r + 1 \xrightarrow{\text{square root}} \sqrt{r+1} \xrightarrow{\text{multiply by 2}} 2\sqrt{r+1} \xrightarrow{\text{add 4}} 2\sqrt{r+1} + 4$

76. 1-1, $f^{-1}(x) = \sqrt{x-1} + 3$ **77.** 1-1, $f^{-1}(x) = \frac{1}{2}(x + 3)$ **78.** $f^{-1}(x) = 3 - \frac{4}{x}$ **79.** not 1-1

80. $S = kA$ **81.** $T = kV$ **82.** $M = kmv$ **83.** $I = \frac{k}{R}$

84. $T = 0.0425A$, where T is the amount of the tax and A is the amount of the purchase. \$2.13
85. $T = 0.0124V$, where T is the amount of the tax and V is the value of the property. \$892.80
86. $C = 4.95 + 1.15\,q$, where q is the number of quarts used. Filter \$4.95, quart of oil \$1.15
87. $C = 3{,}630 - 30t$, where C is the caloric need and t is the temperature in ° Celsius. At 16 degrees Celsius, $C = 3{,}150$.

Chapter Test, page 115

1. Domain $\{x: -2 \le x \le 2\}$, range $\{y: 0 \le y \le 2\}$

2. $f(3) = 5$, $g(-2) = 8$, $f\left(\frac{1}{x}\right) = \frac{2}{x} - 1$, $g(2 - x) = 2(2 - x)^2$

3. $f \cdot g(x) = 4x^2 - 1$, $g \cdot f(x) = 2(2x - 1)^2$ **4.** $V = 2w^3$

5. $d = \sqrt{157}$, $m = -\frac{11}{6}$, $6y + 11x = -14$ **6.** $d = \sqrt{97}$, $m = \frac{9}{4}$, $4y - 9x = 5$

7. $2x - 3y = 6$; $m = \frac{2}{3}$; intercepts $(0, -2)$, $(3, 0)$

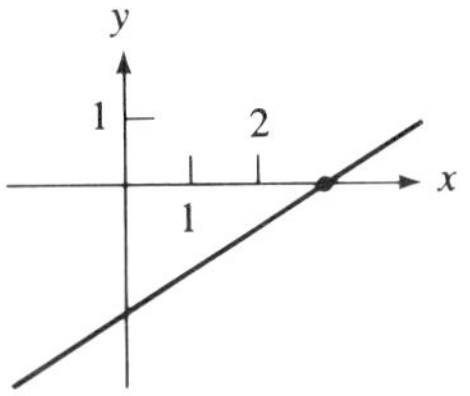

8. $5x + y = 9$; $m = -5$; intercepts $(0, 9)$, $\left(\frac{9}{5}, 0\right)$

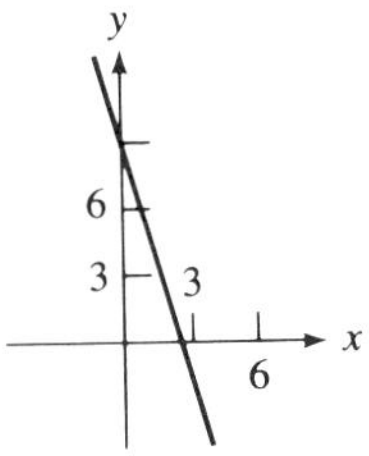

9. $5x = 2y - 23$

10. $3y + 2x = -7$ **11.** symmetric with respect to the y axis

12. symmetric with respect to the x axis

13. $y = 2 - (x + 1)^2$

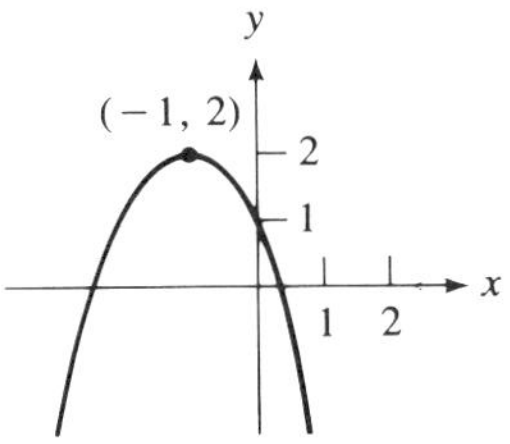

14. no inverse **15.** $g^{-1}(x) = \frac{1}{4}(x + 1)$

16. $H^{-1}(x) = -\frac{2}{x} + \frac{3}{2}$ **17.** daily charge \$25, 19¢/mi

Exercises 4.1, page 122

1. $y = -\frac{1}{2}x^2$; vertex (0, 0)

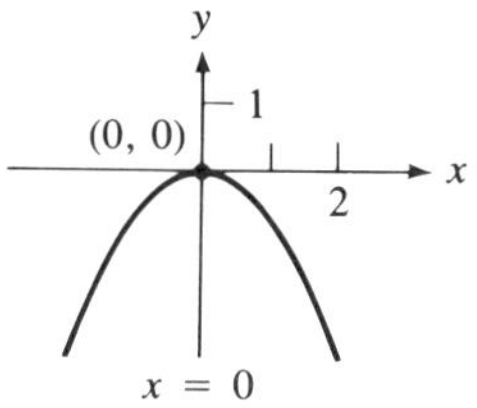

3. $y = 3x^2$; vertex (0, 0)

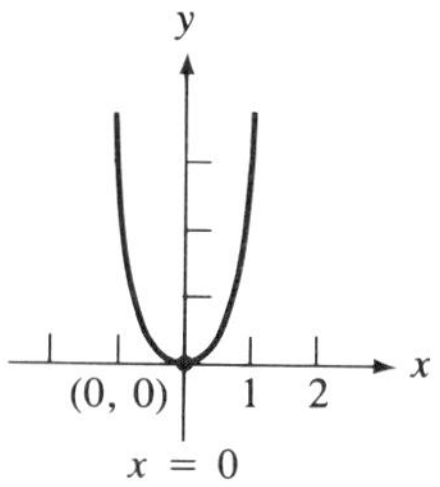

5. $y = 1 - x^2$; vertex (0, 1)

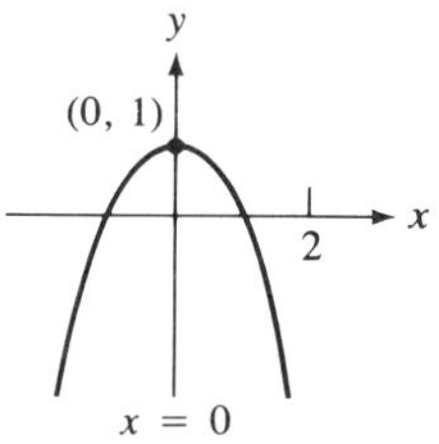

7. $y = 2(x + 1)^2$; vertex $(-1, 0)$

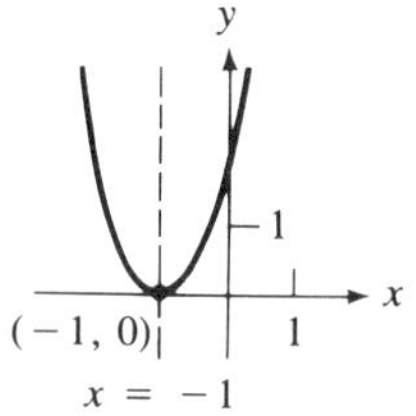

9. $y = (x - 1)^2 - 1$; vertex $(1, -1)$

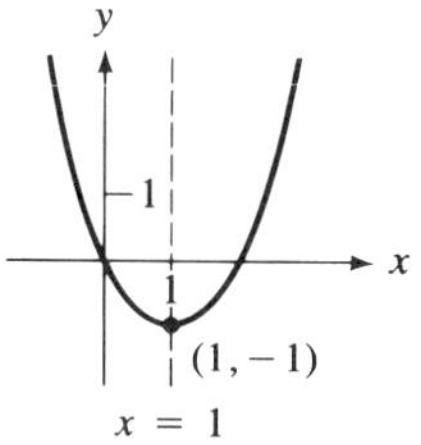

11. $y = 4 - (x + 1)^2$; vertex $(-1, 4)$

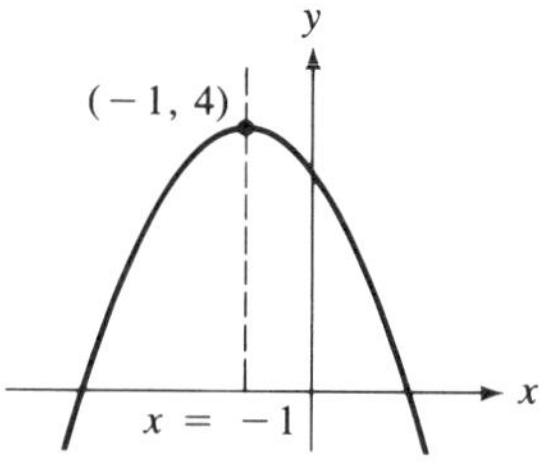

13. $y = x^2 - 2x + 4$; vertex (1, 3)

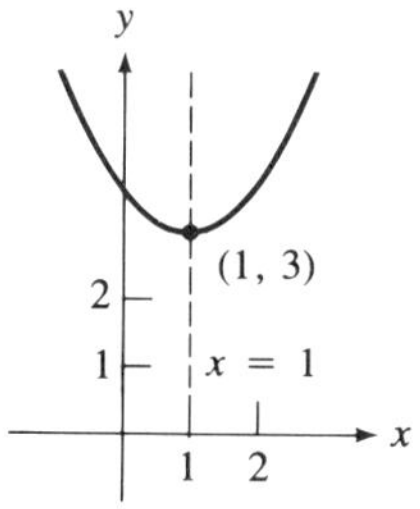

15. $f(x) = -x^2 + 2x - 4$; vertex $(1, -3)$

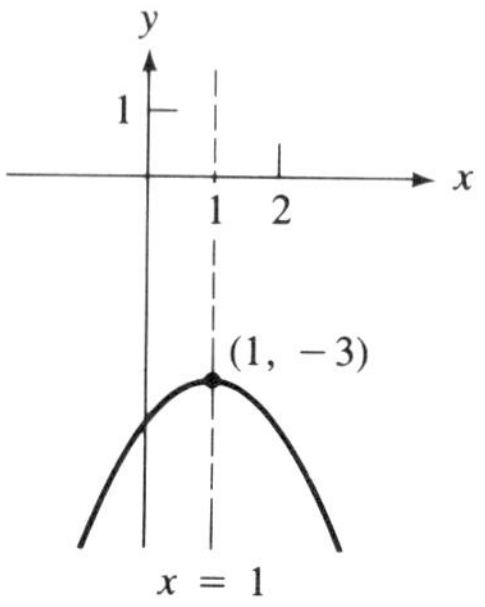

17. $g(x) = -x^2 + 2x$; vertex (1, 1)

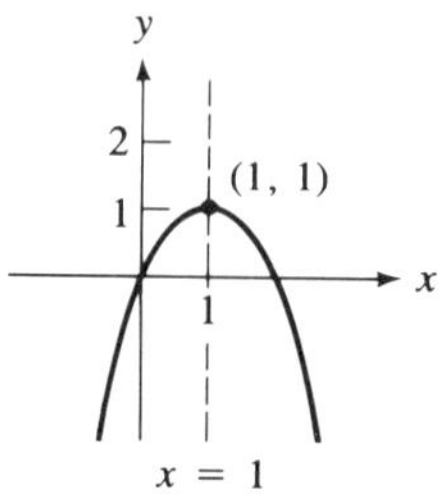

19. $y = 2x^2 - 12x + 24$; vertex (3, 6)

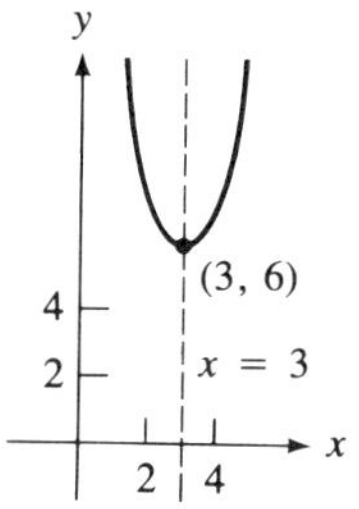

21. $f(x) = 6x - 3x^2 - 4$; vertex $(1, -1)$

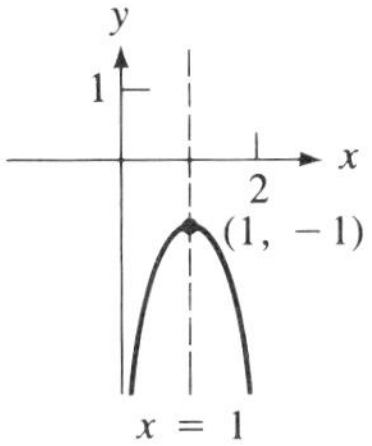

23. $y - 3 = 2(x + 2)^2$ **25.** $y - 3 = -\frac{1}{3}(x - 1)^2$ **27.** $y = \frac{16}{9}\left(x - \frac{7}{2}\right)^2$ **29.** $y + 4 = (x + 2)^2$

31. 1,024 at $t = 8$ **33.** 30 cars **37.** 36 trees per acre **39.** 100 m long and $\frac{200}{\pi}$ m wide

41. $\frac{32}{4 + \pi}$ in. for the square

Exercises 4.2, page 129

1. center $(-7, 3)$, radius $\sqrt{61}$ **3.** center $\left(-\frac{7}{4}, 1\right)$, radius $\frac{1}{4}$ **5.** $(x - 3)^2 + (y - 2)^2 = 13$

7. $(x - 1)^2 + (y - 3)^2 = 10$ **9.** $(x - 3)^2 + \left(y - \frac{11}{2}\right)^2 = \frac{13}{4}$

11. $y^2 = 8x$ **13.** $x^2 = 12y$ **15.** $(x + 2)^2 = -8(y - 6)$ **17.** $12(y - 6) = (x - 2)^2$

19. $x = \frac{1}{8}y^2$; vertex $(0, 0)$, focus $(2, 0)$

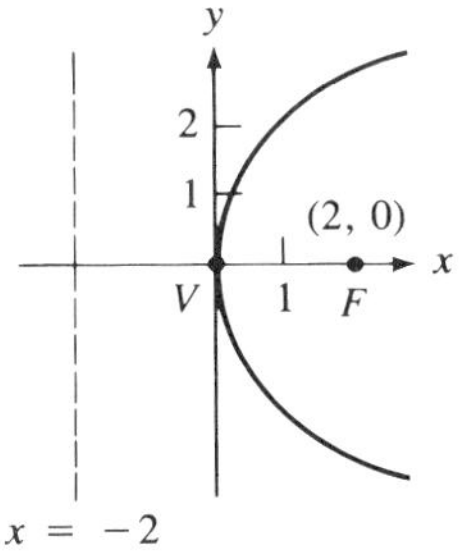

21. $y^2 - 4y + 12 = 8x$; vertex $(1, 2)$, focus $(3, 2)$

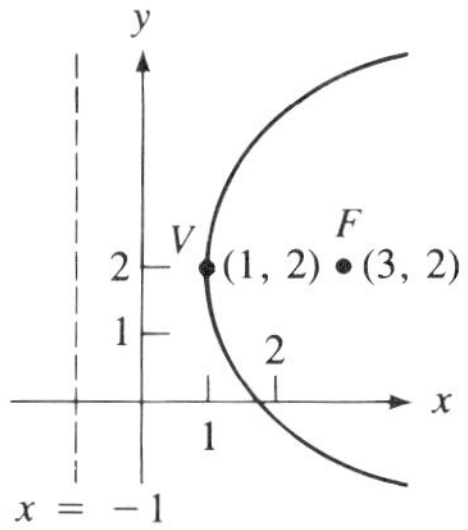

23. $x^2 + 2x + 8y = 15$; vertex $(-1, 2)$, focus $(-1, 0)$

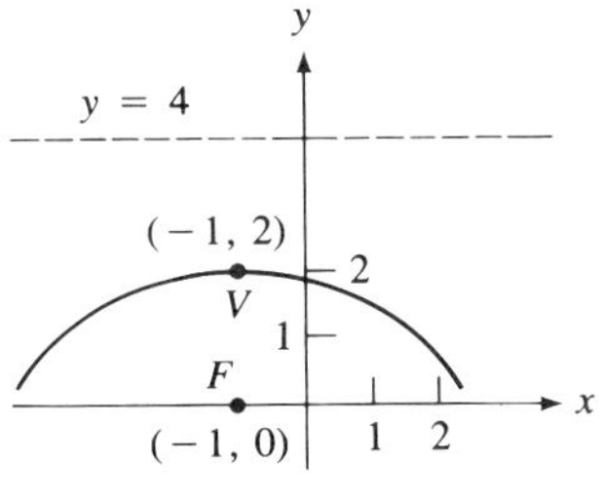

25. 32 in. **27.** maximum height 25,000 ft, distance 100,000 ft

Exercises 4.3, page 134

1. vertices (6, 0), (−6, 0), (0, 2), (0, −2); foci $(\sqrt{32}, 0)$, $(-\sqrt{32}, 0)$

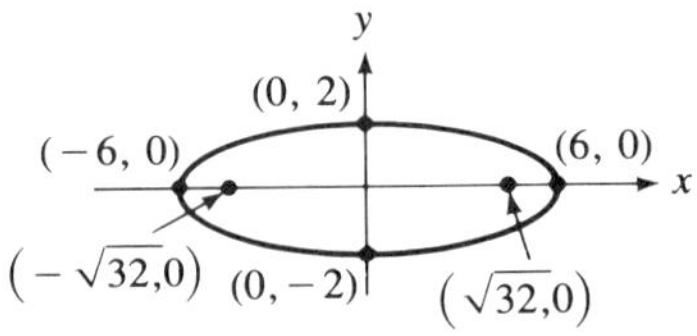

3. vertices (6, 0), (−6, 0), (0, 3), (0, −3); foci $(\sqrt{27}, 0)$, $(-\sqrt{27}, 0)$

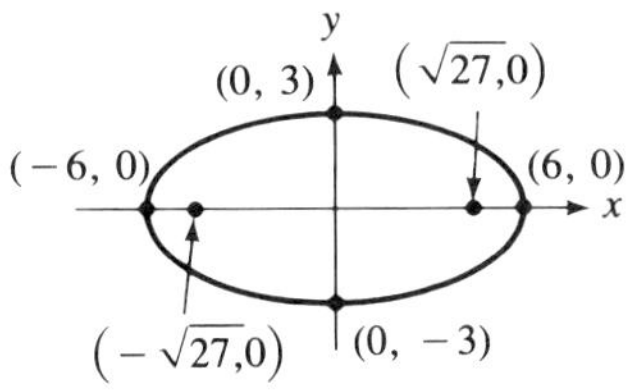

5. vertices $\left(\frac{5}{2}, 0\right)$, $\left(-\frac{5}{2}, 0\right)$, $\left(0, \frac{5}{3}\right)$, $\left(0, -\frac{5}{3}\right)$; foci $\left(\frac{5\sqrt{5}}{6}, 0\right)$, $\left(-\frac{5\sqrt{5}}{6}, 0\right)$

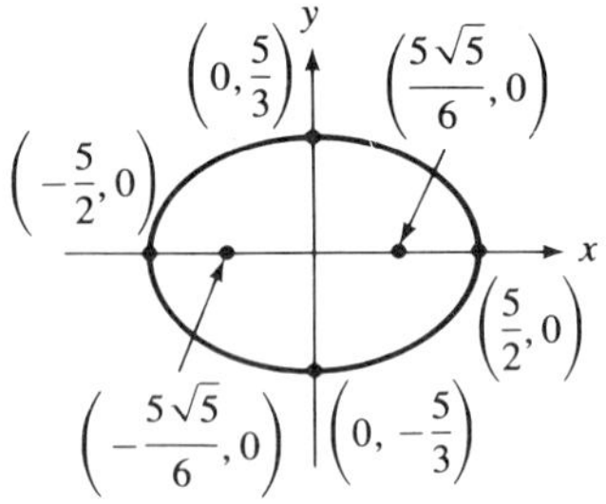

7. vertices (3, 2), (−7, 2), (−2, 12), (−2, −8) foci $(-2, 2 - 5\sqrt{3})$, $(-2, 2 + 5\sqrt{3})$

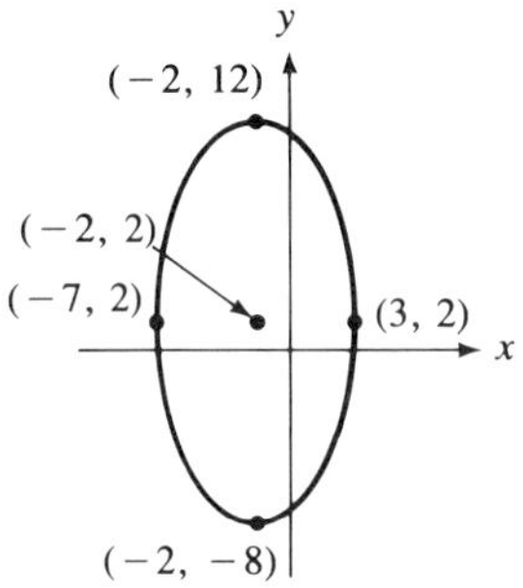

9. vertices (7, −5), (1, −5), (4, 0), (4, −10); foci (4, −1), (4, −9)

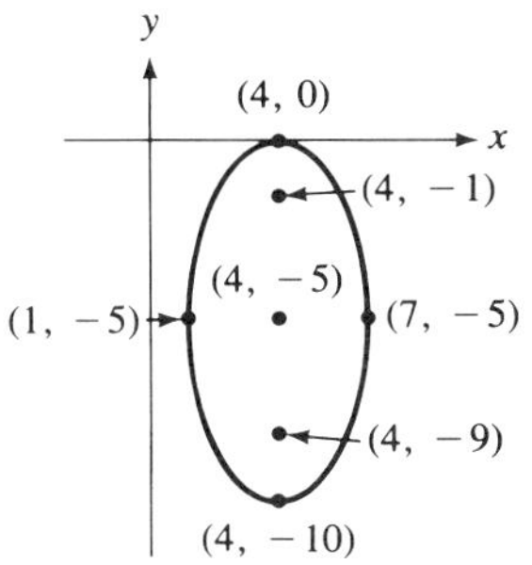

11. $\frac{x^2}{36} + \frac{y^2}{27} = 1$ **13.** $\frac{x^2}{16} + \frac{y^2}{25} = 1$ **15.** $\frac{x^2}{25} + \frac{y^2}{16} = 1$ **17.** $\frac{x^2}{24} + \frac{y^2}{49} = 1$

19. $\frac{(x+3)^2}{20} + \frac{y^2}{36} = 1$ **21.** $\frac{(x-2)^2}{9} + \frac{(y+1)^2}{25} = 1$ **23.** $\frac{2}{3}$ units **25.** $12\sqrt{3}$

Exercises 4.4, page 138

1. $\frac{x^2}{16} - \frac{y^2}{25} = 1$; vertices (4, 0), (−4, 0); foci $(\sqrt{41}, 0)$; $(-\sqrt{41}, 0)$

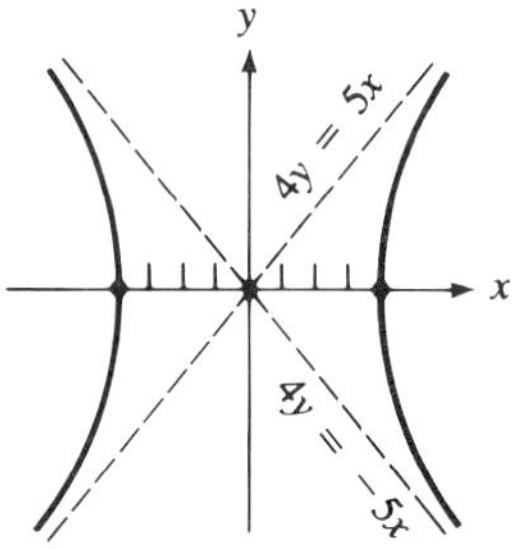

3. $\frac{y^2}{49} - \frac{x^2}{16} = 1$; vertices (0, 7), (0, −7); foci $(0, \sqrt{65})$, $(0, -\sqrt{65})$

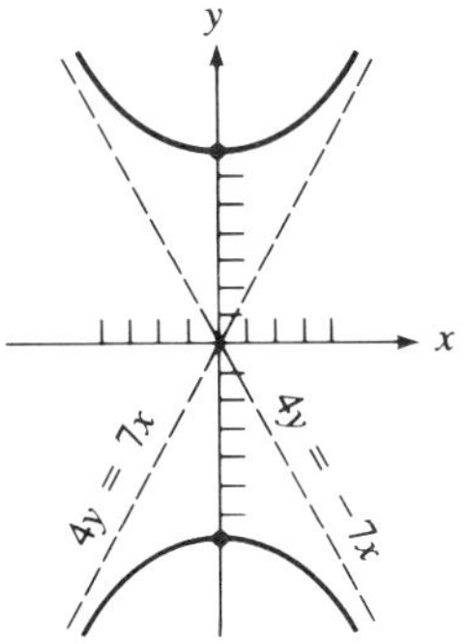

5. $9x^2 - 4y^2 = 36$; vertices $(2, 0)$, $(-2, 0)$; foci $(\sqrt{13}, 0)$, $(-\sqrt{13}, 0)$

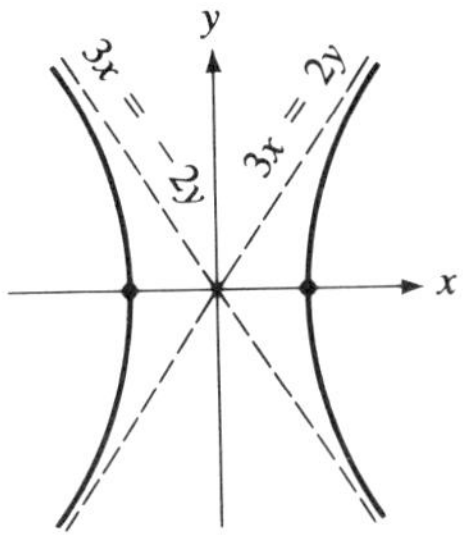

7. $4x^2 - y^2 = 1$; vertices $\left(\frac{1}{2}, 0\right)$, $\left(-\frac{1}{2}, 0\right)$; foci $\left(\frac{\sqrt{5}}{2}, 0\right)$, $\left(-\frac{\sqrt{5}}{2}, 0\right)$

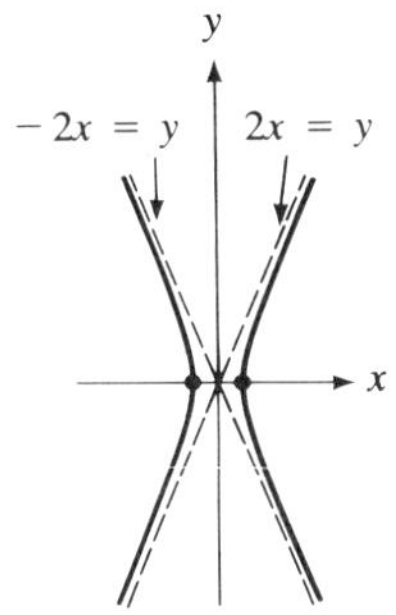

9. $4x^2 - y^2 + 16x - 4y = 88$; vertices $(3, -2)$, $(-7, -2)$; foci $(-2 + 5\sqrt{5}, -2)$, $(-2 - 5\sqrt{5}, -2)$

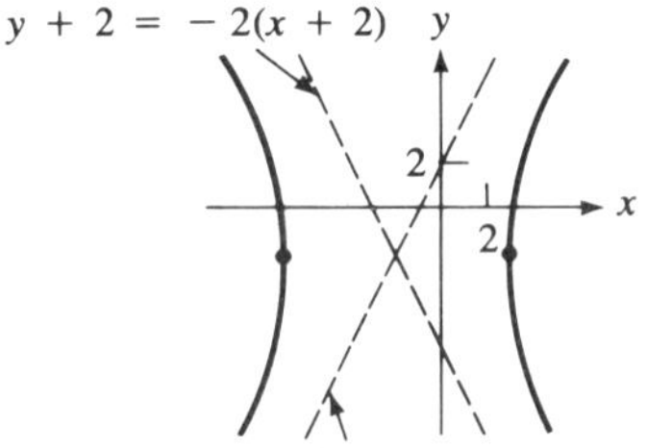

11. $9x^2 - y^2 + 36x - 2y = 1$; vertices $(0, -1)$, $(-4, -1)$; foci $(-2 + \sqrt{40}, -1)$, $(-2 - \sqrt{40}, -1)$

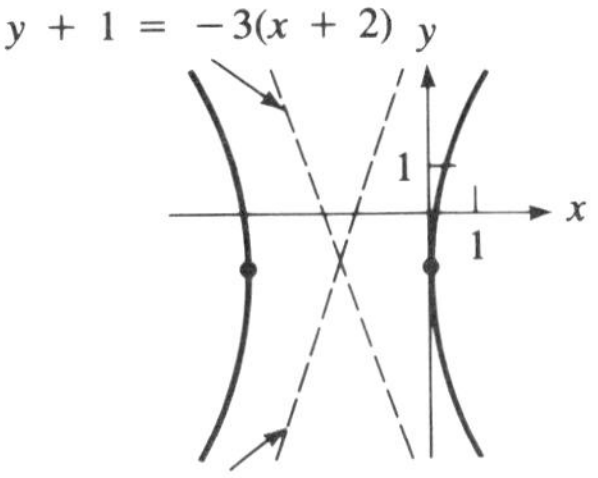

13. $\frac{y^2}{9} - \frac{x^2}{7} = 1$ **15.** $\frac{x^2}{9} - \frac{y^2}{16} = 1$ **17.** $\frac{x^2}{16} - \frac{y^2}{9} = 1$ **19.** $\frac{y^2}{20} - \frac{x^2}{5} = 1$

21. $20(x - 4)^2 - 80(y - 3)^2 = 64$

Exercises 4.5, page 142

1. $y = -\frac{1}{2}x^3$

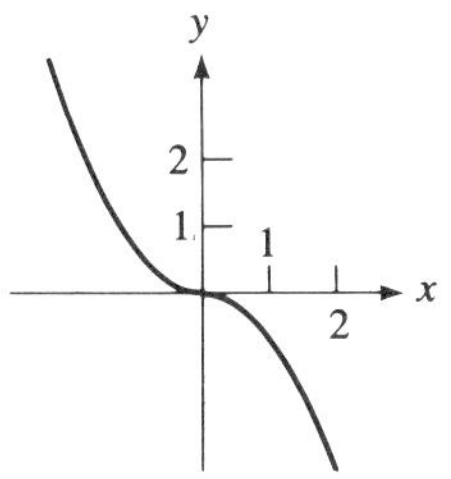

3. $y = \frac{1}{3}x^3$

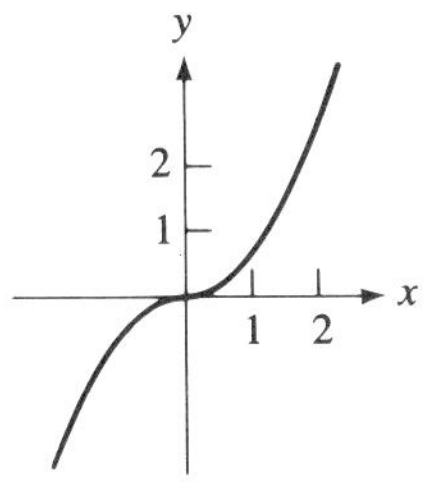

5. $y = \frac{1}{2}(x - 1)^3$

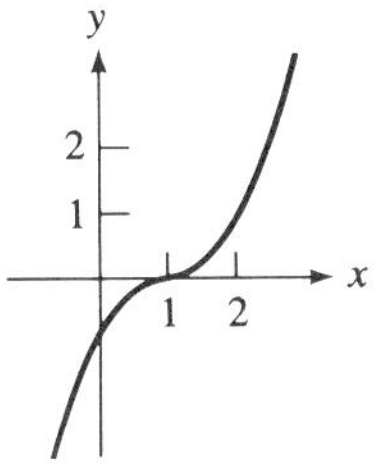

7. $y = 2(x + 2)^3 - 3$

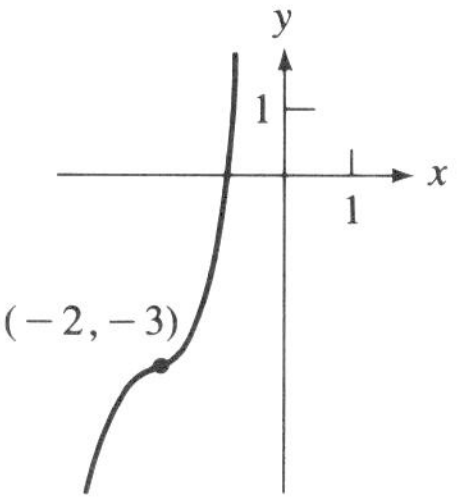

9. $y = -(2 - x)^3 + 1$

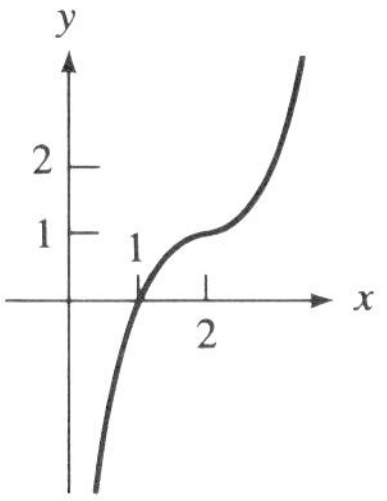

11. $y = -\frac{1}{2}x^4$

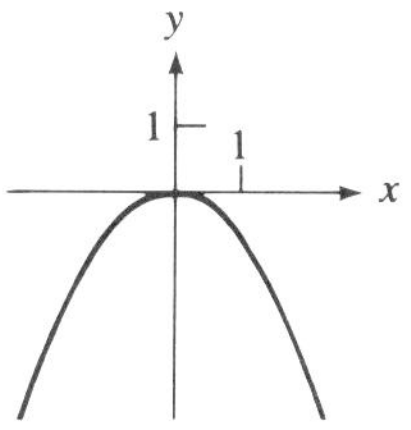

13. $y = -\frac{1}{2}(x - 4)^4 + 1$

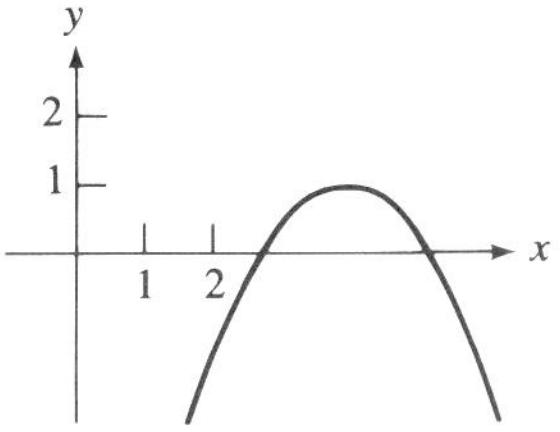

15. $y = x^5 - 1$

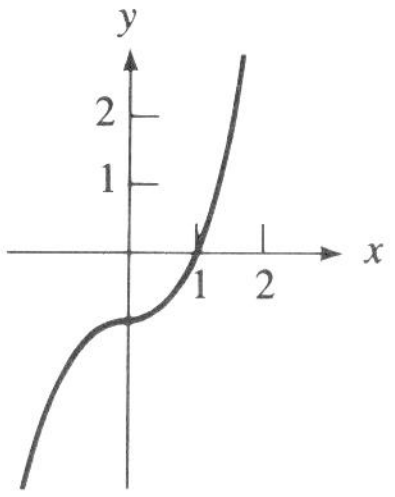

The vertical scale in the graphs in Exercises 17–23 has been distorted.

17. $f(x) = (x - 4)(x - 1)(x + 4)$

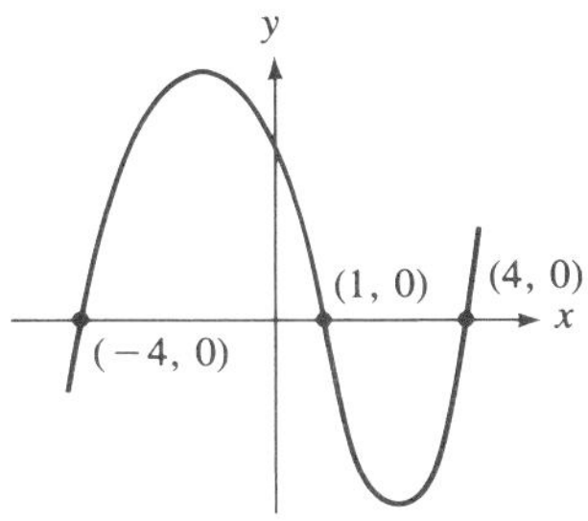

19. $h(x) = (x + 2)(x - 5)x^2$

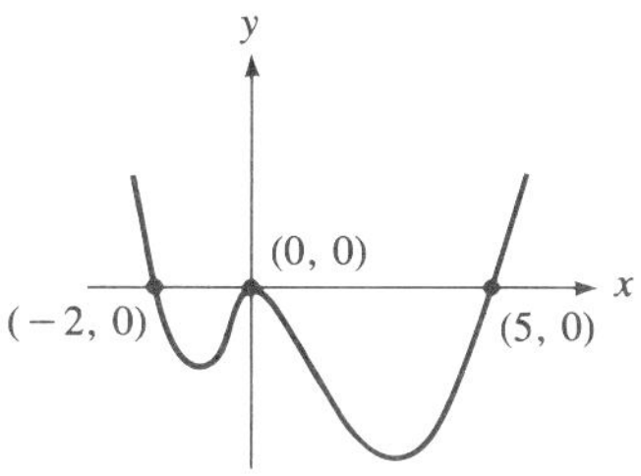

21. $f(x) = x^3 + 3x^2 - 9x - 27$

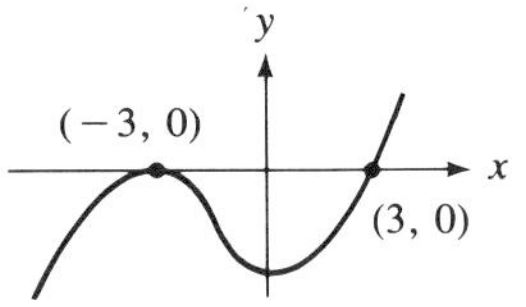

23. $y = x^3 - x^2 - 12x$

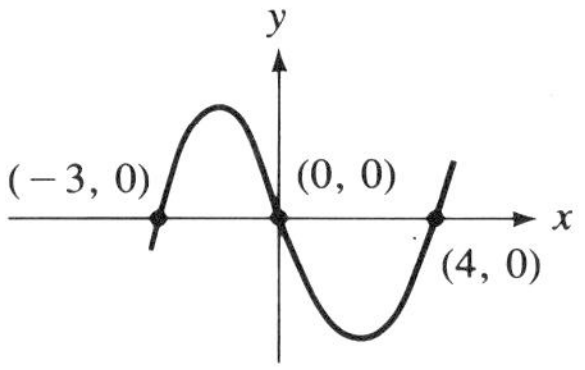

Exercises 4.6, page 147

1. $x = -3, y = 0$ **3.** $x = \frac{3}{4}, y = -\frac{3}{4}$ **5.** $x = \frac{2}{3}, y = \frac{16}{3}$ **7.** $x = 2, x = -2, y = 0$

9. $x = 6$ **11.** $x = 1, x = -1, y = \frac{2}{3}$ **13.** $x = 0, y = \frac{5}{6}$ **15.** $x = 0, y = -\frac{3}{7}$

17. $y = \frac{1}{x + 4}$

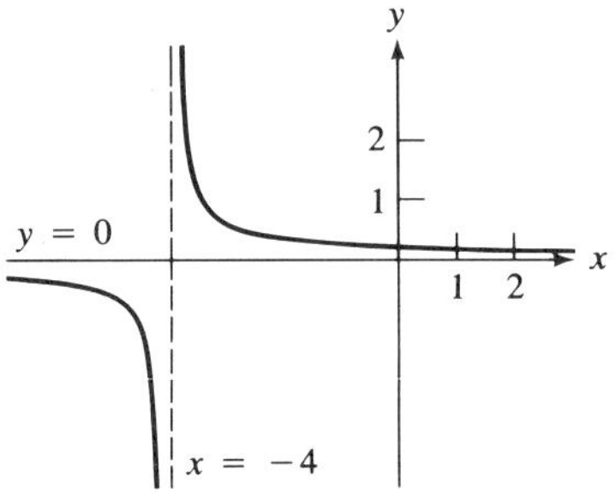

19. $y = \frac{-2}{x - 3} + 4$

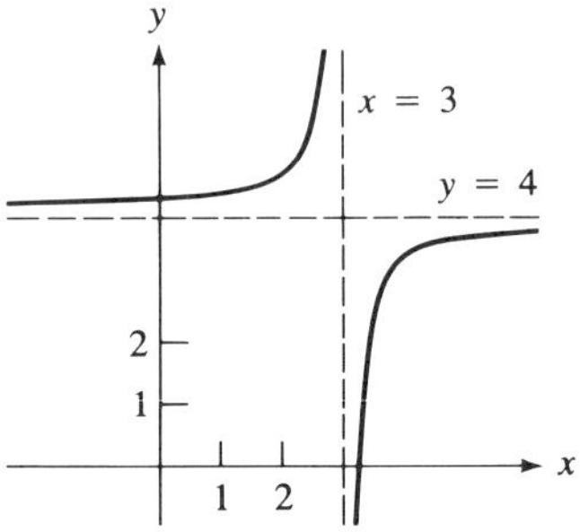

21. $y = \frac{(x - 1)^2}{2(x^2 - 2x - 3)}$

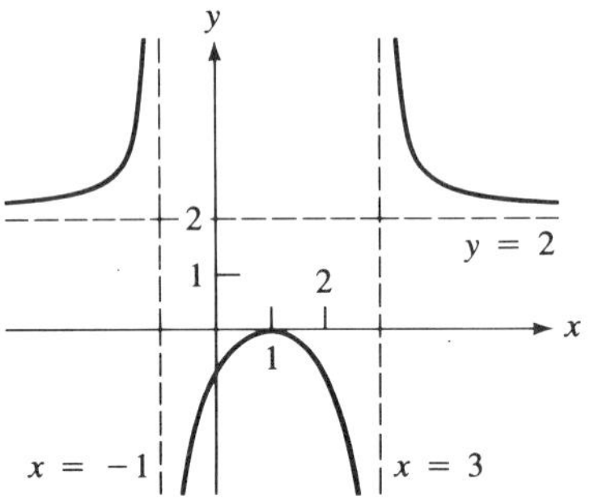

23. $y = \frac{x + 1}{3(x^2 + 2x - 8)}$

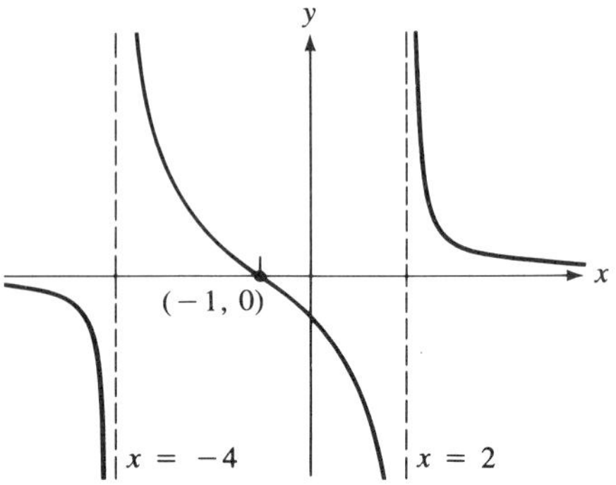

Exercises 4.7, page 151

1. $\frac{2}{x - 3} + \frac{3}{x + 2}$ **3.** $\frac{-2}{x} + \frac{3}{2x - 3}$ **5.** $\frac{3}{x + 4} + \frac{-2}{2x - 1}$ **7.** $\frac{1}{x} + \frac{-2}{x - 1} + \frac{3}{x + 1}$

9. $\frac{2}{x - 1} + \frac{3}{x + 1} + \frac{-4}{x - 2}$ **11.** $\frac{-2}{x - 4} + \frac{1}{x + 1} + \frac{-3}{x + 2}$ **13.** $\frac{1}{x} + \frac{1}{x^2} + \frac{3}{x - 1}$

15. $\frac{3}{x - 1} + \frac{4}{(x - 1)^2} + \frac{1}{x + 2}$ **17.** $\frac{2}{x - 1} + \frac{2x - 1}{x^2 + x + 3}$ **19.** $\frac{4}{x - 2} + \frac{-3x + 2}{2x^2 + x + 2}$

Exercises 4.8, page 156

1.

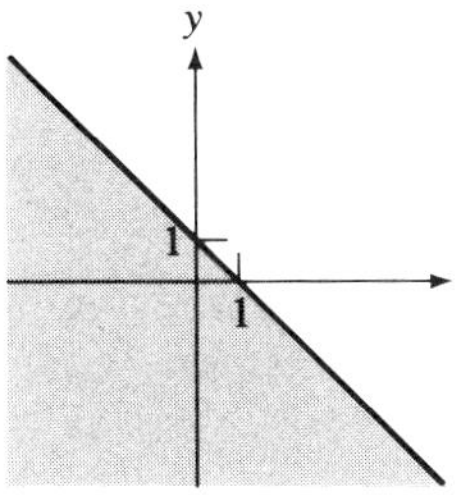

3.

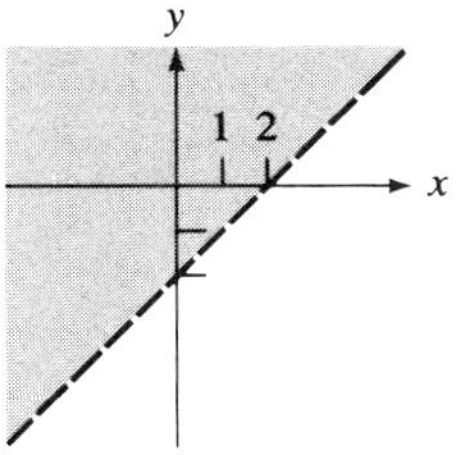

5.

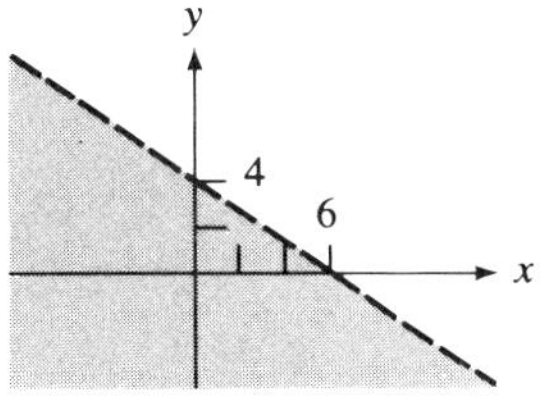

7.

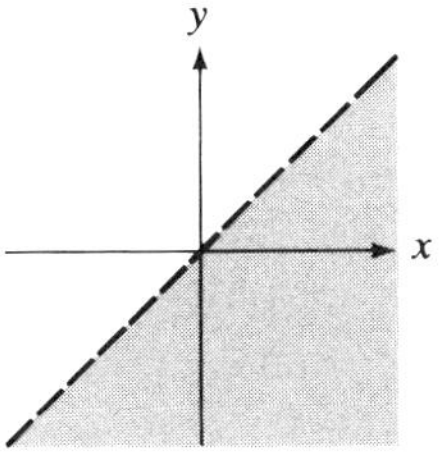

9.

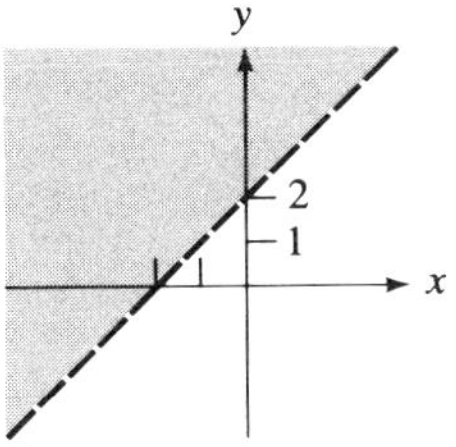

11.

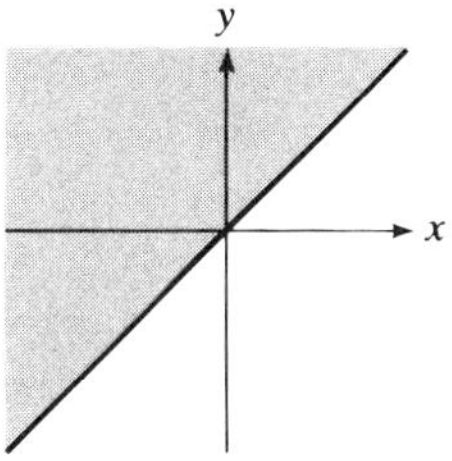

13.

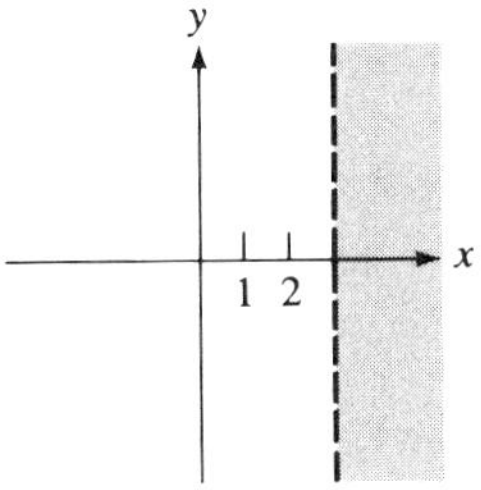

15.

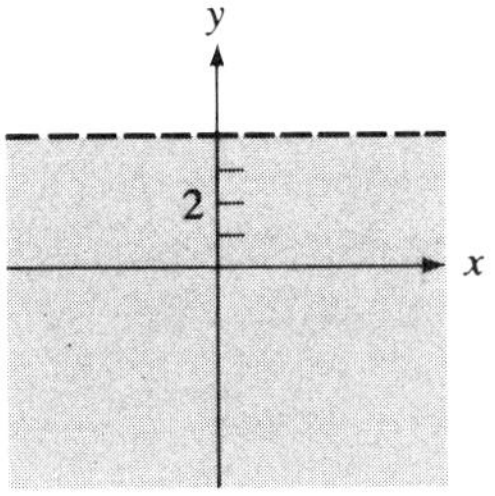

17.

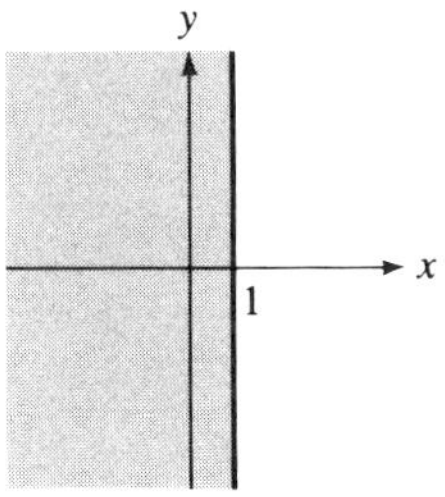

19.

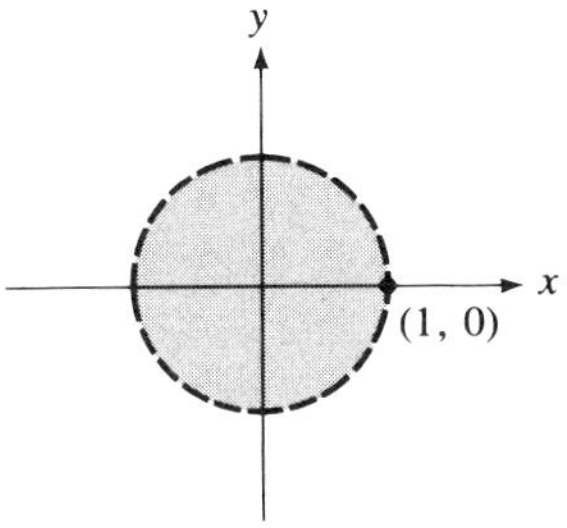

21.

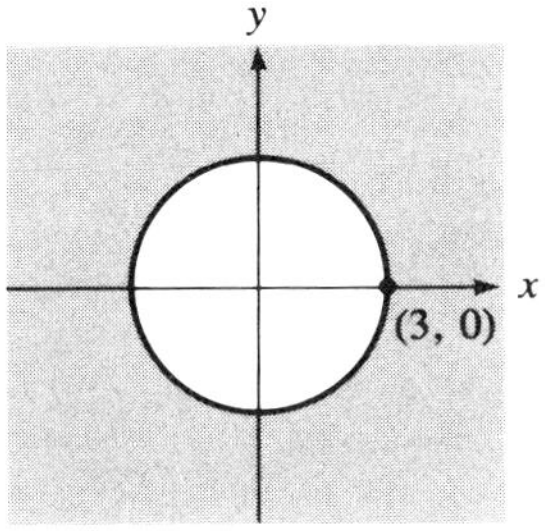

23.

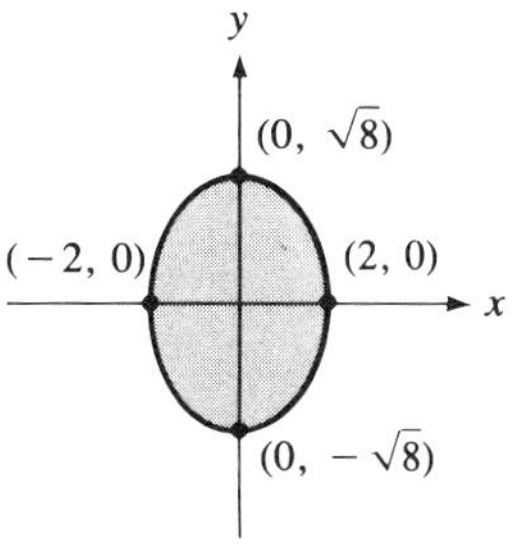

25.

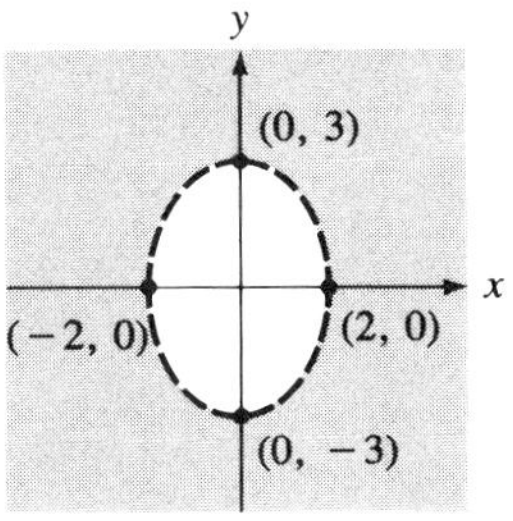

27.

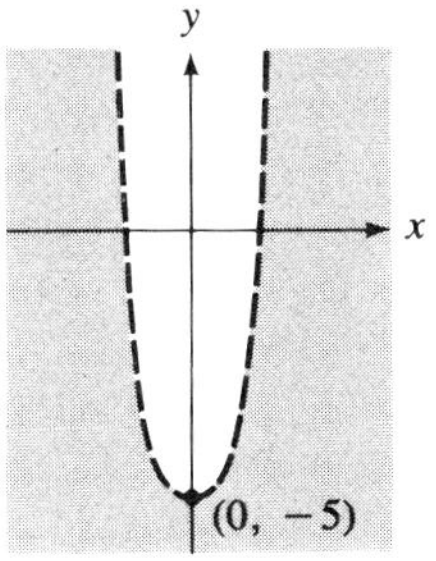

29.

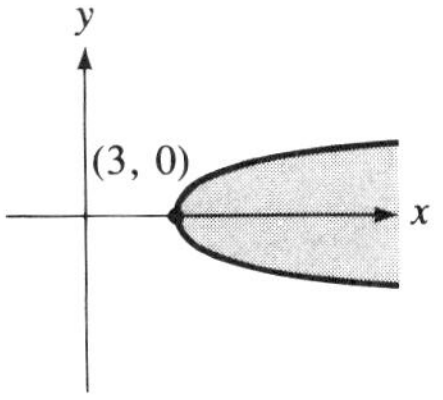

31.

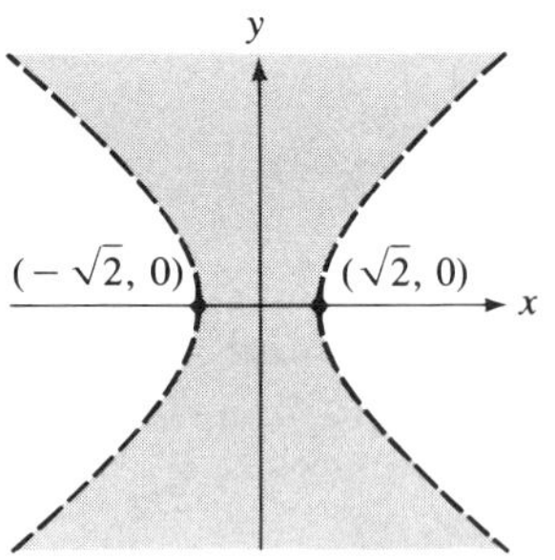

33.

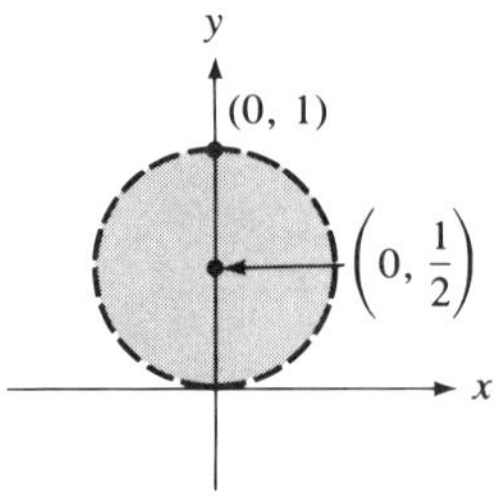

35.

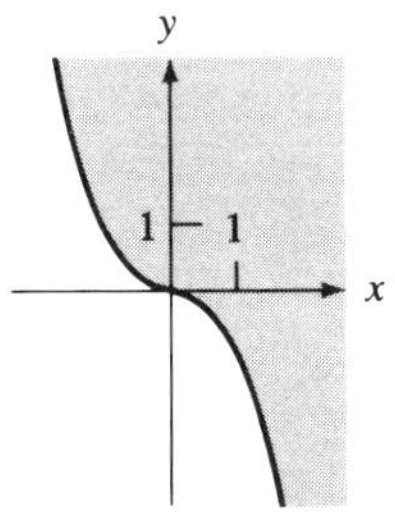

37.

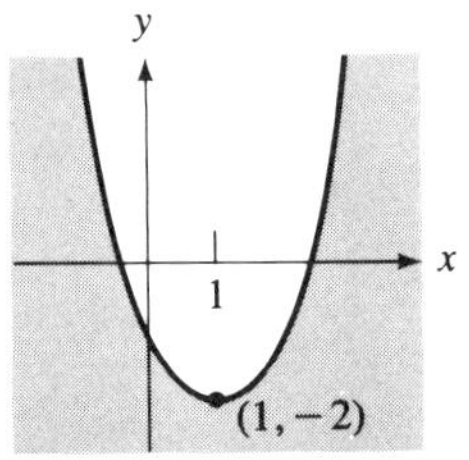

39.

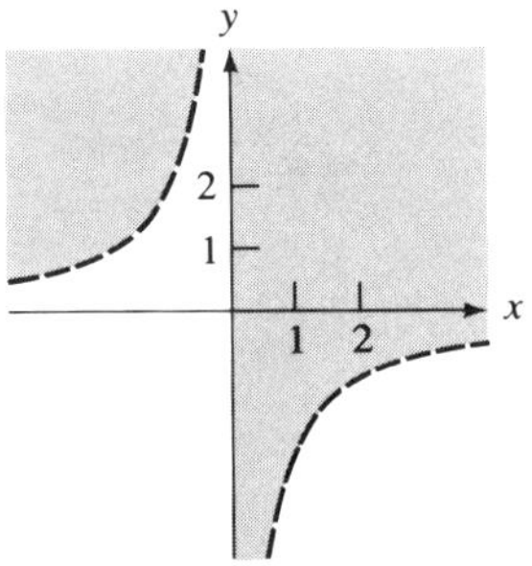

41.

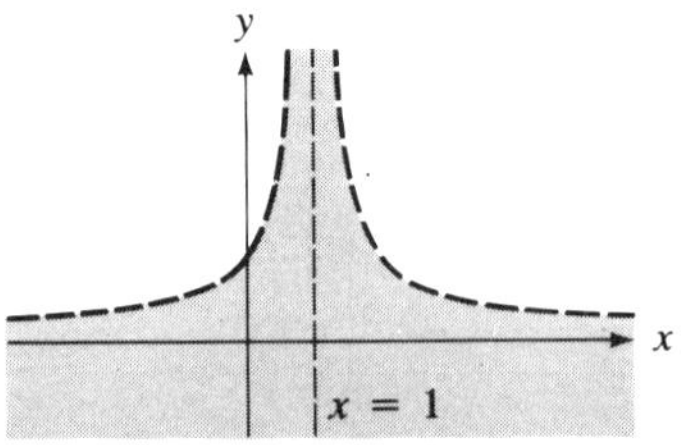

Chapter Review, page 157

1. vertex (0, 0)

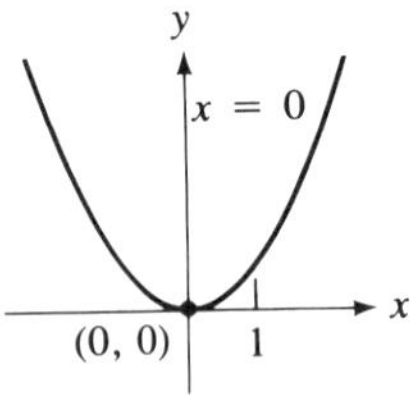

2. vertex (−1, 2)

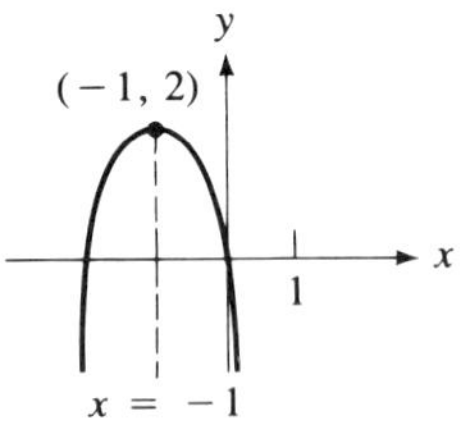

3. vertex (0, 1)

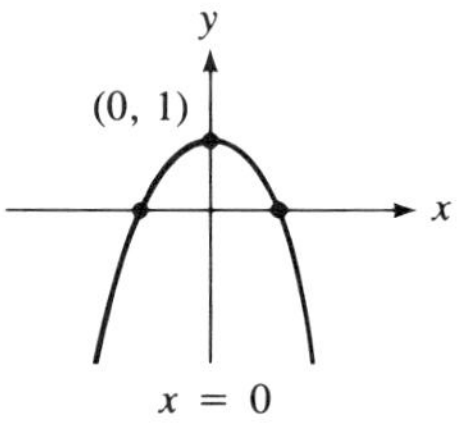

4. vertex (3, 6)

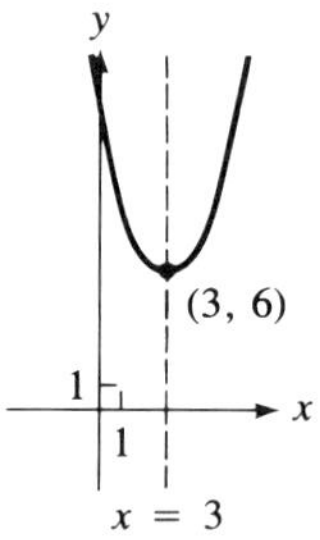

5. vertex $\left(-\frac{1}{4}, -\frac{47}{8}\right)$

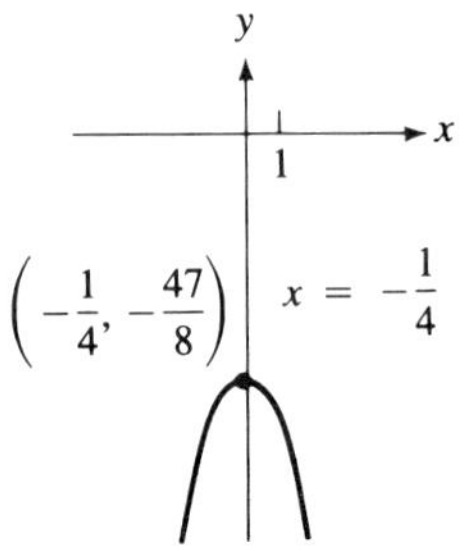

6. $v = 1$ at $t = 1$ **7.** 200 **8.** center $\left(-3, \frac{9}{2}\right)$, radius $\frac{\sqrt{117}}{2}$

9. $y^2 = -12x$

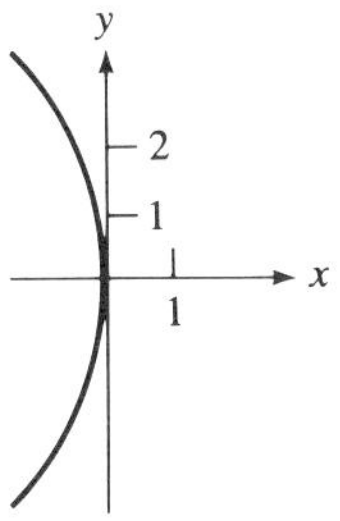

10. $8(y - 6) = (x - 3)^2$

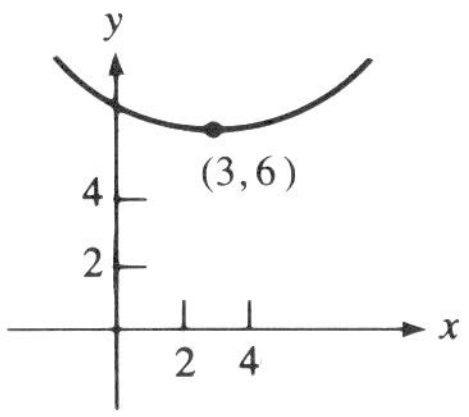

11. vertices (1, 3), (1, −9), (−1, −3), (3, −3); foci $(1, -3 + \sqrt{32})$, $(1, -3 - \sqrt{32})$

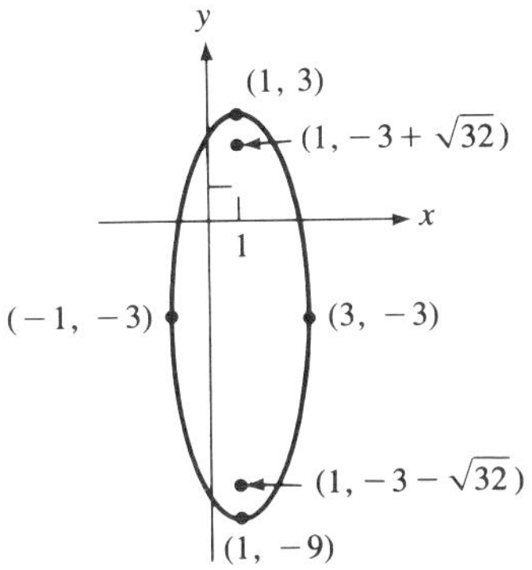

12. $\frac{4(x + 2)^2}{39} + \frac{\left(y - \frac{11}{2}\right)^2}{16} = 1$

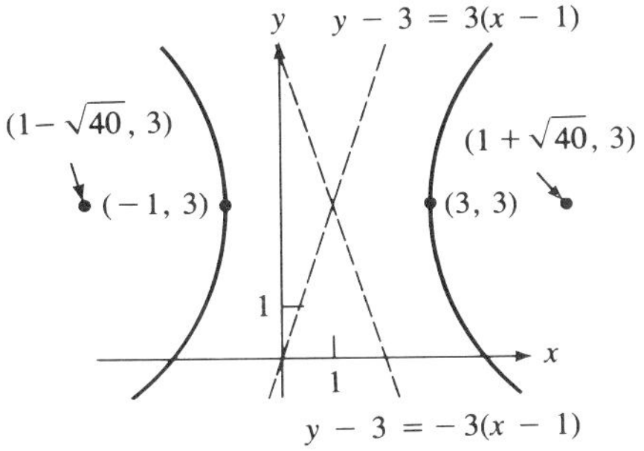

13. vertices (3, 3), (−1, 3); foci $(1 + \sqrt{40}, 3)$, $(1 - \sqrt{40}, 3)$

14. $\frac{4\left(y - \frac{11}{2}\right)^2}{9} - \frac{(x + 2)^2}{4} = 1$ **15.** $\frac{17y^2}{400} - \frac{17x^2}{25} = 1$ **16.** $\frac{(y + 2)^2}{36} - \frac{(x - 3)^2}{13} = 1$

17.

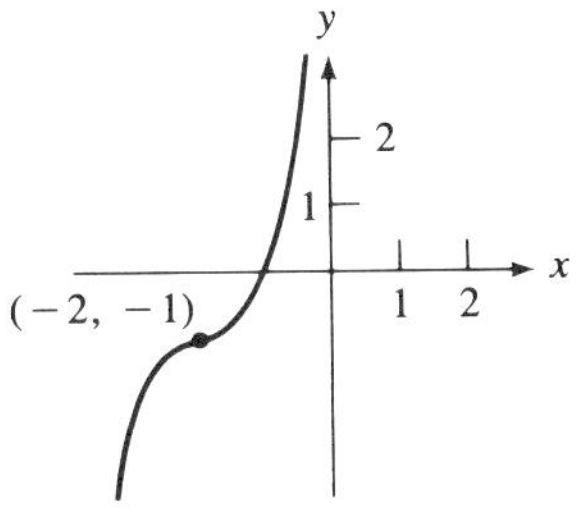

18.

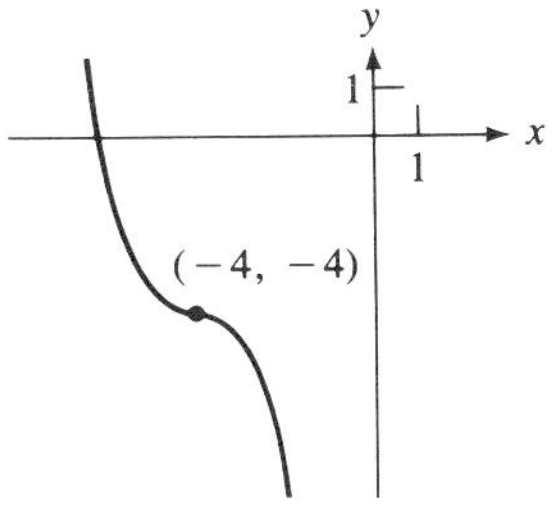

19.

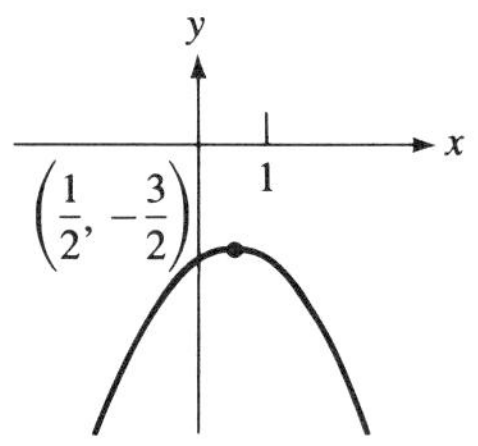

20.

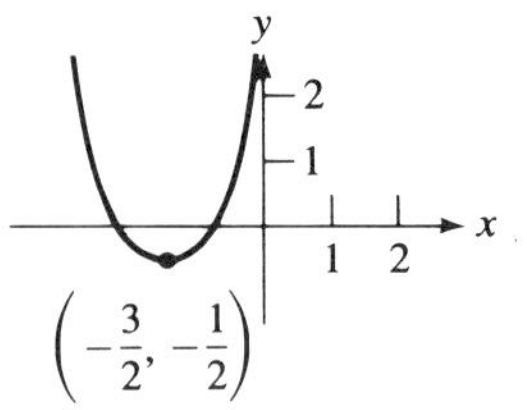

The vertical scale in Exercises 21–24 is distorted.

21.

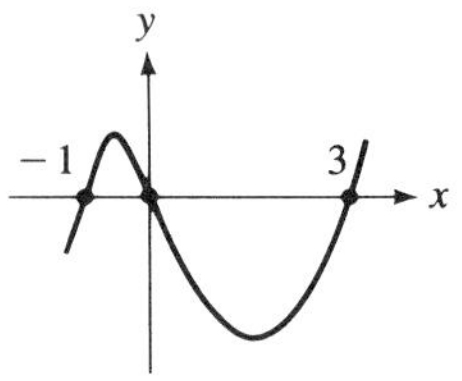

22.

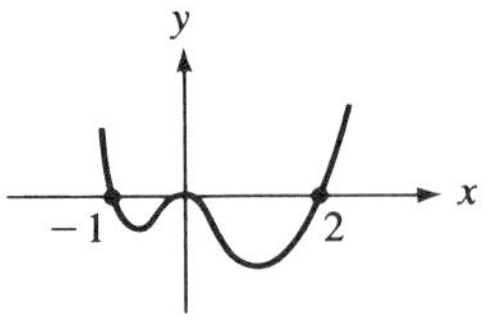

23.

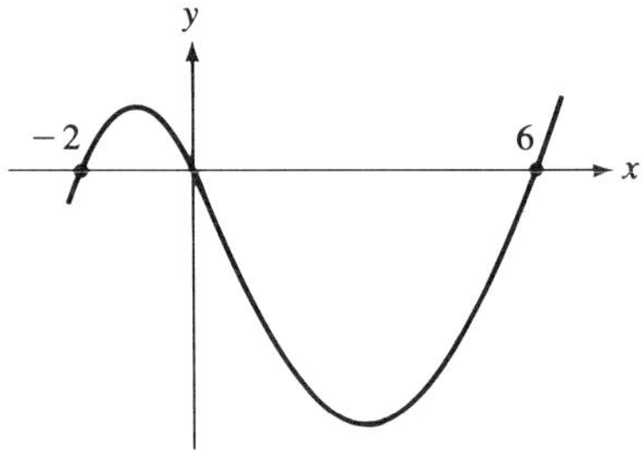

24.

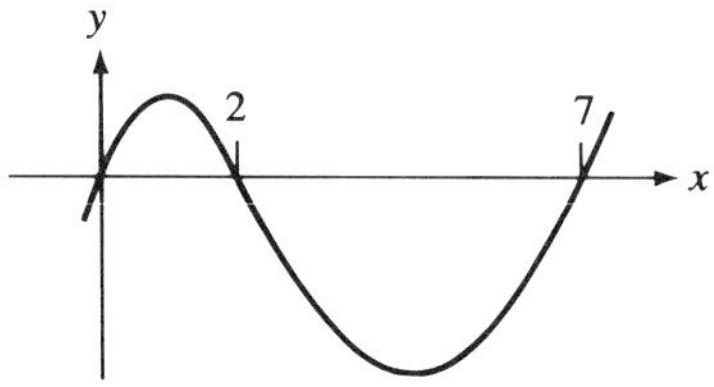

25.

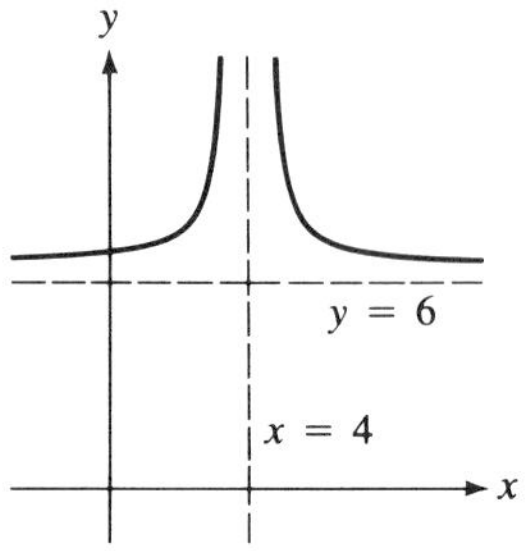

26.

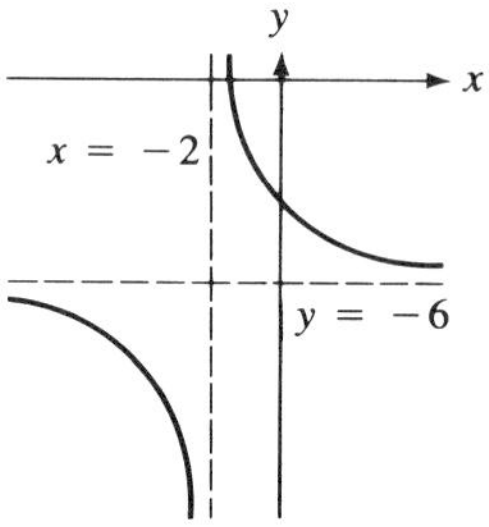

27.

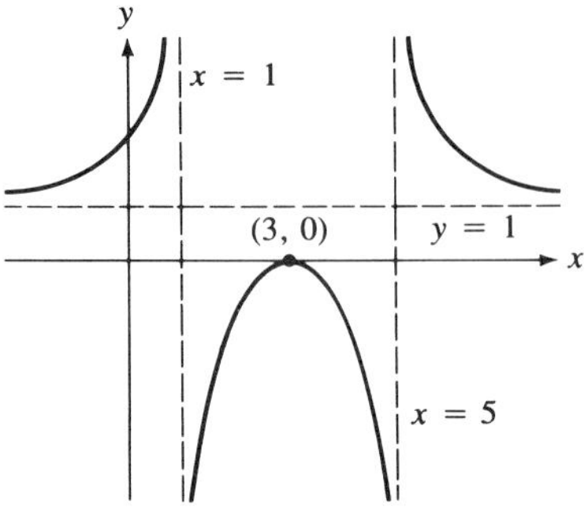

28.

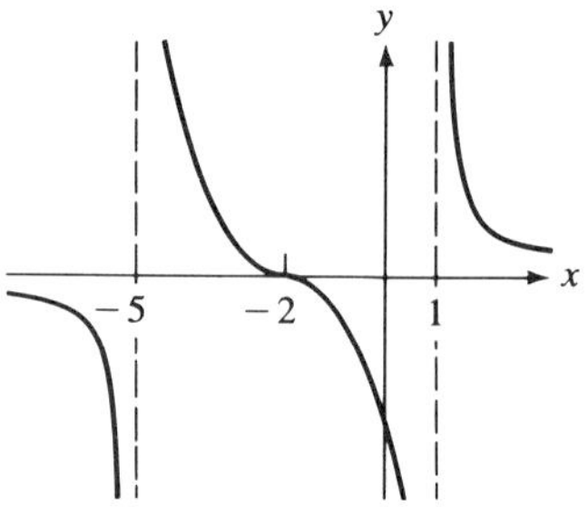

29. $\dfrac{-4}{2x-1}+\dfrac{3}{x+3}$ **30.** $\dfrac{-1}{x-3}+\dfrac{4}{3x+2}$ **31.** $\dfrac{1}{x}+\dfrac{1}{x^2}+\dfrac{2}{x-5}$ **32.** $\dfrac{-2}{x}+\dfrac{3x+1}{x^2-3x+5}$

33.

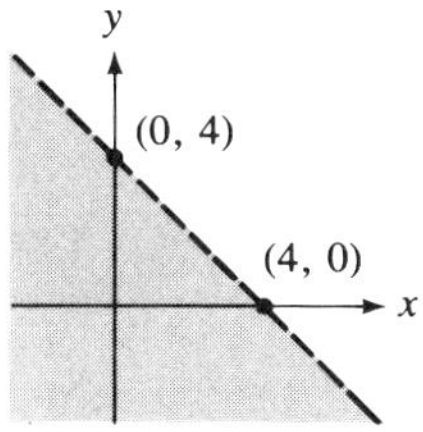

34.

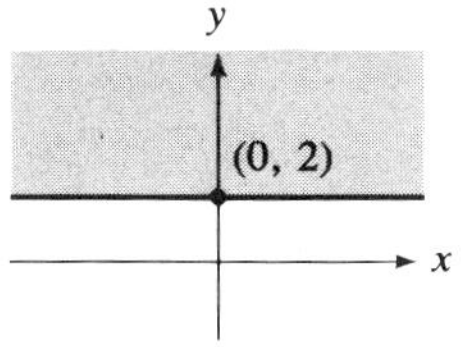

35.

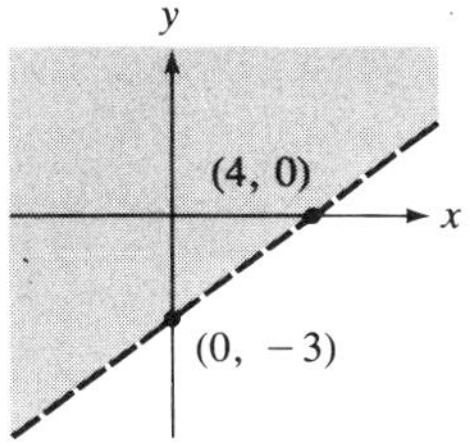

36.

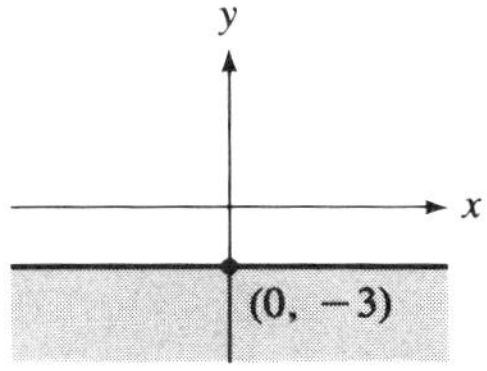

37.

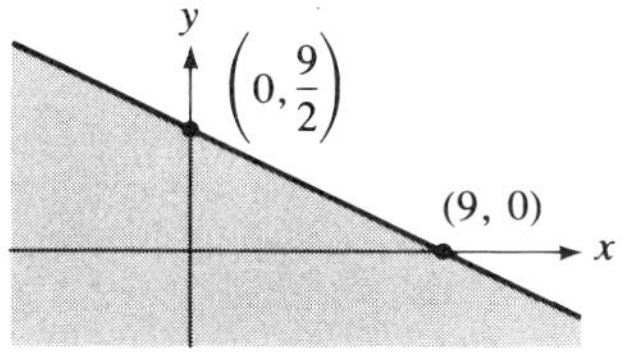

38.

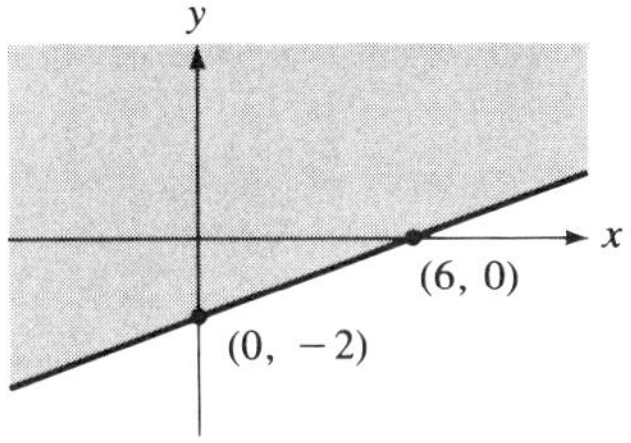

39.

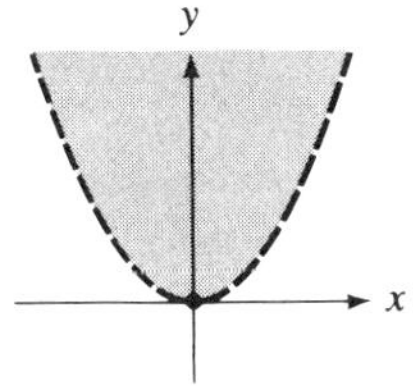

40.

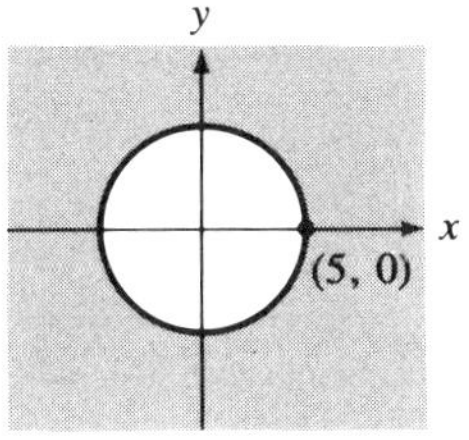

41.

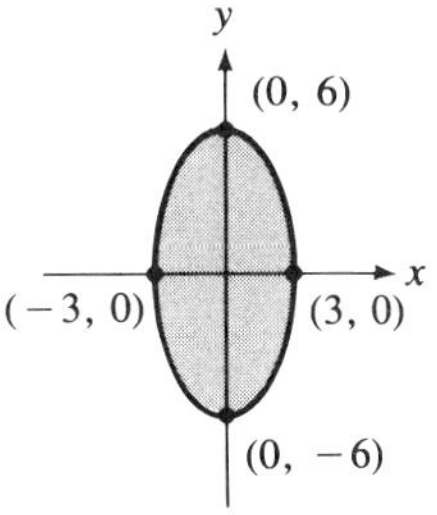

42.

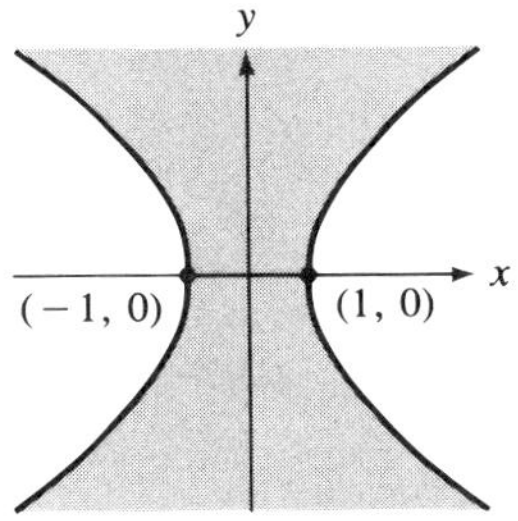

43.

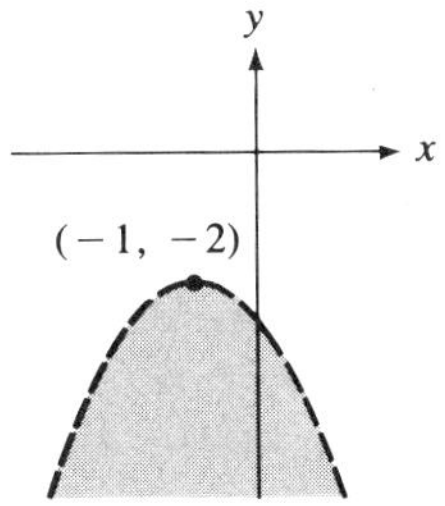

44.

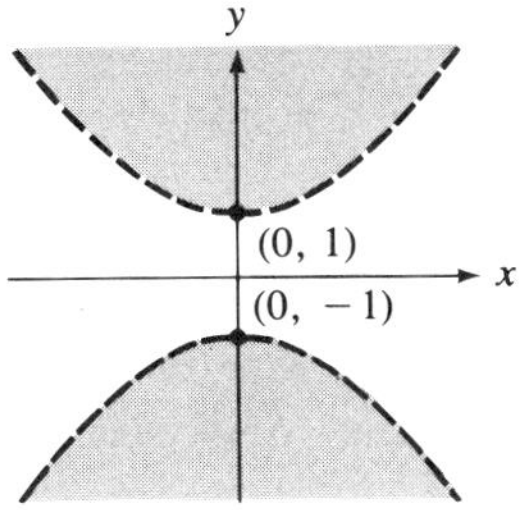

45.

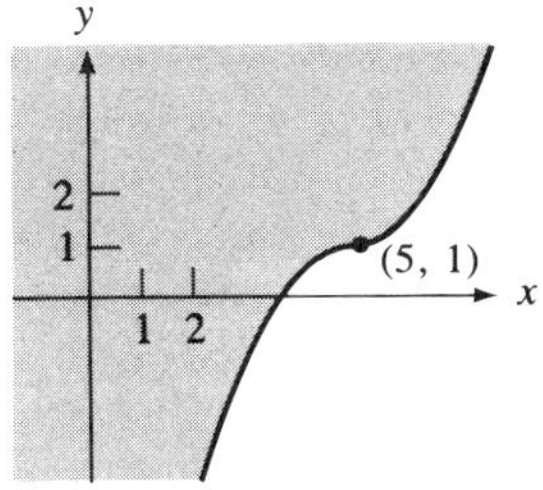

46.

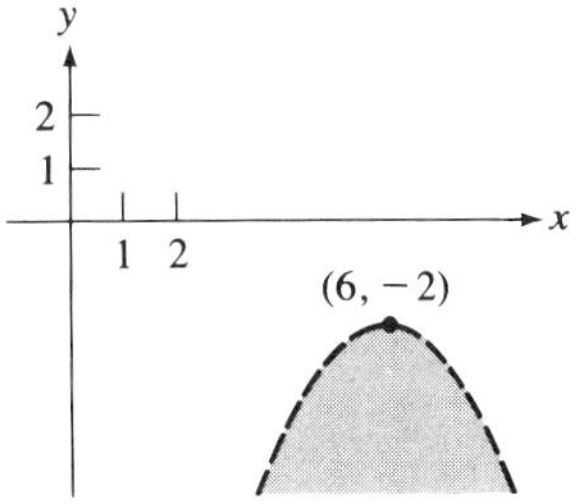

47.

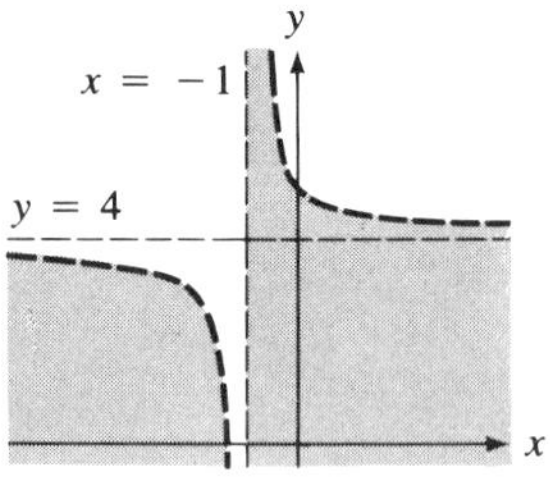

48.

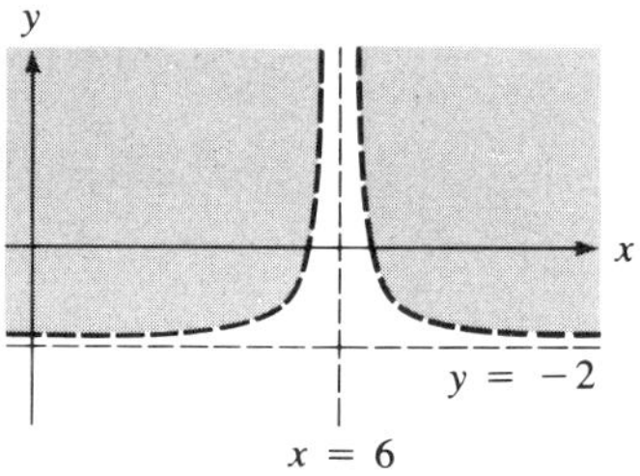

Chapter Test, page 158

1. vertex (2, −4)

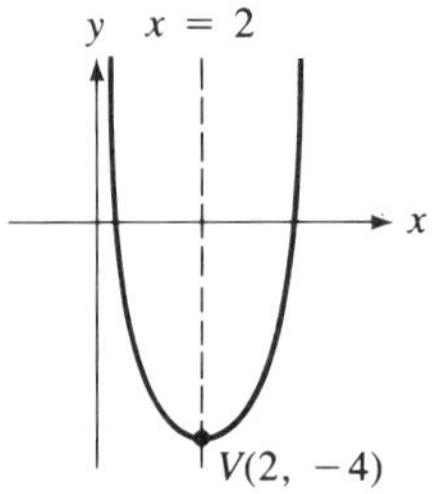

2. center (1, −2), radius 3 **3.** $(x - 2)^2 = 16y$

4. $\dfrac{x^2}{12} + \dfrac{y^2}{16} = 1$ **5.** vertices (7, −1), (−1, −1), (3, 1), (3, −3); foci $(3 + \sqrt{12}, -1)$, $(3 - \sqrt{12}, -1)$

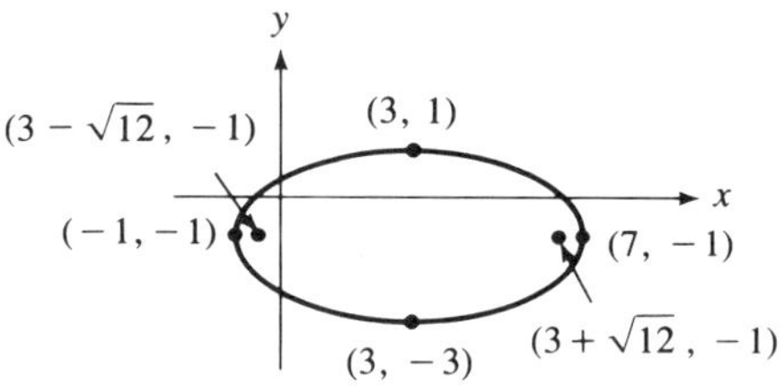

6. $(y - 6)^2 - \dfrac{(x + 2)^2}{8} = 1$ **7.** vertices (1, −1), (−7, −1); foci $(-3 - \sqrt{20}, -1)$, $(-3 + \sqrt{20}, -1)$

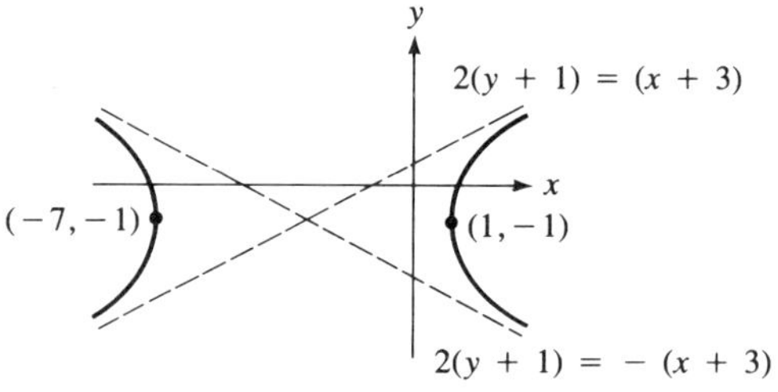

8. hyperbola, circle, parabola

9.

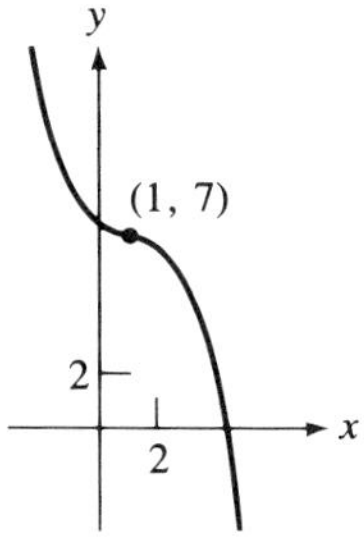

10.

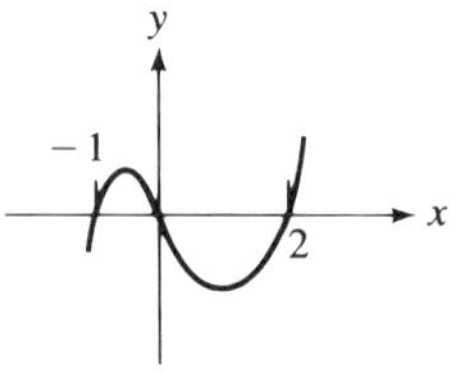

11.

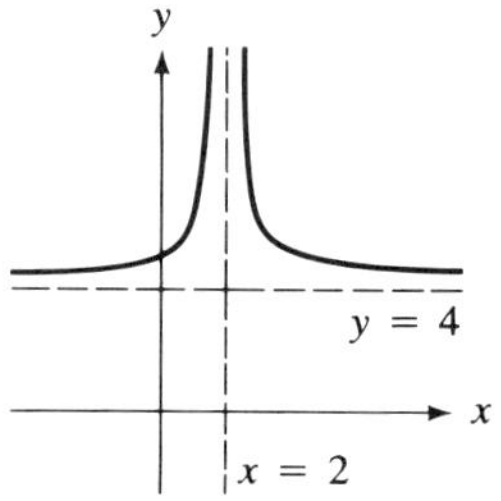

12. $\dfrac{3}{x-2} + \dfrac{-2}{x+3}$

13.

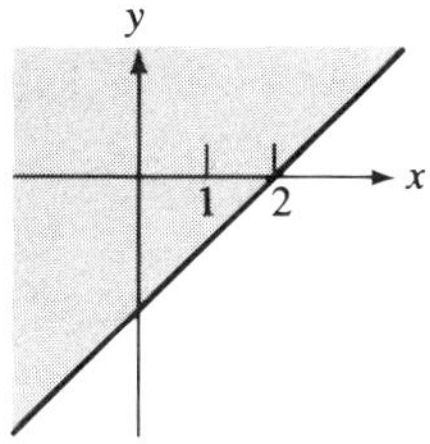

14.

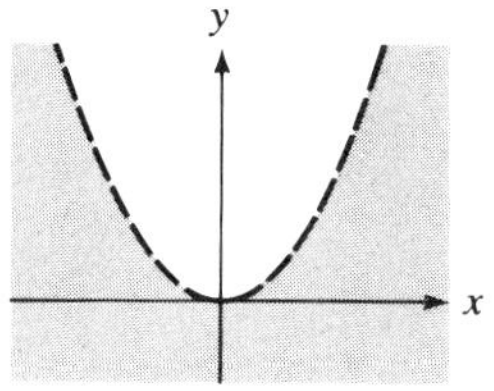

15.

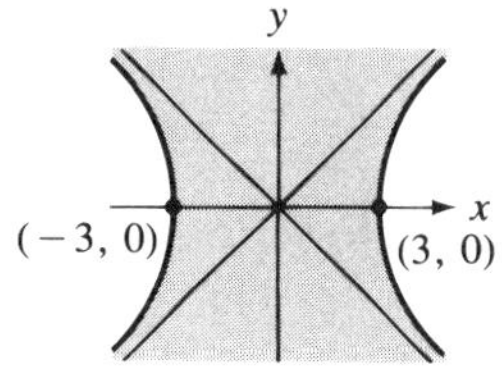

16.

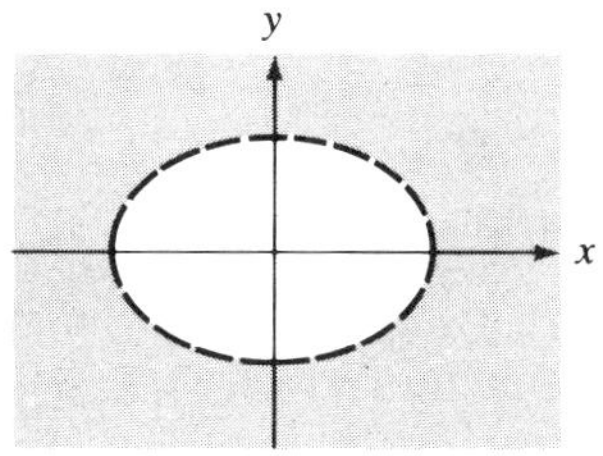

17.

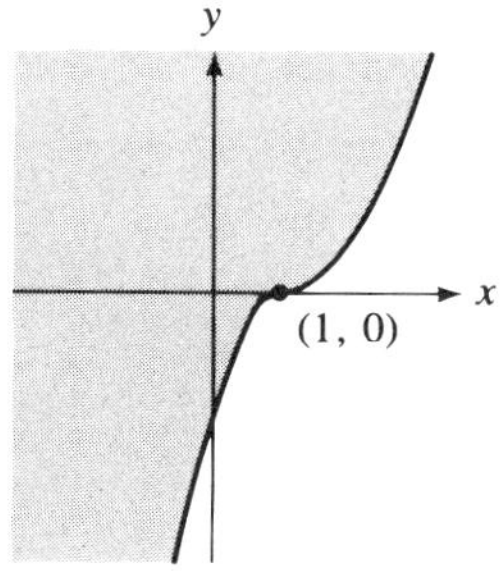

18.

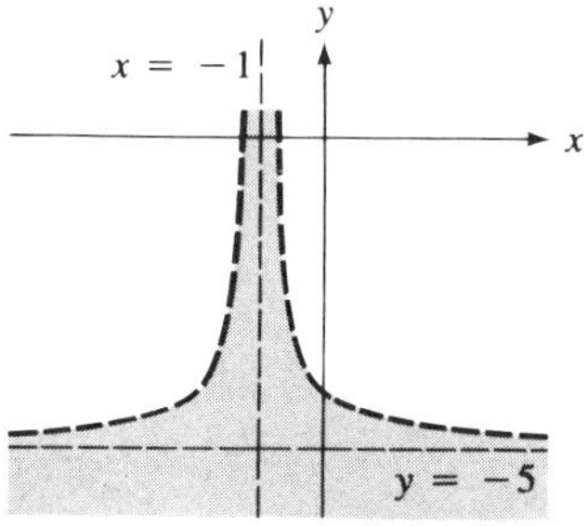

Exercises 5.1, page 164

1. $\frac{1}{8}$ **3.** $\frac{1}{256}$ **5.** 49 **7.** $\frac{64}{729}$ **9.** $\frac{1}{2}$ **11.** $\frac{3}{4}$ **13.** $\frac{1}{81}$ **15.** $\frac{2}{27}$ **17.** $-\frac{1}{2}$ **19.** $\sqrt{2}$

21.

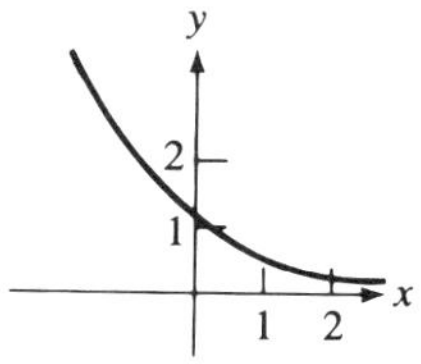

23.

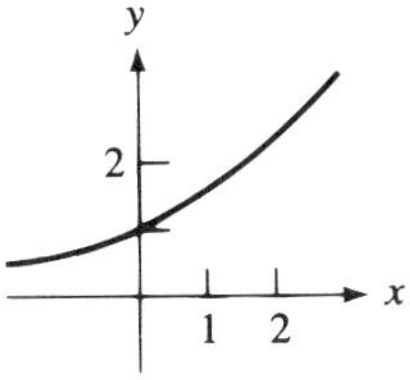

25.

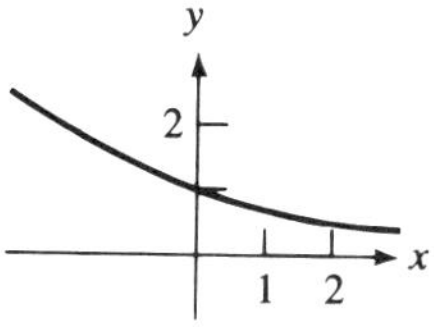

27.

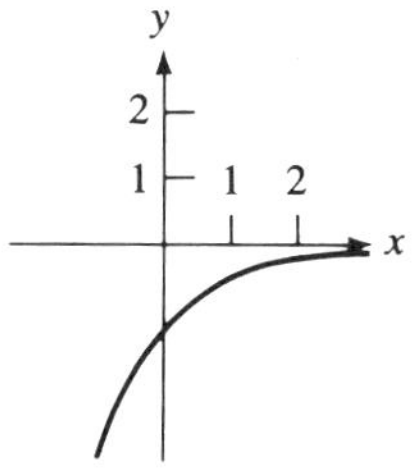

29.

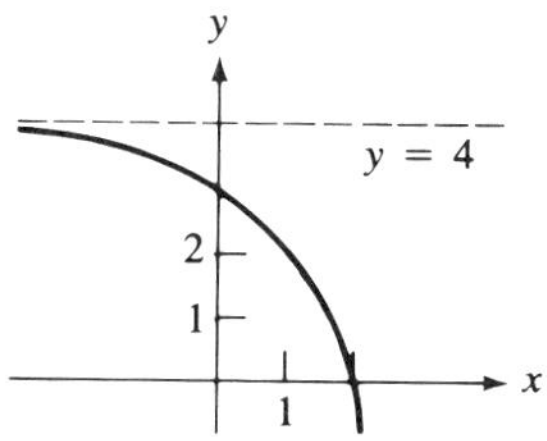

31.

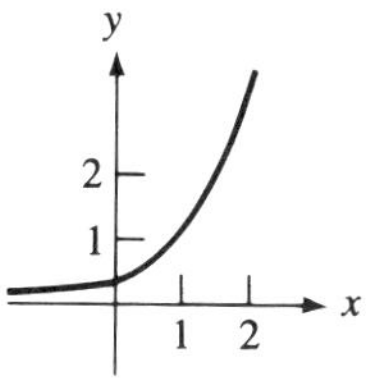

33.

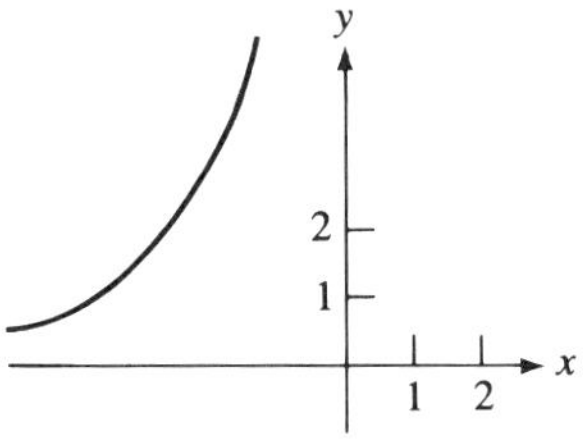

Exercises 5.2, page 168

1.

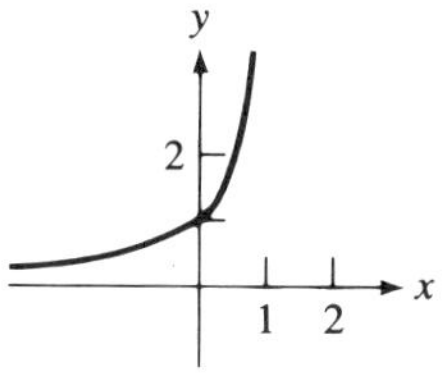

3.

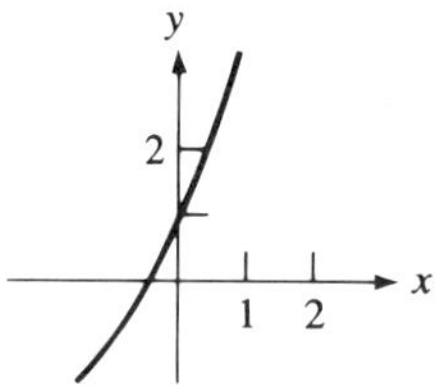

5. \$316 **7.** \$471 **9.** \$63 **11.** between 5 and 6 years **13.** \$6,475 **15.** \$4,049.70
17. \$340 **19.** 321,000 **21.** about 5% **25.** \$10,795 **27.** \$6,728

Exercises 5.3, page 173

1.

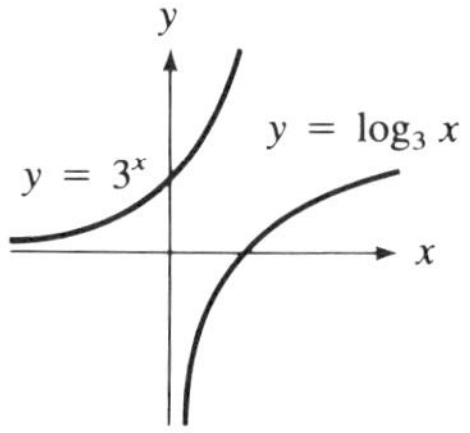

3.

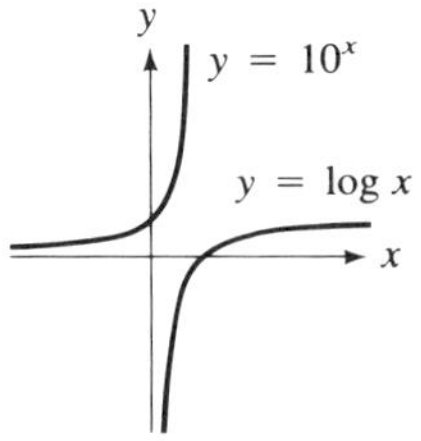

5. -2 **7.** $\frac{1}{2}$ **9.** 3 **11.** $-\frac{1}{2}$ **13.** 5 **15.** 36 **17.** 0.398 **19.** 0.875
21. -0.477 **23.** 1.505 **25.** 1.477 **27.** -2.699 **29.** 2.477 **31.** -1.921
33. $-\log x - \log y$ **35.** $2 \log z - 3 \log x - \log y$ **37.** $\frac{1}{3} \log x + \frac{2}{3} \log y - \frac{1}{2} \log z$
39. $10^6 = 1{,}000{,}000$ times **41.** 0.09531 **43.** 0.02956 **45.** 0.13975 **47.** -0.22307
49. -0.28743 **51.** 64 **53.** $\sqrt[4]{5}$

Exercises 5.4, page 177

1. 0.7543 **3.** -3.2457 **5.** 4.6096 **7.** 0.0043 **9.** $0.9996 - 1$ **11.** $0.9941 - 2$
13. 5.04 **15.** 504 **17.** 0.419 **19.** 631 **21.** 0.000631 **23.** 0.00393 **25.** 2.58
27. 10.7 **29.** 0.535 **31.** 2.5 **33.** 2.77

Exercises 5.5, page 181

1. 4 **3.** $\frac{5}{4}$ **5.** 1 **7.** -2.2 **9.** 2.64 **11.** 6 **13.** 50 **15.** 3, -1 **17.** 4, 3
19. 2 **21.** 16 **23.** no solutions **25.** \$2,458, about 19 years **27.** 35 years **29.** 2.9%
31. \$60.13 **33.** 249,215,000 (rounded off to the nearest thousand) **35.** \$119 (nearest dollar)
37. $P(t) = 450{,}000\, e^{0.019t}$, 544,000

Exercises 5.6, page 185

1. 0.760 **3.** $0.8839 - 4$ **5.** 4.0022 **7.** 0.3021 **9.** $0.5742 - 1$ **11.** 2.3702
13. 5.065 **15.** 3.999 **17.** 0.4566 **19.** 6,467 **21.** 0.003532 **23.** 51,320
25. 0.9647 **27.** 2.643 **29.** 1.962

Chapter Review, page 186

1. 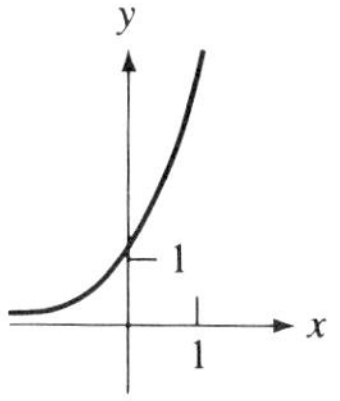

2. 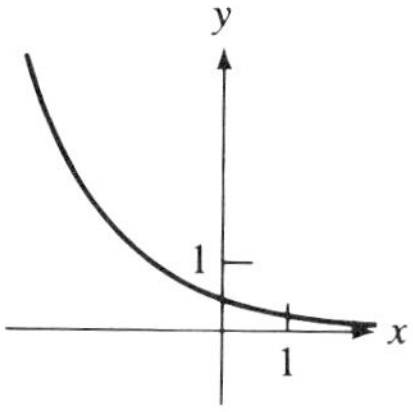

3. 3,280 **4.** 6,298 **5.** −3 **6.** 5 **7.** −6 **8.** 5 **9.** 16 **10.** 64 **11.** 0.778
12. 1.255 **13.** −0.176 **14.** 1.079 **15.** −0.125 **16.** 0.150 **17.** 1.657 **18.** 1.105
19. 5.367 **20.** 1.71 **21.** 2 **22.** 4, −1 **23.** 25 **24.** 2, 4 **25.** 4 **26.** 0.39 g
27. 7.2% **28.** 1,217 million

Chapter Test, page 187

1. 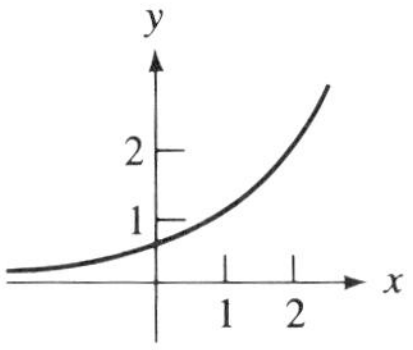

2. 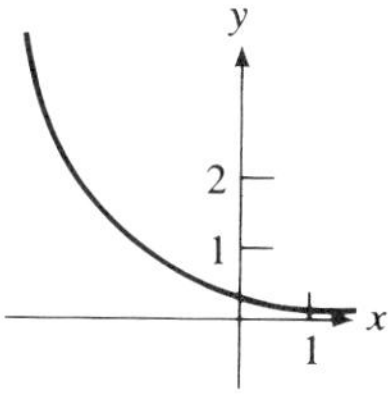

3. 187.50 **4.** 5,474 **5.** 7 **6.** 4 **7.** −3 **8.** −3 **9.** 6 **10.** 5 **11.** 16
12. 2,401 **13.** 1.793 **14.** 2.486 **15.** −0.407 **16.** 2.283 **17.** 0.407 **18.** 0.717
19. 11.822 **20.** 3.615 **21.** 3,352.5 **22.** $-\frac{1}{4}$ **23.** 2, −1 **24.** 1.209
25. no solutions **26.** 7 **27.** $2, \frac{1}{2}$ **28.** 25 g

Exercises 6.1, page 193

1. 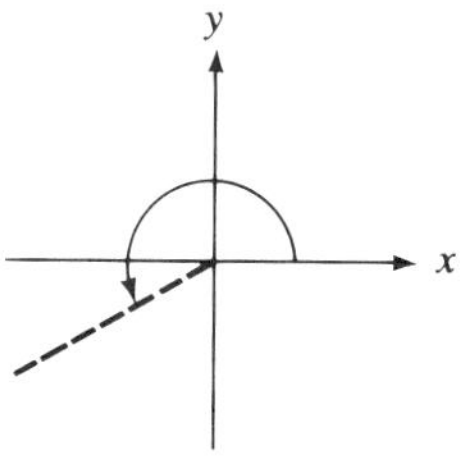

3. 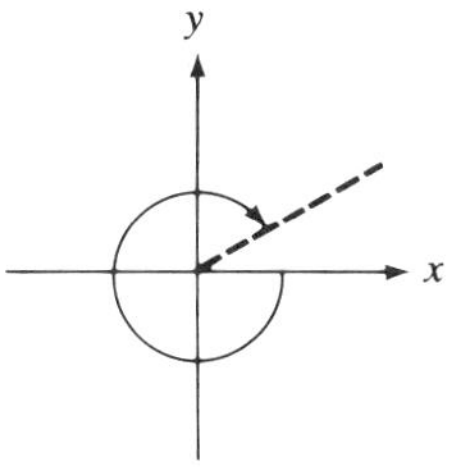

5. 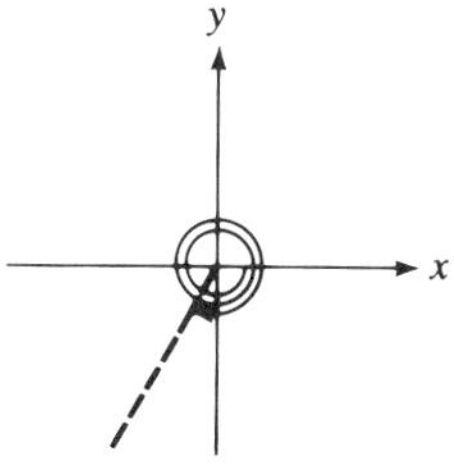

7. 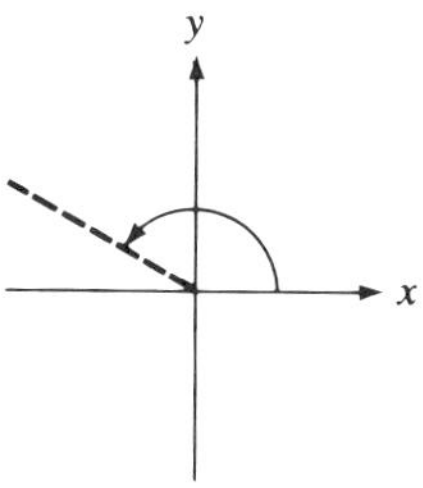

9. 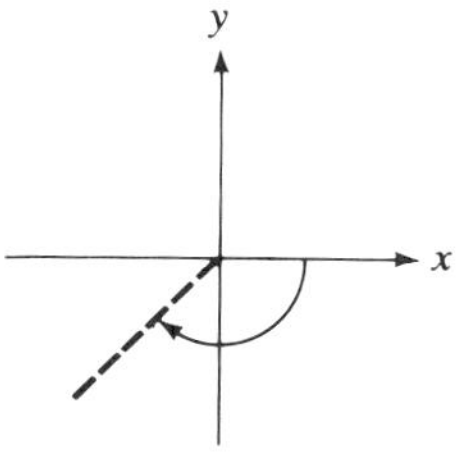

11. 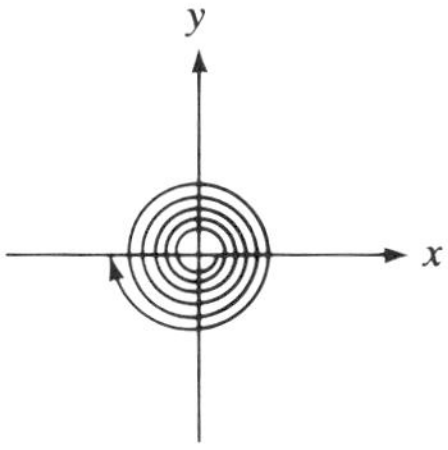

13. 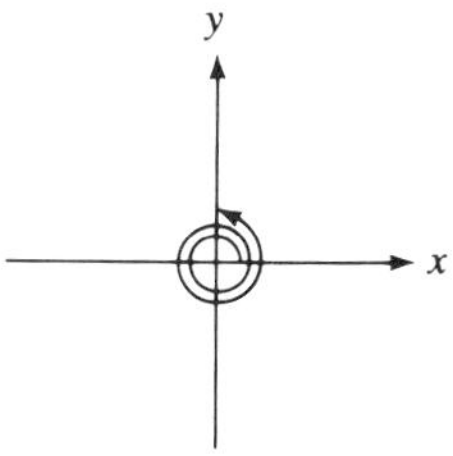

	Coterminal	Positive	Negative	Acute	Right	Quadrantal	Obtuse
15.	720° and 0°	all	none	none	none	720°, 0°, 7π, $\frac{5\pi}{2}$	none
17.	$7\pi/6$ and $-5\pi/6$, $13\pi/6$ and $\pi/6$	$7\pi/6$, $13\pi/6$, $\pi/6$	$-5\pi/6$, $-20\pi/6$	$\pi/6$	none	none	none
19.	−90°, 450°, and $3\pi/2$	450°, $3\pi/2$, 4π	−90°, −450°	none	none	all	none
21.	−12° and 348°, 168° and $14\pi/15$	168°, $14\pi/15$, 348°	−12°, $-14\pi/15$	none	none	none	168°, $14\pi/15$
23.	$3\pi/10$, −306°, and $-17\pi/10$	$3\pi/10$, 18°	−306°, $-17\pi/10$, −54°	$3\pi/10$, 18°	none	none	none

25. $\frac{\pi}{4}$ **27.** $\frac{7\pi}{6}$ **29.** $-\frac{2\pi}{3}$ **31.** 10π **33.** $-\frac{5\pi}{6}$ **35.** $\frac{5\pi}{3}$ **37.** 210° **39.** 480°

41. −315° **43.** $\left(\frac{72}{\pi}\right)^\circ$ **45.** $\left(\frac{540}{\pi}\right)^\circ$ **47.** $\left(\frac{1{,}620}{\pi}\right)^\circ$ **49.** 35°7′30″ **51.** 250°24′36″

53. 0.76 **55.** 3.76 **57.** 171°53′14.4″ **59.** 515°39′43.3″ **61.** $s = 10$ cm, $A = 200$ cm^2

63. 10π cm **65.** $1 \text{ mil} = \frac{2\pi}{6{,}400}$ radians. If $r = 1{,}000$, $s = r\theta = (1{,}000)\left(\frac{2\pi}{6{,}400}\right) = 0.3125\pi \simeq 0.982$ yards.

67. 138.46 radians $\simeq$ 7,933.17° **69.** 108,385 in./min **71.** 28.0625 **73.** 178°52′30″
75. 0.8351 **77.** 2,506.35°

Exercises 6.2, page 201

	$\sin\theta$	$\cos\theta$	$\tan\theta$	$\csc\theta$	$\sec\theta$	$\cot\theta$
1.	$\frac{4}{5}$	$\frac{3}{5}$	$\frac{4}{3}$	$\frac{5}{4}$	$\frac{5}{3}$	$\frac{3}{4}$
3.	$\frac{-5}{\sqrt{29}}$	$\frac{-2}{\sqrt{29}}$	$\frac{5}{2}$	$\frac{-\sqrt{29}}{5}$	$\frac{-\sqrt{29}}{2}$	$\frac{2}{5}$
5.	$\frac{2}{\sqrt{13}}$	$\frac{-3}{\sqrt{13}}$	$-\frac{2}{3}$	$\frac{\sqrt{13}}{2}$	$-\frac{\sqrt{13}}{3}$	$-\frac{3}{2}$
7.	$-\frac{1}{2}$	$\frac{\sqrt{3}}{2}$	$-\frac{1}{\sqrt{3}}$	-2	$\frac{2}{\sqrt{3}}$	$-\sqrt{3}$
9.	1	0	*	1	*	0
11.	$-\frac{1}{\sqrt{2}}$	$\frac{1}{\sqrt{2}}$	-1	$-\sqrt{2}$	$\sqrt{2}$	-1
13.	$\frac{1}{\sqrt{2}}$	$\frac{1}{\sqrt{2}}$	1	$\sqrt{2}$	$\sqrt{2}$	1
15.	$\frac{3}{\sqrt{10}}$	$\frac{-1}{\sqrt{10}}$	-3	$\frac{\sqrt{10}}{3}$	$-\sqrt{10}$	$-\frac{1}{3}$
17.	$-\frac{\sqrt{3}}{2}$	$-\frac{1}{2}$	$\sqrt{3}$	$-\frac{2}{\sqrt{3}}$	-2	$\frac{1}{\sqrt{3}}$
19.	$-\frac{2}{\sqrt{13}}$	$\frac{3}{\sqrt{13}}$	$-\frac{2}{3}$	$-\frac{\sqrt{13}}{2}$	$\frac{\sqrt{13}}{3}$	$-\frac{3}{2}$

	$\sin\theta$	$\cos\theta$	$\tan\theta$	$\csc\theta$	$\sec\theta$	$\cot\theta$
21.	$\frac{3}{\sqrt{13}}$	$-\frac{2}{\sqrt{13}}$	$-\frac{3}{2}$	$\frac{\sqrt{13}}{3}$	$-\frac{\sqrt{13}}{2}$	$-\frac{2}{3}$
23.	$-\frac{2}{\sqrt{5}}$	$-\frac{1}{\sqrt{5}}$	2	$-\frac{\sqrt{5}}{2}$	$-\sqrt{5}$	$\frac{1}{2}$
25.	$-\frac{2}{\sqrt{5}}$	$\frac{1}{\sqrt{5}}$	-2	$-\frac{\sqrt{5}}{2}$	$\sqrt{5}$	$-\frac{1}{2}$

27. $(0, 5)$ **29.** $\left(-\frac{3}{2}, -\frac{3\sqrt{3}}{2}\right)$ **31.** $(-2\sqrt{2}, 2\sqrt{2})$ **33.** $(-24, -10)$

	$\sin\theta$	$\cos\theta$	$\tan\theta$	$\csc\theta$	$\sec\theta$	$\cot\theta$
35.	$\frac{7}{25}$	$-\frac{24}{25}$	$-\frac{7}{24}$	$\frac{25}{7}$	$-\frac{25}{24}$	$-\frac{24}{7}$
37.	$-\frac{12}{13}$	$-\frac{5}{13}$	$\frac{12}{5}$	$-\frac{13}{12}$	$-\frac{13}{5}$	$\frac{5}{12}$
39.	$\frac{2}{\sqrt{13}}$	$\frac{3}{\sqrt{13}}$	$\frac{2}{3}$	$\frac{\sqrt{13}}{2}$	$\frac{\sqrt{13}}{3}$	$\frac{3}{2}$
41.	$-\frac{\sqrt{6}}{3}$	$\frac{\sqrt{3}}{3}$	$-\sqrt{2}$	$-\frac{3}{\sqrt{6}}$	$\frac{3}{\sqrt{3}}$	$-\frac{1}{\sqrt{2}}$
43.	$-\frac{1}{4}$	$-\frac{\sqrt{15}}{4}$	$\frac{1}{\sqrt{15}}$	-4	$\frac{-4}{\sqrt{15}}$	$\sqrt{15}$

45. $\frac{1}{2}$ **47.** $-\frac{1}{2}$ **49.** $-\frac{\sqrt{3}}{2}$ **51.** $\frac{1}{\sqrt{2}}$ **53.** -1 **55.** -1 **57.** 0 **59.** -1 **61.** 0
63. 1

	$\sin\theta$	$\cos\theta$	$\tan\theta$	$\csc\theta$	$\sec\theta$	$\cot\theta$
65.	0	-1	0	*	-1	*
67.	$-\frac{4}{5}$	$-\frac{3}{5}$	$\frac{4}{3}$	$-\frac{5}{4}$	$-\frac{5}{3}$	$\frac{3}{4}$
69.	$-\frac{3}{\sqrt{13}}$	$-\frac{2}{\sqrt{13}}$	$\frac{3}{2}$	$-\frac{\sqrt{13}}{3}$	$-\frac{\sqrt{13}}{2}$	$\frac{2}{3}$
71.	$-\frac{\sqrt{6}}{3}$	$\frac{\sqrt{3}}{3}$	$-\sqrt{2}$	$-\frac{3}{\sqrt{6}}$	$\frac{3}{\sqrt{3}}$	$-\frac{1}{\sqrt{2}}$
73.	$\frac{2}{\sqrt{13}}$	$-\frac{3}{\sqrt{13}}$	$-\frac{2}{3}$	$\frac{\sqrt{13}}{2}$	$-\frac{\sqrt{13}}{3}$	$-\frac{3}{2}$

	$\sin\theta$	$\cos\theta$	$\tan\theta$	$\csc\theta$	$\sec\theta$	$\cot\theta$
75.	+	−	−	+	−	−
77.	+	−	−	+	−	−
79.	−	+	−	−	+	−
81.	−	+	−	−	+	−
83.	−	−	+	−	−	+

85. $\pm\frac{k\pi}{2}$, $k \varepsilon Z$, k odd **87.** $\pm\frac{k\pi}{2}$, $k \varepsilon Z$, k odd **89.** $\frac{16}{81}$ **91.** any α such that $\sin\alpha\cos\alpha \neq 0$

93. 4 **95.** $y - 2 = -\frac{4}{3}(x + 1)$ **97.** 0.5, −0.9524 **99.** −0.1736, 0.8623 **101.** 0.1049, −0.2602

Exercises 6.3, page 209

Values of the trigonometric functions are listed in the order sine, cosine, and tangent. The cosecant, secant, and cotangent are the reciprocals of these, respectively, when defined.

1. $a = 2, b = 2\sqrt{3}$ **3.** $b = \frac{\sqrt{3}}{2}, c = 1$ **5.** $a = 3, c = 6$ **7.** $a = \frac{5}{\sqrt{3}}, c = \frac{10}{\sqrt{3}}$

9. $a = 2, c = 2\sqrt{2}$ **11.** $a = \frac{5}{\sqrt{2}}, b = \frac{5}{\sqrt{2}}$ **13.** $b = \frac{1}{2}, c = \frac{\sqrt{2}}{2}$ **15.** $\frac{1}{\sqrt{2}}, -\frac{1}{\sqrt{2}}, -1$

17. $\frac{1}{2}, -\frac{\sqrt{3}}{2}, -\frac{1}{\sqrt{3}}$ **19.** $-\frac{1}{\sqrt{2}}, \frac{1}{\sqrt{2}}, -1$ **21.** $-\frac{\sqrt{3}}{2}, \frac{1}{2}, -\sqrt{3}$ **23.** $-\frac{1}{2}, -\frac{\sqrt{3}}{2}, \frac{1}{\sqrt{3}}$

25. $-\frac{1}{2}, -\frac{\sqrt{3}}{2}, \frac{1}{\sqrt{3}}$ **27.** $\frac{1}{\sqrt{2}}, \frac{1}{\sqrt{2}}, 1$ **29.** α: $\frac{5}{13}, \frac{12}{13}, \frac{5}{12}$; β: $\frac{12}{13}, \frac{5}{13}, \frac{12}{5}$ **31.** α: $\frac{7}{25}, \frac{24}{25}, \frac{7}{24}$; β: $\frac{24}{25}, \frac{7}{25}, \frac{24}{7}$

33. α: $\sqrt{\frac{2}{11}}, \frac{3}{\sqrt{11}}, \frac{\sqrt{2}}{3}$; β: $\frac{3}{\sqrt{11}}, \sqrt{\frac{2}{11}}, \frac{3}{\sqrt{2}}$ **35.** α: $\frac{2}{\sqrt{7}}, \sqrt{\frac{3}{7}}, \frac{2}{\sqrt{3}}$; β: $\sqrt{\frac{3}{7}}, \frac{2}{\sqrt{7}}, \frac{\sqrt{3}}{2}$

37. α: $\frac{\sqrt{7}}{4}, \frac{3}{4}, \frac{\sqrt{7}}{3}$; β: $\frac{3}{4}, \frac{\sqrt{7}}{4}, \frac{3}{\sqrt{7}}$ **39.** $\cos\theta$ **41.** $\cot\theta$ **43.** $\sec\theta$ **45.** $\sin\frac{\pi}{14}$ **47.** $\cos\frac{\pi}{8}$

49. $\cot 10°$ **51.** $\tan 40°$ **53.** $\csc\frac{\pi}{24}$ **55.** $a = \frac{3}{\sqrt{8}}, c = \frac{9}{\sqrt{8}}$ **57.** $b = \frac{8}{5}, c = \frac{8\sqrt{26}}{5}$

59. $b = \frac{2}{\sqrt{3}}, c = \frac{4}{\sqrt{3}}$ **61.** $a = \frac{1}{8}, b = \frac{\sqrt{15}}{8}$

Exercises 6.4, page 215

Values of the trigonometric functions are listed in the order sine, cosine, and tangent. The cosecant, secant, and cotangent are the reciprocals of these, respectively, when defined.

1. $\frac{1}{2}, \frac{\sqrt{3}}{2}, \frac{1}{\sqrt{3}}$ **3.** $-\frac{1}{\sqrt{2}}, -\frac{1}{\sqrt{2}}, 1$ **5.** 0, −1, 0 **7.** $-\frac{\sqrt{3}}{2}, \frac{1}{2}, -\sqrt{3}$ **9.** −1, 0, *

11. +, −, − **13.** −, +, − **15.** +, −, − **17.** +, −, − **19.** −, +, −

21. $-\frac{2}{5}, -\frac{\sqrt{21}}{5}, \frac{2}{\sqrt{21}}$ **23.** $-\frac{1}{\sqrt{2}}, \frac{1}{\sqrt{2}}, -1$ **25.** $-\frac{1}{\sqrt{2}}, -\frac{1}{\sqrt{2}}, 1$ **27.** 0, −1, 0

29. $\frac{1}{2}, \frac{\sqrt{3}}{2}, \frac{1}{\sqrt{3}}$ **31.** $-\frac{1}{2}, \frac{\sqrt{3}}{2}, -\frac{1}{\sqrt{3}}$ **33.** 0, 1, 0 **35.** $\frac{\sqrt{3}}{2}, -\frac{1}{2}, -\sqrt{3}$ **37.** $-\frac{\sqrt{2}}{2}, \frac{\sqrt{2}}{2}, -1$

39. $\frac{3}{\sqrt{13}}, \frac{2}{\sqrt{13}}, \frac{3}{2}$ **41.** $-\frac{\sqrt{15}}{4}, \frac{1}{4}, -\sqrt{15}$ **43.** $\frac{2\pi}{3}$

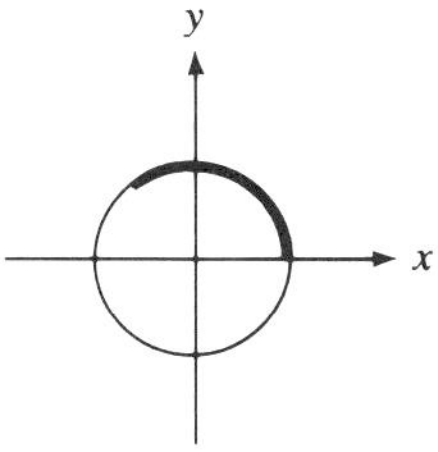

45. $\frac{-\pi}{4}$

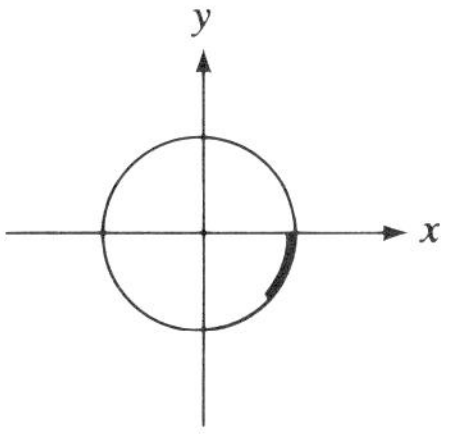

47. $\frac{\pi}{6}$

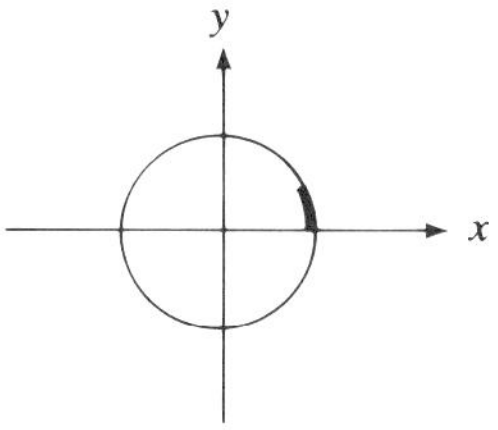

49. $\frac{5\pi}{4}$

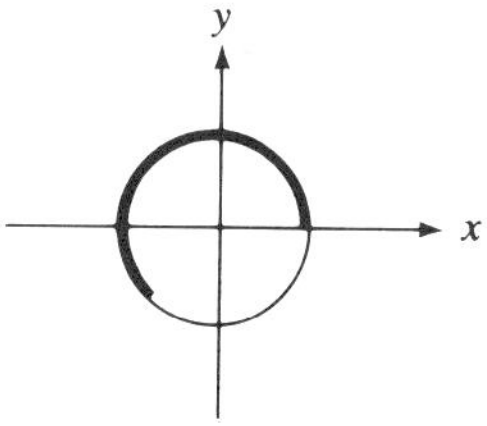

51. $T = \frac{12}{5}$ sec, $f = \frac{5}{12}$ oscillations/sec **53.** 0.9605, 0.2783, 3.4513 **55.** −0.2345, −0.9721, 0.2412
57. −0.9213, −0.3888, 2.3695

Exercise 6.5, page 223

Values of the trigonometric functions are listed in the order sine, cosine, tangent, cosecant, secant, and cotangent.

1. 75° **3.** 80° **5.** $\frac{3\pi}{8}$ **7.** 80° **9.** $\frac{\pi}{4}$ **11.** $\pi - 2$ **13.** $\frac{\pi}{12}$ **15.** $\pi^2 - 3\pi$

17. $4 - \pi$ **19.** $\frac{\pi}{3}$ **21.** 0.6428, −0.6428, −0.6428 **23.** −0.2125, 0.2125, −0.2125

25. $-\frac{5\pi}{4}, -\frac{\pi}{4}, \frac{3\pi}{4}, \frac{5\pi}{4}$ **27.** $-\frac{11\pi}{6}, -\frac{7\pi}{6}, \frac{\pi}{6}, \frac{5\pi}{6}$ **29.** $-\frac{5\pi}{4}, -\frac{3\pi}{4}, \frac{3\pi}{4}, \frac{5\pi}{4}$
31. 0.2644, 0.9644, 0.2742, 3.782, 1.037, 3.647 **33.** 0.7679, 0.6406, 1.199, 1.302, 1.561, 0.8342
35. 0.7951, 0.6065, 1.311, 1.258, 1.649, 0.7627 **37.** 30°40′ **39.** 76°40′ **41.** 53°50′
43. −0.9377, 0.3475, −2.699, −1.066, 2.878, −0.3706
45. −0.7451, −0.6670, 1.117, −1.342, −1.499, 0.8952
47. 0.9659, −0.2588, −3.732, 1.035, −3.864, −0.2679 **49.** 0.4348 **51.** 23°4.8′
53. $b = 4.45$, $c = 6.69$, $A = 11.13$ **55.** $a = 2.22$, $b = 9.75$, $A = 10.82$
57. $a = 5.45$, $c = 6.22$, $A = 8.18$ **59.** 0.7444
61. 0.8770 (interpolation), 0.8770064 (calculator)
63. −3.381 (interpolation), −3.380515 (calculator) **65.** $a = 7.75$, $b = 8.43$, $A = 32.67$

Exercises 6.6, page 230

1.

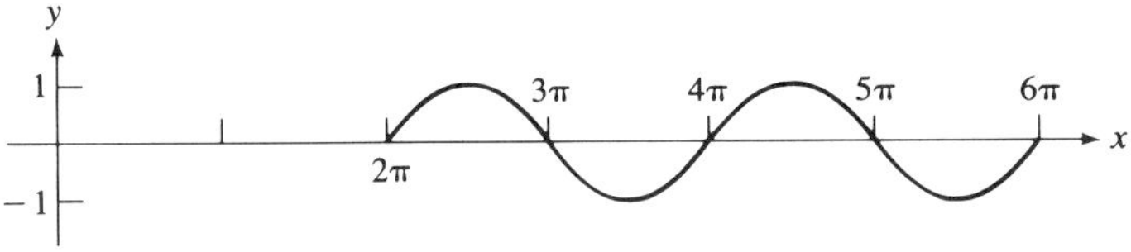

3.

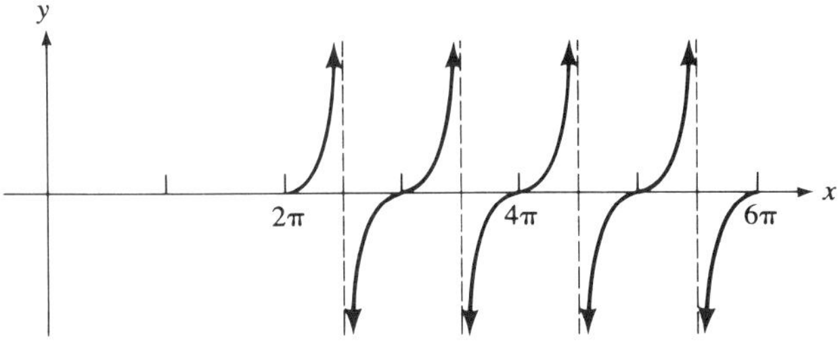

5.

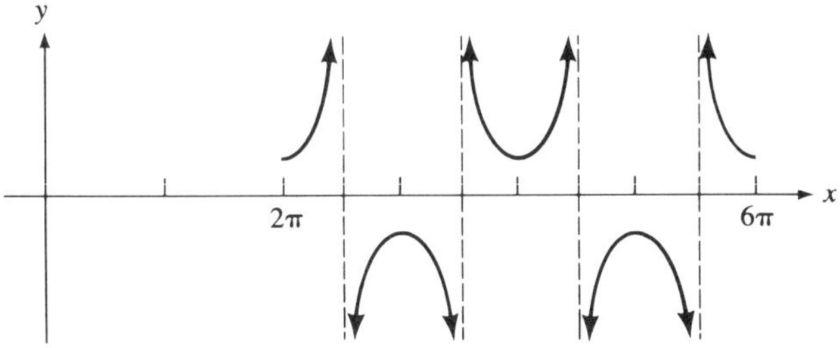

7.

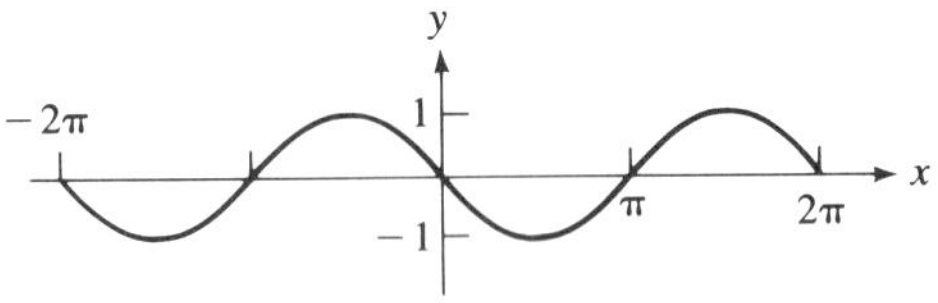

9.

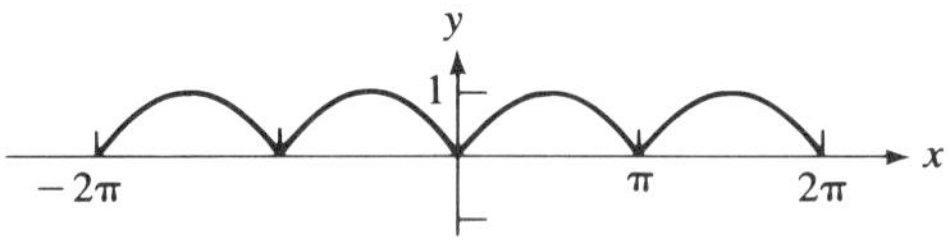

11.

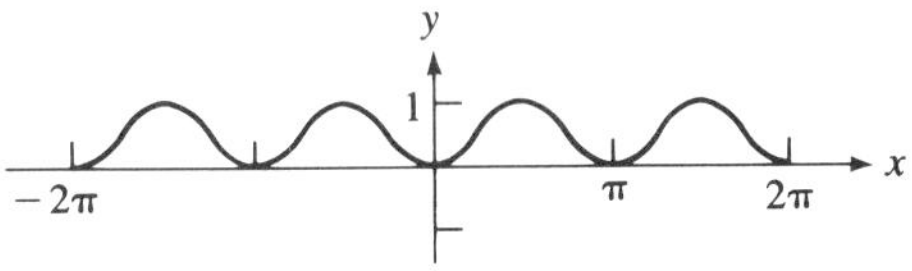

13.

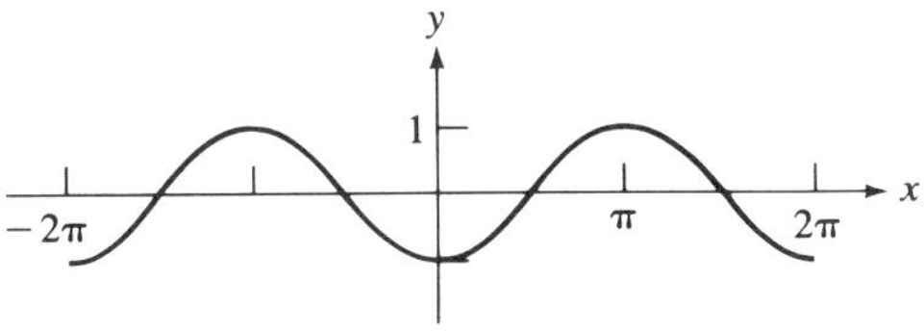

15.

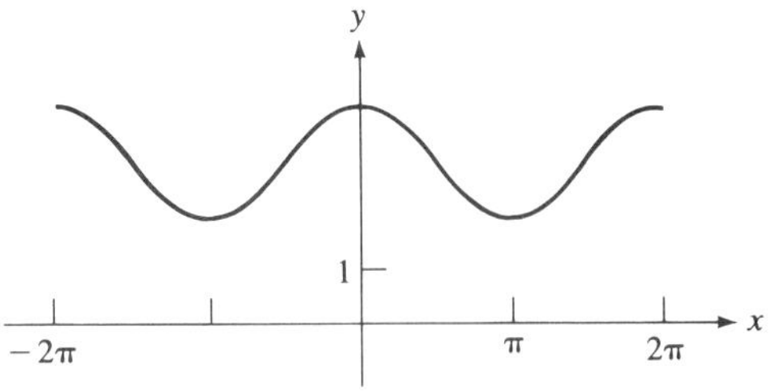

17.

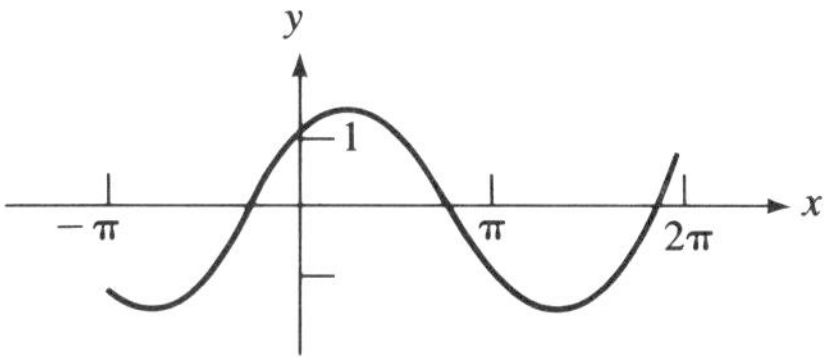

19. $\left[-\pi, -\frac{3\pi}{4}\right), \left(\frac{\pi}{4}, \frac{5\pi}{4}\right)$ **21.** $\left[-\frac{\pi}{4}, -\frac{\pi}{2}\right), \left[\frac{\pi}{4}, \frac{\pi}{2}\right), \left[\frac{5\pi}{4}, \frac{3\pi}{2}\right)$

23. $\left[-\pi, -\frac{3\pi}{4}\right], \left[-\frac{\pi}{4}, \frac{\pi}{4}\right], \left[\frac{3\pi}{4}, \frac{5\pi}{4}\right], \left[\frac{7\pi}{4}, 2\pi\right]$

25. $\left(-\frac{3\pi}{2}, -\frac{\pi}{2}\right), \left(\frac{\pi}{2}, \frac{3\pi}{2}\right)$ **27.** $\left(-2\pi, -\frac{3\pi}{2}\right), \left(-\frac{\pi}{2}, \frac{\pi}{2}\right), \left(\frac{3\pi}{2}, 2\pi\right)$

29. $\left(-\pi, -\frac{\pi}{2}\right), \left(-\frac{\pi}{2}, 0\right), \left(\pi, \frac{3\pi}{2}\right), \left(\frac{3\pi}{2}, 2\pi\right)$ **31.** $-\sin 175$ **33.** $-\cot \frac{78\pi}{3}$

35. a. 92.6, 172.5, 1,255.76, 3,374.6, 10,381.3, 21,585.5, 158,040.4
b. −52.1, −70.4, −108.6, −4,909.8, −9,645.7, −13,573.5, −73,136.5

Exercises 6.7, page 236

1. 1, 2π, $\frac{\pi}{2}$ units to right **3.** $\frac{1}{2}$, $\frac{2\pi}{3}$, $\frac{\pi}{18}$ units to right **5.** π, 1, $\frac{1}{2}$ units to left

7. $\frac{1}{3}$, 6, $\frac{9}{\pi}$ units to left

9.

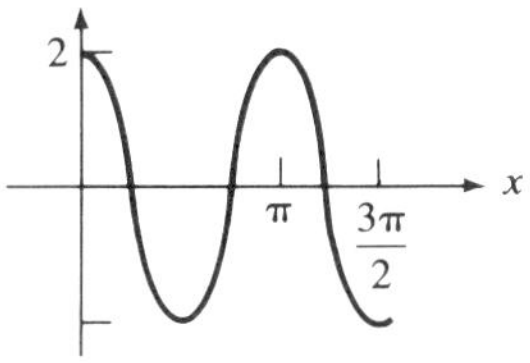

11.

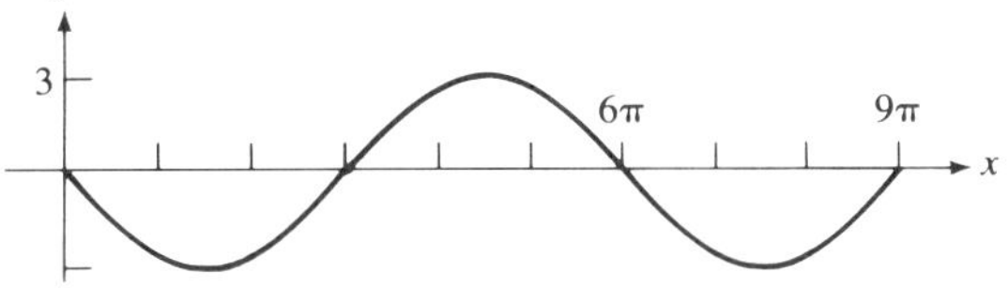

13.

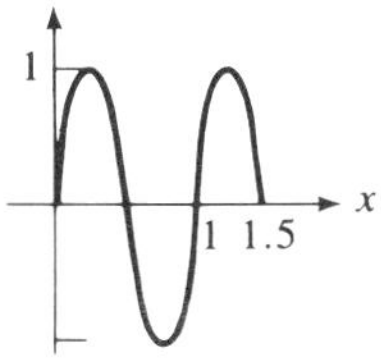

15.

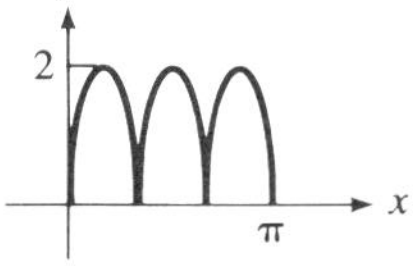

17.

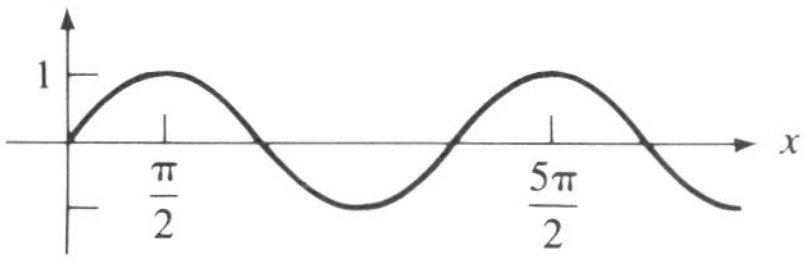

19.

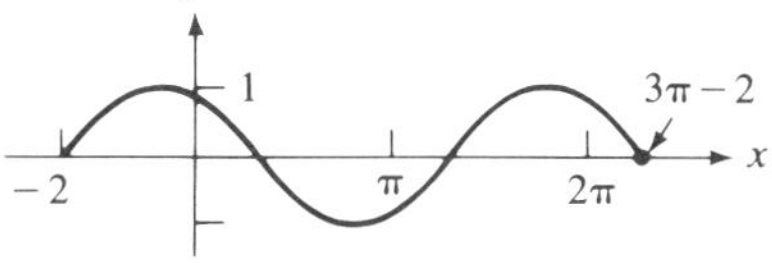

21.

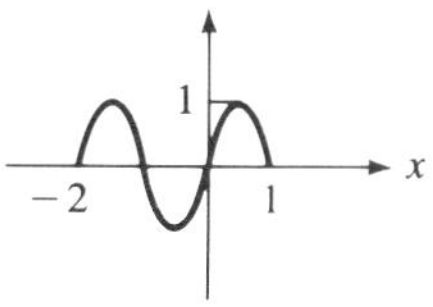

23.

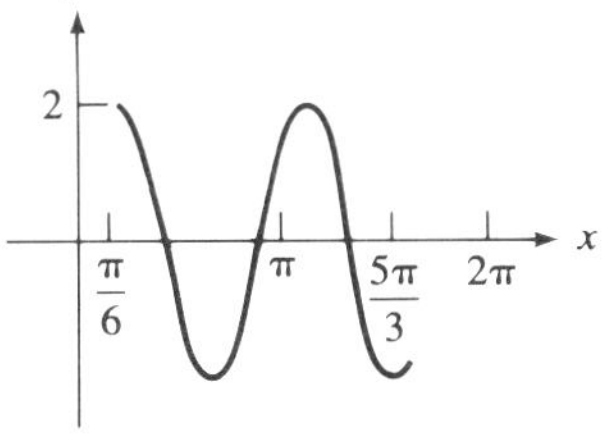

25.

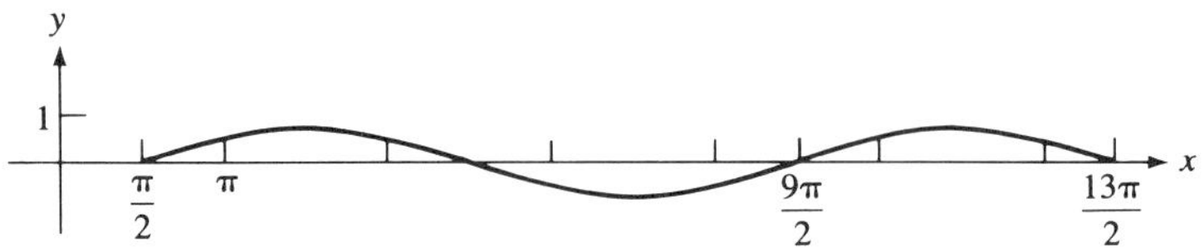

27.

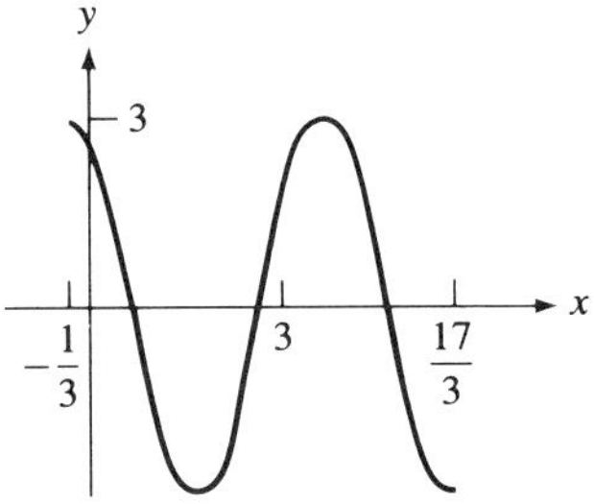

29.

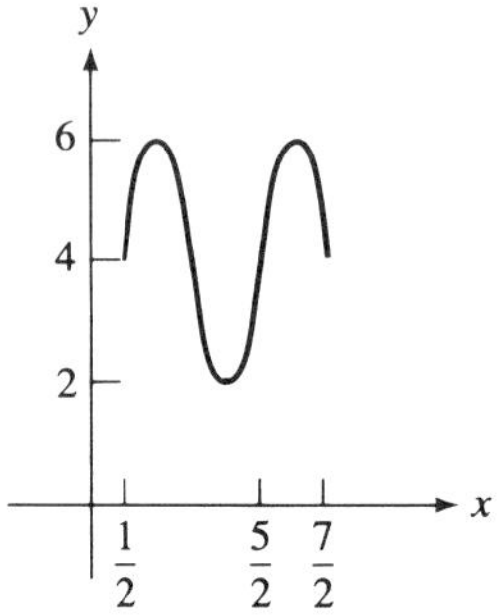

31.

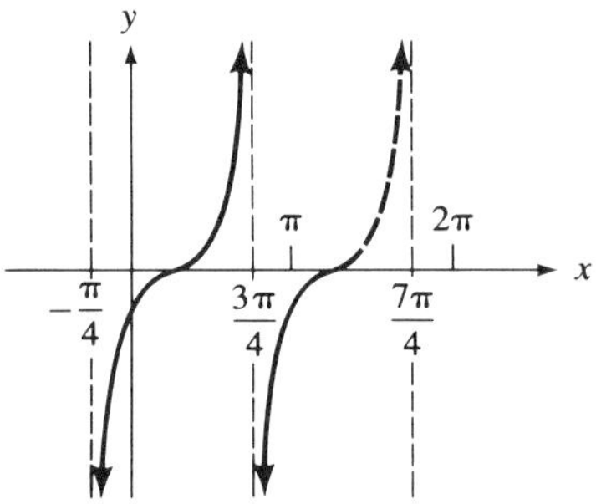

33.

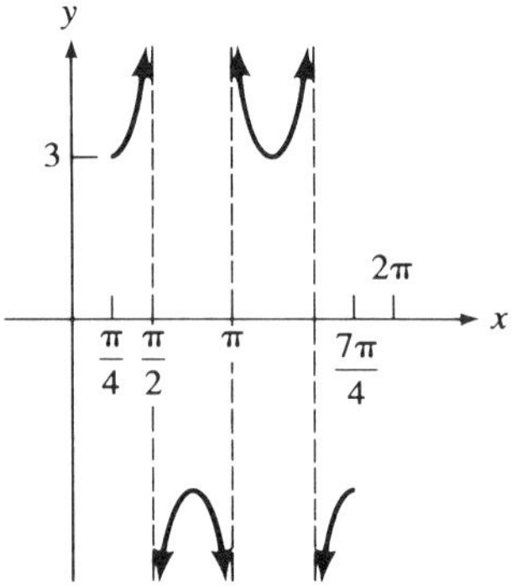

35.

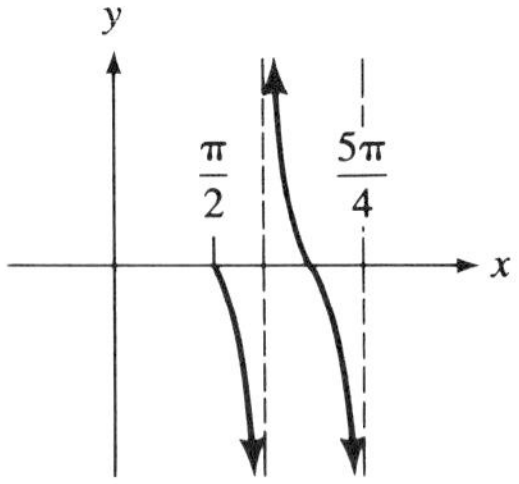

37.

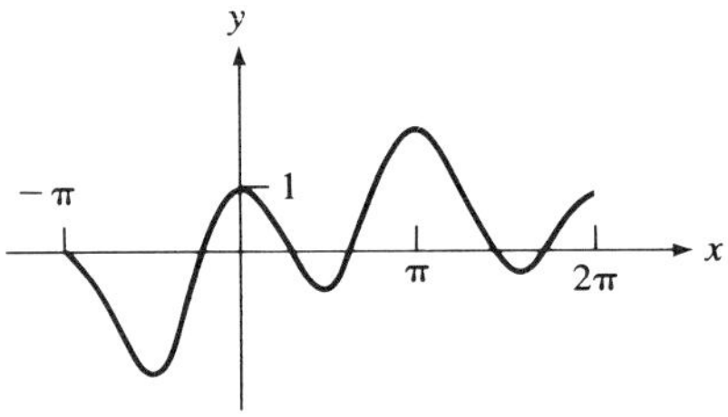

39.

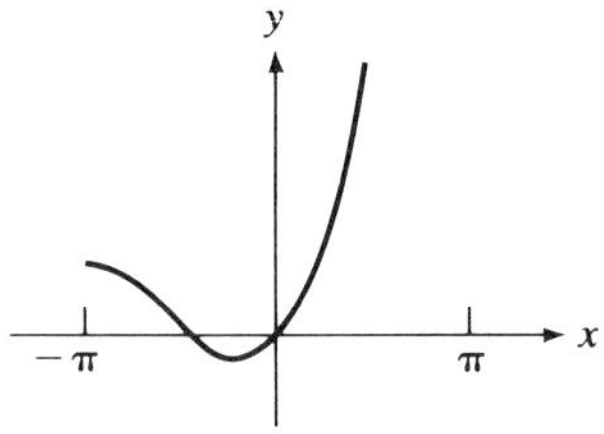

41. The function values should be plotted at the points $x = \dfrac{\pi(n-2)}{12}$, $n = 0, 1, 2, \ldots, 12$.

Exercises 6.8, page 242

1. $\dfrac{\pi}{3}$ **3.** π **5.** $-\dfrac{\pi}{3}$ **7.** $\dfrac{2\pi}{3}$ **9.** 0 **11.** $-\dfrac{\pi}{6}$ **13.** $\dfrac{\pi}{2}$ **15.** $-\dfrac{\pi}{6}$ **17.** $-\dfrac{\pi}{4}$

19. 0.5003 **21.** 2.537 **23.** −1.437 **25.** 2.379 **27.** $\dfrac{\pi}{4}$ **29.** $\dfrac{\pi}{6}$ **31.** 3 **33.** $\dfrac{12}{5}$

35. $6 - 2\pi$ **37.** $-\dfrac{7\pi}{12}$ **39.** $\sqrt{1 - x^2}$ **41.** $\dfrac{\sqrt{x^2 + 4x + 20}}{x + 2}$ **43.** $\dfrac{\sqrt{1 - x^2}}{x}$

45.

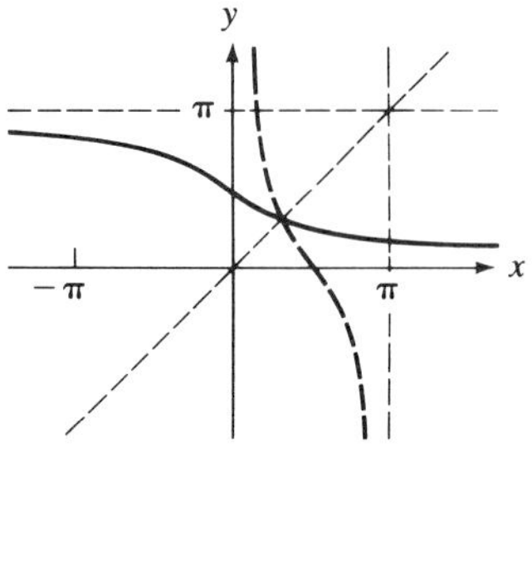

In Exercises 47–53 degree mode is listed first.

47. 0.5687, 0.5687 **49.** 3°, 0.1416; arcsin $x \,\varepsilon \left[-\frac{\pi}{2}, \frac{\pi}{2}\right]$ and $3 > \frac{\pi}{2}$

51. 2°, −1.142; arctan $x \,\varepsilon \left(-\frac{\pi}{2}, \frac{\pi}{2}\right)$ **53.** 100°, 0.5310; range restriction

Chapter Review, page 244

Values of the trigonometric functions are listed in the order sine, cosine, and tangent. The cosecant, secant, and cotangent are the reciprocals of these, respectively, when defined.

1.

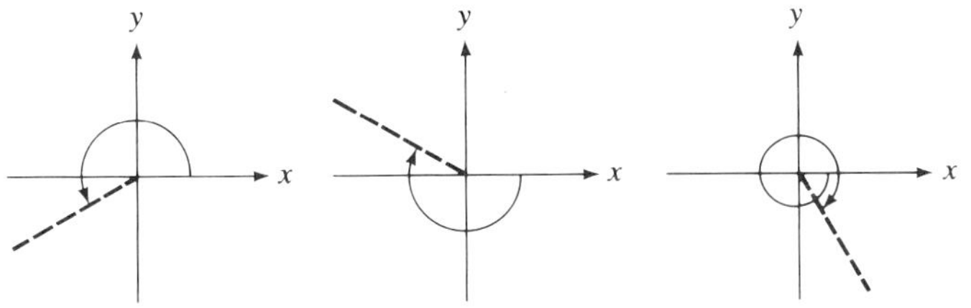

2.

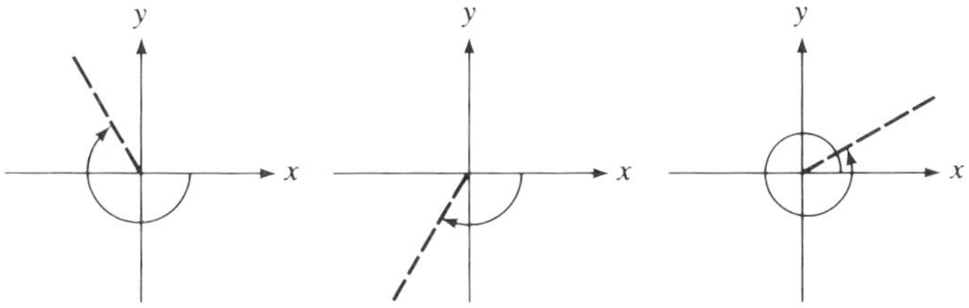

3. $\frac{2\pi}{15}$ **4.** $-\frac{7\pi}{12}$ **5.** −330° **6.** 225° **7.** 31°10′12″ **8.** 83°31′48″ **9.** 0.7224

10. 0.2031 **11.** $\frac{9}{\sqrt{82}}, -\frac{1}{\sqrt{82}}, -9$ **12.** $-\frac{1}{\sqrt{10}}, -\frac{3}{\sqrt{10}}, \frac{1}{3}$ **13.** $-\frac{4}{5}, \frac{3}{5}, -\frac{4}{3}$

14. $-\frac{2}{\sqrt{13}}, -\frac{3}{\sqrt{13}}, \frac{2}{3}$ **15.** $-\frac{5}{\sqrt{61}}, -\frac{6}{\sqrt{61}}, \frac{5}{6}$ **16.** $-\frac{2}{\sqrt{7}}, \frac{\sqrt{3}}{\sqrt{7}}, -\frac{2}{\sqrt{3}}$ **17.** $\frac{10}{11}, -\frac{\sqrt{21}}{11}, -\frac{10}{\sqrt{21}}$

18. $-\frac{24}{25}, \frac{7}{25}, -\frac{24}{7}$ **19.** $\left(\frac{15}{\sqrt{61}}, -\frac{18}{\sqrt{61}}\right)$ **20.** $\left(-\frac{18}{41}, -\frac{80}{41}\right)$ **21.** $a = 2, b = 2\sqrt{3}$

22. $b = 2\sqrt{3}, c = 4\sqrt{3}$ **23.** $a = 4, c = 4\sqrt{2}$ **24.** $a = \sqrt{2}, b = \sqrt{2}$ **25.** $b = 30, c = 3\sqrt{101}$

26. $a = \frac{5}{2}, b = \frac{5\sqrt{3}}{2}$

27. α: $\frac{40}{41}, \frac{9}{41}, \frac{40}{9}$
β: $\frac{9}{41}, \frac{40}{41}, \frac{9}{40}$

28. α: $\frac{5}{\sqrt{61}}, \frac{6}{\sqrt{61}}, \frac{5}{6}$
β: $\frac{6}{\sqrt{61}}, \frac{5}{\sqrt{61}}, \frac{6}{5}$

29. −135°: $-\frac{1}{\sqrt{2}}, -\frac{1}{\sqrt{2}}, 1$
$\frac{5\pi}{3}$: $-\frac{\sqrt{3}}{2}, \frac{1}{2}, -\sqrt{3}$
-3π: 0, −1, 0

30. −150°: $-\frac{1}{2}, -\frac{\sqrt{3}}{2}, \frac{1}{\sqrt{3}}$
$\frac{7\pi}{4}$: $-\frac{1}{\sqrt{2}}, \frac{1}{\sqrt{2}}, -1$
$-\frac{5\pi}{2}$: −1, 0, *

31. $\cot \frac{\pi}{20}$ **32.** $\sin 19°41'25''$ **33.** $-\frac{\sqrt{3}}{2}, -\frac{1}{2}, \sqrt{3}$ **34.** $\frac{1}{\sqrt{2}}, -\frac{1}{\sqrt{2}}, -1$ **35.** $-1, 0,$ *

36. $0, -1, 0$ **37.** $\frac{\sqrt{65}}{9}, -\frac{4}{9}, -\frac{\sqrt{65}}{4}$ **38.** $\frac{4}{5}, -\frac{3}{5}, -\frac{4}{3}$ **39.** $-\frac{\sqrt{3}}{2}, -\frac{1}{2}, \sqrt{3}$

40. $-\frac{3}{\sqrt{13}}, \frac{2}{\sqrt{13}}, -\frac{3}{2}$ **41.** $-\frac{\sqrt{15}}{4}, \frac{1}{4}, -\sqrt{15}$ **42.** $\frac{\sqrt{3}}{2}, -\frac{1}{2}, -\sqrt{3}$ **43.** $-\frac{\pi}{6}$ **44.** $\frac{3\pi}{4}$

45. $\frac{3\pi}{8}, 10°, 2\pi - 6$ **46.** $\frac{\pi}{6}, 80°, 1$ **47.** $51°20'$: 0.7808, 0.6248, 1.250
$-125°30'$: −0.8141, −0.5807, 1.402

48. $170°40'$: 0.1622, −0.9868, −0.1644
$-190°10'$: 0.1765, −0.9843, −0.1793

49. $\frac{11\pi}{12}$: 0.2588, −0.9659, −0.2679
$-\frac{19\pi}{15}$: 0.7431, −0.6691, −1.111

50. $\frac{5\pi}{9}$: 0.9848, −0.1736, −5.671
$-\frac{4\pi}{5}$: −0.5878, −0.8090, 0.7265

51. 0.5184, 0.8551, 0.6064 **52.** 0.7939, 0.6081, 1.305

53. $47°40'$ **54.** $56°20'$ **55.** $55°45'$ **56.** $31°43'$

57.

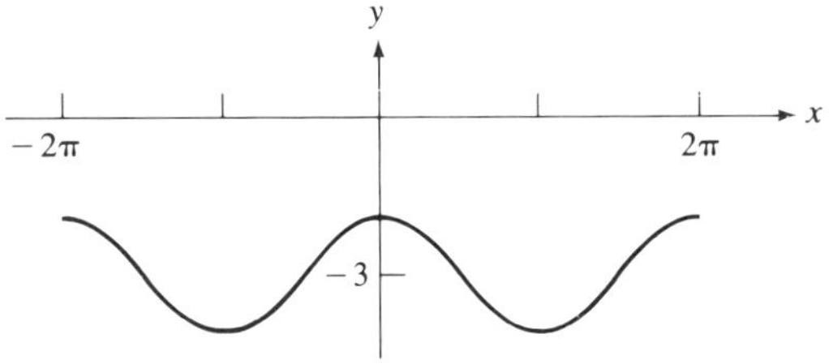

58.

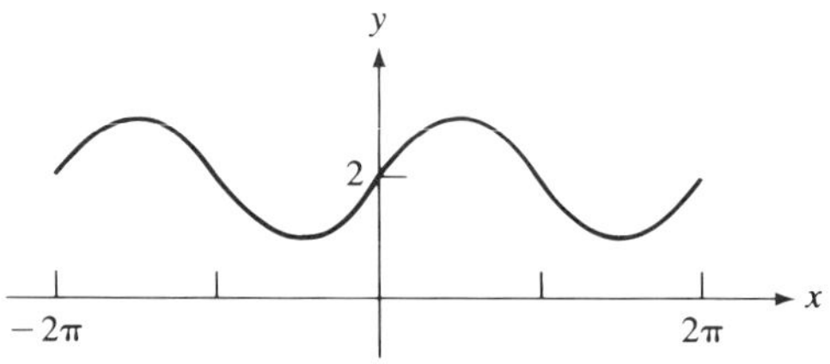

59.

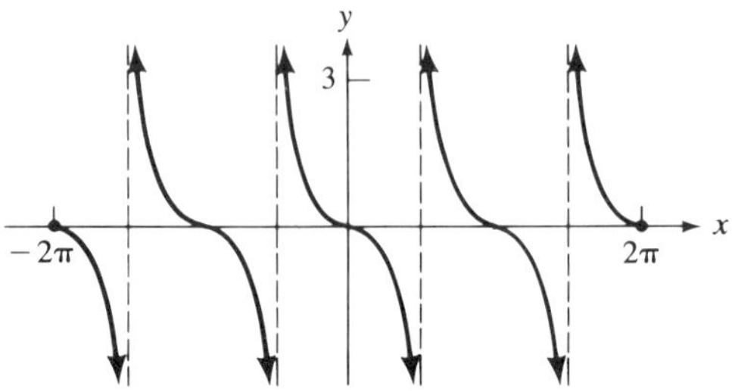

60.

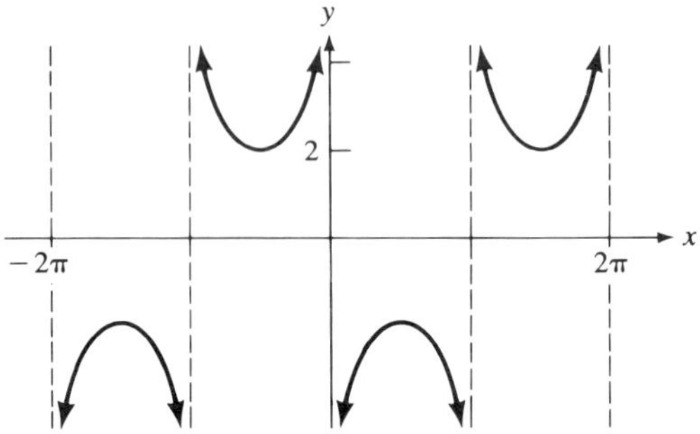

61. $\left(\frac{\pi}{4}, \frac{\pi}{2}\right), \left(\frac{9\pi}{4}, \frac{5\pi}{2}\right)$ **62.** $\left(\frac{\pi}{4}, \pi\right), \left(\frac{9\pi}{4}, 3\pi\right)$

63. amplitude = 2, period = 4π, shift = $\frac{2\pi}{3}$ right

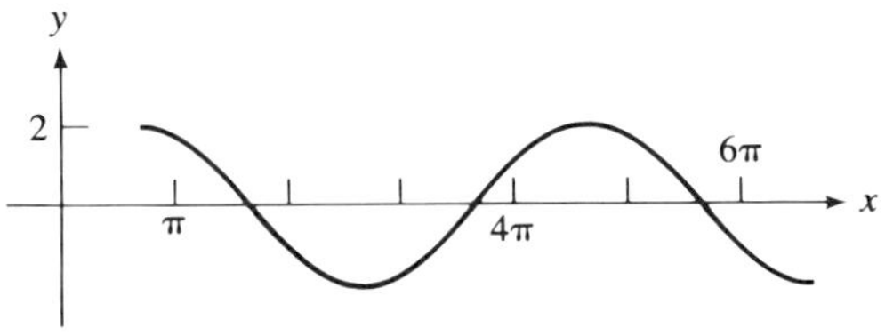

64. amplitude = 2, period = π, shift = $\frac{\pi}{2}$ right

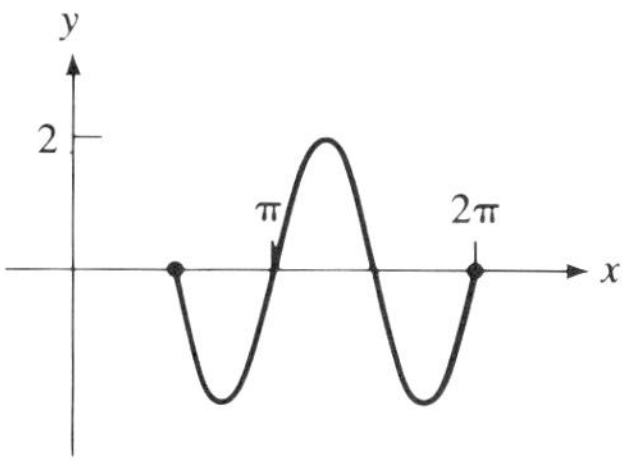

65. period = 2, shift = 2 right

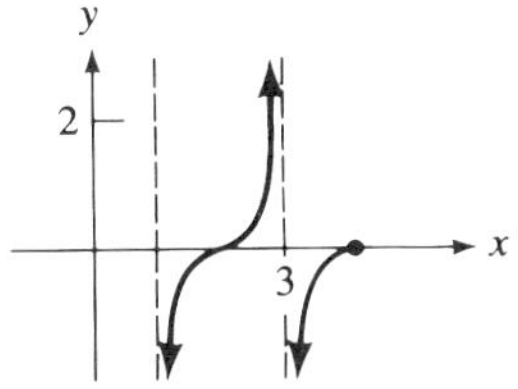

66. period = 4, shift = 1 right

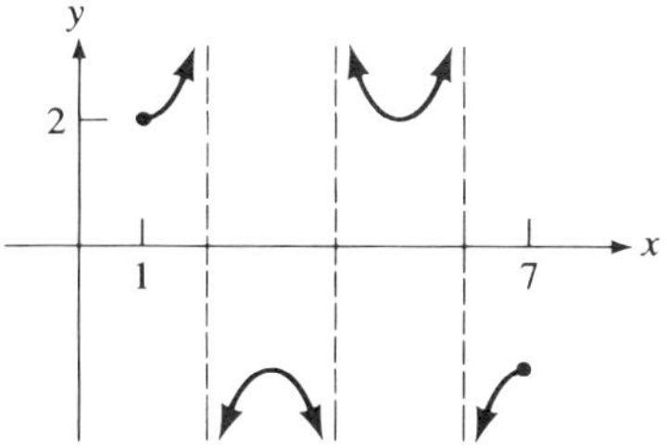

67.

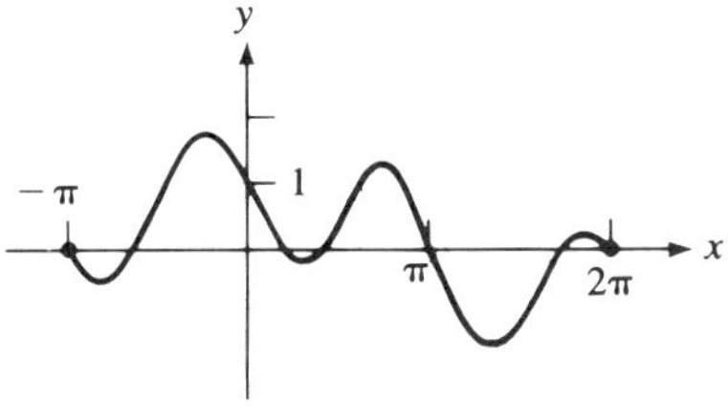

68.

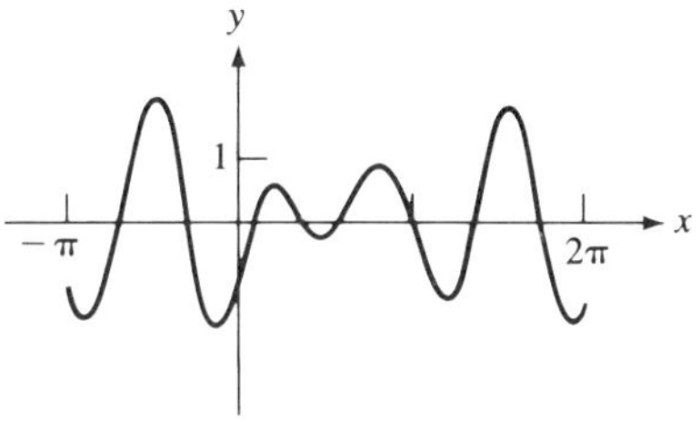

69. $-\frac{\pi}{6}$ **70.** $\frac{11\pi}{12}$ **71.** $\frac{2\pi}{3}$ **72.** -1 **73.** -1.1869 **74.** -0.9076

75. $\frac{x-2}{\sqrt{x^2-4x+20}}$ **76.** $\frac{3x}{2}$ **77.** $\theta = \frac{13}{3}$ radians, $A = 487.5\text{ cm}^2$ **78.** 3.2π cm

Chapter Test, page 247

Values of trigonometric functions are listed in the order sine, cosine, tangent, cosecant, secant, and cotangent.

1. $\frac{16\pi}{5}$ **2.** $-480°$ **3.** $-\frac{5}{13}, -\frac{12}{13}, \frac{5}{12}, -\frac{13}{5}, -\frac{13}{12}, \frac{12}{5}$ **4.** $-\frac{4}{5}, \frac{3}{5}, -\frac{4}{3}, -\frac{5}{4}, \frac{5}{3}, -\frac{3}{4}$

5. $a = 2$, $b = \sqrt{2}$, $\sin\alpha = \frac{2}{\sqrt{6}}$, $\tan\beta = \frac{\sqrt{2}}{2}$, $\sec\beta = \frac{\sqrt{6}}{2}$ **6.** $\frac{1}{\sqrt{2}}, -\frac{1}{\sqrt{2}}, -1, \sqrt{2}, -\sqrt{2}, -1$

7. $-\frac{1}{2}, -\frac{\sqrt{3}}{2}, \frac{1}{\sqrt{3}}, -2, -\frac{2}{\sqrt{3}}, \sqrt{3}$ **8.** $-\frac{\sqrt{3}}{2}, \frac{1}{2}, -\sqrt{3}, -\frac{2}{\sqrt{3}}, 2, -\frac{1}{\sqrt{3}}$ **9.** $-\frac{5\pi}{6}$

10. $-0.9805, -0.1965, 4.989$ **11.** $0.2588, -0.9659, -0.2679$ **12.** $100°40'$ **13.** $-88°40'$

14. $\frac{\pi}{2}$ **15.** $\frac{\pi}{3}$ **16.** $\frac{5}{\sqrt{4x^2+25}}$ **17.** $s = \frac{2\pi}{3}$ cm, $A = \frac{10\pi}{3}\text{ cm}^2$

18.

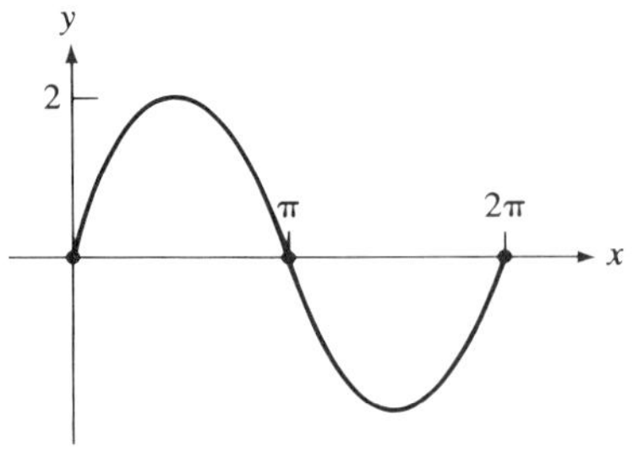

19.

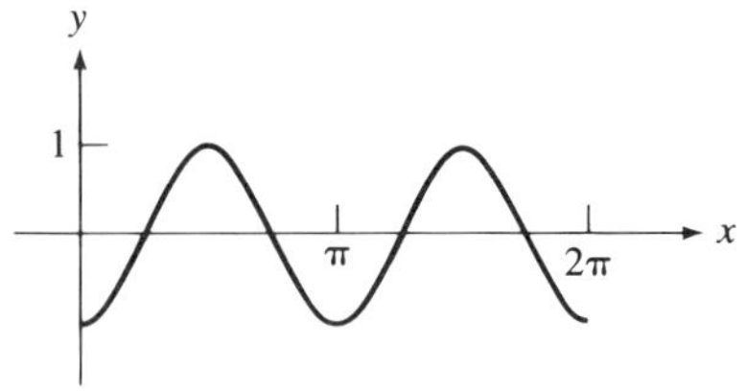

Exercises 7.1, page 252

In Exercises 1–7 one counterexample is given. There are many others.

1. $A = 9, B = 16$ **3.** $x = \frac{\pi}{2}$ **5.** $x = 0$ **7.** $x = \pi$ **27.** $x = \frac{3\pi}{2} + 2k\pi, k \varepsilon Z$

29. $x = \frac{\pi}{4} + \frac{k\pi}{2}, k \varepsilon Z$ **31.** $x = \frac{\pi}{6} + 2k\pi, k \varepsilon Z$ or $x = \frac{5\pi}{6} + 2k\pi, k \varepsilon Z$ **33.** $x = 2 - \sqrt{3}$

35. $\frac{\pi}{2}, \frac{3\pi}{2}$ **37.** $\frac{\pi}{4}, \frac{3\pi}{4}, \frac{5\pi}{4}, \frac{7\pi}{4}$ **39.** $\frac{\pi}{8}, \frac{5\pi}{8}, \frac{9\pi}{8}, \frac{13\pi}{8}$ **41.** $0, \frac{\pi}{3}, \frac{5\pi}{3}, 2\pi$ **43.** $0, \frac{\pi}{4}, \pi, \frac{5\pi}{4}, 2\pi$

45. $\frac{7\pi}{6}, \frac{11\pi}{6}$ **47.** $\frac{\pi}{3}, \frac{\pi}{2}, \frac{5\pi}{3}$ **49.** $x = \arcsin\left(\frac{-1 + \sqrt{5}}{2}\right)$ **51.** 1.4009, 0.1699

53. Each side is approximately 1.5708. **55.** 1 **57.** Each side is approximately 1.5399.

59. 0.3907 (left), 0.2372 (right)

Exercises 7.2, page 258

27. $\tan x = 3$, $\sec x = -\sqrt{10}$, $\cos x = -\frac{1}{\sqrt{10}}$, $\sin x = -\frac{3}{\sqrt{10}}$, $\csc x = -\frac{\sqrt{10}}{3}$, $\cot x = \frac{1}{3}$ **29.** $0, \pi, 2\pi$

31. $\frac{\pi}{4}, \frac{\pi}{2}, \frac{5\pi}{4}, \frac{3\pi}{2}$ **33.** $\frac{\pi}{2}, \pi$ **35.** $\frac{\pi}{4}$ **37.** $\frac{\pi}{4}, \frac{3\pi}{4}, \frac{5\pi}{4}, \frac{7\pi}{4}$ **39.** $0, \frac{\pi}{6}, \frac{5\pi}{6}, \pi, \frac{7\pi}{6}, \frac{11\pi}{6}, 2\pi$

41. π **43.** $0, 2\pi$ **45.** $\frac{\pi}{2}, \frac{3\pi}{2}$ **47.** $0, \frac{\pi}{3}, \pi, \frac{5\pi}{3}, 2\pi$ **49.** $\frac{\pi}{6}, \frac{5\pi}{6}$ **51.** $0, \frac{\pi}{2}, \frac{3\pi}{2}, 2\pi$

53. $\arcsin\left(\frac{-1 + \sqrt{5}}{2}\right)$ **55.** $3 \cos \theta$ **57.** $6 \sec^3 \theta \quad \cot^2 \theta$

Exercises 7.3, page 264

1. No. Take $x = \frac{\pi}{2}, y = -\frac{\pi}{2}$. **3.** No. Take $x = \pi$. **5.** $\frac{\sqrt{2}}{4}(\sqrt{3} - 1)$ **7.** $\frac{\sqrt{3} - 1}{1 + \sqrt{3}}$

9. $\frac{1 - \sqrt{3}}{1 + \sqrt{3}}$ **11.** $\frac{\sqrt{2}}{4}(\sqrt{3} - 1)$ **13.** $\frac{1}{15}(3\sqrt{6} - 4\sqrt{3})$ **15.** II **17.** $-\frac{323}{36}$

19.
$$\begin{aligned} \sin(x - y) = \sin[x + (-y)] &= \sin x \cos(-y) + \cos x \sin(-y) \quad \text{(using Example 1)} \\ &= \sin x \cos y + \cos x(-\sin y) \\ &= \sin x \cos y - \cos x \sin y \end{aligned}$$

37. $-\cos x$ **39.** $-\sin x$ **41.** 0 **43.** $\frac{\sqrt{3}}{2}$ **45.** $\tan \frac{\pi}{2}$ is undefined **47.** $-\frac{\sqrt{2}}{4}(\sqrt{3} + 1)$

49. If $f(x) = mx + b$ is additive, then

$$b = f(0) = f[x + (-x)] = f(x) + f(-x) = (mx + b) + (-mx + b) = 2b$$

so that $b = 0$. If $b = 0$, then

$$f(x + y) = m(x + y) = mx + my = f(x) + f(y)$$

51. $\cos(x + k\pi) = (-1)^k \cos x$

Exercises 7.4, page 272

1. $\frac{24}{25}$ **3.** $-\frac{24}{7}$ **5.** $-\frac{1}{\sqrt{5}}$ **7.** $\frac{3}{\sqrt{13}}$ **9.** $\frac{3}{2}$ **11.** $-\frac{119}{169}$ **13.** $\frac{1}{2}\sqrt{2 + \sqrt{2}}$

15. $-\frac{1}{2}\sqrt{2 - \sqrt{2}}$ **17.** $-\frac{1}{2}\sqrt{2 - \sqrt{3}}$ **19.** $\sin\theta = \frac{1}{\sqrt{10}}, \cos\theta = \frac{3}{\sqrt{10}}$ **21.** $\sin\theta = \frac{4}{5}, \cos\theta = \frac{3}{5}$

23. 1 **25.** $\frac{24}{7}$ **27.** $\frac{2}{\sqrt{5}}$ **43.** $\frac{\pi}{6}, \frac{5\pi}{6}, \frac{7\pi}{6}, \frac{11\pi}{6}$ **45.** $0, \frac{\pi}{2}, \pi, \frac{3\pi}{2}$ **47.** $\frac{\pi}{6}, \frac{\pi}{2}, \frac{5\pi}{6}, \frac{3\pi}{2}$

49. $\frac{\pi}{8}, \frac{5\pi}{8}, \frac{9\pi}{8}, \frac{13\pi}{8}$ **51.** $0, \frac{\pi}{3}, \frac{\pi}{2}, \frac{2\pi}{3}, \pi, \frac{4\pi}{3}, \frac{3\pi}{2}, \frac{5\pi}{3}$ **53.** 0 **54.** π **55.** $0, \frac{2\pi}{3}, \frac{4\pi}{3}$

57. $\frac{1}{8}(3 + 4\cos 2x + \cos 4x)$

59. $(\sqrt{3} + 1)^2 = 4 + 2\sqrt{3} = 2(2 + \sqrt{3})$, so that $\sqrt{3} + 1 = \sqrt{2(2 + \sqrt{3})} = \sqrt{2}\sqrt{2 + \sqrt{3}}$. Divide both sides of this last equation by $2\sqrt{2}$.

Exercises 7.5, page 276

1. $2\cos 4x \sin 2x$ **3.** $2\sin x \sin\frac{1}{2}x$ **5.** $\frac{1}{2}(\sin 2x + \sin x)$ **7.** $\frac{1}{2}\left(\cos\frac{\pi}{10} - \cos\frac{3\pi}{10}\right)$ **9.** $\frac{1}{2}\sqrt{6}$

11. $-\frac{1}{2}\sqrt{6}$ **13.** $\frac{1}{4}(\sqrt{3} - 2)$

15. Add the two identities of Identity 9 to obtain $\cos(x + y) + \cos(x - y) = 2\cos x \cos y$.

17. Let $x + y = u$, $x - y = v$, and write Identity 19 as

$$2\cos\left(\frac{u + v}{2}\right)\cos\left(\frac{u - v}{2}\right) = \cos u + \cos v$$

29. $\frac{\pi}{6}, \frac{\pi}{2}, \frac{5\pi}{6}$ **31.** $0, \frac{\pi}{2}$ **33.** $0, \frac{\pi}{4}, \frac{3\pi}{4}$ **35.** $\frac{\pi}{4}, \frac{3\pi}{4}$ **37.** $0, \frac{\pi}{12}, \frac{5\pi}{12}, \frac{7\pi}{12}, \frac{11\pi}{12}$

Exercises 7.6, page 282

1. Square the left side and apply Identities 6 and 12.

3. Replace $\tan x$ by $\sin x/\cos x$ on the left side, simplify the fraction, then multiply numerator and denominator by $(\cos x - \sin x)$. For the numerator apply Identities 6 and 12. Use Identity 13a for the denominator.

5. Use Identity 13b twice and collect terms.

7. Write $\sin 4x - \sin 2x = \sin 4x + \sin(-2x)$ and apply Identities 21 and 22 to the numerator and denominator of the left side.

9. Arrange the left side and replace $\sin x + \sin 5x$ by $2\sin 3x \cos 2x$ and $(\cos x + \cos 5x)$ by $2\cos 3x \cos 2x$. Then factor.

11. On the left side use Identity 21 on the numerator and Identity 9 on the denominator.

13. Multiply numerator and denominator on the left by $\sin x$, then replace $\sin x \cos x$ by $\frac{1}{2}\sin 2x$ (Identity 12). Continue in this manner, replacing $\sin 2x \cos 2x$ by $\frac{1}{2}\sin 4x$, then $\sin 4x \cos 4x$ by $\frac{1}{2}\sin 8x$.

15. $3\sin x - 4\sin^3 x$ **17.** $0, \frac{3\pi}{2}$ **19.** $\frac{2\pi}{3}, \frac{5\pi}{3}$ **21.** $\frac{\pi}{3}, \frac{5\pi}{3}$ **23.** $\frac{2\pi}{3}, \frac{4\pi}{3}, \frac{3\pi}{2}$

25. $\frac{\pi}{8}, \frac{\pi}{2}, \frac{5\pi}{8}, \frac{9\pi}{8}, \frac{3\pi}{2}, \frac{13\pi}{8}$ **27.** $\frac{\pi}{2}, \frac{7\pi}{6}, \frac{11\pi}{6}$ **29.** $0, \frac{\pi}{6}, \frac{\pi}{2}, \frac{5\pi}{6}, \pi, \frac{3\pi}{2}$

Chapter Review, page 282

Values of trigonometric functions are listed in the order sine, cosine, tangent, cosecant, secant, and cotangent.

1. $-\frac{2\sqrt{2}}{3}, -\frac{1}{3}, 2\sqrt{2}, -\frac{3}{2\sqrt{2}}, -3, \frac{1}{2\sqrt{2}}$ **2.** $\frac{3}{\sqrt{10}}, -\frac{1}{\sqrt{10}}, -3, \frac{\sqrt{10}}{3}, -\sqrt{10}, -\frac{1}{3}$ **3.** $\frac{\sqrt{2}}{4}(\sqrt{3}+1)$

4. $-\frac{\sqrt{2}}{4}(\sqrt{3}+1)$ **5.** $\frac{3-\sqrt{3}}{3+\sqrt{3}}$ **6.** $\frac{1}{2}\sqrt{2-\sqrt{2}}$ **7.** $\frac{1}{2}\sqrt{2-\sqrt{3}}$ **8.** $\frac{\sqrt{6}}{2}$

9. $\frac{1}{4}(1+\sqrt{3})$ **10.** $-\frac{\sqrt{3}}{2}$ **11.** $\frac{7}{25}$ **12.** $\frac{4}{3}$ **13.** $\frac{\sqrt{2}}{10}$ **14.** $-\frac{\sqrt{3}}{3}$ **15.** -0.8 **16.** $\frac{117}{125}$

17. $\frac{7\sqrt{2}}{10}$ **18.** $\frac{24}{7}$ **19.** $\frac{2}{\sqrt{5}}, \frac{1}{\sqrt{5}}$ **20.** $2\cos 4x \sin x$ **21.** $\frac{1}{2}(\sin 4x - \sin x)$ **22.** $4\cos\theta$

23. $2{,}401\tan\theta\sec^3\theta$ **24.** π **25.** $\frac{\pi}{6}$ **26.** π

27. This is an identity. Replace $\tan\left(\frac{3\pi}{2}+x\right)$ by $\dfrac{\sin\left(\frac{3\pi}{2}+x\right)}{\cos\left(\frac{3\pi}{2}+x\right)}$

and use Identities 10 and 9.

28. This is Identity 24 (see Section 7.4 also).

29. $\frac{\pi}{2}$ **30.** $x = \frac{\pi}{4}, y = \frac{\pi}{4}$

31. Replace $\cot x$ by $\cos x/\sin x$. Combine the fractions and use Identity 2.

32. Expand the left side using Identity 9. Recall that $\cos 3\pi/2 = 0$ and $\sin 3\pi/2 = -1$.

33. Replace $\tan^2 x$ and $\cot^2 x$ using Identities 4 and 5. Simplify and use Identity 6.

34. Combine the fractions on the left, simplify using Identity 6, then cancel the common factor $(1 + \cos x)$.

35. On the right side, replace $\tan x$ by $\sin x/\cos x$ and $\cot x$ by $\cos x/\sin x$, multiply numerator and denominator of the complex fraction by $\cos x \sin x$ and use Identity 6.

36. On the left side, replace $\cos^2 x$ by $1 - \sin^2 x$, factor $1 - \sin^2 x$, reduce the fraction, and collect terms.

37. On the left side, replace $\sin(-x)$ by $-\sin x$, $\sec(-x)$ by $\sec x$, and $\cot(-x)$ by $-\cot x$. Then express all functions in terms of sine and cosine and simplify.

38. Expand both terms on the left using Identity 9 and collect terms.

39. On the left, factor $\csc^2 x$ from the first two terms and -1 from the second two terms. Use Identity 6 to obtain $\csc^2 x - 1$ and then use Identity 8.

40. On the left, factor the numerator, reduce the fraction, and use Identity 6 to obtain $1 - \sin x \cos x$. Replace 1 in this expression by $\cos x/\cos x$, factor out $\cos x$, and use Identity 1.

41. On the left, expand both terms using Identity 11. Replace $\tan \pi/4$ by 1, and combine the two fractions.

42. On the left, use Identity 12, multiply by $\sin x/\sin x$, and replace $\cos x/\sin x$ by $\cot x$.

43. On the left, replace $\tan\frac{1}{2}x$ by $\sin\frac{1}{2}x/\cos\frac{1}{2}x$, simplify the complex fraction, and multiply numerator and denominator by $\left(\cos\frac{1}{2}x - \sin\frac{1}{2}x\right)$. Then use Identity 13a on numerator, Identities 6 and 12 on denominator.

44. On the left, express cot x and tan x in terms of sin x and cos x, combine the fractions, multiply by $\frac{2}{2}$, and use Identities 6 and 12. Follow with Identity 2.
45. On the left, combine the fractions, apply Identity 10 to write numerator as sin $4x$, then apply Identity 12 to obtain a factor of sin $2x$ in both numerator and denominator.
46. Combine $\sin 3x + \sin x$ using Identity 21, then factor sin $2x$ from the left side.
47. On the left, write $\sin x - \sin 3x$ as $\sin x + \sin(-3x)$, then use Identities 23 and 21.
48. On the left, write $\sin 3x - \sin x$ as $\sin 3x + \sin(-x)$, then use Identities 21 and 22.
49. $\pm\sqrt{\frac{17}{2}}$ **50.** $\frac{(2k+1)\pi}{2}$, $2k\pi$, $k \varepsilon Z$ **51.** $(1+4k)\frac{\pi}{8}$, $k \varepsilon Z$ **52.** $\frac{\pi}{6}, \frac{5\pi}{6}, \frac{3\pi}{2}$
53. $\arccos\left(\frac{-1+\sqrt{5}}{2}\right)$ **54.** $\pm\frac{\pi}{6}$ **55.** $\frac{\pi}{3}$ **56.** $0, \arccos\left(-\frac{3}{5}\right)$ **57.** $\frac{\pi}{2}$
58. $0, \frac{\pi}{6}, \frac{\pi}{2}, \frac{5\pi}{6}, \pi, \frac{7\pi}{6}, \frac{3\pi}{2}, \frac{11\pi}{6}$ **59.** $\frac{\pi}{2}, \frac{2\pi}{3}, \frac{4\pi}{3}, \frac{3\pi}{2}$ **60.** $0, \frac{\pi}{3}, \frac{5\pi}{3}$ **61.** $\frac{\pi}{2}, \frac{\pi}{4}, \frac{3\pi}{4}, \frac{5\pi}{4}, \frac{3\pi}{2}, \frac{7\pi}{4}$
62. $0, \frac{\pi}{2}, \pi, \frac{3\pi}{2}$ **63.** $0, \frac{\pi}{2}, \frac{2\pi}{3}, \pi, \frac{4\pi}{3}, \frac{3\pi}{2}$ **64.** $\frac{\pi}{6}, \frac{\pi}{3}, \frac{\pi}{2}, \frac{2\pi}{3}, \frac{5\pi}{6}, \frac{7\pi}{6}, \frac{4\pi}{3}, \frac{3\pi}{2}, \frac{5\pi}{3}, \frac{11\pi}{6}$
65. $\frac{\pi}{4}, \frac{3\pi}{4}, \frac{7\pi}{6}, \frac{11\pi}{6}$ **66.** $\frac{2\pi}{3}, \frac{4\pi}{3}, \frac{3\pi}{2}$

Chapter Test, page 284

1. $-\frac{5}{12}$ **2. a.** $\frac{\sqrt{2}}{4}(\sqrt{3}+1)$ **b.** $\frac{1}{2}\sqrt{2+\sqrt{3}}$ **3.** $\frac{1}{4}$ **4.** $\frac{\sqrt{2}}{2}$ **5.** $-\frac{1}{3}$ **6.** $-\frac{\sqrt{3}+2\sqrt{2}}{6}$
7. $\frac{3}{5}$ **8.** $\frac{\pi}{2}$ is a counterexample. **9.** $x = 0$, $y = \frac{\pi}{2}$ is a counterexample.
10. $\sin\left(x - \frac{3\pi}{2}\right) \overset{10}{=} \sin x \cos\frac{3\pi}{2} - \cos x \sin\frac{3\pi}{2} = (\sin x)(0) - (\cos x)(-1)$
$= \cos x = \cos(-x)$
11. $\frac{\sin 2x}{1+\cos 2x} \overset{12}{\underset{13b}{=}} \frac{2\sin x\cos x}{1+(2\cos^2 x - 1)} = \frac{2\sin x\cos x}{2\cos^2 x} = \frac{\sin x}{\cos x} \overset{4}{=} \tan x$
Or: Identity 17, with x replaced by $2x$.
12. $\left(\sin\frac{1}{2}x - \cos\frac{1}{2}x\right)^2 = \sin^2\frac{1}{2}x - 2\sin\frac{1}{2}x\cos\frac{1}{2}x + \cos^2\frac{1}{2}x$
$\overset{6}{=} 1 - 2\sin\frac{1}{2}x\cos\frac{1}{2}x \overset{12}{=} 1 - \sin\left(2\cdot\frac{1}{2}x\right) = 1 - \sin x$
13. $\frac{\pi}{6}, \frac{5\pi}{6}, \frac{7\pi}{6}, \frac{11\pi}{6}$ **14.** $0, \frac{\pi}{2}, \pi, \frac{3\pi}{2}$ **15.** $\frac{3\pi}{2}, \pi \pm \arccos\frac{1}{4}$

Exercises 8.1, page 290

In this chapter answers will vary somewhat, depending on functions and triangular parts used to obtain solutions.
1. $\beta = 24°$, $a = 38$, $c = 42$ **3.** $\alpha = 70°$, $a = 19$, $b = 6.8$ **5.** $\alpha = 21°$, $\beta = 69°$, $c = 14$
7. $\beta = 57°20'$, $b = 105$, $c = 125$ **9.** $\alpha = 24°17'$, $\beta = 65°43'$, $b = 39.27$ **11.** 8.6 ft
13. 6,823 ft **15.** 22°40′ **17.** 209.5 ft **19.** 26 ft **21.** 292 ft, N55°W **23.** 2.8 mi, 318°

Exercises 8.2, page 295

1. $\alpha = 68°$, $b = 14.9$, $c = 15.2$ **3.** no triangle **5.** $\beta = 83°30'$, $a = 2.95$, $c = 5.92$
7. $\alpha = 138°10'$, $\beta = 13°30'$, $a = 6.80$
9. $\alpha = 3°10'$, $\gamma = 155°30'$, $a = 19.4$, or $\alpha = 134°10'$, $\gamma = 24°30'$, $a = 252$
11. $\alpha = 60°$, $\gamma = 90°$, $a = 16\sqrt{3}$ **13.** $\beta = 10°50'$, $\gamma = 41°50'$, $b = 100.9$ **15.** 16 mi
17. 200 m **19.** 39 ft **21.** 7 mi **23.** 4,300 yards

Exercises 8.3, page 300

1. $\alpha = 48°$, $\gamma = 85°$, $b = 23$ **3.** $\alpha = 35°30'$, $\beta = 53°$, $\gamma = 91°30'$ **5.** $\gamma = 74°$, $b = 36$, $c = 36$
7. $\alpha = 30°$, $\beta = 45°$, $c = 31$ **9.** $\beta = 79°20'$, $\gamma = 59°50'$, $a = 363$ **11.** $\alpha = 59°$, $\beta = 67°$, $\gamma = 54°$
13. 88° **15.** 42° **17.** 0.3 mi **19.** 410 mi **21.** S82°W
23. Notice that $\frac{1}{2}(\alpha + \gamma) = \frac{1}{2}(180° - 47°)$. Compute the ratio $(a - c)/(a + c)$, then use the Law of Tangents to find $\frac{1}{2}(\alpha - \gamma) \simeq -18.8°$. Then $\alpha = \frac{1}{2}(\alpha + \gamma) + \frac{1}{2}(\alpha - \gamma)$ and $\gamma = \frac{1}{2}(\alpha + \gamma) - \frac{1}{2}(\alpha - \gamma)$. Find b using the Law of Sines.

Exercises 8.4, page 307

1.
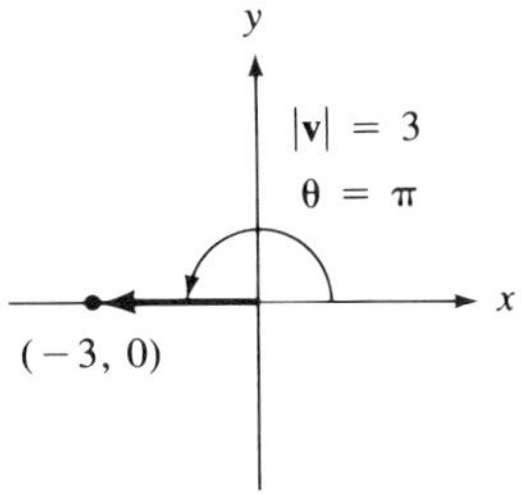

3.
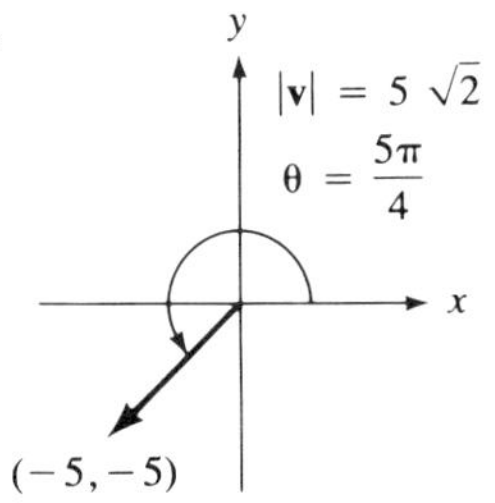

5.
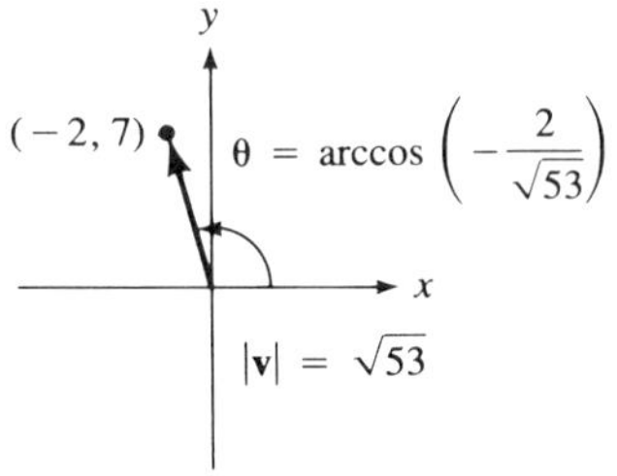

7.
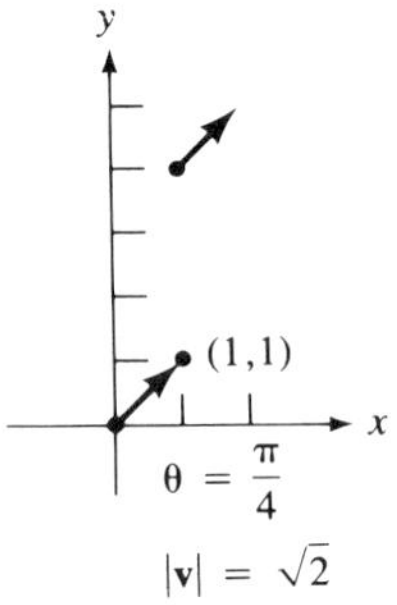

9.
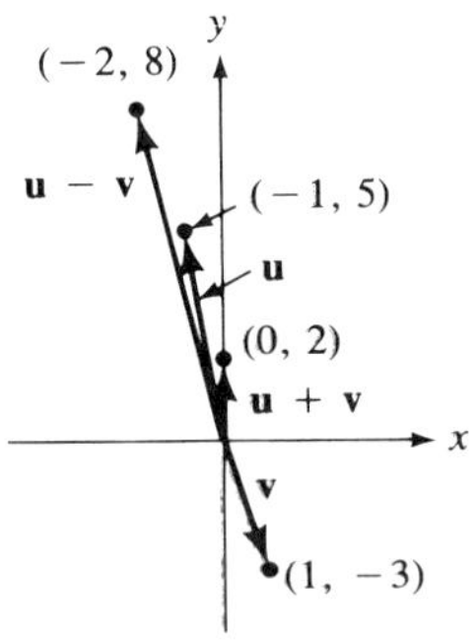

11. $\langle 12, -13\rangle$ **13.** $\langle 3, 1\rangle$ **15.** $\langle -4, -7\rangle$ **17.** $\langle 14, 3\rangle$ **19.** $3\sqrt{13}$ **21.** $-3\mathbf{i} - 5\mathbf{j}$
23. $\frac{21}{2}\mathbf{i} + 4\mathbf{j}$ **25.** $x = 10$, $y = 7$ **27.** $x = \frac{1}{2}$, $y = 1$ **29.** horizontal $\frac{3}{2}$, vertical $-\frac{3\sqrt{3}}{2}$
31. horizontal -1, vertical 1 **33.** $\theta = \pi \pm \arccos \frac{12}{13}$ **35.** $(12, 10)$

In Exercises 37–45 let $\mathbf{u} = \langle a, b\rangle$, $\mathbf{v} = \langle c, d\rangle$.
37. $\mathbf{u} + \mathbf{v} = \langle a, b\rangle + \langle c, d\rangle = \langle a + c, b + d\rangle = \langle c + a, d + b\rangle = \langle c, d\rangle + \langle a, b\rangle = \mathbf{v} + \mathbf{u}$
39. $\mathbf{u} + \mathbf{0} = \langle a, b\rangle + \langle 0, 0\rangle = \langle a + 0, b + 0\rangle = \langle a, b\rangle = \mathbf{u}$
41. $r(s\mathbf{u}) = r\langle sa, sb\rangle = \langle r(sa), r(sb)\rangle$ and $(rs)\mathbf{u} = \langle (rs)a, (rs)b\rangle$
43. $(-1)\mathbf{u} = (-1)\langle a, b\rangle = \langle (-1)a, (-1)b\rangle = \langle -a, -b\rangle = -\langle a, b\rangle$
45. $r\mathbf{0} = r\langle 0, 0\rangle = \langle r0, r0\rangle = \langle 0, 0\rangle = \mathbf{0}$ **47.** $|\mathbf{u}| = \sqrt{\cos^2\alpha + \sin^2\alpha} = \sqrt{1} = 1$

Exercises 8.5, page 313

1. -5 **3.** -24 **5.** -24 **7.** 126

9.

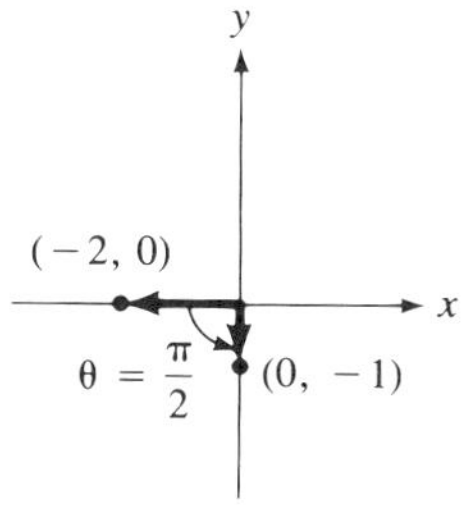

11.

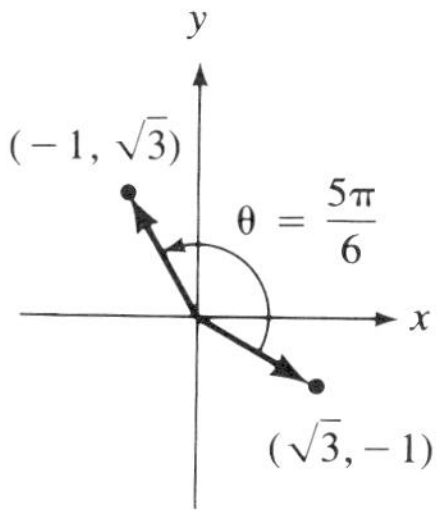

13.

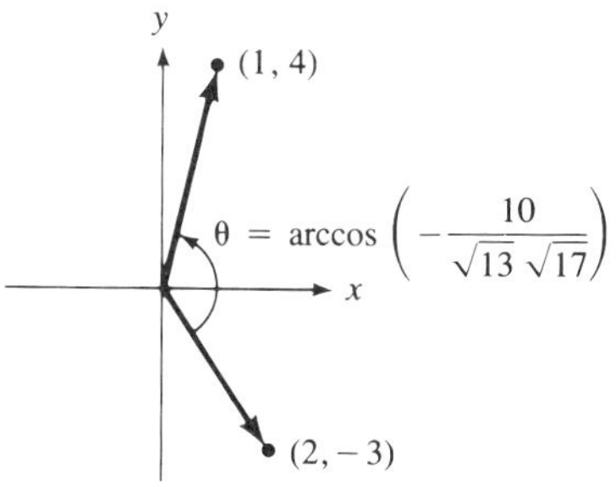

15. $\{\langle 3r, -2r\rangle : r \in R, r \neq 0\}$ **17.** $x = 3$

In Exercises 19–23 let $\mathbf{u} = \langle a, b\rangle$, $\mathbf{v} = \langle c, d\rangle$, $\mathbf{w} = \langle e, f\rangle$.

19. $\mathbf{u} \cdot (\mathbf{v} + \mathbf{w}) = \langle a, b\rangle \cdot \langle c + e, d + f\rangle = a(c + e) + b(d + f)$ and
$\mathbf{u} \cdot \mathbf{v} + \mathbf{u} \cdot \mathbf{w} = (ac + bd) + (ae + bf)$

21. $r(\mathbf{u} \cdot \mathbf{v}) = r(ac + bd)$ and $\mathbf{u} \cdot (r\mathbf{v}) = \langle a, b\rangle \cdot \langle rc, rd\rangle = a(rc) + b(rd)$

23. $|\mathbf{u}|^2 = (\sqrt{a^2 + b^2})^2 = a^2 + b^2 = \langle a, b\rangle \cdot \langle a, b\rangle = \mathbf{u} \cdot \mathbf{u}$

25. $\arccos\left(-\dfrac{6}{5\sqrt{13}}\right) \simeq 109°30'$, $\arccos\left(\dfrac{31}{25\sqrt{2}}\right) \simeq 28°40'$, and $\arccos\left(\dfrac{19}{5\sqrt{26}}\right) \simeq 41°50'$

27. $\pm\left\langle -\dfrac{2}{\sqrt{5}}, \dfrac{1}{\sqrt{5}}\right\rangle$ **29.** 53 ft-lb

31. Let $\mathbf{u} = \langle a, b\rangle$. If $\theta = 0$, there is $r \in R$, $r > 0$ such that $\mathbf{v} = r\mathbf{u} = \langle ra, rb\rangle$. Then

$$|\mathbf{u}|\,|\mathbf{v}| \cos\theta = |\mathbf{u}|\,|\mathbf{v}|\,(1) = \sqrt{a^2 + b^2}\,\sqrt{r^2(a^2 + b^2)} = \sqrt{r^2}\,(a^2 + b^2) = r(a^2 + b^2) = \mathbf{u} \cdot \mathbf{v}$$

Similarly, if $\theta = \pi$, there is $r \in R$, $r < 0$ such that $\mathbf{v} = \langle ra, rb\rangle$. Then

$$|\mathbf{u}|\,|\mathbf{v}| \cos\theta = |\mathbf{u}|\,|\mathbf{v}|\,(-1) = -\sqrt{r^2}\,(a^2 + b^2) = -(-r)(a^2 + b^2) = \mathbf{u} \cdot \mathbf{v}$$

33. $\arctan 2\sqrt{6} \simeq 78°30'$ with the riverbank (N78°30′E). Speed $10\sqrt{6} \simeq 24.5$ mi/hr.

35. $\langle -9\sqrt{3}, -19\rangle$

37. The resultant is the vector $\langle 3 + 5\sqrt{2}, -2\sqrt{2} + 2\sqrt{3}\rangle \simeq \langle 10, 0.64\rangle$. This is about 10 mi in the direction N86°20′E.

Chapter Review, page 314

1. $\gamma = 84°30'$, $a = 9.59$, $b = 18.5$ **2.** $\alpha = 81°$, $\beta = 59°$, $a = 14$ or $\alpha = 19°$, $\beta = 121°$, $c = 5$

3. $\alpha = 35°$, $\beta = 55°$, $c = 12$ **4.** $\gamma = 67°40'$, $a = 3.8$, $c = 9.2$ **5.** $\alpha = 104°$, $\beta = 47°$, $\gamma = 29°$

6. $\alpha = 111°40'$, $\beta = 37°50'$, $\gamma = 30°30'$ **7.** $\beta = 14°$, $\gamma = 26°$, $a = 32$ **8.** $\alpha = 37°$, $\beta = 123°$, $c = 3$

9. no solution **10.** no solution **11.** $\langle -5, 13\rangle$ **12.** $\left\langle -\dfrac{11}{2}, -\dfrac{5}{2}\right\rangle$ **13.** $\langle -12, 13\rangle$

14. $\dfrac{1}{3}\sqrt{833}$ **15.** 22 **16.** 9 **17.** $\arccos\left(-\dfrac{3}{\sqrt{205}}\right) \simeq 102°10'$ **18.** $\arccos\left(-\dfrac{27}{\sqrt{730}}\right) \simeq 177°50'$

19. $\left\langle -\dfrac{5}{\sqrt{41}}, -\dfrac{4}{\sqrt{41}}\right\rangle$ **20.** $x = -8$, $y = \dfrac{5}{2}$ **21.** $x = -3$, $y = 8$ **22.** $x = 6$

23. $x = \dfrac{3}{\sqrt{13}}$, $y = \dfrac{2}{\sqrt{13}}$ **24.** $\mathbf{v} = -\mathbf{i} + \sqrt{3}\,\mathbf{j}$, horizontal component is -1, vertical is $\sqrt{3}$.

25. $\pi + \arcsin\dfrac{3}{5}$, $2\pi - \arcsin\dfrac{3}{5}$ **26.** $(0, -5)$

27. $\arccos\left(\dfrac{49}{5\sqrt{170}}\right) \simeq 41°20'$, $\arccos\left(\dfrac{1}{5\sqrt{74}}\right) \simeq 88°40'$, and $\arccos\left(\dfrac{36}{\sqrt{85}\,\sqrt{37}}\right) \simeq 50°$

28. 55 ft-lb **29.** 31 mi **30.** 17 mi **31.** 112°
32. Actual course is 39°, ground speed is 620 mi/hr. **33.** 42° **34.** S32°W, 640 mi
35. The resultant is the vector $\mathbf{F} = \langle 6 + 10\sqrt{2}, 6\sqrt{3} + 10\sqrt{2}\rangle \approx \langle 20.14, 24.53\rangle$. The magnitude of $\mathbf{F}$ is about 31.7 lb in the direction N39°20′E. The force $-\mathbf{F}$ will balance. The magnitude of $-\mathbf{F}$ is the same as $|\mathbf{F}|$. The direction of $-\mathbf{F}$ is S39°20′W.
36. 32 and 29

Chapter Test, page 315

1. a. $\langle 14, -7\rangle$ **b.** -5 **c.** $\sqrt{13}$ **d.** $\pm\left\langle \frac{3}{\sqrt{13}}, \frac{2}{\sqrt{13}}\right\rangle$ **e.** $-\frac{5}{\sqrt{13}\sqrt{17}}$

f.

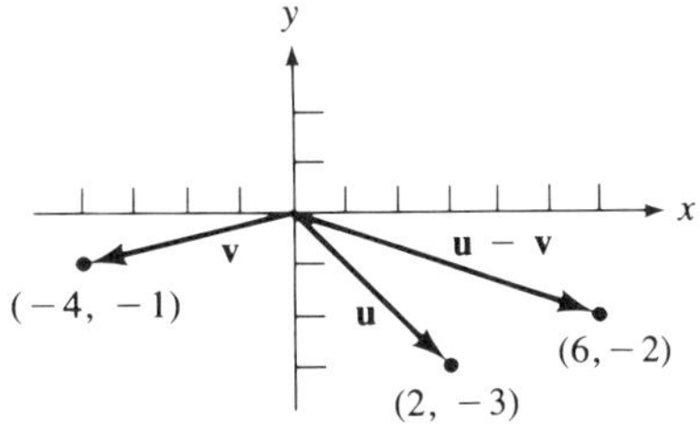

2. $2\sqrt{7}$ **3.** $\alpha = \arcsin 0.8, \pi - \arcsin 0.8$

4. west: 75 sin 50°, south: 75 cos 50° (mi) **5.** $\arccos\left(\frac{23}{5\sqrt{29}}\right)$ **6.** 100 tan 80° m **7.** $\langle 20, 10\sqrt{3}\rangle$

Exercises 9.1, page 320

1, 3, 5, 7.

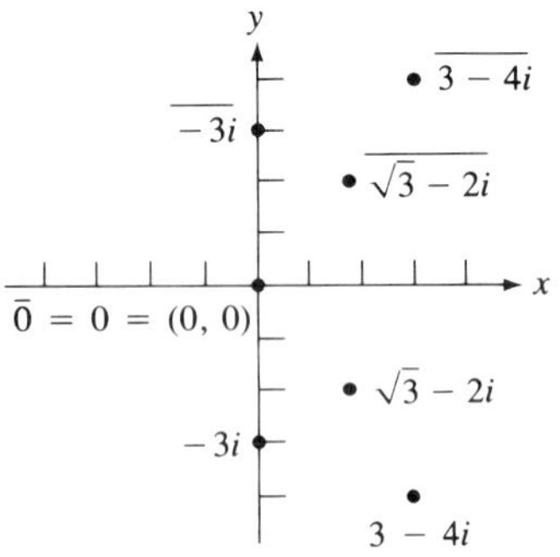

9.

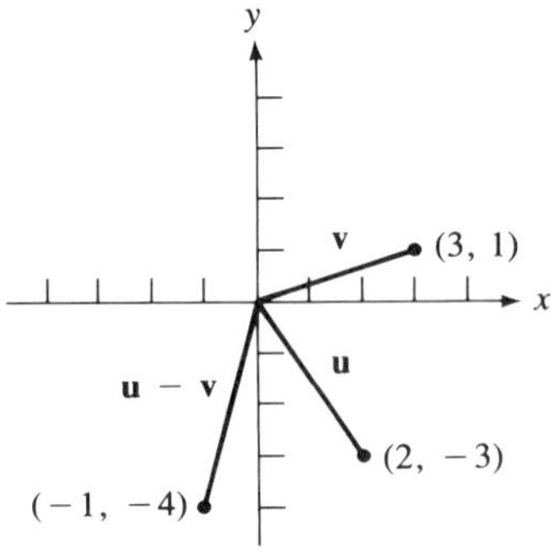

11. $\sqrt{2}$ **13.** 4 **15.** 0 **17.** $\sqrt{a^2 + 4}$

In Exercises 19–23 the principal argument is given in radian measure.

19. 0 **21.** $\frac{5\pi}{4}$ **23.** $\frac{11\pi}{6}$ **25.** $\sqrt{2} + \sqrt{2}\,i$ **27.** $-\frac{1}{2}i$ **29.** $-\frac{3}{2} + \frac{\sqrt{3}}{2}i$

31. $\sqrt{2}\left(\cos\frac{3\pi}{4} + i\sin\frac{3\pi}{4}\right)$ **33.** $3(\cos 0 + i\sin 0)$ **35.** $2\left(\cos\frac{2\pi}{3} + i\sin\frac{2\pi}{3}\right)$

37. $\pi\left(\cos\frac{\pi}{2} + i\sin\frac{\pi}{2}\right)$ **39.** $7\sqrt{2}\left(\cos\frac{7\pi}{4} + i\sin\frac{7\pi}{4}\right)$ **41.** $3\left(\cos\frac{\pi}{3} + i\sin\frac{\pi}{3}\right)$

43. $\sqrt{65}(\cos\theta + i\sin\theta)$, where $\theta = \pi - \arctan\frac{7}{4}$

45. Let $u = a + bi$. Then $|\bar{u}| = |a - bi| = \sqrt{a^2 + (-b)^2} = \sqrt{a^2 + b^2} = |u|$.
47. Let $u = a + bi$, $v = c + di$. Then

$$\frac{u}{v} = \frac{a + bi}{c + di} = \frac{(a + bi)(c - di)}{(c + di)(c - di)} = \frac{u\bar{v}}{c^2 + d^2} = \frac{u\bar{v}}{|v|^2}$$

Exercises 9.2, page 326

1. $uv = 6\sqrt{2}\left(\cos\frac{3\pi}{4} + i\sin\frac{3\pi}{4}\right)$

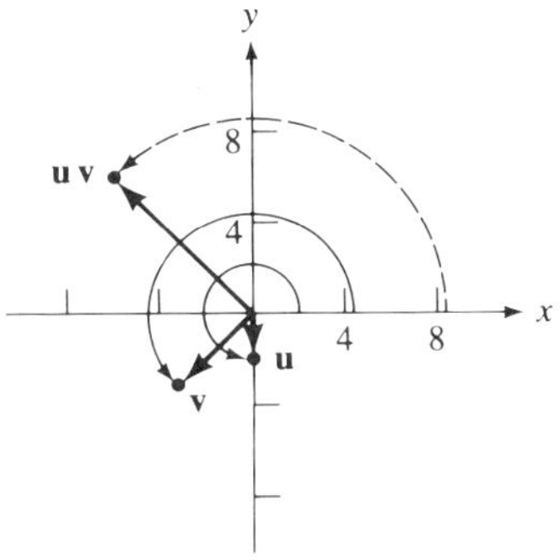

3. $\frac{u}{v} = \left(\cos\frac{\pi}{2} + i\sin\frac{\pi}{2}\right)$

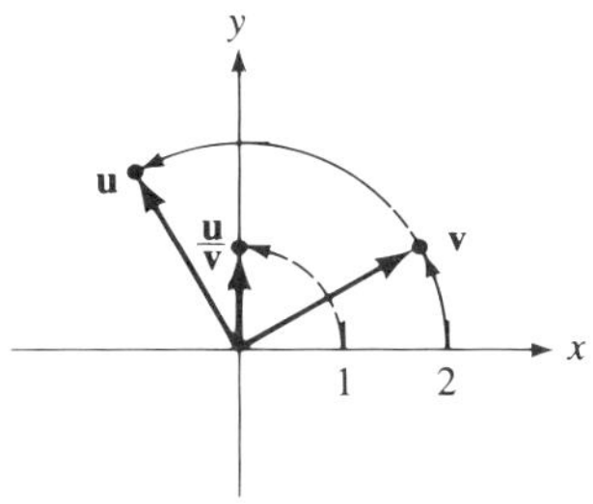

5. $uv = 6(\cos 0 + i\sin 0) = 6$
$u \div v = 1.5(\cos\pi + i\sin\pi) = -1.5$

7. $uv = 2(\cos 0 + i\sin 0) = 2$
$u \div v = \cos\frac{3\pi}{2} + i\sin\frac{3\pi}{2} = -i$

9. $uv = 16\left(\cos\frac{\pi}{4} + i\sin\frac{\pi}{4}\right) = 8\sqrt{2} + 8\sqrt{2}\,i$

$u \div v = 4\left(\cos\frac{7\pi}{4} + i\sin\frac{7\pi}{4}\right) = 2\sqrt{2} - 2\sqrt{2}\,i$

11. $uv = 8\left(\cos\frac{5\pi}{6} + i\sin\frac{5\pi}{6}\right) = -4\sqrt{3} + 4i$

$u \div v = \frac{1}{2}\left(\cos\frac{\pi}{2} + i\sin\frac{\pi}{2}\right) = \frac{1}{2}i$

13. $uv = 2\sqrt{2}\left(\cos\frac{\pi}{12} + i\sin\frac{\pi}{12}\right) = (1 + \sqrt{3}) + (\sqrt{3} - 1)i$

$u \div v = \sqrt{2}\left(\cos\frac{7\pi}{12} + i\sin\frac{7\pi}{12}\right) = \frac{1 - \sqrt{3}}{2} + \frac{1 + \sqrt{3}}{2}i$

15. $13[\cos(\alpha + \beta) + i\sin(\alpha + \beta)]$, where $\alpha = \arctan 1.5$ and $\beta = \arccos\left(-\frac{3}{\sqrt{13}}\right)$

17. $\sqrt{2.6}\left[\cos\left(\frac{\pi}{4} + \alpha - \beta\right) + i\sin\left(\frac{\pi}{4} + \alpha - \beta\right)\right]$, where $\alpha = \arccos\left(-\frac{2}{\sqrt{13}}\right)$ and $\beta = \arccos\left(-\frac{3}{\sqrt{13}}\right)$

19. $2 + 2i$ **21.** 1 **23.** $\frac{1}{8} - \frac{\sqrt{3}}{8}i$ **25.** $15i$

27. $[r(\cos\theta + i\sin\theta)]^{-1} = \frac{\cos 0 + i\sin 0}{r(\cos\theta + i\sin\theta)} = \frac{1}{r}[\cos(0 - \theta) + i\sin(0 - \theta)]$
$= \frac{1}{r}[\cos(-\theta) + i\sin(-\theta)] = \frac{1}{r}(\cos\theta - i\sin\theta)$

29. $u^2 = [r(\cos\theta + i\sin\theta)][r(\cos\theta + i\sin\theta)]$
$= r^2[\cos(\theta + \theta) + i\sin(\theta + \theta)] = r^2(\cos 2\theta + i\sin 2\theta)$

31. It suffices to show that $(-1 + \sqrt{3}\,i)^3 = 8$. Since

$$-1 + \sqrt{3}\,i = 2\left(\cos\frac{2\pi}{3} + i\sin\frac{2\pi}{3}\right)$$

we have

$$(-1 + \sqrt{3}\,i)^3 = 2\cdot 2\cdot 2\left[\cos\left(\frac{2\pi}{3} + \frac{2\pi}{3} + \frac{2\pi}{3}\right) + i\sin\left(\frac{2\pi}{3} + \frac{2\pi}{3} + \frac{2\pi}{3}\right)\right]$$
$$= 8(\cos 2\pi + i\sin 2\pi) = 8(1 + 0) = 8$$

Exercises 9.3, page 332

1. $-\frac{1}{2} + \frac{\sqrt{3}}{2}i$ **3.** i **5.** 16 **7.** 64 **9.** $-8 - 8\sqrt{3}\,i$ **11.** $-\frac{1}{2} + \frac{\sqrt{3}}{2}i$

13. $u_0 = \frac{\sqrt{3}}{2} + \frac{1}{2}i$

$u_1 = -\frac{\sqrt{3}}{2} + \frac{1}{2}i$

$u_2 = -i$

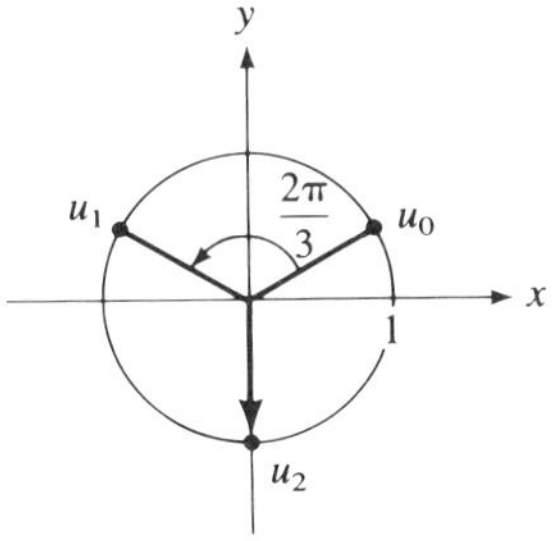

15. $u_0 = 1$, $u_1 = \frac{1}{\sqrt{2}} + \frac{1}{\sqrt{2}}i$

$u_2 = i$

$u_3 = -\frac{1}{\sqrt{2}} + \frac{1}{\sqrt{2}}i$

$u_4 = -1$

$u_5 = -\frac{1}{\sqrt{2}} - \frac{1}{\sqrt{2}}i$

$u_6 = -i$

$u_7 = \frac{1}{\sqrt{2}} - \frac{1}{\sqrt{2}}i$

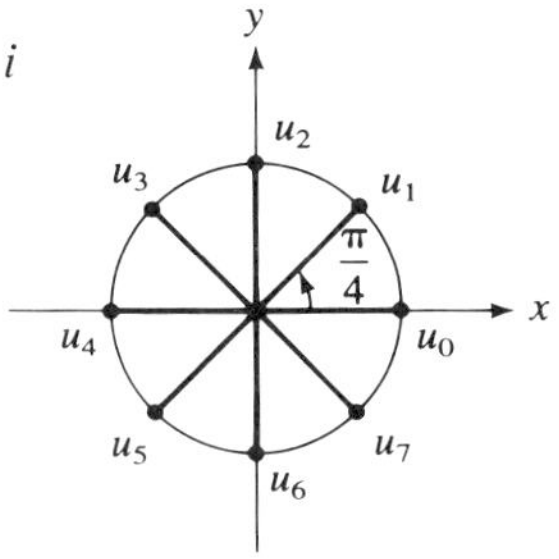

17. $u_0 = \sqrt{3} + i$
$u_1 = -1 + \sqrt{3}\,i$
$u_2 = -\sqrt{3} - i$
$u_3 = 1 - \sqrt{3}\,i$

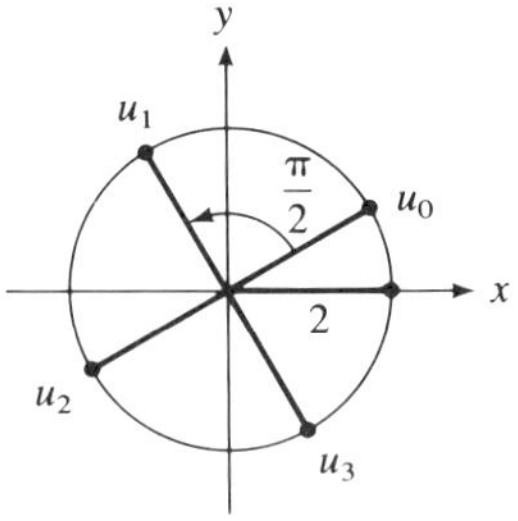

19. $u_0 = \sqrt{2}(\cos 63° + i \sin 63°)$
$u_1 = \sqrt{2}(\cos 135° + i \sin 135°)$
$u_2 = \sqrt{2}(\cos 207° + i \sin 207°)$
$u_3 = \sqrt{2}(\cos 279° + i \sin 279°)$
$u_4 = \sqrt{2}(\cos 351° + i \sin 351°)$

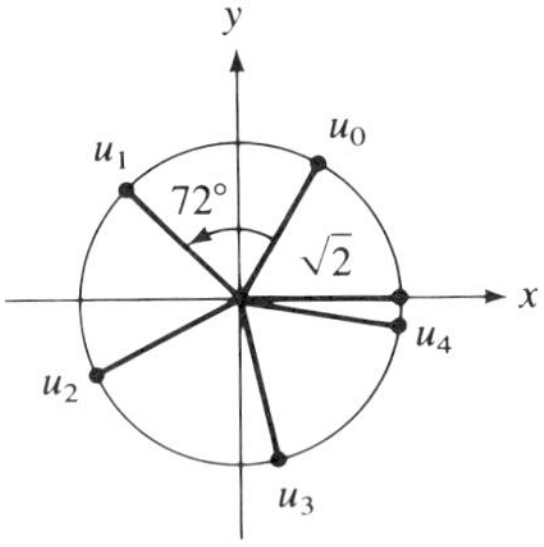

21. $u_0 = 2(\cos 110° + i \sin 110°)$
$u_1 = 2(\cos 230° + i \sin 230°)$
$u_2 = 2(\cos 350° + i \sin 350°)$

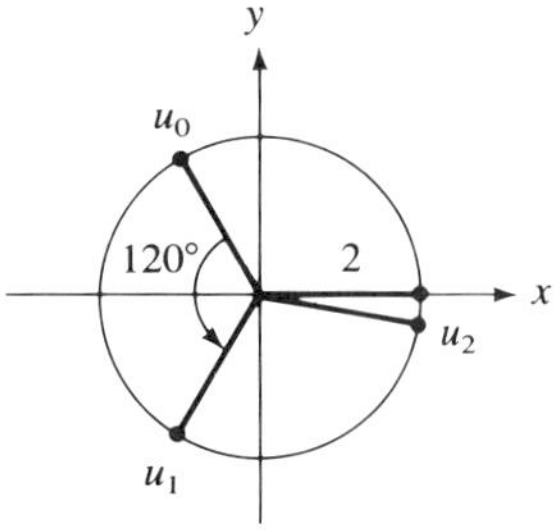

23. $u_0 = \sqrt{2}(\cos 15° + i \sin 15°)$
$u_1 = \sqrt{2}(\cos 105° + i \sin 105°)$
$u_2 = \sqrt{2}(\cos 195° + i \sin 195°)$
$u_3 = \sqrt{2}(\cos 285° + i \sin 285°)$

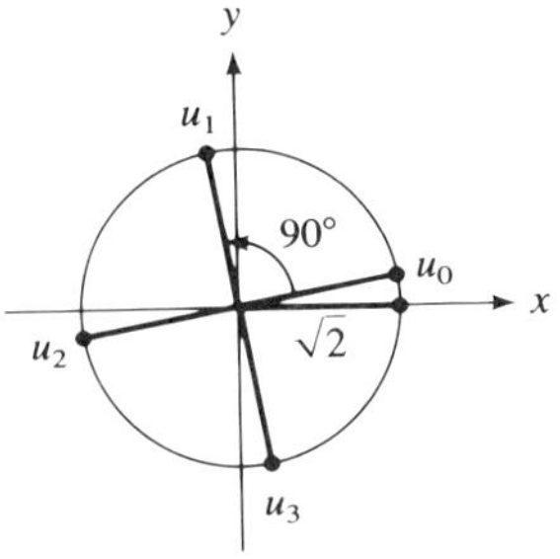

25. $u_0 = 3(\cos 15° + i \sin 15°)$
$u_1 = 3(\cos 105° + i \sin 105°)$
$u_2 = 3(\cos 195° + i \sin 195°)$
$u_3 = 3(\cos 285° + i \sin 285°)$

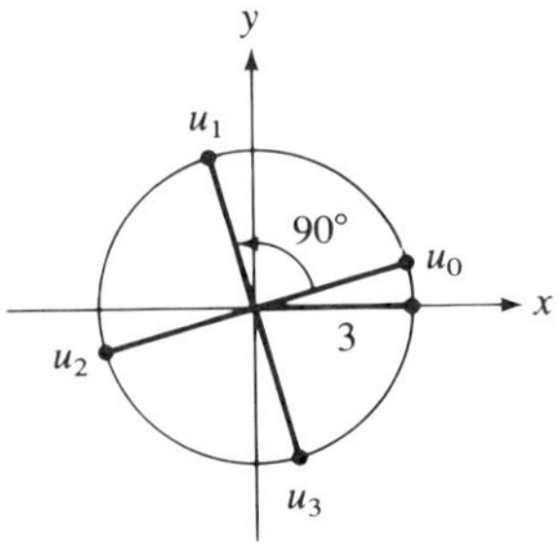

27. $i, -\frac{\sqrt{3}}{2} - \frac{1}{2}i, \frac{\sqrt{3}}{2} - \frac{1}{2}i$

29. $(\cos 75° + i \sin 75°), (\cos 165° + i \sin 165°), (\cos 255° + i \sin 255°), (\cos 345° + i \sin 345°)$

31. $-\frac{1}{2} + \frac{1}{2}i$ **33.** $\frac{\sqrt{2}}{2}(\cos 255° + i \sin 255°)$ **35.** $4 \sin\theta \cos\theta (\cos^2\theta - \sin^2\theta)$

Exercises 9.4, page 338

1.

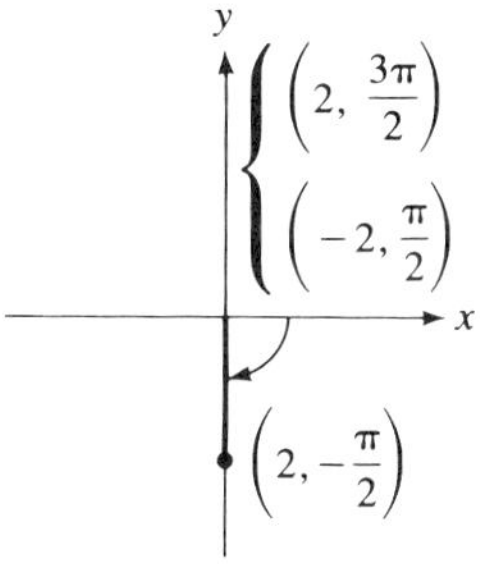

3.

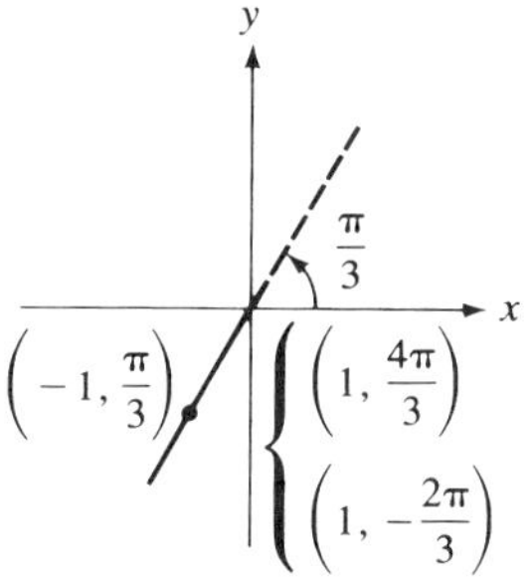

5.

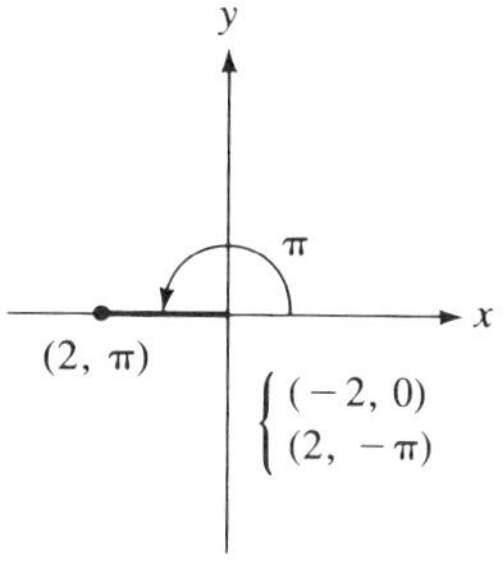

7.

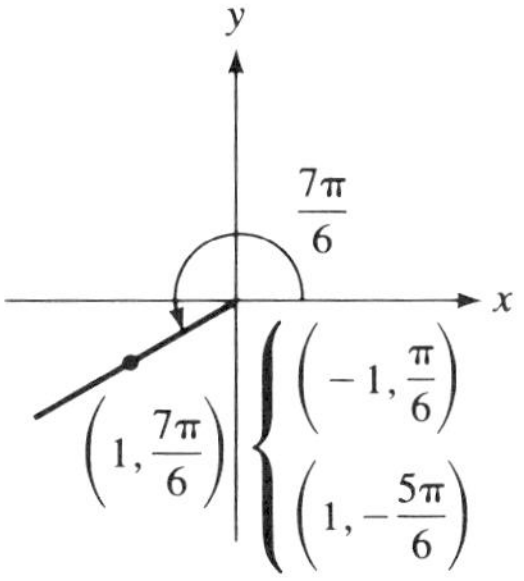

9.

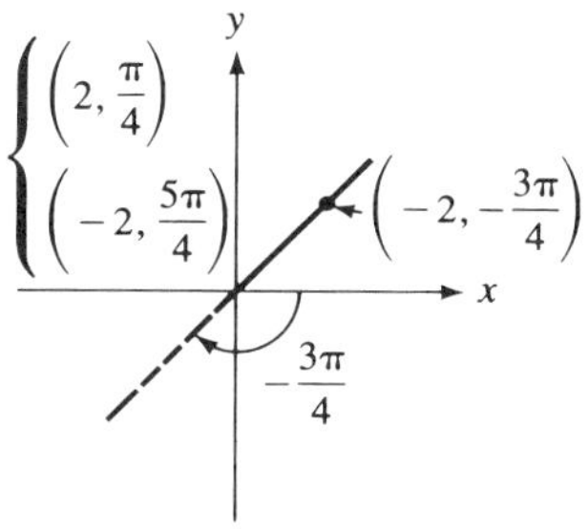

11.

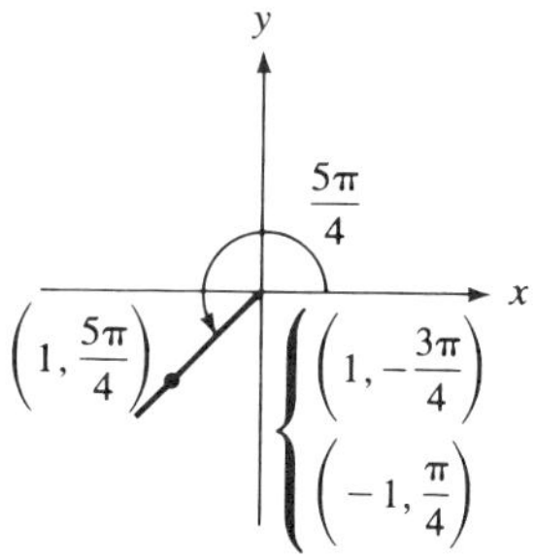

13. $\left(\frac{3\sqrt{3}}{2}, \frac{3}{2}\right)$ **15.** $(-2\sqrt{2}, -2\sqrt{2})$ **17.** $(-2, 0)$ **19.** $(0, 1)$ **21.** $\left(2, \frac{3\pi}{2}\right)$ **23.** $(2, \pi)$

25. $\left(2\sqrt{2}, \frac{3\pi}{4}\right)$ **27.** $\left(2, -\frac{\pi}{6}\right)$ **29.** $r + 2(\cos\theta - \sin\theta) = 0$ **31.** $r^2(\cos^2\theta - \sin^2\theta) = 1$

33. $\tan\theta = 5$ **35.** $r^2 + 2r\cos\theta + 4r\sin\theta + 4 = 0$ **37.** $y = 1$ **39.** $x^2 + y^2 = 25$

41. $3y = -x$ **43.** $9(x^2 + y^2) = (x^2 + y^2 - 2x)^2$ **45.** $x^2 + y^2 = 2x + y$

47. $x^2 + y^2 = x + \sqrt{3}\,y$ **49.** $xy = 0$ **51.** $(x^2 + y^2)^2 = 2xy$

Exercises 9.5, page 342

1.

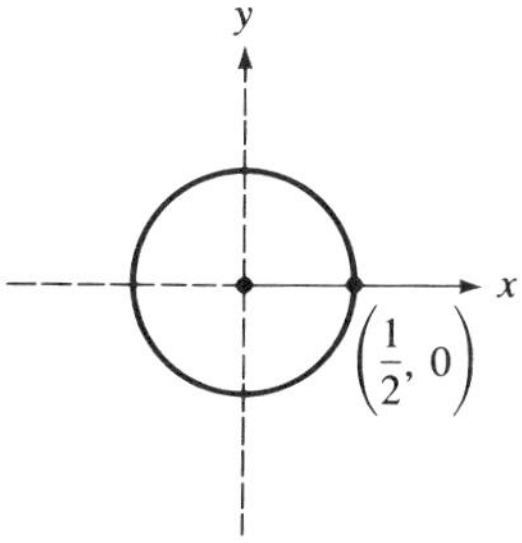

3.

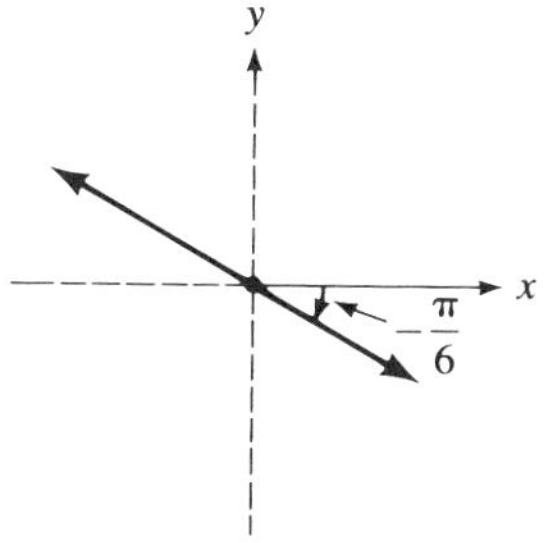

5.

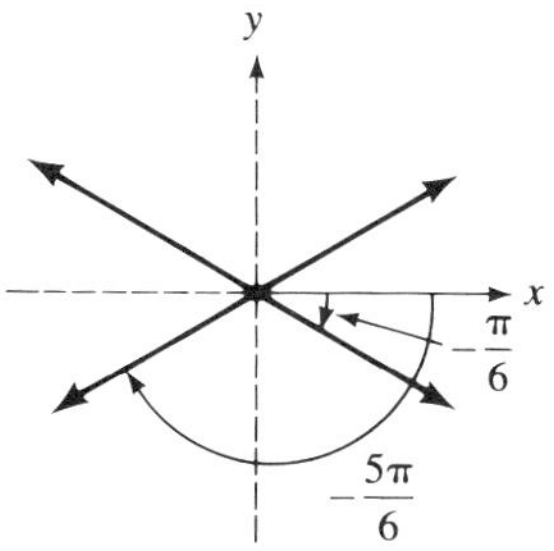

7.

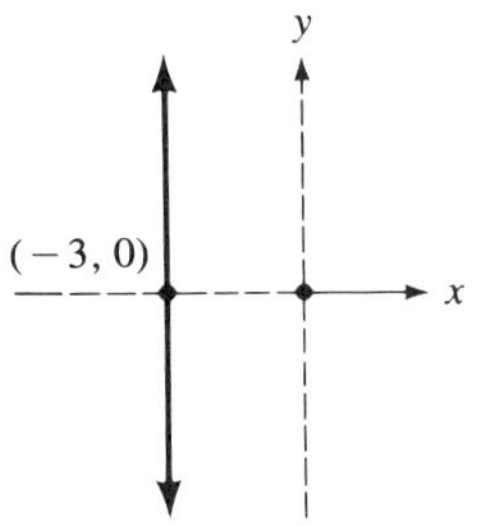

9.

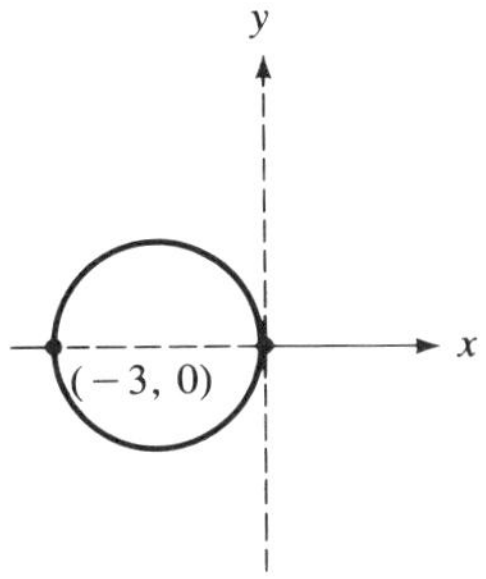

11.

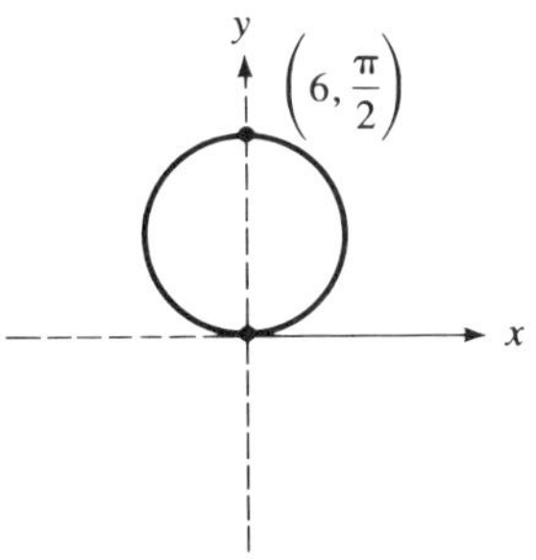

13.

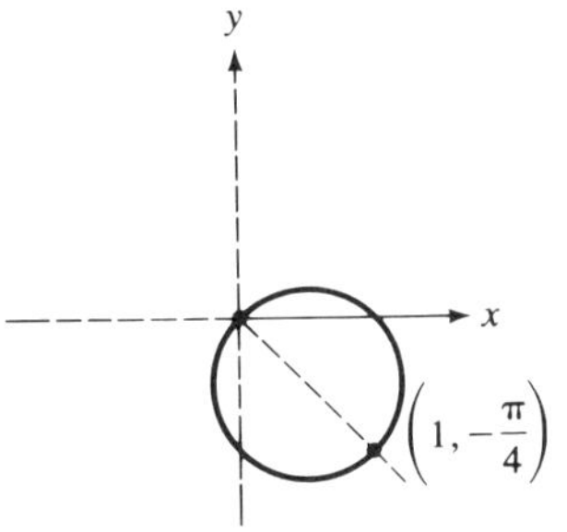

15.

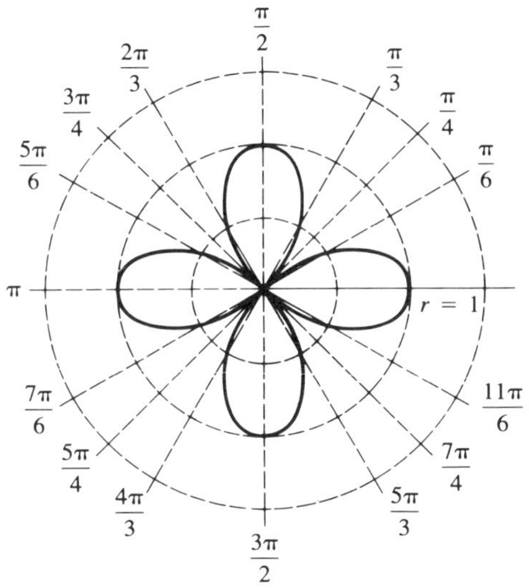

17.

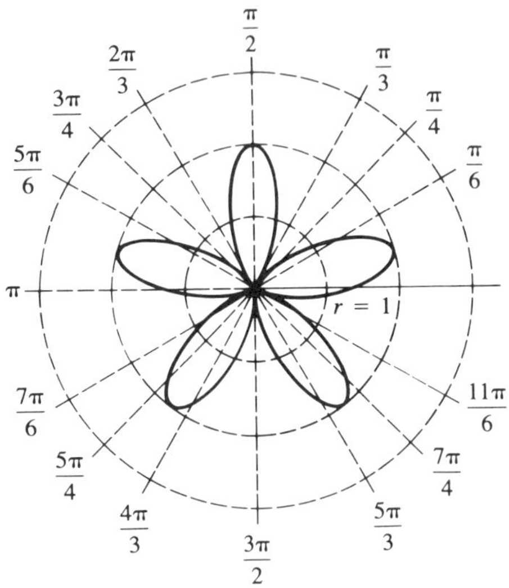

19.

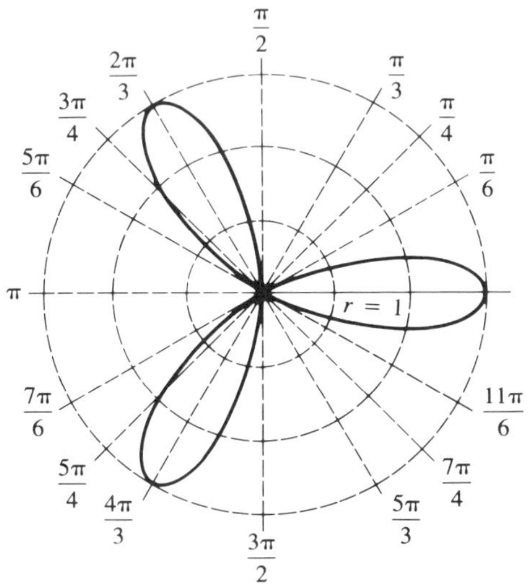

21.

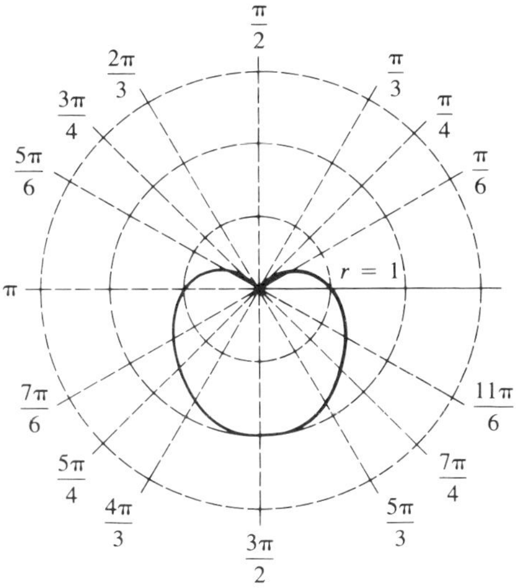
π/2
2π/3
π/3
3π/4
π/4
5π/6
π/6
π
r = 1
7π/6
11π/6
5π/4
7π/4
4π/3
5π/3
3π/2

23.

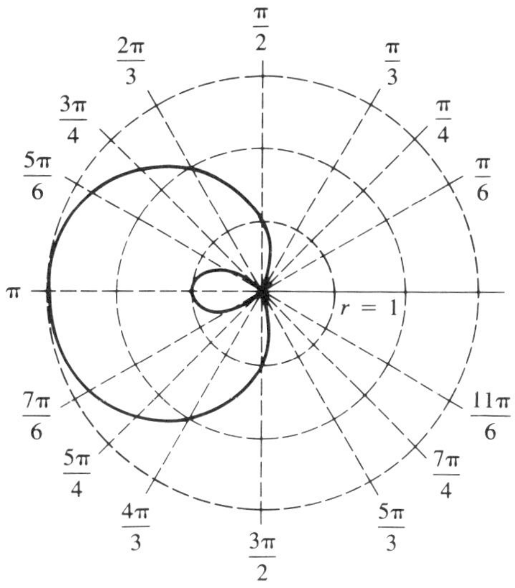
π/2
2π/3
π/3
3π/4
π/4
5π/6
π/6
π
r = 1
7π/6
11π/6
5π/4
7π/4
4π/3
5π/3
3π/2

25.

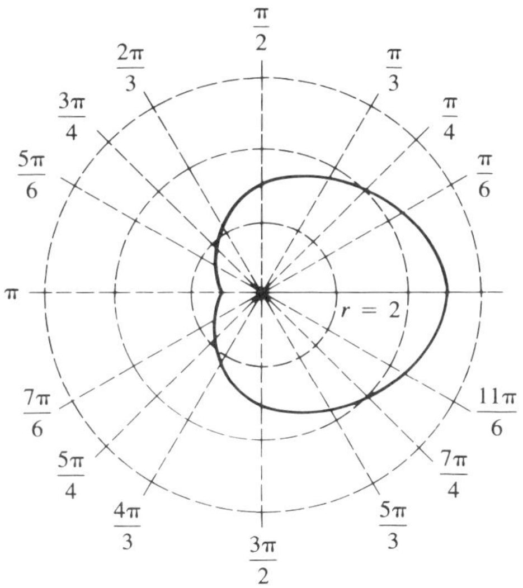
π/2
2π/3
π/3
3π/4
π/4
5π/6
π/6
π
r = 2
7π/6
11π/6
5π/4
7π/4
4π/3
5π/3
3π/2

27.

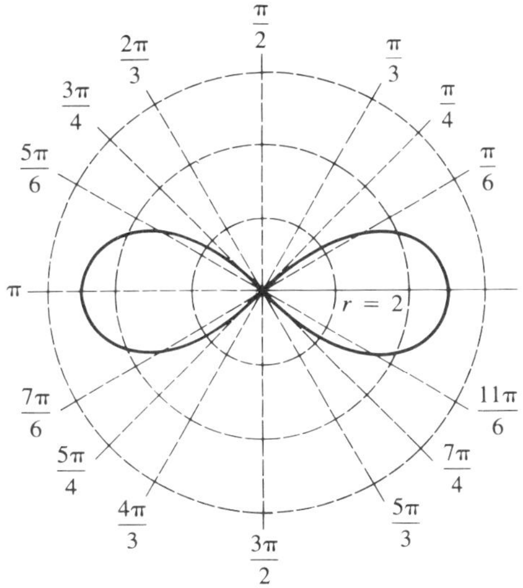
π/2
2π/3
π/3
3π/4
π/4
5π/6
π/6
π
r = 2
7π/6
11π/6
5π/4
7π/4
4π/3
5π/3
3π/2

29.

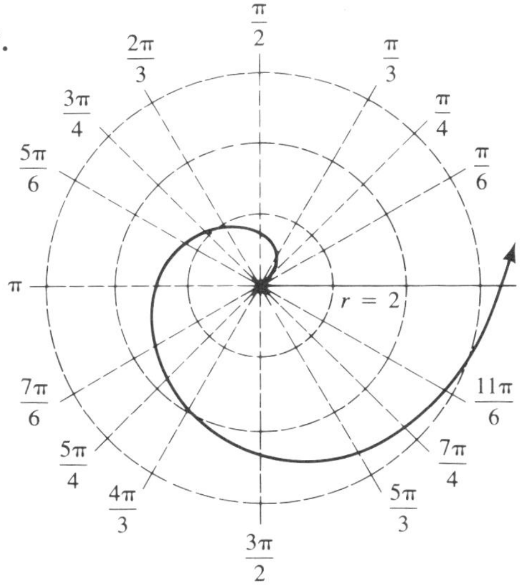
π/2
2π/3
π/3
3π/4
π/4
5π/6
π/6
π
r = 2
7π/6
11π/6
5π/4
7π/4
4π/3
5π/3
3π/2

31. In rectangular coordinates: center $\left(0, \frac{a}{2}\right)$, radius $\frac{|a|}{2}$ **33.** $r = \frac{2}{1 - \cos\theta}$

Chapter Review, page 343

1.

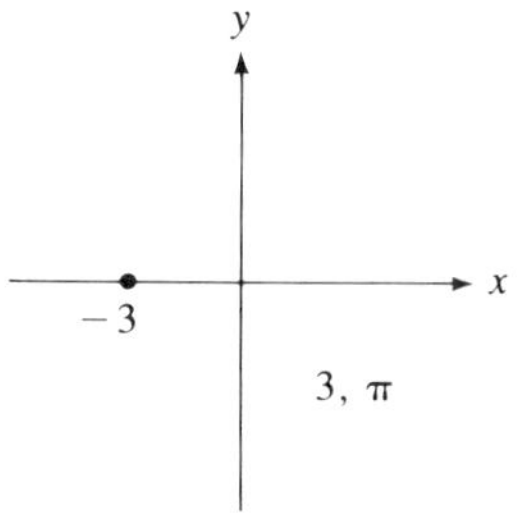

2.

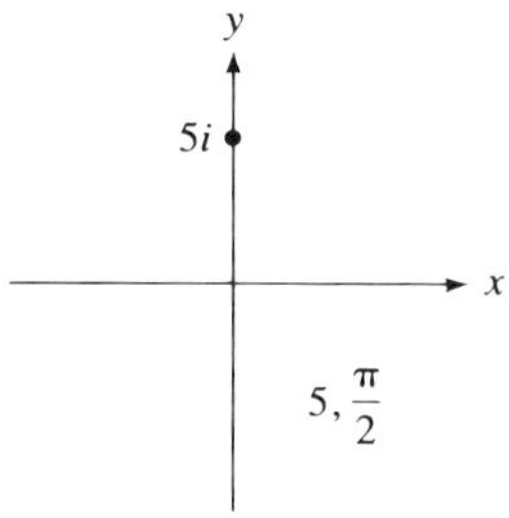

3.

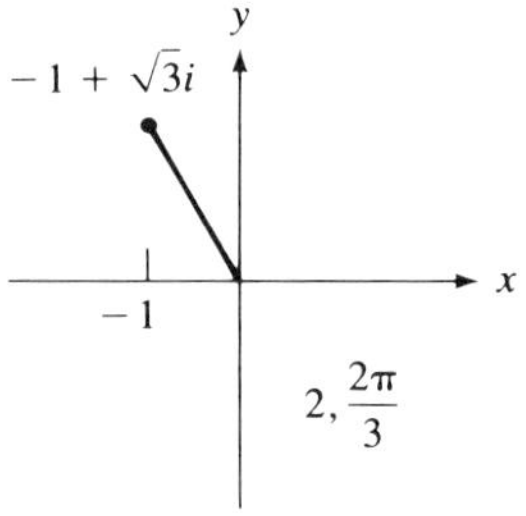

4.

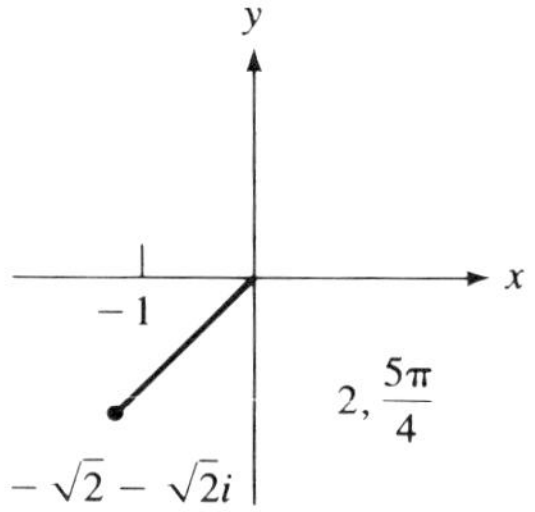

5. $-4\sqrt{2} + 4\sqrt{2}\,i$ **6.** $-\frac{\sqrt{3}}{2} - \frac{1}{2}i$ **7.** $7\left(\cos\frac{3\pi}{2} + i\sin\frac{3\pi}{2}\right)$

8. $12\left(\cos\frac{11\pi}{6} + i\sin\frac{11\pi}{6}\right)$ **9.** $\pi\sqrt{2}\left(\cos\frac{7\pi}{4} + i\sin\frac{7\pi}{4}\right)$

10. $\sqrt{29}(\cos\theta + i\sin\theta)$, where $\theta = \arccos\left(-\frac{5}{\sqrt{29}}\right)$ **11.** $uv = 2\left(\cos\frac{\pi}{2} + i\sin\frac{\pi}{2}\right) = 2i$

$u \div v = (\cos\pi + i\sin\pi) = -1$

12. $uv = 20\left(\cos\frac{3\pi}{2} + i\sin\frac{3\pi}{2}\right) = -20i$

$u \div v = \frac{5}{4}\left(\cos\frac{3\pi}{2} + i\sin\frac{3\pi}{2}\right) = -\frac{5}{4}i$

13. $uv = 4\left(\cos\frac{11\pi}{6} + i\sin\frac{11\pi}{6}\right) = 2\sqrt{3} - 2i$

$u \div v = \left(\cos\frac{3\pi}{2} + i\sin\frac{3\pi}{2}\right) = -i$

14. $uv = 32\sqrt{2}\left(\cos\frac{5\pi}{4} + i\sin\frac{5\pi}{4}\right) = -32 - 32i$

$u \div v = \sqrt{2}\left(\cos\frac{7\pi}{4} + i\sin\frac{7\pi}{4}\right) = 1 - i$

15. $\sqrt{1{,}537}[\cos(\alpha + \beta) + i\sin(\alpha + \beta)]$, where $\alpha = \arctan\left(-\frac{2}{5}\right)$ and $\beta = \arccos\left(-\frac{2}{\sqrt{53}}\right)$

16. $\frac{1}{5}\sqrt{10}[\cos(\alpha - \beta) + i\sin(\alpha - \beta)]$, where $\alpha = \arctan(-3)$ and $\beta = \arctan\left(-\frac{4}{3}\right)$

17. $12(\cos 270° + i\sin 270°) = -12i$ **18.** $2\left(\cos\frac{5\pi}{3} + i\sin\frac{5\pi}{3}\right) = 1 - \sqrt{3}\,i$ **19.** -64

20. $-\frac{1}{\sqrt{2}} + \frac{1}{\sqrt{2}}i$ **21.** 1 **22.** $-648 + 648\sqrt{3}\,i$

23. $u_0 = 2 + 2\sqrt{3}\, i$
$u_1 = -4$
$u_2 = 2 - 2\sqrt{3}\, i$

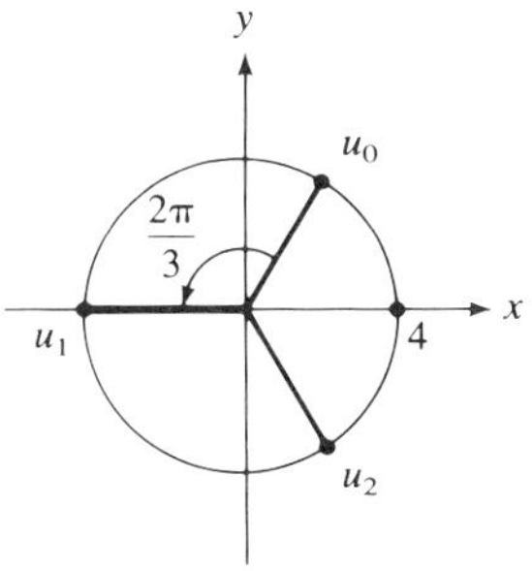

24. $u_0 = 1$
$u_1 = i$
$u_2 = -1$
$u_3 = -i$

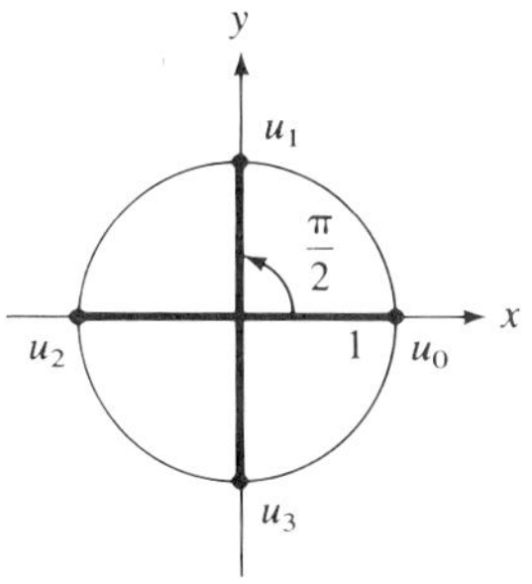

25. $u_0 = \sqrt[5]{2}(\cos 30^\circ + i \sin 30^\circ)$
$u_1 = \sqrt[5]{2}(\cos 102^\circ + i \sin 102^\circ)$
$u_2 = \sqrt[5]{2}(\cos 174^\circ + i \sin 174^\circ)$
$u_3 = \sqrt[5]{2}(\cos 246^\circ + i \sin 246^\circ)$
$u_4 = \sqrt[5]{2}(\cos 318^\circ + i \sin 318^\circ)$

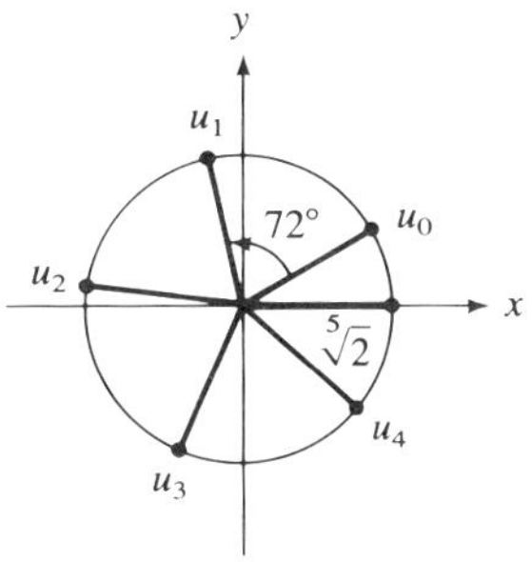

26. $u_0 = \sqrt[6]{2}(\cos 75^\circ + i \sin 75^\circ)$
$u_1 = \sqrt[6]{2}(\cos 195^\circ + i \sin 195^\circ)$
$u_3 = \sqrt[6]{2}(\cos 315^\circ + i \sin 315^\circ)$

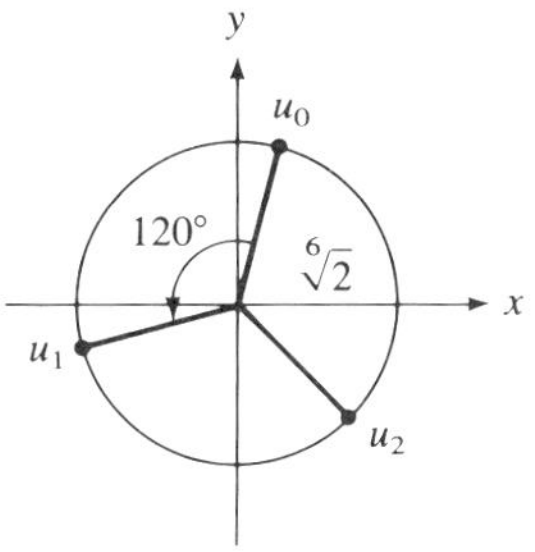

27. $\frac{\sqrt{3}}{2} + \frac{1}{2}i,\ -\frac{\sqrt{3}}{2} + \frac{1}{2}i,\ -i$

28. $4(\cos\theta + i\sin\theta)$, $\theta = 0^\circ, 72^\circ, 144^\circ, 216^\circ, 288^\circ$

29.

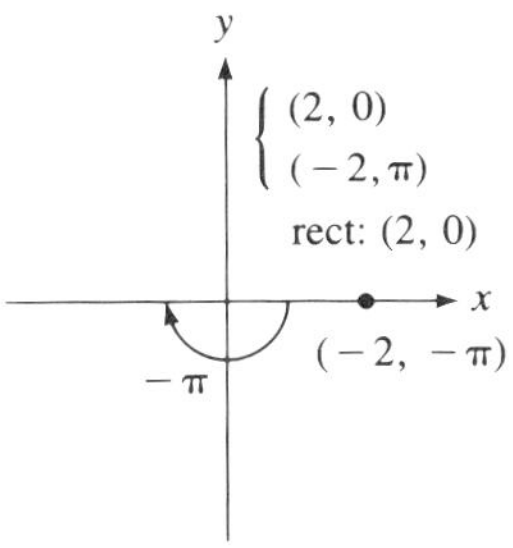

30.

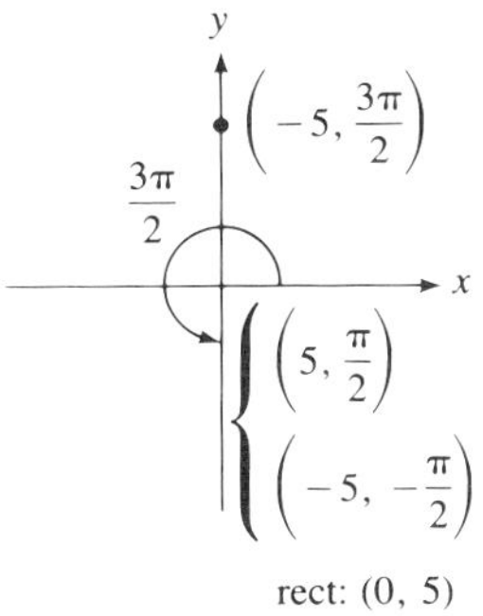

31.

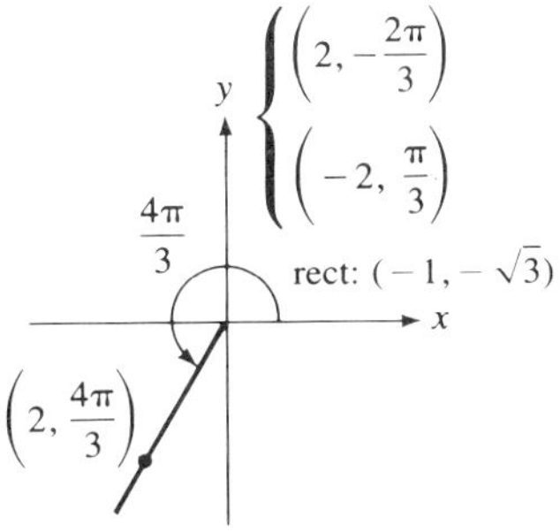

32.

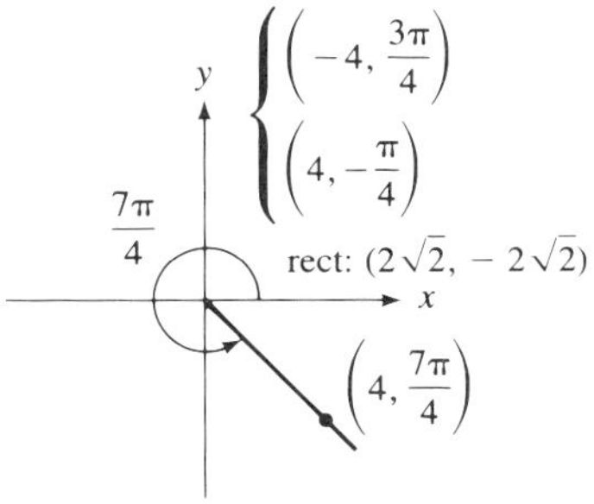

33. $\left(4\sqrt{2}, \dfrac{3\pi}{4}\right)$ **34.** $\left(6, \dfrac{11\pi}{6}\right)$ **35.** $r^2 \cos^2 \theta = 4r \sin \theta + 1$ **36.** $r = 4 \sin \theta$

37. $y = -4$ **38.** $x^2 + y^2 = 2x$ **39.** $(x^2 + y^2)y = 3$ **40.** $9(x^2 + y^2) = (x^2 + y^2 + y)^2$

41.

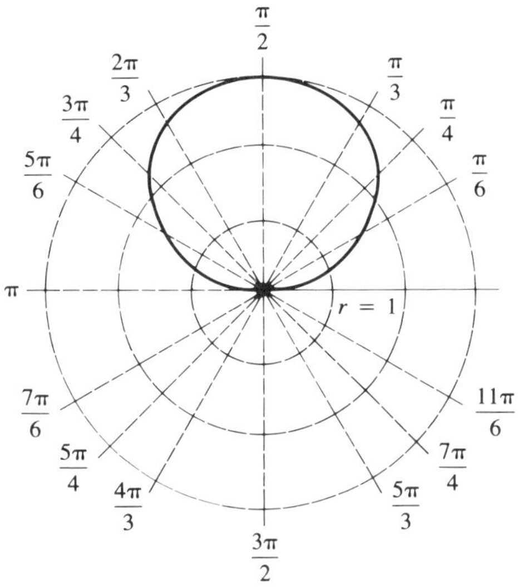

42.

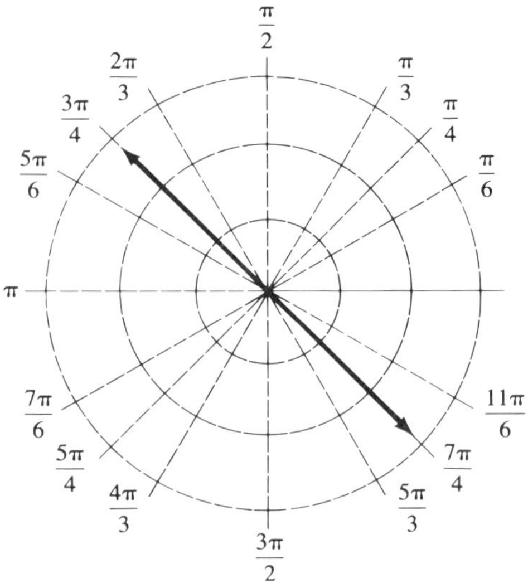

43.

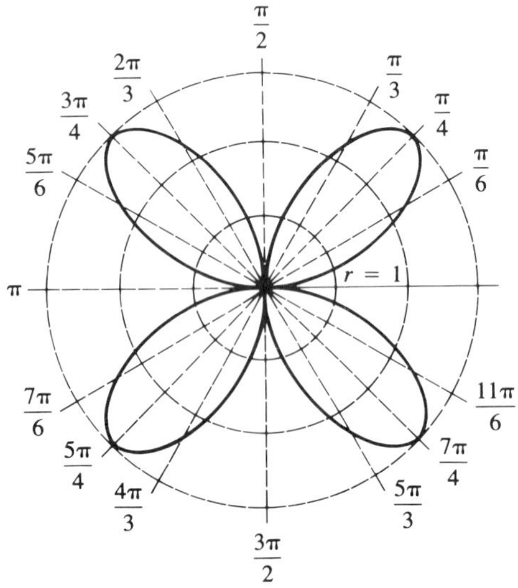

44.

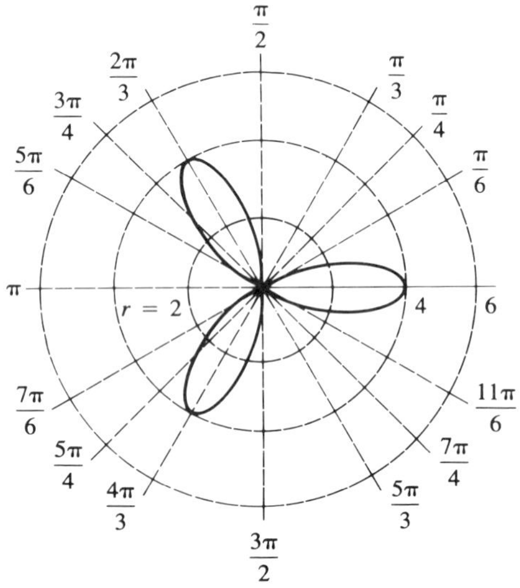

45.

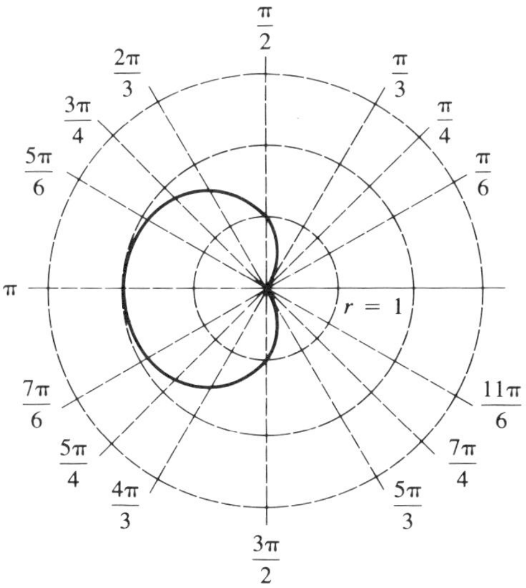

46.

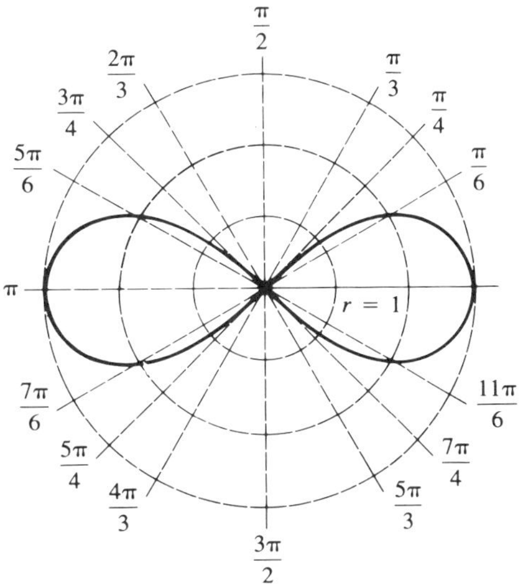

47.

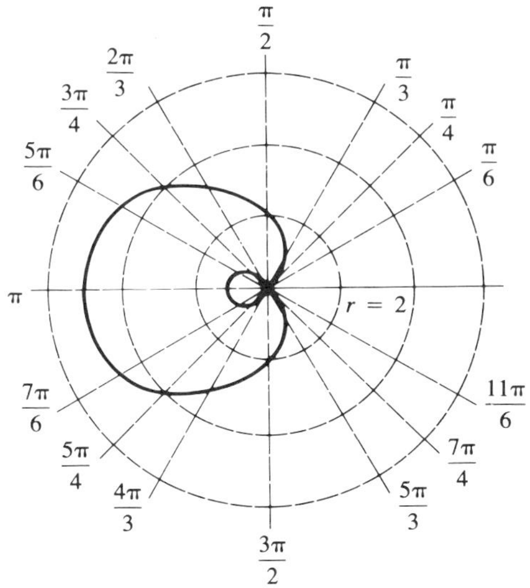

48.

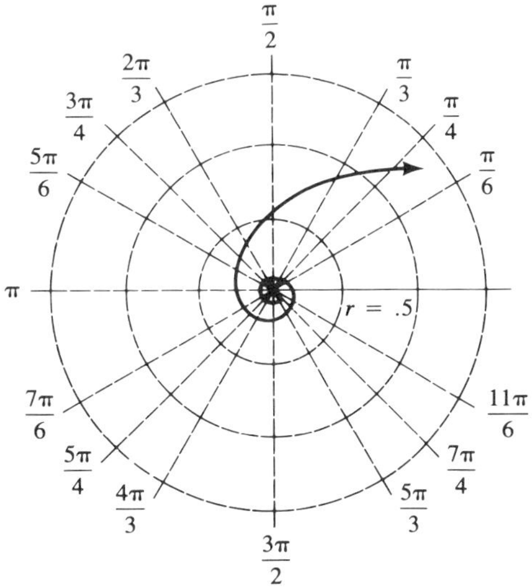

49. in rectangular coordinates: center $\left(\frac{b}{2}, \frac{a}{2}\right)$, radius $\frac{\sqrt{a^2 + b^2}}{2}$

Chapter Test, page 345

1. a.

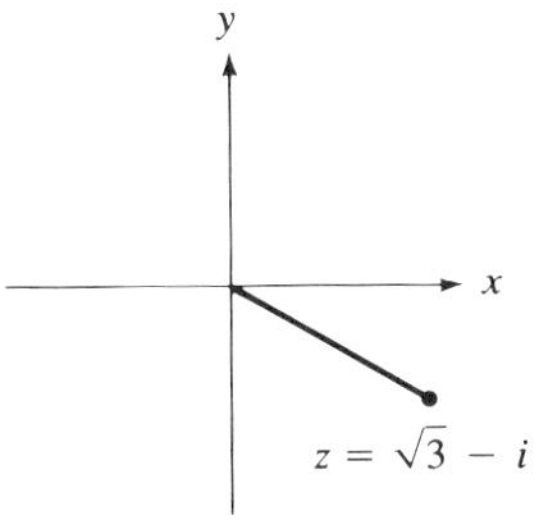

b. 2 **c.** $\frac{11\pi}{6}$ **d.** $2\left(\cos \frac{11\pi}{6} + i \sin \frac{11\pi}{6}\right)$

e. $8\left(\cos\frac{3\pi}{2} + i\sin\frac{3\pi}{2}\right)$ **f.** $-8i$ **2.** $\cos\frac{\pi}{6} + i\sin\frac{\pi}{6} = \frac{\sqrt{3}}{2} + \frac{1}{2}i$ **3.** $\cos\pi + i\sin\pi = -1$

4. $i, -\frac{\sqrt{3}}{2} - \frac{1}{2}i, \frac{\sqrt{3}}{2} - \frac{1}{2}i$ **5. a.** $\left(-2\sqrt{2}, \frac{7\pi}{4}\right)$ **b.** $\left(\frac{1}{2}, \frac{\sqrt{3}}{2}\right)$ **6.** $r = -\frac{3}{2}\cos\theta$

7. $2x^2 + 2y^2 - y = 0$ **8.** $\tan\theta = -\frac{1}{3}\sqrt{3}$

9.

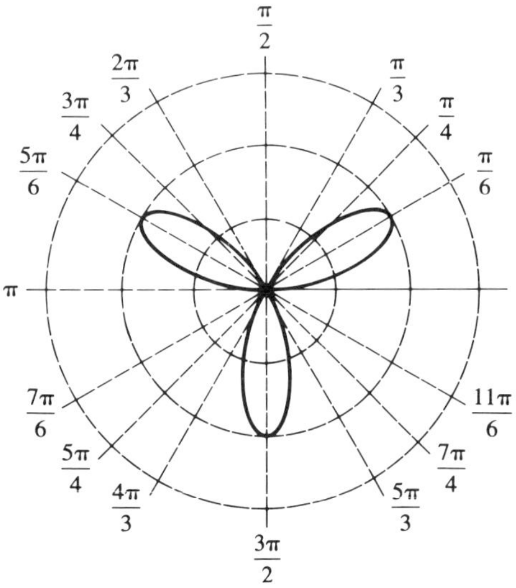

Exercises 10.1, page 353

1. inconsistent **3.** $(1, -1)$ **5.** $(2, -1)$ **7.** lines coincide **9.** $(-1, 2)$ **11.** $(1, 2)$

13. $\left(\frac{3}{4}, \frac{1}{2}\right)$ **15.** $\left(\frac{5}{11}, -\frac{25}{11}\right)$ **17.** lines coincide **19.** $\left(-\frac{9}{5}, -\frac{66}{5}\right)$ **21.** $(3, 2), (-2, -3)$

23. $(6, 1), (-1, -6)$ **25.** $(-3, 0), (3, 0)$ **27.** $(2, 2), (-2, -2), (4, 1), (-4, -1)$

29. $\left(-\sqrt{5}, -\frac{\sqrt{5}}{2}\right), \left(\sqrt{5}, \frac{\sqrt{5}}{2}\right), \left(-\sqrt{2}, -\frac{5\sqrt{2}}{4}\right), \left(\sqrt{2}, \frac{5\sqrt{2}}{4}\right)$

31. $(0, \sqrt{8}), (0, -\sqrt{8}), (-2, -4), (2, 4)$ **33.** 25 acres of onions, 50 acres of lettuce **35.** 14¢

37. Vena has 5, Jack has 5, and Bob has 0. **39.** $\frac{1}{2}$ liter of dry milk and $1\frac{1}{2}$ liters of whole milk

41. 4 liters of water **43.** \$1,500 at 5% **45.** 1,500 **47.** 50

Exercises 10.2, page 358

1.

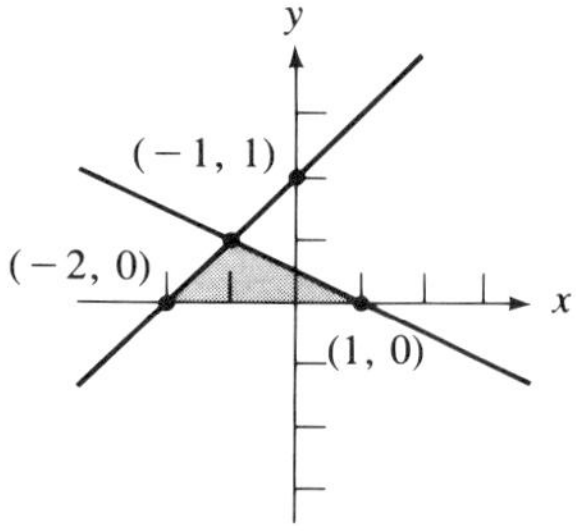

3.

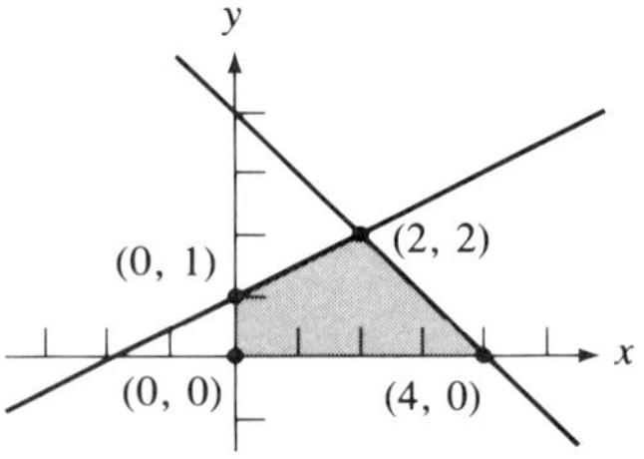

5.

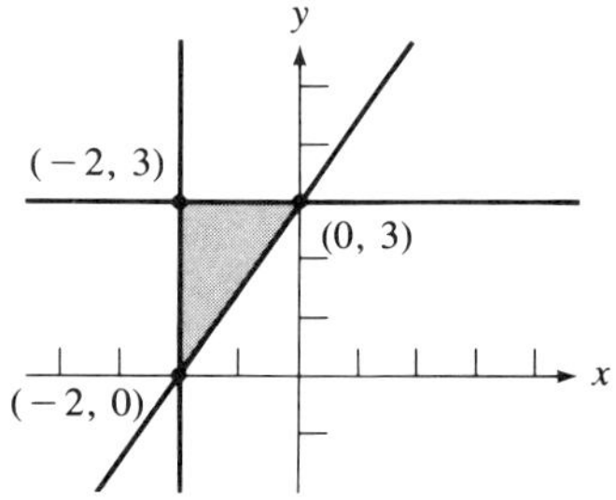

7.

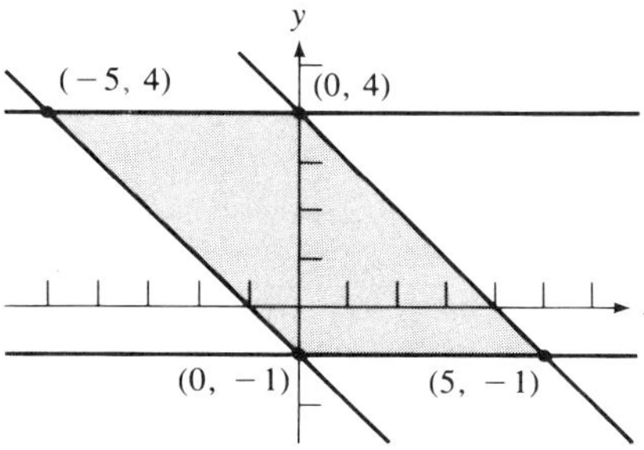

9.

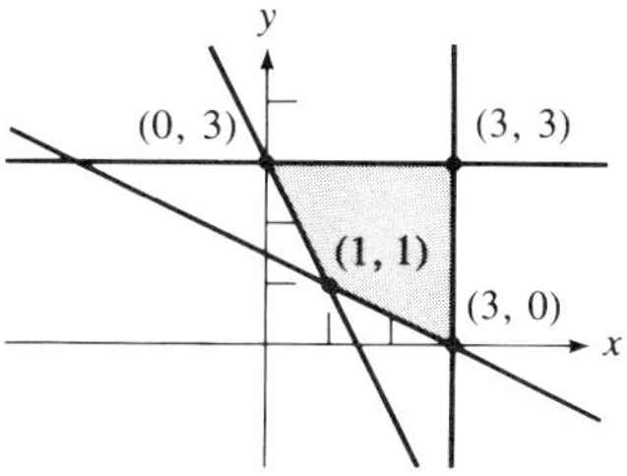

11.

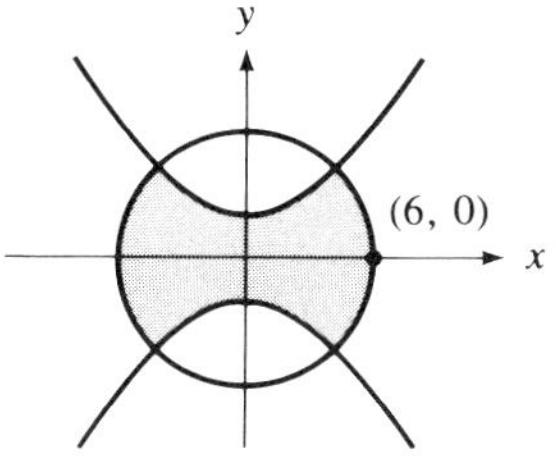

13.

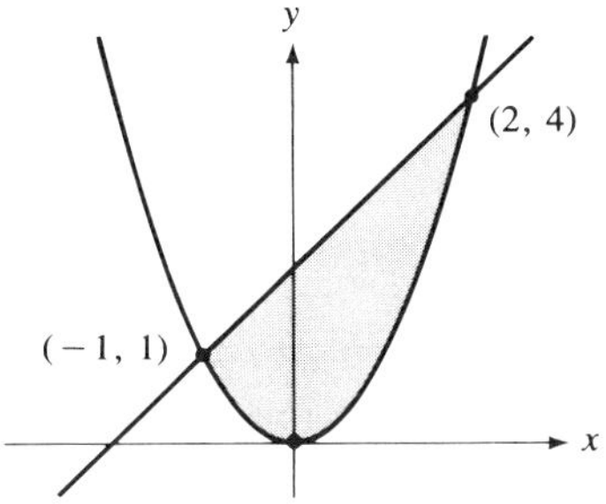

15.

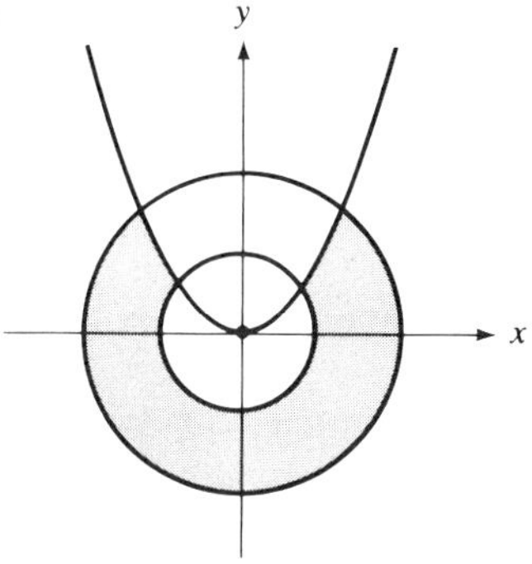

17. $2y \le x + 6$
$y + 3x \le 10$
$5y + x \ge 8$

19. $x \le 3$
$x \ge -2$
$5y - x \le 12$
$5y - x \ge -3$

21. $2y \le x + 2$
$3y + 2x \le 3$
$x + y \ge -2$
$y \ge x - 4$

23. $y \le x + 1$
$3y + x \le 15$
$y \le -2x + 15$
$3y - 2x \ge -11$
$x + y \ge 3$

25. any point in the region shown, where x = number of acres in onions and y = number of acres in chile

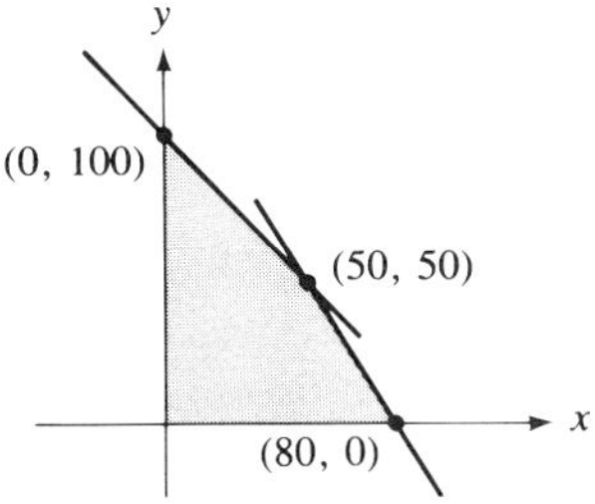

27. any point in the region shown, where x = number of acres of milo and y = number of acres of cotton

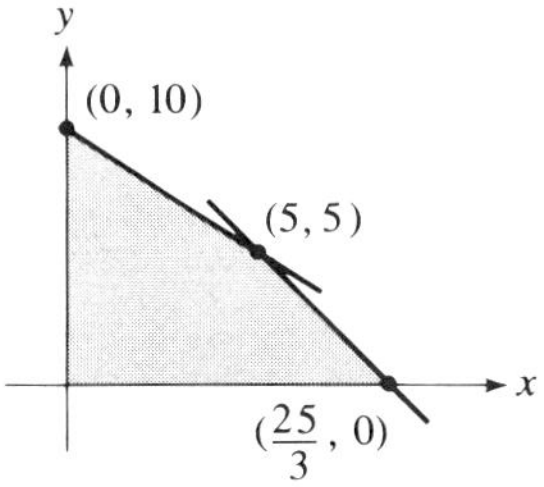

Exercises 10.3, page 363

1. maximum 3 at (1, 0), minimum −6 at (−2, 0)
3. maximum 8 at (4, 0), minimum −6 at (2, 2)
5. maximum 6 at (0, 3), minimum −6 at (−2, 0)
7. maximum 8 at (5, −1), minimum −27 at (−5, 4)
9. maximum −2 at (1, 1), minimum −16 at (3, 3)
11. 80 acres of onions, no chile

13. 4 lb of food B, none of food A **15.** 30 copies on each machine
17. 16 hardback and 30 paperback **19.** 6 hours on machine A, 16 hours on machine B

Exercises 10.4, page 369

1. $y = x,\ z = 1 - 2x,\ (-1, -1, 3)$ **3.** $y = -2x,\ z = 2 - x,\ (-1, 2, 3)$
5. $z = 2y - 6,\ x = y - 9,\ (-7, 2, -2)$ **7.** $z = 2y + 2,\ x = 5 - y,\ (3, 2, 6)$
9. $x = \frac{3}{4}z,\ y = -\frac{1}{4}z,\ (-3, 1, -4)$ **11.** $y = z - 4,\ x = z - 8,\ (-12, -8, -4)$ **13.** $(-2, -1, 3)$
15. $(4, 2, -1)$ **17.** $(-1, 4, 2)$ **19.** $\left(-\frac{1}{3}, 1, \frac{1}{2}\right)$ **21.** $(1, 6, 0)$ **23.** $s = 4 + t,\ r = 2t - 3$
25. $x = y - 5,\ z = 2y - 10$ **27.** $\left(-\frac{1}{8}, \frac{1}{7}, -\frac{1}{4}\right)$ **29.** $\left(-4, \frac{1}{3}, 2\right)$
31. 10 of bag A, 0 of bag B, 5 of bag C **33.** $y = 2x^2 - x - 1$
35. Use x 4-person tables, y 6-person tables, and z 8-person tables, where $y = 20 - 2z$ and $x = 25 + z$.
37. 24 chickens

Exercises 10.5, page 376

1. $x = 0,\ y = 4$ **3.** $x = 0,\ y = -2$ **5.** not possible **7.** 0 **9.** -5 **11.** $\begin{bmatrix} 0 & 0 \\ 60 & -30 \end{bmatrix}$

13. $\begin{bmatrix} 6 & 9 \\ -4 & -6 \end{bmatrix}$ **15.** $\begin{bmatrix} -5 & 8 \\ 7 & -12 \end{bmatrix}$ **17.** $\begin{bmatrix} 5 & -26 \\ 5 & -51 \end{bmatrix}$ **19.** not defined **21.** $\begin{bmatrix} -5 & 11 \\ 9 & -15 \\ -11 & 21 \end{bmatrix}$

23. not defined **25.** $[16 \quad -20]$ **27.** $\begin{bmatrix} -16 & 24 & -8 \\ 4 & -6 & 2 \\ 6 & -9 & 3 \end{bmatrix}$ **29.** $\begin{bmatrix} 1 & 0 \\ 0 & 1 \end{bmatrix}$

31. $\begin{bmatrix} 2 & -3 \\ -1 & 4 \end{bmatrix}\begin{bmatrix} x \\ y \end{bmatrix} = \begin{bmatrix} 7 \\ 0 \end{bmatrix}$ **33.** $\begin{bmatrix} 5 & -9 & 2 \\ 1 & 14 & -2 \end{bmatrix}\begin{bmatrix} r \\ s \\ t \end{bmatrix} = \begin{bmatrix} 0 \\ 0 \end{bmatrix}$ **35.** $\begin{bmatrix} 4 & 5 & -9 \\ -1 & 3 & 8 \\ -3 & 0 & 1 \end{bmatrix}\begin{bmatrix} r \\ s \\ t \end{bmatrix} = \begin{bmatrix} 0 \\ 0 \\ 9 \end{bmatrix}$

37. $\begin{bmatrix} 1 & -7 & 4 \\ 0 & 2 & -1 \\ 4 & 2 & -6 \end{bmatrix}\begin{bmatrix} x \\ y \\ z \end{bmatrix} = \begin{bmatrix} 9 \\ 1 \\ 9 \end{bmatrix}$ **39.** $\begin{aligned} 2a - 5b + 6c &= 7 \\ -6b + 4c &= 9 \end{aligned}$ **41.** $s = 8,\ t = 7$

43. $\begin{aligned} 7a - 6b &= 9 \\ 4a - 5b &= 4 \\ -2a + b &= 3 \end{aligned}$ **45.** \$401.40 **47.** \$203.46 **49.** \$34.20

Exercises 10.6, page 382

1. $(46, 33)$ **3.** no solutions **5.** $(-45, 66)$ **7.** $a = 12 - 2c,\ b = 2c - 2$
9. $y = 2z + 4,\ x = -z - 3$ **11.** $x = -\frac{1}{17}(11z + 9),\ y = \frac{1}{17}(45 + 38z)$ **13.** inconsistent
15. $a = 6 - c,\ b = \frac{1}{2}(c - 3)$ **17.** $(2, -3, 8)$ **19.** $(2, -1, -3)$ **21.** $(0, 1, 1)$ **23.** $\left(3, \frac{1}{2}, -2\right)$
25. $\left(\frac{1}{2}, \frac{1}{3}, \frac{1}{4}\right)$ **27.** $x = \dfrac{z}{1 - 4z},\ y = \dfrac{z}{3z - 1}$
29. A: radius = 0.475 cm; B: radius = 0.125 cm; C: radius = 0.275 cm **31.** Q: 2.6, P: 3.9, R: 3.9

Exercises 10.7, page 388

1. $\begin{bmatrix} 1 & 4 \\ 2 & 9 \end{bmatrix}$ **3.** $\begin{bmatrix} \frac{5}{6} & \frac{2}{3} \\ -1 & -1 \end{bmatrix}$ **5.** no inverse **7.** $\begin{bmatrix} 3 & 1 & 1 \\ 5 & 3 & 2 \\ -3 & 0 & -1 \end{bmatrix}$ **9.** no inverse

11. no inverse **13.** $\begin{bmatrix} 1 & -3 & 2 \\ -3 & 10 & -7 \\ 2 & -4 & 3 \end{bmatrix}$ **15.** $(-35, 117, -46)$ **21.** $\begin{bmatrix} -\frac{1}{3} & -\frac{1}{3} & \frac{5}{3} \\ \frac{1}{3} & \frac{4}{3} & -\frac{11}{3} \\ \frac{4}{3} & \frac{10}{3} & -\frac{29}{3} \end{bmatrix}$

23. $a = -\dfrac{28}{3}, b = \dfrac{148}{3}, c = \dfrac{55}{3}$

25. $-\dfrac{7}{2}$ **27.** $a = \dfrac{3}{4}, b = \dfrac{5}{4}$ **29.** $a = 9.5, b = -1$ **31.** $a = -3, b = 2$

33. 2 of bag A, 2 of bag B, and 3 of bag C **35.** 1 of bag A, 3 of bag B, and 1 of bag C

Exercises 10.8, page 395

1. -9 **3.** 10 **5.** 0 **7.** 11 **9.** -286 **11.** 11 **13.** 1 **15.** -31 **17.** 21 **19.** -3 **21.** -186

Exercises 10.9, page 398

1. $(-11, 17)$ **3.** $(-3, -18)$ **5.** $(-4, 0)$ **7.** $(1, -1, 2)$ **9.** $\left(\dfrac{11}{3}, \dfrac{17}{3}, -\dfrac{67}{3}\right)$

11. $(24, -18, 18, -10)$ **15.** $\begin{bmatrix} 3 & -2 \\ 10 & -7 \end{bmatrix}, \begin{bmatrix} \frac{5}{7} & -\frac{3}{7} \\ -\frac{1}{7} & \frac{2}{7} \end{bmatrix}$

Chapter Review, page 398

1. $\left(\dfrac{1}{2}, \dfrac{5}{2}\right)$ **2.** $(3, 2)$ **3.** $\left(-\dfrac{1}{2}, \dfrac{10}{3}\right)$ **4.** $(-2, -3)$ **5.** $(4, 2), (2, -2)$

6. $(3, 4), (-3, -4), (4, -3), (-4, 3)$ **7.** $(\sqrt{2}, 5), (-\sqrt{2}, 5), (-\sqrt{2}, -5), (\sqrt{2}, -5)$

8. 7% and 8% **9.** 20 persons at \$7 or 28 persons at \$5

10.

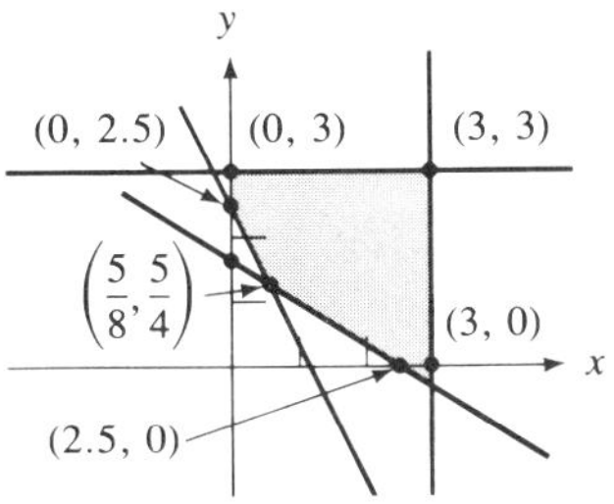

11.

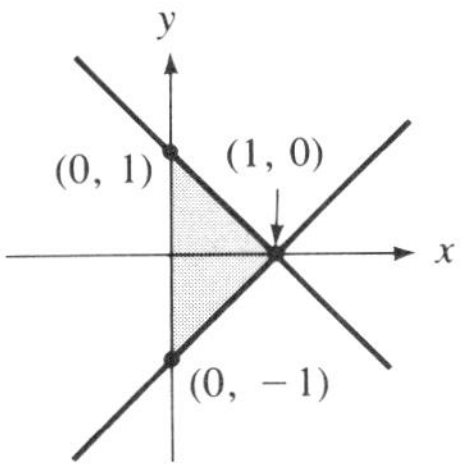

12.

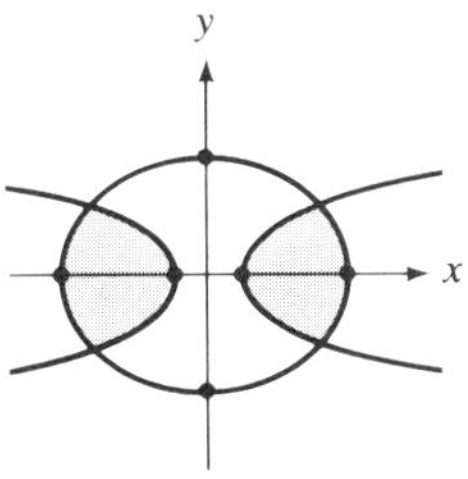

13.

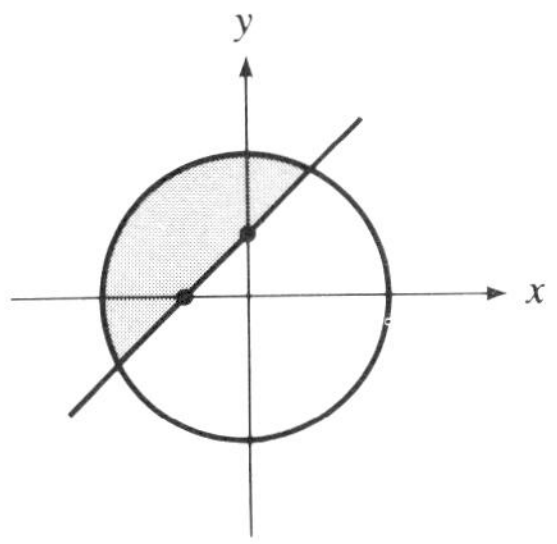

14. machine A, 5 hr; machine B, 10 hr **15.** $x = \frac{1}{19}(54 - 2z),\ y = \frac{1}{19}(5z - 2);\ \left(\frac{58}{19}, -\frac{12}{19}, -2\right)$

16. $x = \frac{1}{5}(13z + 7),\ y = -\frac{1}{5}(2 - 7z);\ \left(-\frac{19}{5}, -\frac{16}{5}, -2\right)$

17. $x = \frac{1}{3}(6z - 2),\ y = \frac{1}{3}(12z - 2);\ \left(-\frac{14}{3}, -\frac{26}{3}, -2\right)$

18. $x = \frac{1}{2}(z - 5),\ y = -2z - 4;\ \left(-\frac{7}{2}, 0, -2\right)$ **19.** $(1, -2, 2)$ **20.** inconsistent

21. $x = 2y - 1,\ z = -y + 3$ **22.** inconsistent **23.** 24°, 72°, and 84°

24. $x^2 + y^2 - 2x - 8y - 8 = 0$ **25.** $\begin{bmatrix} 17 \\ 18 \end{bmatrix}$ **26.** $\begin{bmatrix} 10 & 12 \\ 34 & 60 \end{bmatrix}$ **27.** $\begin{bmatrix} 35 & 27 \\ 9 & 4 \end{bmatrix}$ **28.** $\begin{bmatrix} 4 \\ 1 \\ 2 \end{bmatrix}$

29. $\begin{bmatrix} -6 & 33 & -45 \\ -4 & 14 & -20 \\ 24 & -10 & 18 \end{bmatrix}$ **30.** $\begin{bmatrix} -15 & 228 \\ -36 & 270 \end{bmatrix}$ **31.** $\begin{bmatrix} 3 & 7 \\ 1 & -5 \end{bmatrix}\begin{bmatrix} x \\ y \end{bmatrix} = \begin{bmatrix} 16 \\ 17 \end{bmatrix}$ **32.** $\begin{bmatrix} 4 & -17 \\ -1 & 6 \end{bmatrix}\begin{bmatrix} x \\ y \end{bmatrix} = \begin{bmatrix} 11 \\ 0 \end{bmatrix}$

33. $\begin{bmatrix} 3 & -2 & 1 \\ 0 & 1 & -1 \end{bmatrix}\begin{bmatrix} x \\ y \\ z \end{bmatrix} = \begin{bmatrix} 4 \\ 9 \end{bmatrix}$ **34.** $\begin{bmatrix} 4 & 6 & -11 \\ 2 & -1 & 0 \end{bmatrix}\begin{bmatrix} x \\ y \\ z \end{bmatrix} = \begin{bmatrix} 19 \\ 8 \end{bmatrix}$

35.
$$\begin{aligned} 5x + 4y &= 1 \\ 2x + 3y &= -1 \\ -x \qquad &= 6 \end{aligned}$$

36.
$$\begin{aligned} x - 2y + 3z &= 0 \\ 4x \qquad + 2z &= 1 \\ 2x - 3y \qquad &= 8 \end{aligned}$$

37. $(1, -3, 3)$ **38.** $(3, 1, -2)$ **39.** $(27, -3, -3)$ **40.** $(-8, 26, -17)$ **41.** $\left(\frac{199}{22}, -\frac{35}{22}\right)$

42. $(21, -2, -11)$

Chapter Test, page 400

1. $(2, -2)$ **2.** $\left(\frac{54}{31}, -\frac{11}{31}\right)$ **3.** Clarence 10, Luis 40

4. $x = \frac{1}{13}(31z - 4),\ y = \frac{1}{13}(-20z - 10)$ **5.** $\left(\frac{27}{13}, -\frac{30}{13}, 1\right)$ **6.** $\left(-\frac{74}{17}, -\frac{21}{17}\right), (2, 3)$

7.

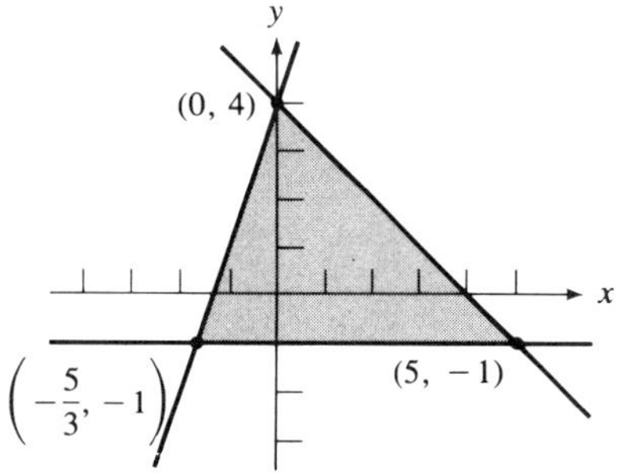

8. maximum of 15 at $(5, -1)$, minimum of -20 at $(0, 4)$ **9.** $(1, -1, -1)$ **10.** $(1, -4, -7)$

11. 10 **12.** $(-21, -16)$

Exercises 11.1, page 406

1. $(x - 3)(x^3 + 2x^2 + 5x + 16) + 47$ **3.** $(x - 1)(x^2 - 2x + 1)$ **5.** $x(4x^4 - 3x + 2) - 9$

7. $(x + 1)(x^3 - 7x^2 + 7x - 2) + 4$ **9.** $\left(x - \frac{1}{4}\right)\left(\frac{1}{2}x - \frac{5}{24}\right) + \frac{11}{96}$

11. $(x-3)(2x^3+3x^2-3x-2)$ **13.** $(x-i)(4x^3+4ix^2-7x+2-7i)+7-2i$
15. $(x-2i)(x^2-3ix+10)+24i$ **17.** -5 **19.** 121 **21.** -107 **23.** 28 **25.** yes
27. yes **29.** no **31.** yes

Exercises 11.2, page 409

1. $(x+3)(x^3+3x^2-16x+54)-163$ **3.** $(x-2)(2x^4+x^3+2x^2+4x+9)+19$
5. $(x-1)(x^3+x^2+x+1)$ **7.** $(x+1)(x^2-x+1)$ **9.** $(x-1)(x^3+2x^2+3x+4)+5$
11. $(x-\frac{1}{2})(x^2+\frac{7}{2}x+\frac{3}{4})+\frac{51}{8}$ **13.** 133 **15.** -154 **17.** 0 **19.** $-21-14i$ **21.** $-\dfrac{3}{16}$
23. $-4+5i$ **25.** 4 **27.** 47 **29.** 283

Exercises 11.3, page 413

1. $(x-2)(x+1)(x-7)$ **3.** $4(x+1)\left(x-\frac{1}{2}-\frac{3\sqrt{2}}{2}i\right)\left(x-\frac{1}{2}+\frac{3\sqrt{2}}{2}i\right)$

5. $(x+1)(x-2)\left(x+\frac{1}{2}-\frac{\sqrt{3}}{2}i\right)\left(x+\frac{1}{2}+\frac{\sqrt{3}}{2}i\right)$ **7.** $2(x-2)(x-3)(x+1)\left(x+\frac{3}{2}\right)$

9. $(x-3-i)(x-3+i)(x-5)$ **11.** $2\left(x-\frac{3}{2}\right)(x-3i)(x+3i)$ **13.** $(y-2i)(y+2i)(y-1)(y+4)$

15. $(x-4)\left(x-\frac{5}{2}-\frac{\sqrt{13}}{2}\right)\left(x-\frac{5}{2}+\frac{\sqrt{13}}{2}\right)$ **17.** $(x-2)(x-2)(x-4)(x+1)$

19. $(2x-1)(x-1)(x+4)$ **21.** $(x-1)(x-1)(x+1)(x+1)$ **23.** $(x+6)(x+6)(x-1)(x-4)$
25. $(2x-1)(3x+2)$ **27.** $(2x-1)(3x-1)(2x+3)(3x+2)$

Exercises 11.4, page 418

1. $-1, 0; 1, 2; 3, 4$ **3.** $-1, 0; 1, 2$ **5.** $-1, 0$ **7.** 2, 3 **9.** 1, 2 **11.** $-4, -3; 2, 3$
13. $-2, 0$ **15.** $-1, 5$ **17.** $-1, 4$ **19.** $-1, 2$
21. two or zero positive roots, one negative root, four, two or zero complex roots
23. two or zero positive roots; two or zero negative roots; zero, two, or four complex roots
25. one positive root, two or zero negative roots, two or zero complex roots
27. one positive root, one negative root, two complex roots
29. one positive root, one negative root, two complex roots

Exercises 11.5, page 422

1. 2, 2, and -3 **3.** 1, -2, and 3 **5.** $-\frac{4}{3}, \frac{1}{2}$, and 5 **7.** $\frac{1}{2}$, 1, 2, and $\frac{5}{3}$ **9.** $5, -1+i, -1-i$

11. $5, -2-\sqrt{2}$, and $-2+\sqrt{2}$ **13.** $2+\sqrt{7}, 2-\sqrt{7}, \frac{1}{2}$ **15.** $\frac{3}{2}, \frac{-1+\sqrt{23}\,i}{2}, \frac{-1-\sqrt{23}\,i}{2}$

17. $\frac{5}{2}, -1, \frac{-3+\sqrt{5}}{2}, \frac{-3-\sqrt{5}}{2}$ **19.** $1, -\frac{1}{2}, 1-\sqrt{3}, 1+\sqrt{3}$ **21.** 2 in. **23.** 2, 3, and 6
25. 3 hr

Exercises 11.6, page 425

1. 2.35 **3.** 1.26 **5.** 0.13 **7.** 0.61 **9.** 4.79 **11.** 2.92 **13.** 7 in.

Chapter Review, page 425

1. $(x-2)(3x^2+17)+33$ **2.** $(x+1)(4x^4-4x^3-2x^2+2x-1)-18$
3. $(x+2)(x^5-2x^4+3x^3-5x^2+13x-27)+55$ **4.** $(x-1)(3x^3-3x^2-2x-4)$ **5.** 64

6. $-8-10i$ **7.** 0 **8.** $-3-60i$ **9.** yes **10.** no **11.** yes **12.** yes **13.** $\frac{1}{2}, 2, -\frac{3}{2}$

14. $i, -i, 3+\sqrt{6}, 3-\sqrt{6}$ **15.** $1, 1, \frac{3+\sqrt{17}}{2}, \frac{3-\sqrt{17}}{2}$ **16.** $2(x-1+\sqrt{5})(x-1-\sqrt{5})$
17. $(x-1)(x-2)(x-2-2i)(x-2+2i)$ **18.** 1, 2; −1, 0; −4, −3 **19.** 1, 2 **20.** 2, 3; −4, −3
21. −1, 0 **22.** −1, 1 **23.** 2 or zero positive roots, 1 negative root, 2 or zero complex roots
24. 2 or zero positive roots, no negative roots, zero is a root, 2 or zero complex roots.
25. one positive root, 1 or three negative roots, 2 or zero complex roots **26.** $\frac{3}{2}, \frac{1}{3}, -2$
27. $-2, -3, \frac{1}{5}$ **28.** $-1, -\frac{1}{2}, 5, 3$ **29.** $\frac{1}{3}, \frac{1}{5}, \frac{1}{2}$ **30.** 1.16 **31.** 0.75

Chapter Test, page 426

1. $(x-3)(5x^2+12x+38)+123$ **2.** $(x+2)(6x^2-12x+23)-48$ **3.** 123 **4.** −48 **5.** no
6. no **7.** $3, 3, -1, \frac{1}{2}$ **8.** $3, \frac{1+\sqrt{7}\,i}{2}, \frac{1-\sqrt{7}\,i}{2}$ **9.** $3, -\frac{1}{2}, \frac{2}{3}$ **10.** 1.41

Exercises 12.1, page 432

1. 0, 2, 6, 12 **3.** 1, −5, −15, −29 **5.** −2, 3, −4, 5 **7.** 1, −1, 1, −1 **9.** 1, 7, 9, 63
11. 6, 4, 5, 9/2 **13.** 5, 2, −1, −4, $a_n = 5 + (n-1)(-3)$
15. −3, −1, 1, 3, $a_n = -3 + (n-1)(2)$ **17.** 7, 6, 5, 4, $a_n = 7 + (n-1)(-1)$
19. −19, −11, −3, 5, $a_n = -19 + (n-1)(8)$ **21.** 5, 1, .2, .04, $a_n = 5(.2)^{n-1}$
23. −8, 8, −8, 8, $a_n = 8(-1)^n$ **25.** 1/3, 2/3, 4/3, 8/3, $a_n = (1/3)(2)^{n-1}$
27. 3/2, 1, 2/3, 4/9, $a_n = (3/2)(2/3)^{n-1}$ **29.** 1/6, 1/2, 3/2, 9/2, $a_n = (1/6)(3)^{n-1}$
31. nth payment = 435 + $(n-1)(-6.75)$, last payment = 306.75 **33.** 734.66 **35.** 365.35

Exercises 12.2, page 437

1. 30 **3.** 45 **5.** −5/16 **7.** 199 **9.** 287/30 **11.** 49/36 **13.** $\Sigma_{n=0}^{4}(4+3n)$
15. $\Sigma_{n=1}^{5}(-1)^{n+1}(2)^n$ **17.** $\Sigma_{n=0}^{3}(-1)^n(2)^{3-n}$ **19.** −138 **21.** 88 **23.** 11.625 **25.** 3
27. 1,417.50 **29.** 8079.96

Exercises 12.4, page 445

1. $a^3 - 9a^2b + 27ab^2 - 27b^3$ **3.** $a^6 - 6a^5b + 15a^4b^2 - 20a^3b^3 + 15a^2b^4 - 6ab^5 + b^6$
5. $32x^5 + 80x^4y + 80x^3y^2 + 40x^2y^3 + 10xy^4 + y^5$ **7.** $u^4 - 2u^3 + (3/2)u^2 - (1/2)u + 1/16$
9. $1 + 4(.05) + 6(.05)^2 + 4(.05)^3 + (.05)^4$ **11.** $x^6 - 3x^4y^2 + 3x^2y^4 - y^6$
13. $16a^8 - 32a^6 + 24a^4 - 8a^2 + 1$ **15.** $x^9 - 3x^6 + 3x^3 - 1$ **17.** $40x^3y^2$ **19.** $15p^4q^2$
21. 1, 10, 45, 120, 210, 252, 210, 120, 45, 10, 1

Chapter Review, page 445

1. 1, 5, 9, 13, 17 **2.** 0, 3/4, −4/3, −15/8, −12/5 **3.** −41, −39, −35, −29, −21
4. 4, 4/3, 4/9, 4/27, 4/81 **5.** 4, 7, 10, 13, 16 **6.** 12, −2, 5, 3/2, 13/4 **7.** −20 **8.** 15
9. 16 **10.** 162 **11.** 1,480, 1,360, 1,240, 1,120 **12.** 8,162.93 **13.** 396.05 **14.** 70
15. 85 **16.** 48 **17.** 363/140 **18.** 33 **19.** 16 **20.** 5.93 **21.** −180
22. $\Sigma_{n=2}^{5} 2n$ **23.** $\Sigma_{n=1}^{5}(4n-1)$ **24.** $\Sigma_{n=0}^{3} 4(3/2)^n$ **25.** $\Sigma_{n=3}^{6}(-1)^{n-1}(2n)$
26. $\Sigma_{n=2}^{6}(n^2-1)$ **27.** $\Sigma_{n=2}^{5}(n^3+1)$ **28.** 820 **29.** 1277.96
30. $a^5 - 25a^4b + 250a^3b^2 - 1250a^2b^3 + 3125ab^4 - 3125b^5$ **31.** $-1792x^5$
32. $128x^7 - 1344x^6y + 6048x^5y^2 - 15{,}120x^4y^3 + 22{,}680x^3y^4 - 20{,}412x^2y^5 + 10{,}206xy^6 - 2{,}187y^7$

Chapter Test, page 447

1. 1, 0, −3, −8, −15 **2.** 6, 4, 3, 2.5, 2.25 **3.** −2 **4.** 3 **5.** −16 **6.** −8
7. 1400, 1300, 1200, 1100 **8.** 50 **9.** $\Sigma_{n=0}^{4}(-2+2n)$ **10.** −5 **11.** 124
13. $32 - 80ab + 80a^2b^2 - 40a^3b^3 + 10a^4b^4 - a^5b^5$ **14.** $(21)(32)x^5y^2$

Tables

Table I Squares, square roots, and prime factors
Table II Common logarithms
Table III Exponential functions
Table IV Natural logarithms of numbers
Table V Values of the trigonometric functions

Table I Squares, square roots, and prime factors

No.	Sq.	Sq. Root	Prime Factors	No.	Sq.	Sq. Root	Prime Factors
1	1	1.000		51	2,601	7.141	$3 \cdot 17$
2	4	1.414	2	52	2,704	7.211	$2^2 \cdot 13$
3	9	1.732	3	53	2,809	7.280	53
4	16	2.000	2^2	54	2,916	7.348	$2 \cdot 3^3$
5	25	2.236	5	55	3,025	7.416	$5 \cdot 11$
6	36	2.449	$2 \cdot 3$	56	3,136	7.483	$2^3 \cdot 7$
7	49	2.646	7	57	3,249	7.550	$3 \cdot 19$
8	64	2.828	2^3	58	3,364	7.616	$2 \cdot 29$
9	81	3.000	3^2	59	3,481	7.681	59
10	100	3.162	$2 \cdot 5$	60	3,600	7.746	$2^2 \cdot 3 \cdot 5$
11	121	3.317	11	61	3,721	7.810	61
12	144	3.464	$2^2 \cdot 3$	62	3,844	7.874	$2 \cdot 31$
13	169	3.606	13	63	3,969	7.937	$3^2 \cdot 7$
14	196	3.742	$2 \cdot 7$	64	4,096	8.000	2^6
15	225	3.873	$3 \cdot 5$	65	4,225	8.062	$5 \cdot 13$
16	256	4.000	2^4	66	4,356	8.124	$2 \cdot 3 \cdot 11$
17	289	4.123	17	67	4,489	8.185	67
18	324	4.243	$2 \cdot 3^2$	68	4,624	8.246	$2^2 \cdot 17$
19	361	4.359	19	69	4,761	8.307	$3 \cdot 23$
20	400	4.472	$2^2 \cdot 5$	70	4,900	8.367	$2 \cdot 5 \cdot 7$
21	441	4.583	$3 \cdot 7$	71	5,041	8.426	71
22	484	4.690	$2 \cdot 11$	72	5,184	8.485	$2^3 \cdot 3^2$
23	529	4.796	23	73	5,329	8.544	73
24	576	4.899	$2^3 \cdot 3$	74	5,476	8.602	$2 \cdot 37$
25	625	5.000	5^2	75	5,625	8.660	$3 \cdot 5^2$
26	676	5.099	$2 \cdot 13$	76	5,776	8.718	$2^2 \cdot 19$
27	729	5.196	3^3	77	5,929	8.775	$7 \cdot 11$
28	784	5.292	$2^2 \cdot 7$	78	6,084	8.832	$2 \cdot 3 \cdot 13$
29	841	5.385	29	79	6,241	8.888	79
30	900	5.477	$2 \cdot 3 \cdot 5$	80	6,400	8.944	$2^4 \cdot 5$
31	961	5.568	31	81	6,561	9.000	3^4
32	1,024	5.657	2^5	82	6,724	9.055	$2 \cdot 41$
33	1,089	5.745	$3 \cdot 11$	83	6,889	9.110	83
34	1,156	5.831	$2 \cdot 17$	84	7,056	9.165	$2^2 \cdot 3 \cdot 7$
35	1,225	5.916	$5 \cdot 7$	85	7,225	9.220	$5 \cdot 17$
36	1,296	6.000	$2^2 \cdot 3^2$	86	7,396	9.274	$2 \cdot 43$
37	1,369	6.083	37	87	7,569	9.327	$3 \cdot 29$
38	1,444	6.164	$2 \cdot 19$	88	7,744	9.381	$2^3 \cdot 11$
39	1,521	6.245	$3 \cdot 13$	89	7,921	9.434	89
40	1,600	6.325	$2^3 \cdot 5$	90	8,100	9.487	$2 \cdot 3^2 \cdot 5$
41	1,681	6.403	41	91	8,281	9.539	$7 \cdot 13$
42	1,764	6.481	$2 \cdot 3 \cdot 7$	92	8,464	9.592	$2^2 \cdot 23$
43	1,849	6.557	43	93	8,649	9.644	$3 \cdot 31$
44	1,936	6.633	$2^2 \cdot 11$	94	8,836	9.695	$2 \cdot 47$
45	2,025	6.708	$3^2 \cdot 5$	95	9,025	9.747	$5 \cdot 19$
46	2,116	6.782	$2 \cdot 23$	96	9,216	9.798	$2^5 \cdot 3$
47	2,209	6.856	47	97	9,409	9.849	97
48	2,304	6.928	$2^4 \cdot 3$	98	9,604	9.899	$2 \cdot 7^2$
49	2,401	7.000	7^2	99	9,801	9.950	$3^2 \cdot 11$
50	2,500	7.071	$2 \cdot 5^2$	100	10,000	10.000	$2^2 \cdot 5^2$

Table II Common logarithms

x	0	1	2	3	4	5	6	7	8	9
1.0	.0000	.0043	.0086	.0128	.0170	.0212	.0253	.0294	.0334	.0374
1.1	.0414	.0453	.0492	.0531	.0569	.0607	.0645	.0682	.0719	.0755
1.2	.0792	.0828	.0864	.0899	.0934	.0969	.1004	.1038	.1072	.1106
1.3	.1139	.1173	.1206	.1239	.1271	.1303	.1335	.1367	.1399	.1430
1.4	.1461	.1492	.1523	.1553	.1584	.1614	.1644	.1673	.1703	.1732
1.5	.1761	.1790	.1818	.1847	.1875	.1903	.1931	.1959	.1987	.2014
1.6	.2041	.2068	.2095	.2122	.2148	.2175	.2201	.2227	.2253	.2279
1.7	.2304	.2330	.2355	.2380	.2405	.2430	.2455	.2480	.2504	.2529
1.8	.2553	.2577	.2601	.2625	.2648	.2672	.2695	.2718	.2742	.2765
1.9	.2788	.2810	.2833	.2856	.2878	.2900	.2923	.2945	.2967	.2989
2.0	.3010	.3032	.3054	.3075	.3096	.3118	.3139	.3160	.3181	.3201
2.1	.3222	.3243	.3263	.3284	.3304	.3324	.3345	.3365	.3385	.3404
2.2	.3424	.3444	.3464	.3483	.3502	.3522	.3541	.3560	.3579	.3598
2.3	.3617	.3636	.3655	.3674	.3692	.3711	.3729	.3747	.3766	.3784
2.4	.3802	.3820	.3838	.3856	.3874	.3892	.3909	.3927	.3945	.3962
2.5	.3979	.3997	.4014	.4031	.4048	.4065	.4082	.4099	.4116	.4133
2.6	.4150	.4166	.4183	.4200	.4216	.4232	.4249	.4265	.4281	.4298
2.7	.4314	.4330	.4346	.4362	.4378	.4393	.4409	.4425	.4440	.4456
2.8	.4472	.4487	.4502	.4518	.4533	.4548	.4564	.4579	.4594	.4609
2.9	.4624	.4639	.4654	.4669	.4683	.4698	.4713	.4728	.4742	.4757
3.0	.4771	.4786	.4800	.4814	.4829	.4843	.4857	.4871	.4886	.4900
3.1	.4914	.4928	.4942	.4955	.4969	.4983	.4997	.5011	.5024	.5038
3.2	.5051	.5065	.5079	.5092	.5105	.5119	.5132	.5145	.5159	.5172
3.3	.5185	.5198	.5211	.5224	.5237	.5250	.5263	.5276	.5289	.5302
3.4	.5315	.5328	.5340	.5353	.5366	.5378	.5391	.5403	.5416	.5428
3.5	.5441	.5453	.5465	.5478	.5490	.5502	.5514	.5527	.5539	.5551
3.6	.5563	.5575	.5587	.5599	.5611	.5623	.5635	.5647	.5658	.5670
3.7	.5682	.5694	.5705	.5717	.5729	.5740	.5752	.5763	.5775	.5786
3.8	.5798	.5809	.5821	.5832	.5843	.5855	.5866	.5877	.5888	.5899
3.9	.5911	.5922	.5933	.5944	.5955	.5966	.5977	.5988	.5999	.6010
4.0	.6021	.6031	.6042	.6053	.6064	.6075	.6085	.6096	.6107	.6117
4.1	.6128	.6138	.6149	.6160	.6170	.6180	.6191	.6201	.6212	.6222
4.2	.6232	.6243	.6253	.6263	.6274	.6284	.6294	.6304	.6314	.6325
4.3	.6335	.6345	.6355	.6365	.6375	.6385	.6395	.6405	.6415	.6425
4.4	.6435	.6444	.6454	.6464	.6474	.6484	.6493	.6503	.6513	.6522
4.5	.6532	.6542	.6551	.6561	.6571	.6580	.6590	.6599	.6609	.6618
4.6	.6628	.6637	.6646	.6656	.6665	.6675	.6684	.6693	.6702	.6712
4.7	.6721	.6730	.6739	.6749	.6758	.6767	.6776	.6785	.6794	.6803
4.8	.6812	.6821	.6830	.6839	.6848	.6857	.6866	.6875	.6884	.6893
4.9	.6902	.6911	.6920	.6928	.6937	.6946	.6955	.6964	.6972	.6981
5.0	.6990	.6998	.7007	.7016	.7024	.7033	.7042	.7050	.7059	.7067
5.1	.7076	.7084	.7093	.7101	.7110	.7118	.7126	.7135	.7143	.7152
5.2	.7160	.7168	.7177	.7185	.7193	.7202	.7210	.7218	.7226	.7235
5.3	.7243	.7251	.7259	.7267	.7275	.7284	.7292	.7300	.7308	.7316
5.4	.7324	.7332	.7340	.7348	.7356	.7364	.7372	.7380	.7388	.7396
x	0	1	2	3	4	5	6	7	8	9

Table II Common logarithms *(Continued)*

x	0	1	2	3	4	5	6	7	8	9
5.5	.7404	.7412	.7419	.7427	.7435	.7443	.7451	.7459	.7466	.7474
5.6	.7482	.7490	.7497	.7505	.7513	.7520	.7528	.7536	.7543	.7551
5.7	.7559	.7566	.7574	.7582	.7589	.7597	.7604	.7612	.7619	.7627
5.8	.7634	.7642	.7649	.7657	.7664	.7672	.7679	.7686	.7694	.7701
5.9	.7709	.7716	.7723	.7731	.7738	.7745	.7752	.7760	.7767	.7774
6.0	.7782	.7789	.7796	.7803	.7810	.7818	.7825	.7832	.7839	.7846
6.1	.7853	.7860	.7868	.7875	.7882	.7889	.7896	.7903	.7910	.7917
6.2	.7924	.7931	.7938	.7945	.7952	.7959	.7966	.7973	.7980	.7987
6.3	.7993	.8000	.8007	.8014	.8021	.8028	.8035	.8041	.8048	.8055
6.4	.8062	.8069	.8075	.8082	.8089	.8096	.8102	.8109	.8116	.8122
6.5	.8129	.8136	.8142	.8149	.8156	.8162	.8169	.8176	.8182	.8189
6.6	.8195	.8202	.8209	.8215	.8222	.8228	.8235	.8241	.8248	.8254
6.7	.8261	.8267	.8274	.8280	.8287	.8293	.8299	.8306	.8312	.8319
6.8	.8325	.8331	.8338	.8344	.8351	.8357	.8363	.8370	.8376	.8382
6.9	.8388	.8395	.8401	.8407	.8414	.8420	.8426	.8432	.8439	.8445
7.0	.8451	.8457	.8463	.8470	.8476	.8482	.8488	.8494	.8500	.8506
7.1	.8513	.8519	.8525	.8531	.8537	.8543	.8549	.8555	.8561	.8567
7.2	.8573	.8579	.8585	.8591	.8597	.8603	.8609	.8615	.8621	.8627
7.3	.8633	.8639	.8645	.8651	.8657	.8663	.8669	.8675	.8681	.8686
7.4	.8692	.8698	.8704	.8710	.8716	.8722	.8727	.8733	.8739	.8745
7.5	.8751	.8756	.8762	.8768	.8774	.8779	.8785	.8791	.8797	.8802
7.6	.8808	.8814	.8820	.8825	.8831	.8837	.8842	.8848	.8854	.8859
7.7	.8865	.8871	.8876	.8882	.8887	.8893	.8899	.8904	.8910	.8915
7.8	.8921	.8927	.8932	.8938	.8943	.8949	.8954	.8960	.8965	.8971
7.9	.8976	.8982	.8987	.8993	.8998	.9004	.9009	.9015	.9020	.9025
8.0	.9031	.9036	.9042	.9047	.9053	.9058	.9063	.9069	.9074	.9079
8.1	.9085	.9090	.9096	.9101	.9106	.9112	.9117	.9122	.9128	.9133
8.2	.9138	.9143	.9149	.9154	.9159	.9165	.9170	.9175	.9180	.9186
8.3	.9191	.9196	.9201	.9206	.9212	.9217	.9222	.9227	.9232	.9238
8.4	.9243	.9248	.9253	.9258	.9263	.9269	.9274	.9279	.9284	.9289
8.5	.9294	.9299	.9304	.9309	.9315	.9320	.9325	.9330	.9335	.9340
8.6	.9345	.9350	.9355	.9360	.9365	.9370	.9375	.9380	.9385	.9390
8.7	.9395	.9400	.9405	.9410	.9415	.9420	.9425	.9430	.9435	.9440
8.8	.9445	.9450	.9455	.9460	.9465	.9469	.9474	.9479	.9484	.9489
8.9	.9494	.9499	.9504	.9509	.9513	.9518	.9523	.9528	.9533	.9538
9.0	.9542	.9547	.9552	.9557	.9562	.9566	.9571	.9576	.9581	.9586
9.1	.9590	.9595	.9600	.9605	.9609	.9614	.9619	.9624	.9628	.9633
9.2	.9638	.9643	.9647	.9652	.9657	.9661	.9666	.9671	.9675	.9680
9.3	.9685	.9689	.9694	.9699	.9703	.9708	.9713	.9717	.9722	.9727
9.4	.9731	.9736	.9741	.9745	.9750	.9754	.9759	.9763	.9768	.9773
9.5	.9777	.9782	.9786	.9791	.9795	.9800	.9805	.9809	.9814	.9818
9.6	.9823	.9827	.9832	.9836	.9841	.9845	.9850	.9854	.9859	.9863
9.7	.9868	.9872	.9877	.9881	.9886	.9890	.9894	.9899	.9903	.9908
9.8	.9912	.9917	.9921	.9926	.9930	.9934	.9939	.9943	.9948	.9952
9.9	.9956	.9961	.9965	.9969	.9974	.9978	.9983	.9987	.9991	.9996
x	**0**	**1**	**2**	**3**	**4**	**5**	**6**	**7**	**8**	**9**

Table III Exponential functions

x	e^x	e^{-x}	x	e^x	e^{-x}
0.00	1.0000	1.0000	1.5	4.4817	0.2231
0.01	1.0101	0.9901	1.6	4.9530	0.2019
0.02	1.0202	0.9802	1.7	5.4739	0.1827
0.03	1.0305	0.9705	1.8	6.0496	0.1653
0.04	1.0408	0.9608	1.9	6.6859	0.1496
0.05	1.0513	0.9512	2.0	7.3891	0.1353
0.06	1.0618	0.9418	2.1	8.1662	0.1225
0.07	1.0725	0.9324	2.2	9.0250	0.1108
0.08	1.0833	0.9331	2.3	9.9742	0.1003
0.09	1.0942	0.9139	2.4	11.023	0.0907
0.10	1.1052	0.9048	2.5	12.182	0.0821
0.11	1.1163	0.8958	2.6	13.464	0.0743
0.12	1.1275	0.8869	2.7	14.880	0.0672
0.13	1.1388	0.8781	2.8	16.445	0.0608
0.14	1.1503	0.8694	2.9	18.174	0.0550
0.15	1.1618	0.8607	3.0	20.086	0.0498
0.16	1.1735	0.8521	3.1	22.198	0.0450
0.17	1.1853	0.8437	3.2	24.533	0.0408
0.18	1.1972	0.8353	3.3	27.113	0.0369
0.19	1.2092	0.8270	3.4	29.964	0.0334
0.20	1.2214	0.8187	3.5	33.115	0.0302
0.21	1.2337	0.8106	3.6	36.598	0.0273
0.22	1.2461	0.8025	3.7	40.447	0.0247
0.23	1.2586	0.7945	3.8	44.701	0.0224
0.24	1.2712	0.7866	3.9	49.402	0.0202
0.25	1.2840	0.7788	4.0	54.598	0.0183
0.30	1.3499	0.7408	4.1	60.340	0.0166
0.35	1.4191	0.7047	4.2	66.686	0.0150
0.40	1.4918	0.6703	4.3	73.700	0.0136
0.45	1.5683	0.6376	4.4	81.451	0.0123
0.50	1.6487	0.6065	4.5	90.017	0.0111
0.55	1.7333	0.5769	4.6	99.484	0.0101
0.60	1.8221	0.5488	4.7	109.95	0.0091
0.65	1.9155	0.5220	4.8	121.51	0.0082
0.70	2.0138	0.4966	4.9	134.29	0.0074
0.75	2.1170	0.4724	5.0	148.41	0.0067
0.80	2.2255	0.4493	5.5	244.69	0.0041
0.85	2.3396	0.4274	6.0	403.43	0.0025
0.90	2.4596	0.4066	6.5	665.14	0.0015
0.95	2.5857	0.3867	7.0	1096.6	0.0009
1.0	2.7183	0.3679	7.5	1808.0	0.0006
1.1	3.0042	0.3329	8.0	2981.0	0.0003
1.2	3.3201	0.3012	8.5	4914.8	0.0002
1.3	3.6693	0.2725	9.0	8103.1	0.0001
1.4	4.0552	0.2466	10.0	22026	0.00005

Table IV Natural logarithms of numbers

n	$\ln n$	n	$\ln n$	n	$\ln n$
		4.5	1.5041	9.0	2.1972
0.1	−2.3026	4.6	1.5261	9.1	2.2083
0.2	−1.6094	4.7	1.5476	9.2	2.2192
0.3	−1.2040	4.8	1.5686	9.3	2.2300
0.4	−0.9163	4.9	1.5892	9.4	2.2407
0.5	−0.6931	5.0	1.6094	9.5	2.2513
0.6	−0.5108	5.1	1.6292	9.6	2.2618
0.7	−0.3567	5.2	1.6487	9.7	2.2721
0.8	−0.2231	5.3	1.6677	9.8	2.2824
0.9	−0.1054	5.4	1.6864	9.9	2.2925
1.0	0.0000	5.5	1.7047	10	2.3026
1.1	0.0953	5.6	1.7228	11	2.3979
1.2	0.1823	5.7	1.7405	12	2.4849
1.3	0.2624	5.8	1.7579	13	2.5649
1.4	0.3365	5.9	1.7750	14	2.6391
1.5	0.4055	6.0	1.7918	15	2.7081
1.6	0.4700	6.1	1.8083	16	2.7726
1.7	0.5306	6.2	1.8245	17	2.8332
1.8	0.5878	6.3	1.8405	18	2.8904
1.9	0.6419	6.4	1.8563	19	2.9444
2.0	0.6931	6.5	1.8718	20	2.9957
2.1	0.7419	6.6	1.8871	25	3.2189
2.2	0.7885	6.7	1.9021	30	3.4012
2.3	0.8329	6.8	1.9169	35	3.5553
2.4	0.8755	6.9	1.9315	40	3.6889
2.5	0.9163	7.0	1.9459	45	3.8067
2.6	0.9555	7.1	1.9601	50	3.9120
2.7	0.9933	7.2	1.9741	55	4.0073
2.8	1.0296	7.3	1.9879	60	4.0943
2.9	1.0647	7.4	2.0015	65	4.1744
3.0	1.0986	7.5	2.0149	70	4.2485
3.1	1.1314	7.6	2.0281	75	4.3175
3.2	1.1632	7.7	2.0412	80	4.3820
3.3	1.1939	7.8	2.0541	85	4.4427
3.4	1.2238	7.9	2.0669	90	4.4998
3.5	1.2528	8.0	2.0794	100	4.6052
3.6	1.2809	8.1	2.0919	110	4.7005
3.7	1.3083	8.2	2.1041	120	4.7875
3.8	1.3350	8.3	2.1163	130	4.8676
3.9	1.3610	8.4	2.1282	140	4.9416
4.0	1.3863	8.5	2.1401	150	5.0106
4.1	1.4110	8.6	2.1518	160	5.0752
4.2	1.4351	8.7	2.1633	170	5.1358
4.3	1.4586	8.8	2.1748	180	5.1930
4.4	1.4816	8.9	2.1861	190	5.2470

Table V Values of the trigonometric functions

t	t degrees	$\sin t$	$\cos t$	$\tan t$	$\cot t$	$\sec t$	$\csc t$		
.0000	**0° 00′**	.0000	1.0000	.0000	—	1.000	—	**90 00′**	1.5708
.0029	10	.0029	1.0000	.0029	343.8	1.000	343.8	50	1.5679
.0058	20	.0058	1.0000	.0058	171.9	1.000	171.9	40	1.5650
.0087	30	.0087	1.0000	.0087	114.6	1.000	114.6	30	1.5621
.0116	40	.0116	.9999	.0116	85.94	1.000	85.95	20	1.5592
.0145	50	.0145	.9999	.0145	68.75	1.000	68.76	10	1.5563
.0175	**1° 00′**	.0175	.9998	.0175	57.29	1.000	57.30	**89 00′**	1.5533
.0204	10	.0204	.9998	.0204	49.10	1.000	49.11	50	1.5504
.0233	20	.0233	.9997	.0233	42.96	1.000	42.98	40	1.5475
.0262	30	.0262	.9997	.0262	38.19	1.000	38.20	30	1.5446
.0291	40	.0291	.9996	.0291	34.37	1.000	34.38	20	1.5417
.0320	50	.0320	.9995	.0320	31.24	1.001	31.26	10	1.5388
.0349	**2° 00′**	.0349	.9994	.0349	28.64	1.001	28.65	**88 00′**	1.5359
.0378	10	.0378	.9993	.0378	26.43	1.001	26.45	50	1.5330
.0407	20	.0407	.9992	.0407	24.54	1.001	24.56	40	1.5301
.0436	30	.0436	.9990	.0437	22.90	1.001	22.93	30	1.5272
.0465	40	.0465	.9989	.0466	21.47	1.001	21.49	20	1.5243
.0495	50	.0494	.9988	.0495	20.21	1.001	20.23	10	1.5213
.0524	**3° 00′**	.0523	.9986	.0524	19.08	1.001	19.11	**87 00′**	1.5184
.0553	10	.0552	.9985	.0553	18.07	1.002	18.10	50	1.5155
.0582	20	.0581	.9983	.0582	17.17	1.002	17.20	40	1.5126
.0611	30	.0610	.9981	.0612	16.35	1.002	16.38	30	1.5097
.0640	40	.0640	.9980	.0641	15.60	1.002	15.64	20	1.5068
.0669	50	.0669	.9978	.0670	14.92	1.002	14.96	10	1.5039
.0698	**4° 00′**	.0698	.9976	.0699	14.30	1.002	14.34	**86 00′**	1.5010
.0727	10	.0727	.9974	.0729	13.73	1.003	13.76	50	1.4981
.0756	20	.0756	.9971	.0758	13.20	1.003	13.23	40	1.4952
.0785	30	.0785	.9969	.0787	12.71	1.003	12.75	30	1.4923
.0814	40	.0814	.9967	.0816	12.25	1.003	12.29	20	1.4893
.0844	50	.0843	.9964	.0846	11.83	1.004	11.87	10	1.4864
.0873	**5° 00′**	.0872	.9962	.0875	11.43	1.004	11.47	**85 00′**	1.4835
.0902	10	.0901	.9959	.0904	11.06	1.004	11.10	50	1.4806
.0931	20	.0929	.9957	.0934	10.71	1.004	10.76	40	1.4777
.0960	30	.0958	.9954	.0963	10.39	1.005	10.43	30	1.4748
.0989	40	.0987	.9951	.0992	10.08	1.005	10.13	20	1.4719
.1018	50	.1016	.9948	.1022	9.788	1.005	9.839	10	1.4690
.1047	**6° 00′**	.1045	.9945	.1051	9.514	1.006	9.567	**84 00′**	1.4661
.1076	10	.1074	.9942	.1080	9.255	1.006	9.309	50	1.4632
.1105	20	.1103	.9939	.1110	9.010	1.006	9.065	40	1.4603
.1134	30	.1132	.9936	.1139	8.777	1.006	8.834	30	1.4573
.1164	40	.1161	.9932	.1169	8.556	1.007	8.614	20	1.4544
.1193	50	.1190	.9929	.1198	8.345	1.007	8.405	10	1.4515
.1222	**7° 00′**	.1219	.9925	.1228	8.144	1.008	8.206	**83 00′**	1.4486
		$\cos t$	$\sin t$	$\cot t$	$\tan t$	$\csc t$	$\sec t$	t degrees	t

Table V Values of the trigonometric functions *(Continued)*

t	t degrees	sin t	cos t	tan t	cot t	sec t	csc t		
.1222	**7° 00′**	.1219	.9925	.1228	8.144	1.008	8.206	**83 00′**	1.4486
.1251	10	.1248	.9922	.1257	7.953	1.008	8.016	50	1.4457
.1280	20	.1276	.9918	.1287	7.770	1.008	7.834	40	1.4428
.1309	30	.1305	.9914	.1317	7.596	1.009	7.661	30	1.4399
.1338	40	.1334	.9911	.1346	7.429	1.009	7.496	20	1.4370
.1367	50	.1363	.9907	.1376	7.269	1.009	7.337	10	1.4341
.1396	**8° 00′**	.1392	.9903	.1405	7.115	1.010	7.185	**82 00′**	1.4312
.1425	10	.1421	.9899	.1435	6.968	1.010	7.040	50	1.4283
.1454	20	.1449	.9894	.1465	6.827	1.011	6.900	40	1.4254
.1484	30	.1478	.9890	.1495	6.691	1.011	6.765	30	1.4224
.1513	40	.1507	.9886	.1524	6.561	1.012	6.636	20	1.4195
.1542	50	.1536	.9881	.1554	6.435	1.012	6.512	10	1.4166
.1571	**9° 00′**	.1564	.9877	.1584	6.314	1.012	6.392	**81 00′**	1.4137
.1600	10	.1593	.9872	.1614	6.197	1.013	6.277	50	1.4108
.1629	20	.1622	.9868	.1644	6.084	1.013	6.166	40	1.4079
.1658	30	.1650	.9863	.1673	5.976	1.014	6.059	30	1.4050
.1687	40	.1679	.9858	.1703	5.871	1.014	5.955	20	1.4021
.1716	50	.1708	.9853	.1733	5.769	1.015	5.855	10	1.3992
.1745	**10° 00′**	.1736	.9848	.1763	5.671	1.015	5.759	**80 00′**	1.3963
.1774	10	.1765	.9843	.1793	5.576	1.016	5.665	50	1.3934
.1804	20	.1794	.9838	.1823	5.485	1.016	5.575	40	1.3904
.1833	30	.1822	.9833	.1853	5.396	1.017	5.487	30	1.3875
.1862	40	.1851	.9827	.1883	5.309	1.018	5.403	20	1.3846
.1891	50	.1880	.9822	.1914	5.226	1.018	5.320	10	1.3817
.1920	**11° 00′**	.1908	.9816	.1944	5.145	1.019	5.241	**79 00′**	1.3788
.1949	10	.1937	.9811	.1974	5.066	1.019	5.164	50	1.3759
.1978	20	.1965	.9805	.2004	4.989	1.020	5.089	40	1.3730
.2007	30	.1994	.9799	.2035	4.915	1.020	5.016	30	1.3701
.2036	40	.2022	.9793	.2065	4.843	1.021	4.945	20	1.3672
.2065	50	.2051	.9787	.2095	4.773	1.022	4.876	10	1.3643
.2094	**12° 00′**	.2079	.9781	.2126	4.705	1.022	4.810	**78 00′**	1.3614
.2123	10	.2108	.9775	.2156	4.638	1.023	4.745	50	1.3584
.2153	20	.2136	.9769	.2186	4.574	1.024	4.682	40	1.3555
.2182	30	.2164	.9763	.2217	4.511	1.024	4.620	30	1.3526
.2211	40	.2193	.9757	.2247	4.449	1.025	4.560	20	1.3497
.2240	50	.2221	.9750	.2278	4.390	1.026	4.502	10	1.3468
.2269	**13 00′**	.2250	.9744	.2309	4.331	1.026	4.445	**77 00′**	1.3439
.2298	10	.2278	.9737	.2339	4.275	1.027	4.390	50	1.3410
.2327	20	.2306	.9730	.2370	4.219	1.028	4.336	40	1.3381
.2356	30	.2334	.9724	.2401	4.165	1.028	4.284	30	1.3352
.2385	40	.2363	.9717	.2432	4.113	1.029	4.232	20	1.3323
.2414	50	.2391	.9710	.2462	4.061	1.030	4.182	10	1.3294
.2443	**14 00′**	.2419	.9703	.2493	4.011	1.031	4.134	**76 00′**	1.3265
		cos t	sin t	cot t	tan t	csc t	sec t	t degrees	t

Table V Values of the trigonometric functions *(Continued)*

t	t degrees	$\sin t$	$\cos t$	$\tan t$	$\cot t$	$\sec t$	$\csc t$		
.2443	**14 00′**	.2419	.9703	.2493	4.011	1.031	4.134	**76 00′**	1.3265
.2473	10	.2447	.9696	.2524	3.962	1.031	4.086	50	1.3235
.2502	20	.2476	.9689	.2555	3.914	1.032	4.039	40	1.3206
.2531	30	.2504	.9681	.2586	3.867	1.033	3.994	30	1.3177
.2560	40	.2532	.9674	.2617	3.821	1.034	3.950	20	1.3148
.2589	50	.2560	.9667	.2648	3.776	1.034	3.906	10	1.3119
.2618	**15 00′**	.2588	.9659	.2679	3.732	1.035	3.864	**75 00′**	1.3090
.2647	10	.2616	.9652	.2711	3.689	1.036	3.822	50	1.3061
.2676	20	.2644	.9644	.2742	3.647	1.037	3.782	40	1.3032
.2705	30	.2672	.9636	.2773	3.606	1.038	3.742	30	1.3003
.2734	40	.2700	.9628	.2.805	3.566	1.039	3.703	20	1.2974
.2763	50	.2728	.9621	.2836	3.526	1.039	3.665	10	1.2945
.2793	**16 00′**	.2756	.9613	.2867	3.487	1.040	3.628	**74 00′**	1.2915
.2822	10	.2784	.9605	.2899	3.450	1.041	3.592	50	1.2886
.2851	20	.2812	.9596	.2931	3.412	1.042	3.556	40	1.2857
.2880	30	.2840	.9588	.2962	3.376	1.043	3.521	30	1.2828
.2909	40	.2868	.9580	.2994	3.340	1.044	3.487	20	1.2799
.2938	50	.2896	.9572	.3026	3.305	1.045	3.453	10	1.2770
.2967	**17 00′**	.2924	.9563	.3057	3.271	1.046	3.420	**73 00′**	1.2741
.2996	10	.2952	.9555	.3089	3.237	1.047	3.388	50	1.2712
.3025	20	.2979	.9546	.3121	3.204	1.048	3.356	40	1.2683
.3054	30	.3007	.9537	.3153	3.172	1.049	3.326	30	1.2654
.3083	40	.3035	.9528	.3185	3.140	1.049	3.295	20	1.2625
.3113	50	.3062	.9520	.3217	3.108	1.050	3.265	10	1.2595
.3142	**18 00′**	.3090	.9511	.3249	3.078	1.051	3.236	**72 00′**	1.2566
.3171	10	.3118	.9502	.3281	3.047	1.052	3.207	50	1.2537
.3200	20	.3145	.9492	.3314	3.018	1.053	3.179	40	1.2508
.3229	30	.3173	.9483	.3346	2.989	1.054	3.152	30	1.2479
.3258	40	.3201	.9474	.3378	2.960	1.056	3.124	20	1.2450
.3287	50	.3228	.9465	.3411	2.932	1.057	3.098	10	1.2421
.3316	**19 00′**	.3256	.9455	.3443	2.904	1.058	3.072	**71 00′**	1.2392
.3345	10	.3283	.9446	.3476	2.877	1.059	3.046	50	1.2363
.3374	20	.3311	.9436	.3508	2.850	1.060	3.021	40	1.2334
.3403	30	.3338	.9426	.3541	2.824	1.061	2.996	30	1.2305
.3432	40	.3365	.9417	.3574	2.798	1.062	2.971	20	1.2275
.3462	50	.3393	.9407	.3607	2.773	1.063	2.947	10	1.2246
.3491	**20 00′**	.3420	.9397	.3640	2.747	1.064	2.924	**70 00′**	1.2217
.3520	10	.3448	.9387	.3673	2.723	1.065	2.901	50	1.2188
.3549	20	.3475	.9377	.3706	2.699	1.066	2.878	40	1.2159
.3578	30	.3502	.9367	.3739	2.675	1.068	2.855	30	1.2130
.3607	40	.3529	.9356	.3772	2.651	1.069	2.833	20	1.2101
.3636	50	.3557	.9346	.3805	2.628	1.070	2.812	10	1.2072
.3665	**21 00′**	.3584	.9336	.3839	2.605	1.071	2.790	**69 00′**	1.2043
		$\cos t$	$\sin t$	$\cot t$	$\tan t$	$\csc t$	$\sec t$	t degrees	t

Table V Values of the trigonometric functions *(Continued)*

t	t degrees	sin t	cos t	tan t	cot t	sec t	csc t		
.3665	**21° 00′**	.3584	.9336	.3839	2.605	1.071	2.790	**69° 00′**	1.2043
.3694	10	.3611	.9325	.3872	2.583	1.072	2.769	50	1.2014
.3723	20	.3638	.9315	.3906	2.560	1.074	2.749	40	1.1985
.3752	30	.3665	.9304	.3939	2.539	1.075	2.729	30	1.1956
.3782	40	.3692	.9293	.3973	2.517	1.076	2.709	20	1.1926
.3811	50	.3719	.9283	.4006	2.496	1.077	2.689	10	1.1897
.3840	**22° 00′**	.3746	.9272	.4040	2.475	1.079	2.669	**68° 00′**	1.1868
.3869	10	.3773	.9261	.4074	2.455	1.080	2.650	50	1.1839
.3898	20	.3800	.9250	.4108	2.434	1.081	2.632	40	1.1810
.3927	30	.3827	.9239	.4142	2.414	1.082	2.613	30	1.1781
.3956	40	.3854	.9228	.4176	2.394	1.084	2.595	20	1.1752
.3985	50	.3881	.9216	.4210	2.375	1.085	2.577	10	1.1723
.4014	**23° 00′**	.3907	.9205	.4245	2.356	1.086	2.559	**67° 00′**	1.1694
.4043	10	.3934	.9194	.4279	2.337	1.088	2.542	50	1.1665
.4072	20	.3961	.9182	.4314	2.318	1.089	2.525	40	1.1636
.4102	30	.3987	.9171	.4348	2.300	1.090	2.508	30	1.1606
.4131	40	.4014	.9159	.4383	2.282	1.092	2.491	20	1.1577
.4160	50	.4041	.9147	.4417	2.264	1.093	2.475	10	1.1548
.4189	**24° 00′**	.4067	.9135	.4452	2.246	1.095	2.459	**66° 00′**	1.1519
.4218	10	.4094	.9124	.4487	2.229	1.096	2.443	50	1.1490
.4247	20	.4120	.9112	.4522	2.211	1.097	2.427	40	1.1461
.4276	30	.4147	.9100	.4557	2.194	1.099	2.411	30	1.1432
.4305	40	.4173	.9088	.4592	2.177	1.100	2.396	20	1.1403
.4334	50	.4200	.9075	.4628	2.161	1.102	2.381	10	1.1374
.4363	**25° 00′**	.4226	.9063	.4663	2.145	1.103	2.366	**65° 00′**	1.1345
.4392	10	.4253	.9051	.4699	2.128	1.105	2.352	50	1.1316
.4422	20	.4279	.9038	.4734	2.112	1.106	2.337	40	1.1286
.4451	30	.4305	.9026	.4770	2.097	1.108	2.323	30	1.1257
.4480	40	.4331	.9013	.4806	2.081	1.109	2.309	20	1.1228
.4509	50	.4358	.9001	.4841	2.066	1.111	2.295	10	1.1199
.4538	**26° 00′**	.4384	.8988	.4877	2.050	1.113	2.281	**64° 00′**	1.1170
.4567	10	.4410	.8975	.4913	2.035	1.114	2.268	50	1.1141
.4596	20	.4436	.8962	.4950	2.020	1.116	2.254	40	1.1112
.4625	30	.4462	.8949	.4986	2.006	1.117	2.241	30	1.1083
.4654	40	.4488	.8936	.5022	1.991	1.119	2.228	20	1.1054
.4683	50	.4514	.8923	.5059	1.977	1.121	2.215	10	1.1025
.4712	**27° 00′**	.4540	.8910	.5095	1.963	1.122	2.203	**63° 00′**	1.0996
.4741	10	.4566	.8897	.5132	1.949	1.124	2.190	50	1.0966
.4771	20	.4592	.8884	.5169	1.935	1.126	2.178	40	1.0937
.4800	30	.4617	.8870	.5206	1.921	1.127	2.166	30	1.0908
.4829	40	.4643	.8857	.5243	1.907	1.129	2.154	20	1.0879
.4858	50	.4669	.8843	.5280	1.894	1.131	2.142	10	1.0850
.4887	**28° 00′**	.4695	.8829	.5317	1.881	1.133	2.130	**62° 00′**	1.0821
		cos t	sin t	cot t	tan t	csc t	sec t	t degrees	t

Table V Values of the trigonometric functions *(Continued)*

t	t degrees	$\sin t$	$\cos t$	$\tan t$	$\cot t$	$\sec t$	$\csc t$		
.4887	**28 00′**	.4695	.8829	.5317	1.881	1.133	2.130	**62 00**	1.0821
.4916	10	.4720	.8816	.5354	1.868	1.134	2.118	50	1.0792
.4945	20	.4746	.8802	.5392	1.855	1.136	2.107	40	1.0763
.4974	30	.4772	.8788	.5430	1.842	1.138	2.096	30	1.0734
.5003	40	.4797	.8774	.5467	1.829	1.140	2.085	20	1.0705
.5032	50	.4823	.8760	.5505	1.816	1.142	2.074	10	1.0676
.5061	**29 00′**	.4848	.8746	.5543	1.804	1.143	2.063	**61 00′**	1.0647
.5091	10	.4874	.8732	.5581	1.792	1.145	2.052	50	1.0617
.5120	20	.4899	.8718	.5619	1.780	1.147	2.041	40	1.0588
.5149	30	.4924	.8704	.5658	1.767	1.149	2.031	30	1.0559
.5178	40	.4950	.8689	.5696	1.756	1.151	2.020	20	1.0530
.5207	50	.4975	.8675	.5735	1.744	1.153	2.010	10	1.0501
.5236	**30 00′**	.5000	.8660	.5774	1.732	1.155	2.000	**60 00**	1.0472
.5265	10	.5025	.8646	.5812	1.720	1.157	1.990	50	1.0443
.5294	20	.5050	.8631	.5851	1.709	1.159	1.980	40	1.0414
.5323	30	.5075	.8616	.5890	1.698	1.161	1.970	30	1.0385
.5352	40	.5100	.8601	.5930	1.686	1.163	1.961	20	1.0356
.5381	50	.5125	.8587	.5969	1.675	1.165	1.951	10	1.0327
.5411	**31 00′**	.5150	.8572	.6009	1.664	1.167	1.942	**59 00**	1.0297
.5440	10	.5175	.8557	.6048	1.653	1.169	1.932	50	1.0268
.5469	20	.5200	.8542	.6088	1.643	1.171	1.923	40	1.0239
.5498	30	.5225	.8526	.6128	1.632	1.173	1.914	30	1.0210
.5527	40	.5250	.8511	.6168	.1621	1.175	1.905	20	1.0181
.5556	50	.5275	.8496	.6208	1.611	1.177	1.896	10	1.0152
.5585	**32 00′**	.5299	.8480	.6249	1.600	1.179	1.887	**58°00′**	1.0123
.5614	10	.5324	.8465	.6289	1.590	1.181	1.878	50	1.0094
.5643	20	.5348	.8450	.6330	1.580	1.184	1.870	40	1.0065
.5672	30	.5373	.8434	.6371	1.570	1.186	1.861	30	1.0036
.5701	40	.5398	.8418	.6412	1.560	1.188	1.853	20	1.0007
.5730	50	.5422	.8403	.6453	1.550	1.190	1.844	10	.9977
.5760	**33 00′**	.5446	.8387	.6494	1.540	1.192	1.836	**57°00′**	.9948
.5789	10	.5471	.8371	.6536	1.530	1.195	1.828	50	.9919
.5818	20	.5495	.8355	.6577	1.520	1.197	1.820	40	.9890
.5847	30	.5519	.8339	.6619	1.511	1.199	1.812	30	.9861
.5876	40	.5544	.8323	.6661	1.501	1.202	1.804	20	.9832
.5905	50	.5568	.8307	.6703	1.492	1.204	1.796	10	.9803
.5934	**34 00′**	.5592	.8290	.6745	1.483	1.206	1.788	**56 00′**	.9774
.5963	10	.5616	.8274	.6787	1.473	1.209	1.781	50	.9745
.5992	20	.5640	.8258	.6830	1.464	1.211	1.733	40	.9716
.6021	30	.5664	.8241	.6873	1.455	1.213	1.766	30	.9687
.6050	40	.5688	.8225	.6916	1.446	1.216	1.758	20	.9657
.6080	50	.5712	.8208	.6959	1.437	1.218	1.751	10	.9628
.6109	**35 00′**	.5736	.8192	.7002	1.428	1.221	1.743	**55 00′**	.9599
		$\cos t$	$\sin t$	$\cot t$	$\tan t$	$\csc t$	$\sec t$	t degrees	t

Table V Values of the trigonometric functions *(Continued)*

t	t degrees	$\sin t$	$\cos t$	$\tan t$	$\cot t$	$\sec t$	$\csc t$		
.6109	**35 00′**	.5736	.8192	.7002	1.428	1.221	1.743	**55 00′**	.9599
.6138	10	.5760	.8175	.7046	1.419	1.223	1.736	50	.9570
.6167	20	.5783	.8158	.7089	1.411	1.226	1.729	40	.9541
.6196	30	.5807	.8141	.7133	1.402	1.228	1.722	30	.9512
.6225	40	.5831	.8124	.7177	1.393	1.231	1.715	20	.9483
.6254	50	.5854	.8107	.7221	1.385	1.233	1.708	10	.9454
.6283	**36 00′**	.5878	.8090	.7265	1.376	1.236	1.701	**54°00′**	.9425
.6312	10	.5901	.8073	.7310	1.368	1.239	1.695	50	.9396
.6341	20	.5925	.8056	.7355	1.360	1.241	1.688	40	.9367
.6370	30	.5948	.8039	.7400	1.351	1.244	1.681	30	.9338
.6400	40	.5972	.8021	.7445	1.343	1.247	1.675	20	.9308
.6429	50	.5995	.8004	.7490	1.335	1.249	1.668	10	.9279
.6458	**37 00′**	.6018	.7986	.7536	1.327	1.252	1.662	**53 00′**	.9250
.6487	10	.6041	.7969	.7581	1.319	1.255	1.655	50	.9221
.6516	20	.6065	.7951	.7627	1.311	1.258	1.649	40	.9192
.6545	30	.6088	.7934	.7673	1.303	1.260	1.643	30	.9163
.6574	40	.6111	.7916	.7720	1.295	1.263	1.636	20	.9134
.6603	50	.6134	.7898	.7766	1.288	1.266	1.630	10	.9105
.6632	**38 00′**	.6157	.7880	.7813	1.280	1.269	1.624	**52 00′**	.9076
.6661	10	.6180	.7862	.7860	1.272	1.272	1.618	50	.9047
.6690	20	.6202	.7844	.7907	1.265	1.275	1.612	40	.0918
.6720	30	.6225	.7826	.7954	1.257	1.278	1.606	30	.8988
.6749	40	.6248	.7808	.8002	1.250	1.281	1.601	20	.8959
.6778	50	.6271	.7790	.8050	1.242	1.284	1.595	10	.8930
.6807	**39 00′**	.6293	.7771	.8098	1.235	1.287	1.589	**51 00′**	.8901
.6836	10	.6316	.7753	.8146	1.228	1.290	1.583	50	.8872
.6865	20	.6338	.7735	.8195	1.220	1.293	1.578	40	.8843
.6894	30	.6361	.7716	.8243	1.213	1.296	1.572	30	.8814
.6923	40	.6383	.7698	.8292	1.206	1.299	1.567	20	.8785
.6952	50	.6406	.7679	.8342	1.199	1.302	1.561	10	.8756
.6981	**40 00′**	.6428	.7660	.8391	1.192	1.305	1.556	**50 00′**	.8727
.7010	10	.6450	.7642	.8441	1.185	1.309	1.550	50	.8698
.7039	20	.6472	.7623	.8491	1.178	1.312	1.545	40	.8668
.7069	30	.6494	.7604	.8541	1.171	1.315	1.540	30	.8639
.7098	40	.6517	.7585	.8591	1.164	1.318	1.535	20	.8610
.7127	50	.6539	.7566	.8642	1.157	1.322	1.529	10	.8581
.7156	**41 00′**	.6561	.7547	.8693	1.150	1.325	1.524	**49 00′**	8.552
.7185	10	.6583	.7528	.8744	1.144	1.328	1.519	50	.8523
.7214	20	.6604	.7509	.8796	1.137	1.332	1.514	40	.8494
.7243	30	.6626	.7490	.8847	1.130	1.335	1.509	30	.8465
.7272	40	.6648	.7470	.8899	1.124	1.339	1.504	20	.8436
.7301	50	.6670	.7451	.8952	1.117	1.342	1.499	10	.8407
.7330	**42 00′**	.6691	.7431	.9004	1.111	1.346	1.494	**48 00′**	.8378
		$\cos t$	$\sin t$	$\cot t$	$\tan t$	$\csc t$	$\sec t$	t degrees	t

Table V Values of the trigonometric functions *(Continued)*

t	t degrees	sin t	cos t	tan t	cot t	sec t	csc t		
.7330	**42 00′**	.6691	.7431	.9004	1.111	1.346	1.494	**48 00′**	.8378
.7359	10	.6713	.7412	.9057	1.104	1.349	1.490	50	.8348
.7389	20	.6734	.7392	.9110	1.098	1.353	1.485	40	.8319
.7418	30	.6756	.7373	.9163	1.091	1.356	1.480	30	.8290
.7447	40	.6777	.7353	.9217	1.085	1.360	1.476	20	.8261
.7476	50	.6799	.7333	.9271	1.079	1.364	1.471	10	.8232
.7505	**43 00′**	.6820	.7314	.9325	1.072	1.367	1.466	**47 00′**	.8203
.7534	10	.6841	.7294	.9380	1.066	1.371	1.462	50	.8174
.7563	20	.6862	.7274	.9435	1.060	1.375	1.457	40	.8145
.7592	30	.6884	.7254	.9490	1.054	1.379	1.453	30	.8116
.7621	40	.6905	.7234	.9545	1.048	1.382	1.448	20	.8087
.7650	50	.6926	.7214	.9601	1.042	1.386	1.444	10	.8058
.7679	**44 00′**	.6947	.7193	.9657	1.036	1.390	1.440	**46 00′**	.8029
.7709	10	.6967	.7173	.9713	1.030	1.394	1.435	50	.7999
.7738	20	.6988	.7153	.9770	1.024	1.398	1.431	40	.7970
.7767	30	.7009	.7133	.9827	1.018	1.402	1.427	30	.7941
.7796	40	.7030	.7112	.9884	1.012	1.406	1.423	20	.7912
.7825	50	.7050	.7092	.9942	1.006	1.410	1.418	10	.7883
.7854	**45°00′**	.7071	.7071	1.0000	1.0000	1.414	1.414	**45 00′**	.7854
		cos t	sin t	cot t	tan t	csc t	sec t	t degrees	t

Index

Abcissa, 79
Absolute value:
 of a complex number, 317
 of a real number, 9
Accuracy, 287–288
Addition:
 of complex numbers, 54
 formula, 260
 of fractions, 36
 of ordinates, 100, 235
 of real numbers, 1
 of vectors, 303
Additive:
 function, 260
 identity, 1
 inverse, 1
Affine function, 93
Algebra, fundamental theorem of, 410
Ambiguous case, 293
Amplitude, 232
Angle:
 acute, 189
 between two vectors, 309
 of depression, 288
 of elevation, 287
 initial side, 188
 measure, 189–190, 192
 negative, 188
 obtuse, 189
 positive, 188
 quadrantal, 188
 reference, 218
 right, 189
 standard position, 188
 terminal side, 188
 vertex, 188
Angles:
 complementary, 209
 coterminal, 188
Arccosecant x ($\csc^{-1}x$), 240
Arccosine x ($\cos^{-1}x$), 240
Arccotangent x ($\cot^{-1}x$), 240
Arcsecant x ($\sec^{-1}x$), 240
Arcsine x ($\sin^{-1}x$), 238
Arctangent x ($\tan^{-1}x$), 240
Area of a sector, 192
Argument:
 of a complex number, 318
 principal, 318
Arithmetic:
 progression, 430
 sequence, 429
 sum, 434
Associative axiom:
 for addition, 1
 for multiplication, 1
Asymptotes:
 definition, 136
 horizontal, 143
 hyperbola, 136
 vertical, 143, 228
Augmented matrix, 378
Axis:
 imaginary, 317
 major, 131
 minor, 131
 polar, 333
 real, 317

Basic formulas of analytic trigonometry, 278
Bearing, 289
Binomial, 23
 coefficient, 441
 theorem, 441–445
Bounds for roots, 416
Break-even point, 352

Calculator exercises, 6, 31, 47, 78, 85, 98, 103, 139, 147, 169, 174, 194, 203, 216, 225, 231, 237, 243, 253, 409
Cancellation law:
 for addition, 6
 for multiplication, 6
Cardioid, 341
Cartesian coordinate system, 78
Center:
 ellipse, 131
 circle, 123
Characteristic, 175
China, Texas, 188
Circle, 81, 123–125
 unit, 190

Circular sections, 212
Coefficient, 23
  matrix, 375
Cofactor of a matrix, 391
Cofunctions, 209
Collecting like terms, 5
Column matrix, 372
Common:
  denominator, 36
  logarithm, 171
Commutative axioms:
  for addition, 1
  for multiplication, 1
Complementary angles, 209
Completing the square, 59
Complex numbers:
  absolute value, 317
  addition, 54
  argument of, 318
  argument, principal, 318
  conjugate of, 56
  graphing, 316
  magnitude, 317
  modulus, 317
  multiplication, 55, 322
  *n*th root, 329–330
  polar form, 318
  quotient of, 56, 323
  standard form of, 56
  trigonometric form, 318
  as vectors, 317
Complex plane, 316
Components of a vector, 306
Composition of functions, 101
Compound interest, 165
Conditional equation, 248
Conic sections, 123–139
Conjugate of a complex number, 56
Constant of variation, 108
Convex polygonal region, 356
Coordinate system:
  polar, 333
  rectangular, 78
Cosecant x (csc x), 196
Cosines, law of, 297
Cosine x (cos x), 196
Cotangent x (cot x), 196
Coterminal angles, 188
Cramer's rule, 395–398
Cube of a binomial, 25

Declining balances, 165
Decreasing function, 162
Degree:
  of a monomial, 23
  of a polynomial, 23
  unit of angle measure, 189
Degree-radian conversion, 190
De Moivre's theorem, 327
Denominator:
  common, 36
  rationalizing, 51
Depreciation, 165
Depression, angle of, 288
Descartes' rule of signs, 417
Determinant:
  definition, 390
  properties of, 392–393
Difference:
  of two cubes, 28
  of two squares, 24, 28
  of two vectors, 304
Direct variation, 108
Directly proportional, 108
Directrix, 125
Discriminant, 61
Distance formula, 80
Distributive axiom, 1
Division:
  of complex numbers, 56, 323
  of fractions, 38
  of polynomials, 402
  of real numbers, 4
  synthetic 406, 408
Domain, 74
  of a function, 74
  of exponential function, 160
  of inverse trigonometric functions, 240
  of logarithmic function, 170
  of trigonometric functions, 197, 213
Dot product, 308
  properties, 308
Double angle formulas, 266

e, 167
Effective speed, 287
Elementary row operation, 380
Elevation, angle of, 287
Elimination, Gaussian, 380
Ellipse, 130, 134
Equations:
  of a circle, 81
  conditional, 248
  definition, 11
  equivalent, 11
  exponential, 178–182
  linear, *see* Linear equations
  logarithmic, 178–182
  polar, 337
  polynomial, *see* Polynomial equations
  quadratic, *see* Quadratic equations
  radical, 65
  root, 11
  of rose curves, 342
  solution set, 11
  systems, 346–351, 365, 371
  trigonometric, 248
Equivalent:
  equations, 8
  inequalities, 14
Expansion by cofactors, 263
Exponential:
  equations, 178–182
  functions, 159–169
Exponents:
  negative, 42
  properties of, 26, 43
  rational, 49
  zero, 42
Extraneous roots, 257
Extreme point, 361

Factorial notation, 441
Factoring:
  common factors, 28
  difference of two cubes, 28
  difference of two squares, 30
  formulas (trigonometric), 275
  by grouping, 30
  perfect squares, 28
  solution by, 116
  sum of cubes, 30
  trinomials, 29
Factor theorem, 405
Field, 2
Focus:
  ellipse, 130
  hyperbola, 134
  parabola, 125
FOIL method, 25
Formulas of analytic trigonometry, 278
Four-leafed rose, 342
Fractions:
  addition, 36
  cancellation law, 35
  common denominator, 36
  division, 38
  equality, 35
  partial, 148–152
  products, 37
  quotients, 35
Frequency, 215
Functions:
  additive, 260
  circular, 212
  composition of, 101
  constant, 75
  cosecant, 196
  cosine, 196
  cotangent, 196
  decreasing, 162
  definition, 74, 82
  difference of two, 99
  domain, 74
  even, 94
  exponential, 159–169
  graph, 82
  identity, 75
  image, 74
  increasing, 161
  inverse, 104

Functions: (*continued*)
inverse trigonometric, 238–240
linear, 83, 86
logarithmic, 169–174
odd, 95
one-to-one, 75
periodic, 199
polynomial, 139–142
product of, 99
quadratic, 116–123
quotient of, 99
range, 74
rational, 142–148
secant, 196
sine, 196
sum, 99
tangent, 196
value, 74
zero of, 120
Fundamental:
identities, 254
Theorem of Algebra, 410
Future value, 431

Gaussian elimination, 378–382
Generalized Pythagorean theorem, 298
General solution, 366
Geometric:
progression, 431
sequence, 431
sum, 435
Graph(s):
of cardiod, 341
of cosecant function, 229,
of cosine function, 227,
of cotangent function, 228,
of ellipse, 130
of exponential function, 161
of four-leafed rose, 341
of a function, 82
of hyperbola, 135
of inequalities, 15, 153–157
of inverse trigonometric functions, 239–240
of lemniscate, 342
of limacon, 340
of logarithmic function, 170
of parabola, 127
polar, 339
of secant function, 229
of sine function, 226
of tangent function, 228
Graphing by addition of ordinates, 100
Graph-points, 226, 232
Greater than, 8

Half-angle formulas, 269
Half-life, 180
Heading, 290
Horizontal component, 306
Hyperbola, 134–139
Hypotenuse, 207
Hypothesis, induction, 438

Inconsistent system, 347
Increasing function, 161
Identities:
addition formulas, 261
basic formulas, 278
double-angle formulas, 266
factoring formulas, 275
fundamental, 254
half-angle formulas, 269
product formulas, 274
Identity, 248
Imaginary axis, 317
Induction, 437–441
hypothesis, 438
Inequalities:
equivalent, 14
linear, 152
quadratic, 67–68, 154
systems, 354–360
Infinity, 143
Initial:
side, 188
point, 301
Inner product, 308
Integers, set of, 1
Intercepts, 89
form, 92
Interpolation, linear, 182–185, 222
Interval notation, 16
Inverse:
function, 104
matrix, 384–389
trigonometric functions, 238–240
variation, 109
Inversely proportional, 109
Irrational:
numbers, 1
roots, 422–425

Joint variation, 110

Law of:
cosines, 297
sines, 291
tangents, 301
Laws, cancellation:
for addition, 6
for fractions, 35
for multiplication, 6
Lemniscate, 342
Less than, 8
Limacon, 340
Line:
equation, 88–93
point-slope form, 88
slope, 87
slope-intercept form, 89
of symmetry, 118
Linear:
combination, 305
functions, 83, 86–93
interpolation, 182–185, 322
programming, 360–364
Linear equations:
in one unknown, 13
in three unknowns, 365–371
in two unknowns, 86
Logarithmic:
equations, 178–182
functions, 169–174
Logarithms:
common, 171
natural, 171
properties, 171
Long division, polynomials, 402

Magnitude:
of complex number, 317
of a force, 310
of a vector, 301
Major axis, 131
Mantissa, 175
Mathematical induction, 438
Matrix:
augmented, 378
coefficient, 375
cofactor of, 391
column, 372
definition, 371
determinant of, 390–395
elementary row operations, 380
entry, 371
equality, 371
equations, 376
identity, 375
inversion, 384–389
minor of, 391
multiplicative inverse, 384–389
$n \times m$, 371
product, 372–376
row, 371
scalar product, 372
Midpoint formula, 80
Mil, 192
Minor, 391
axis, 131
Minute, 189
Modulus of a complex number, 317
Mollweide's formula, 296
Monomial:
definition, 23
degree, 23
Multiple roots, 411
Multiplication:
complex numbers, 55, 322, 327
fractions, 37
matrices, 373
real numbers, 2
of two vectors, 308
of a vector by a scalar, 303

Multiplicative identity:
 for matrices, 375
 for real numbers, 2
Multiplicative inverse:
 for matrices, 384
 for real numbers, 2

Natural:
 logarithms, 171
 numbers, 1
Negative:
 angle, 188
 numbers, 8
Nested form, 31
Newton's formula, 296
Notation,:
 factorial, 441
 scientific, 45
*n*th root(s):
 complex number, 329–330
 principal, 48
 of unity, 331
Number:
 complex, 316
 real, 1
 reference, 218
Numerator, 35

Oblique triangle, 291
Odd function, 95
Ordered:
 pair, 79
 triple, 365
Order relation, 8–11
Ordinate(s), 79
 addition of, 100, 235
Origin, 78
Orthogonal vectors, 310

Parabola, 116, 125–130
Paralellogram law, 304
Partial fractions, 148–152
Particular solution, 366
Pascal's triangle, 444
Period:
 of cosecant function, 199
 of cosine function, 199, 233
 of cotangent function, 199
 of secant function, 199
 of sine function, 199, 232
 of tangent function, 199, 235
 of vibrating mass, 215
Perpendicular vectors, 310
Phase shift, 233
Point-slope form, 88
Polar:
 axis, 333
 coordinate system, 335
 equations, 337
 form of complex numbers, 318
 graphs, 339
 to rectangular conversion formulas, 335
Pole, 333
Polynomial equations:
 conjugate pairs of roots, 412
 irrational roots, 422–425
 locating roots, 414–419
 multiple roots, 411
 number of roots, 411
 rational roots, 419–425
Polynomials:
 definition, 23
 degree, 23
 division, 402
 equations, *see* Polynomial equations
 factoring, 28–31
 functions, 139–142
 value of, 23
Position vector, 301
Positive:
 angle, 188
 number, 8
Present value, 431
Principal argument, 318
Products:
 complex numbers, 55, 322
 formulas, 274
 fractions, 37
 matrices, 373
 radicals, 50
 scalar, 372
 special, 24–26
Progression:
 arithmetic, 429
 geometric, 431
Pythagorean theorem, 80
 generalized, 298

Quadrant, 79
Quadrantal angle, 188
Quadratic:
 formula, 60
 function, 116–120
 inequalities, 67–68
Quadratic equations:
 definition, 58
 discriminant, 61
 nature of roots, 61
 solution by completing the square, 59
 solution by factoring, 61
 solution by formula, 60
Quotients:
 of complex numbers, 56, 323
 of fractions, 35
 of functions, 99
 of polynomials, 35–42, 402
 or real numbers, 4

Radian, 190
Radian-degree conversion, 190
Radicals:
 definition, 48
 equations, 65–66
Radicals (*continued*)
 product, 50
 quotients, 50
 sums, 52
Range:
 of exponential function, 161
 of a function, 74
 of inverse trigonometric functions, 240
 of logarithmic function, 170
 of a relation, 80
 of trigonometric functions, 198, 214
Rational:
 functions, 142–148
 number, 1
 roots, 419–422
Rationalizing the denominator, 51
Real axis, 317
Reciprocal, 2
Rectangular to polar coordinate formulas, 335
Reference:
 angle, 218
 number, 218
Relations:
 definition, 81
 domain, 81
 range, 81
Remainder theorem, 404
Resultant of forces, 311
Roots:
 bounds for, 416
 for a complex number, 329–330
 in conjugate pairs, 412
 of equations, 11
 irrational, 422–425
 locating, 414–419
 multiple, 411
 *n*th, 48, 329–330
 of a polynomial equation, 411
 principal *n*th, 48
 rational, 419–425
 of unity, 331
Rose curves, 342
Row operations, 380
Rule of signs, 417

Scalar, 303
 product, 308, 372
Scientific notation, 45
Secant function (sec *x*), 196
Second, 189
Sector of a circle, 192
Sequence:
 arithmetic, 429
 definition, 428
 geometric, 431
 terms, 428
Series:
 arithmetic, 434
 definition, 433
 geometric, 435

Set builder notation, 15
Side:
  adjacent, 207
  opposite, 207
Side-angle-angle (SAA), 292
Side-angle-side (SAS), 299
Side-side-angle (SSA), 293
Side-side-side (SSS), 298
Sigma notation, 433
Significant digits, 287–288
Sine function (sin *x*), 196
Sines, law of, 291
Slope:
  of line, 87
  of parallel lines, 90
  of perpendicular lines, 91
Slope-intercept form, 89
Solution:
  extraneous, 66, 257
  set, 11
  of triangles, 286
Special angles, 204–207
Spiral, 343, 344
Square of a binomial, 24
Standard:
  form for complex numbers, 56
  lettering for triangles, 286
  position of an angle, 188
Sum of two cubes, 30
Symmetric graphs:
  with respect to origin, 94
  with respect to *x*-axis, 95
  with respect to *y*-axis, 93
Synthetic:
  division, 406–408
  substitution, 409
Systems:
  inconsistent, 347
  inequalities, 354–360
  linear equations, 346–351
  nonlinear equations, 349–351
  solution by elimination, 348

Tangent function (tan *x*), 196
Tangents, law of, 301
Term, 23
Terminal:
  point, 301
  side, 188
Translation, 96–98
Triangle:
  30°-60°-90°, 204
  45°-45°-90°, 206
  oblique, 291
  solution of right, 286
Trigonometric:
  equations, 248
  form of complex numbers, 318
  functions of complementary angles, 209
  functions, defined, 196
  functions of differences of two angles, 261
  functions, domain, 197, 213
  functions, range, 198, 214
  functions of real numbers, 210
  functions, signs of, 200
  functions of special angles, 204–207
  functions of sums of two angles, 261
  identities, 248, 278
Trigonometry, basic formulas of, 278
Trinomial, 23

Unit circle, 190
Unity, roots of, 331
Unknown, 11
Upper bounds for roots, 416

Value:
  absolute, 9, 317
  of a function, 74
Variable, 11
Variation:
  constant of, 108
  direct, 108
  inverse, 109
  joint, 110
Vector:
  addition, 303
  addition and scalar multiplication properties, 305
  angle between, 309
  column, 372
  definition, 301–302
  difference of two, 304
  equality, 301
  i, 305
  j, 305
  magnitude of, 301
  multiplication of, 308
  orthogonal, 310
  position, 301
  row, 371
  unit, 305
Vertical line test, 83
Vertex:
  of an angle, 188
  of an ellipse, 131
  of a hyperbola, 137
  of a parabola, 118
Vertical component, 306

Work, 310

Zero:
  of a function, 120
  vector, 303